MERRILL INTEGRATED MATHEMATICS

Klutch • Bumby • Collins • Egbers

Course 3

MERRILL
PUBLISHING COMPANY

The logo for *Merrill Integrated Mathematics* is emblematic of the blend of topics presented.
The equals sign (=) at the top indicates the inclusion of equations and algebra. The radical symbol ($\sqrt{}$) indicates that all numbers through the real numbers are treated. The angle symbol ($\angle$) and triangle ($\triangle$) symbolize geometry. The summation symbol (Σ) stands for probability and statistics.

Imprint 1997

Send all inquiries to:
Glencoe/McGraw-Hill
936 Eastwind Drive
Westerville, OH 43081-3374

ISBN 0-675-05548-2

6 7 8 9 10 11 12 13 14 15 16 026/043 02 01 00 99 98 97

Authors

Mr. Richard J. Klutch taught mathematics at Freeport High School in New York and now teaches mathematics at the Hunter College Campus Schools of the City University of New York. He has participated in the development of major mathematics curriculum projects and played a major role in developing such programs at Hunter College High School and elsewhere. He has served as a consultant to the Bureau of Mathematics Education of the State of New York. He is one of the authors of the new syllabus for the three-year sequence for high school mathematics for the state of New York.

Mr. Donald W. Collins is a lecturer in the Department of Mathematics and Informational Sciences at Sam Houston State University in Huntsville, Texas. Mr. Collins taught mathematics at Elmhurst Public Schools in Elmhurst, Illinois and Abilene Public Schools in Abilene, Texas. He has had many years of experience in the field of mathematical education, including over 20 years of work in the mathematics textbook publishing industry.

Dr. Douglas R. Bumby is chairman of the mathematics department at Scarsdale High School, Scarsdale, New York. Dr. Bumby taught mathematics at Hunter College High School and was Clinical Associate in Mathematics Education at Teachers College, Columbia University, where he taught graduate courses and advised doctoral candidates. Dr. Bumby's research interests are in the areas of mathematics curriculum development and readability of mathematics textbooks. He was instrumental in developing a model integrated mathematics program at Hunter College High School in New York City.

Mr. Elden B. Egbers is the Supervisor of Mathematics for the state of Washington. Mr. Egbers taught mathematics and was the department chairman at Queen Anne High School in Seattle, Washington. Most recently, he directed the writing of the Washington State Curriculum Guidelines. Mr. Egbers has also given workshops on integrated mathematics to mathematics teachers and supervisors.

Reviewers

Robert Drajem
Mathematics Teacher
Buffalo Academy for Visual and Performing Arts
Buffalo, New York

Tim Frawley
Mathematics Department Chair
Corning-Painted Post West High School
Painted Post, New York

Dr. Felix Labaki
Coordinator of Mathematics Instruction
Williamsville Central Schools
Williamsville, New York

Dick Moore
Mathematics Teacher
Penfield High School
Penfield, New York

Oystein Ostebo
Mathematics Teacher
Corning-Painted Post East High School
Painted Post, New York

Ken Sohmer
Academic Supervisor for Mathematics
Jamestown High School
Jamestown, New York

William Wallace Stout
Teacher
Peninsula High School
Gig Harbor, Washington

George Turk
Mathematics Department Chairperson
Franklin High School
Seattle, Washington

Preface

Merrill Integrated Mathematics, Course 3, is the third textbook in a three-textbook series that integrates the study of algebra and geometry. It continues the development of the useful topics of logic, statistics, and probability presented in *Merrill Integrated Mathematics, Courses 1 and 2*. The goals of the text are to develop proficiency with mathematical skills, to expand understanding of mathematical concepts, to improve logical thinking, and to promote success. To achieve these goals the following strategies are used.

Build upon a Solid Foundation. The spiraled nature of the topics in *Merrill Integrated Mathematics* provides an on-going review of previously learned mathematical concepts throughout the text.

Utilize Sound Pedagogy. Concepts are introduced when they are needed and in a logical sequence. Each concept presented is then used both within that lesson and in later lessons.

Provide a Variety of Topics. A wide variety of topics increases student interest and motivation. Students are exposed to a more representative selection of mathematical ideas than would be possible in a traditional curriculum.

Facilitate Learning. A clear, concise format aids the student in understanding the mathematical concepts. Furthermore, many photographs, illustrations, graphs, and tables provide help for the student in visualizing the ideas presented.

Use Relevant Real-Life Applications. Applications provide a practical approach to mathematics, relating it to other disciplines and to everyday life.

The text offers a variety of special features to aid the student.

Student Annotations	Help students identify important concepts as they study.
Selected Answers	Allows students to check their progress as they work. These answers are provided at the back of the text.
Mixed Review	Provides students with a quick review of skills and concepts previously taught.
Chapter Review	Permits students to review each chapter by working sample problems from each section.
Chapter Test	Enables students to check their own progress.
Mathematical Excursions	Enliven and help maintain student interest by providing interesting side trips. Topics are varied and include glimpses into the development and uses of mathematics.
Problem Solving Application	Instructs students in the uses of different problem-solving techniques and strategies as tools for solving problems.

Contents

Working with Polynomial and Rational Expressions

Working with Complex Numbers

Relations and Functions

An Introduction to Transformation Geometry

Exponential and Logarithmic Functions

Circles

Chapter 6

An Introduction to Circular Functions

Chapter 7

Applications of Circular Functions

Chapter 8

Trigonometric Identities and Equations

Chapter 9

More Work with Transformations

Chapter 10

Probability and the Binomial Theorem

Chapter 11

Statistics

Chapter 12

Sequences and Series

Chapter 13

An Introduction to Matrices

Chapter 14

Polynomials, Complex Numbers, and DeMoivre's Theorem

Chapter 15

Mathematical Induction

Chapter 16

Problem Solving Applications

Staff

Project Editor: Donald T. Porzio
Project Artist: Dick Smith
Project Designer: Larry W. Collins
Photo Editor: Barbara Buchholz

Photo Credits

1, 29, Gerard Photography; **47,** First Image; **72,** Historical Pictures Service, Inc.; **91,** Janeart Ltd./The Image Bank; **131,** Gerard Photography; **135,** file photo; **146,** Gerard Photography; **167,** Kessel/Shih © 1976, Springer-Verlag; **188,** Ruth Dixon; **209,** Tim Courlas; **249,** Wolfgang Filser/G&J Images/The Image Bank; **289,** Doug Martin; **329,** Hickson-Bender Photography; **336,** Tom Stack/Tom Stack & Assoc.; **345,** Courtesy Westinghouse Electric Corporation; **363,** M. C. Escher © Beeldrecht, Amsterdam, New York Collection Haags Gemeate Museum; **401,** Alan Carey/The Image Works; **425,** Gerard Photography; **437,** Alan Carey/The Image Works; **438,** First Image; **449,** Gerard Photography; **469,** David M. Dennis; **505,** Gerard Photography; **543,** Cincinnati Convention & Visitor's Bureau; **583,** Steve Marquart; **586,** Gerard Photography.

Working with Polynomial and Rational Expressions

Chapter 1

A musical note exactly one harmonic octave lower than another note can be produced by plucking a guitar string exactly twice as long as the string that produced the higher note.

In fact, the notes of the scale downward are produced by increasing the length of the string according to rational numbers. For example, the note B is produced by a string $\frac{16}{15}$ of the length of a C-string.

The ancient Greek mathematicians first discovered that musical intervals are governed by rational numbers.

1.1 The Reals

Much of your work in mathematics so far has dealt with the real number system. Although this number system is now taken for granted, it was not until the late nineteenth century that mathematicians gained a full understanding of real numbers. This section reviews some important properties of the real number system.

First, let us review some important subsets of the reals.

natural numbers $\mathcal{N} = \{1, 2, 3, \ldots\}$ *also called counting numbers*

whole numbers $\mathcal{W} = \{0, 1, 2, 3, \ldots\}$

integers $\mathcal{Z} = \{\ldots, -3, -2, -1, 0, 1, 2, 3, \ldots\}$

rational numbers $\mathcal{Q} = \left\{\text{All numbers in the form of } \frac{a}{b}\text{, where } a \text{ and } b \text{ are integers and } b \neq 0.\right\}$

Therefore, all integers are rational numbers. So are numbers such as $\frac{-2}{3}$, $\frac{7}{8}$, 0.75 or $\frac{75}{100}$, and $0.\overline{3}$ or $\frac{1}{3}$. *Recall that* $0.\overline{3} = 0.3333\ldots$

The decimal representation of a **rational number** is always a repeating decimal. Note that a decimal such as 0.5 is considered a repeating decimal since 0.5 is equivalent to $0.5\overline{0}$.

Examples

Give the decimal representations for $\frac{11}{8}$ and $\frac{1}{11}$.

To find the decimal representation, divide the numerator by the denominator.

1 $\frac{11}{8} \longrightarrow 8\overline{)11.0000}$ = 1.3750 . . . or 1.375

2 $\frac{1}{11} \longrightarrow 11\overline{)1.0000}$ = 0.0909 . . . or $0.\overline{09}$

Conversely, it is also the case that every repeating decimal represents a rational number. Example 3 illustrates a procedure for expressing repeating decimals in $\frac{a}{b}$ form. Note that the number you multiply n by depends on how many digits are in the repeating part of the decimal. For example, $0.0\overline{987}$ would call for n to be multiplied by 1000 because there are three digits in the repeating part of that decimal. Likewise, $2.0\overline{34782}$ would require multiplication by 100,000. The following statement summarizes how to determine what the multiplier of n will be.

The multiplier of n, where n is a repeating decimal, is 10^m where m represents the number of digits in the repeating part of the decimal.

Example

3 Show that $0.2\overline{54}$ represents a rational number.

$$\text{Let } n = 0.2\overline{54}.$$

$$100n = 25.4\overline{54}$$ *Multiply both sides by 100 because the repeating part of the decimal occupies two decimal places.*

Now subtract the first equation from the second equation.

$$\begin{aligned} 100n &= 25.4\overline{54} \\ n &= 0.2\overline{54} \\ \hline 99n &= 25.2 \\ n &= \frac{25.2}{99} \text{ or } \frac{252}{990} \text{ or } \frac{14}{55} \end{aligned}$$

Therefore, $0.2\overline{54} = \frac{252}{990}$ or $\frac{14}{55}$, which are both in the required $\frac{a}{b}$ form.

Examples 1 and 2 illustrate the following important equivalence.

A number is *rational* if and only if it can be represented as a repeating decimal.

Some numbers cannot be represented by repeating decimals. For example, $\sqrt{2}$, $\sqrt{7}$, and π have nonrepeating decimal representations. They are examples of *irrational numbers.* A number which is represented by a nonrepeating decimal is an **irrational number.**

The set consisting of all rational numbers together with all irrational numbers is called the set of **real numbers.** The symbol $\mathcal{R}$ names the set of real numbers. Unless otherwise specified, the replacement set for all variables in this chapter is $\mathcal{R}$.

The following examples review some important properties and applications of the set of real numbers.

Examples

4 Evaluate $3x^2 - 2y$ if $x = -4$ and $y = -5$.

$$\begin{aligned} 3x^2 - 2y &= 3(-4)^2 - 2(-5) \\ &= 3(16) - (-10) \quad \textit{Use the order of operations.} \\ &= 48 + 10 \\ &= 58 \end{aligned}$$

5 **Simplify $15x + 12y - 10x + 7y$.**

$$\begin{aligned} 15x + 12y - 10x + 7y &= 15x - 10x + 12y + 7y && \textit{Commutative property} \\ &= (15 - 10)x + (12 + 7)y && \textit{Distributive property is used to} \\ &= 5x + 19y && \textit{combine like terms.} \end{aligned}$$

Since $15x + 12y - 10x + 7y = 5x + 19y$ for all values of the variables, this equation is called an *identity*.

6 **Write an algebraic expression in simplest form to represent the perimeter of the rectangle.**

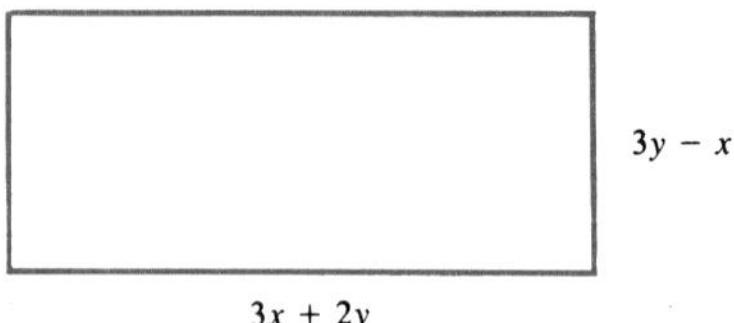

Recall that the formula for the perimeter, P, is $P = 2 \times \text{length} + 2 \times \text{width}$.

$$\begin{aligned} P &= 2(3x + 2y) + 2(3y - x) \\ &= 6x + 4y + 6y - 2x \\ &= 6x - 2x + 4y + 6y \\ &= 4x + 10y \end{aligned}$$

Exercises

Exploratory **Tell whether each statement is *true* or *false*. Justify your answer.**

1. Any integer is a whole number.
2. Any natural number is a whole number.
3. Any whole number is an integer.
4. Any integer is a rational number.
5. Any real number is rational.
6. Any irrational number is real.

Name the property of the real number system that justifies each statement.

7. $6 + \sqrt{2} = \sqrt{2} + 6$
8. $\sqrt{3} + (-\sqrt{3}) = 0$
9. $17 \cdot 1 = 17$
10. $\frac{3}{4} \cdot \frac{4}{3} = 1$
11. $\sqrt{3} \cdot \frac{1}{2} = \frac{1}{2} \cdot \sqrt{3}$
12. $\pi + 6 = 6 + \pi$
13. $4(2 + \pi) = 4(2) + 4(\pi)$
14. $8 + (3 + \sqrt{5}) = (8 + 3) + \sqrt{5}$

Written **Express each of the following as a decimal.**

1. $\frac{4}{5}$
2. $\frac{5}{12}$
3. $\frac{4}{9}$
4. $\frac{3}{11}$
5. $\frac{7}{4}$
6. $-2\frac{1}{3}$
7. $\frac{2}{15}$
8. $\frac{2}{7}$
9. 4%
10. 0.5%

Express each of the following in $\frac{a}{b}$ form.

11. -18
12. 0
13. $2\frac{1}{4}$
14. 0.8
15. 0.635
16. -3.4
17. $0.\overline{6}$
18. 0.4
19. $0.\overline{4}$
20. $0.\overline{25}$
21. $0.\overline{03}$
22. $0.\overline{41}$
23. $0.8\overline{43}$
24. $0.14\overline{7}$
25. $4.0\overline{3}$

Evaluate these expressions if $a = 2$, $b = -3$, and $c = \frac{1}{2}$.

26. $2a + b$
27. $3b - 4a$
28. $4c + a$
29. $a^2 + b$
30. $ab + b^2$
31. $a^3 + b^2$
32. $b^3 + a^2$
33. $ab - c$
34. $c(a + b)$
35. $ab(a + c)$
36. $c^2 + \frac{1}{a^2}$
37. $\frac{ab}{c}$
38. $4b^2 - 3b + 1$
39. $c^2 + \frac{5}{ab}$
40. $\frac{b - a}{c}$
41. $\frac{1 + \frac{1}{a}}{1 + c}$

Simplify.

42. $3x + 2 + 4x + 5$
43. $4y + 6 + 8y - 2$
44. $6a - 2 + 8a + 1$
45. $2 + 3m - 1 + m$
46. $-8 + 3x - 2 + 4x$
47. $2z - 10 - 4z + 10$
48. $-6x + 1 - 2x + 1$
49. $10x - 3y + 2x + 1$
50. $-4x + 3 + 2x - 4$
51. $2(3x - y) + 3(2x + y)$
52. $3(5a + 4b) + 8(2a - 3b)$
53. $4(2x - y) - 5(x - 2y)$
54. $2(7c - 3d) - 6(2d - 4c)$
55. $\frac{1}{4}(12 + 8a) - \frac{1}{4}(16a - 20)$
56. $\frac{2}{3}\left(\frac{1}{2}x + 3y\right) - \frac{1}{2}\left(\frac{2}{3}x - 4y\right)$
57. $\frac{1}{5}(10 - 15x) - \frac{5}{8}(-16 - 24x)$
58. $-\frac{5}{6}(-12 - 18y) - \frac{4}{5}(-20y + 5)$
59. $7(0.2m + 0.3n) - 5(0.6m - n)$
60. $9(0.6x - 0.2y) - 3(0.2x + 1.1y)$

Write an algebraic expression in simplest form for each phrase.

61. x decreased by 5
62. y increased by 11
63. twice the sum of x and 7
64. five times the sum of y and 4
65. eight more than two times x
66. six less than the product of 5 and x
67. the total cost of t tickets if one ticket costs \$1.68
68. the value in cents of d dimes and q quarters
69. the distance a car travels in h hours if its rate is 75 mph
70. the distance a car travels in 5 hours if its rate is x mph
71. the time it takes a car traveling 80 mph to go y miles
72. the perimeter of a rectangle with length $2x - y$ and width $3x - 4y$.
73. the area of a triangle if its altitude is 4 units longer than its base b
74. the perimeter of a rectangle if its length is 3 units more than twice its width w
75. the price in cents of y apples if x apples cost d dollars
76. the price in dollars of x pears if y pears cost c cents

Challenge **Simplify. Write your answers using decimals.**

1. $7\%x + \frac{1}{2}x + 2x$
2. $3\%a + 6\%b + \frac{3}{4}a$
3. $2\%\left(\frac{3}{4}x + \frac{1}{2}y\right) + y$

Which of the following systems are groups? Justify your answer.

4. $(\mathcal{W}, +)$
5. $(\mathcal{Z}, +)$
6. $(\mathcal{Q}, \cdot)$
7. $(\mathcal{R}, \cdot)$

Which of the following systems are fields? Justify your answer.

8. $(\mathcal{W}, +, \cdot)$
9. $(\mathcal{Z}, +, \cdot)$
10. $(\mathcal{Q}, +, \cdot)$
11. $(\mathcal{R}, +, \cdot)$

1.2 Equations and Inequalities

The following examples review methods of solving linear equations and inequalities. In each case, the domain of the variable is $\mathcal{R}$.

Examples

1 **Solve $4(2x - 3) + 3 = 4x - 29$.**

$4(2x - 3) + 3 = 4x - 29$
$8x - 12 + 3 = 4x - 29$ *Distributive Property*
$8x - 9 = 4x - 29$ *Simplify the left side.*
$8x = 4x - 20$ *Add 9 to each side.*
$4x = -20$ *Add $-4x$, or subtract $4x$.*
$x = -5$ *Multiply by $\frac{1}{4}$, or divide by 4.*

The solution set is $\{-5\}$.

2 **Solve and graph the solution set of $-2(y - 3) + 2 > 14$.**

$-2(y - 3) + 2 > 14$
$-2y + 6 + 2 > 14$ *Distributive Property*
$-2y + 8 > 14$ *Simplify.*
$-2y > 6$ *Subtract 8.*
$y < -3$ *Divide by -2.*

Now graph $y < -3$.

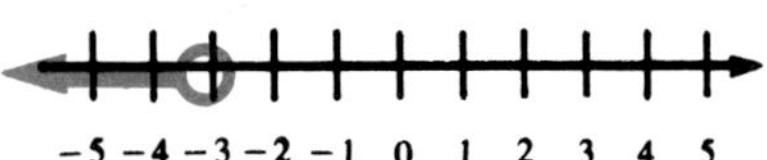

Remember that when you multiply or divide each side of an inequality by the same negative number, you must reverse the order of the inequality.

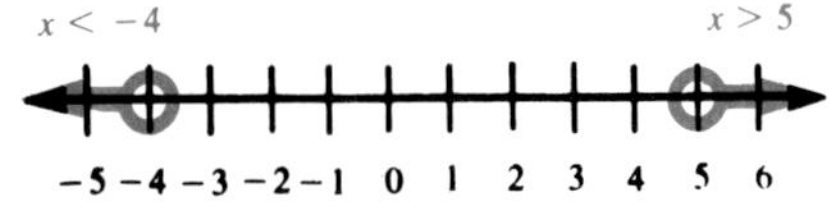

Recall that a **disjunction** ($\vee$) is an *or* sentence. It is true if one or both of its disjuncts are true. The graph of the solution set of $(x < -4) \vee (x > 5)$ is shown at the left.

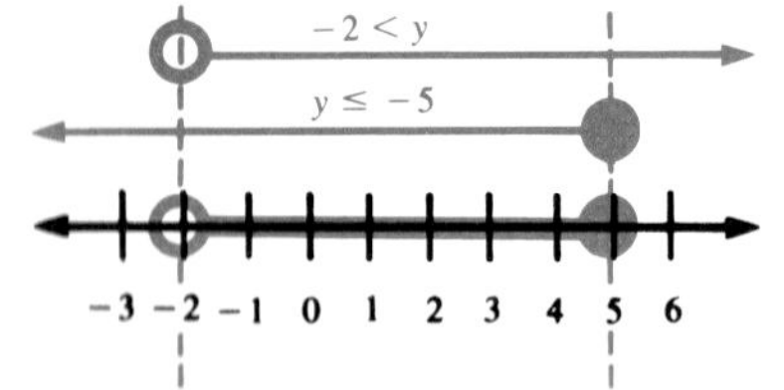

A **conjunction** ($\wedge$) is an *and* sentence. It is true if, and only if, both conjuncts are true. A sentence such as $-2 < y \leq 5$ is really a conjunction since it is equivalent to $(-2 < y) \wedge (y \leq 5)$. This graph is shown at the left.

Examples 3 and 4 demonstrate other graphs of disjunctions and conjunctions. Look for a correlation between the appearance of the graph and its type of connective.

Examples

3 **Solve and graph the solution set of $(2x - 1 < -7) \vee (-3x + 1 < -11)$.**

First solve each disjunct.

$$2x - 1 < -7 \quad \vee \quad -3x + 1 < -11$$
$$2x < -6 \quad \vee \quad -3x < -12$$
$$x < -3 \quad \vee \quad x > 4$$

Then graph the solution set.

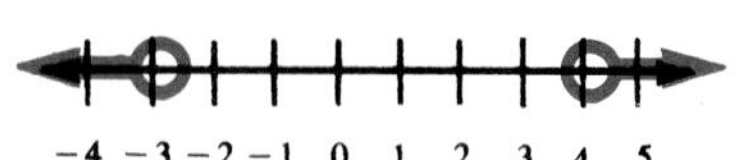

The solution set may be written in either of the following ways.

$\{x \in \mathcal{R}: (x < -3) \vee (x > 4)\}$ or {all real numbers less than -3 or greater than 4}

4 **Solve and graph the solution set of $4 \leq 2x - 6 < 8$.**

First, write the expression as a conjunction and solve each conjunct.

$$4 \leq 2x - 6 \quad \wedge \quad 2x - 6 < 8$$
$$10 \leq 2x \quad \wedge \quad 2x < 14$$
$$5 \leq x \quad \wedge \quad x < 7$$

Then graph the solution set.

The solution set may be written as {all real numbers greater than or equal to 5 and less than 7} or $\{x \in \mathcal{R}: 5 \leq x < 7\}$.

5 **In driving from Pine City to Eldonville, Sue drove at 70 km/h for 3 hours and then at 80 km/h for the rest of the trip. If the total distance was 322 km, how long did she drive at 80 km/h?**

Explore

We know that Sue drove a total distance of 322 km.
Since rate · time = distance, we also know that Sue drove (70 km/h) · (3 h) or 210 km at 70 km/h.
We want to know how long she drove at 80 km/h.

Let t = the number of hours driven at 80 km/h.
Then $80t$ = the distance driven at 80 km/h.

Plan

We can write an equation to represent the situation described in this problem.

$$\underbrace{\text{distance at 70 km/h}}_{210} + \underbrace{\text{distance at 80 km/h}}_{80t} = \underbrace{\text{total distance}}_{322}$$

Solve

$$80t = 112$$
$$t = 1.4$$

Sue drove 1.4 hours, or 1 hour and 24 minutes, at 80 km/h.

Examine

If Sue drove for 1.4 hours at 80 km/h, she would have traveled (80 km/h) · (1.4 hours) or 112 kilometers at 80 km/h.
Since $210 + 112 = 322$, the answer is correct.

Exercises

Exploratory **Tell whether each statement is *true* or *false*. Justify your answer.**

1. If $a = b$, then $ac = bc$.
2. If $a > b$, then $a + c > b + c$.
3. If $c < d$, then $ac < ad$.
4. If $a > b$ and $c > 0$, then $\frac{a}{c} > \frac{b}{c}$.
5. $(5 > 1) \vee (-3 < 0)$
6. $(5 > 1) \wedge (-3 < 0)$
7. $(4 > -1) \wedge (-5 > -2)$
8. $-4 \leq x < 7$ is equivalent to $(x \geq -4) \wedge (x < 7)$.

Graph each set on a number line.

9. $\{x \in \mathcal{R}: (x < -2) \vee (x > 5)\}$
10. $\{r \in \mathcal{Z}: (r \leq -1) \vee (r > 4)\}$

Written **Solve. Graph the solution sets of all inequalities.**

1. $3x - 1 = 5$
2. $7 + 4y = 19$
3. $1 + 2x = -3$
4. $4 + 6x = 4$
5. $16 = 3x + 1$
6. $-11 = 2y + 1$
7. $-3y - 1 = 20$
8. $-3 - 8y = -11$
9. $-12 = -11x + 10$
10. $\frac{2}{3}x = 14$
11. $\frac{4}{5}y + 1 = 9$
12. $-4 + \frac{3}{4}x = 5$
13. $5x + 7 = 2x + 13$
14. $3x + 5 = 9x + 2$
15. $3y - 4 = 7y + 12$
16. $2y - 8 = 14 - 9y$
17. $3x - 4 = 13x + 6$
18. $5m - 1 = 13 - 2m$
19. $32 + 5t = 8 - t$
20. $-4 - 3b = 12 - b$
21. $8x - 3 = 5(2x + 1)$
22. $-9 + 6b - b = 2(6 + b)$
23. $2(1 + 2y) = 5 - 2y$
24. $2(3m - 5) + 4 = 44$
25. $2(x + 10) = 55 - 3x$
26. $3m - 3(2m - 3) = 29 + 2m$
27. $\frac{3}{8} - \frac{1}{4}x = \frac{1}{16}$
28. $3x + 1 > 16$
29. $\frac{3}{4}x \leq \frac{3}{16}$
30. $-5y < 60$
31. $-7m + 2 < 23$
32. $-3x + 7 > 43$
33. $7 - 6m \leq 29$
34. $3 - 2x \geq 0$
35. $3y + 1 < y + 5$
36. $9(x + 2) \geq 72$
37. $-\frac{4}{5}y \geq -\frac{2}{15}$
38. $2(x + 10) \geq 55 - 3x$
39. $14 - 2x \leq 4 - x$
40. $5p - (4p - 5) = 3(2p - 6)$
41. $24 - 2(6 - 2x) < 4(x - 1)$
42. $4(2x - 6) = 8(x - 5)$
43. $6 - (2x - 4) = 2(5 - x)$
44. $a - 2(3a - 1) < 4 - 5a$
45. $25x - 2(12x - 6) \leq 16 + x - 4$
46. $\frac{4}{3}x - \frac{5}{9} > \frac{x}{2} + \frac{5}{6}$
47. $2 - 3(2x - 6) = 4\left(2x - \frac{1}{4}\right)$
48. $(x < -4) \vee (x + 1 > 3)$
49. $(x \geq -3) \wedge (x - 2 < 6)$
50. $(x - 4 \leq 2) \wedge (x - 2 > -5)$
51. $(y - 4 \geq 2) \vee (y + 1 < 1)$
52. $(8 < -y) \vee (1 - y > 2)$
53. $-3 < x + 4 < 5$
54. $4 \leq t + 6 < 10$
55. $(2x + 1 < -3) \vee (3x - 2 > 4)$
56. $(2y + 2 > 5 - y) \vee (2y - 2 < 6)$
57. $3 \leq 2w - 3 \leq 5$
58. $3 \leq 6 + 3y < 12$
59. $(x - 4 < 1) \vee (x + 2 > 1)$
60. $(y + 6 > -1) \vee (y - 2 < 4)$
61. $(5b < 9 + 2b) \vee (9 - 2b > 11)$
62. $0 \leq 4 - 3m \leq 12$
63. $2y - 2 \leq 3y \leq 4y - 1$

Solve algebraically.

64. A number decreased by 89 is 29. Find the number.

65. The sum of twice a number and 3 is 49. Find the number.

66. Eighty-seven increased by three times a number is 165. Find the number.

67. Thirty-two decreased by twice a number is 18. Find the number.

68. The sum of three consecutive integers is -30. Find the greatest integer.

69. The sum of three consecutive integers is 0. Find the integers.

70. The length of a rectangle is 7 cm less than twice its width. The perimeter is 148 cm. Find the length and width.

71. The width of a rectangle is 5 m more than half its length. The perimeter is 286 m. Find the length and width.

72. The Pine Car Rental Plan is \$18.95 a day plus 16¢ a mile. The Cone Car Rental Plan is \$15.95 a day plus 19¢ a mile. How many miles would you have to drive for the Pine Plan to cost less than the Cone Plan?

73. Harry's Rent-A-Car offers two plans. The first plan costs \$111 for 6 days with unlimited mileage. The second plan costs \$11.95 a day plus 15¢ a mile. When is the second plan the better buy?

74. Two trains leave Pine City Central Station and travel in opposite directions. After 11 hours they are 1485 km apart. The rate of one train is 15 km more per hour than the rate of the other. Find both rates.

75. Zeb drove at 50 mph for 1.5 hours. He drove 55 mph for the rest of the trip. If he drove a total of 174 miles, how long did he drive at 55 mph?

76. Taka is driving to Pine City, a distance of 662 miles. If she drives 55 mph for 5 hours, at what speed must she travel to complete the total trip in 14 hours?

77. Two hours after a truck leaves Pine City traveling at 45 mph, a car leaves to overtake the truck. If it takes the car 10 hours to reach the truck, what was the car's speed?

78. Can 90 stamps of 15¢ and 25¢ denominations cost \$14.75? Explain your answer.

79. Can two consecutive positive integers have a difference of 28? Explain.

80. Is it possible to find two positive integers whose sum is even and whose difference is odd? Explain.

81. Find the greatest possible length of the side of a square whose perimeter is at most 64 cm.

82. Is it possible to have a total of \$4.25 from a group of 12 coins composed solely of nickels and quarters? Explain.

83. Is it possible to have a total of \$3.45 from a group of 25 coins composed solely of nickels and quarters? Explain.

84. Find four consecutive integers such that the sum of the second and twice the third is 68.

85. Four times a number decreased by twice the number is 100. Find the number.

86. Twice a number increased by 12 is 31 less than three times the number. Find the number.

87. Find two consecutive integers such that twice the lesser integer, increased by the greater integer, is 50.

88. Shan Wong's car used 25 gallons of gasoline to travel 350 miles. How much gasoline will his car use to travel 462 miles?

89. Andrea Jonas drove 244 kilometers in four hours. At that rate, how long will it take her to drive 366 kilometers?

90. Ron Dorris is on his way to San Diego, 300 miles away. He drives 45 mph for three hours. He drives 55 mph for the rest of the trip. How long does Mr. Dorris drive at 55 mph?

91. An express train leaves Cedarville at the same time a passenger train leaves Pine City, 470 miles away. The express train, traveling 10 mph faster, passes the other train in 2.5 hours. Find the speed of each train.

1.3 Absolute Value

The idea of absolute value is important in many applications of mathematics. The absolute value of x is written in symbols as $|x|$.

Geometric Definition of Absolute Value

> The **absolute value** of a real number is the number of units it is from 0 on the number line.

Examples

Find the absolute value of 3 and of −4.

1

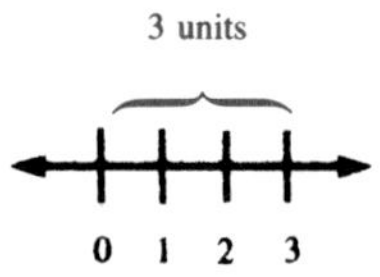

Since the distance from 0 to 3 is 3 units, $|3| = 3$.

2

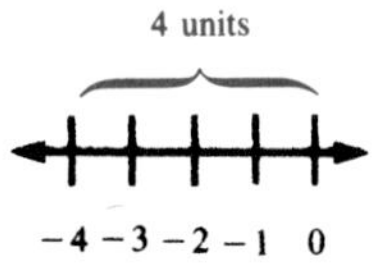

Since the distance from 0 to −4 is 4 units, $|-4| = 4$.

The absolute value can also be found using the following definition.

Algebraic Definition of Absolute Value

> **For any real number x,**
> **if x is positive or zero, then $|x| = x$;**
> **if x is negative, then $|x| = -x$.**

Examples

3 Solve $|x| = 7$.

The equation states that the absolute value of x is 7. The only possible replacements for x are 7 and −7.

The solution can be written in the following manner.

$|x| = 7$	*Note that $|x| = 7$ is equivalent to a disjunction.*
$x = 7 \quad \vee \quad x = -7$	*The solution set is $\{7, -7\}$.*

4 **Solve $|3x - 4| = 8$.**

$$|3x - 4| = 8$$

$$3x - 4 = 8 \quad \vee \quad 3x - 4 = -8 \qquad \textit{Definition of Absolute Value}$$

$$3x = 12 \quad \vee \quad 3x = -4$$

$$x = 4 \quad \vee \quad x = -\frac{4}{3}$$

Check $x = 4$.

$$|3x - 4| = 8$$
$$|3(4) - 4| \stackrel{?}{=} 8$$
$$|12 - 4| \stackrel{?}{=} 8$$
$$8 = 8$$

Check $x = -\frac{4}{3}$.

$$\left|3\left(-\frac{4}{3}\right) - 4\right| \stackrel{?}{=} 8$$
$$|-4 - 4| \stackrel{?}{=} 8$$
$$8 = 8$$

The solutions are 4 or $-\frac{4}{3}$. *The solution set is $\left\{4, -\frac{4}{3}\right\}$.*

5 **Solve $|x + 1| = -9$.**

The equation states that the absolute value of $x + 1$ is *negative* 9. No value of x can make this equation true. Thus, the equation has no solution. *The solution set is { } or $\varnothing$.*

6 **Solve and graph the solution set of $|x| < 3$.**

$|x| < 3$ means that the distance from 0 to x is less than 3 units. Thus, a solution for $|x| < 3$ would be any number that lies between -3 and 3. This is shown in the graph.

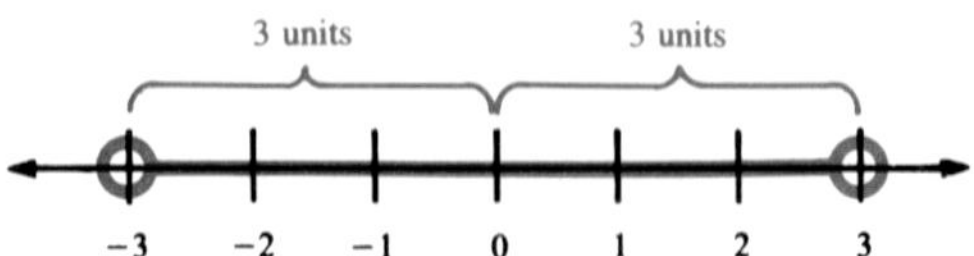

The solution set is $\{x \in \mathcal{R}: -3 < x < 3\}$.

Note that $|x| < 3$ is equivalent to the conjunction $-3 < x < 3$.

7 **Solve and graph the solution set of $|x| > 3$.**

$|x| > 3$ means that the distance from 0 to x is greater than 3 units. Thus, a solution for $|x| > 3$ would be any number that lies more than 3 units from 0, as shown in the graph.

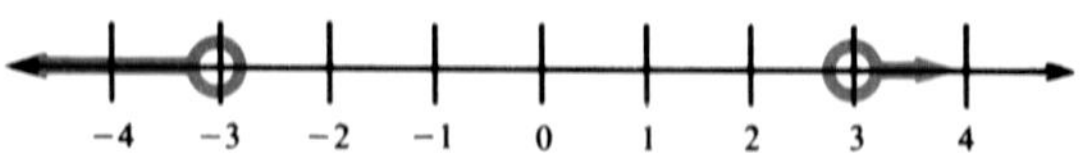

The solution set is $\{x \in \mathcal{R}: (x < -3) \vee (x > 3)\}$.

Note that $|x| > 3$ is equivalent to the disjunction $(x < -3) \vee (x > 3)$.

8 **Solve and graph the solution set of $|2x + 5| \geq 11$.**

$|2x + 5| \geq 11$ means that $2x + 5$ is 11 or more units from 0.

$$2x + 5 \leq -11 \quad \vee \quad 2x + 5 \geq 11$$
$$2x \leq -16 \quad \vee \quad 2x \geq 6$$
$$x \leq -8 \quad \vee \quad x \geq 3$$

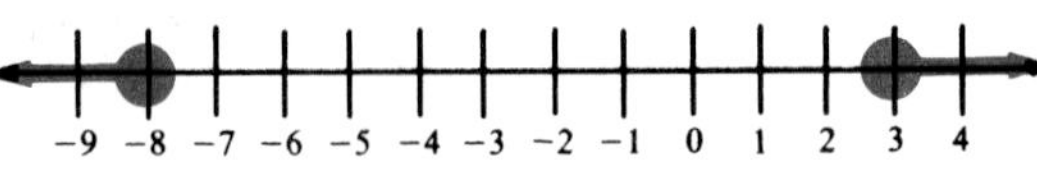

The solution set is $\{x \in \mathcal{R}: (x \leq -8) \vee (x \geq 3)\}$.

9 **Solve and graph the solution set of $|2x - 5| < 9$.**

$|2x - 5| < 9$ means that $2x - 5$ is less than 9 units from 0.

$$-9 < 2x - 5 < 9$$
$$-4 < 2x < 14$$
$$-2 < x < 7$$

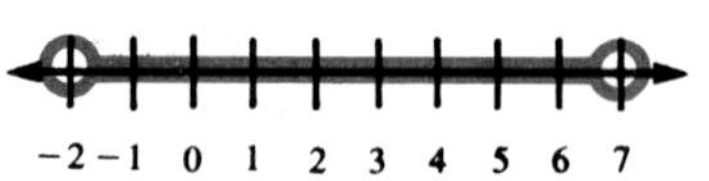

The solution set is $\{x \in \mathcal{R}: -2 < x < 7\}$.

Exercises

Exploratory **Write each of the following as a disjunction or conjunction.**

1. $|x| = 7$
2. $|x| > 5$
3. $|x| < 10$
4. $|x| < 4$
5. $|x| = 11$
6. $|x| \leq 9$
7. $|x| \geq 15$
8. $|x - 1| > 8$

Written **Solve each of the following. Graph solution sets of inequalities.**

1. $|y| = 7$
2. $|x| = \frac{3}{4}$
3. $|y| = -6$
4. $|m| + 1 = 3$
5. $|y| - 3 = 5$
6. $6 + |x| = 5$
7. $|x + 1| = 4$
8. $|y - 4| = 7$
9. $|3 - m| = 4$
10. $|m - 4| = -2$
11. $|m + 6| = 19$
12. $|y| = 0$
13. $|x - 4| = -3$
14. $|2y + 7| = 19$
15. $|3x + 12| = 48$
16. $|5x + 30| = 65$
17. $|2x - 37| = 15$
18. $5|x + 4| = 45$
19. $8|x - 3| = 88$
20. $3|x + 7| = 36$
21. $|2x - 3| = 0$
22. $5|x - 2| = 10$
23. $|x + 1| + 4 = 0$
24. $|2m + 3| - 9 = 0$
25. $|x| > 9$
26. $|y| > 5$
27. $|t| < 7$
28. $|y| < 4$
29. $|m| \geq 3$
30. $|n| \geq 4$
31. $|2y| < 8$
32. $|5x| \leq 15$
33. $|3t| \geq 27$
34. $|y| > -5$
35. $|t| \geq -1$
36. $|x + 1| > 4$
37. $|x + 1| \geq 4$
38. $|x + 2| > 5$
39. $|y - 6| < 4$
40. $|x + 1| \leq 3$
41. $|y - 9| > 5$
42. $|y - 4| \leq -12$
43. $|2x - 9| \leq 27$
44. $|2x - 5| \leq 7$
45. $6 + |3x| > 0$
46. $|3x + 7| \leq 26$
47. $|5 - 2p| > 3$
48. $|7 - 3y| \leq 26$
49. $|9 - 2n| \leq 1$
50. $|3 - 3s| < 27$
51. $|1 + 2(x - 3)| > 7$
52. $|2 - 3(2x - 1)| \leq 7$
53. $|6x + 25| + 14 < 6$
54. $2 + |3 - 2x| > 0$

Challenge **Solve.**

1. $|y| - 4 = 3y$
2. $|x| = -x$
3. $|x| < x$
4. $|m - 1| = 3m$
5. $|5 + 2y| = 1 + y$
6. $|x + 3| + |x - 3| > 8$

1.4 Polynomials

To solve more complicated equations, it is necessary to review some skills involving polynomials. The building blocks of polynomials are monomials.

A **monomial** or term is a numeral, a variable, or a product of a numerical factor and one or more variables. Examples of monomials are 15, y, $-8xy$, $5x^3$, and $-7x^2y^3$.

The numerical factor is called the **coefficient.** The **degree** of a monomial is the sum of the exponents of its variables. Two or more monomials are called *like monomials,* or *like terms,* if they are the same or they differ only in their coefficients.

Examples

1 The following demonstrate the vocabulary associated with monomials.

a. The coefficient of $-8xy$ is -8.

b. $-8xy$ has degree 2 since x and y each have exponent 1.

c. The degree of $7x^2y^3$ is 5.

d. $-9x^3y$ and $12x^3y$ are like monomials, or like terms.

e. $7c^2d$ and $9cd^2$ are unlike monomials, or unlike terms.

The distributive property allows us to simplify the sum or difference of like monomials. Sometimes this is called combining like terms.

2 Simplify $2a^2b - 5a + 6a^2b + 7a$.

$$\begin{aligned} 2a^2b - 5a + 6a^2b + 7a &= 2a^2b + 6a^2b - 5a + 7a \\ &= (2 + 6)a^2b + (-5 + 7)a \\ &= 8a^2b + 2a \end{aligned}$$

To multiply monomials, recall the laws of exponents.

Laws of Exponents

For all non-zero real numbers x and y and for all integers m and n, the following are true.

$$\begin{aligned} x^m \cdot x^n &= x^{m+n} \\ (x^m)^n &= x^{mn} \\ (xy)^m &= x^m y^m \\ \frac{x^m}{x^n} &= x^{m-n} \\ x^0 &= 1 \end{aligned}$$

0^0 is not defined.

Examples

3 **Simplify $(2x^3)^4(y^3)^2$.**

$$(2x^3)^4(y^3)^2 = 2^4 \cdot (x^3)^4 \cdot (y^3)^2$$
$$= 16x^{12}y^6$$

4 **Simplify $\frac{(x^2y^4)^5}{(x^3y)^6}$.**

$$\frac{(x^2y^4)^5}{(x^3y)^6} = \frac{x^{10}y^{20}}{x^{18}y^6}$$
$$= x^{10-18}y^{20-6}$$
$$= x^{-8}y^{14} \text{ or } \frac{y^{14}}{x^8}$$

Recall that $x^{-n} = \frac{1}{x^n}$.

A **polynomial** is a monomial or the sum of two or more monomials. The monomials are called the *terms* of the polynomial. A polynomial is *simplified* or in *simplest* form when no two of its terms are like terms. A polynomial in simplest form can be classified by the number of its terms. A polynomial containing two terms is a **binomial;** one with three terms is a **trinomial.** The **degree of a polynomial** is the degree of the monomial with the greatest degree. When a polynomial is arranged in descending order of degree, the polynomial is in *standard form.*

Examples

5 **These expressions demonstrate binomials, trinomials, and degree.**

a. $x + 5$ is a binomial.

b. $x^2 + 4x - 6$ is a trinomial.

c. The degree of $5xy^2 + 2x - 6$ is 3. *The degree of* $5xy^2$ *is 3.*

d. $3y^2 - 6y + 5$ is in standard form. $-6y + 3y^2 + 5$ *is not in standard form.*

6 **Use the distributive property to multiply $4x(2x - 5xy)$.**

$$4x(2x - 5xy) = 4x \cdot 2x - 4x \cdot 5xy$$
$$= 8x^2 - 20x^2y$$

7 **Simplify $(2x + 5)(x + 3)$.**

Method 1: Horizontal Arrangement

$$(2x + 5)(x + 3) = 2x(x + 3) + 5(x + 3)$$
$$= 2x^2 + 6x + 5x + 15$$
$$= 2x^2 + 11x + 15$$

Method 2: Vertical Arrangement

Rewrite $(2x + 5)(x + 3)$ as a multiplicand and a multiplier. Multiply $2x + 5$ by 3 and then by x, aligning like terms for easier addition.

$$\begin{array}{rrrl} 2x & + \ 5 & & \\ x & + \ 3 & & \\ \hline 2x^2 & + \ 5x & & \text{Product of } x(2x+5) \\ & + \ 6x & + \ 15 & \text{Product of } 3(2x+5) \\ \hline 2x^2 & + \ 11x & + \ 15 & \text{Sum of the two products} \end{array}$$

Method 3: FOIL *This method is most efficient with products of binomials.*

First *Last*

$$(2x + 5)(x + 3) = \underbrace{2x \cdot x}_{\text{First}} + \underbrace{2x \cdot 3}_{\text{Outer}} + \underbrace{5 \cdot x}_{\text{Inner}} + \underbrace{5 \cdot 3}_{\text{Last}}$$

Inner

Outer

$$= 2x^2 + 6x + 5x + 15$$
$$= 2x^2 + 11x + 15$$

F *Multiply the first terms of each binomial.*
O *Multiply the outer terms.*
I *Multiply the inner terms.*
L *Multiply the last terms.*

8 **Simplify $(3x - 4)(x^2 - 5x + 3)$.**

Method 1

$$\begin{aligned}(3x - 4)(x^2 - 5x + 3) &= 3x(x^2 - 5x + 3) - 4(x^2 - 5x + 3) \\ &= 3x^3 - 15x^2 + 9x - 4x^2 + 20x - 12 \\ &= 3x^3 - 19x^2 + 29x - 12\end{aligned}$$

Method 2

$$(3x - 4)(x^2 - 5x + 3) \longrightarrow \begin{array}{rrrr} & x^2 & - \ 5x & + \ 3 \\ & & 3x & - \ 4 \\ \hline 3x^3 & - \ 15x^2 & + \ 9x & \\ & - \ 4x^2 & + \ 20x & - \ 12 \\ \hline 3x^3 & - \ 19x^2 & + \ 29x & - \ 12 \end{array}$$

A polynomial in one variable of the form $ax + b$, where $a \neq 0$, is called a **linear polynomial.** A polynomial of the form $ax^2 + bx + c$, where $a \neq 0$, is called a **quadratic** polynomial.

Exercises

Exploratory **Tell whether each of the following is a polynomial. If it is, classify it as a monomial, binomial, or trinomial; write it in standard form; and give its degree.**

1. $15m^2n^2 + 3mn^2$ **2.** $r + 6r^2 - 1$ **3.** $4 - 16t^2$ **4.** $\sqrt{x}$
5. $20mn^2t^3$ **6.** $\frac{x}{3} + x^5$ **7.** $|x| + 4x^2$ **8.** $4y - 3y^5 + 2$

Written **Simplify.**

1. $7x - 2y + 9x + 4y$
2. $5xy + 6x - 8xy$
3. $4a^2 - 2a - 11a^2$
4. $5m^2 - 2n + 3m^2$
5. $2a - 4ab + a$
6. $8 - 9x^2 + 2$
7. $-6a + 2b^2 + a$
8. $4xy^2 - 2 + xy^2$
9. $4 - 2(x + 4)$
10. $5a + 2b - 3a + 7b$
11. $2x^2 - 3x^2$
12. $(x^4)(x^3)$
13. $5a^2 - 11b^2 + 7a^2$
14. $(x^5)^2$
15. $(xy)^2$
16. $5xy - y + 3xy$
17. $(2a)^3$
18. $(-3y)^2$
19. $4a^2b + 4 - 5a^2b$
20. $(4mn)(-3m)$
21. $(-5m^2n)^3$
22. $\frac{m^5}{m^2}$
23. $\frac{x^{10}y^3}{x^4y}$
24. $\frac{y^{12}z^8}{y^7z^7}$
25. $\frac{6a^2}{12a}$
26. $\frac{56x^2y^2}{72x^5}$
27. $\frac{40a^3x^4}{60a^2x^5}$
28. $\frac{x^4y^3z^5}{x^3y^2z^5}$
29. $\frac{9abx^2}{12abx^3}$
30. $\frac{-48ab^2c^3}{64ab^3c^2}$
31. $(-x^2y^3)^3(-2xy)^5$
32. $(-2xy)^2(x^3y)$
33. $(2a + 3)^2$
34. $-3 + t^2 - 5t + 7 - 3t^3 + 4t^2$
35. $q^2 - 2q(q + 4) + 7$
36. $8x^3 - 2x^2 + 5 + 11x^3 + 7 - 3x^2$
37. $y^2 + 3y(y - 2) + 2$
38. $5m^2 + 2(m^2 - 1) - 2(m^2 - 3)$
39. $(2a^2 + 4a) - (a^2 + 5a)$
40. $3x^2(5 - 3x) - 4(x^2 - 7x^3)$
41. $(2x + 1)(x + 3)$
42. $(7m - 3n)(-2m + 6n)$
43. $(6a + 2b)(6a - 2b)$
44. $(3m - 1)(m + 2)$
45. $(y - 6)(4y - 7)$
46. $(2t - 5)(3t + 4)$
47. $(2a + b)(a - b)$
48. $(3x - 1)^2$
49. $(2q - 7)^2$
50. $(x - 1)(x + 1)$
51. $(3m + 2)(3m - 2)$
52. $(2a + 3)^3$
53. $(x + 1)(x^2 + 2x + 3)$
54. $(2x - 1)(x^2 - 3x + 5)$
55. $(2y - 3)(y^2 - 3y - 8)$
56. $(3t^2 - 4t - 6)(t + 4)$
57. $(2s + 1)(s - 1)(s + 4)$
58. $(x - 1)(x + 1)(2x - 3)$
59. $(a + 3)(a^2 - 3a + 9)$
60. $(m + 4)(m^2 - 4m + 16)$
61. $(a + b)(a^2 - ab + b^2)$
62. $(x + y)(x^2 - xy + y^2)$
63. $(a - 3)(a^2 + 3a + 9)$
64. $(m - 4)(m^2 + 4m + 16)$
65. $(a - b)(a^2 + ab + b^2)$
66. $(x - y)(x^2 + xy + y^2)$

Challenge **Simplify.**

1. $[(a + b)(a - b)](a^2 + b^2)$
2. $(m + 3)^2(m - 1)$
3. $(2x - 1)(x + 3)(x - 3)$
4. $(a + b + c)(a - b - c)$

1.5 Factoring

Factoring is a common method used to solve polynomial equations. Later in this chapter other important uses of factoring will be presented.

Examples

1 **Factor $6x^2 + 8x$.**

First, find the greatest common factor (GCF) of the terms of the polynomial. The GCF of $6x^2$ and $8x$ is $2x$. Then, use the distributive property.

$$6x^2 + 8x = 2x(3x + 4)$$ *$2x(3x + 4)$ is the factored form of $6x^2 + 8x$.*

2 **Factor $18a^3bc^2 - 27a^2b^3 + 24a^2bc$.**

The GCF of the terms is $3a^2b$.

$$18a^3bc^2 - 27a^2b^3 + 24a^2bc = 3a^2b(6ac^2 - 9b^2 + 8c)$$

Sometimes a polynomial can be recognized as one of the following special products.

Difference of Two Squares

$$(a + b)(a - b) = a^2 - b^2$$

Perfect Trinomial Squares

$$(a + b)^2 = a^2 + 2ab + b^2$$
$$(a - b)^2 = a^2 - 2ab + b^2$$

Examples

3 **Factor $16y^2 - 1$**

$$16y^2 - 1 = (4y + 1)(4y - 1)$$ *Difference of Two Squares*

4 **Factor $9m^2 - 12m + 4$.**

$$9m^2 - 12m + 4 = (3m - 2)^2$$ *Perfect Trinomial Square*

Often a trinomial can be factored into a product of binomials by reversing the FOIL method.

Examples

5 **Factor $2m^2 - 5m - 3$.**

We must fill the parentheses. $2m^2 - 5m - 3 = (\quad)(\quad)$

The first terms must be $2m$ and m. The last terms must be either 1 and -3 or -1 and 3. These are the possible factors.

$(2m + 1)(m - 3)$
$(2m - 3)(m + 1)$
$(2m - 1)(m + 3)$
$(2m + 3)(m - 1)$

The only pair of factors that will produce the correct middle term $-5m$ is

$(2m + 1)(m - 3)$.

6 **Factor $2x^2 - 8$.**

First, factor out the GCF, which is 2.

$2x^2 - 8 = 2(x^2 - 4)$ *Note that $x^2 - 4$ is the difference of two squares so we must factor again.*
$2x^2 - 8 = 2(x + 2)(x - 2)$

7 **Factor $6y^3 + 14y^2 - 12y$.**

$6y^3 + 14y^2 - 12y = 2y(3y^2 + 7y - 6)$ *Factor out the GCF.*
$= 2y(3y - 2)(y + 3)$ *Factor the trinomial.*

8 **Use factoring to solve $x^2 = 3x$.**

$x^2 = 3x$
$x^2 - 3x = 0$ *Write equation in standard form.*
$x(x - 3) = 0$ *Factor.*
$x = 0 \lor x - 3 = 0$ *If $ab = 0$, then $a = 0$ or $b = 0$.*
$x = 0 \lor x = 3$ *The solution set is $\{0, 3\}$.*

9 **Solve $m^2 + 4m = -4$.**

$m^2 + 4m + 4 = 0$ *Express in standard form by adding 4.*
$(m + 2)(m + 2) = 0$ *Factor the trinomial.*
$m + 2 = 0$ *Because both factors are the same, only one equation is needed.*
$m = -2$ *The solution set is $\{-2\}$.*

10 **Solve $2x^3 + 4x^2 = 6x$.**

$2x^3 + 4x^2 = 6x$
$2x^3 + 4x^2 - 6x = 0$ *Express in standard form by subtracting 6x.*
$2x(x^2 + 2x - 3) = 0$ *Factor out the GCF.*
$2x(x + 3)(x - 1) = 0$ *Factor the trinomial.*
$2x = 0 \lor x + 3 = 0 \lor x - 1 = 0$ *Set each factor equal to 0.*
$x = 0 \lor x = -3 \lor x = 1$ *The solution set is $\{-3, 0, 1\}$.*

Of course, any quadratic equation can be solved by using the quadratic formula.

Quadratic Formula

The solution of any quadratic equation in the form of $ax^2 + bx + c = 0$, $a \neq 0$, is given by

$$x = \frac{-b \pm \sqrt{b^2 - 4ac}}{2a}.$$

For instance, if the equation in example 8 is used, the quadratic equation becomes $x^2 - 3x = 0$. When the values for a, b, and c are substituted into the formula, the solutions 0 and 3 are easily obtained.

Exercises

Exploratory **Factor.**

1. $8x + 24$
2. $6x - 18$
3. $3x^2 + x$
4. $5y - y^2$
5. $5x^2 - 25$
6. $14b - 2b^2$
7. $x^2 - 9$
8. $1 - t^2$
9. $m^2 - 25$
10. $y^3 - y^2$
11. $r^2 - 49$
12. $4t^2 - 16$
13. $x^2 + 6x + 9$
14. $t^2 - 8t + 16$
15. $y^2 - 7y + 10$
16. $4 - 4c + c^2$

Written **Factor.**

1. $4y + 12y^2$
2. $2x^2 - 2$
3. $3s^3 - 27s$
4. $rs - ts$
5. $12 - 4y - 3x$
6. $16xy^2 - 8x^2y$
7. $b^2 - 144$
8. $3d^2 - 48$
9. $7pm + 2p^2 - 14px$
10. $5x^2y - 10xy^2$
11. $r^4 + r^3s + r^2s^2$
12. $36y^2 - 100$
13. $12x^2 + 3xy - 9$
14. $m^2 - 7m + 10$
15. $p^2 - 5p + 4$
16. $b^2 + 7b + 6$
17. $2a^2 + 10a + 12$
18. $3y^2 + 5y + 2$
19. $4x^2 + 11x + 6$
20. $4m^2 - 20m + 21$
21. $xy^3 - 9xy$
22. $2x^2 + 24x + 72$
23. $x^2y + 6xy + 9y$
24. $6a^2 + 27a - 15$
25. $r^3 - r^2 - 30r$
26. $y^4 - 1$
27. $16 - 4m^2$
28. $m^5 - 25m^7$
29. $2x^3 - 9x^2 - 5x$
30. $80 - 5y^2$
31. $64 - 4t^2$
32. $m^4 - 16$
33. $9x^2 - 9y^2$
34. $6x^2 - 5 - 13x$
35. $6y^2 + 10y - 4$
36. $3x^4 - x^2 - 2$

Solve each equation.

37. $x^2 = 4x$
38. $4x^2 - 1 = 0$
39. $x^2 - x = 0$
40. $x^3 - x^2 = 0$
41. $x^2 + x = 30$
42. $m^2 + 6m = 27$
43. $y^2 - y = 12$
44. $2x^2 + 9x + 4 = 0$
45. $2x^2 + 5x + 3 = 0$
46. $4x^2 + 9 = 12x$
47. $9x^4 - x^2 = 0$
48. $12x^3 + 10x^2 - 8x = 0$

Solve each equation using the quadratic formula.

49. $x^2 - 3x - 2 = 0$
50. $x^2 - 5x = 2$
51. $x^2 - 8x = -14$
52. $x^2 + 12x = -4$
53. $8x^2 = 60$
54. $2x^2 = 7 - 3x$

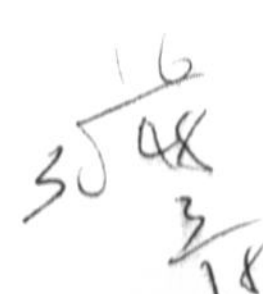

Solve.

55. Find two consecutive integers whose product is 182.

56. The area of a parallelogram is 110 m^2. Find the lengths of a base and altitude to it if the base is 2 m longer than 4 times the altitude.

57. The sum of the squares of three consecutive integers is 194. Find the integers.

58. The sum of the squares of two consecutive even integers is 244. Find the integers.

Find the dimensions of the rectangles that satisfy each set of conditions.

59. The area is 65 cm^2. The length is 3 cm more than twice the width.

60. The area is 195 m^2. The measures of the dimensions are consecutive odd integers.

61. A rectangular garden, 20 m by 40 m, is surrounded by a walk of uniform width. The area of the walk is 1216 m^2. Find the width of the walk.

62. The walk around the Pine City duck pond is 3 m wide. The area of the walk is 450 m^2. If the pond is 9 m longer than wide, what are its dimensions?

63. A local park is 30 meters long by 20 meters wide. Plans are being made to double the area by adding a strip at one end and another of the same width on one side. Find the width of the strips.

64. The length of Hillcrest Garden is 6 feet more than its width. A walkway 3 feet wide surrounds the outside of the garden. The total area of the walkway is 288 square feet. Find the dimensions of the garden.

65. The Hillside Garden Club wants to double the area of its rectangular display of roses. If it is now 6 meters by 4 meters, by what equal amount must each dimension be increased?

66. A rectangular garden 25 feet by 50 feet is increased on all sides by the same amount. Its area increases 400 square feet. By how much is each dimension increased?

Challenge **Factor.**

1. $x^{2b+1} - x^b$ **2.** $y^{b+3} + 4y^b$ **3.** $x^{2b} - 1$ **4.** $x^{2b+2} - x^{2b}$

5. Recall the meaning of the symbol ${}_nC_r$. Find n if ${}_nC_2 = 28$.

Mathematical Excursions

Sums and Differences of Cubes

Two special products involving cubes are shown below.

Sum of Cubes $(a + b)(a^2 - ab + b^2) = a^3 + b^3$

Difference of Cubes $(a - b)(a^2 + ab + b^2) = a^3 - b^3$

Exercises **Using the example above, factor the following.**

1. $x^3 + 27$ **2.** $x^3 + 8$ **3.** $y^3 + 64$ **4.** $y^3 - 8$

5. $8a^3 + 1$ **6.** $8 - x^3$ **7.** $27a^3 - b^3$ **8.** $7n^3 - 7$

1.6 Simplifying Rational Expressions

An equation such as $\frac{x}{6} = \frac{6}{x + 5}$ can be solved by using the Cross Multiplication Principle. But it is not so easy to solve equations such as $y + \frac{y}{y - 1} = \frac{4y - 3}{y - 1}$. Before you learn how to solve such equations, it is necessary to become very familiar with a new type of algebraic expression.

A *rational expression* is a fraction whose numerator and denominator are polynomials. Here are some rational expressions.

$$\frac{x}{6} \qquad \frac{6}{x + 5} \qquad \frac{t^2}{t - 1} \qquad \frac{m^2 - 36}{2m^2 + 3m + 1}$$

Just as with rational numbers, the denominator of a rational expression cannot be zero. Therefore, we often must restrict the values of the variables that appear in the denominator.

Examples

1 For what values of m is $\frac{m^2 - 36}{2m^2 + 3m + 1}$ undefined?

The expression is undefined for those values of m that make the denominator, $2m^2 + 3m + 1$, equal to zero.

$$2m^2 + 3m + 1 = 0$$
$$(2m + 1)(m + 1) = 0$$
$$2m + 1 = 0 \quad \vee \quad m + 1 = 0$$
$$m = -\frac{1}{2} \quad \vee \quad m = -1$$

Thus, the expression $\frac{m^2 - 36}{2m^2 + 3m + 1}$ is undefined when $m = -\frac{1}{2}$ or $m = -1$.

2 For what values of x is $\frac{x^2 + 8x + 7}{3x^3 - 12x}$ undefined?

$$3x^3 - 12x = 0$$
$$3x(x^2 - 4) = 0$$
$$3x(x + 2)(x - 2) = 0$$
$$3x = 0 \quad \vee \quad x + 2 = 0 \quad \vee \quad x - 2 = 0$$
$$x = 0 \quad \vee \quad x = -2 \quad \vee \quad x = 2$$

The values for which the expression is undefined are 0, −2, and 2.

From now on, the domain of all variables will be restricted so that the denominators are not zero.

Since rational expressions are fractions, the key to working with them is to recall your work with fractions in arithmetic.

One of the first things you learned about fractions was that different fractions can represent the same number. Such fractions are called *equivalent fractions*.

Example

3 Write two fractions equivalent to $\frac{3}{5}$.

Multiplying the numerator and denominator of a fraction by the same nonzero number produces an equivalent fraction.

$$\frac{3}{5} = \frac{3 \times 2}{5 \times 2} = \frac{6}{10} \qquad \text{Thus, } \frac{3}{5} = \frac{6}{10}.$$

$$\frac{3}{5} = \frac{3 \times (-4)}{5 \times (-4)} = \frac{-12}{-20} = \frac{12}{20} \qquad \text{Thus, } \frac{3}{5} = \frac{12}{20}.$$

Example 3 illustrates the following principle.

Principle of Equivalent Fractions

If n and b are not 0, then $\frac{a \cdot n}{b \cdot n} = \frac{a}{b}$.

This principle allows us to *reduce a fraction to lower terms* by dividing the numerator and denominator by the same nonzero number. Dividing the numerator and denominator by their greatest common factor (GCF) produces a fraction in *lowest terms,* or *simplest form.*

Examples

4 Simplify $\frac{114}{102}$.

First, factor the numerator and denominator. $\frac{114}{102} = \frac{19 \cdot 2 \cdot 3}{17 \cdot 2 \cdot 3}$

Then, divide each by their GCF, 6 ⟶ $= \frac{19}{17}$

5 Simplify $\frac{2x - 4}{3x - 6}$.

$\frac{2x - 4}{3x - 6} = \frac{2(x - 2)}{3(x - 2)}$ *Factor the numerator and denominator.*

$= \frac{2}{3}$ *Divide the numerator and denominator by their GCF, $x - 2$.*

Therefore, $\frac{2x - 4}{3x - 6} = \frac{2}{3}$.

6 **Simplify** $\frac{y+1}{y^2-1}$.

$$\frac{y+1}{y^2-1} = \frac{y+1}{(y+1)(y-1)}$$ *Factor the denominator.*

$$= \frac{(y+1)\cdot 1}{(y+1)(y-1)}$$

$$= \frac{1}{y-1}$$ *Divide the numerator and denominator by their GCF, (y + 1).*

Therefore, $\frac{y+1}{y^2-1} = \frac{1}{y-1}$.

7 **Simplify** $\frac{t^2-25}{2t^2-9t-5}$.

$$\frac{t^2-25}{2t^2-9t-5} = \frac{(t+5)(t-5)}{(2t+1)(t-5)}$$ *Factor the numerator and denominator.*

$$= \frac{t+5}{2t+1}$$ *Divide by GCF, (t − 5).*

8 **Simplify** $\frac{3-y}{y-3}$.

$$\frac{3-y}{y-3} = \frac{-1(-3+y)}{y-3}$$ *Note that 3 − y and y − 3 are additive inverses.*

$$= -1 \cdot \frac{y-3}{y-3}$$ *Use the commutative property.*

$$= -1$$

9 **Simplify** $\frac{4-4r}{2r^2+r-3}$.

$$\frac{4-4r}{2r^2+r-3} = \frac{4(1-r)}{(2r+3)(r-1)}$$

$$= \frac{4(-1)(-1+r)}{(2r+3)(r-1)}$$ *Factor out −1 to create a common factor in the numerator and denominator.*

$$= \frac{4(-1)(r-1)}{(2r+3)(r-1)}$$ *Use the commutative property.*

$$= \frac{-4}{2r+3} \text{ or } -\frac{4}{2r+3}$$ *Divide by the GCF, (r − 1).*

Exercises

Exploratory **Find all values of the variable for which the expression is undefined.**

1. $\frac{3}{x}$
2. $\frac{4y}{x-3}$
3. $\frac{2x}{5y}$
4. $\frac{7-x}{x}$
5. $\frac{7}{7-x}$
6. $\frac{4x}{2x-1}$
7. $\frac{6y}{y^2+3y}$
8. $\frac{7}{x^2-4}$
9. $\frac{4x+12}{x^3-9x}$
10. $\frac{2y}{y^2-6y-7}$

Complete each of the following. Justify your answer.

11. $\frac{4}{9} = \frac{8}{?}$
12. $\frac{5}{x} = \frac{20}{?}$
13. $\frac{2y}{3m} = \frac{?}{12m^2}$
14. $\frac{4}{x-2} = \frac{8}{?}$
15. $\frac{y-3}{y+4} = \frac{3y-9}{?}$
16. $\frac{x}{7x} = \frac{1}{?}$
17. $\frac{3x^2}{x^5} = \frac{3}{?}$
18. $\frac{4x}{4x+16} = \frac{?}{x+4}$
19. $\frac{1}{x-3} = \frac{?}{x^2-9}$

Written **Simplify if possible. Be prepared to state any restrictions on variables.**

1. $\frac{24}{72}$
2. $\frac{33}{303}$
3. $\frac{4}{9}$
4. $\frac{625}{25}$
5. $\frac{37}{81}$
6. $\frac{7x}{x}$
7. $\frac{x^5}{x^2}$
8. $\frac{m}{mn}$
9. $\frac{y^3}{y^2}$
10. $\frac{10r}{2t}$
11. $\frac{x^{10}}{x^4}$
12. $\frac{x^9}{x^2}$
13. $\frac{34x^2}{42x^5}$
14. $\frac{x^2y}{xy}$
15. $\frac{a^3b^2}{a^4b}$
16. $\frac{x^2y^3}{xy^2}$
17. $\frac{42y^3x}{18y^7}$
18. $\frac{12n^9}{2n^3}$
19. $\frac{-3x^2y^5}{18x^5y^2}$
20. $\frac{14y^2z}{49yz^3}$
21. $\frac{42y^3x^7}{18y^7}$
22. $\frac{(2xy)^4}{(x^2y)^2}$
23. $\frac{(-2x^2y)^3}{4x^5y}$
24. $\frac{a^3b^2}{(-ab)^3}$
25. $\frac{(-3t^2u)^3}{(6tu^2)^2}$
26. $\frac{m+5}{5}$
27. $\frac{m+5}{2m+10}$
28. $\frac{y+3}{y+4}$
29. $\frac{y+3}{3y+9}$
30. $\frac{4x}{x^2-x}$
31. $\frac{6y+3}{9y+6}$
32. $\frac{m-3}{m^2-9}$
33. $\frac{t+2}{t^2-4}$
34. $\frac{u-5}{u^2-25}$
35. $\frac{x-6}{x^2-6x}$
36. $\frac{1-x^2}{x+1}$
37. $\frac{a+b}{a^2-b^2}$
38. $\frac{r^2-36}{(r+6)^2}$
39. $\frac{m^2-m}{m-1}$
40. $\frac{ab+a}{a^2-a}$
41. $\frac{u^2-100}{u^2+20u+100}$
42. $\frac{3t^2+9}{3t+9}$
43. $\frac{2y^2-18}{2y-6}$
44. $\frac{8m^2}{2m^2-4m^3}$
45. $\frac{2x^2}{x^3+6x^2y+9xy}$
46. $\frac{x^2+7}{x}$
47. $\frac{a^2-4}{2a^2-3a-2}$
48. $\frac{x^2+12x+36}{x^2+7x+6}$
49. $\frac{y-3}{y^2+y-12}$
50. $\frac{t^2-25}{3t^2-16t+5}$
51. $\frac{2x^2+8x}{x^2+x-12}$
52. $\frac{6y^3-9y^2}{2y^2+5y-12}$
53. $\frac{8a^2-16a}{a^2-4a+4}$
54. $\frac{9s^2-25}{9s^2+30s+25}$
55. $\frac{3-m}{6-17m+5m^2}$
56. $\frac{5t^2+10t+5}{3t^2+6t+3}$
57. $\frac{r-1}{1-r}$
58. $\frac{s-t}{t-s}$
59. $\frac{s^3-s^2}{1-s}$
60. $\frac{t^2-t^3}{t^2-2t+1}$
61. $\frac{1-x}{x+1}$
62. $\frac{16-4x^2}{2x^2-8}$
63. $\frac{2r^2-6r-8}{32-2r^2}$
64. $\frac{x^4-13x^2+36}{x^2+5x+6}$

Solve each proportion.

65. $\frac{m}{4} = \frac{4}{m}$

66. $\frac{t}{4} = \frac{8}{11}$

67. $\frac{3}{t-5} = \frac{1}{6}$

68. $\frac{5}{3} = \frac{v+1}{v}$

69. $\frac{1}{4} = \frac{x-3}{8x}$

70. $\frac{4}{y-3} = \frac{7}{y-2}$

71. $\frac{y}{9} = \frac{1}{y-8}$

72. $\frac{x}{x+1} = \frac{2x}{7}$

73. $\frac{y}{3} = \frac{5}{8-y}$

74. $\frac{5}{t} = \frac{10-t}{5}$

75. $\frac{m}{3} = \frac{5}{8-m}$

76. $\frac{3-w}{14} = \frac{-2}{w}$

Challenge **Solve.**

1. In a right triangle, the altitude to the hypotenuse separates the hypotenuse into two segments. The length of one segment is 5 cm more than the length of the other. If the altitude's length is 6 cm, find the length of the hypotenuse.

Use the figure at the right. Find each of the following.

2. s, if $h = 9$ and $r = 4$

3. s, if $x = 8$ and $z = 4$

4. h, if $r = 24$ and $s = 6$

5. x, if $s = 4$ and $z = 8$

6. y, if $x = 20$ and $r = 4$

7. h, if $y = 20$ and $x = 50$

8. x, if $z = 8$ and $r = 12$

9. r and s, if $h = 3$ and $s = r + 8$

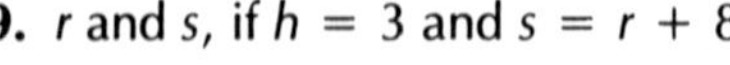

Find the value of x for each triangle.

10.

11.

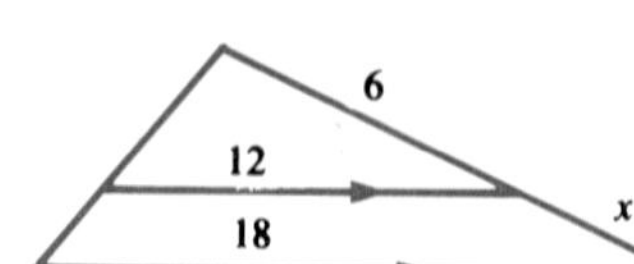

12.

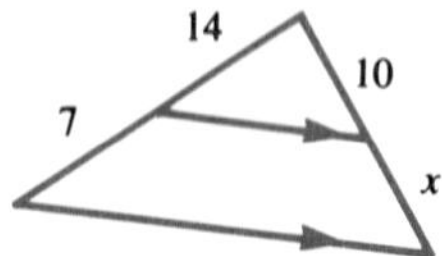

Mixed Review

Solve each equation or inequality.

1. $4a - 3(1 - 2a) = 7a$

2. $2(7 - n) > 4 - 7n$

3. $-3 \leq 5 - 4x < 6$

4. $|2k - 11| = 11$

5. $|3b + 17| \geq 4$

6. $|5 - 7x| - 9 < 0$

7. The measure of the supplement of an angle is 8 less than three times the measure of its complement. Find the measure of the angle.

8. Lita is driving to Niagara Falls, a distance of 651 miles. If she drives 5 hours at 55 mph, at what speed must she travel to complete the total trip in 13 hours?

9. Jean bought some peaches for 89¢ per pound and some bananas for 39¢ per pound. She bought 14 pounds of fruit. If her total bill was \$8.46, how many pounds of each fruit did she buy?

10. A strip of uniform width is mowed around the outer edge of a rectangular lawn with dimensions 100 feet by 120 feet. How wide is the strip if one-third of the lawn has been mowed?

1.7 Multiplying and Dividing Rational Expressions

To multiply rational expressions, we use a familiar rule from arithmetic.

Multiplication of Rational Expressions

If b and d are not 0, then $\frac{a}{b} \cdot \frac{c}{d} = \frac{ac}{bd}$.

In the following examples, *simplify* means to perform the indicated operations and then express the result in lowest terms.

Examples

1 **Simplify** $\frac{5}{y} \cdot \frac{7}{y+1}$.

$$\frac{5}{y} \cdot \frac{7}{y+1} = \frac{35}{y(y+1)} \text{ or } \frac{35}{y^2+y}$$

2 **Simplify** $\frac{3}{4} \cdot \frac{16}{21}$.

Simplify the product as you multiply.

Method 1: Factor to reveal GCF, then divide both numerator and denominator by GCF.

$$\frac{3}{4} \cdot \frac{16}{21} = \frac{3 \cdot 4 \cdot 4}{4 \cdot 3 \cdot 7} = \frac{4}{7}$$

Method 2: Divide numerator and denominator by common factors shortening the procedure.

$$\frac{\overset{1}{\cancel{3}}}{\underset{1}{\cancel{4}}} \cdot \frac{\overset{4}{\cancel{16}}}{\underset{7}{\cancel{21}}} = \frac{4}{7}$$

3 **Simplify** $\frac{y}{4} \cdot \frac{8}{y-1}$.

$$\frac{y}{4} \cdot \frac{8}{y-1} = \frac{8y}{4(y-1)} = \frac{4(2y)}{4(y-1)} = \frac{2y}{y-1}$$

or

$$\frac{y}{\underset{1}{\cancel{4}}} \cdot \frac{\overset{2}{\cancel{8}}}{y-1} = \frac{2y}{y-1}$$

In both methods the numerator and denominator are divided by 4.

4 Simplify $\frac{x^2 - 16}{x^2 - x - 20} \cdot \frac{x + 4}{x - 4}$.

$$\frac{x^2 - 16}{x^2 - x - 20} \cdot \frac{x + 4}{x - 4} = \frac{(x^2 - 16)(x + 4)}{(x^2 - x - 20)(x - 4)}$$

$$= \frac{(x + 4)\overset{1}{\cancel{(x - 4)}}\overset{1}{\cancel{(x + 4)}}}{(x - 5)\underset{1}{\cancel{(x + 4)}}\underset{1}{\cancel{(x - 4)}}} \quad \text{Factor. Divide by the GCF.}$$

$$= \frac{x + 4}{x - 5}$$

To divide rational expressions, use the familiar rule from arithmetic.

Division of Rational Expressions

If b, c, and d are not 0, then $\frac{a}{b} \div \frac{c}{d} = \frac{a}{b} \cdot \frac{d}{c}$ or $\frac{ad}{bc}$.

Examples

5 Simplify $\frac{12}{a} \div \frac{bc}{5}$.

$$\frac{12}{a} \div \frac{bc}{5} = \frac{12}{a} \cdot \frac{5}{bc}$$

$$= \frac{60}{abc}$$

6 Simplify $\frac{y^2}{2y^2 - 3y} \div \frac{y}{4y^2 - 9}$.

$$\frac{y^2}{2y^2 - 3y} \div \frac{y}{4y^2 - 9} = \frac{y^2}{2y^2 - 3y} \cdot \frac{4y^2 - 9}{y}$$

$$= \frac{\overset{1}{\cancel{y^2}}(2y + 3)\overset{1}{\cancel{(2y - 3)}}}{\underset{1}{\cancel{y}}\underset{1}{\cancel{(2y - 3)}} \cdot \underset{1}{\cancel{y}}} \quad \text{Factor and divide by the common factors.}$$

$$= 2y + 3$$

Exercises

Exploratory **Simplify each product or quotient.**

1. $\frac{5}{2} \cdot \frac{3}{4}$ **2.** $\frac{2}{3} \cdot \frac{5}{7}$ **3.** $\frac{4}{7} \div \frac{3}{5}$ **4.** $\frac{5}{8} \div \frac{1}{3}$ **5.** $7 \cdot \frac{1}{5}$

6. $6 \div \frac{1}{4}$ **7.** $\frac{1}{5} \div 10$ **8.** $3\frac{4}{5} \cdot 1\frac{1}{4}$ **9.** $-2\frac{1}{3} \div 6$ **10.** $7\frac{5}{8} \div (-4)$

11. $\left(-2\frac{5}{8}\right) \div \left(-1\frac{5}{3}\right)$ **12.** $\frac{5y}{2} \cdot \frac{1}{3}$ **13.** $\frac{x}{3} \cdot \frac{4}{8}$ **14.** $\frac{2ab}{c} \cdot \frac{3a}{4}$ **15.** $\frac{3c}{5y} \cdot \frac{5}{c}$

Written **Express each product in simplest form.**

1. $\frac{1}{18y} \cdot 2y$ **2.** $y \cdot \frac{-6}{y}$ **3.** $\frac{4t}{3} \cdot \frac{9}{t}$ **4.** $4x^2 \cdot \frac{2}{x}$ **5.** $\frac{2y^3}{3} \cdot \frac{9}{2y}$

6. $\frac{xy}{xz} \cdot \frac{z}{x}$ **7.** $\frac{m^2n}{n^2p} \cdot \frac{p}{q}$ **8.** $\frac{10x^2}{15y^2} \cdot \frac{5y}{2x}$ **9.** $\frac{27a^3}{18b^3} \cdot \frac{12a^4b^5}{9ac}$ **10.** $\left(\frac{x^2}{y}\right)^2 \cdot \frac{5}{3x}$

11. $\frac{2}{y} \cdot \frac{x+3}{2}$ **12.** $\frac{b}{5} \cdot \frac{b+1}{3b}$ **13.** $\frac{x+1}{3} \cdot \frac{4}{x}$ **14.** $\frac{3}{y^2+2} \cdot \frac{x}{3}$

15. $\frac{a+b}{4} \cdot \frac{5}{a+b}$ **16.** $\frac{x+5}{3} \cdot \frac{9}{x}$ **17.** $\frac{5a}{2} \cdot \frac{4}{5a+7}$ **18.** $\frac{y-7}{6} \cdot \frac{12}{y-7}$

19. $\frac{2n-4}{5} \cdot \frac{10}{n-2}$ **20.** $\frac{3a-3b}{b} \cdot \frac{b^2}{a-b}$ **21.** $\frac{-(1-a)}{3} \cdot \frac{-9}{a-1}$ **22.** $\frac{-(2-b)}{x} \cdot \frac{x^2}{b-2}$

23. $\frac{4x+40}{3x} \cdot \frac{9x}{3x+30}$ **24.** $\frac{x^2-y^2}{y^2} \cdot \frac{y^3}{x-y}$ **25.** $\frac{a+b}{14} \cdot \frac{7}{b+a}$ **26.** $\frac{x^2-y^2}{x+y} \cdot \frac{11}{x-y}$

27. $\frac{4y^2-9}{4y^2} \cdot \frac{8y}{2y-3}$ **28.** $\frac{a^2+b^2}{x^2-y^2} \cdot \frac{x-y}{x+y}$ **29.** $\frac{3x+15}{2} \cdot \frac{8}{x^2+4x-5}$

30. $\frac{3m^2-6m}{9m} \cdot \frac{2}{m-2}$ **31.** $\frac{y^2+8y+15}{y} \cdot \frac{y^2}{2y+10}$ **32.** $\frac{w^2-11w+24}{w^2-18w+80} \cdot \frac{w^2-15w+50}{w^2-9w+20}$

Express each quotient in simplest form.

33. $\frac{a}{b} \div \frac{1}{b}$ **34.** $\frac{x^2}{y^2} \div \frac{x^3}{a}$ **35.** $\frac{-b^3}{2a} \div \frac{b}{a^2}$ **36.** $\frac{r^3}{2s} \div \frac{-r^4}{8s}$

37. $\frac{x}{y+1} \div \frac{1}{y+1}$ **38.** $\frac{r^2}{r-5} \div (r-5)$ **39.** $\frac{t^2-36}{4t} \div (6-t)$ **40.** $\frac{m^2}{m+2} \div \frac{m}{m+2}$

41. $\frac{c^2+8c+16}{b^2} \div (c+4)$ **42.** $\frac{1-r}{3} \div \frac{r-1}{2}$ **43.** $\frac{y^2-25}{25-y^2} \div \frac{3}{y}$

44. $\frac{u+1}{u-1} \div \frac{u^2+2u+1}{3}$ **45.** $\frac{16-x^2}{10} \div \frac{x-4}{5}$ **46.** $\frac{2y}{y^2-4} \div \frac{4}{y^2-4y+4}$

47. $\frac{c^2+2c-3}{3c+3} \div \frac{c^2+5c+6}{2c+2}$ **48.** $\frac{p^2+7p}{3p} \div \frac{49-p^2}{3p-21}$

49. $\frac{9-4t^2}{t^2+6t+9} \div \frac{8t-12}{2t^2+5t-3}$ **50.** $\frac{3+10t^2-17t}{5t^2+4t-1} \div \frac{4t^2-9}{3+5t+2t^2}$

51. $\left(1+\frac{1}{x}\right) \div \left(2+\frac{3}{x}\right)$ **52.** $\left(x+\frac{1}{x}\right) \div \left(x+\frac{2}{x}\right)$

53. $\left(\frac{5a^2-20}{2a+2}\right) \div \left(\frac{5}{2}+\frac{5}{a}\right)$ **54.** $\frac{2y^2-4y}{y^2-4} \div \frac{8}{y^2+4y+4}$

Challenge **Find the values of the variables for which the expression is undefined.**

1. exercise 40 **2.** exercise 50 **3.** exercise 28

Mathematical Excursions

Rational Numbers and Photography

To take good pictures, photographers must make sure the correct amount of light enters the camera. The light enters the camera through an adjustable opening. The sizes of the opening are called *f-stops*. The *f*-stops are determined by dividing the focal length of the lens by the diameter of the opening. For example, a camera lens has a focal length of 50 mm. The diameter of the opening is 6.25 mm.

$$\begin{aligned} f\text{-stop} &= \frac{\text{focal length}}{\text{diameter of opening}} \\ &= \frac{50 \text{ mm}}{6.25 \text{ mm}} \\ &= 8 \quad \text{The } f\text{-stop is 8.} \end{aligned}$$

The dial that shows *f*-stops has the numbers 1.4, 2, 2.8, 4, 5.6, 8, 11, and 16. Each *f*-stop lets twice as much light into the camera as the next. For example, *f*/4 lets in *twice* as much light as *f*/5.6.

Another dial marked 1, 2, 4, 8, 15, 30, 60, 125, 250, 500, and 1000 shows the shutter speed. The shutter speed in seconds is the multiplicative inverse of the number on the dial. For example, a shutter speed marked 30 means light enters the camera for $\frac{1}{30}$ of a second. Each shutter speed lets light into the camera for twice as long as the next setting.

Suppose you want to photograph a parked sports car on a sunny day. The setting *f*/11 at $\frac{1}{60}$ would allow the correct amount of light into the camera. For a moving car, you need a faster shutter speed, such as $\frac{1}{125}$. At $\frac{1}{125}$, only half the amount of light is allowed into the camera. For proper exposure, the *f*-stop must be increased to *f*/8.

Exercises **A lens has a focal length of 200 mm. Find the *f*-stop for each diameter opening.**

1. 50 mm **2.** 25 mm **3.** 12.5 mm **4.** 36 mm

For each exposure combination, state two others that give the same exposure.

5. *f*/8 at $\frac{1}{60}$ **6.** *f*/5.6 at $\frac{1}{60}$ **7.** *f*/2.8 at $\frac{1}{125}$ **8.** *f*/4 at $\frac{1}{30}$

1.8 Adding and Subtracting Rational Expressions

Rational expressions are added and subtracted in the same way that rational numbers are in arithmetic.

Examples

Simplify.

1 $\frac{7}{11} + \frac{3}{11} - \frac{5}{11} = \frac{7 + 3 - 5}{11}$ or $\frac{5}{11}$

2 $\frac{8y}{15x^2} + \frac{9y}{15x^2} = \frac{8y + 9y}{15x^2}$ or $\frac{17y}{15x^2}$

To add or subtract fractions with different denominators, first find fractions with common denominators that are equivalent to the original fractions. Then add or subtract the equivalent fractions. Any common denominator may be used, but it is usually more efficient to use the least common denominator (LCD). The LCD is the least common multiple of the denominators.

Examples

3 **Compute** $\mathbf{\frac{2}{3} + \frac{1}{6} - \frac{1}{2}}$**.**

$$\frac{2}{3} + \frac{1}{6} - \frac{1}{2} = \frac{2 \cdot 2}{3 \cdot 2} + \frac{1}{6} - \frac{1 \cdot 3}{2 \cdot 3}$$

The least common denominator (LCD) of 3, 6, and 2 is 6.

$$= \frac{4}{6} + \frac{1}{6} - \frac{3}{6}$$

$$= \frac{4 + 1 - 3}{6}$$

$$= \frac{2}{6} \quad \text{or} \quad \frac{1}{3}$$

4 **Simplify** $\mathbf{\frac{4x + 1}{3} - \frac{x - 5}{2}}$**.**

$$\frac{4x + 1}{3} - \frac{x - 5}{2} = \frac{2(4x + 1)}{2 \cdot 3} - \frac{3(x - 5)}{3 \cdot 2}$$

Express each fraction as an equivalent fraction with denominator 6.

$$= \frac{8x + 2}{6} - \frac{3x - 15}{6}$$

$$= \frac{8x + 2 - (3x - 15)}{6}$$

$$= \frac{8x + 2 - 3x + 15}{6}$$

$$= \frac{5x + 17}{6}$$

Sometimes the LCD is not obvious. For instance, consider adding $\frac{7}{120}$ and $\frac{1}{36}$. To find the LCD, first factor each denominator into prime factors. The LCD is the number that contains each prime factor the greatest number of times it appears.

$$120 = \textcircled{2} \cdot \textcircled{2} \cdot \textcircled{2} \cdot 3 \cdot \triangle\!\!\!\!5 \qquad 36 = 2 \cdot 2 \cdot \boxed{3} \cdot \boxed{3}$$

$$\textcircled{2} \cdot \textcircled{2} \cdot \textcircled{2} \cdot \boxed{3} \cdot \boxed{3} \cdot \triangle\!\!\!\!5 = 360$$

$$\frac{7}{120} + \frac{1}{36} = \frac{7 \cdot 3}{120 \cdot 3} + \frac{1 \cdot 5 \cdot 2}{36 \cdot 5 \cdot 2}$$

Express $\frac{7}{120}$ and $\frac{1}{36}$ as equivalent fractions with denominator 360.

$$= \frac{21}{360} + \frac{10}{360} \text{ or } \frac{31}{360}$$

The following example uses this procedure with rational expressions.

Example

5 **Simplify $\frac{x + 1}{x^2 - 2x} - \frac{6}{x^2 - 4}$.**

$$\frac{x + 1}{x^2 - 2x} - \frac{6}{x^2 - 4} = \frac{x + 1}{x(x - 2)} - \frac{6}{(x - 2)(x + 2)}$$

Factor denominators.

$$= \frac{(x + 1)(x + 2)}{x(x - 2)(x + 2)} - \frac{6(x)}{x(x - 2)(x + 2)}$$

The LCD is $x(x - 2)(x + 2)$.

$$= \frac{x^2 + 3x + 2 - 6x}{x(x - 2)(x + 2)} \text{ or } \frac{x^2 - 3x + 2}{x(x - 2)(x + 2)}$$

$$= \frac{\overset{1}{\cancel{(x - 2)}}(x - 1)}{x\underset{1}{\cancel{(x - 2)}}(x + 2)}$$

Factor the numerator and divide both numerator and denominator by $x - 2$.

$$= \frac{x - 1}{x(x + 2)} \text{ or } \frac{x - 1}{x^2 - 2x}$$

Exercises

Exploratory **Express each sum or difference in simplest form.**

1. $\frac{2x^2}{y^2} - \frac{x}{y^2}$
2. $\frac{5r}{s^3} - \frac{r^2}{s^3}$
3. $\frac{x + 6}{x + 3} + \frac{4 - x}{x + 3}$

4. $\frac{x+3}{4} - \frac{x+5}{4}$

5. $\frac{2y+6}{8} - \frac{3y-1}{8}$

6. $\frac{3+x}{y} - \frac{2(x-3)}{y}$

7. $\frac{2x}{x+1} - \frac{x-1}{x+1}$

8. $\frac{3s}{s-4} - \frac{4+2s}{s-4}$

9. $\frac{x^2+7}{x^3-x} - \frac{7-x}{x^3-x}$

For each pair of denominators, find the LCD.

Sample: 18, 27 The LCD for 18 and 27 is 54.

10. 2, 5
11. 8, 12
12. x, 15
13. a, a^2
14. x^3, x^2
15. xy, y^2
16. $(x+3)$, $(x-3)$
17. y^2-4, $y+2$
18. $36x^2y$, $20xyz$
19. $x+5$, $x^2+3x-15$
20. x^2-8x, y^2-8y
21. $96x^2$, $16x+144$

Written **Express each sum or difference in simplest form.**

1. $\frac{3}{7} + \frac{2}{14}$

2. $\frac{5}{18} - \frac{2}{27}$

3. $\frac{2}{3} - \frac{1}{4} + \frac{5}{6}$

4. $4\frac{2}{3} + 1\frac{5}{8} - 7\frac{1}{6}$

5. $\frac{8}{9} + \frac{y}{3}$

6. $\frac{x}{4} + \frac{3x}{8}$

7. $\frac{r}{7} - \frac{2}{3}$

8. $\frac{2y}{7} - \frac{3x}{4}$

9. $\frac{3}{2x} + \frac{5}{x}$

10. $\frac{5}{2b} - \frac{3}{4}$

11. $\frac{2x}{3} + \frac{x+1}{6}$

12. $\frac{2y+3}{10} - \frac{3}{5}$

13. $\frac{4x}{7y} + \frac{3y}{21y}$

14. $7 + \frac{1}{x}$

15. $a + \frac{1}{b}$

16. $\frac{x}{x+y} + \frac{y}{x+y}$

17. $\frac{4a}{7b^2} - \frac{2a}{b}$

18. $\frac{x-2}{8} + \frac{x+1}{4}$

19. $\frac{2-a}{3} + \frac{2a+1}{5}$

20. $\frac{x}{3} - \frac{x+1}{9}$

21. $\frac{y-7}{7} - \frac{3y-2}{14}$

22. $\frac{5y}{6} - \frac{y+2}{2}$

23. $\frac{2a}{5} - \frac{4-3a}{4}$

24. $\frac{3}{r} + \frac{r+1}{s}$

25. $\frac{x}{x-2} - \frac{8}{x^2-4}$

26. $\frac{y}{y-4} - \frac{32}{y^2-16}$

27. $\frac{b+2}{b-3} - \frac{15}{b^2-3b}$

28. $\frac{2x}{2x+6} + \frac{5x}{x+3}$

29. $\frac{y+4}{y^2-16} + \frac{2}{y-4}$

30. $\frac{4}{x+y} - \frac{7}{x-y}$

31. $\frac{x^2}{x-y} + \frac{y^2}{y-x}$

32. $y - 1 + \frac{1}{y+1}$

33. $\frac{2}{x^2-9} - \frac{1}{x^2-3x}$

34. $2y + 1 + \frac{3}{2y-1}$

35. $\frac{b-2}{b-3} - \frac{b-4}{b-5}$

36. $\frac{t}{t+3} + \frac{18}{9-t^2}$

37. $3m + 1 - \frac{2m}{3m+1}$

38. $\frac{-4y}{y^2-4} - \frac{y}{y+2}$

39. $\frac{m}{m-3} + \frac{6m}{9-m^2}$

40. $\frac{-a}{5-a} - \frac{10a}{a^2-25}$

41. $\frac{1}{x^2-x-2} + \frac{1}{x^2+2x+1}$

42. $\frac{y+3}{3y-3} - \frac{2y-2}{3y^2-6y+3}$

43. $\frac{2}{3a-1} - \frac{1}{3a+1} - \frac{2}{9a^2-1}$

44. $\frac{x+1}{x-1} + \frac{x+2}{x-2} + \frac{2x}{x^2-3x+2}$

45. $\frac{a+3}{a^2-9} - \frac{4a}{6a^2-17a-3}$

46. $\frac{2y}{x-y} - \frac{2xy}{y^2-x^2} + \frac{x}{x+y}$

Challenge

Evaluate your answer for exercise 46 for $x = 3$ and $y = -3$. Then evaluate the original expression for those same values. Explain your findings.

1.9 Complex Fractions

A **complex fraction** is a fraction whose numerator or denominator contains one or more fractions. Some complex fractions are shown below.

$$\frac{1}{2 + \frac{1}{3}} \qquad \frac{\frac{1}{r} + 3}{\frac{4}{5} - 6} \qquad \frac{\frac{x^2 - y^2}{r^2 + 2rs + s^2} + s}{\frac{x^2 + y^2}{x + y} + s}$$

One way to simplify a complex fraction is to think of it as a division problem and use the procedures learned in Section 1.7. The complex fractions in examples 1 and 2 are simplified in this way.

Examples

Simplify $\frac{\frac{2}{3}}{\frac{5}{8}}$ and $\frac{\frac{x+3}{4}}{\frac{2}{x-1}}$ by rewriting them as division exercises.

1 $$\frac{\frac{2}{3}}{\frac{5}{8}} = \frac{2}{3} \div \frac{5}{8}$$
$$= \frac{2}{3} \cdot \frac{8}{5}$$
$$= \frac{16}{15}$$

2 $$\frac{\frac{x+3}{4}}{\frac{2}{x-1}} = \frac{x+3}{4} \div \frac{2}{x-1}$$
$$= \frac{x+3}{4} \cdot \frac{x-1}{2}$$
$$= \frac{x^2 + 2x - 3}{8}$$

3 **Simplify $\frac{x + \frac{1}{4}}{y + \frac{3}{8}}$ by using common denominators.**

Method 1: $$\frac{x + \frac{1}{4}}{y + \frac{3}{8}} = \frac{\frac{4x}{4} + \frac{1}{4}}{\frac{8y}{8} + \frac{3}{8}} \text{ or } \frac{\frac{4x+1}{4}}{\frac{8y+3}{8}}$$

Write the numerator and denominator as single fractions.

$$\frac{4x+1}{4} \div \frac{8y+3}{8} = \frac{4x+1}{4} \cdot \frac{8}{8y+3}$$

Then proceed as in example 1.

$$= \frac{2(4x+1)}{8y+3} \text{ or } \frac{8x+2}{8y+3}$$

Method 2: $$\frac{x + \frac{1}{4}}{y + \frac{3}{8}} = \frac{8\left(x + \frac{1}{4}\right)}{8\left(y + \frac{3}{8}\right)}$$

Multiply the numerator and denominator by the LCD of all fractions contained in the complex fraction.

$$= \frac{8x+2}{8y+3}$$

4 **Simplify** $\dfrac{\frac{1}{x} - \frac{1}{y}}{1 + \frac{1}{x}}$.

Method 1

$$\frac{\frac{1}{x} - \frac{1}{y}}{1 + \frac{1}{x}} = \frac{\frac{y}{xy} - \frac{x}{xy}}{\frac{x}{x} + \frac{1}{x}}$$

$$= \frac{\frac{y - x}{xy}}{\frac{x + 1}{x}}$$

$$= \frac{y - x}{xy} \div \frac{x + 1}{x}$$

$$= \frac{y - x}{xy} \cdot \frac{x}{x + 1}$$

$$= \frac{y - x}{y(x + 1)} \text{ or } \frac{y - x}{xy + y}$$

Method 2

$$\frac{\frac{1}{x} - \frac{1}{y}}{1 + \frac{1}{x}} = \frac{\left(\frac{1}{x} - \frac{1}{y}\right) \cdot xy}{\left(1 + \frac{1}{x}\right) \cdot xy}$$

The LCD of all the fractions is xy.

$$= \frac{\frac{xy}{x} - \frac{xy}{y}}{xy + \frac{xy}{x}}$$

$$= \frac{y - x}{xy + y}$$

Exercises

Exploratory **Simplify each of the following.**

1. $\dfrac{\frac{2}{3}}{\frac{4}{5}}$
2. $\dfrac{\frac{4}{11}}{\frac{5}{8}}$
3. $\dfrac{1 + \frac{1}{3}}{5}$
4. $\dfrac{1}{2 + \frac{1}{5}}$
5. $\dfrac{\frac{2}{5}}{\frac{1}{3} - \frac{2}{5}}$
6. $6 + \dfrac{\frac{1}{5}}{2 - \frac{3}{5}}$
7. $\dfrac{x - 1}{\frac{3}{x}}$
8. $\dfrac{\frac{x}{4}}{x}$
9. $\dfrac{\frac{y}{2}}{\frac{2y}{7}}$
10. $\dfrac{\frac{1}{a}}{\frac{2}{b - 1}}$

Written **Simplify.**

1. $\dfrac{\frac{x}{y}}{\frac{x^2}{y}}$
2. $\dfrac{\frac{8}{b^2}}{\frac{6}{b}}$
3. $\dfrac{\frac{1}{x}}{\frac{-3}{x^2}}$
4. $\dfrac{\frac{y + 3}{2}}{\frac{y + 2}{3}}$
5. $\dfrac{\frac{2x - 6}{5}}{\frac{3x - 9}{5}}$

6. $\dfrac{\frac{x+y}{b}}{\frac{x+y}{b^2}}$

7. $\dfrac{\frac{2}{3}+\frac{3}{5}}{\frac{4}{5}-\frac{7}{10}}$

8. $\dfrac{1+\frac{1}{a}}{1-\frac{1}{a^2}}$

9. $\dfrac{\frac{x^2}{9}-1}{\frac{x}{9}-\frac{1}{3}}$

10. $\dfrac{\frac{x}{2}+\frac{x}{4}}{\frac{x}{8}-\frac{x}{2}}$

11. $\dfrac{\frac{a}{5}-\frac{a}{4}}{\frac{a}{2}+\frac{a}{3}}$

12. $\dfrac{1-\frac{1}{y^2}}{1+\frac{1}{y}}$

13. $\dfrac{y-\frac{1}{y}}{\frac{1-y^2}{y}}$

14. $\dfrac{x-\frac{1}{x}}{1-\frac{1}{x^2}}$

15. $\dfrac{9-\frac{1}{x^2}}{3-\frac{1}{x}}$

16. $\dfrac{1-\frac{1}{4x^2}}{2+\frac{1}{x}}$

17. $\dfrac{\frac{1}{a}-\frac{1}{b}}{\frac{1}{a}+\frac{1}{b}}$

18. $\dfrac{c+\frac{c}{d}}{1+\frac{1}{d}}$

19. $\dfrac{\frac{2a+6}{a^3}}{1+\frac{3}{a}}$

20. $\dfrac{\frac{1}{a}}{1+\frac{1}{a}}$

21. $\dfrac{a-\frac{1}{4}}{a+\frac{1}{4}}$

22. $\dfrac{\frac{x+y}{x}}{\frac{1}{x}+\frac{1}{y}}$

23. $\dfrac{\frac{1}{5}+\frac{1}{y}}{y-\frac{25}{y}}$

24. $\dfrac{\frac{6}{5}-\frac{1}{y}}{36y-\frac{25}{y}}$

25. $\dfrac{t+\frac{1}{2}}{t^2-\frac{1}{4}}$

26. $\dfrac{a-1-2a^{-1}}{1-2a^{-1}}$

27. $\dfrac{x+2-3x^{-1}}{1-x^{-1}}$

28. $\dfrac{\frac{c}{4}-1}{\frac{c^2}{16}-1}$

29. $\dfrac{\frac{1}{3}+x}{x^2-\frac{1}{9}}$

30. $\dfrac{2x-5-\frac{3}{x}}{2+\frac{1}{x}}$

31. $\dfrac{\frac{y}{3}-\frac{3}{y}}{\frac{1}{3}-\frac{1}{y}}$

32. $\dfrac{\frac{c}{9}-4c^{-1}}{\frac{1}{2}-3c^{-1}}$

33. $\dfrac{\frac{m}{2}-\frac{m}{8}}{\frac{m}{2}+\frac{m}{3}}$

34. $\dfrac{\frac{1}{x}-\frac{2}{x^2}-\frac{3}{x^3}}{\frac{1}{x}+\frac{1}{x^2}-\frac{12}{x^3}}$

35. $\dfrac{\frac{1}{a}-\frac{1}{a^2}-\frac{20}{a^3}}{\frac{1}{a}+\frac{8}{a^2}+\frac{16}{a^3}}$

36. $\dfrac{2x^{-1}+9x^{-2}-5x^{-3}}{2x-\frac{1}{2}x^{-1}}$

37. $\dfrac{1-\frac{x}{7}}{\frac{1}{7}-\frac{1}{x}}$

38. $a+\dfrac{1}{a-\frac{1}{a}}$

39. $\dfrac{1-\frac{3}{y}}{1-\frac{2}{y}-\frac{3}{y^2}}$

40. $\dfrac{\frac{x}{y}-\frac{y}{x}}{\frac{1}{x}+\frac{1}{y}}$

41. $\dfrac{\frac{x}{y}-\frac{y}{x}}{\frac{y}{x}-1}$

42. $\dfrac{1+\frac{2}{y+1}}{1+\frac{4}{y-1}}$

43. $\dfrac{3+\frac{5}{a+2}}{3-\frac{10}{a+7}}$

44. $\dfrac{\frac{2x}{2x-1}-1}{1+\frac{2x}{1-2x}}$

45. $\dfrac{4+\frac{8}{m}}{2m-\frac{8}{m}}$

Challenge **Simplify.**

1. $\dfrac{\frac{x^2+7x}{3x}}{\frac{49-x^2}{3x-21}} \div \dfrac{\frac{10x^2-17x+3}{5x^2+4x-1}}{\frac{4x^2-9}{2x^2+5x+3}}$

2. $\dfrac{(3x^2+15x)(x-5)^{-2}}{(x^2-3x-10)^{-1}} \div \dfrac{(x+2)(x^2-25)^{-1}}{5x^{-2}(x^2+3x-10)^{-1}}$

1.10 Fractions and Equations

A fractional equation contains one or more fractions. Consider the following fractional equation.

$$\frac{y}{3} + \frac{y-1}{2} = -3$$

One way to solve this equation is to multiply both sides of the equation by the LCD of the fractions. In this case, the LCD is 6.

$$6\left(\frac{y}{3} + \frac{y-1}{2}\right) = 6(-3)$$

$$\frac{6y}{3} + \frac{6(y-1)}{2} = -18 \quad \textit{Distributive Property}$$

$$2y + 3(y-1) = -18 \quad \textit{Simplify each fraction on the left side.}$$

$$2y + 3y - 3 = -18 \quad \textit{Distributive Property}$$

$$5y = -15$$

$$y = -3$$

The solution set is $\{-3\}$. *You should check this result in the original equation.*

Example

1 **Solve $\frac{x}{x-2} + \frac{x+2}{x} = 2$.**

The LCD is $x(x-2)$.
Multiply both sides of the original equation by $x(x-2)$. Then solve.

$$x(x-2)\left(\frac{x}{x-2} + \frac{x+2}{x}\right) = 2x(x-2)$$

$$x \cdot x + (x-2)(x+2) = 2x^2 - 4x$$

$$x^2 + x^2 - 4 = 2x^2 - 4x$$

$$-4 = -4x$$

$$1 = x$$

The solution set is $\{1\}$. *The check is left for you.*

Multiplying both sides of an equation by an algebraic expression may produce an equation that is not equivalent to the original equation. Here is an example of this situation.

Example

2 Solve $x - \frac{15}{x+3} = \frac{5x}{x+3}$ and check.

$$(x+3)\left(x - \frac{15}{x+3}\right) = \frac{5x}{x+3}(x+3) \quad \textit{Multiply both sides by } x + 3.$$
$$x^2 + 3x - 15 = 5x$$
$$x^2 - 2x - 15 = 0$$
$$(x+3)(x-5) = 0$$
$$x + 3 = 0 \quad \vee \quad x - 5 = 0$$
$$x = -3 \quad \vee \quad x = 5$$

Notice what happens when the solutions are checked in the original equation.

Check for $x = -3$.

$$-3 - \frac{15}{-3+3} \stackrel{?}{=} \frac{5(-3)}{-3+3}$$
$$-3 - \frac{15}{0} \stackrel{?}{=} \frac{-15}{0}$$

Check for $x = 5$.

$$5 - \frac{15}{5+3} \stackrel{?}{=} \frac{5 \cdot 5}{5+3}$$
$$5 - \frac{15}{8} \stackrel{?}{=} \frac{25}{8}$$
$$\frac{25}{8} \stackrel{?}{=} \frac{25}{8}$$

For $x = -3$, we get denominators of 0 which are not allowed! Clearly -3 is not a solution of the original equation. We have no such problems in the check for $x = 5$. Therefore, the solution set is $\{5\}$.

Consider that "extra" root, -3, in example 2. Recall that multiplying both sides of an equation by 0 does not necessarily produce an equivalent equation. In example 2, both sides of the original equation were multiplied by $x + 3$. But $x + 3$ is zero when $x = -3$. Therefore, the second equation, $x^2 + 3x - 15 = 5x$, is not equivalent to the original one. Fortunately, the solution set of the second equation includes the solution of the original equation. For this reason, it is very important that the solution set of the second equation be carefully checked. Thus, the following principle always applies.

Whenever you multiply both sides of an equation by a polynomial, the roots of the new equation include—but may not be limited to—the roots of the original equation. Therefore, each root of the new equation should be checked in the *original* equation.

Exercises

Exploratory **Explain your answers to each of the following.**

1. Is every proportion a fractional equation? Explain.
2. Is every fractional equation a proportion? Explain.

Explain how the second equation is obtained from the first equation.

3. $\frac{x}{5} + 4 = \frac{2}{5}$
   $x + 20 = 2$
4. $\frac{1}{y} + 2 = \frac{3}{y}$
   $1 + 2y = 3$
5. $\frac{1}{x} + 3 = \frac{2}{x}$
   $1 + 3x = 2$
6. $\frac{x-1}{3} + \frac{1}{2} = 5$
   $\frac{2x-2}{3} + 1 = 10$
7. $\frac{x-1}{3} + \frac{1}{2} = 5$
   $x - 1 + \frac{3}{2} = 15$
8. $\frac{x-1}{3} + \frac{1}{2} = 5$
   $2x - 2 + 3 = 30$
9. Look at exercises 6–8. In each case, a different equation was obtained from $\frac{x-1}{3} + \frac{1}{2} = 5$. Which would be easiest to solve? Why?
10. Under what circumstances is it essential that the roots obtained in solving a fractional equation be checked? Give an example.
11. Do you agree with the solution at the right that Clara wrote for the equation $x^2 - x = 0$? Explain.

$$x^2 - x = 0$$
$$x^2 = x$$
$$x = 1 \quad \textit{Multiply each side by } \frac{1}{x}.$$

The solution set is $\{1\}$.

Written **Solve and check.**

1. $\frac{3}{x} = 10$
2. $\frac{6}{y-1} = 3$
3. $\frac{2}{x} + 3 = \frac{29}{x}$
4. $4 + \frac{2}{t} = \frac{-2}{t}$
5. $\frac{y}{2} = 7 + \frac{y}{3}$
6. $\frac{1}{x} = x$
7. $\frac{x+4}{2} + \frac{x+6}{3} = \frac{x}{6}$
8. $\frac{y-2}{4} + \frac{y}{6} = 2$
9. $\frac{7}{y+1} = 7$
10. $\frac{11}{2y} - \frac{2}{3y} = \frac{1}{6}$
11. $\frac{1}{4} = \frac{x-3}{8x}$
12. $\frac{3}{y+2} = \frac{4}{y-1}$
13. $\frac{y+4}{3} = \frac{4}{y}$
14. $\frac{x-1}{x+5} = \frac{x+1}{x+4}$
15. $\frac{x}{x^2-3} = \frac{5}{x+4}$
16. $\frac{1}{x-1} + \frac{2}{x} = 0$
17. $\frac{x+2}{3} + \frac{x+1}{9} = \frac{1}{3}$
18. $\frac{x^2}{6} - \frac{x}{3} = \frac{1}{2}$
19. $r^2 + \frac{17r}{6} = \frac{1}{2}$
20. $\frac{2y}{3} - \frac{y+3}{6} = 2$
21. $\frac{2y+1}{5} - \frac{2+7y}{15} = \frac{2}{3}$
22. $\frac{2y-5}{6} - \frac{y-5}{4} = \frac{3}{4}$
23. $\frac{4t-3}{5} - \frac{4-2t}{3} = 1$
24. $\frac{5+7p}{8} - \frac{3(5+p)}{10} = 2$
25. $\frac{2q-1}{3} - \frac{4q+5}{8} = -\frac{19}{24}$
26. $8 - \frac{2-5x}{4} = \frac{4x+9}{3}$
27. $\frac{3x-1}{4} - \frac{x-5}{5} = -2$
28. $0.8(y - 9) = 0.6(4 + y)$
29. $0.08y + 0.06(2000 - y) = 150$
30. $0.25(60) = 0.20(60 + r)$
31. $0.03y + 0.08(1000 - y) = 77.75$

32. $x + 5 = \frac{6}{x}$

33. $a + 1 = \frac{6}{a}$

34. $x + \frac{12}{x} = 8$

35. $\frac{2}{x} + \frac{1}{4} = \frac{11}{12}$

36. $\frac{1}{9} + \frac{1}{2a} = \frac{1}{a^2}$

37. $\frac{1}{x - 1} + \frac{2}{x} = 0$

38. $\frac{5}{6} - \frac{m}{m + 4} = \frac{7}{6}$

39. $\frac{4t}{3t - 2} + \frac{2t}{3t + 2} = 2$

40. $\frac{1 - m}{1 + m} + \frac{3m}{m - 1} = -1$

41. $\frac{q}{q - 5} + \frac{q}{q - 5} = 3$

42. $\frac{2p}{2p + 3} - \frac{3p}{2p - 3} = 1$

43. $\frac{12}{x^2 - 16} - \frac{24}{x - 4} = 3$

44. $\frac{2y - 1}{y - 4} + 2 = \frac{-5y}{y - 4}$

45. $\frac{x}{x - 2} - \frac{x + 7}{x + 2} = 1$

46. $\frac{a}{a - 1} + a = \frac{4a - 3}{a - 1}$

47. $\frac{1}{b^2 - 1} = \frac{2}{b^2 + b - 2}$

48. $\frac{1}{1 - y} = 1 - \frac{y}{y - 1}$

49. $\frac{t - 3}{7} = \frac{t - 3}{4 + t}$

50. $x + \frac{x}{x - 1} = \frac{4x - 3}{x - 1}$

51. $\frac{t}{t^2 - 1} + \frac{2}{t + 1} = \frac{1}{2t - 2}$

52. $\frac{2}{x + 2} - \frac{x}{2 - x} = \frac{x^2 + 4}{x^2 - 4}$

53. $\frac{y}{y - 5} + \frac{17}{25 - y^2} = \frac{1}{y + 5}$

54. $\frac{t + 4}{t} + \frac{3}{t - 4} = \frac{-16}{t^2 - 4t}$

55. $\frac{x + 3}{x + 2} = 2 - \frac{3}{x^2 + 5x + 6}$

56. If $x + y = 1$ and $\frac{x}{y} = 1$, find the value of xy.

57. Find the measure of an angle that is $12\frac{1}{2}\%$ of its supplement's measure.

Find the values of *x* that satisfy the following conditions.

58. $2x$ is 60% of $x + 3$.

59. $2x$ is 40% of $2x - 7$.

60. The ratio of $x + 2$ to $x - 5$ is 30%.

61. $7x - 3.7$ is 3% of $5x + 2$.

Challenge

1. If $y \neq 0$ and $y = 2 + 2m$, show that $\frac{1}{y} + \frac{1}{y} + \frac{1}{y} + \frac{1}{y} = \frac{2}{1 + m}$.

2. If $\frac{3x}{5y} = 11$, find the value of $\frac{3x - 5y}{5y}$.

Mixed Review

Perform the indicated operations and simplify.

1. $\frac{r^2 - 6r}{2r + 12} - \frac{36 - r^2}{10r}$

2. $\frac{x^2 - x - 6}{x^2 + x - 6} \cdot \frac{x^2 - 9}{x^2 - 4}$

3. $\left(t - \frac{1}{t}\right) \div \left(t + 1 - \frac{2}{t}\right)$

4. $\frac{m^2}{m - 3} - \frac{5m - 24}{3 - m}$

5. $\frac{x}{x^2 + 3x + 2} + \frac{6}{x^2 + x - 2}$

6. $\frac{y}{y + 2} - \frac{3}{y - 2} + \frac{12}{y^2 - 4}$

7. $\frac{x - 4x^{-1}}{4x^{-2} - 1}$

8. $\frac{xy^{-1} - 1}{x^{-1} - y^{-1}}$

9. $\frac{2 - 9x^{-1} - 18x^{-2}}{2x + 27(2x)^{-2}}$

Solve each equation.

10. $\frac{-3}{y - 4} = \frac{1}{y + 4}$

11. $\frac{w - 3}{w + 2} = \frac{6 - w}{w - 12}$

12. $r + \frac{r^2 - 5}{r^2 - 1} = \frac{r^2 + r + 2}{r + 1}$

1.11 Fractional Equations in Problem Solving

Many problems can be solved using fractional equations.

Example

1 **Working at a steady pace, it takes Gil 3 hours to mow a lawn. Ali can mow the same lawn in 2 hours. If they work together, how long will it take them to mow the lawn?**

Explore We know that Gil can mow one-third of the lawn in one hour and Ali can mow one-half of the lawn in one hour.

Let t = the number of hours it will take them to mow the lawn.

Then $\frac{1}{3}t$ = the amount of lawn Gil will mow, and $\frac{1}{2}t$ = the amount of lawn Ali will mow.

Plan *amount mowed by Gil* + *amount mowed by Ali* = *entire lawn mowed*

$$\frac{1}{3}t + \frac{1}{2}t = 1$$

Solve

$$\begin{aligned} 6\left(\frac{1}{3}t + \frac{1}{2}t\right) &= 6(1) \\ 2t + 3t &= 6 \\ 5t &= 6 \\ t &= \frac{6}{5} \text{ or } 1\frac{1}{5} \end{aligned}$$

Working together, Gil and Ali will take $1\frac{1}{5}$ hours, or 1 hour and 12 minutes, to mow the lawn. *Examine this solution by checking that the answer satisfies the conditions of the original problem.*

Exercises

Written **Solve.**

1. A number divided by 9 is equal to the number decreased by 80. Find the number.
2. The denominator of a fraction is one more than the numerator. If 1 is added to both the numerator and the denominator, the new fraction is $\frac{2}{3}$. Find the original fraction.
3. The ratio of 4 less than a number to 26 more than that number is 1 to 3. Find the number.
4. Five times the multiplicative inverse of a number is added to the number. The result is $10\frac{1}{2}$. Find the number.

5. The denominator of a fraction is 1 less than twice the numerator. If 7 is added to both the numerator and the denominator the result is $\frac{7}{10}$. Find the original fraction.

6. Two numbers are in a ratio of 6 to 7. If the first is increased by 2 and the second is increased by 1, the resulting numbers are in the ratio of 4 to 5. Find the original numbers.

7. Farmer Gray can plow his field in 10 hours with his old tractor. The same field can be plowed in 6 hours with his new tractor. If Farmer Gray and his daughter plow the field using both tractors how long will it take?

8. The Pine City Municipal Pool can be filled by a pipe in 10 hours. The pool can be emptied by a drain pipe in 20 hours. If the drain is left open while the pool is filling, how long will it take to fill?

9. Susie can rake all the leaves on her lawn in 12 hours. Lori takes 18 hours to do the same job. Susie and Lori work together for a certain time and Lori finishes the raking in twice as many hours more. Find the number of hours each girl works.

10. The Pine City Express Bus can travel 300 km in the same time that the Pine City Commuter Train travels 200 km. If the speed of the bus is 20 km/h greater than the speed of the train, find the speeds of the bus and the train.

11. At Pine City High School, it takes 5 hours for one computer to schedule classes. Another computer can do the job in 4 hours. If the computers work together, how long will it take to do the job?

12. Last year Michael painted his room in 3 hours. This year Michael and John were able to paint the room together in 2 hours. How long would it have taken John to paint the room working alone?

13. The Pine City Seal Pond can be filled in 6 hours using a hose. A second larger hose can fill the pond in 4 hours. If both hoses are used, how long will it take to fill the pond?

14. The Pine City baseball team has won 25 of 36 games. How many of the remaining 12 games must be won for the team to end up winning 75% of all their games?

15. Increasing the average speed of the Pine City Express Bus by 13 km/h resulted in the 260 km trip taking an hour less than before. What was the original average speed of the bus?

16. Paul can jog to Zelda's house in 10 minutes and Zelda takes 6 minutes to ride her bike to Paul's house. If they start from their houses at the same time, in how many minutes do they meet?

17. How much pure acid should be added to 30 ounces of a solution which is now 40% acid to produce a solution which will be 50%?

18. How many liters of water need to be added to 9 liters of a 50% alcohol solution to produce a solution containing 30% alcohol?

19. How much pure sulfuric acid should be mixed with 10 liters of water to produce a mixture that is 10% sulfuric acid?

20. How many liters of pure acid should be added to 40 liters of a solution which is 5% pure acid to produce a solution which will be 24% pure acid?

21. Sea water contains 3.5% dissolved solids. Drinking water contains 0.05% dissolved solids. Water for a certain irrigation project is obtained by mixing sea water and drinking water. How many gallons of sea water should be mixed with drinking water to obtain 1000 gallons of irrigation water containing 0.15% dissolved solids?

22. Collinsburg Music makes a profit of $66\frac{2}{3}$% on the sale of a Blaze-O Stereo. For one week only, the price of the stereo is reduced to \$540. If the store makes a profit of $33\frac{1}{3}$% selling the stereo at this price, then \$540 is what percent of the original selling price?

Problem Solving Application: Using Quadratic Equations

When solving problems involving quadratic equations, the solutions of the equation may not satisfy some of the conditions of the problem. For example, a negative number cannot be a length or width. It is important to check all solutions for reasonableness.

Example

1 Greenwood School hoped to asphalt a parking lot with dimensions 30 meters by 60 meters. However, the budget for this project was cut so only 1000 square meters could be asphalted. It was decided to reduce both the length and the width of the lot by the same amount. Find the dimensions of the new lot.

Explore Make a diagram of the lot as originally planned. Since both the length and width are to be reduced by the same amount, let x represent the amount of reduction.

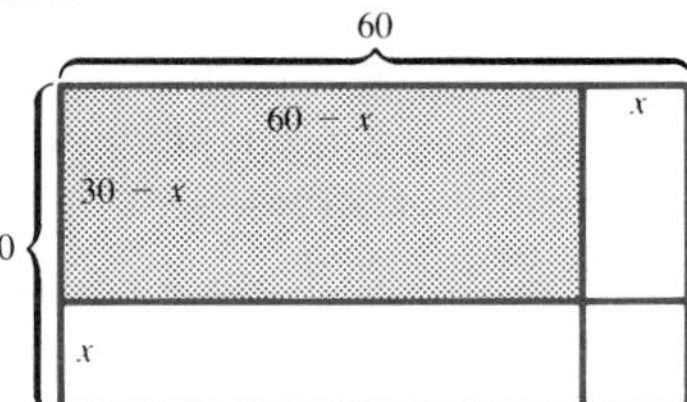

Plan The length of the new parking lot is $60 - x$. The width of the new parking lot is $30 - x$.

length	*times*	*width*	*is equal to*	*area*
$(60 - x)$	$\times$	$(30 - x)$	$=$	1000

Solve

$$
\begin{aligned}
(60 - x)(30 - x) &= 1000 \\
1800 - 90x + x^2 &= 1000 \\
800 - 90x + x^2 &= 0 \\
(10 - x)(80 - x) &= 0
\end{aligned}
$$

$$
\begin{aligned}
10 - x = 0 \quad &\text{or} \quad 80 - x = 0 \\
10 = x \quad & \quad\quad 80 = x
\end{aligned}
$$

Examine If $x = 80$, then $60 - x = -20$ and $30 - x = -50$. This is not a reasonable answer for the dimensions of a rectangle.
If $x = 10$, then $60 - x = 50$ and $30 - x = 20$. This solution is reasonable.

The dimensions of the new parking lot are 50 meters by 20 meters.

Exercises

Written **Solve each problem.**

1. Find two consecutive even integers whose product is 360.

2. The sum of a number and its reciprocal is $\frac{34}{15}$. Find the number.

3. A garden is 40 feet by 35 feet. A gravel path of equal width surrounds it. If the area of the garden and the path is 1554 square feet, what is the width of the path?

4. A recreational field is 100 meters by 120 meters. A strip of uniform width is paved to create a jogging track. If the track is one-third of the area of the field, what is the width of the track?

5. The length of a rectangular pool is 6 feet more than its width. A walkway 3 feet wide surrounds the outside of the pool. The total area of the walkway is 288 square feet. Find the dimensions of the pool.

6. If each side of a square is increased by 4 inches, the area of the square is increased by 176 square inches. What is the area of the original square?

7. The measure of the hypotenuse of a right triangle is 40. The measure of one leg is 3 times the measure of the other. Find the measures of both legs.

8. Triangle ABC is similar to triangle XYZ. AB is three times BC, $BC = XY$, and $YZ = 3$. Find the length of $\overline{AB}$.

9. Chords $\overline{AB}$ and $\overline{CD}$ of a circle intersect at point P. $AP = 2$, $BP = 16$, and $CD = 12$. Find the length of the longer segment $\overline{DP}$.

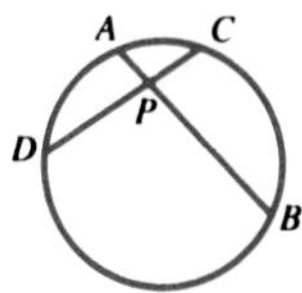

10. A photo has been enlarged so that the length of the new photo is twice the length of the original. The width of the original photo was 3 inches less than its length. The width of the enlarged photo is 2 inches greater than the length of the original. What are the dimensions of the enlarged photo?

11. Three less than a number is the geometric mean between twice the number and three times the number. Find the number.

12. The square of a number increased by 21 is equal to 7 times the number increased by 21. Find the number.

13. Wendell Hooks is a professional photographer. He has a photo 8 centimeters long and 6 centimeters wide. A customer wants a print of the photo. The print is to have half the area of the original. Jim plans to reduce the length and width of the photo by the same amount. What are the dimensions of the print?

14. Kyle and Renee's family room has a rug that is 9 feet by 12 feet. A strip of floor of equal width is uncovered along all edges of the rug. If the area of the uncovered floor is 270 square feet, how wide is the strip?

15. In the diagram at the right, the red rectangle is similar to the whole rectangle. The value of x represents the "golden ratio." Write a proportion and solve for x to find the golden ratio. Round to the nearest thousandth.

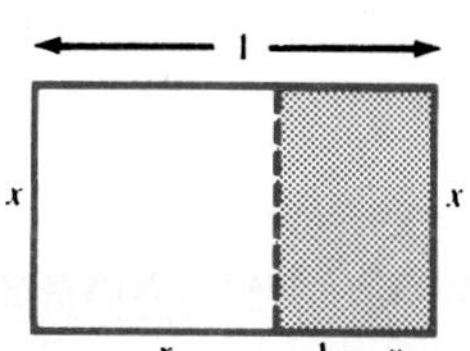

Vocabulary

rational number
real number
irrational number
disjunction
conjunction
absolute value
monomial
coefficient
degree of a monomial
laws of exponents
polynomial
binomial
trinomial
degree of a polynomial
linear polynomial
quadratic polynomial
difference of two squares
perfect trinomial square
quadratic formula
equivalent fractions
complex fractions

Chapter Summary

1. The set of real numbers contains the following subsets: natural numbers, whole numbers, integers, rational numbers, irrational numbers.
2. Absolute value equations or inequalities can be written as a conjunction or disjunction.
3. If b and d are not 0, then $\frac{a}{b} \cdot \frac{c}{d} = \frac{ac}{bd}$.
4. If b, c, and d are not 0, then $\frac{a}{b} \div \frac{c}{d} = \frac{a}{b} \cdot \frac{d}{c}$ or $\frac{ad}{bc}$.
5. To add or subtract rational expressions, a common denominator must be found.
6. When solving fractional equations, all possible solutions must be checked in the original equation to avoid the possibility of 0 as a denominator.

Chapter Review

1.1 **Express each decimal as a rational number in $\frac{a}{b}$ form.**

1. 3.25
2. 0.033
3. $0.\overline{7}$
4. $0.\overline{53}$

Simplify.

5. $13xy - 4y - 17xy$
6. $3(ab + 2b) - 4b(2a - 6)$

Solve. Graph the solution set of each inequality in exercises 7–17.

1.2 7. $2 + 3(4x - 5) = 5(3x - 2)$
8. $2y - 3(2y - 1) = 5(2y - 5)$
9. $3 - \frac{x}{4} < 5$
10. $x + 2 \geq 4x - 4$
11. $(x \leq -4) \vee (x > 1)$
12. $-3 < 2y - 1 \leq 3$
13. It costs $600 per truck plus $8 for each kilogram over 1100 kg to use an Acme Moving Company vehicle. Zeppo Movers charge $500 per truck plus $10 for each kilogram over 1100 kg. When is it less expensive to use Zeppo Movers?

1.3 **14.** $|3y + 7| = 26$ **15.** $|1 + 5x| + 3 = 1$

16. $|x + 1| > 5$ **17.** $|y - 2| < 4$

1.4 **Simplify.**

18. $3a^2 - 2a - 5a^2 + a$ **19.** $(3m^2n^3)(8m^3n)$ **20.** $(3y - 1)(5y + 4)$

21. $5x(x - 7) - 6x(2x + 8)$ **22.** $(3y - 5)^2$ **23.** $(x - 2)(x^2 - 3x + 4)$

1.5 **Factor.**

24. $3xy^2 - 27x^2y$ **25.** $n^2 - 16$ **26.** $w^2 - 6w - 55$

27. $5p^2 - 20$ **28.** $a^2x - b^2x$ **29.** $5h^2 - 66h + 72$

Solve.

30. $2y^2 - y = 0$ **31.** $2y^2 + 9y = 5$ **32.** $2x^3 - x^2 = 0$

1.6 **Find all values of *y* for which the expression is undefined.**

33. $\frac{y - 3}{4y - 20}$ **34.** $\frac{4y}{y^2 - 2y}$ **35.** $\frac{y - 2}{y^2 - 2y - 15}$

Simplify.

1.7 **36.** $\frac{x - 4}{16x^2} \cdot \frac{20x^5}{x - 4}$ **37.** $\frac{x^2 - 25}{x - 6} \cdot \frac{x^2 - 6x}{x^2 + 5x}$

38. $\frac{x^2 - y^2}{10x^2} \div \frac{x - y}{25xy}$ **39.** $\frac{4 - 3x}{x + 2} \div \frac{3x - 4}{2x + 4}$

1.8 **40.** $\frac{4}{ab} - \frac{6}{bc}$ **41.** $\frac{y + 2}{y} + \frac{y - 8}{4y}$

42. $\frac{m}{m - 3} - \frac{18}{m^2 - 9}$ **43.** $\frac{4x + 8}{x^2 + 4x + 4} + \frac{x + 1}{x + 2}$

1.9 **44.** $\frac{1 - \frac{4}{x}}{1 - \frac{16}{x^2}}$ **45.** $\frac{\frac{2a + 4}{a}}{3 - \frac{12}{a^2}}$

Solve.

1.10 **46.** $\frac{8}{x - 4} = \frac{x + 4}{x - 4}$ **47.** $\frac{5}{x} + \frac{1}{2} = \frac{16}{2x}$

48. $\frac{3x + 2}{4} = \frac{9}{4} - \frac{3 - 2x}{6}$ **49.** $\frac{x}{x^2 - 1} + \frac{2}{x + 1} = 1 + \frac{1}{2x - 2}$

1.11 **50.** One integer is 2 less than another integer. Three times the reciprocal of the lesser integer plus five times the reciprocal of the greater integer is $\frac{7}{8}$. Find the integers.

51. Theater tickets cost \$2.50 for children and \$3.50 for adults. The price of eight tickets was \$23.00. How many adult tickets were purchased?

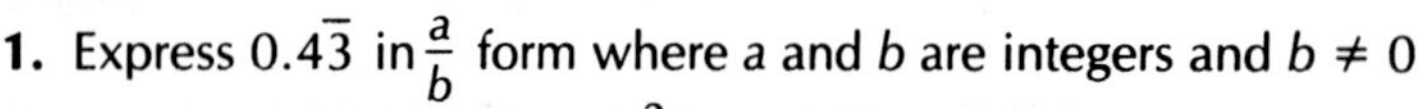

Chapter Test

1. Express $0.4\overline{3}$ in $\frac{a}{b}$ form where a and b are integers and $b \neq 0$.
2. For what values of x is $\frac{2x}{3x - 15}$ undefined?
3. Graph the solution set of $(y \geq 3) \vee (y > -1)$.
4. Find the degree of $8x - 5x^2 + 3$.
5. If $x^2 \cdot x = x^m$, find the value of m.

Factor.

6. $8x^2 + 10x$
7. $4y^2 - 25$
8. $x^2 + 12x + 36$
9. $3x^2 + 11x - 4$
10. $3a^3b^2 + 6ab^3$
11. $7t^2 - 28$

Simplify.

12. $2y - 5y^2 + 3y + 6y^2$
13. $(-2xy^2)(3x^3y^4)$
14. $(4m - 3)(5m + 2)$
15. $\frac{x^2 - y^2}{a^2 - b^2} \cdot \frac{a + b}{x - y}$
16. $\frac{7}{5a} - \frac{10}{3ab}$
17. $\frac{1}{m - 4} + \frac{2}{m - 3}$
18. $\frac{x - 5}{5x} + \frac{x - 1}{x}$
19. $\frac{x + 3}{x - 1} - \frac{4}{x^2 - x}$
20. $\frac{2}{1 - x} + \frac{5x}{x^2 - 1}$
21. $\frac{x^2 - 2x + 1}{y - 5} \div \frac{x - 1}{y^2 - 25}$
22. $\left(\frac{5}{y} - 1\right) \div \left(\frac{25}{y^2} - 1\right)$

Solve. Graph the solution sets of the inequalities.

23. $3x + 4(x - 2) > 2(3x - 2)$
24. $-4 < 3y - 1 \leq 5$
25. $|2x - 1| = 7$
26. $|2 + y| \leq 6$
27. $|x - 4| > 5$
28. $5x^2 - x = 0$
29. $a - \frac{5}{a} = 4$
30. $\frac{3}{x} + \frac{x}{x + 2} = \frac{-2}{x + 2}$
31. A fraction is equivalent to $\frac{3}{4}$. If 5 is subtracted from its numerator, the new fraction is equivalent to $\frac{4}{7}$. Find the original fraction.

Tell whether each statement is true or false.

32. Any irrational number is a real number.
33. $x + 3$ is a factor of $x^2 + 9$.
34. If $x \neq y$ then $\frac{x - y}{y - x} = -1$.
35. If $a > b$ and $c \neq 0$, then $ac > bc$.

Working with Complex Numbers

Chapter 2

Early mathematicians like Girolano Cardono (1501 – 1576) were somewhat puzzled by the complex numbers. Their investigation of these numbers that, at that time, seemed to have no practical value, laid the groundwork for later advances in electrical engineering, electrostatics, quantum mechanics, aerodynamics, mapmaking, and other fields.

2.1 Radicals

Expressions like $\sqrt{3}$, $\sqrt{\pi}$, and $\sqrt{x + 7}$ are examples of *square root radicals*. The symbol $\sqrt{\ }$ is called a **radical sign.** The expression written underneath the radical sign is called the **radicand.**

The radical $\sqrt{9}$ names the nonnegative real number whose square is 9. The nonnegative square root of a number is often called its *principal square root*. The radical $-\sqrt{9}$ names the negative real number whose square is 9. Therefore,

$$\sqrt{9} = 3 \quad \text{and} \quad -\sqrt{9} = -3.$$

Both square roots can be denoted by $\pm\sqrt{9}$ or ± 3.

Definition of Square Root

A number x is called a **square root** of a number y if $x^2 = y$.

Example

1 **Solve $x^2 = \frac{9}{4}$.**

$$x^2 = \frac{9}{4}$$

$$x = \sqrt{\frac{9}{4}} \text{ or } x = -\sqrt{\frac{9}{4}}$$

Every positive real number has two square roots.

$$x = \frac{3}{2} \quad \text{or } x = -\frac{3}{2}$$

The solution set is $\left\{\frac{3}{2}, -\frac{3}{2}\right\}$.

You should be careful not to confuse the idea of the *negative square root* of a number with the idea of a *square root of a negative number.* For example, $-\sqrt{9} = -3$ since $(-3)(-3) = 9$. But $\sqrt{-9}$ cannot name a real number since no real number squared is -9.

In general, if x is negative, $\sqrt{x}$ does not name a real number.

Example

2 **For what values of x does $\sqrt{x + 5}$ name a real number?**

$\sqrt{x + 5}$ will name a real number when $x + 5$ is *not* negative.

$$x + 5 \geq 0$$
$$x \geq -5$$

Therefore $\sqrt{x + 5}$ names a real number when $x \geq -5$.

Rational numbers like 144, 64, and $\frac{4}{9}$, whose square roots are rational numbers, are called **perfect squares.**

Since numbers like $\pm\sqrt{2}$, $\pm\sqrt{3}$, $\pm\sqrt{6}$, $\pm\sqrt{7}$, $\pm\sqrt{18}$, and so forth are irrational numbers, their decimal representations are non-repeating decimals. Rational approximations for these non-repeating decimals can be found by using the table in the back of the book, a calculator or computer, or the *divide-and-average* procedure illustrated in Example 3.

Example

3 Give a rational approximation to the nearest tenth for $\sqrt{7}$.

$$4 < 7 < 9$$
$$(2)^2 < 7 < (3)^2$$
$$2 < \sqrt{7} < 3$$

Step 1 *Estimate.* Since 7 is closer to 9, choose a number a little less than 3 for your first estimate, for example, 2.8.

Step 2 *Divide.*

$$2.8\overline{)7.00} = 2.5$$

Step 3 *Average.*

$$\frac{2.8 + 2.5}{2} = 2.65$$

Therefore, $\sqrt{7} \approx 2.7$ to the nearest tenth.

If a more precise estimation is required, repeat steps 1–3 using 2.65 as the second estimate.

Frequently, it is necessary to work with other roots. We call x a *cube root* of y if $x^3 = y$. For example, 4 is a cube root of 64 since $4^3 = 64$. We call x a *fourth root* of y if $x^4 = y$. Therefore, 2 is a fourth root of 16, since $2^4 = 16$. In a similar manner we can define fifth roots, sixth roots, and so on.

Definition of *n*th root

If n is a positive integer and if $x^n = y$, then x is called an nth root of y.

As was the case with square roots, a number may have more than one root. For example, -2 and 2 are both fourth roots of 16. On the other hand, -3 is the only real cube root of -27, and -9 has no real square root. The following chart shows the various possibilities for real roots of a number y.

The Real nth Roots of y

	$y > 0$	$y < 0$	$y = 0$
***n* even**	one positive root one negative root	no real roots	one real root, namely 0
***n* odd**	one positive root no negative roots	no positive roots one negative root	one real root, namely 0

Since a number may have more than one root, mathematicians generally make the following agreement.

The principal nth root of a number y is a nonnegative number unless n is odd and y is negative.

The symbol $\sqrt[n]{y}$ is read "the principal nth root of y." n is called the **index.** When n is 2, we write the familiar symbol $\sqrt{y}$.

Examples

4 **The following demonstrate principal roots.**

$\sqrt[3]{-27} = -3$ $\qquad$ $\sqrt[4]{16} = 2$

$\sqrt[4]{81} = 3$ $\qquad$ $-\sqrt[4]{16} = -2$

$\sqrt[4]{-16}$ is not a real number. $\qquad$ $(\sqrt[3]{-8})^3 = -8$

If the variables represent positive numbers, simplify the following.

5 $\sqrt{25y^4}$

Since $25y^4 = (5y^2)(5y^2)$,
$\sqrt{25y^4} = 5y^2$.

6 $\sqrt[3]{-8y^3}$

Since $-8y^3 = (-2y)(-2y)(-2y)$,
$\sqrt[3]{-8y^3} = -2y$.

7 **Solve $2x^4 = 162$.**

$$2x^4 = 162$$
$$x^4 = 81$$
$$x = \sqrt[4]{81} \quad \text{or} \quad x = -\sqrt[4]{81}$$
$$x = 3 \quad \text{or} \quad x = -3$$

The solution set is $\{3, -3\}$.

Exercises

Exploratory Simplify.

1. $\sqrt{9}$ 2. $-\sqrt{49}$ 3. $-\sqrt{4}$ 4. $\sqrt{144}$ 5. $-\sqrt{121}$
6. $-\sqrt{64}$ 7. $\sqrt{0.01}$ 8. $-\sqrt{0.04}$ 9. $\sqrt{\frac{1}{4}}$ 10. $-\sqrt{\frac{4}{9}}$
11. $\sqrt{\frac{36}{49}}$ 12. $\sqrt[3]{27}$ 13. $-\sqrt[3]{-8}$ 14. $\sqrt[3]{64}$ 15. $-\sqrt[5]{-32}$
16. $\sqrt[3]{-27}$ 17. $-\sqrt[3]{27}$ 18. $\sqrt[6]{64}$ 19. $\sqrt[3]{-125}$ 20. $\sqrt[11]{-1}$

Tell whether each of the following is rational, irrational, or neither.

21. $\sqrt{2}$ 22. $-\sqrt{100}$ 23. $\sqrt{-4}$ 24. $\sqrt[3]{-27}$ 25. $-\sqrt{0.09}$
26. $\sqrt[3]{6}$ 27. $\sqrt[5]{-32}$ 28. $-\sqrt{24}$ 29. $-\sqrt[4]{256}$ 30. $\sqrt[6]{-64}$

Written Simplify. Assume all variables represent positive numbers.

1. $-\sqrt{81}$ 2. $\sqrt{169}$ 3. $-\sqrt{225}$ 4. $-\sqrt[4]{81}$ 5. $\sqrt{x^2}$
6. $\sqrt{16y^2}$ 7. $\sqrt{9a^2}$ 8. $\sqrt{49y^4}$ 9. $\sqrt{x^8}$ 10. $-\sqrt{c^{10}}$
11. $\sqrt{4m^{12}}$ 12. $\sqrt[3]{8m^3}$ 13. $\sqrt{\frac{y^2}{4}}$ 14. $-\sqrt{\frac{a^4}{16}}$ 15. $-\sqrt{169y^{14}}$
16. $\sqrt[5]{32x^5}$ 17. $\sqrt[4]{16m^8}$ 18. $-\sqrt[5]{x^5y^{10}}$ 19. $-\sqrt[9]{-y^9}$ 20. $\sqrt{25x^4y^2}$
21. $\sqrt{64a^2b^8}$ 22. $\sqrt[3]{-27r^3s^6}$ 23. $\sqrt[3]{-8m^9n^3}$ 24. $-\sqrt{121b^2c^6}$ 25. $\sqrt[4]{16x^8y^{12}z^4}$

Find the smallest integral value of x for which each expression represents a real number.

26. $\sqrt{x+3}$ 27. $\sqrt{x-9}$ 28. $\sqrt{x+10}$ 29. $\sqrt{2x-4}$
30. $\sqrt{3x-12}$ 31. $\sqrt{2x-1}$ 32. $\sqrt{5x+2}$ 33. $\sqrt{4x-30}$

Use the divide-and-average method to find a rational approximation correct to the nearest tenth.

34. $\sqrt{11}$ 35. $\sqrt{21}$ 36. $\sqrt{12}$ 37. $\sqrt{27}$ 38. $\sqrt{30}$ 39. $\sqrt{286}$

Solve each equation if the domain of the variable is $\mathcal{R}$.

40. $x^2 = 49$ 41. $t^2 = 625$ 42. $36y^2 = 1$
43. $3a^2 = 48$ 44. $m^2 + 9 = 25$ 45. $x^2 + x^2 = 50$
46. $125m^3 + 1 = 0$ 47. $8y^3 - 1 = 0$ 48. $16x^4 + 81 = 0$

Challenge

1. Does $(\sqrt{x})^2 = x$ no matter what value x represents? Explain.
2. Does $\sqrt{x^2} = x$ no matter what value x represents? Explain.
3. Is it ever the case that $\sqrt{x^2} = -x$? Explain.
4. Under what circumstances is $\sqrt[n]{y^n} = y$?

Solve each of the following if the domain is $\mathcal{R}$.

5. $x^2 < x$ 6. $x > \sqrt{x}$ 7. $x < \sqrt{x}$ 8. $x^2 > x^3$

2.2 Simplifying Radical Expressions

The following examples illustrate two familiar properties of square roots.

$$\sqrt{36} = \sqrt{4} \cdot \sqrt{9} \qquad \sqrt{\frac{4}{9}} = \frac{\sqrt{4}}{\sqrt{9}}$$

Multiplication Property of Square Roots

If a and b are not negative, then $\sqrt{ab} = \sqrt{a} \cdot \sqrt{b}$.

Division Property of Square Roots

If a is not negative and b is positive, then $\sqrt{\frac{a}{b}} = \frac{\sqrt{a}}{\sqrt{b}}$.

These properties often help in simplifying computations involving radicals.

Example

1 Find a rational approximation for $\sqrt{300}$.

Instead of using the divide-and-average method, note that $300 = 100 \cdot 3$ and that 100 is a perfect square.

$$\begin{aligned}\sqrt{300} &= \sqrt{100 \cdot 3} \\ &= \sqrt{100} \cdot \sqrt{3} && \textit{Multiplication Property of Square Roots} \\ &= 10\sqrt{3} \\ &\approx 10(1.732) && \textit{Recall } \sqrt{3} \approx 1.732. \\ &= 17.32\end{aligned}$$

Thus, 17.32 is a rational approximation for $\sqrt{300}$.

To find a rational approximation for $\frac{1}{\sqrt{2}}$, we could proceed this way.

$$\frac{1}{\sqrt{2}} \approx \frac{1}{1.414} \qquad 1.414\overline{)1.000000} = 0.707$$

Dividing 1 by 1.414 is time consuming. The procedure illustrated below enables us to express $\frac{1}{\sqrt{2}}$ as an equivalent fraction with a rational denominator.

$$\frac{1}{\sqrt{2}} = \frac{1}{\sqrt{2}} \cdot \frac{\sqrt{2}}{\sqrt{2}} \qquad \textit{Note that } \frac{\sqrt{2}}{\sqrt{2}} = 1.$$

$$= \frac{\sqrt{2}}{2} \approx \frac{1.414}{2} = 0.707 \qquad \textit{Dividing 1.414 by 2 is easier than dividing 1 by 1.414.}$$

The process of eliminating radicals from the denominator of a fraction is called *rationalizing the denominator.*

Definition of Simplest Radical Form

A radical expression is said to be in simplified form if the following conditions are satisfied.

1. The radicand does not contain any perfect square factors other than 1.
2. The radicand does not contain a fraction.
3. A radical does not appear in a denominator; that is, the denominator is rationalized.

Examples

2 **Simplify $\sqrt{\frac{7}{8}}$.**

Method 1

$$\sqrt{\frac{7}{8}} = \frac{\sqrt{7}}{\sqrt{8}} = \frac{\sqrt{7}}{\sqrt{8}} \cdot \frac{\sqrt{8}}{\sqrt{8}} = \frac{\sqrt{56}}{8} = \frac{\sqrt{4} \cdot \sqrt{14}}{8} = \frac{\sqrt{14}}{4}$$

Method 2

$$\sqrt{\frac{7}{8}} = \frac{\sqrt{7}}{\sqrt{8}} = \frac{\sqrt{7}}{2\sqrt{2}} = \frac{\sqrt{7}}{2\sqrt{2}} \cdot \frac{\sqrt{2}}{\sqrt{2}} = \frac{\sqrt{14}}{4}$$

3 **If x and y represent positive numbers, simplify $\sqrt{72x^2y^2}$.**

$$\sqrt{72x^2y^2} = \sqrt{36 \cdot 2 \cdot x^2 \cdot y^2} = \sqrt{36} \cdot \sqrt{2} \cdot \sqrt{x^2} \cdot \sqrt{y^2} = 6 \cdot \sqrt{2} \cdot x \cdot y = 6xy\sqrt{2}$$

To simplify nth roots, we use the following properties of radicals.

Multiplication Property of Radicals

Division Property of Radicals

If $\sqrt[n]{x}$ and $\sqrt[n]{y}$ represent real numbers, then

$$\sqrt[n]{xy} = \sqrt[n]{x} \cdot \sqrt[n]{y}$$

and if $y \neq 0$,

$$\sqrt[n]{\frac{x}{y}} = \frac{\sqrt[n]{x}}{\sqrt[n]{y}}.$$

To multiply two or more radicals of the form $a\sqrt[n]{b}$, we use the multiplication property of radicals and the commutative and associative properties of multiplication.

Examples

4 **Simplify $\sqrt[3]{54x^3y^4}$.**

Find factors of 54 that are powers of 3. Then use the *Multiplication Property of Radicals.*

$$\begin{aligned}\sqrt[3]{54x^3y^4} &= \sqrt[3]{3^3 \cdot 2 \cdot x^3 \cdot y^3 \cdot y}\\ &= \sqrt[3]{3^3} \cdot \sqrt[3]{2} \cdot \sqrt[3]{x^3} \cdot \sqrt[3]{y^3} \cdot \sqrt[3]{y}\\ &= 3 \cdot \sqrt[3]{2} \cdot x \cdot y \cdot \sqrt[3]{y}\\ &= 3xy\sqrt[3]{2y}\end{aligned}$$

Multiply the following expressions.

5 $2\sqrt{2} \cdot 4\sqrt{6}$

$$\begin{aligned}2\sqrt{2} \cdot 4\sqrt{6} &= 2 \cdot 4 \cdot \sqrt{2} \cdot \sqrt{6}\\ &= 8\sqrt{12}\\ &= 8 \cdot 2\sqrt{3}\\ &= 16\sqrt{3}\end{aligned}$$

6 $5\sqrt[3]{9} \cdot \frac{1}{3}\sqrt[3]{3}$

$$\begin{aligned}5\sqrt[3]{9} \cdot \frac{1}{3}\sqrt[3]{3} &= 5 \cdot \frac{1}{3} \cdot \sqrt[3]{9} \cdot \sqrt[3]{3}\\ &= \frac{5}{3}\sqrt[3]{27}\\ &= \frac{5}{3} \cdot 3\\ &= 5\end{aligned}$$

Exercises

Exploratory **Simplify. Assume all variables represent positive numbers.**

1. $\sqrt{2} \cdot \sqrt{5}$
2. $\sqrt{3} \cdot \sqrt{10}$
3. $\sqrt{3} \cdot \sqrt{7}$
4. $3\sqrt{2} \cdot \sqrt{2}$
5. $4\sqrt{2} \cdot \sqrt{5}$
6. $3\sqrt{6} \cdot 4\sqrt{6}$
7. $\sqrt[3]{2} \cdot \sqrt[3]{4}$
8. $2\sqrt[3]{16} \cdot 3\sqrt[3]{4}$
9. $\sqrt{8}$
10. $\sqrt{24}$
11. $\sqrt{75}$
12. $\sqrt{27}$
13. $\sqrt{20}$
14. $\sqrt{72}$
15. $\sqrt{50}$
16. $\sqrt{48}$
17. $5\sqrt{28}$
18. $6\sqrt{32}$
19. $\sqrt{m^3}$
20. $7\sqrt{12x^2}$

Written **Simplify. Assume all variables represent positive numbers.**

1. $5\sqrt{54}$
2. $-3\sqrt{50}$
3. $\sqrt{81y}$
4. $\sqrt{y^5}$
5. $\sqrt{t^7}$
6. $\sqrt{8a^2b}$
7. $\sqrt{12x^2y}$
8. $\sqrt{9xy^2}$
9. $\sqrt{x^2y^5}$
10. $\sqrt{20a^3b}$
11. $\sqrt{48x^3y^2}$
12. $5\sqrt{40m^2}$
13. $2\sqrt{5} \cdot 3\sqrt{10}$
14. $6\sqrt{15} \cdot 2\sqrt{3}$
15. $3\sqrt{5} \cdot 2\sqrt{20}$
16. $3\sqrt{14} \cdot 6\sqrt{2}$
17. $\sqrt[3]{16}$
18. $\sqrt[3]{54}$
19. $\sqrt[3]{128}$
20. $\sqrt[4]{32}$
21. $\sqrt[5]{x^7}$
22. $\sqrt[3]{y^4}$
23. $\sqrt[3]{16y^3}$
24. $\sqrt[3]{56x^6}$
25. $\sqrt[3]{9} \cdot \sqrt[3]{6}$
26. $\sqrt[4]{2} \cdot \sqrt[4]{24}$
27. $5\sqrt[3]{y} \cdot \sqrt[3]{16y^4}$
28. $\sqrt[3]{6x^4} \cdot \sqrt[3]{9x}$
29. $\frac{1}{\sqrt{3}}$
30. $\frac{2}{\sqrt{5}}$

31. $\frac{5}{\sqrt{3}}$ **32.** $\frac{6}{\sqrt{5}}$ **33.** $\frac{7}{\sqrt{3}}$ **34.** $\frac{10}{\sqrt{10}}$ **35.** $\frac{6}{\sqrt{6}}$

36. $\frac{3}{\sqrt{7}}$ **37.** $\sqrt{\frac{2}{9}}$ **38.** $\sqrt{\frac{5}{9}}$ **39.** $\sqrt{\frac{3}{4}}$ **40.** $\sqrt{\frac{7}{36}}$

41. $8\sqrt{\frac{5}{16}}$ **42.** $14\sqrt{\frac{3}{49}}$ **43.** $\sqrt{\frac{8}{9}}$ **44.** $\sqrt{\frac{24}{25}}$ **45.** $\sqrt{\frac{3}{5}}$

46. $\sqrt{\frac{2}{3}}$ **47.** $\sqrt{\frac{5}{3}}$ **48.** $18\sqrt{\frac{3}{2}}$ **49.** $\frac{6}{\sqrt{18}}$ **50.** $\frac{7\sqrt{2}}{\sqrt{7}}$

51. $\frac{\sqrt{10}}{4\sqrt{2}}$ **52.** $\frac{4\sqrt{7}}{3\sqrt{2}}$ **53.** $\sqrt{\frac{5}{9}} \cdot \sqrt{5}$ **54.** $\sqrt{\frac{1}{2}} \cdot \sqrt{72}$ **55.** $\sqrt{\frac{5}{36}} \cdot \sqrt{45}$

56. $3\sqrt{\frac{1}{6}} \cdot \sqrt{24}$ **57.** $\frac{6}{5\sqrt{24}}$ **58.** $\frac{2}{3\sqrt{18}}$ **59.** $\frac{\sqrt{40}}{\sqrt{5}}$ **60.** $\frac{18\sqrt{48}}{36\sqrt{6}}$

61. $\frac{5\sqrt{98}}{7\sqrt{2}}$ **62.** $\frac{7\sqrt{300}}{10\sqrt{3}}$ **63.** $\frac{7\sqrt{75}}{10\sqrt{3}}$ **64.** $\frac{2\sqrt{48x^3}}{8\sqrt{2x}}$ **65.** $\frac{24\sqrt{96}}{18\sqrt{3}}$

66. $\frac{3}{\sqrt[3]{4}}$ **67.** $\frac{4}{\sqrt[3]{2}}$ **68.** $\frac{7}{\sqrt[3]{9}}$ **69.** $\sqrt[4]{\frac{2}{3}}$ **70.** $\frac{\sqrt{30y}}{\sqrt{5y}}$

71. $\sqrt{1 + \frac{1}{7}}$ **72.** $\sqrt{4\frac{4}{5}}$ **73.** $\sqrt[3]{\frac{3}{8}}$ **74.** $\sqrt[3]{\frac{7}{64}}$ **75.** $3\sqrt[3]{\frac{8}{25}}$

76. In the Multiplication Property of Square Roots, why is it required that $a \geq 0$ and $b \geq 0$?

77. In the Division Property of Square Roots, why is it required that $a \geq 0$ and $b > 0$?

78. Do you think $\sqrt{a + b} = \sqrt{a} + \sqrt{b}$ for $a \geq 0$ and $b \geq 0$? Try the rule for $a = 16$ and $b = 9$. What conclusion do you reach?

Mathematical Excursions — Properties of Roots

Study the following examples and look for a pattern.

Examples

1. $\sqrt[3]{(-3)^3} = \sqrt[3]{-27} = -3$
2. $\sqrt{(-5)^2} = \sqrt{25} = 5 = |-5|$
3. $\sqrt[4]{(-2)^4} = \sqrt[4]{16} = 2 = |-2|$

These examples suggest the following principle.

If n is odd then $\sqrt[n]{y^n} = y$.
If n is even then $\sqrt[n]{y^n} = |y|$.

Exercises Simplify each of the following if the variables represent real numbers.

1. $\sqrt{x^2}$ **2.** $\sqrt{121a^2}$ **3.** $\sqrt{169a^4b^2}$ **4.** $\sqrt[3]{27x^3}$

5. $-\sqrt{121b^2c^6}$ **6.** $\sqrt{(r + s)^2}$ **7.** $\sqrt[4]{(x + y)^4}$ **8.** $\sqrt{x^2 - 6xy + 9y^2}$

2.3 Operating with Radicals

$\sqrt[3]{9}$ and $\sqrt[4]{9}$ are not like radicals (different indexes). Also, $3\sqrt{5a}$ and $3\sqrt{5b}$ are not like radicals (different radicands).

Radicals that have the same index and the same radicand are called *like radicals*. For example, $5\sqrt{2}$ and $3\sqrt{2}$ are like radicals. So are $5\sqrt[3]{7}$ and $\sqrt[3]{7}$.

Like radicals can be added or subtracted using the distributive property. For example, $5\sqrt{2} + 4\sqrt{2} = (5 + 4)\sqrt{2} = 9\sqrt{2}$.

Sometimes, the radicals in a radical expression must be simplified in order to find like radicals that can be combined.

Examples

1 Simplify $4\sqrt{27} + 5\sqrt{3} - 8\sqrt{48}$.

$$
\begin{aligned}
4\sqrt{27} + 5\sqrt{3} - 8\sqrt{48} &= 4\sqrt{9 \cdot 3} + 5\sqrt{3} - 8\sqrt{16 \cdot 3} \\
&= 4 \cdot 3\sqrt{3} + 5\sqrt{3} - 8 \cdot 4\sqrt{3} && \textit{Simplify each radical.} \\
&= 12\sqrt{3} + 5\sqrt{3} - 32\sqrt{3} && \textit{Combine like radicals.} \\
&= -15\sqrt{3}
\end{aligned}
$$

2 Simplify $\frac{1}{\sqrt{8}} + \sqrt{32}$.

$$
\begin{aligned}
\frac{1}{\sqrt{8}} + \sqrt{32} &= \frac{1}{2\sqrt{2}} + 4\sqrt{2} \\
&= \frac{1}{2\sqrt{2}} \cdot \frac{\sqrt{2}}{\sqrt{2}} + 4\sqrt{2} \\
&= \frac{\sqrt{2}}{4} + 4\sqrt{2} \\
&= \frac{\sqrt{2}}{4} + \frac{16\sqrt{2}}{4} \\
&= \frac{17\sqrt{2}}{4}
\end{aligned}
$$

The distributive property can also be used to multiply radical expressions.

Examples

3 Multiply $\sqrt{6}(\sqrt{3} + 2\sqrt{15})$ and express the product in simplest form.

$$
\begin{aligned}
\sqrt{6}(\sqrt{3} + 2\sqrt{15}) &= \sqrt{6} \cdot \sqrt{3} + \sqrt{6} \cdot 2\sqrt{15} && \textit{Distributive Property} \\
&= \sqrt{18} + 2\sqrt{90} \\
&= 3\sqrt{2} + 6\sqrt{10}
\end{aligned}
$$

4 **Simplify $(3 + 5\sqrt{2})(2 + 3\sqrt{2})$.**

$$\begin{aligned}(3 + 5\sqrt{2})(2 + 3\sqrt{2}) &= 3 \cdot 2 + 3 \cdot 3\sqrt{2} + 5\sqrt{2} \cdot 2 + 5\sqrt{2} \cdot 3\sqrt{2} \\ &= 6 + 9\sqrt{2} + 10\sqrt{2} + 30 \\ &= 36 + 19\sqrt{2}\end{aligned}$$

Simplify the following products.

5 $(3 + \sqrt{2})(3 - \sqrt{2})$

6 $(4 + 2\sqrt{3})(4 - 2\sqrt{3})$

Each product is of the form $(a + b)(a - b)$, which is the difference of two squares, $a^2 - b^2$.

$(3 + \sqrt{2})(3 - \sqrt{2}) = 9 - 2 = 7$

$(4 + 2\sqrt{3})(4 - 2\sqrt{3}) = 16 - 12 = 4$

Look carefully at examples 5 and 6. The product of two irrational numbers is a rational number. *Can you explain why this happens?*

Binomial pairs such as $3 + \sqrt{2}$ and $3 - \sqrt{2}$ are called **conjugates** of each other. The conjugate of $4 - 2\sqrt{3}$ is $4 + 2\sqrt{3}$.

To rationalize the denominator of a fraction, such as $\frac{2}{3 + \sqrt{5}}$, we multiply both numerator and denominator by the conjugate, $3 - \sqrt{5}$.

Examples

Simplify the following fractions.

7 $\dfrac{2}{3 + \sqrt{5}}$

$$\begin{aligned}\frac{2}{3 + \sqrt{5}} &= \frac{2}{3 + \sqrt{5}} \cdot \frac{3 - \sqrt{5}}{3 - \sqrt{5}} \\ &= \frac{2(3 - \sqrt{5})}{(3 + \sqrt{5})(3 - \sqrt{5})} \\ &= \frac{6 - 2\sqrt{5}}{9 - 5} \\ &= \frac{6 - 2\sqrt{5}}{4} \\ &= \frac{2(3 - \sqrt{5})}{2 \cdot 2} \\ &= \frac{3 - \sqrt{5}}{2}\end{aligned}$$

8 $\dfrac{1 + \sqrt{2}}{3 - \sqrt{2}}$

$$\begin{aligned}\frac{1 + \sqrt{2}}{3 - \sqrt{2}} &= \frac{1 + \sqrt{2}}{3 - \sqrt{2}} \cdot \frac{3 + \sqrt{2}}{3 + \sqrt{2}} \\ &= \frac{(1 + \sqrt{2})(3 + \sqrt{2})}{(3 - \sqrt{2})(3 + \sqrt{2})} \\ &= \frac{3 + \sqrt{2} + 3\sqrt{2} + 2}{9 - 2} \\ &= \frac{5 + 4\sqrt{2}}{7}\end{aligned}$$

9 **Solve $x^2 - 8x + 8 = 0$.**

$x^2 - 8x + 8 = 0$ *Use the quadratic formula,* $x = \frac{-b \pm \sqrt{b^2 - 4ac}}{2a}$.

$$x = \frac{-(-8) \pm \sqrt{64 - 32}}{2}$$

$$= \frac{8 \pm \sqrt{32}}{2}$$

$$= \frac{8 \pm 4\sqrt{2}}{2}$$

$$= \frac{\overset{2}{\cancel{4}}(2 \pm \sqrt{2})}{\underset{1}{\cancel{2}}} \text{ or } 4 \pm 2\sqrt{2}$$

The roots are $4 \pm 2\sqrt{2}$. *The check is left for you.*

Exercises

Exploratory **Simplify, if possible.**

1. $\sqrt{3} + 5\sqrt{3}$
2. $2\sqrt{7} + 4\sqrt{7}$
3. $8\sqrt{2} - 6\sqrt{2}$
4. $3\sqrt{2} + \sqrt{8}$
5. $2\sqrt{3} + 7\sqrt{27}$
6. $2\sqrt{45} - 6\sqrt{5}$
7. $\sqrt{5} + \sqrt{10}$
8. $\sqrt{24} + \sqrt{8}$
9. $\sqrt{48} - \sqrt{12}$
10. $\sqrt{\frac{1}{3}} + \sqrt{3}$
11. $5\sqrt{\frac{2}{5}} + 3\sqrt{10}$
12. $3\sqrt{\frac{2}{3}} + \sqrt{24}$

Written **Simplify, if possible. Assume all variables represent positive numbers.**

1. $3\sqrt{72} - 5\sqrt{32}$
2. $7\sqrt{40} - 2\sqrt{5}$
3. $8\sqrt{27} - 10\sqrt{75}$
4. $\sqrt{6x} + \sqrt{24x}$
5. $8\sqrt{3y} - 3\sqrt{75y}$
6. $\sqrt{16m} - \sqrt{4m}$
7. $\sqrt{\frac{1}{2}} + \frac{\sqrt{8}}{2}$
8. $5\sqrt{8} - \sqrt{\frac{1}{2}}$
9. $3\sqrt{12} - 2\sqrt{75} + \sqrt{27}$
10. $\frac{1}{2}\sqrt{32} - \frac{1}{5}\sqrt{50} - \frac{1}{4}\sqrt{8}$
11. $\frac{2\sqrt{5}}{5} - \sqrt{\frac{1}{5}}$
12. $15\sqrt{\frac{1}{5}} - 3\sqrt{20}$
13. $4\sqrt{\frac{1}{8}} - \sqrt{\frac{1}{2}}$
14. $\sqrt{\frac{1}{5}} - \sqrt{\frac{75}{5}}$
15. $15\sqrt{\frac{1}{3}} + \frac{2}{3}\sqrt{27} - \sqrt{3}$
16. $\sqrt{\frac{2}{5}} + \sqrt{40} + \sqrt{10}$
17. $7\sqrt[3]{3} - \sqrt[3]{24}$
18. $\sqrt[3]{48} - \sqrt[3]{6}$
19. $7\sqrt[4]{2} + 8\sqrt[4]{32}$
20. $\sqrt{2x^2y^4} + \sqrt{8x^2y^4}$
21. $\sqrt{100m^3n} - \sqrt{64mn^3}$
22. $\sqrt[3]{27m^5n^6} + \sqrt[3]{8m^8n^3}$
23. $3\sqrt{2}(\sqrt{2} - 1)$
24. $\sqrt{3}(3 + \sqrt{3})$
25. $\sqrt{3}(2\sqrt{3} + \sqrt{12})$
26. $\sqrt{5}(\sqrt{20} + 3\sqrt{2})$

27. $-\sqrt{2}(4\sqrt{3} + \sqrt{2})$
28. $2\sqrt{6}\left(8 - \sqrt{\frac{1}{2}}\right)$
29. $(2 + \sqrt{5})(3 + \sqrt{5})$
30. $(8 - \sqrt{2})(6 + \sqrt{2})$
31. $(4 + \sqrt{7})(1 - \sqrt{7})$
32. $(3 - \sqrt{3})(5 - \sqrt{3})$
33. $(2\sqrt{7} - 1)(2\sqrt{7} - 3)$
34. $(4 + 2\sqrt{3})(1 - \sqrt{3})$
35. $(3\sqrt{5} - 2)(2\sqrt{5} + 1)$
36. $(2 + \sqrt{3})^2$
37. $(5 + \sqrt{2})^2$
38. $(\sqrt{7} - 2)^2$
39. $(\sqrt{10} - 5)^2$
40. $(2 + \sqrt{3})(2 - \sqrt{3})$
41. $(5 + \sqrt{7})(5 - \sqrt{7})$
42. $(5 - \sqrt{6})(5 + \sqrt{6})$
43. $(8 - \sqrt{2})(8 + \sqrt{2})$
44. $(2 + 3\sqrt{5})(2 - 3\sqrt{5})$
45. $(3\sqrt{6} - 2)(3\sqrt{6} + 2)$
46. $(2\sqrt{3} - \sqrt{2})(2\sqrt{3} + \sqrt{2})$
47. $(6\sqrt{2} + 2\sqrt{7})(6\sqrt{2} - 2\sqrt{7})$

48. $\frac{7}{2 + \sqrt{3}}$
49. $\frac{2}{1 + \sqrt{3}}$
50. $\frac{3}{1 - \sqrt{7}}$
51. $\frac{2}{2 - \sqrt{2}}$
52. $\frac{8}{4 - \sqrt{6}}$
53. $\frac{3}{4 - \sqrt{7}}$
54. $\frac{2}{3 - \sqrt{5}}$
55. $\frac{1}{\sqrt{5} + 3}$
56. $\frac{-2}{\sqrt{2} - 2}$
57. $\frac{8}{\sqrt{15} + 5}$
58. $\frac{\sqrt{5}}{\sqrt{5} + 1}$
59. $\frac{2\sqrt{3}}{1 - \sqrt{3}}$
60. $\frac{4\sqrt{5}}{\sqrt{5} - 1}$
61. $\frac{1 - \sqrt{2}}{3 - \sqrt{2}}$
62. $\frac{2 - \sqrt{3}}{5 + \sqrt{3}}$
63. $\frac{1 + \sqrt{5}}{4 - \sqrt{5}}$
64. $\frac{1 + \sqrt{5}}{3 - \sqrt{5}}$
65. $\frac{1 + \sqrt{7}}{1 - \sqrt{7}}$
66. $\frac{\sqrt{6} + 4}{\sqrt{6} + 2}$
67. $\frac{1 + 3\sqrt{2}}{2\sqrt{2} - 3}$

Solve and check. Express roots in simplest radical form.

68. $x^2 - 4x - 2 = 0$
69. $x^2 - 4x + 1 = 0$
70. $x^2 - 5x - 5 = 0$
71. $x^2 - 4x = 8$
72. $x^2 - 10x = -5$
73. $15y = 15 - 3y^2$
74. $8r^2 = 60$
75. $12x^2 - 11x = 3$
76. $7y^2 = y + 2$

Write an equation having the given roots.

77. $\sqrt{3}, 2\sqrt{3}$
78. $\sqrt{2}, -5\sqrt{2}$
79. $2 + \sqrt{3}, 2 - \sqrt{3}$
80. $5 - \sqrt{2}, 5 + \sqrt{2}$
81. $4 + \sqrt{3}, 4 - \sqrt{3}$
82. $\frac{1 + \sqrt{7}}{2}, \frac{1 - \sqrt{7}}{2}$

Challenge Simplify.

1. $\dfrac{\frac{2}{\sqrt{2}} - 2}{\frac{1 - \sqrt{2}}{\sqrt{2}}}$
2. $\dfrac{\frac{5}{\sqrt{3}} + \sqrt{3}}{\frac{2}{\sqrt{3}} + 1}$
3. $\dfrac{\frac{1}{\sqrt{7}} + 4}{\frac{4}{\sqrt{7}} - 1}$

4. $\frac{3}{\sqrt{2} + \sqrt{3} + \sqrt{5}}$
5. $\frac{2}{\sqrt{3} + \sqrt{7} + \sqrt{10}}$
6. $\frac{4}{\sqrt{3} + \sqrt{5} - \sqrt{2}}$

Hint: Use the associative property to group the denominators.

2.4 Radical Equations

Equations in which a variable appears in the radicand of a radical are called **radical equations.** For example, $\sqrt{y-2} - 3 = 0$ is a radical equation.

Example

1 Solve $\sqrt{y-2} - 3 = 0$. Check your answer.

$$\sqrt{y-2} - 3 = 0$$

$$\sqrt{y-2} = 3 \quad \textit{Isolate the radical.}$$

$$(\sqrt{y-2})^2 = (3)^2 \quad \textit{Square both sides.}$$

$$y - 2 = 9$$

$$y = 11$$

The solution set is {11}.

Check $y = 11$.

$$\sqrt{11-2} - 3 \stackrel{?}{=} 0$$

$$\sqrt{9} - 3 \stackrel{?}{=} 0$$

$$3 - 3 = 0$$

Squaring both sides of an equation may produce an equation that is not equivalent to the original. For example, squaring both sides of $x = 1$ produces $x^2 = 1$. The solution sets are {1} and {1, −1}, respectively. Therefore, when we square both sides of an equation, it is necessary to check possible solutions in the original equation.

Example

2 Solve $\sqrt{2x+1} + 1 = x$.

$$\sqrt{2x+1} + 1 = x$$

$$\sqrt{2x+1} = x - 1 \quad \textit{Isolate the radical.}$$

$$2x + 1 = (x-1)^2 \quad \textit{Square both sides.}$$

$$2x + 1 = x^2 - 2x + 1$$

$$0 = x^2 - 4x$$

$$0 = x(x-4)$$

$$x = 0 \lor x = 4$$

Check $x = 0$.

$$\sqrt{2(0)+1} + 1 \stackrel{?}{=} 0$$

$$\sqrt{1} + 1 \stackrel{?}{=} 0$$

0 does not check.
Reject $x = 0$

Check $x = 4$.

$$\sqrt{2(4)+1} + 1 \stackrel{?}{=} 4$$

$$\sqrt{9} + 1 \stackrel{?}{=} 4$$

$$3 + 1 \stackrel{?}{=} 4$$

$$4 = 4$$

The solution set is {4}.

Exercises

Exploratory **Solve each equation and check your solutions.**

1. $\sqrt{x} = 9$
2. $\sqrt{y} = 16$
3. $\sqrt{y} - 2 = 0$
4. $\sqrt{2y} = 8$
5. $2\sqrt{y} = 8$
6. $\sqrt{5x} = 2$
7. $\sqrt{3y} + 1 = 0$
8. $1 - \sqrt{x} = 10$
9. $3 - \sqrt{x} = 5$

Written **Solve each equation and check your solutions.**

1. $\sqrt{3y} = 6$
2. $5 + \sqrt{x} = 8$
3. $\sqrt{5 + x} = 3$
4. $\sqrt{2 - y} = 1$
5. $\sqrt{y + 2} = 3$
6. $\sqrt{m + 5} = 4$
7. $\sqrt{2y + 7} = 5$
8. $\sqrt{1 - 7y} = 6$
9. $\sqrt{11 - 10x} = 9$
10. $5 - \sqrt{y} = -1$
11. $\sqrt{3m - 2} = -1$
12. $y = \sqrt{4y + 12}$
13. $x = \sqrt{2x + 15}$
14. $\sqrt{5m - 6} = m$
15. $y = 3\sqrt{4 - y}$
16. $x - 2 = \sqrt{x}$
17. $t = 2\sqrt{3t - 8}$
18. $x - 4 = 3\sqrt{x}$
19. $4\sqrt{x} + 5 = x$
20. $\sqrt{2y} + 4 = y$
21. $\sqrt{y + 2} + 10 = y$
22. $\sqrt{5m + 1} + 6 = 10$
23. $7 + \sqrt{y - 3} = 1$
24. $\sqrt{y - 6} + 8 = y$
25. $\sqrt{x - 5} + 7 = x$
26. $u - 6 = \sqrt{u^2 - 3u}$
27. $y - 2\sqrt{4y - 3} = 3$
28. $y + 3\sqrt{y + 7} = 3$
29. $\sqrt{x^2 - 21} - x = -3$
30. $m - 2\sqrt{3m - 5} = -1$
31. $\sqrt{2x + 2} = \sqrt{3x - 5}$
32. $3y + \sqrt{4y - 7} = 7$
33. $\sqrt[3]{2 + y} = 2$
34. $\sqrt{x^2 - 3x} = 3x - 8$
35. $2 + \sqrt{y} = \sqrt{4 + y}$
36. $2\sqrt{x} = \sqrt{3x - 11} + 2$

Challenge **Solve each equation and check your solutions.**

1. $\sqrt{x + 8} + 3 = \sqrt{x + 35}$
2. $\sqrt{y + 12} + 1 = \sqrt{y + 21}$
3. $\sqrt{2y + 5} + \sqrt{y - 1} = 4$

Mixed Review

Solve each equation.

1. $x + 2\sqrt{3} = 5$
2. $2 + 5y\sqrt{10} = 0$
3. $x\sqrt{2} + 3x = 4$
4. $x - x\sqrt{5} = 2$
5. $x^2 - x\sqrt{2} - 4 = 0$
6. $x^2\sqrt{3} - 2x + \sqrt{6} = 0$
7. $\frac{3}{t} + \frac{5t}{t - 2} = 5$
8. $\frac{x}{x + 3} - \frac{27}{x^2 + x - 6} = \frac{x + 1}{2 - x}$
9. Don Mattingfield has 183 hits in 534 at bats. How many consecutive hits would he need to raise his batting average (hits ÷ at bats) to over 0.360?
10. Carita needs to make 1000 milliliters of a 30% alcohol solution by mixing 25% and 50% alcohol solutions. How much of each solution should she use?
11. The load capacity of two trucks are in the ratio of 5 to 2. The smaller truck has a capacity 3 tons less than that of the larger truck. What is the capacity of the larger truck?
12. Pipe A can fill a tank in 4 hours and pipe B can fill the tank in 3 hours. With the tank empty, pipe A is turned on, and one hour later, pipe B is turned on. How long will pipe B run before the tank is full?

2.5 Imaginary Numbers

Through the years, mathematicians have developed new number systems in order to create roots for equations that had none in previously existing systems.

For example, the equation $x^2 - 2 = 0$ has no solution within the set of rational numbers. When the real numbers were developed, the equation $x^2 - 2 = 0$ acquired two solutions, $\sqrt{2}$ and $-\sqrt{2}$.

The equation $x^2 + 1 = 0$ has no solution within the set of real numbers. To "create" a solution for this equation, it seems reasonable to proceed in the following manner.

$$x^2 = -1$$
$$x = \sqrt{-1} \text{ or } x = -\sqrt{-1}$$

However, the expressions $\sqrt{-1}$ and $-\sqrt{-1}$ do *not* name real numbers.

Early mathematicians were reluctant to accept $\sqrt{-1}$ and $-\sqrt{-1}$ as numbers. For this reason, they referred to them as "imaginary" numbers, in contrast to the more comfortable "real" numbers.

To simplify notation, we agree that

$$i = \sqrt{-1}. \qquad i^2 = -1$$

The number i, which is *not a real number*, is called the **imaginary unit.**

The laws of exponents continue to apply. For example, $(i\sqrt{2})^2 = i^2(\sqrt{2})^2 = (-1)(2) = -2$. Therefore, $(i\sqrt{2})^2 = -2$. Taking the square root of both sides, we have $i\sqrt{2} = \sqrt{-2}$.

Simplifying $\sqrt{-r}$

If r is a positive number, then $\sqrt{-r} = i\sqrt{r}$.

To avoid confusion, we write $i\sqrt{2}$ instead of $\sqrt{2}\, i$.

Examples

Simplify the following expressions.

1 $\sqrt{-32}$

$$\sqrt{-32} = i\sqrt{32}$$
$$= i\sqrt{16 \cdot 2}$$
$$= 4i\sqrt{2}$$

2 $\sqrt{-\frac{8}{9}}$

$$\sqrt{-\frac{8}{9}} = i\sqrt{\frac{8}{9}}$$
$$= \frac{i\sqrt{8}}{3}$$
$$= \frac{2i\sqrt{2}}{3}$$

Expressions such as $4i\sqrt{2}$ and $\frac{2i\sqrt{2}}{3}$ are of the form bi, where b is a real number such that $b \neq 0$. Such numbers are called **pure imaginary numbers.**

In working with real numbers, we used the property $\sqrt{a} \cdot \sqrt{b} = \sqrt{ab}$, where $a \geq 0$ and $b \geq 0$. This property does *not* hold if a and b are both negative.

Examples

Simplify the following products.

3 $\sqrt{-4}\sqrt{-9}$

$$\begin{aligned}\sqrt{-4}\sqrt{-9} &= i\sqrt{4} \cdot i\sqrt{9} \\ &= 2i \cdot 3i \\ &= 6i^2 \\ &= -6\end{aligned}$$

Note that $\sqrt{-4} \cdot \sqrt{-9} \neq \sqrt{36}$.

4 $\sqrt{-6}\sqrt{-18}$

$$\begin{aligned}\sqrt{-6}\sqrt{-18} &= i\sqrt{6} \cdot i\sqrt{18} \\ &= i\sqrt{6} \cdot 3i\sqrt{2} \\ &= 3i^2\sqrt{12} \\ &= -3 \cdot 2\sqrt{3} \\ &= -6\sqrt{3}\end{aligned}$$

An interesting pattern occurs when the powers of i are simplified. Look carefully at the list below and discover the pattern. Note that $i^0 = 1$ by definition.

$i^0 = 1$	$i^4 = i^2 \cdot i^2 = 1$	$i^8 = i^4 \cdot i^4 = 1$
$i^1 = i$	$i^5 = i^4 \cdot i = i$	$i^9 = i^8 \cdot i = i$
$i^2 = -1$	$i^6 = i^5 \cdot i = -1$	$i^{10} = i^9 \cdot i = -1$
$i^3 = i^2 \cdot i = -i$	$i^7 = i^6 \cdot i = -i$	$i^{11} = i^{10} \cdot i = -i$

What would i^{12} *equal?*

The powers of i repeat their values in cycles of four. Any power of i has the same value as i^0, i^1, i^2, or i^3. This can be expressed as a clock 4 system.

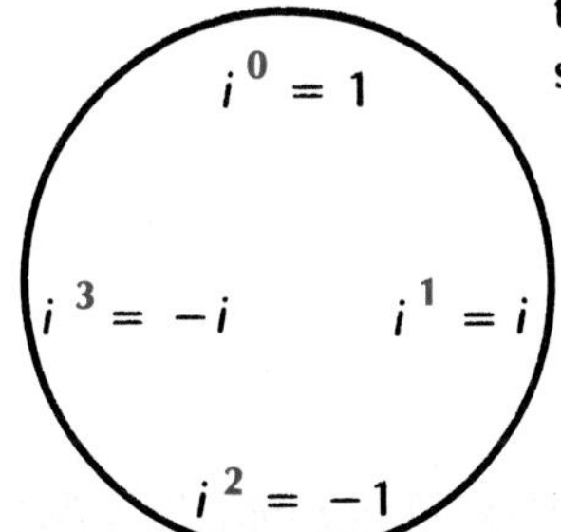

To find the position of i^{18}, divide 18 by 4. The remainder, 2, means that i^{18} is equivalent to i^2 or -1.

Examples

Simplify the following expressions.

5 i^{29}

Divide. $29 \div 4 = 7R1$

$i^{29} = i^1 = i$

6 i^{64}

Divide. $64 \div 4 = 16\ R0$

$i^{64} = i^0 = 1$

Exercises

Exploratory **Simplify.**

1. $\sqrt{-9}$
2. $\sqrt{-16}$
3. $\sqrt{-36}$
4. $\sqrt{-25}$
5. $3\sqrt{-64}$
6. $\frac{1}{2}\sqrt{-4}$
7. $-\frac{2}{3}\sqrt{-36}$
8. $\sqrt{4}\sqrt{-1}$
9. $\sqrt{-2}\sqrt{-2}$
10. $\sqrt{-3}\sqrt{3}$

Written **Simplify.**

1. $\sqrt{-49}$
2. $\sqrt{-144}$
3. $\sqrt{-3}$
4. $4\sqrt{-2}$
5. $\sqrt{-\frac{1}{9}}$
6. $\sqrt{-\frac{1}{4}}$
7. $-8\sqrt{-\frac{3}{16}}$
8. $-10\sqrt{-\frac{7}{25}}$
9. $\sqrt{-18}$
10. $\sqrt{-8}$
11. $\sqrt{-12}$
12. $\sqrt{-24}$
13. $-2\sqrt{-2}$
14. $-3\sqrt{-3}$
15. $-\sqrt{-28}$
16. $-3\sqrt{-20}$
17. $-4\sqrt{-12}$
18. $2\sqrt{-50}$
19. $2\sqrt{-98}$
20. $-\sqrt{-125}$
21. $\frac{3}{4}\sqrt{-64}$
22. $\sqrt{-\frac{1}{3}}$
23. $\sqrt{-\frac{1}{2}}$
24. $-5\sqrt{-\frac{1}{5}}$
25. $-6\sqrt{-\frac{1}{8}}$
26. $\sqrt{-3}\sqrt{-3}$
27. $\sqrt{-4}\sqrt{-1}$
28. $\sqrt{-9}\sqrt{-1}$
29. $\sqrt{-8}\sqrt{-2}$
30. $\sqrt{-3}\sqrt{-4}$
31. $\sqrt{-15}\sqrt{-5}$
32. $\sqrt{-10}\sqrt{-9}$
33. $\sqrt{5}\sqrt{-20}$
34. $\sqrt{-3}\sqrt{48}$
35. $\sqrt{-2}\sqrt{-72}$
36. $(-\sqrt{-27})(-\sqrt{-3})$
37. $(-2\sqrt{-5})(\sqrt{-80})$
38. $(\sqrt{-15})(2\sqrt{-20})$
39. $(\sqrt{-45})(-\sqrt{-5})$
40. $\frac{\sqrt{-16}}{\sqrt{-9}}$
41. $\frac{\sqrt{-25}}{\sqrt{-4}}$
42. $\frac{\sqrt{-81}}{\sqrt{-9}}$
43. $\frac{\sqrt{-12}}{\sqrt{3}}$
44. $\frac{\sqrt{-24}}{\sqrt{6}}$
45. $\frac{\sqrt{-48}}{\sqrt{-3}}$
46. $\frac{\sqrt{-50}}{\sqrt{-8}}$
47. $\frac{\sqrt{-60}}{\sqrt{-75}}$
48. $\frac{3\sqrt{-6}}{6\sqrt{-2}}$
49. $\frac{2\sqrt{-8}}{4\sqrt{-3}}$
50. i^{13}
51. i^{16}
52. i^{26}
53. i^{21}
54. i^{72}
55. $i^{15} \cdot i^3$
56. $i^{91} \cdot i^{14}$
57. $i^{38} + i^{100}$
58. $i^3\sqrt{-8}$
59. $i^7\sqrt{-12}$

Challenge Let $S = \{1, i, -1, -i\}$. Show that $(S, \cdot)$ is a commutative group.

Suppose a mathematical system consists of a set of numbers, S, together with the operations of addition and multiplication. If the following properties hold for all elements a, b, and c in S, then the number system is called a **field.**

Field Properties	Addition	Multiplication
closure	$a + b$ is in S.	ab is in S.
commutative	$a + b = b + a$	$a \cdot b = b \cdot a$
associative	$(a + b) + c = a + (b + c)$	$(a \cdot b) \cdot c = a \cdot (b \cdot c)$
identity	0 is in S. $a + 0 = a = 0 + a$	1 is in S. $a \cdot 1 = a = 1 \cdot a$
inverse	$-a$ is in S. $a + -a = 0 = -a + a$	$\frac{1}{a}$ is in S. $a \cdot \frac{1}{a} = 1 = \frac{1}{a} \cdot a$ if $a \neq 0$
distributive of multiplication over addition	$a(b + c) = ab + ac$ and $(b + c)a = ba + ca$	

You know that these properties hold for all real numbers. So, the real number system is a field.

To determine if a number system is a field, decide whether the closure property holds for addition and multiplication. Then check to see if the commutative, associative, and distributive properties hold. Also, check whether identity elements and inverses exist. If you can find a counterexample, the system is *not* a field.

Example: **Is the whole number system, $(\mathcal{W}, +, \cdot)$, a field?**

The additive inverse of 5 is -5. Since -5 is not a whole number, the whole number system is not a field.

Example: **Is $(J, +, \cdot)$, where J = {odd integers}, a field?**

The sum of 1 and 3 is 4. Since 4 is not an odd integer, $(J, +, \cdot)$ is not closed under addition. Thus, $(J, +, \cdot)$ is not a field.

Exercises **Determine whether each number system is a field. If the system is not a field, give a counterexample for each field property that is not satisfied.**

1. $(\mathcal{Z}, +, \cdot)$
2. $(\mathcal{Q}, +, \cdot)$
3. $(I, +, \cdot)$ where I = {irrational numbers}
4. $(\mathcal{R}, +, \div)$
5. $(K, +, \cdot)$ where K = {even integers}
6. $(V, +, \cdot)$ where $V = \{x \in \mathcal{R}: -1 \leq x \leq 1\}$
7. $(P, +, \cdot)$ where P = {pure imaginary numbers}
8. $(\mathcal{R}, +, !)$ where $a!b = a + b - ab$ for all real numbers a and b
9. $(A, +, \cdot)$ where A = {all numbers of the form $a\sqrt{2}$ where a is a rational number}
10. $(B, +, \cdot)$ where B = {all numbers of the form $a + b\sqrt{3}$ where a and b are rational numbers}

2.6 The Complex Numbers

In the preceding section you were introduced to pure imaginary numbers. These numbers, together with all real numbers, give rise to a new set of numbers.

Definition of Complex Numbers

A complex number is a number that can be written in the form a + bi, where a and b are real numbers and $i = \sqrt{-1}$. The set of complex numbers is denoted by the symbol $\mathscr{C}$.

In $a + bi$, a is called the real part and b is called the imaginary part.

The following are examples of complex numbers.

$$3i \quad 7 \quad -5i \quad 4 + 6i \quad i\sqrt{7} \quad \sqrt{10} - i\sqrt{11}$$

When $b = 0$, the complex number $a + bi$ is simply the real number a. Therefore, every real number is also a complex number.

When $b \neq 0$, the complex number $a + bi$ is called an **imaginary number.** Both $5 + 6i$ and $5i$ are examples of imaginary numbers.

When $a = 0$ and $b \neq 0$, the complex number bi is called a **pure imaginary number.** Thus, $4i$, $-\pi i$, and $5i$ are examples of pure imaginary numbers.

The preceding relationships are summarized in the following diagram.

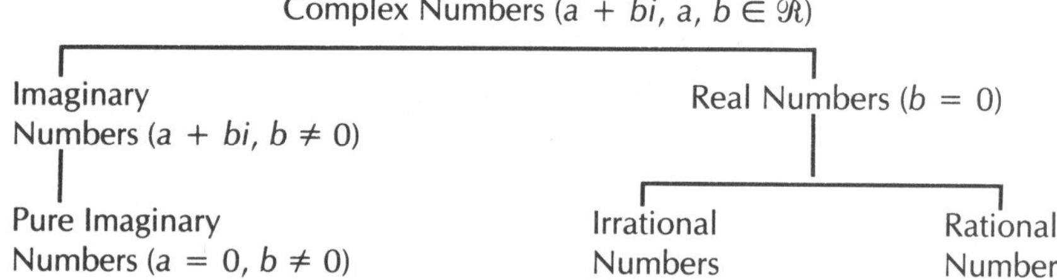

Two complex numbers are equal if and only if their real parts are equal and their imaginary parts are equal. That is, $a + bi = c + di$, if and only if $a = c$ and $b = d$.

Example

1 **Solve for x and y if $2x + 3yi = 6 + 2i$.**

$$2x + 3yi = 6 + 2i$$
$$2x = 6 \quad \text{and} \quad 3y = 2$$
$$x = 3 \quad \text{and} \quad y = \frac{2}{3}$$

To guarantee that the complex number system is a field, mathematicians have defined addition and multiplication of complex numbers so that all the requirements for a field are satisfied. To add or multiply complex numbers, treat i as a variable and use the commutative, associative, and distributive properties.

Examples

2 Express the sum of $3 + 5i$ and $4 + 7i$ in $a + bi$ form.

$$\begin{aligned} 3 + 5i + 4 + 7i &= 3 + 4 + 5i + 7i \\ &= 7 + 12i \end{aligned}$$

3 Express $(4 - 2i\sqrt{3}) - 2(3 - 5i\sqrt{3})$ in $a + bi$ form.

$$\begin{aligned} (4 - 2i\sqrt{3}) - 2(3 - 5i\sqrt{3}) &= 4 - 2i\sqrt{3} - 6 + 10i\sqrt{3} \\ &= 4 - 6 + 10i\sqrt{3} - 2i\sqrt{3} \\ &= -2 + 8i\sqrt{3} \end{aligned}$$

4 Express $(2 - 3i)(3 + 4i)$ in $a + bi$ form.

$$\begin{aligned} (2 - 3i)(3 + 4i) &= 2 \cdot 3 + 2 \cdot 4i - 3i \cdot 3 - 3i \cdot 4i \\ &= 6 + 8i - 9i - 12i^2 \\ &= 6 - i - 12(-1) \\ &= 6 - i + 12 \\ &= 18 - i \end{aligned}$$

5 Find the additive inverse of $-4 + 3i$.

The number $0 + 0i$, or 0, is the *additive identity*. Let $a + bi$ be the additive inverse of $-4 + 3i$.

Then, $(-4 + 3i) + (a + bi) = 0 + 0i$ *Definition of additive inverse.*

$$(-4 + a) + (3 + b)i = 0 + 0i$$

$$-4 + a = 0 \quad \text{and} \quad 3 + b = 0$$

$$a = 4 \quad \text{and} \quad b = -3$$

The additive inverse of $-4 + 3i$ is $4 - 3i$.

The preceding examples suggest the following definitions.

Addition of Complex Numbers

Multiplication of Complex Numbers

For any complex numbers $a + bi$ and $c + di$,

$$(a + bi) + (c + di) = (a + c) + (b + d)i,$$

and

$$(a + bi)(c + di) = (ac - bd) + (ad + bc)i.$$

Using these definitions, it can be shown that $(\mathscr{C}, +)$ is a commutative group. To show that $(\mathscr{C}, +, \cdot)$ is a field, we must also show that $(\mathscr{C}/\{0\}, \cdot)$ is a commutative group. This will be investigated in the next section.

Exercises

Exploratory **Tell whether each statement is *true* or *false*.**

1. A real number is a complex number.
2. A complex number is an imaginary number.
3. The sum of two complex numbers is a complex number.
4. An irrational number is a complex number.
5. A pure imaginary number is an imaginary number.
6. An imaginary number is a pure imaginary number.
7. The additive inverse of $3 + 2i$ is $3 - 2i$.
8. The number 0 is not a complex number.

Solve for x and y.

9. $2x + 5yi = 4 + 15i$
10. $x + yi = 2 - 3i$
11. $x - yi = 4 + 5i$
12. $3x + 2yi = 18 + 7i$
13. $x + 2yi = 3$
14. $2x + yi = 5i$

Find the additive inverse of each number.

15. $1 + i$
16. $-3 + 2i$
17. 12
18. $-5i$

Written **Express in simplest $a + bi$ form.**

1. $(2 + 3i) + (4 + 5i)$
2. $(5 + 2i) + (4 + 6i)$
3. $(1 + i) + (-2 + 3i)$
4. $(2 + 5i) + (5 - 2i)$
5. $(2 - 10i) - (3 + i)$
6. $(-1 + 8i) - (-2 - 3i)$
7. $(-5 + 7i) + (+3 + 4i)$
8. $(-8 - 17i) + (+12 + 27i)$
9. $(2 + 3i) - (1 + i\sqrt{3})$
10. $(\sqrt{5} - i) + (\sqrt{20} - 3i)$
11. $\left(\frac{3}{4} - \frac{1}{3}i\right) - \left(\frac{5}{8} + \frac{5}{6}i\right)$
12. $(1.001 - 0.98i) - (0.7 - 0.3i)$
13. $(1 + 2i)(2 + i)$
14. $(3 + i)(1 + 3i)$
15. $(2 + 3i)(1 + 5i)$
16. $(3 + 4i)(2 - 5i)$
17. $(-2 - 5i)(3 - 4i)$
18. $(-2 + 5i)(6 - i)$
19. $(1 + 3i)(2 - 4i)$
20. $(-1 + 2i)(-3 - 4i)$
21. $(2 - 5i)^2$
22. $(-2 + 7i)^2$
23. $(\sqrt{2} - i)^2$
24. $(\sqrt{3} + 2i)^2$

25. $\left(\frac{1}{2} + \frac{1}{2}i\right)\left(\frac{1}{2} - \frac{1}{2}i\right)$

26. $(2 + i\sqrt{3})^2$

27. $(2 - i\sqrt{2})(3 + i\sqrt{2})$

28. $(8 + \sqrt{-4})(2 + \sqrt{-1})$

29. $(3 - \sqrt{-16})(2 + \sqrt{-25})$

30. $(2 + i\sqrt{5})(2 - i\sqrt{5})$

31. $(-1 + i\sqrt{3})(-1 - i\sqrt{3})$

32. $2(3 + i) + 5(1 + 5i)$

33. $3(2 - i) + 2(4 - 7i)$

34. $4(3 - 5i) - 2(1 - 3i)$

35. $5(-2 + 7i) - 4(3 - 2i)$

36. $(1 - i)^2 + (2 + i)^2$

37. $(\sqrt{5} - i)(\sqrt{5} + 4i)$

38. $(1 + i)^2 - 2(7 - 3i)$

39. $(-2 + i)^2 - 5(2 - i)$

40. $i + i^2$

41. $(3 - 4i)(3 - 4i)$

42. $-i(-i)(-i)$

43. $33 - (2 + i)(3 - 5i)$

44. $(a - bi)(c - di)$

45. $(6 - i\sqrt{2}) + (6 - i\sqrt{2})$

46. $(6 - i\sqrt{2})(6 + i\sqrt{2})$

47. $(-i - 1)(2i - 2)$

48. $(i - 2)(i + 2)(2 - i)$

49. $(4 - i)^3$

50. $(1 - i)(1 + i)(1 + 2i)$

51. $(2 + 3i)(3i - 2)$

52. $\frac{8 + 4i}{6} - \frac{2 - 3i}{4} + i$

53. $3(7i - 3) + 4(9 - 9i)$

54. $(3 - 2i) + (7 + 4i)$

55. $i + i^2 + 5$

56. $(3 - 4i)(6 + i)$

57. $(5i + 2)(5i - 2)$

58. $i(a + bi)$

59. $i^2 + i^3$

60. $i + i^2 + i^3 + i^4$

61. $i^{11} + i^3$

62. $i^{13} + i^{12}$

63. $i^8 + 2i^9 - 5i^{17}$

64. $-i^{16} - i^{10} + (2i)^5$

65. $i^7 + 2i^6 - i^{10}$

66. $i^6 + 3i^9 - 2i^{16}$

67. $i^{11} + 5i + (2i)(i^3)$

68. $(i\sqrt{8} + i\sqrt{12})(i\sqrt{2} - i\sqrt{3})$

69. $(\sqrt{2} + \sqrt{-12})(1 - \sqrt{3})$

70. $(2 - 2i)^2 - (-1 + 4i)$

71. $3(-1 + 5i)^2 - 3(-2 + 7i)^2$

72. $\left(-\frac{1}{2} + \frac{i\sqrt{3}}{2}\right)^3$

73. $\left(-\frac{1}{2} - \frac{i\sqrt{3}}{2}\right)^3$

74. $i^{11} + 5i + (2i)(i^3)$

75. $(\sqrt{-8} + \sqrt{-12})^2$

76. $\left(\frac{1}{2} + \frac{i\sqrt{2}}{2}\right)^2$

77. $\left(\frac{1}{3} + \frac{i\sqrt{3}}{2}\right)^3$

78. $\left(-\frac{1}{3} + i\sqrt{2}\right)^2$

79. $i^9 + i + (2i)(i^2)$

80. $(\sqrt{2} - \sqrt{-12})(\sqrt{1} - \sqrt{3})$

81. $i^9 + 2i^6 - i^{12}$

82. $(3 - 3i)^3 - (-1 + i^2)$

83. $3(3 - 4i)^2 - (-2 + i)$

84. $i^8 + 3i^9 - 2i^{14}$

85. $(\sqrt{3} + i\sqrt{10})(i\sqrt{2} - i\sqrt{3})$

86. $\left(\frac{1}{2}\right)(i^3) + i^{11}$

87. $3(-2 + i)^2 - 2(-1 + 3i)$

Find the values of x and y for which each sentence is true.

88. $x + xi - y + yi = 2 - 4i$

89. $2x + xi + y - yi = 7 - i$

90. $x + 2xi + 2y - yi = 5 + 5i$

91. $x + 2xi + 4y - 3yi = 13 + 7i$

92. Show that $0 + 0i$ is the additive identity in $(\mathscr{C}, +, \cdot)$.

93. What is the additive inverse of $a + bi$? Justify your answer.

94. Show that $(a + bi)(a - bi)$ is a real number.

Challenge

1. Show that $2i$ and $-2i$ are solutions of $x^4 - 16 = 0$.

2. If the domain of x is $\mathscr{C}$, show that $-\frac{1}{2} + \frac{i\sqrt{3}}{2}$ is a solution of $x^3 - 1 = 0$.

3. Explain why $(\mathscr{C}, +)$ is a commutative group.

2.7 Conjugates and Division of Complex Numbers

In earlier work, we saw that the product of irrational numbers such as $2 + \sqrt{3}$ and $2 - \sqrt{3}$ is a rational number. This fact is used to rationalize irrational denominators.

There is a similar relationship among complex numbers. Two complex numbers of the form $a + bi$ and $a - bi$ are called **conjugates.** For example, $2 - 3i$ and $2 + 3i$ are each conjugates of the other. Study the following examples carefully.

$$(2 + 3i)(2 - 3i) = 2^2 - (3i)^2 = 4 + 9 = 13$$
$$(4 - 2i)(4 + 2i) = 4^2 - (2i)^2 = 16 + 4 = 20$$
$$(\sqrt{3} + i\sqrt{2})(\sqrt{3} - i\sqrt{2}) = (\sqrt{3})^2 - (i\sqrt{2})^2 = 3 + 2 = 5$$

In each case, the product is a real number.

Product of Conjugates

> The product of two complex numbers that are conjugates is a real number.

Now we can change expressions, such as $\frac{1}{2 + 3i}$, to $a + bi$ form. We use a procedure similar to the one used to rationalize denominators containing radicals.

Examples

1 **Express $\frac{1}{2 + 3i}$ in $a + bi$ form.**

$$\frac{1}{2 + 3i} = \frac{1}{2 + 3i} \cdot \frac{2 - 3i}{2 - 3i}$$

Multiply the numerator and denominator by $2 - 3i$, the conjugate of $2 + 3i$.

$$= \frac{2 - 3i}{4 - 9i^2}$$
$$= \frac{2 - 3i}{13} \text{ or } \frac{2}{13} - \frac{3}{13}i$$

2 **Express $(1 + 3i) \div (2 + i)$ in $a + bi$ form.**

$$(1 + 3i) \div (2 + i) = \frac{1 + 3i}{2 + i}$$
$$= \frac{1 + 3i}{2 + i} \cdot \frac{2 - i}{2 - i}$$
$$= \frac{2 - i + 6i - 3i^2}{4 - i^2}$$
$$= \frac{5 + 5i}{5} = \frac{5(1 + i)}{5} = 1 + i$$

Therefore, $(1 + 3i) \div (2 + i) = 1 + i$.

Now we can complete our investigation of the field properties of the complex numbers. For any complex number $a + bi$, the product $(a + bi)(1 + 0i)$ equals $a + bi$. Thus, $1 + 0i$ is the *multiplicative identity*.

The *multiplicative inverse* of a number such as $2 + 3i$ is $\frac{1}{2 + 3i}$, since $(2 + 3i)\left(\frac{1}{2 + 3i}\right) = 1$.

In example 1, we saw that $\frac{1}{2 + 3i}$ is equal to the complex number $\frac{2}{13} - \frac{3}{13}i$. You should check that $(2 + 3i)\left(\frac{2}{13} - \frac{3}{13}i\right) = 1$.

Example

3 **Express the multiplicative inverse of $3 - 4i$ in $a + bi$ form.**

The multiplicative inverse of $3 - 4i$ is $\frac{1}{3 - 4i}$.

$$\begin{aligned} \frac{1}{3 - 4i} &= \frac{1}{3 - 4i} \cdot \frac{3 + 4i}{3 + 4i} \\ &= \frac{3 + 4i}{9 - 16i^2} \\ &= \frac{3 + 4i}{25} \text{ or } \frac{3}{25} + \frac{4}{25}i \end{aligned}$$

The multiplicative inverse of $3 - 4i$ is $\frac{3}{25} + \frac{4}{25}i$.

The procedure demonstrated in example 3 can be used to find the multiplicative inverse of any nonzero complex number. You will be asked to verify the following statement in the exercises.

$$(\mathscr{C}, +, \cdot) \text{ is a field.}$$

Exercises

Exploratory **State the conjugate of each number.**

1. $1 + i$ **2.** $2 - 3i$ **3.** $-2 + 4i$ **4.** i **5.** $-i$

6. $2 + 3i$ **7.** 3 **8.** $1 + i\sqrt{2}$ **9.** $-2 + i\sqrt{3}$ **10.** $6 - i\sqrt{7}$

Find the product of each number and its conjugate for exercises 1–10.

11. exercise 1 **12.** exercise 2 **13.** exercise 3 **14.** exercise 4 **15.** exercise 5

16. exercise 6 **17.** exercise 7 **18.** exercise 8 **19.** exercise 9 **20.** exercise 10

21. What is the conjugate of $a - bi$?

22. Explain why the product of a complex number and its conjugate must be a real number.

Written **Express the additive inverse and multiplicative inverse in $a + bi$ form.**

1. $1 + i$ **2.** $-1 + i$ **3.** $2 + i$ **4.** $-2 + i$ **5.** $-1 - i$

6. $-3 + 2i$ **7.** $1 - 4i$ **8.** $-1 + 2i$ **9.** $-2 + 4i$ **10.** $4i$

11. $-2i$ **12.** $\sqrt{3} - i$ **13.** $1 + i\sqrt{3}$ **14.** π **15.** $-\pi - \pi i$

Express each of the following in $a + bi$ form.

16. $\frac{2}{1 + i}$ **17.** $\frac{5}{2 + i}$ **18.** $\frac{4}{1 - i}$ **19.** $\frac{-1}{2 + i}$ **20.** $\frac{2i}{1 - i}$

21. $\frac{3i}{2 + i}$ **22.** $\frac{1 + i}{2 + i}$ **23.** $\frac{2 + i}{3 - i}$ **24.** $\frac{1 + 3i}{2 - i}$ **25.** $\frac{5 + i}{2 - 3i}$

26. $\frac{2 + 3i}{3 - 2i}$ **27.** $\frac{3 - 2i}{1 - i}$ **28.** $\frac{1 + i}{3 + 2i}$ **29.** $\frac{1 - i}{4 - 5i}$ **30.** $\frac{4 - 2i}{-2i}$

31. $\frac{\sqrt{2} + i}{\sqrt{2} - i}$ **32.** $\frac{7}{\sqrt{2} - 3i}$ **33.** $\frac{4}{\sqrt{3} + 2i}$ **34.** $\frac{1 + i\sqrt{3}}{1 - i\sqrt{3}}$ **35.** $\frac{2 - i\sqrt{7}}{2 + i\sqrt{7}}$

36. $\frac{1}{1 + i} + \frac{1}{2 + i}$ **37.** $\frac{(2 + 3i)^2}{(3 + i)^2}$ **38.** $\frac{3}{1 - i} + \frac{4}{1 + i}$ **39.** $\frac{1}{i} \div (1 + i)$

40. $\frac{3 - 2i}{1 - i} - \frac{4 + 5i}{1 + i}$ **41.** $\frac{1 - i}{(1 + i)^2}$ **42.** $\left(\frac{3 + 3i}{1 + i}\right)^2$ **43.** $(4 + 3i)^2 \div (3 - i)^2$

44. Show that the sum of two conjugates is a real number.

45. Explain why $(\mathscr{C}, +, \cdot)$ is a field.

Challenge

If $a + bi \neq 0$, show that its multiplicative inverse is $\frac{a}{a^2 + b^2} - \frac{b}{a^2 + b^2}i$.

Mathematical Excursions

The History of Complex Number Terminology

Carl Friedrich Gauss (1777–1855), pictured at the left, was one of the greatest mathematicians of Germany. He is credited with coining the term complex number. The use of the letter i for $\sqrt{-1}$ is due to the Swiss mathematician **Leonard Euler** (1707–1783). **Augustin-Louis Cauchy** (1789–1857), a French mathematician, suggested the name conjugates to describe pairs of complex numbers such as $a + bi$ and $a - bi$.

Can the Complex Numbers be Ordered?

Whenever you are given two different real numbers, you can determine which of the numbers is less than the other. We say that the real numbers can be *ordered*.

A field $(F, +, \cdot)$ is ordered if it is possible to define an order relation, $<$, that satisfies the following properties. Assume a, b, and c are elements of F.

Trichotomy Property:	Exactly one of the following is true. $a < b, \quad a = b, \quad \text{or} \quad b < a$
Transitive Property:	If $a < b$ and $b < c$, then $a < c$.
Addition Property:	If $a < b$, then $a + c < b + c$.
Multiplication Property:	If $a < b$ and $0 < c$, then $ac < bc$.

Some familiar ordered fields are $(\mathcal{R}, +, \cdot)$ and $(\mathcal{Q}, +, \cdot)$.

One property of an ordered field is that the square of any nonzero element is greater than zero. Using symbols, this result can be stated as follows.

If a is an element of an ordered field and $a \neq 0$, then $a^2 > 0$.

Since the complex number system is a field, it is natural to inquire whether it is also an ordered field.

If it *is* an ordered field, the property given above must hold for all complex numbers. In particular, since $i \neq 0$, the property says that $i^2 > 0$. But $i^2 = -1$. Therefore, we find that $-1 > 0$. Clearly this contradicts the fact that in the real numbers $-1 < 0$. If the trichotomy property is to hold, $-1 > 0$ *and* $-1 < 0$ cannot both be true.

The assumption that the complex numbers form an ordered field leads to a contradiction. We have just proven the following statement.

The complex number system is *not* an ordered field.

It follows that imaginary numbers cannot be classified as positive or negative. For example, even though the real number 3 is positive, it is meaningless to say $3i$ is positive or negative. Likewise, a statement such as "$5i$ is larger than $2i$" is meaningless. This means that imaginary numbers cannot appear in inequalities.

Exercise Ted offers the following solution to $x^2 + 9 > 8$. Criticize Ted's work.

$$x^2 > -1$$
$$x > \pm i$$

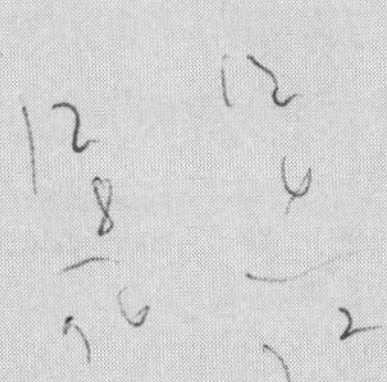

2.8 Complex Roots

We have seen that the solution set of $x^2 = -1$ is $\{i, -i\}$ when the domain of x is the set of complex numbers. Here is another example of solving an equation with complex roots.

Example

1 **Solve $x^2 = 2x - 8$. Express the roots in simplest $a + bi$ form.**

$x^2 - 2x + 8 = 0$ *Standard Form*

$x = \frac{-(-2) \pm \sqrt{4 - 4(1)(8)}}{2}$ *Use the quadratic formula,* $x = \frac{-b \pm \sqrt{b^2 - 4ac}}{2a}$.

$= \frac{2 \pm \sqrt{4 - 32}}{2}$

$= \frac{2 \pm \sqrt{-28}}{2}$

$= \frac{2 \pm i\sqrt{28}}{2} = \frac{2 \pm 2i\sqrt{7}}{2} = 1 \pm i\sqrt{7}$

The solution set is $\{1 + i\sqrt{7}, 1 - i\sqrt{7}\}$. *You should check the roots in the original equation.*

You may recall that quadratic equations can also be solved by graphing the related parabola. The following example shows different graphing situations and the solution set determined in each situation.

Examples

2 **Solve $x^2 - 2x - 3 = 0$ by graphing.**

To solve the equation graphically, graph the associated equation $y = x^2 - 2x - 3$.

This graph is a parabola with the line $x = 1$ as its axis of symmetry.

A table of values with three points on either side of the axis of symmetry is shown below, next to the graph of $y = x^2 - 2x - 3$.

x	$y = x^2 - 2x - 3$
−2	5
−1	0
0	−3
1	−4
2	−3
3	0
4	5

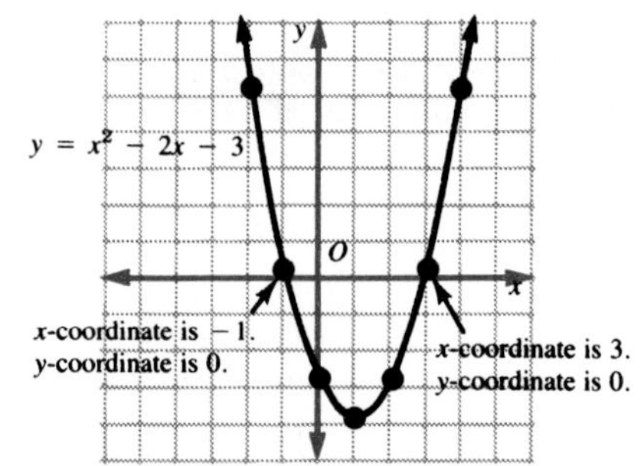

Look carefully at the points where the graph crosses the x-axis. Clearly the y-coordinate of these points is 0. The x-coordinates are the values that make $x^2 - 2x - 3 = 0$. Therefore, they are the roots of our original equation, $x^2 - 2x - 3 = 0$. We conclude that the solution set is $\{-1, 3\}$.

3 **Solve $x^2 + 6x + 9 = 0$ by graphing.**

The graph of the associated equation $y = x^2 + 6x + 9$ and a table of values appears below.

x	y
−5	4
−4	1
−3	0
−2	1
−1	4

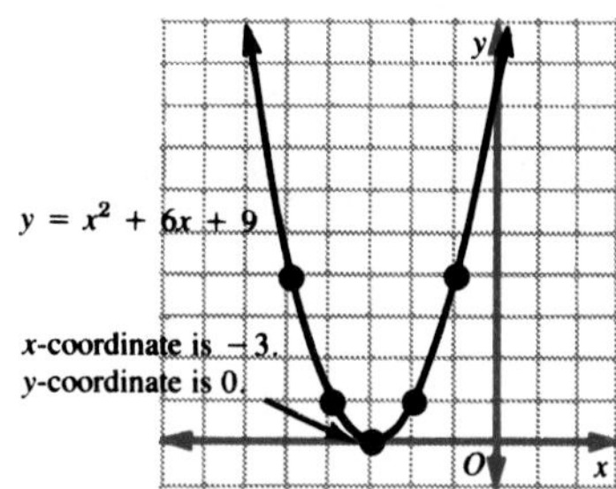

In this case, the graph crosses the x-axis in only one place. We conclude that the solution set is $\{-3\}$.

4 **Solve the equation in example 1 by graphing.**

The graph of the associated equation, $y = x^2 - 2x + 8$, and a table of values appears below.

x	y
−1	11
0	8
1	7
2	8
3	11

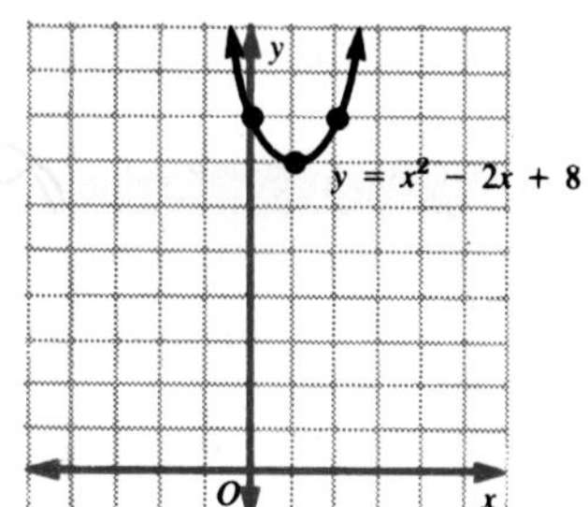

The graph does not cross the x-axis. Therefore there are no real roots. We could have predicted this from the imaginary roots we found in example 1.

From examples 2–4 we see that a parabola of the form $y = ax^2 + bx + c$ may intersect the x-axis in two, one or no points. This means that a quadratic equation of the form $ax^2 + bx + c = 0$ where a, b, and c are real numbers, may have two, one, or no real roots. When there are no real roots, there are two conjugate imaginary roots.

Is there a way of determining the nature of the roots of a quadratic equation without solving the equation or sketching the associated parabola? When using the quadratic formula to solve an equation, we can tell that the roots will be imaginary if the radicand is negative. The expression $b^2 - 4ac$ is called the **discriminant** of a quadratic equation. It gives us some information about the nature of the roots.

If $b^2 - 4ac$ is negative, there are two conjugate imaginary roots.

To see what happens when the discriminant is zero or positive, study the algebraic solutions given in exercises 5 and 6.

Examples

5 **Solve $x^2 - 2x - 3 = 0$.**

$$x = \frac{2 \pm \sqrt{(-2)^2 - 4(1)(-3)}}{2}$$

$$= \frac{2 \pm \sqrt{4 + 12}}{2}$$

$$= \frac{2 \pm \sqrt{16}}{2}$$

$x = 3$ or $x = -1$

The discriminant is positive. There are *two* real roots.

6 **Solve $x^2 + 6x + 9 = 0$.**

$$x = \frac{-6 \pm \sqrt{36 - 4(1)(9)}}{2}$$

$$= \frac{-6 \pm \sqrt{36 - 36}}{2}$$

$$= \frac{-6 \pm \sqrt{0}}{2}$$

$x = -3$

The discriminant is zero. There is *one* real root.

The following chart summarizes the results for the quadratic equation $ax^2 + bx + c = 0$ when $a \neq 0$ and a, b, c are real numbers.

Value of $b^2 - 4ac$	**Nature of Roots**
negative	two conjugate imaginary roots
zero	one real root (or two equal roots)
positive	two real roots

Suppose the quadratic equation $ax^2 + bx + c = 0$ has *integral coefficients*. The expression $\frac{-b \pm \sqrt{b^2 - 4ac}}{2a}$ will be *rational* if, and only if, $\sqrt{b^2 - 4ac}$ is rational. But $\sqrt{b^2 - 4ac}$ is rational if, and only if, $b^2 - 4ac$ is the square of an integer.

When a, b, and c are integers, the equation $ax^2 + bx + c = 0$ has rational roots if, and only if, $b^2 - 4ac$ is the square of an integer.

Examples

7 **Describe the nature of the roots of $y^2 + \frac{5y}{2} - \frac{3}{2} = 0$.**

This equation is equivalent to the equation $2y^2 + 5y - 3 = 0$. The discriminant is $(5)^2 - 4(2)(-3) = 25 + 24 = 49 = 7^2 > 0$. Since the coefficients are *integers* and the discriminant is the square of an integer, there are two rational roots.

8 **Describe the nature of the roots of $m^2 + 2m\sqrt{5} + 1 = 0$.**

The discriminant is $(2\sqrt{5})^2 - 4(1)(1) = 20 - 4 = 16 = 4^2 > 0$.

There are two real roots since the discriminant is positive. But the equation does *not* have integral coefficients, and it is not equivalent to one that does. Therefore *no* conclusion regarding rational roots is possible. You should verify that the roots are, in fact, irrational.

9 **Find all the values of k for which $x^2 + 2x + k = 0$ has two real roots.**

For the roots to be real, the discriminant must be positive. Therefore, we must solve the following inequality.

$$\begin{aligned} b^2 - 4ac &> 0 \\ 2^2 - 4(1)(k) &> 0 \\ 4 - 4k &> 0 \\ -4k &> -4 \\ k &< 1 \end{aligned}$$

When k is a real number such that $k < 1$, the equation $x^2 + 2x + k = 0$ will have two real roots.

Exercises

Exploratory **Solve each equation. Express the roots in simplest $a + bi$ form.**

1. $x^2 + 4 = 0$
2. $x^2 + 16 = 0$
3. $x^2 + 49 = 0$
4. $x^2 + 3 = 0$
5. $x^2 - 4x + 5 = 0$
6. $x^2 - 2x + 5 = 0$
7. $x^2 - 6x + 10 = 0$
8. $x^2 + 17 = 8x$
9. $2x^2 - 2x + 1 = 0$
10. $9x^2 = 6x - 2$
11. $6x(3x - 1) = -1$
12. $2z^2 = -3(z + 3)$
13. $3m^2 - 2m = -2$
14. $m - 1 = 2m^2$
15. $1 - 2t = 6t^2$
16. $p^2 = \frac{3p - 9}{4}$
17. $\frac{2}{x} - \frac{3}{x - 4} = 1$
18. $\frac{x + 5}{5} = \frac{2}{1 - x}$

Written **Solve by graphing.**

1. $x^2 - 4 = 0$
2. $x^2 + 2x + 1 = 0$
3. $x^2 + x - 6 = 0$
4. $x^2 - 4x = -4$
5. $x^2 + 3x = 4$
6. $x^2 + 6x = -9$

For each of the following find the value of the discriminant. Then describe the nature of the roots as completely as possible.

7. $x^2 - 2x - 1 = 0$
8. $x^2 - 2x + 1 = 0$
9. $x^2 + 5x - 2 = 0$
10. $y^2 + 6y + 9 = 0$
11. $x^2 - 3x - 10 = 0$
12. $x^2 - 20x = -100$
13. $t^2 = 25$
14. $x^2 + 3(2x + 3) = 0$
15. $x^2 - x + 5 = 0$
16. $2m^2 - 5m + 3 = 0$
17. $3x^2 + x + 1 = 0$
18. $2y^2 = 10 - y$
19. $x^2 - 2x\sqrt{2} + 2 = 0$
20. $y^2 + y = 1$
21. $3m^2 = m + 5$
22. $x^2 + 3x + \frac{5}{2} = 0$
23. $y^2 - \frac{1}{4}y = \frac{1}{8}$
24. $x^2 + x + \frac{1}{4} = 0$
25. $x^2 + 2 = 2(x - 1)$
26. $t = 4 - 3t^2$
27. $x\sqrt{5} = 2x^2 - 3$
28. $m^2 + 3m\sqrt{5} - 1 = 0$
29. $3x^2 - 4x + \sqrt{3} = 0$
30. $y^2\sqrt{5} = 6y - \sqrt{5}$
31. $4y^2 - 11y = 2$
32. $3t^2 = -(4 + 8t)$
33. $y^2 - y\sqrt{5} + \frac{21}{2} = 0$

Determine all real values of *k* for which the given equation has the indicated type of roots.

34. $x^2 + 4x + k = 0$, two real roots
35. $x^2 + 2x + 2k = 0$, two real roots
36. $x^2 + 8x + k = 0$, one real root
37. $x^2 + kx - 9 = 0$, one real root
38. $kx^2 + 2x + 3 = 0$, two imaginary roots
39. $3x^2 + 4x + 2k = 0$, two imaginary roots
40. $x^2 - kx + 8 = 0$, one real root
41. $2x^2 - kx + 2 = 0$, two imaginary roots

42. Explain how the graph of $y = ax^2 + bx + c$ shows the nature of the roots of $ax^2 + bx + c = 0$.
43. If one root of a quadratic equation is of the form $m + ni$, where $n \neq 0$, what is the form of the other root?
44. Ted computes the discriminant of a quadratic equation and finds that it is the square of an integer. He concludes that the roots of the equation must be rational. Do you agree? Explain.

Challenge **Find two complex numbers with the following sums and products.**

1. sum 6; product 10
2. sum 8; product 17
3. sum 8; product 24
4. sum 6; product 17
5. sum 10; product 34
6. sum 12; product 37

2.9 Roots and Coefficients

Suppose we want to find a quadratic equation whose roots are 5 and 7. One way to proceed is to reverse the usual factoring process.

$$x = 5 \quad \text{or} \quad x = 7$$
$$x - 5 = 0 \quad \text{or} \quad x - 7 = 0$$
$$(x - 5)(x - 7) = 0$$
$$x^2 - 12x + 35 = 0$$

Therefore, $x^2 - 12x + 35 = 0$ is an equation with roots 5 and 7.

Although this procedure will always work, it is sometimes more convenient to use an interesting relationship between the roots and the coefficients of a quadratic equation. In the equation above, notice that the sum of the roots, 5 and 7, is 12 and the product of the roots is 35. Compare these to the equation.

$$x^2 - \underbrace{12}_{5+7}x + \underbrace{35}_{5 \cdot 7} = 0$$

It appears that the sum and product of the roots have something to do with the coefficients of the original equation.

In general, suppose r_1 and r_2 are the roots of the quadratic equation $ax^2 + bx + c = 0$. Using the quadratic formula, let

$$r_1 = \frac{-b + \sqrt{b^2 - 4ac}}{2a} \text{ and } r_2 = \frac{-b - \sqrt{b^2 - 4ac}}{2a}.$$

$$r_1 + r_2 = \frac{-b + \sqrt{b^2 - 4ac} - b - \sqrt{b^2 - 4ac}}{2a}$$

$$r_1 + r_2 = \frac{-2b}{2a} = -\frac{b}{a} \qquad \textit{The sum of the roots is } -\frac{b}{a}.$$

The product of the roots is in the form of the difference of two squares.

$$r_1 r_2 = \left(\frac{-b + \sqrt{b^2 - 4ac}}{2a}\right)\left(\frac{-b - \sqrt{b^2 - 4ac}}{2a}\right)$$

$$r_1 r_2 = \frac{(-b)^2 - (b^2 - 4ac)}{4a^2}$$

$$r_1 r_2 = \frac{b^2 - b^2 + 4ac}{4a^2} = \frac{4ac}{4a^2} = \frac{c}{a}$$

The preceding developments are summarized below.

Sum and Product of the Roots of a Quadratic Equation

For any quadratic equation $ax^2 + bx + c = 0$, where $a \neq 0$, the sum of the roots is $-\frac{b}{a}$ and the product of the roots is $\frac{c}{a}$.

Examples

1 **Find the sum and product of the roots of $2x^2 - 6x + 7 = 0$.**

In this case $a = 2$, $b = -6$, and $c = 7$.

Therefore, the sum of the roots is $-\frac{b}{a} = -\frac{-6}{2} = 3$.

The product of the roots is $\frac{c}{a} = \frac{7}{2}$.

2 **Write a quadratic equation whose roots are $1 + 2i$ and $1 - 2i$.**

Method 1 Use the sum and product of the roots relationship.

Sum of the roots: $(1 + 2i) + (1 - 2i) = 2$ $\quad -\frac{b}{a} = 2$ or $\frac{2}{1}$

Product of the roots: $(1 + 2i)(1 - 2i) = 1 - 4i^2 = 5$ $\quad \frac{c}{a} = 5$ or $\frac{5}{1}$

Therefore, $a = 1$, $b = -2$, and $c = 5$, producing the equation

$$x^2 - 2x + 5 = 0.$$

Method 2 Use the method of factoring.

If $1 + 2i$ and $1 - 2i$ are roots, then

$$[x - (1 + 2i)][x - (1 - 2i)] = 0$$
$$[(x - 1) - 2i][(x - 1) + 2i] = 0 \quad \textit{Associative Property}$$
$$(x - 1)^2 - 4i^2 = 0 \quad \textit{Difference of Two Squares}$$
$$x^2 - 2x + 1 + 4 = 0$$
$$x^2 - 2x + 5 = 0$$

Exercises

Exploratory **Without solving, find the sum and product of the roots of each equation.**

1. $x^2 + 6x - 4 = 0$
2. $x^2 - 8x + 7 = 0$
3. $x^2 + 5x - 3 = 0$
4. $x^2 - 10x + 1 = 0$
5. $2x + 5 = x^2$
6. $2x^2 - x + 5 = 0$
7. $5x^2 + 4x - 7 = 0$
8. $3x^2 - 7x + 4 = 0$
9. $-4t^2 + 2t + 1 = 0$
10. $-10m^2 - 5m + 2 = 0$
11. $3m + m^2 - 6 = 0$
12. $4x - x^2 = 0$
13. $8 + 2z - z^2 = 0$
14. $x^2 - \frac{2x}{3} + 4 = 0$
15. $\frac{3t^2}{4} + t = 1$
16. $\frac{2y^2}{3} - \frac{1y}{4} = \frac{5}{6}$
17. $y^2\sqrt{3} + y + \sqrt{7} = 0$
18. $p^2\sqrt{2} - 3p\sqrt{5} = -1$
19. $(1 - \sqrt{2})x^2 + x - 3 = 0$
20. $(2 + \sqrt{3})x^2 = 4x + 2$
21. $3x^2 = \frac{x + 4}{5}$

Written **Write a quadratic equation whose roots, r_1 and r_2, have the following sums and products.**

1. $r_1 + r_2 = 5; r_1r_2 = 6$
2. $r_1 + r_2 = -3; r_1r_2 = -4$
3. $r_1 + r_2 = -7; r_1r_2 = 5$
4. $r_1 + r_2 = -11; r_1r_2 = -10$
5. $r_1 + r_2 = -9; r_1r_2 = 10$
6. $r_1 + r_2 = -\frac{1}{2}; r_1r_2 = 2$
7. $r_1 + r_2 = \frac{2}{3}; r_1r_2 = -4$
8. $r_1 + r_2 = \frac{3}{4}; r_1r_2 = 1$

Write a quadratic equation in standard form with the following roots.

9. $-2, 7$
10. $1, -6$
11. $-5, -4$
12. $-3, -10$
13. $3, \frac{1}{2}$
14. $6, \frac{2}{3}$
15. $-\frac{1}{4}, 12$
16. $\frac{1}{2}, \frac{1}{2}$
17. $-\frac{1}{2}, \frac{1}{2}$
18. $-\frac{3}{5}, \frac{3}{5}$
19. $\frac{3}{4}, -4$
20. $1 + \sqrt{2}, 1 - \sqrt{2}$
21. $3 + \sqrt{5}, 3 - \sqrt{5}$
22. $-6 + \sqrt{7}, -6 - \sqrt{7}$
23. $2i, -2i$
24. $3i, -3i$
25. $1 + i, 1 - i$
26. $2 + i, 2 - i$
27. $4 + 3i, 4 - 3i$
28. $-2 + 5i, -2 - 5i$
29. $3 + 4i, 3 - 4i$
30. $2 + i\sqrt{2}, 2 - i\sqrt{2}$
31. $4 + 2i\sqrt{3}, 4 - 2i\sqrt{3}$
32. $-8 + 3i\sqrt{2}, -8 - 3i\sqrt{2}$

In each case, one root r_1, of a quadratic equation is given. Find the other root and the missing coefficient.

33. $x^2 - 5x + c = 0; r_1 = 2$
34. $2x^2 - 6x + c = 0; r_1 = -1$
35. $2x^2 + bx + 8 = 0; r_1 = 2$
36. $x^2 + bx - 7 = 0; r_1 = -1$
37. $3t^2 - 12t + c = 0; r_1 = 2 - \sqrt{5}$
38. $2x^2 - 12x + c = 0; r_1 = 3 + \sqrt{2}$

Challenge **If r_1 and r_2 are roots of $ax^2 + bx + c = 0$, find expressions in simplest form for the following in terms of a, b, and c.**

1. $(r_1 + r_2)^2 + (r_1)^2(r_2)^2$
2. $\dfrac{(r_1)^2 + (r_2)^2}{(r_1)^2(r_2)^2}$
3. $\dfrac{1}{r_1} + \dfrac{1}{r_2}$

Mixed Review

Express in simplest $a + bi$ form.

1. $(\sqrt{-12} - \sqrt{-8})(\sqrt{-3} - \sqrt{-2})$
2. $\dfrac{4 - 3i}{4 + 3i} - \dfrac{4 + 3i}{4 - 3i}$
3. Solve $n + \sqrt{4 - 3n} = -2$.
4. Write a quadratic equation in simplest form whose roots are $\sqrt{5} - 2$ and $\sqrt{5} + 2$.
5. Determine all real values of k for which $x^2 + kx + (3 - k) = 0$ has exactly one real root.
6. The Blockville Garden Club wants to triple the area of its rectangular display of carnations. If the display now has dimensions 8 meters by 4 meters, by what equal amounts must the length and width be increased?

We often graph real numbers on the real number line. It would be helpful to have a way of picturing imaginary numbers. Obviously we cannot put these numbers on the number line, since the reals occupy all positions on the line. To solve the problem, we use the familiar idea of ordered pairs. Each complex number $x + yi$ is associated with the ordered pair (x, y). Then (x, y) is graphed in the usual way.

For example, $2 + 3i$ is associated with the ordered pair (2, 3). This is shown on the graph at the right. Every ordered pair corresponds to a complex number.

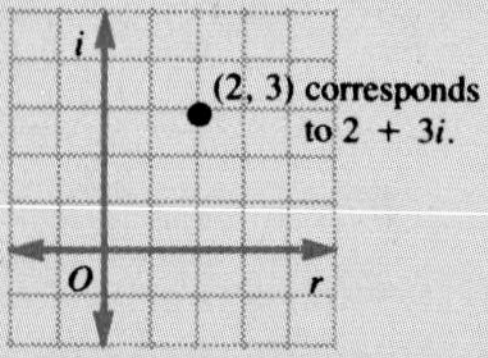

Since the real part of $x + yi$ is x, the horizontal axis is labeled r for the real axis. Likewise, since y is the imaginary part of $x + yi$, the vertical axis is labeled i, for the imaginary axis. When a coordinate plane is used to graph complex numbers, it is called a **complex number plane.**

Example Graph the following points that correspond to $-2 + 3i$, $4i$, 3, and $1 - i$.

$x + yi$	x	y
$-2 + 3i$	-2	3
$4i$	0	4
3	3	0
$1 - i$	1	-1

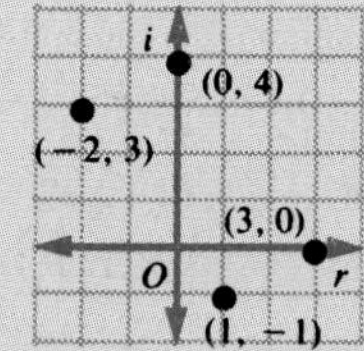

In the graph at the right, points A and B represent $1 + 5i$ and $3 + 2i$, respectively. The sum of $1 + 5i$ and $3 + 2i$ is $4 + 7i$. Point C represents $4 + 7i$. Notice that quadrilateral $OACB$ is a parallelogram.

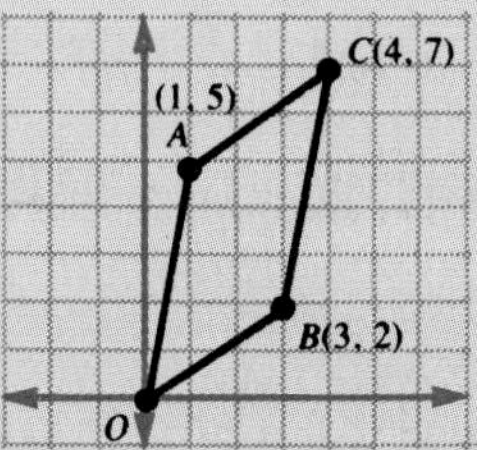

Exercises Graph each complex number.

1. $2 + 5i$ **2.** $1 + 2i$ **3.** $-2 + i$ **4.** $3 - 2i$ **5.** $-3i$

6. -1 **7.** $5i$ **8.** 2 **9.** $-2 - 4i$ **10.** $-1 - 5i$

Demonstrate each sum graphically in the complex plane.

11. $(1 + 2i) + (2 + i)$ **12.** $(3 - i) + (4 - i)$

13. $(7 - 3i) + (2 + 5i)$ **14.** $(6 + 2i) + (-4 + 2i)$

2.10 Quadratic Inequalities

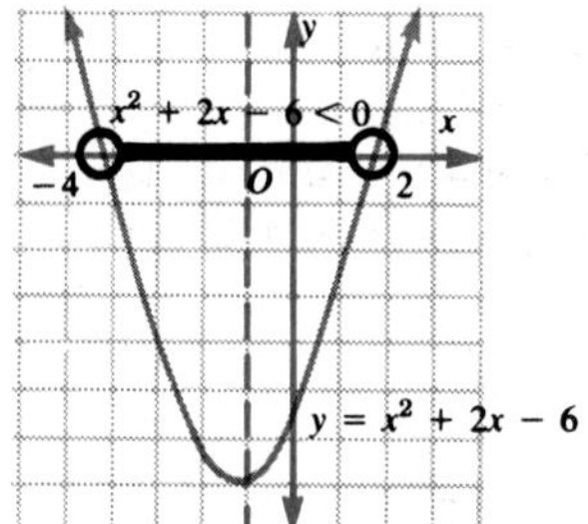

Earlier in this chapter, we solved quadratic equations by graphing. A quadratic inequality, such as $x^2 + 2x - 6 < 0$, can also be solved by graphing. In this section, all variables represent real numbers.

The graph of the parabola associated with $x^2 + 2x - 6 < 0$ is shown. Each point on the x-axis has a y-coordinate of 0. The graph shows that each x-coordinate between −4 and 2 has a corresponding y-coordinate that is less than 0. Thus any value of x between −4 and 2 satisfies the inequality $x^2 + 2x - 6 < 0$. The solution set can be written in the following manner.

$\{x \in \mathcal{R}: -4 < x < 2\}$ *Solution set of $x^2 + 2x - 6 < 0$*

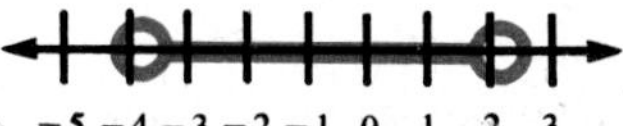

Graph of the solution set

The same graph can be used to solve $x^2 + 2x - 6 > 0$. In this case, the x-coordinates less than −4 or greater than 2 have corresponding y-coordinates that are greater than 0. We write the solution set for this inequality in the following manner.

$\{x \in \mathcal{R}: x < -4 \text{ or } x > 2\}$ *Solution set of $x^2 + 2x - 6 > 0$*

−5 −4 −3 −2 −1 0 1 2 3

Graph of the solution set

Quadratic inequalities can also be solved algebraically. The solutions are based on one of the following two principles.

1. If $ab < 0$, then $a < 0$ and $b > 0$, or $a > 0$ and $b < 0$.
2. If $ab > 0$, then $a < 0$ and $b < 0$, or $a > 0$ and $b > 0$.

Examples

1 **Solve $x^2 + 4x - 5 < 0$ algebraically, and graph the solution set.**

$x^2 + 4x - 5 < 0$

$(x + 5)(x - 1) < 0$ *Using principle 1 we get the following situations.*

$x + 5 < 0 \wedge x - 1 > 0$ or $x + 5 > 0 \wedge x - 1 < 0$

$x < -5 \wedge x > 1$ $\qquad$ $x > -5 \wedge x < 1$

This conjunction is never true for any value of x.

This conjunction is equivalent to $-5 < x < 1$.

−6 −5 −4 −3 −2 −1 0 1 2

The solution set is $\{x \in \mathcal{R}: -5 < x < 1\}$.

2 **Solve $x^2 + 8x \geq -15$ algebraically and graph the solution set.**

$$x^2 + 8x \geq -15$$
$$x^2 + 8x + 15 \geq 0$$
$$(x + 5)(x + 3) \geq 0$$

$$x + 5 \leq 0 \wedge x + 3 \leq 0 \quad \text{or} \quad x + 5 \geq 0 \wedge x + 3 \geq 0$$
$$x \leq -5 \wedge x \leq -3 \qquad\qquad x \geq -5 \wedge x \geq -3$$

This conjunction is equivalent to $x \leq -5$.

This conjunction is equivalent to $x \geq -3$.

The solution set is $\{x \in \mathcal{R}: x \leq -5 \text{ or } x \geq -3\}$.

3 **Solve $x^2 - 7 > 6x$ and graph the solution set.**

An alternate method for solving quadratic inequalities uses three test points. First, solve the equation $x^2 - 7 = 6x$

$$x^2 - 7 = 6x$$
$$x^2 - 6x - 7 = 0$$
$$(x - 7)(x + 1) = 0$$

The solutions to this equation are often called critical points.

$$x - 7 = 0 \quad \text{or} \quad x + 1 = 0$$
$$x = 7 \qquad\qquad x = -1$$

The points 7 and −1 separate the number line into three intervals, $x < -1$, $-1 < x < 7$, and $x > 7$, as shown below.

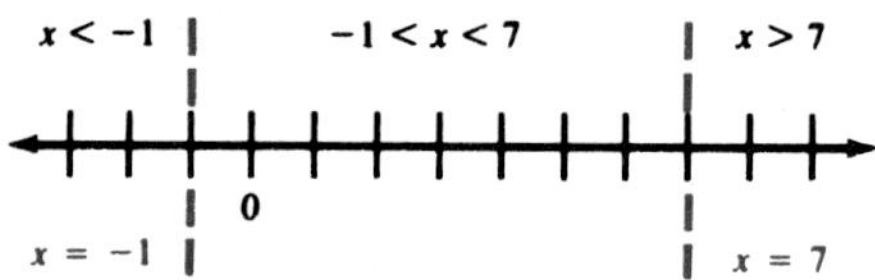

Choose a value of x from each of the three intervals. Then substitute each value into $x^2 - 7 > 6x$ to determine if the values from that interval satisfy the inequality.

For $x < -1$, try −2.

$$x^2 - 7 > 6x$$
$$(-2)^2 - 7 \overset{?}{>} 6(-2)$$
$$-3 > -12 \quad \textit{true}$$

For $-1 < x < 7$, try 0.

$$x^2 - 7 > 6x$$
$$(0)^2 - 7 \overset{?}{>} 6(0)$$
$$-7 > 0 \quad \textit{false!}$$

For $x > 7$, try 10.

$$x^2 - 7 > 6x$$
$$(10)^2 - 7 \overset{?}{>} 6(10)$$
$$93 > 60 \quad \textit{true}$$

The solution set is $\{x \in \mathcal{R}: x < -1 \text{ or } x > 7\}$.

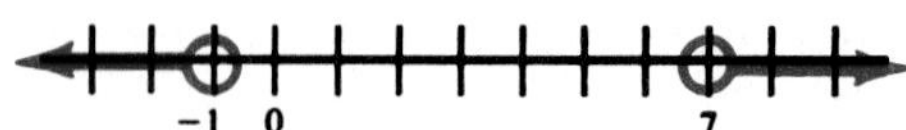

Exercises

Exploratory **Each of the following inequalities is followed by four graphs. Select the graph that shows the solution set of each inequality.**

1. $x^2 - 1 > 0$

a. 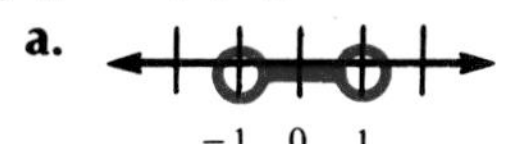

b. −1 0 1

c. 

d. −1 0 1

2. $x^2 - 9 < 0$

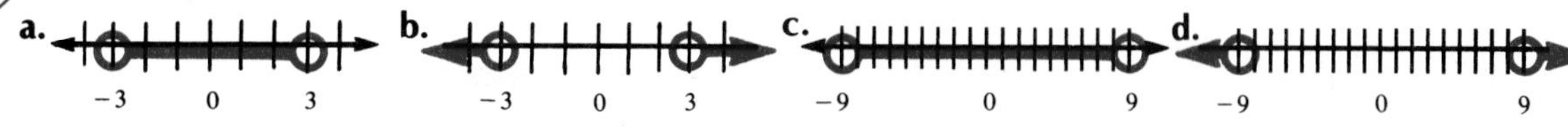

3. $x^2 + 2x - 3 > 0$

a. 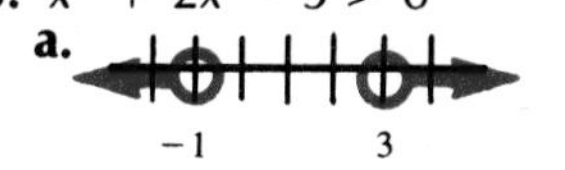

b. −3 1

c. −1 3

d. −3 1

4. $x^2 - 6x + 8 \geq 0$

a. 2 4

b. −2 4

c. 

d. 

Written **Solve each inequality. Then graph the solution set.**

1. $x^2 - 16 > 0$

2. $x^2 < 4$

3. $x^2 - 25 > 0$

4. $3x^2 - 3 > 0$

5. $x^2 + 6x \leq 0$

6. $x^2 + 3x \geq 0$

7. $x^2 + x - 6 < 0$

8. $x^2 + 4x - 21 \geq 0$

9. $x^2 - 4x - 5 \leq 0$

10. $x^2 \geq 3x + 28$

11. $x^2 \leq 36$

12. $x^2 + 2x > 24$

13. $2x^2 - x < 6$

14. $5x - 2x^2 > -3$

15. $-5x - 3x^2 < -2$

16. $x^2 + 8x \geq -16$

17. $x^2 \leq 3$

18. $x^2 + 1 > 0$

Challenge **Solve each inequality. Then graph the solution set.**

1. $(x - 3)(x + 4)(x - 1) > 0$

2. $(x + 2)(x - 3)(x + 6) < 0$

Mixed Review

1. Express $(-1 + i\sqrt{3})^3$ in simplest $a + bi$ form.

2. Find all values of x for which $x - 4$ is 5% of $x^2 + 2x - 3$.

3. The product of two numbers is 30. If the sum of the squares of the two numbers is 64, find the difference of the two numbers.

4. The simple interest for one year on a sum of money is \$108. Suppose the interest rate is increased by 2%. Then \$450 less than the original sum could be invested and yield the same annual interest. What were the original sum of money and the original rate of interest?

5. Janice has made 75 free throws in 103 attempts and Denise has made 82 in 98 attempts. For how many consecutive attempts would Janice have to make her free throws and Denise have to miss in order for Janice to have a higher free throw percentage than Denise?

Problem Solving Application: Guess and Check

Some problems can be solved by using a guess and check strategy. The first step is to guess a solution to the problem. The next step is to check your guess. Even though your guess may be incorrect, it can be used to improve your next guess. This procedure is repeated until you find the correct answer.

Example

1 Find the least prime number greater than 720.

Explore The least possible solution is 721.

Plan To solve this problem, divide each integer greater than 720 by prime divisors until you reach an integer whose only factor is itself. Since 720 is divisible by 3, then 723, 726, 729, . . . are divisible by 3. Therefore, none of these integers are prime. Also, all even integers greater than 2 and all integers whose last digit is 5 are not prime.

Solve Try 721. *A calculator could be used to do these computations.*

$721 \div 7 = 103$ *721 is not prime.*

You do not need to try 722, 723, 724, 725, and 726. *Why?*

Try 727.

$727 \div 7 \approx 103.9$

$727 \div 11 \approx 66.1$

$727 \div 13 \approx 55.9$

$727 \div 17 \approx 42.8$

$727 \div 19 \approx 38.3$

$727 \div 23 \approx 31.6$

$727 \div 29 \approx 25.1$ *Why can you stop at $727 \div 29$?*

Therefore, 727 is the least prime number greater than 720.

Examine The integers 721, 722, 723, 724, 725, and 726 are all divisible by a prime divisor. Therefore, the least prime number greater than 720 is 727.

Exercises

Written Solve each problem.

1. Using $100 bills, $50 bills, and $5 bills, how many of each bill is needed to total $1000 if exactly 100 bills are used?

2. What is the largest amount in United States coins that you can have and still not be able to change a one-dollar bill?

3. Dave purchased several pens for $3.21. If the price of each pen was greater than 50¢, how many pens did Dave purchase? What was the price of each pen?

4. Gail buys 36 donuts for 10¢ each. She wants to eat some of the donuts and sell the rest for 15¢ each to make a profit of $1.50. How many must she sell?

5. Canned peaches can be purchased in cases of 24 or 36. If a shipment of 360 cans was packed in 11 cases, how many were cases of 24?

6. Eleven football players are introduced. Each football player shakes hands with all the other football players on the field. How many handshakes take place?

7. A marching band has 240 members. To perform a new routine, the number of rows in the band's formation must be reduced. The band director decides to decrease the number of rows by 4 and increase the number of members in each row by 10. Find the number of rows in the original formation.

8. On an exam of 120 questions, the final score is based on 1 point for each correct answer and minus $\frac{1}{4}$ point for each incorrect answer. A student answered all of the questions and received a final score of 100. How many correct answers did the student have?

9. A man was born in the 18th century. In a certain year, his age was the square root of the year. How old was he in 1885?

10. For what value(s) of y does 2^{y-1} equal $2y - 1$?

11. Given the diagram at the right, which two lines can be moved to make four squares?

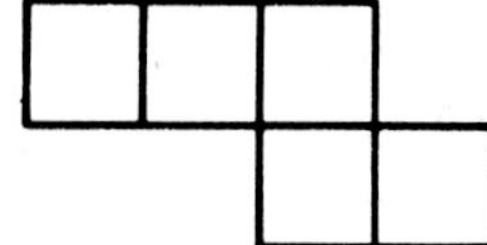

Supply a digit for each letter so that each addition or subtraction problem is correct. Each letter represents a different digit.

12.

```
 FORTY
   TEN
 + TEN
 SIXTY
```

13.

```
  SEVEN
  THREE
  + TWO
 TWELVE
```

14.

```
 GRAPE
 -PEAR
 APPLE
```

15.

```
  WRONG
 +WRONG
  RIGHT
```

16.

```
  MOON
 + SUN
 EARTH
```

17.

```
  HOCUS
 +POCUS
 PRESTO
```

Vocabulary

radical sign	index	pure imaginary number
radicand	conjugate	complex number
square root	radical equation	imaginary number
perfect square	imaginary unit	discriminant

Chapter Summary

1. The principal n^{th} root of a number y is a positive number unless n is odd and y is negative.
2. If $a \geq 0$ and $b \geq 0$, then $\sqrt{ab} = \sqrt{a} \cdot \sqrt{b}$.
3. If $a \geq 0$ and $b > 0$, then $\sqrt{\frac{a}{b}} = \frac{\sqrt{a}}{\sqrt{b}}$.
4. If the denominator of a fraction contains a radical, the denominator must be rationalized to obtain a fraction in simplest form.
5. If $\sqrt[n]{x}$ and $\sqrt[n]{y}$ represent real numbers, then $\sqrt[n]{xy} = \sqrt[n]{x} \cdot \sqrt[n]{y}$, and if $y \neq 0$, $\sqrt[n]{\frac{x}{y}} = \frac{\sqrt[n]{x}}{\sqrt[n]{y}}$.
6. Like radicals are those with the same index and the same radicand.
7. The conjugate of a binomial expression containing radicals is used to rationalize the denominator of fractions with such a binomial expression in the denominator.
8. To solve a radical equation it is necessary to isolate the radical and square both sides of the equation.
9. Possible solutions to any equation should always be checked in the *original* equation.
10. The product of two complex numbers that are conjugates is a real number.
11. The discriminant of a quadratic equation indicates the nature of the roots of the equation.
12. Given the roots of a quadratic equation, the sum and product of those roots can be used to determine the original equation.

Chapter Review

2.1 Simplify. Assume all variables represent positive numbers.

1. $-\sqrt{169}$
2. $-\sqrt[3]{-8}$
3. $\sqrt[5]{\frac{1}{32}}$
4. $\sqrt{0.49}$
5. $\sqrt{36a^2}$
6. $\sqrt{\left(-\frac{1}{4}\right)^2}$
7. $\sqrt{144x^4y^2}$
8. $-\sqrt[3]{-27x^6y^3}$

Find the smallest integral value of x for which each expression represents a real number.

9. $\sqrt{x + 7}$
10. $\sqrt{2x - 11}$
11. $\sqrt{10 + 3x}$

2.2 Simplify. Assume all variables represent positive numbers.

12. $\sqrt{75}$
13. $\sqrt{28x^3}$
14. $-4\sqrt{45y^5}$
15. $2\sqrt{7} \cdot 5\sqrt{14}$
16. $\dfrac{10}{\sqrt{3}}$
17. $\sqrt{\dfrac{1}{2}}\sqrt{32}$
18. $\sqrt{3m^3n^4}\sqrt{24m^6n}$
19. $\dfrac{6\sqrt{3}}{\sqrt{6}}$
20. $\dfrac{5\sqrt{50x^3}}{10\sqrt{2x}}$

2.3

21. $4\sqrt{18} - 3\sqrt{98}$
22. $\dfrac{1}{\sqrt{12}} + \sqrt{48}$
23. $\sqrt{3}(2\sqrt{15} - 5\sqrt{6})$
24. $(4 + 5\sqrt{3})(2 - 7\sqrt{3})$
25. $\dfrac{10}{\sqrt{3} - 2}$
26. $\dfrac{3 - \sqrt{2}}{4 + \sqrt{2}}$

2.4 Solve and check.

27. $x^2 - 8x + 13 = 0$
28. $x^2 = 4x + 14$
29. $\sqrt{5x + 1} = 6$
30. $2 - \sqrt{y} = 8$

2.5 Simplify.

31. $2\sqrt{-10} \cdot 3\sqrt{-5}$
32. $\dfrac{\sqrt{-18}}{\sqrt{-2}}$
33. i^{94}

2.6 Classify each number as completely as possible.

34. 5
35. $\dfrac{3}{4}$
36. $-5i$
37. $\sqrt{3}$
38. $-4 + \sqrt{-6}$

Express in $a + bi$ form.

39. $(3 - 7i) + (5 + 2i)$
40. $(3 - 10i) - (11 - 5i)$
41. $(2 + 7i)(3 - 10i)$
42. $(3 - 2i)^2$
43. $3(2 + i) + i(4 - i)$
44. $3i^9 - 4i^{83}$

2.7

45. the additive inverse of $3 - 4i$
46. the multiplicative inverse of $5 + 2i$
47. $\dfrac{4}{1 + i}$
48. $\dfrac{2i}{1 + i}$
49. $\dfrac{2 + i}{6 - 7i}$

2.8 Find the value of the discriminant and describe the nature of the roots as completely as possible.

50. $x^2 + 10 = 6x$
51. $9x^2 = 12x - 4$
52. Find all values of k for which $x^2 + 4x + 2k = 0$ has two real roots.

2.9 For each set of roots, write a quadratic equation in standard form.

53. $3i, -3i$
54. $-5, 7$
55. $1 + 5i, 1 - 5i$

2.10 Solve and graph the solution set for each inequality.

56. $x^2 - 36 \leq 0$
57. $x^2 + x - 12 > 0$
58. $x^2 + 7x < 0$

Chapter Test

In exercises 1–8, state whether each statement is true or false.

1. The conjugate of $-4 + 2i$ is $4 - 2i$.
2. The additive inverse of $1 + i\sqrt{3}$ is $-1 - i\sqrt{3}$.
3. Every complex number has a multiplicative inverse.
4. If $x + yi = (1 - i)^2$, then $y = 2$.
5. The roots of $x^2 - 10x + 22 = 0$ are imaginary.
6. The product of the roots of $5x^2 = x - 5$ is -1.
7. The sum of two complex numbers is a complex number.
8. The product of $2i^5$ and i^{10} is $-2i$.

9. Express the multiplicative inverse of $2 + 3i$ in $a + bi$ form.
10. Find the product of $4 - 3i$ and its conjugate in simplest terms.
11. Find the sum of $\sqrt{-25}$ and $3\sqrt{-16}$ in simplest form.
12. If the roots of $x^2 - 4x + c = 0$ are $2 + i$ and $2 - i$, find the value of c.
13. Write a quadratic equation in standard form that has roots $3 + i$ and $3 - i$.
14. Classify the roots of $x^2 + x + 1 = 0$ as completely as possible.
15. Find all values of k such that $2x^2 - kx + 2 = 0$ has one real root.

Simplify. Assume all variables represent positive numbers.

16. $2\sqrt{27} + 2\sqrt{3} - 7\sqrt{48}$
17. $\sqrt{\frac{98}{9}} - \sqrt{\frac{25}{2}}$
18. $(\sqrt{3} - \sqrt{5})(\sqrt{15} + 3)$
19. $\sqrt{81x^2y^2}\sqrt{24x^2y}$
20. $15\sqrt{\frac{2}{5}}$
21. $\frac{4}{2 - \sqrt{2}}$
22. $(2 + 3i)(4 - 5i)$
23. $\sqrt{-3}\sqrt{-24}$
24. $5i^{17}$
25. $\sqrt{-2} + \sqrt{-18}$
26. $\frac{2}{1 - i}$
27. $\frac{2 + i}{1 - 2i}$
28. $(2 + \sqrt{-9})(3 - \sqrt{-16})$
29. $\sqrt[3]{54y^4}$
30. $\frac{6}{\sqrt[3]{2}}$

Solve.

31. $\sqrt{2x - 5} = 9$
32. $x^2 - 2x + 17 = 0$
33. $x^2 - 4x = -13$

Solve each of the following inequalities. Then graph the solution set.

34. $x^2 - 3x - 10 < 0$
35. $2x^2 > 3x - 1$
36. $3x^2 \geq 27$

Relations and Functions

Chapter

3

Mathematics is the language of the physical world. This language describes relationships among ideas and systems through such concepts as relations and functions. Motion of all objects can be described and analyzed using mathematical concepts you will become familiar with in this chapter.

3.1 Relations

It is often convenient to display information in a chart like the one shown. This particular one lists students who received the given test scores.

TEST GRADE	STUDENTS
83	*Alex, Mary*
87	*Sara*
92	*Ellen, Tom, Sue*
98	*Maria, Tina*

The information in the chart can also be displayed as a set of ordered pairs.

{(83, Alex), (83, Mary), (87, Sara), (92, Ellen), (92, Tom), (92, Sue), (98, Maria), (98, Tina)}

The first member of each ordered pair is the test score of the second member. In mathematics, any set of ordered pairs, such as the one above, is called a *relation.*

Sometimes the ordered pairs of a relation are illustrated by an arrow diagram like the one shown below.

Definition of Relation

A **relation** is a set of ordered pairs. The set of all the first coordinates is the **domain** of the relation. The set of all the second coordinates is the **range** of the relation.

Thus, in the relation shown above, the domain is {83, 87, 92, 98} and the range is {Alex, Mary, Sara, Ellen, Tom, Sue, Maria, Tina}.

For some relations, there may be a rule or formula that describes the relationship between the coordinates of each ordered pair. In the following relation, each y-coordinate is related to its x-coordinate by an algebraic rule.

$$\{(-2, 4), (-1, 1), (0, 0), (1, 1), (2, 4)\}$$

In this relation, the y-coordinate is the square of the x-coordinate, or, algebraically expressed, $y = x^2$. With the understanding that the domain is $\{-2, -1, 0, 1, 2\}$, we may write this relation in the following symbolic form

$$\{(x, y): y = x^2\}$$

which is read "the set of all ordered pairs (x, y) such that $y = x^2$." The range of this relation is $\{0, 1, 4\}$.

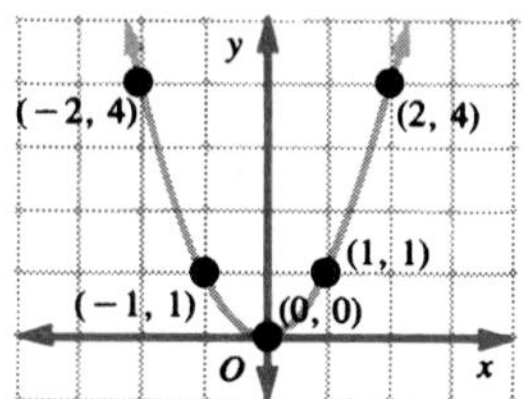

Recall that ordered pairs can also be expressed as a graph in the coordinate plane. When the ordered pairs of the relation $y = x^2$ are graphed, they suggest the familiar shape of a parabola. When the domain is extended to include all real numbers, then the graph of the equation $y = x^2$ is formed. This graph is the parabola shown in gray at the left.

Example

1 If the domain is $\mathscr{R}$ draw the graph of the relation $\{(x, y): y < x + 1\}$.

First, graph the ordered pairs that satisfy $y = x + 1$. These appear on the dashed line shown in the graph.

The points that satisfy $y < x + 1$ lie below the dashed line. The graph of the given relation is the shaded area.

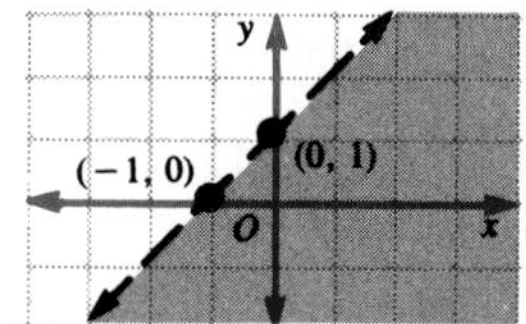

Consider the relation $\left\{(x, y): y = \frac{1}{x - 1}\right\}$. Is the domain of this relation $\mathscr{R}$? Consider what happens when $x = 1$. By substitution, we get $y = \frac{1}{1 - 1}$ or $\frac{1}{0}$, which is undefined. Therefore, 1 cannot be included in the domain. The largest possible domain is {all real numbers except 1}, or $\{x \in \mathscr{R}: x \neq 1\}$, or $\mathscr{R}/\{1\}$.

From now on, unless otherwise stated, if the domain and range of a relation are to consist of real numbers, then the domain will consist of all real numbers x that produce real numbers y in the range.

Example

2 Find the domain of the relation $\left\{(x, y): y = \frac{2}{\sqrt{x - 1}}\right\}$.

In this case, the denominator $\sqrt{x - 1}$ cannot be 0, nor can it be imaginary. $\sqrt{x - 1}$ will be 0 when $x = 1$. Therefore, we exclude 1 from our domain. $\sqrt{x - 1}$ will be real if, and only if, $x - 1 > 0$. This means $x > 1$.

Combining these two restrictions, we conclude that the domain is $\{x: x > 1\}$.

Exercises

Exploratory **For each of the following relations, state the domain and the range.**

1. {(1, 1), (1, 3)}
2. {(2, 3), (0, 3)}
3. {(Mary, Bob), (Alice, Tom)}
4. {(1, 3), (2, 4), (1, 5), (2, 6)}
5. {(a,b), (b,c), (c,a)}
6. $\left\{(1, 7), (3, 7), \left(\frac{8}{3}, 7\right)\right\}$
7. {(△, ⊥), (~, ⊥), (△, ∥)}
8. {(June, 1952), (February, 1966), (May, 1983)}
9. {(Alabama, Montgomery), (Alaska, Juneau), (Arizona, Phoenix), (Arkansas, Little Rock)}

Express the data in each table as a set of ordered pairs. Explain what each number in each ordered pair means.

10.

Celsius	Fahrenheit
100	212
35	95
20	68
0	32

11.

Cost of telegram	Number of words
\$ 5	1–4
\$10	5–8

12.

Table number	Students
1	John, Ted, Alice
2	Sally
3	Joan, Tim

Write the relation indicated by each arrow diagram as a set of ordered pairs.

13.
2 → 3
4 → 3
5 → 1

14.
2 → 4
−3 → 5
7 → 9

15.
9 → 3
9 → −3
4 → 2
4 → −2

Written **Write a rule for each relation and list its domain and range.**

1. {(1, 3), (2, 5), (0, 1), (−2, −3), (10, 21)}
2. {(−3, 3), (−5, 5), (6, −6), (10, −10), (0, 0)}
3. {(3, 9), (−3, 9), (0, 0), (1, 1), (5, 25)}
4. {(−3, 3), (−5, 5), (5, 5), (3, 3), (0, 0)}

In exercises 5–10, refer to the graph of the relation at the right.

5. Write the relation as a set of ordered pairs.
6. Write an arrow diagram to represent the relation.
7. State the domain and the range of the relation.
8. What number in the range corresponds to −1?
9. What numbers in the domain are paired with 4?
10. Try to find a rule that expresses the relationship between corresponding members of the domain and range.

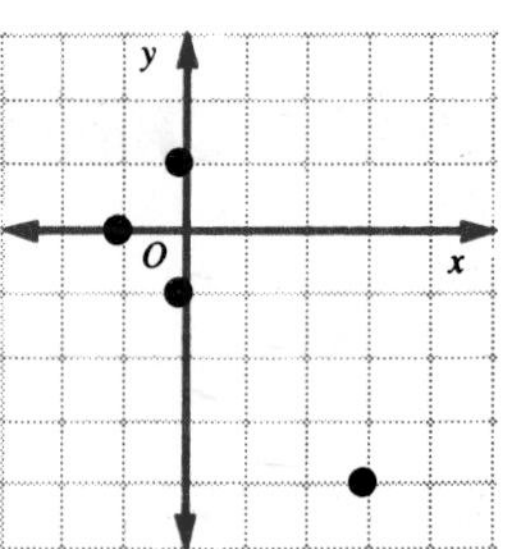

If the domain is {−3, −2, −1, 0, 1, 2, 3}, graph each relation and state the range.

11. $\{(x, y): y = x\}$
12. $\{(x, y): y = |x|\}$
13. $\{(x, y): y = x^2\}$
14. $\{(x, y): y = x^3\}$
15. $\{(x, y): y = \pm\sqrt{x + 3}\}$
16. $\{(x, y): y = |x| - 2\}$

17. $\{(x, y): y = x - 4\}$ **18.** $\{(x, y): y = 3\}$ **19.** $\{(x, y): y = -4\}$

20. $\{(x, y): y \leq -2\}$ **21.** $\{(x, y): x^2 = y\}$ **22.** $\{(x, y): y = x^2 - 4\}$

23. $\{(x, y): y = |x + 1|\}$ **24.** $\{(x, y): y = -x^2\}$ **25.** $\{(x, y): y = -|x|\}$

State the values of *x*, if any, for which each expression does not name a real number.

26. $x + 3$ **27.** $\sqrt{x}$ **28.** $\sqrt{x + 1}$ **29.** $\sqrt{x - 3}$ **30.** $\sqrt{2x - 7}$

31. $\frac{1}{x}$ **32.** $\frac{4}{x - 3}$ **33.** $\frac{2x + 3}{2x - 1}$ **34.** $\frac{x^3}{x^2}$ **35.** $\frac{5}{x^2 - 4}$

36. $\frac{8}{x^2 - 9}$ **37.** $\frac{3x}{x^2 + 4x + 3}$ **38.** $\frac{5}{x^2 - x - 6}$ **39.** $\frac{9}{25 - x^2}$ **40.** $\frac{1 + \frac{1}{x}}{x - 5}$

41. $\frac{3}{|x - 1|}$ **42.** $\frac{1}{|2x + 1|}$ **43.** $\frac{3}{x^2 + 1}$ **44.** $\frac{4}{\sqrt{2x - 1}}$ **45.** $\frac{\sqrt{x - 5}}{7 - x}$

46. $\frac{\sqrt{2x + 6}}{8 - x}$ **47.** $\frac{3}{|x + 2x^2|}$ **48.** $\frac{5}{\sqrt{x^2 - 4}}$ **49.** $\frac{\sqrt{2 + x}}{\sqrt{x - 5}}$ **50.** $\frac{\sqrt{x}}{\sqrt{9 - x^2}}$

If the domain is $\mathcal{R}$, draw a graph of each of the following relations.

51. $\{(x, y): y = x + 3\}$ **52.** $\{(x, y): y = |x + 1|\}$ **53.** $\{(x, y): y \leq -4\}$

54. $\{(x, y): y > x + 2\}$ **55.** $\{(x, y): y \geq 3 - x\}$ **56.** $\{(x, y): y = \sqrt{x}\}$

Give the domain of each relation.

57. $\{(x, y): y < x + 5\}$ **58.** $\{(x, y): x = |y|\}$ **59.** $\{(x, y): y = \sqrt{x}\}$

60. $\left\{(x, y): y = \frac{3}{x - 5}\right\}$ **61.** $\left\{(x, y): y = \frac{3}{|2x - 7|}\right\}$ **62.** $\left\{(x, y): y = \frac{3}{x^2}\right\}$

63. $\{(x, y): 2y = -x\}$ **64.** $\left\{(x, y): y = \frac{4}{3}\right\}$ **65.** $\{(x, y): y \leq 2x\}$

66. $\{(x, y): y > 2x\}$ **67.** $\{(x, y): y \geq x + 3\}$ **68.** $\{(x, y): x = 4\}$

69. $\{(x, y): x \geq 0\}$ **70.** $\{(x, y): x < -1\}$

71. $\{(x, y): x = -1 \text{ or } y = -1\}$ **72.** $\{(x, y): x = 1 \text{ and } y = 1\}$

73. $\{(x, y): y = |3x|\}$ **74.** $\{(x, y): y = -|3x|\}$

75. $\{(x, y): y = |x| - 1\}$ **76.** $\{(x, y): y = 2 - |x|\}$

Challenge **Graph each relation. State the domain and range of the relation.**

1. $\{(x, y): 2x + y = 4\}$ **2.** $\{(x, y): 2x + y \leq 4\}$

3. $\{(x, y): y < |x|\}$ **4.** $\{(x, y): x^2 + y^2 = 4\}$

5. $\{(x, y): x = y^2\}$ **6.** $\{(x, y): y = x^2 + 4x + 4\}$

7. $\{(x, y): 0 < x < 4 \wedge y < 2\}$ **8.** $\{(x, y): y = -|x + 4|\}$

3.2 What Is a Function?

The idea of a function is one of the most important in all mathematics and has many applications in other fields of study.

The table below gives the mileage (miles per gallon) for a particular car traveling at various speeds.

Speed of Car (mph)	Mileage (Miles per gallon)
10	12
20	14
30	15
40	17
50	16
60	14

The information in this table determines a relation whose domain is {10, 20, 30, 40, 50, 60} and whose range is {12, 14, 15, 16, 17}.

As the diagram given below indicates, each speed is paired with exactly one mileage. Each member of the domain is paired with exactly one member of the range.

Domain *(Speed)* **Range** *(Mileage)*

10 → 12
20 → 14
30 → 15
40 → 17
50 → 16
60 → 14

This relation is an example of a *function*.

Definition of Function

A **function** or **mapping** is a relation in which each member of the domain is paired with exactly one member of the range.

Examples

1 The relation indicated by the diagram is *not* a function.

Since 70 is paired with both Jim and Jane, we do not have each member of the domain paired with exactly one member of the range.

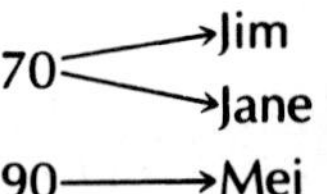

2 **The relation indicated by this diagram *is* a function.**

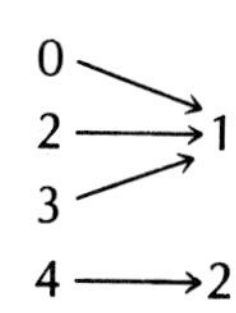

Each member of the domain is paired with exactly one member of the range. Since this relation is a mapping, we may read the symbol $0 \rightarrow 1$ as "0 maps to 1." We call 1 the *image* of 0. Note that the image of 2 and of 3 is also 1. The image of 4 is 2.

3 **The relation $\{(-1, 2), (3, 5), (4, 7), (-1, 3)\}$ is *not* a function.**

The ordered pairs $(-1, 2)$ and $(-1, 3)$ have the same first coordinate. This indicates that -1 is paired with two different members of the range, 2 and 3.

Examples 2 and 3 illustrate the following principle.

If no two ordered pairs in a relation have the same first coordinate, then the relation is a function, or mapping.

In many cases, functions can be specified by an equation. The function in example 4 illustrates an equation in the form, $y = mx + b$. This is known as a **linear function.**

Examples

4 **The relation $\{(x, y): y = 2x - 1\}$ is a function.**

A few ordered pairs are given in the table. If the domain is $\mathcal{R}$, the graph that appears at the right is the graph of all ordered pairs for the function.

x	y
−3	−7
−2	−5
0	−1
1	1

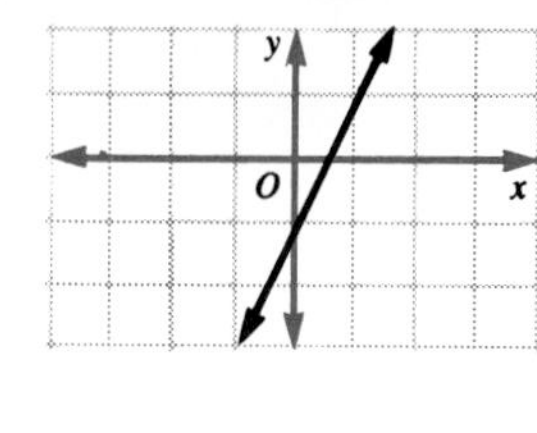

5 **The familiar Fahrenheit-Celsius conversion formula $C = \frac{5}{9}(F - 32)$ determines a linear function.**

The graph shows ordered pairs in which the domain is the set of Fahrenheit measures and the range is the set of Celsius measures.

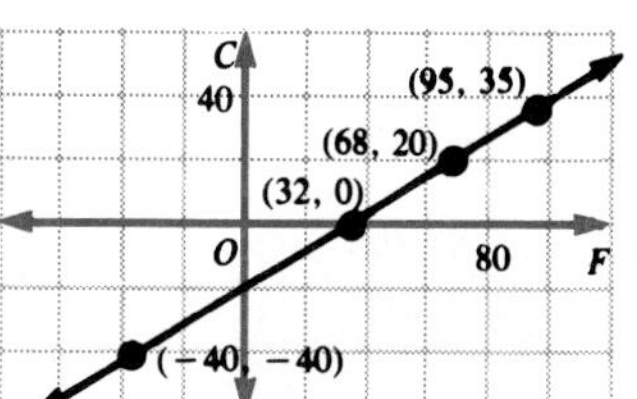

6 **The distance D traveled in two hours depends on the rate R at which you are traveling. This relationship can be expressed by the linear equation, $D = 2R$.**

The linear function $\{(R, D): D = 2R\}$ is graphed at the right. It makes sense to restrict the domain to all nonnegative real numbers since the rate cannot be negative.

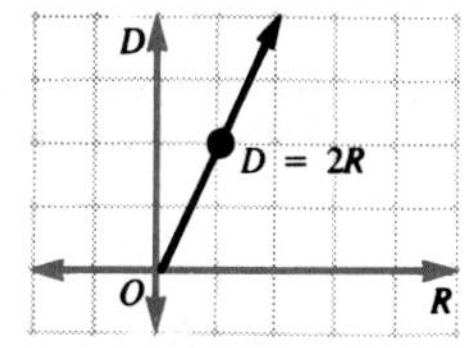

In example 5, the Celsius temperature depends upon the Fahrenheit temperature. Similarly, in example 6, the distance traveled depends upon the rate of speed. Whenever the value of one variable depends on, or is determined by, the value of another variable, we say that the first variable is a *function of the second.* Thus, the Celsius temperature is a function of the Fahrenheit temperature and the distance traveled in two hours is a function of the rate of speed.

Examples

7 **The relation $\{(x, y): y = |x|\}$ is an example of an *absolute value* function.**

Note that the range consists of all nonnegative real numbers.

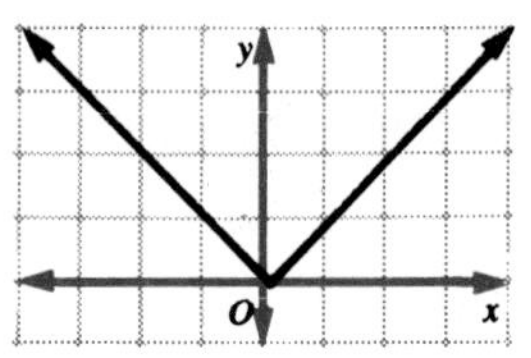

8 **The relation $\{(x, y): x = y^2\}$ is *not* a function.**

As the table shows, there are many cases where more than one y-coordinate is paired with the same x-coordinate. For example, (4, 2) and (4, −2) indicate that 4 is paired with two members of the range.

x	y
0	0
1	1
1	−1
4	2
4	−2

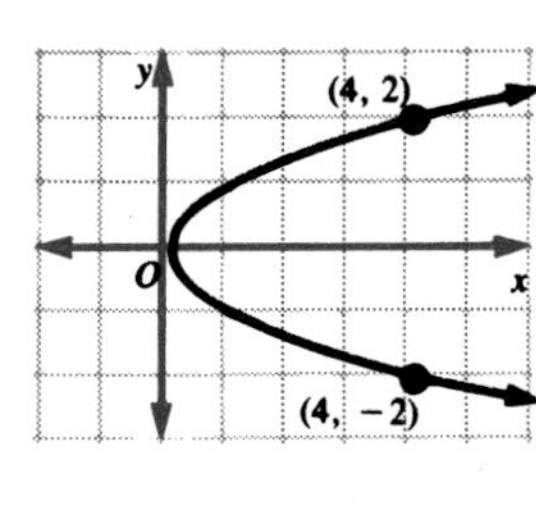

Whenever any graph contains two ordered pairs with the same first coordinate, as in example 8, the graph cannot be the graph of a function. Since all ordered pairs that have the same first coordinate lie on the same vertical line, we can use the following test to tell whether a graph is that of a function.

Vertical Line Test

If no vertical line intersects a graph more than once, the graph is the graph of a function.

Example

9 **The Vertical Line Test verifies that Figure 1 shows a function. Figure 2 does not.**

Since the gray line intersects the graph in Figure 2 twice, the graph is not that of a function.

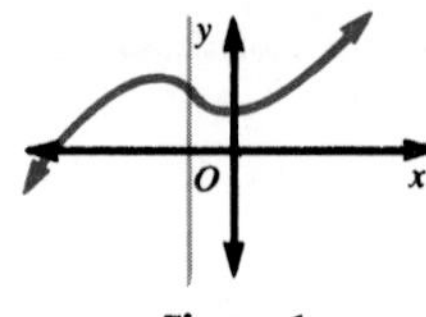

Figure 1

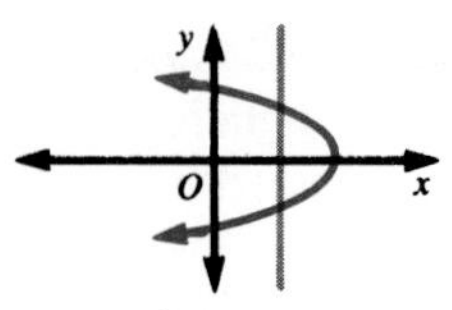

Figure 2

Exercises

Exploratory **Give an example of a relation that is not a function for each of the domains given in exercises 1–3.**

1. {1, 2, 4, 5} **2.** {−1, 0, 1} **3.** {Columbus, Akron, Toledo}

4. Explain why the Vertical Line Test works.

Give the domain and range for each of the following relations. Then decide whether each relation is a function. If it is not a function, explain why.

5. {(1, 4), (2, 4)} **6.** {(−4, 3), (−7, 10)} **7.** {(1, 8), (1, 9)}

8. {(8, 1), (9, 1), (10, 1)} **9.** {(1, 1), (1, 2), (2, 3), (2, 4)} **10.** {(a, b), (b, c), (a, c)}

11. Bernie → 78
Maria → 78
Harry → 78

12.

No. of sides of Polygon	Sum of Interior Angles
3	→ 180
4	→ 360
5	→ 540
6	→ 720

13. 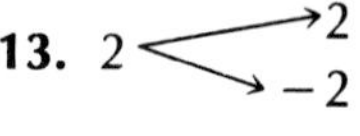

14. $\{(x, y): y = \sqrt{x}\}$ **15.** $\{(x, y): x^2 + y^2 = 4\}$ **16.** $\{(x, y): xy = 10\}$

Using the formula in example 5, find the Celsius measure that corresponds to each of the following Fahrenheit measures.

17. 32 **18.** 212 **19.** 50 **20.** 89 **21.** 59

Tell whether each graph is the graph of a function. If not, explain why. In each case, give the domain and range.

22. 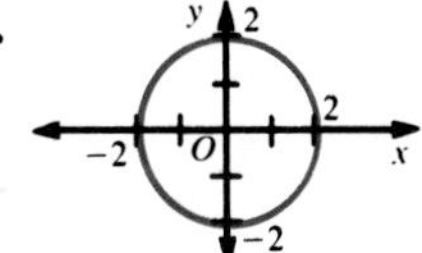

23. 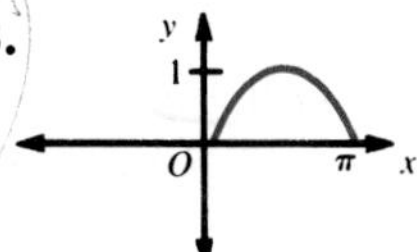

24. 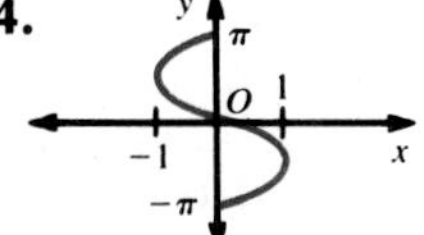

25. 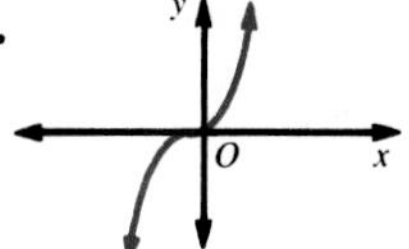

26. 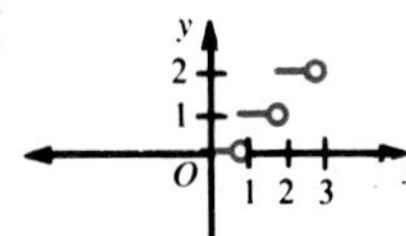

27. 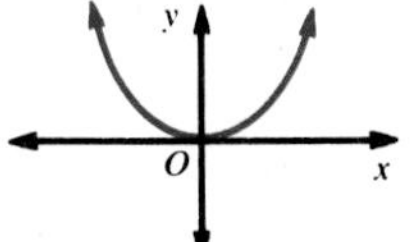

28. 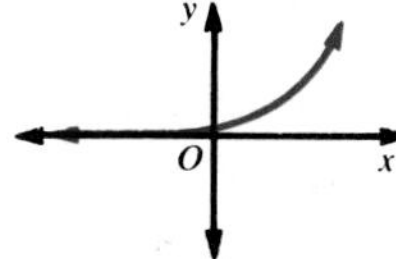

29. 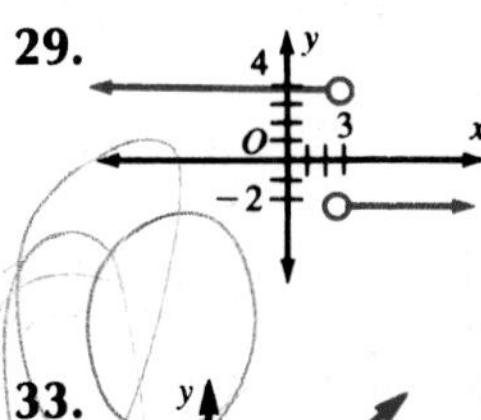

30. 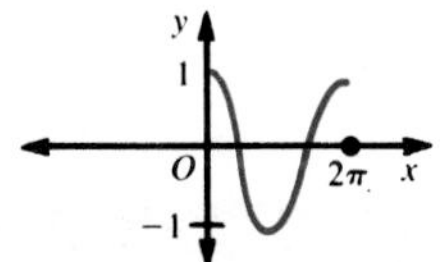

31. 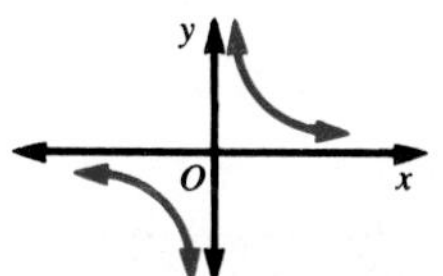

32. 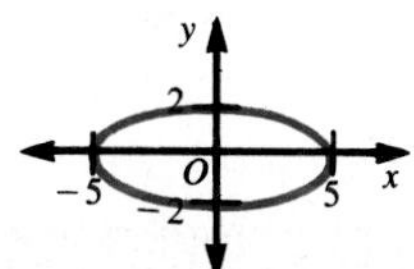

33. 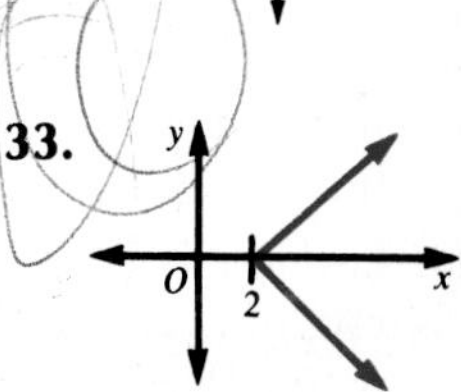

Written **For each of the following functions, find the range if the domain is $\{-1, 0, 3\}$.**

1. $y = 2x - 1$ **2.** $y = 2x - 3$ **3.** $y = x^2 - 3$

4. $y = x^3 + 1$ **5.** $y = x$ **6.** $y = |x - 1|$

7. $y = |2x + 1|$ **8.** $y = 2x^2 - x + 1$ **9.** $y = \frac{x}{x - 1}$

Write an equation for the linear function whose graph has slope *m* and contains the given point.
Recall the point-slope form: $y - y_1 = m(x - x_1)$.

10. $m = 2, (1, 4)$ **11.** $m = -\frac{1}{2}, (-3, 1)$ **12.** $m = -\frac{2}{3}, (-5, 0)$ **13.** $m = -\frac{3}{5}, (-1, -7)$

Write an equation for the linear function that contains the two given ordered pairs.

14. $(0, 0), (2, 5)$ **15.** $(1, 2), (3, 6)$ **16.** $(-2, 1), (3, 4)$ **17.** $\left(\frac{1}{2}, 1\right), \left(\frac{1}{4}, \frac{1}{2}\right)$

In exercises 18–26, graph each relation and tell whether it is a function. If it is a function, tell whether it is a linear function, absolute value function, or neither. Then give the domain and range, and give the image of 5.

18. $\{(x, y): y = 3x - 4\}$ **19.** $\{(x, y): y = x\}$ **20.** $\{(x, y): y = 4\}$

21. $\{(x, y): x = |y|\}$ **22.** $\{(x, y): y = |x|\}$ **23.** $\{(x, y): y = -x^2\}$

24. $\{(x, y): y = 4 - x^2\}$ **25.** $\{(x, y): y = |x| - 2\}$ **26.** $\{(x, y): y = |x - 2|\}$

27. Explain what it means to say that "temperature is a function of the time of day."

For each function, find the domain if the range is $\{-3, 0, 5\}$.

28. $y = 2x + 1$ **29.** $y = 3 - x$ **30.** $y = \frac{x - 3}{3x + 2}$

31. $y = |2x - 1| - 3$ **32.** $y = x^2 - 4$ **33.** $y = x^2 + 4x$

34. The Auto Shoppe has a standard \$25 charge for each job. In addition, the mechanic working on the job charges \$32 per hour. Write an equation for the linear function that describes the relationship between time spent on a job and the total charge.

35. A telephone company charges \$14.50 per month plus 13¢ for each local call. Write an equation for the linear function that describes the relationship between the number of local calls made per month and the monthly telephone charge.

36. The underground temperature of rocks varies with their depth below the surface. The temperature at the surface is about 20° C. At a depth of 2 km, the temperature is about 90° C. Write an equation to describe the relationship between underground temperature and depth, assuming the relationship is linear.

37. The number of times a cricket chirps per minute varies with the temperature. At 23° C, crickets chirp about 68 times per minute. At 7° C, they chirp about 3 times per minute. Write an equation to describe the relationship between the temperature and the number of chirps per minute, assuming the relationship is linear.

38. Emilio earns a salary of \$180 per week for selling records. He also receives an extra 35¢ for each record over 100 that he sells. Write an equation to describe the relationship between Emilio's total weekly income and the number of records he sells per week.

3.3 Function Notation

You are already familiar with several ways of representing a function. For example, the set of ordered pairs and arrow diagram shown below each represent the same function.

$\{(2, 4), (3, 9), (4, 16), (5, 25)\}$

$2 \longrightarrow 4$
$3 \longrightarrow 9$
$4 \longrightarrow 16$
$5 \longrightarrow 25$

This function maps each member of the domain to its square. A convenient way of naming this function is to use the letter f and write

$$f: x \longrightarrow x^2$$

which is read "f is the function that maps x to x^2" or "f maps x to x^2."

Example

1 The domain of g: $x \to 2x + 1$ is $\{1, 2, 3, 4\}$. Represent g with an arrow diagram, and give the range of g. Then represent g as a set of ordered pairs, and draw the graph of g.

Arrow diagram for g

$1 \longrightarrow 3$
$2 \longrightarrow 5$
$3 \longrightarrow 7$
$4 \longrightarrow 9$

The range of g is $\{3, 5, 7, 9\}$.

A set of ordered pairs for g is $\{(1, 3), (2, 5), (3, 7), (4, 9)\}$.

The graph of g appears at the right.

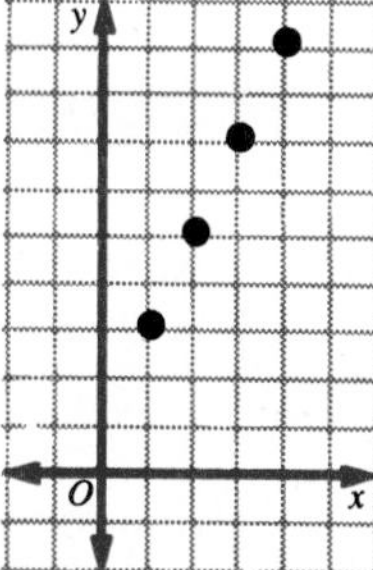

The members of the range of a function are called *values* of the function. In the preceding example, the values of g are 3, 5, 7, and 9.

To symbolize "g maps 1 to 3" we may write

$$g: 1 \to 3 \quad \text{or} \quad 1 \xrightarrow{g} 3.$$

Another common notation is $g(1) = 3$, which may be read in any of the following ways.

"g maps 1 to 3" "g of 1 equals 3"
"1 maps to 3 under g" "the value of g at 1 is 3"

It is important to recognize that the symbol $g(1)$ has nothing to do with multiplication. It represents the member of the range that is paired with 1 under the mapping g. The other members of the range in example 1 can be expressed as

$$g(2) = 5, \quad g(3) = 7, \quad \text{and} \quad g(4) = 9.$$

Therefore, the rule for the function g can be expressed as $g: x \to 2x + 1$ or as $g(x) = 2x + 1$.

In general, if x represents any member of the domain of a function f, then $f(x)$ represents the value of f at x.

Examples

If $f(x) = 2x^2 - 3x + 1$, find each of the following.

2 $f(-2)$

$$f(-2) = 2(-2)^2 - 3(-2) + 1$$
$$= 2 \cdot 4 + 6 + 1$$
$$f(-2) = 15$$

3 **the value of f at 3**

To find the value of f at 3, we compute $f(3)$.

$$f(3) = 2 \cdot 3^2 - 3 \cdot 3 + 1$$
$$= 2 \cdot 9 - 9 + 1$$
$$f(3) = 10$$

Recall that an equation like $y = 2x$ defines a function. The equation $y = 2x$ can also be written $f(x) = 2x$. In general, the expression $y = f(x)$ expresses the fact that the variable y is a function of the variable x.

A common way to define a function f is to state its domain and then give a formula for $f(x)$. Unless otherwise specified, you can assume that the domain for any function f consists of all those real numbers x for which $f(x)$ is also a real number.

Examples

4 **Find the domain of f if $f(x) = \dfrac{3}{x^2 - x - 12}$.**

For $f(x)$ to be a real number, the denominator $x^2 - x - 12$ cannot equal 0. To find the values of x for which the function is undefined, solve the following equation.

$$x^2 - x - 12 = 0$$
$$(x + 3)(x - 4) = 0$$
$$x = -3 \text{ or } x = 4$$

The domain of f is {all real numbers except -3 and 4}. This can be expressed as $\{x \in \mathcal{R}: x \neq -3 \text{ and } x \neq 4\}$ or $\mathcal{R}/\{-3, 4\}$.

5 **Using the function in example 4, find x if $f(x) = -\frac{1}{2}$.**

$$f(x) = \frac{3}{x^2 - x - 12} = -\frac{1}{2} \quad \textit{Substitution}$$
$$x^2 - x - 12 = -6 \quad \textit{Cross Multiplication}$$
$$x^2 - x - 6 = 0$$
$$(x - 3)(x + 2) = 0$$
$$x = 3 \text{ or } x = -2$$

For $f(x) = -\frac{1}{2}$, $x = 3$ or $x = -2$.

Exercises

Exploratory **Explain the meaning of each of the following.**

1. $h: x \rightarrow 3x$
2. $g(4) = 3$
3. $f: 4 \rightarrow -6$
4. value of a function
5. h of 2 equals 7
6. value of k at 2 equals 0

The domain for each of the following functions is $\{-1, 0, 1, 2\}$. Represent each function by an arrow diagram. Then express the function as a set of ordered pairs. Finally, write the range of each function as a set.

7. $f(x) = x^2$
8. $f(x) = x^3$
9. $f(x) = 3x - 1$
10. $f(x) = 2x + 5$
11. $g(x) = \frac{x + 1}{3}$
12. $h(x) = \frac{x + 1}{x + 3}$

Find $f(-3)$, $f(-1)$, $f(0)$, and $f(5)$ for each of the following functions.

13. $f(x) = x + 3$
14. $f(x) = x^2 - 1$
15. $f(x) = 2x^3$
16. $f(z) = z^2 - z - 1$
17. $f(r) = |r + 3|$
18. $f(t) = 32t^2$
19. $f(t) = -16t^2$
20. $f(n) = n^4 - 1$
21. $f(x) = |x| + x^2$

Written **The domain is $\{-2, -1, 0, 1, 2\}$. Represent each function as a set of ordered pairs. Then give the range and graph the function. For exercise 3, exclude 0 from the domain.**

1. $h: x \rightarrow 3x - 1$
2. $k: x \rightarrow |x|$
3. $g(x) = \frac{1}{x}$
4. $t(a) = 2a^2 - a$
5. $p(n) = |4 - n|$
6. $G(v) = -2v^2$
7. $f(x) = x^2 + 3x + 5$
8. $g(x) = 3x^2 - x - 1$

Find $f(-2)$ for each of the following.

9. $f: x \rightarrow x + 2$
10. $f: x \rightarrow x^2$
11. $f(x) = x^2 + 3$
12. $f: x \rightarrow \sqrt{x + 6}$
13. $f(x) = x^3 - 2x^2 + 3$
14. $f(x) = \frac{1}{8x} - \frac{2}{x}$
15. $f(x) = \frac{3}{4 - x}$
16. $f(x) = |5 - x|$

If $f(x) = 2x^3 - 5$, find each of the following.

17. $f(-3)$ **18.** $f(-0.5)$ **19.** $f(\sqrt{2})$ **20.** $f(2n)$ **21.** $f(p + 1)$

For each equation, use arrow notation to define a function.

22. $y = 2x + 1$ **23.** $y = |x|$ **24.** $y = x^2 - 2$ **25.** $y = |x - 3|$

Determine the domain of each function so that the range is in $\mathcal{R}$.

26. $J(x) = \sqrt{2 - 3x}$ **27.** $g(x) = \sqrt{x + 4}$ **28.** $k(x) = \sqrt{4x - 3}$ **29.** $f(x) = \frac{3}{x}$

30. $h(x) = \frac{2}{x - 5}$ **31.** $m(x) = \frac{\sqrt{2x + 4}}{8 - x}$ **32.** $C(x) = \frac{3}{x^2 - 7x + 12}$ **33.** $H(x) = \frac{x + 2}{x^2 - 9}$

Give the domain and range of each function.

34. $h(x) = 3x$ **35.** $f(x) = |x|$ **36.** $g(x) = x^2$ **37.** $k(x) = \frac{1}{x}$

38. $G(x) = -x^2$ **39.** $H(x) = x^2 + 1$ **40.** $f: x \rightarrow x^3$ **41.** $k(x) = |x| - 4$

The graph of the function *g* appears at the right. Use the graph to find each of the following.

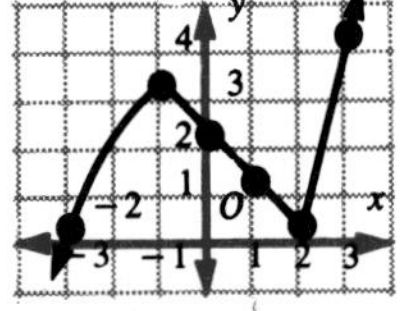

42. $g(0)$ **43.** $g(1)$ **44.** $g(-1)$

45. $g(-3)$ **46.** $g(3) - g(2)$ **47.** $g(2) - g(-3)$

If *f* is a linear function, find $f(x)$ for the following pairs of values.

48. $f(3) = -1$ and $f(-5) = -2$ **49.** $f(5) = -2$ and $f(-2) = 5$

For each function, determine the coordinates of the points where the graph of the function crosses the *x*-axis.

50. $f(x) = x^2 - 4$ **51.** $h(x) = |x - 2|$ **52.** $t(x) = |8x - 10| - 6$

For exercises 53–61, refer to the graph of the linear function $y = f(x)$.

Find the coordinates of each point.

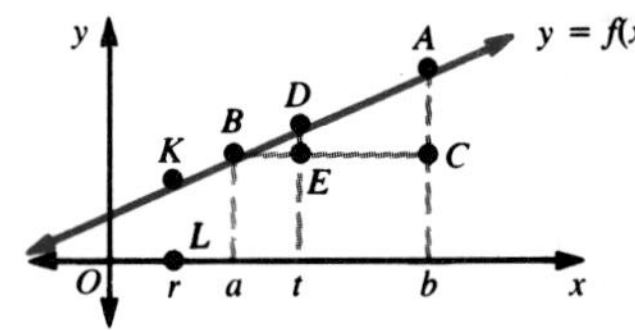

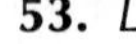

53. L **54.** B **55.** C

56. D **57.** E **58.** A

Use the graph to verify each of the following measures.

59. $BE = t - a$ **60.** $BC = b - a$ **61.** $DE = f(t) - f(a)$

Challenge

Explain how the graphs of each pair of functions differ.

1. $f(x) = |x - 3|$ and $g(x) = |x| - 3$ **2.** $f(x) = |3x|$ and $g(x) = 3|x|$

3. $f(x) = |x^2 - 5|$ and $g(x) = |x^2| - 5$ **4.** $f(x) = -2|x|$ and $g(x) = |-2x|$

3.4 Some Special Functions

In this section we will look at some functions that have a variety of applications in everyday life. Recall that function f is a *linear function* if $f(x) = mx + b$. Depending on the domain, the graph of a linear function is all or part of a line with slope m and y-intercept b.

Example

1 **In 1787, the French scientist Jacques Charles performed a series of experiments in which he recorded the volume of various gases at different temperatures. He observed that gases expand when heated and contract when cooled. By plotting his data on a graph, he also observed a linear relationship between the volume, v, and the temperature, t. That is, he saw that $v = f(t)$ is a linear function.**

Suppose that the volume is 400 cc when a certain gas is at 30°C and that the volume is 600 cc when the gas is at 80°C. Find a rule for $f(t)$.

Since f is linear, $f(t) = mt + b$. We know that (30, 400) and (80, 600) are points on the graph of f. Use the slope formula to find m.

$$m = \frac{y_2 - y_1}{x_2 - x_1} = \frac{600 - 400}{80 - 30} = \frac{200}{50} = 4$$

Using substitution, $f(t) = mt + b$ becomes $f(t) = 4t + b$.

To find b, we substitute 400 for $f(t)$ and 30 for t.

$$400 = 4 \cdot 30 + b$$
$$400 = 120 + b$$
$$b = 280$$

Therefore $f(t) = 4t + 280$.

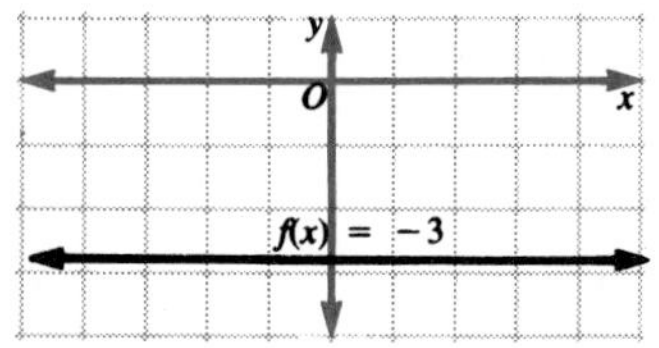

A function f is a **constant function** if $f(x) = b$ where b is some real number. The range of a constant function contains only one element. Its graph will be all or part of a horizontal line. Pictured at the left is the graph of the constant function defined by $f(x) = -3$.

A linear function defined by $f(x) = kx$, where $k \neq 0$, is called a **direct variation.** When a direct variation is written $y = kx$ we say that "y varies directly as x", or "y varies with x", or "y is directly proportional to x." The slope, k, is called the **constant of variation** or **constant of proportionality.**

Example

2 **If the sales tax, t, is 5% of the purchase price, p, then $t = 0.05p$.**

This means that the *sales tax varies directly with the purchase price,* or that the sales tax is directly proportional to the purchase price. That is, the higher the cost, the greater the sales tax will be. The constant of proportionality is 0.05.

A function f is a **quadratic function** if $f(x) = ax^2 + bx + c$, where $a \neq 0$. The graph of a quadratic function is a *parabola*. A function f is called an **exponential function** if $f(x) = a^x$, where a is a positive real number. Exponential functions are often used to study population growth or decline, radioactive decay, and compound interest.

These two types of functions are shown in examples 3 and 4.

Examples

3 **The distance that an object will fall after t seconds at an initial velocity of 32 feet per second is defined by $f(t) = 16t^2 + 32t$.* Find how far the object will fall after 2 seconds has elapsed.**

**$16t^2$ represents the force of gravity on the object.*

Evaluate $f(2)$. *$t = 2$ seconds*

$$f(2) = 16(2)^2 + 32(2) = 64 + 64 = 128$$

Therefore, after 2 seconds, the object has fallen 128 feet.

4 **In a biology experiment, Sara prepares a culture containing a single-celled organism. On each subsequent day she examines the culture and notes that the organism has divided. She records her findings in the following chart.**

t	*number of days (time)*	0	1	2	3	4	5
$f(t)$	*number of organisms*	1	2	4	8	16	32

The data can be described by the function $f(t) = 2^t$.

Another special type of function, the **step function,** is shown in example 5. Its name is derived from the appearance of its graph. One of the more familiar step functions, the **greatest integer function,** is illustrated in example 6.

Examples

5 **Pine City Rent-A-Bike rents bikes for \$5 per day. If you use a bike for any part of a day, you are charged for a full day's rental. This situation suggests a function *f* whose domain is {number of days rental} and whose range is {number of dollars of rental fee}.**

$f: x \rightarrow 5$ whenever $0 < x \leq 1$,
$f: x \rightarrow 10$ whenever $1 < x \leq 2$,
$f: x \rightarrow 15$ whenever $2 < x \leq 3$,
$f: x \rightarrow 20$ whenever $3 < x \leq 4$,
and so on.

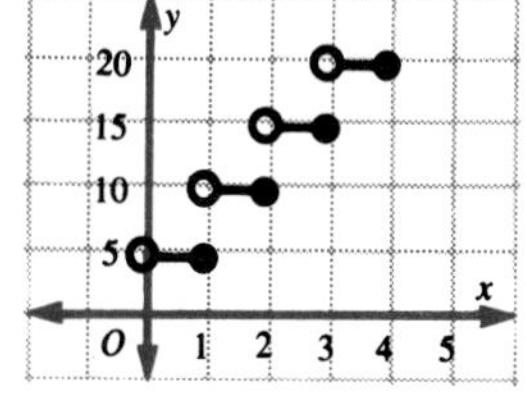

The open circles show that the indicated endpoints are not included in the graph.

6 **The symbol $[x]$ stands for "the greatest integer less than or equal to x."**

$[1\frac{1}{2}] = 1$ 1 is the greatest integer less than or equal to $1\frac{1}{2}$.
$[1] = 1$ 1 is the greatest integer equal to 1.

Some other examples and the graph of $f(x) = [x]$ appear below.

$[\frac{1}{2}] = 0$ $[2.9] = 2$ $[-1] = -1$
$[-1.3] = -2$ $[-4.7] = -5$

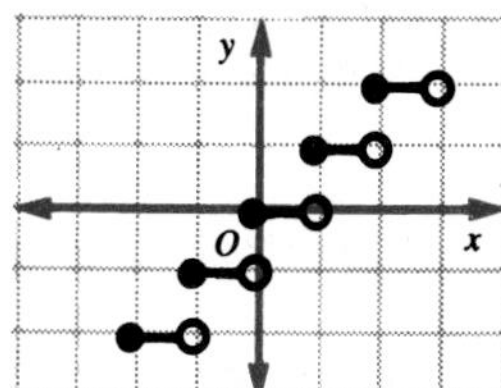

Remember the figure only shows part of the graph of $f(x) = [x]$.

A function defined by $f(x) = \frac{k}{x}$, where $k \neq 0$, is called an **inverse variation.** When an inverse variation is written as $y = \frac{k}{x}$ or $xy = k$, we say that "y varies inversely as x" or "y is inversely proportional to x." Again, k is called the constant of variation or constant of proportionality.

Example

7 **The time it takes to drive a certain distance is inversely proportional to the rate of speed. Marie drives for 4 hours at 55 mph. How long would it take her to make the same trip if she drives at 50 mph?**

Explore The inverse variation that relates time (t) and rate of speed (r) when distance is constant is $rt = d$, or $d = rt$. It should take Marie longer to make the trip at 50 mph than at 55 mph.

Plan First find the distance of the trip using her actual rate of speed (55 mph) and time (4 hours). Then use this distance and the new rate of speed (50 mph) to determine the new time for the trip.

Solve

$$d = rt$$
$$d = (55 \text{ mph}) \cdot (4 \text{ hours})$$
$$d = 220 \text{ miles}$$

$$rt = d$$
$$(50 \text{ mph}) \cdot t = 220 \text{ miles}$$
$$t = \frac{220 \text{ miles}}{50 \text{ mph}} \text{ or } 4.4 \text{ hours}$$

Substitute 220 miles for d and 50 mph for r. Why?

Marie would take 4.4 hours, or 4 hours and 24 minutes, to make the same trip traveling at 50 mph.

Examine Since this is an inverse variation, as the rate decreases, the time should increase. The answer seems reasonable.

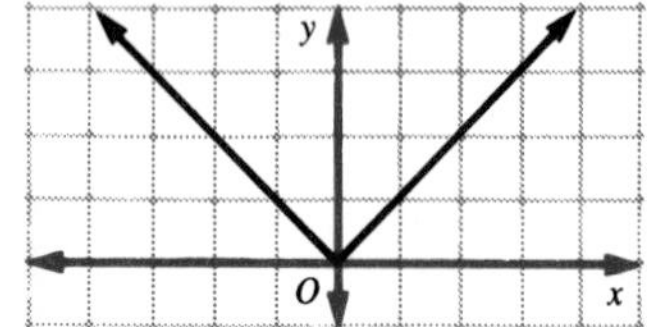

$$f(x) = \begin{cases} x \text{ if } x > 0 \\ 0 \text{ if } x = 0 \\ -x \text{ if } x < 0 \end{cases}$$

Sometimes more than one rule must be used to define a function. Consider the function f, shown at the left, and its graph.

You may have noticed that the rule specified for f corresponds to the definition of the *absolute value* of x. Therefore the graph of f is the same as the graph of the **absolute value function,** $f(x) = |x|$.

Exercises

Exploratory **Classify each function as linear, constant, direct variation, quadratic, exponential, step, inverse variation, absolute value, or none of these.**

1. $f(x) = x$ **2.** $g(x) = 2x - 3$ **3.** $f{:}x \rightarrow 4$ **4.** $g(x) = -x^2 + 1$

5. $f(x) = [x]$ **6.** $f{:}x \rightarrow 1.5x$ **7.** $f(x) = x^3 + 2x^2 - 1$ **8.** $g(x) = \dfrac{2}{x - 3}$

Written **Classify each function as linear, constant, direct variation, quadratic, exponential, step, inverse variation, absolute value, or none of these.**

1. $y = 7x$
2. $y = 7^x$
3. $xy = 7$
4. $y = |-3|$
5. $y = |-3|x + 1$
6. $y = |-3x| + 1$
7. $y = [-3]x + 1$
8. $y = [-3x] + 1$

If f is a linear function, with the given slope and value, find $f(x)$.

9. $m = 3, f(2) = 4$
10. $m = 2, f(1) = 5$
11. $m = -1, f(-2) = 7$
12. $m = 0, f(4) = -4$
13. $m = -5, f(10) = -6$
14. $m = -\frac{2}{3}, f(1) = -6$

If g is a linear function with the given values, find $g(x)$.

15. $g(-2) = 5, g(3) = -6$
16. $g(0) = 4, g(-2) = 6$
17. $g(-1) = -6, g(4) = 0$

Using the formula in example 1, calculate the volume of a certain gas at each temperature.

18. 40°C
19. 60°C
20. 95°C
21. −40°C
22. −60°C

Find each of the following.

23. $[2]$
24. $[3.6]$
25. $[9.8]$
26. $[-0.5]$
27. $[-3.4]$

Graph each of the following functions.

28. $f(x) = [x]$
29. $h(x) = -x^2 + 2$
30. $h(t) = 3t^2$
31. $F(x) = x^2 - 2x$
32. $G(x) = x^2 + 6x - 5$
33. $f(x) = x^2 + 5x + 1$
34. $h(x) = \begin{cases} 3 \text{ if } x < 0 \\ -3 \text{ if } x \geq 0 \end{cases}$
35. $F(x) = \begin{cases} 0 \text{ if } x \geq 1 \\ 1 \text{ if } x < 1 \end{cases}$
36. $G(x) = \begin{cases} 0 \text{ if } x \leq 0 \\ x \text{ if } x > 0 \end{cases}$
37. $g(x) = \begin{cases} 2x \text{ if } 0 < x < 1 \\ 0 \text{ if } x \leq 0 \\ x + 1 \text{ if } x \geq 1 \end{cases}$
38. $m(x) = x^3$
39. $K(x) = (x - 5)^2 + 2$

Each of the following determines a quadratic function. Give the function.

40. The area of a circle is a function of its radius.
41. The area of a square is a function of the length of a side.

In exercises 42–44, write an equation for each direct variation. State the constant of variation.

42. The distance d traveled at 50 km/h varies directly as the time t.
43. The cost, c, of a number of T-shirts at $4.99 per T-shirt varies directly with the number, n, of T-shirts.
44. The perimeter, P, of a square is directly proportional to the length, s, of a side of the square.
45. The distance that a spring stretches is directly proportional to the weight hung on it. When a weight of 64 g is hung on the spring, it stretches 3.4 cm. Find the distance the spring stretches if a weight of 80 g is hung on it.
46. The velocity of a falling object is directly proportional to the number of seconds it has been falling. If the velocity of a certain object is 96 ft/s at the end of three seconds, what is its velocity at the end of 8.5 seconds?

In example 4 on page 106, Sara started with one organism. Suppose she had started with ten organisms. How many would be in the culture at the end of the following time intervals?

47. two days **48.** three days **49.** six days **50.** ten days

On the planet Glorfio, it is found that an organism called a Glorfium divides into three Glorfiums at regular intervals. Use this information to solve exercises 51–52.

51. If a culture begins with one Glorfium, write an equation for the exponential function G that gives the number of Glorfiums after t time intervals.

52. If a culture begins with twelve Glorfiums, how many can be expected to be in the culture at the end of five time intervals?

53. The Pine City Rent-All Agency rents chain saws for a maximum of five days. The rental fee for one day, or portion thereof, is \$10. The rental fee for over one day up to three days is \$15. For anything over three days, the fee is \$20. Graph this rental function.

54. The Pine City High School Pep Club is selling greeting cards to raise money for a project. A customer can buy boxes of cards at three different rates. For 1–4 boxes, the cost is \$10 each. For orders of 5–7 boxes, the cost is \$9 each. Any of 8 or more boxes is \$8 each. Graph this function.

55. A Pine City educational statute states that there be one teacher for every 25 students in the school system. This year the student enrollment is 6000 but a $12\frac{1}{2}\%$ drop is expected for the next year. How many teachers will Pine City Schools need for the next year?

56. Charles' Law states that the volume of a gas is directly proportional to its temperature. If the volume of a gas is 2.5 cubic feet at 150° Kelvin, what is the volume of the same gas when the temperature is raised 50°?

In exercises 57–58, write an equation for each inverse variation. State the constant of variation.

57. The number, n, of T-shirts that can be purchased for \$24.99 varies inversely with the cost, c, of one T-shirt.

58. The length, ℓ, of a rectangle with an area of 128 m^2 is inversely proportional to the width, w, of the rectangle.

59. When air is pumped into an automobile tire, the pressure required is inversely proportional to the volume. If the pressure is 30 pounds when the volume is 140 cubic inches, find the pressure when the volume is 100 cubic inches.

60. In a closed room, the number of hours of safe oxygen level varies inversely as the number of people in the room. If there are 2 hours of safe oxygen level for 400 people, how many hours of safe oxygen level are there for 600 people?

Challenge

1. If y varies directly as the square of x, and $y = 7$ when $x = 9$, find y when $x = 7$.

2. If the square of y varies inversely as x, and $y = 4$ when $x = 2$, find y when $x = 11$.

3. The intensity of illumination on a surface varies inversely as the square of the distance from the light source. A surface is 12 meters from a light source. How far must that surface be from the light source to receive twice as much illumination?

4. Graph $K(x) = [x] + |x|$.

5. Graph $m(x) = [x] + 1$.

3.5 Inverse Functions

Given a function, it is often possible to form a new function in a special way. For example, consider the function f defined by the set

$$\{(2, 6), (3, 7), (4, 8)\}.$$

If we interchange the numbers of each ordered pair, we obtain the function

$$\{(6, 2), (7, 3), (8, 4)\}.$$

This function is the **inverse of f,** or **f-inverse,** symbolized by f^{-1}.

Note that f^{-1} does *not* mean $\frac{1}{f}$.

As another example, consider the function g defined by

$$\{(-2, 4), (2, 4)\}.$$

Then, g^{-1} is represented by $\{(4, -2), (4, 2)\}$.

This time, g^{-1} is *not* a function, since 4 is paired with two elements, -2 and 2. Therefore the inverse of a function may not be a function.

Examples

For each of the following, $f(x) = |x|$ and the domain is $\{-1, 1, 2\}$.

1 **List the ordered pairs of f.**

The ordered pairs of f are $(-1, 1)$, $(1, 1)$, $(2, 2)$.

2 **List the ordered pairs of f^{-1}.**

The ordered pairs of f^{-1} are $(1, -1)$, $(1, 1)$, $(2, 2)$.

3 **Determine if f^{-1} is a function.**

f^{-1} is *not* a function since $(1, -1)$ and $(1, 1)$ have the same first coordinate.

When will the inverse of a function also be a function? Consider the previous examples once again. The function f has two ordered pairs with the same y-coordinate, $(-1, 1)$ and $(1, 1)$. When the coordinates are interchanged, there are two ordered pairs with the same x-coordinate, $(1, -1)$ and $(1, 1)$. Therefore, the inverse, f^{-1} is not a function. If no two ordered pairs of a function have the same y-coordinate, then its inverse will also be a function. A function with this property is called a *one-to-one function*.

Definition of One-to-One Function

A function f is a **one-to-one function** if each member of the range corresponds to exactly one member of the domain.

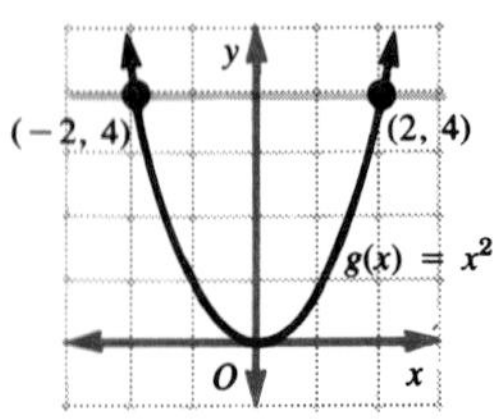

If a function f is one-to-one, then its inverse, f^{-1} is also a function.

Another indicator of a one-to-one function is the graph of that function. If we inspect the graph of f and see that no points lie on the same horizontal line, then f must be one-to-one. The graph of $g(x) = x^2$, shown at the left, does *not* pass this test.

Horizontal-Line Test

If no horizontal line intersects the graph of f in more than one point, then f is one-to-one.

By using the rule for a one-to-one function f, we can find a rule for its inverse function, f^{-1}. For example, suppose f is the linear function

$$\{(x, y) : y = 2x + 1\}.$$

Then interchanging x and y produces f^{-1}.

$$\{(x, y) : x = 2y + 1\}$$

Since it is customary to express y in terms of x, we solve the equation $x = 2y + 1$ for y.

$$x = 2y + 1$$
$$x - 1 = 2y$$
$$\frac{x - 1}{2} = y \quad \text{or} \quad \frac{1}{2}x - \frac{1}{2} = y$$

Thus, f^{-1} can be represented by

$$\left\{(x, y) : y = \frac{1}{2}x - \frac{1}{2}\right\}.$$

Examples

4 If $g(x) = 3x + 5$, determine whether g is one-to-one.

Since the graph of g is a non-horizontal line, g is a one-to-one function.

5 If $g(x) = 3x + 5$, find an equation for $g^{-1}(x)$ in terms of x.

Since $g(x) = 3x + 5$, we know that $y = 3x + 5$ is an equation for g. An equation for g^{-1} is $x = 3y + 5$.

Solving for y, we obtain $\frac{x - 5}{3} = y$. Therefore, $g^{-1}(x) = \frac{x - 5}{3}$.

Exercises

Exploratory **Answer each of the following.**

1. What is meant by the inverse of a function?
2. Under what circumstances will the inverse of a function also be a function?
3. Under what circumstances is a function one-to-one?

For each function draw an arrow diagram for its inverse. Tell if the inverse is also a function.

4. $A \to -1$, $B \to -1$, $C \to 3$

5. $2 \to a$, $5 \to b$, $7 \to c$

6. $a \to 1$, $b \to 1$, $c \to 1$

7. $-7 \to 7$, $0 \to 0$, $7 \to 7$

Write the inverse of each function as a set of ordered pairs. Tell if the inverse is also a function.

8. $\{(-1, 1), (1, -1)\}$

9. $\{(-2, 3), (-1, 5), (4, 3)\}$

10. $\{(\sqrt{2}, 3), (-\sqrt{2}, 4), (-4, 3)\}$

11. $\{(-3, 7), (4, 9), (7, -3), (8, 5)\}$

12. If g is one-to-one, why must g^{-1} be a function?

13. Suppose h is a one-to-one function. Its domain has 18 elements. How many elements are in its range? Explain.

14. What is the horizontal-line test? Why does it work?

15. Explain why the function defined by $y = 3x + 7$ is the inverse of the function defined by $x = 3y + 7$.

16. Is $x = 3y + 7$ equivalent to $y = \frac{x - 7}{3}$? Explain.

Written **Write the inverse of each function as a set of ordered pairs. Determine if the inverse is also a function.**

1. $\{(1, 5), (2, 3), (4, 3), (-1, 5)\}$

2. $\{(-6, 5), (5, -6), (7, 7), (8, -7)\}$

3. $\{(-2, 5), (3, 4), (4, 4), (-1, 4)\}$

4. $\{(10, -3), (11, 10), (-3, 11), (-10, -3)\}$

For each of the following functions the domain is $\{-3, -2, -1, 1, 2\}$.

a. **Write the function as a set of ordered pairs.**
b. **Determine if the function is one-to-one.**
c. **Write the inverse as a set of ordered pairs.**
d. **Determine if the inverse is also a function.**

5. $\{(x, y) : y = x + 1\}$

6. $\{(x, y) : y = x - 4\}$

7. $\{(x, y) : y = 4x\}$

8. $\{(x, y) : y = x\}$

9. $f(x) = x^2$

10. $g(x) = x^3$

11. $h(x) = |x + 3|$

12. $k(x) = |x + 1|$

13. $T(x) = \sqrt{x + 3}$

14. $K(x) = x^2 - 2x + 1$

15. $R(x) = \frac{3}{x}$

16. $A(x) = \frac{x}{|x| + 1}$

Tell whether the functions graphed below are one-to-one. Which ones have inverses that are functions?

17. 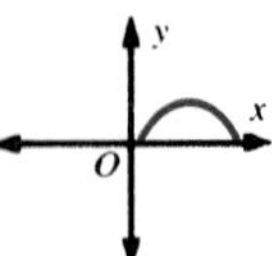

18. 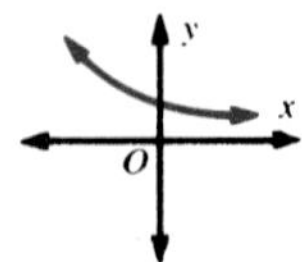

19. 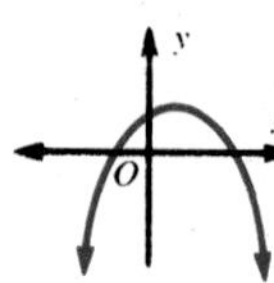

20. 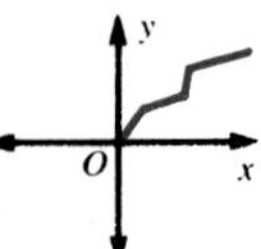

21. 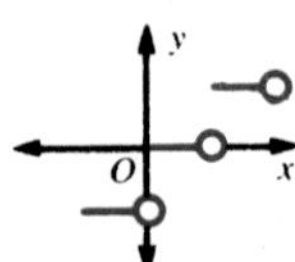

22. 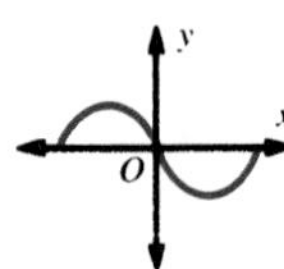

If $g(7) = 6$, $g(8) = 9$, $h(-2) = 3$, and $h(4) = -10$, find each of the following values.

23. $g^{-1}(6)$ **24.** $h^{-1}(3)$ **25.** $g^{-1}(9)$ **26.** $h^{-1}(-10)$

Find an equation for the inverse function of each of the following.

27. $y = 2x$ **28.** $y = -3x$ **29.** $y = \frac{1}{2}x$ **30.** $y = 2x + 1$ **31.** $y = -3x - 5$

Write an equation for $f^{-1}(x)$ in terms of x.

32. $f(x) = 6x$ **33.** $f(x) = 4x$ **34.** $f(x) = \frac{2}{3}x$ **35.** $f(x) = -3x$

36. $f(x) = \frac{1}{4}x$ **37.** $f(x) = -x + 5$ **38.** $f(x) = -3x + 1$ **39.** $f(x) = \frac{1}{2}x + \frac{1}{2}$

40. $f(x) = \frac{2}{5}x$ **41.** $f(x) = \frac{10 - x}{4}$ **42.** $f(x) = \frac{1}{5}x - \frac{1}{3}$ **43.** $f(x) = \frac{4}{3} - \frac{1}{2}x$

Use the graph to answer exercises 44–52.

44. What is the domain of the function?

45. Is the function one-to-one? Explain.

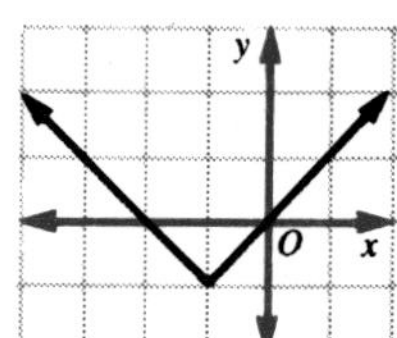

Sketch the graph of the function for each of the following domains. Tell if the function is one-to-one in its new domain.

46. $\mathcal{Z}$ **47.** $\mathcal{W}$ **48.** negative reals **49.** positive reals

50. $\{x \in \mathcal{R} : x \leq -1\}$ **51.** $\{x \in \mathcal{R} : -2 \leq x \leq 0\}$ **52.** $\{x \in \mathcal{R} : -1 \leq x \leq 7\}$

Determine a domain of f such that f^{-1} is also a function.

53. $f(x) = x + 4$ **54.** $f(x) = |x|$ **55.** $f(x) = \sqrt{x}$

56. $f(x) = |x| - 4$ **57.** $f(x) = |x - 3|$ **58.** $f(x) = |x + 5|$

59. $f(x) = (x + 3)^2$ **60.** $f(x) = \sqrt{1 - x}$ **61.** $f(x) = \sqrt{1 + x^2}$

62. $f(x) = |x - 2| - 2$ **63.** $f(x) = 2 - x^2$ **64.** $f(x) = |1 - x^2|$

Mixed Review

Choose the best answer.

1. What is the solution set for $x^2 - 30 < 13x$?
   a. $\{x \in \mathcal{R}: -15 < x < 2\}$ **b.** $\{x \in \mathcal{R}: x < -2 \text{ or } x > 15\}$
   c. $\{x \in \mathcal{R}: -2 < x < 15\}$ **d.** $\{x \in \mathcal{R}: 3 < x < 10\}$
2. If the domain of $y = [1.2x]$ is $\{-2, -1, 0, 1, 2\}$, what is the range?
   a. $\{-2, -1, 0, 2, 3\}$ **b.** $\{-3, -2, 0, 1, 2\}$ **c.** $\{-2, -1, 0, 1, 2\}$ **d.** $\{0, 1.2, 2.4\}$
3. If the domain of $\{(x, y): |y + 1| = x\}$ is $\{0, 1, 3\}$, what is the range?
   a. $\{-1, 0, 2\}$ **b.** $\{1, 2, 4\}$ **c.** $\{-4, -2, -1, 0, 2\}$ **d.** none of these
4. Which equation represents a direct variation?
   a. $xy = -4$ **b.** $y = -4x + 1$ **c.** $y = -4$ **d.** $y = -4x$
5. If g is a linear function with $g(-2) = 7$ and $g(3) = 4$, what is $g(x)$?
   a. $g(x) = -\frac{3}{5}x + \frac{29}{5}$ **b.** $g(x) = -\frac{5}{3}x + \frac{11}{3}$ **c.** $g(x) = \frac{3}{5}x + \frac{41}{5}$ **d.** $g(x) = -\frac{3}{5}x + \frac{17}{5}$
6. Which quadratic equation has roots $2 - 3i$ and $2 + 3i$?
   a. $x^2 + 4x + 13 = 0$ **b.** $x^2 - 4x + 13 = 0$ **c.** $x^2 - 4x - 5 = 0$ **d.** $x^2 - 4x + 7 = 0$

Mathematical Excursions

Inverse Functions and Graphing

$y = 5x - 3$	x	0	1	2	−1
	y	−3	2	7	−8
$y = \frac{x + 3}{5}$	x	−3	2	7	−8
	y	0	1	2	−1

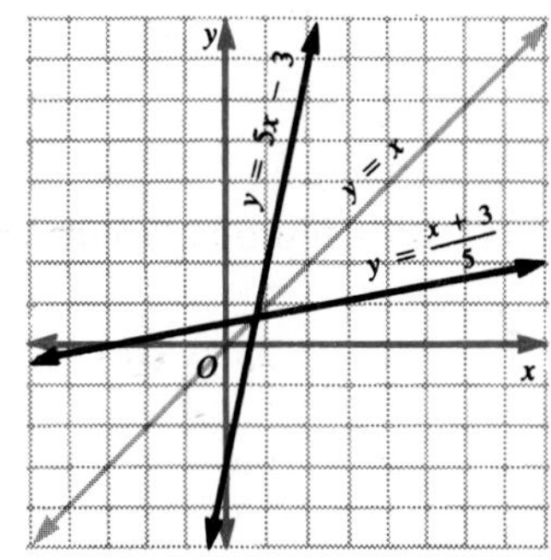

The graphs of $f(x) = 5x - 3$ and its inverse $f^{-1}(x) = \frac{x + 3}{5}$ are shown with a table of values for each function. Notice that for each ordered pair (a, b) in f, there exists an ordered pair (b, a) in f^{-1}. Therefore, if f and f^{-1} are inverse functions, $f(a) = b$ if and only if $f^{-1}(b) = a$.

Another interesting observation involves the graphs of f, f^{-1}, and the line $y = x$. Let P be (a, b) on f and P' be (b, a) on f^{-1}. The line containing P and P' is perpendicular to the line $y = x$, and the line $y = x$ bisects the line segment PP'. Therefore, P and P' are equidistant from the line $y = x$ and $R_{y=x}(P) = P'$.

Exercises **Use graphing to determine whether f and g are inverse functions.**

1. $f(x) = 2x + 1$, $g(x) = x - 0.5$
2. $f(x) = -2x + 3$, $g(x) = 2x - 3$
3. $f(x) = x + 4$, $g(x) = x - 4$
4. $f(x) = x^2$ for $x \geq 0$, $g(x) = \sqrt{x}$

3.6 Conic Sections

In your studies so far, you have graphed many relations involving x and y. Most of these were functions. These included $y = mx + b$ (a *line*), $y = ax^2 + bx + c$ (a *parabola*), and $y = |x| + c$ (an *absolute value function*). Another equation, $x^2 + y^2 = r^2$, is a relation, but *not* a function. Its graph is familiar to you as that of a *circle* with center at the origin.

In this section, two other relations that are not functions will be explored. These are $ax^2 + by^2 = c$, and $ax^2 - by^2 = c$ (or $-ax^2 + by^2 = c$). In both relations a, b, and $c > 0$.

To determine some characteristics of $ax^2 + by^2 = c$, let us graph the relation $4x^2 + 9y^2 = 36$. First, make a table of x-coordinates and find the corresponding y-coordinates. Note that, since x is always squared, for every ordered pair (x, y) with a positive x value, there is a corresponding ordered pair $(-x, y)$. An example of solving the equation for a given value of x is shown below.

x	y
0	± 2
1	$\pm\frac{4\sqrt{2}}{3} \approx \pm 1.9$
2	$\pm\frac{2\sqrt{5}}{3} \approx \pm 1.5$
3	0
4	$\pm\frac{2i\sqrt{7}}{3}$

Corresponding pairs

x	y
-1	$\pm\frac{4\sqrt{2}}{3} \approx \pm 1.9$
-2	$\pm\frac{2\sqrt{5}}{3} \approx \pm 1.5$
-3	0
-4	$\pm\frac{2i\sqrt{7}}{3}$

Rational approximations for irrational expressions make graphing easier.

Solve for y if $x = 2$.

$$4x^2 + 9y^2 = 36$$
$$4(4) + 9y^2 = 36$$
$$9y^2 = 20$$
$$y^2 = \frac{20}{9}$$
$$y = \pm\sqrt{\frac{20}{9}}$$
$$y = \pm\frac{2\sqrt{5}}{3} \approx \pm 1.49$$

Note that for $x = 4$ and $x = -4$, the y-coordinate is not a real number so it cannot be graphed in the real number coordinate plane. If the relation is to be defined in the real number system, it appears that the domain must be restricted to the interval $-3 \leq x \leq 3$. From the y-coordinates in the table, we can see that the range is the interval $-2 \leq y \leq 2$.

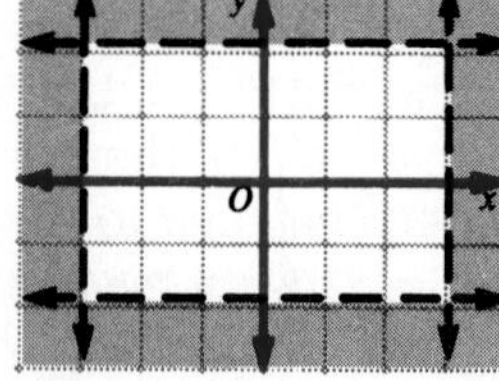

These intervals create a "boundary" for the graph of $4x^2 + 9y^2 = 36$. The gray area on the coordinate plane indicates the ordered pairs outside the intervals for x and y.

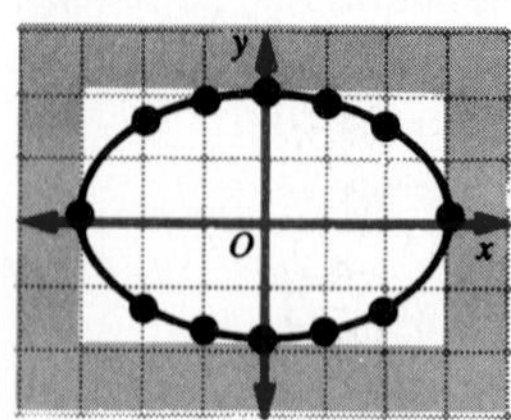

If we graph the points in the table and include other points such as those in which $x = \pm\frac{1}{2}$ and $\pm 2\frac{1}{2}$, the graph takes on a definite shape. This curve is an example of an **ellipse.** Its *x-intercepts* are ± 3 and its *y-intercepts* are ± 2.

There is a more efficient way of determining the graph of this equation than randomly plotting points. Solving the equation for y enables us to see more important facts about the curve and to complete a table of values more quickly.

$$\begin{aligned} 4x^2 + 9y^2 &= 36 \\ 9y^2 &= 36 - 4x^2 \\ y^2 &= \frac{36 - 4x^2}{9} \\ y &= \pm\sqrt{\frac{36 - 4x^2}{9}} \end{aligned}$$

For $y \in \mathcal{R}$, $4x^2$ must be less than or equal to 36. Why? Solving this inequality, we find that $|x| \leq 3$, or $-3 \leq x \leq 3$. This domain indicates that the numerator of the fraction under the radical sign can never be more than 36. So $|y|$ is never more than $\sqrt{\frac{36}{9}}$ or 2. Thus, $-2 \leq y \leq 2$.

These boundaries create the rectangle in which the curve must be contained. These intervals also narrow the choices for the x-values used in finding ordered pairs to graph.

Example

1 Solve $25x^2 + 4y^2 = 100$ for y. Then graph the equation.

$$\begin{aligned} 25x^2 + 4y^2 &= 100 \\ 4y^2 &= 100 - 25x^2 \\ y^2 &= \frac{100 - 25x^2}{4} \\ y &= \pm\sqrt{\frac{100 - 25x^2}{4}} \end{aligned}$$

To find the interval for the domain, solve $100 - 25x^2 \geq 0$.

$$|x| \leq 2 \text{ or } -2 \leq x \leq 2$$

The interval for the range is

$$-5 \leq y \leq 5.$$

x	y
0	± 5
$\pm\frac{1}{2}$	$\approx \pm 4.8$
± 1	$\approx \pm 4.3$
$\pm 1\frac{1}{2}$	$\approx \pm 3.3$
± 2	0

Recall that $(\pm\frac{1}{2}, \pm 4.8)$ symbolizes four ordered pairs: $(\frac{1}{2}, 4.8)$ $(\frac{1}{2}, -4.8)$ $(-\frac{1}{2}, 4.8)$ $(-\frac{1}{2}, 4.8)$.

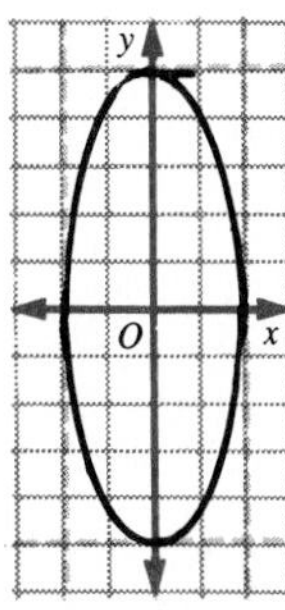

The x-intercepts are ± 2.
The y-intercepts are ± 5.

Suppose you wish to graph $x^2 - 9y^2 = 9$. This time the coefficient of y^2 is negative. Proceed as before, solving the equation for y.

$$x^2 - 9y^2 = 9$$
$$-9y^2 = 9 - x^2$$
$$y^2 = \frac{9 - x^2}{-9} \text{ or } \frac{x^2 - 9}{9}$$
$$y = \pm\sqrt{\frac{x^2 - 9}{9}}$$

For $y \in \mathcal{R}$, we see that $|x|$ cannot be less than 3. That is, $x \geq 3$ or $x \leq -3$. For this domain, y may be any real number, so there is no restricted interval for the range.

Using this information, let us complete a table of values and graph the relation.

x	y
± 3	0
± 4	$\approx \pm 0.9$
± 5	± 1.5
± 6	$\approx \pm 1.7$

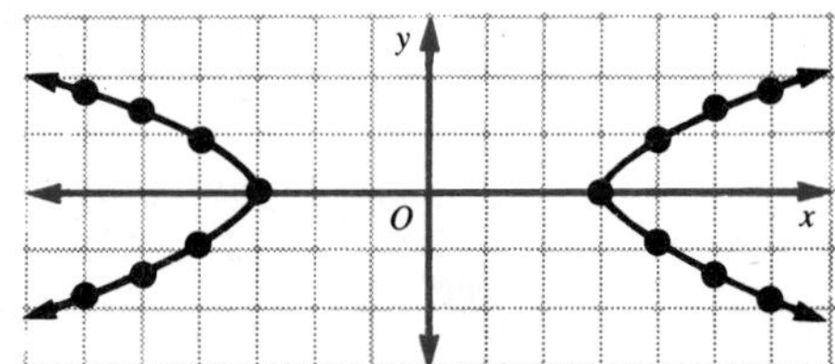

This graph is called a **hyperbola.** This type of curve can have different appearances depending upon the equation graphed. Example 2 shows another hyperbola.

Example

2 Sketch the graph of $xy = 6$.

By solving for y or for x, we see that neither x nor y can equal 0.

x	y
6	1
3	2
2	3
1	6
-1	-6
-2	-3
-3	-2
-6	-1

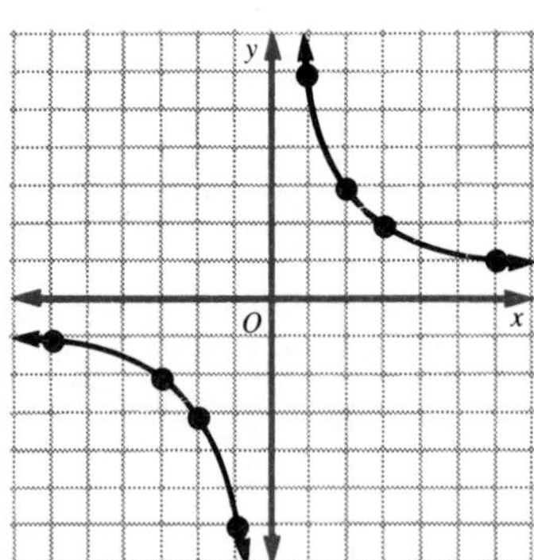

Note that this hyperbola is the graph of an inverse variation. What are the domain and range of this function?

The ellipse and hyperbola, as well as the circle and parabola, are known as **conic sections.** They have this name because each curve can be formed by the intersection of a plane with a conical surface. The following chart summarizes the conic sections we have studied.

Graph	Center	General Equation
Circle	(0, 0) (h, k)	$x^2 + y^2 = r^2$, (radius r) $(x - h)^2 + (y - k)^2 = r^2$, (radius r)
Parabola		$y = ax^2 + bx + c$, $(a \neq 0)$
Ellipse	(0, 0)	$ax^2 + by^2 = c$, $(a, b, c > 0)$
Hyperbola	(0, 0)	$ax^2 - by^2 = c$, $(a, b, c > 0)$ $-ax^2 + by^2 = c$, $(a, b, c > 0)$ $xy = c$, $(c > 0)$

Exercises

Exploratory **Identify the graphs of each of the following.**

1. $x^2 + y^2 = 25$
2. $x^2 + 4y^2 = 16$
3. $x + 4y = 10$
4. $4x^2 + 4y^2 = 20$
5. $-9x^2 + 12y^2 = 144$
6. $x^2 + 4y^2 = 4$
7. $x^2 - 4y^2 = 4$
8. $y = 4x^2 - 2x + 1$
9. $-x^2 + x = 4$
10. $2x + y = 4$
11. $16x^2 - y^2 = 400$
12. $x^2 + 4 = y$
13. $xy = 28$
14. $8x^2 - 20 = 8y^2$
15. $xy = -1.44$

16. Explain how to find the x- or y-intercepts of any curve.

Written **Find the *x*- and *y*-intercepts, if they exist, of the graph of each of the following.**

1. $2x^2 + 3y^2 = 12$
2. $y = x^2 - 7x - 18$
3. $y = x^2 + 6x + 9$
4. $xy = 8$
5. $x^2 + y^2 = 25$
6. $16x^2 - 4y^2 = 400$
7. $2x^2 + 2y^2 = 14$
8. $-x^2 + 4y^2 = 100$
9. $y = x^2 + 2x + 6$
10. $x^2 - 10x = y$
11. $16x^2 + y^2 = 16$
12. $16x^2 + 25y^2 = 400$

Identify the graph of each of the following equations.

13. $y + 2x = x^2 + 9$
14. $\frac{x^2}{9} + \frac{y^2}{4} = 1$
15. $\frac{x^2}{16} - \frac{y^2}{4} = 1$
16. $\frac{y^2}{9} - \frac{x^2}{16} = 1$
17. $2x^2 + 2y^2 = 4\sqrt{3}$
18. $\frac{x^2}{9} + \frac{y^2}{16} = 1$

Identify the graph of each of the following equations. Then graph each conic section.

19. $y = -x^2 + 2x - 5$
20. $xy = 12$
21. $xy = -6$
22. $x^2 + 4y^2 = 16$
23. $4x^2 + 25y^2 = 100$
24. $4x^2 - 25y^2 = 100$

25. $9x^2 + y^2 = 36$
26. $16x^2 + 4y^2 = 64$
27. $16x^2 + 9y^2 = 144$
28. $-x^2 + 9y^2 = 144$
29. $-2x^2 + y^2 = 25$
30. $4x^2 + y^2 = 144$
31. $(x - 3)^2 = y$
32. $x - 16y^2 = 16$
33. $\frac{x^2}{9} + \frac{y^2}{16} = 1$
34. $\frac{x^2}{4} + \frac{y^2}{9} = 1$
35. $\frac{x^2}{9} - y^2 = 1$
36. $\frac{x^2}{25} + \frac{y^2}{16} = 1$

Challenge **The equation $x^2 + y^2 = r^2$, when graphed, is a circle with center at (0, 0). A circle, located elsewhere on the plane, has the equation $(x - h)^2 + (y - k)^2 = r^2$ with (h, k) as its center. Using that analogy, describe the graphs of each of the following equations. Then verify your answer by graphing.**

1. $4(x - 1)^2 + 9(y + 3)^2 = 36$
2. $x^2 + 4y^2 + 2x + 40y + 85 = 0$
3. $25x^2 + 4y^2 - 100x - 40y + 100 = 0$
4. $9y^2 - 16x^2 - 72y - 64x = 64$

Mathematical Excursions

The Origin of Conics

The name *conic sections* is derived from the intersection of a plane with a conical surface. The illustrations below show the formation of each of the conic sections you have studied.

circle

ellipse

parabola

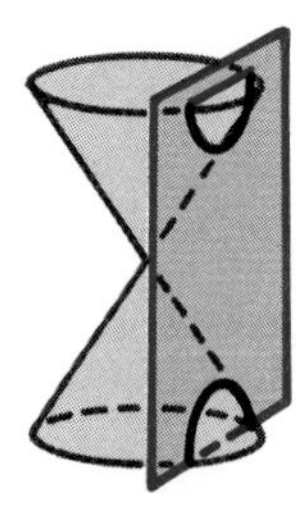
hyperbola

When a plane intersects a conical surface at the vertex, a special set of conics is formed. These are called *degenerate conics* and are shown below.

point

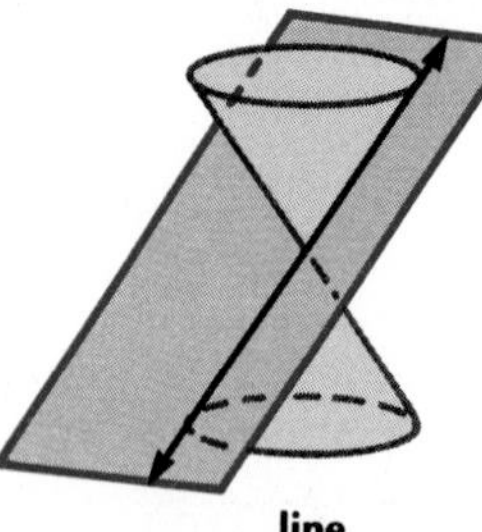
line

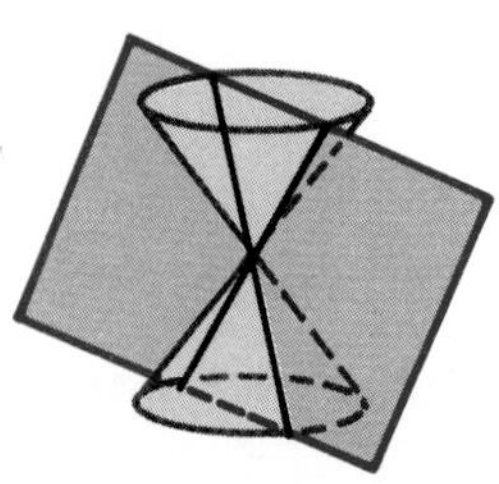
intersecting lines

3.7 Composing Functions

In many applications of functions it is necessary to combine the action of two functions to produce a new function. For example, suppose we have the functions $f(x) = x^2$ and $g(x) = 2x$ shown below.

$$\begin{array}{ccc} & g & \\ 1 & \longrightarrow & 2 \\ 2 & \longrightarrow & 4 \\ 3 & \longrightarrow & 6 \\ 4 & \longrightarrow & 8 \end{array} \qquad \begin{array}{ccc} & f & \\ 2 & \longrightarrow & 4 \\ 4 & \longrightarrow & 16 \\ 6 & \longrightarrow & 36 \\ 8 & \longrightarrow & 64 \end{array}$$

The situation pictured above suggests that it is possible to combine the action of f and g as shown below.

$$\begin{array}{ccccc} & g & & f & \\ 1 & \longrightarrow & 2 & \longrightarrow & 4 \\ 2 & \longrightarrow & 4 & \longrightarrow & 16 \\ 3 & \longrightarrow & 6 & \longrightarrow & 36 \\ 4 & \longrightarrow & 8 & \longrightarrow & 64 \end{array}$$

By combining f and g in this way, we produce a new function with domain $\{1, 2, 3, 4\}$ and range $\{4, 16, 36, 64\}$. This function is called the **composite function** of f and g. The composite of f and g is symbolized "$f \circ g$", which is read, "f following g." Here is a diagram for $f \circ g$.

$$\begin{array}{ccc} & f \circ g & \\ 1 & \longrightarrow & 4 \\ 2 & \longrightarrow & 16 \\ 3 & \longrightarrow & 36 \\ 4 & \longrightarrow & 64 \end{array}$$

Examples

1 **If f and g are the functions illustrated below, draw an arrow diagram for $f \circ g$.**

$$\begin{array}{ccc} & f & \\ 0 & \longrightarrow & 5 \\ 1 & \longrightarrow & 6 \\ 2 & \longrightarrow & 7 \end{array} \qquad \begin{array}{ccc} & g & \\ 3 & \longrightarrow & 0 \\ 4 & \longrightarrow & 1 \\ 5 & \longrightarrow & 2 \end{array}$$

$f \circ g$ means "f following g." Perform function g first, and then function f.

$$\begin{array}{ccccc} & g & & f & \\ 3 & \longrightarrow & 0 & \longrightarrow & 5 \\ 4 & \longrightarrow & 1 & \longrightarrow & 6 \\ 5 & \longrightarrow & 2 & \longrightarrow & 7 \end{array} \quad \text{or} \quad \begin{array}{ccc} & f \circ g & \\ 3 & \longrightarrow & 5 \\ 4 & \longrightarrow & 6 \\ 5 & \longrightarrow & 7 \end{array}$$

In function notation, these values can be expressed as

$$(f \circ g)(3) = 5$$
$$(f \circ g)(4) = 6$$
$$\text{and } (f \circ g)(5) = 7.$$

Sometimes it is not possible to form the composite of two functions.

2 **Use the functions f and g from example 1 to show that $g \circ f$ does not exist.**
$g \circ f$ *means "g following f."*

Perform function f first.

$$\begin{array}{c} f \\ 0 \longrightarrow 5 \\ 1 \longrightarrow 6 \\ 2 \longrightarrow 7 \end{array}$$

Now, perform g.

But this is impossible because the elements of the range, 5, 6, and 7, are not in the domain of g. Therefore $g \circ f$ does not exist.

Limitations of a Composite Function

In order for the composite $f \circ g$ to exist, each member of the range of g must be a member of the domain of f.

Suppose that g: $3 \rightarrow 5$ and h: $5 \rightarrow 6$. This means that $g(3) = 5$ and $h(5) = 6$. It follows that $(h \circ g)(3) = h[g(3)]$
$= h(5)$ or 6.

In general, $(f \circ g)(x) = f[g(x)]$.

Examples

3 **If $f(x) = 2x + 3$ and $g(x) = x^2 + 1$ find $(f \circ g)(x)$.**

Method 1

$$\begin{aligned}(f \circ g)(x) &= f[g(x)] \\ &= f(x^2 + 1) \\ &= 2(x^2 + 1) + 3 \\ &= 2x^2 + 2 + 3 \\ &= 2x^2 + 5\end{aligned}$$

Method 2

$$x \xrightarrow{g} (x^2 + 1) \xrightarrow{f} 2(x^2 + 1) + 3$$

$$\begin{aligned}\text{Therefore } [f \circ g](x) &= 2(x^2 + 1) + 3 \\ &= 2x^2 + 2 + 3 \\ &= 2x^2 + 5\end{aligned}$$

4 **Using f and g from example 3, find $(g \circ f)(x)$.**

$$\begin{aligned}(g \circ f)(x) &= g[f(x)] \\ &= g(2x + 3) \\ &= (2x + 3)^2 + 1 \\ &= 4x^2 + 12x + 9 + 1 \\ &= 4x^2 + 12x + 10\end{aligned}$$

Examples 3 and 4 show that $(f \circ g)(x) \neq (g \circ f)(x)$. Therefore, composition of functions is not a commutative operation.

Examples

If $I(x) = x$ and $h(x) = x^2 + 2x + 1$, find each of the following.

5 $(h \circ I)(x)$

$$\begin{aligned}(h \circ I)(x) &= h[I(x)] \\ &= h(x) \\ &= x^2 + 2x + 1 \\ &= h(x)\end{aligned}$$

6 $(I \circ h)(x)$

$$\begin{aligned}(I \circ h)(x) &= I[h(x)] \\ &= I(x^2 + 2x + 1) \\ &= x^2 + 2x + 1 \\ &= h(x)\end{aligned}$$

In examples 5 and 6, composing $I(x)$ with $h(x)$ produces $h(x)$. Perhaps this reminds you of familiar situations involving the real numbers 0 and 1, which are the identities for addition and multiplication, respectively. The function $I(x) = x$ behaves in much the same way. The function $I(x) = x$ is called the **identity function for composition,** since, for any function f, $(f \circ I)(x) = f(x)$ and $(I \circ f)(x) = f(x)$.

Once we have an identity, it is natural to inquire about inverses. For example, you know that in the reals, 2 and -2 are additive inverses because $2 + (-2) = 0$ and $-2 + 2 = 0$. Similarly, 3 and $\frac{1}{3}$ are multiplicative inverses because $3 \cdot \frac{1}{3} = 1$ and $\frac{1}{3} \cdot 3 = 1$. In the preceding section, you learned that a one-to-one function f has an inverse function f^{-1}. You can see that the name "inverse" is used because *composition of a function and its inverse produces the identity function*. That is, the inverse of a function "undoes" the function.

Examples

If $f(x) = 2x - 1$, show that each of the following is true.

7 $(f \circ f^{-1})(x) = I(x)$

Since $f(x) = 2x - 1$, $f^{-1}(x) = \frac{x+1}{2}$.

$$\begin{aligned}(f \circ f^{-1})(x) &= f[f^{-1}(x)] \\ &= f\left[\frac{x+1}{2}\right] \\ &= 2 \cdot \frac{x+1}{2} - 1 \\ &= x \\ &= I(x)\end{aligned}$$

8 $(f^{-1} \circ f)(x) = I(x)$

$$\begin{aligned}(f^{-1} \circ f)(x) &= f^{-1}[f(x)] \\ &= f^{-1}(2x - 1) \\ &= \frac{(2x-1)+1}{2} \\ &= \frac{2x}{2} \\ &= x \\ &= I(x)\end{aligned}$$

Exercises

Exploratory For exercises 1–18, refer to the mappings f and g shown at the right.

f		g	
3	→ 7	7	→ 3
4	→ 9	13	→ 6
5	→ 11	11	→ 4
6	→ 13	9	→ 5

Find each of the following.

1. domain of f
2. domain of g
3. range of f
4. range of g
5. $f(3)$
6. $g(9)$
7. $f(5)$
8. $g(11)$

Construct an arrow diagram for each composition. Then write the result as a set of ordered pairs.

9. $f \circ g$
10. $g \circ f$

Find each of the following.

11. domain of $f \circ g$
12. domain of $g \circ f$
13. range of $f \circ g$
14. range of $g \circ f$
15. $(g \circ f)(3)$
16. $(g \circ f)(4)$
17. $(g \circ f)(5)$
18. $(g \circ f)(6)$

Written Find each of the following if $f(x) = x^2$, $g(x) = 3x$, and $h(x) = x - 1$.

1. $(f \circ g)(1)$
2. $(g \circ f)(1)$
3. $(h \circ f)(3)$
4. $(f \circ h)(3)$
5. $g(f(-2))$
6. $f(h(-3))$
7. $g(h(-2))$
8. $h(g(-2))$
9. $f\left[h\left(-\frac{1}{4}\right)\right]$
10. $g\left[f\left(-\frac{1}{4}\right)\right]$

11. $h\left[f\left(-\frac{1}{5}\right)\right]$ **12.** $f[h(\sqrt{2} + 3)]$ **13.** $f[h(5 - \sqrt{3})]$ **14.** $f[g(1 + \sqrt{2})]$

15. $h[f(\sqrt{5} - 2)]$ **16.** $f[g(x)]$ **17.** $h[g(x)]$ **18.** $f[h(x)]$

19. Find $[f \circ (g \circ h)](x)$ and $[(f \circ g) \circ h](x)$. From your results would you say that composition is an association operation? Explain.

For each of the following pairs of functions, find $f[g(x)]$ and $g[f(x)]$.

20. $f(x) = 2x + 1$
$g(x) = x - 4$

21. $f(x) = 2x - 5$
$g(x) = 3x + 7$

22. $f(x) = x^2 + 4$
$g(x) = 2x$

23. $f(x) = x^3$
$g(x) = x - 1$

24. $f(x) = |x|$
$g(x) = x + 4$

25. $f(x) = \sqrt{x}$
$g(x) = x^2$

Determine whether f and g are inverse functions. Justify your answers.

26. $f(x) = 2x + 1,\ g(x) = x - \frac{1}{2}$

27. $f(x) = -2x + 3,\ g(x) = 2x - 3$

28. $f(x) = x + 4,\ g(x) = x - 4$

29. $f(x) = x^2$ for $x \geq 0$, $g(x) = \sqrt{x}$

If $f(x) = \frac{1}{x}$ and $g(x) = x^2$, find each of the following.

30. domain of f **31.** domain of g **32.** range of g **33.** domain of $(f \circ g)$

34. range of f **35.** domain of $(g \circ f)$ **36.** range of $(f \circ g)$ **37.** range of $(g \circ f)$

Mixed Review

Solve each equation.

1. $r + \frac{r^2 - 5}{r^2 - 1} = \frac{r^2 + r + 2}{r + 1}$

2. $n - 3\sqrt{3n - 5} = 6$

If $f = \{(3, 8), (4, 0), (6, 3), (7, -1)\}$ and $g = \{(3, 6), (8, 6), (-1, -8), (0, 4)\}$, write each of the following as sets of ordered pairs.

3. the inverse of f

4. the inverse of g

5. $f \circ g$, if it exists

6. $g \circ f$, if it exists

7. Identify the graph of the equation $(x + 2y)^2 - 6 = (x + y)(x + 4y)$.

8. Express the additive and multiplicative inverses of $5 - 3i$ in $a + bi$ form.

9. The commission received on the sale of a house varies directly as the sale price of the house. If a real estate agent receives a commission of \$7200 on the sale of a \$135,000 house, how much commission would she receive on a \$198,000 home?

10. If 2 crates have the same capacity and depth, the length is inversely proportional to the width. Crate A is 60 cm long and 40 cm wide. Crate B is 5 cm long. What is crate B's width if the crates have the same depth and capacity?

Problem Solving Application: Using Joint Variation

Consider the formula for the area of a triangle, $A = \frac{1}{2}bh$. Notice that the area depends directly on *both* the base and the height. Such an equation is called a **joint variation.** We say that the area *varies jointly* as the base and the height.

Joint Variation

A joint variation is described by an equation of the form $z = kxy$, where $k \neq 0$. The variable k is called the *constant of variation.*

Examples

1 **Suppose z varies jointly as x and y, and $z = 42$ when $x = 18$ and $y = 7$. Find z when $x = 15$ and $y = 8$.**

First find the value of k.

$$z = kxy$$
$$42 = k(18)(7) \quad \textit{Substitute.}$$
$$\frac{1}{3} = k$$

The equation for this joint variation is $z = \frac{1}{3}xy$. Use the equation to find z.

$$z = \frac{1}{3}(15)(8) \quad \textit{Substitute 15 for x and 8 for y in } z = \frac{1}{3}xy.$$
$$z = 40$$

Thus, $z = 40$ when $x = 15$ and $y = 8$.

2 **The hourly heat loss (H) through a pane of glass of specific thickness varies jointly as the area (A) of the glass (in square inches) and the difference (d) in the temperatures (in degrees Fahrenheit) on each side of the glass. The equation that represents this joint variation is $H = kAD$, where $k \neq 0$. For a certain 30 inch square pane of glass, the hourly heat loss is 6000 calories when the difference in the temperatures is 40°F. What is the hourly heat loss for a 25 inch by 35 inch pane of glass with the same thickness if the difference in temperatures is 30°F?**

Explore Since this is a joint variation, the hourly heat loss depends directly on the area of glass and the temperature difference.

Plan First determine the value of k using the information on hourly heat loss for the 30 inch square pane of glass. Then use this value to determine the hourly heat loss for the 25 inch by 35 inch pane.

Solve

$$H = k(\ell w)d$$
$$6000 = k[(30)(30)]40$$
$$6000 = 36000k$$
$$k = \frac{1}{6}$$

Substitute ℓw for A in $H = kAd$. Substitute 6000 for H, 30 for ℓ, 30 for w, and 40 for d.

$$H = \frac{1}{6}(\ell w)d$$
$$= \frac{1}{6}[(25)(35)]30$$
$$= 5250$$

Now substitute 25 for ℓ, 35 for w, and 30 for d.

The hourly heat loss would be 5250 calories. *Examine this solution.*

Exercises

Use the formula from example 2 to answer the following. Assume that the panes of glass all have the same thickness.

1. What is the hourly heat loss for a 40 inch by 21 inch pane of glass when the difference in temperatures is 27°F?
2. What is the area of a pane of glass that has an hourly heat loss of 8800 calories when the difference in temperatures is 48°F?
3. What is the difference in the temperatures on each side of a 26 inch by 52 inch pane of glass if its hourly heat loss is 4056 calories?
4. Suppose the 30 inch square pane of glass in example 2 is replaced by a thicker pane of glass, reducing the hourly heat loss to 5400 calories when the difference in temperatures is 40°F. What is the hourly heat loss for a 24 inch by 32 inch pane of this glass when the difference in temperatures is 25°F?

The equation that relates simple interest earned (*I*), amount of money invested (*P*), annual rate of interest (*r*), and amount of time, in years, for the investment (*t*) is $I = Prt$. If the amount of the investment is constant, then this equation represents a joint variation. Suppose an investment earns \$900 in simple interest after 4 years at an annual rate of 7.5%.

5. How much interest will the investment earn after 10 years at a rate of 6.8%?
6. How long will it take the investment to earn \$1107 in interest at a rate of 8.2%?
7. If the investment is to earn \$1848 in interest after 8 years, what must be the annual rate of interest?

Suppose some money is invested for a specific amount of time at a given annual rate of interest. Describe what happens to the amount of simple interest earned from the investment if the following changes are made while all other things remain the same.

8. The investment is doubled.
9. The annual rate of interest is tripled.
10. The investment is halved, but it is invested for five times as long as before.

Vocabulary

relation
domain
range
function
mapping
linear function
vertical line test
constant function
direct variation
constant of variation
constant of proportionality
quadratic function
exponential function
step function
greatest integer function
absolute value function
inverse of a function
one-to-one function
horizontal line test
ellipse
hyperbola
conic sections
composite function
identity function for composition

Chapter Summary

1. Algebraic relations can be written in symbolic form, such as $\{(x, y): y = x^2\}$, which is read "all ordered pairs (x, y) such that $y = x^2$."
2. If no two ordered pairs in a relation have the same first coordinate, then the relation is a function.
3. Whenever the value of one variable depends on the value of another variable, the first variable is a function of the second variable.
4. A function can be represented by a set of ordered pairs, an arrow diagram, mapping notation, or special function notation.
5. The graph of a quadratic function is a parabola.
6. The inverse of a function f is symbolized by f^{-1}, which does not mean $\frac{1}{f}$.
7. The inverse of a function is not necessarily a function.
8. Each conic section can be written in a general equation form.
9. Composition of functions is not a commutative operation.

Chapter Review

3.1 **If the domain is $\{x: -3 \le x \le 3\}$, graph each relation and state the range.**

1. $\{(x, y) : y = x\}$
2. $\{(x, y) : y = x^2\}$
3. $\{(x, y) : y = x^4\}$
4. $\{(x, y) : y = x^2 + 4\}$
5. $\{(x, y) : y = |x + 1|\}$
6. $\{(x, y) : y = -x^3\}$

State the values of x for which the expression is not a real number.

7. $\frac{2x + 3}{2x - 1}$ 8. $\sqrt{x + 1}$ 9. $\sqrt{2x - 7}$ 10. $\frac{4}{\sqrt{2x + 1}}$

3.2 **Tell whether each of the following is the graph of a function. If not, explain why. State the domain and range.**

11.
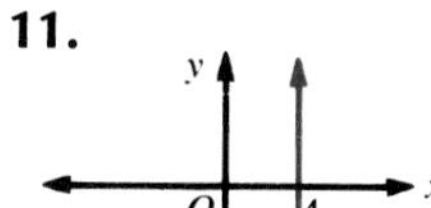

12.
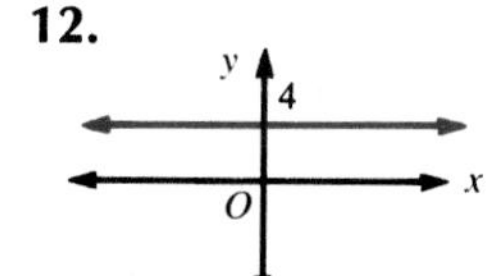

13.
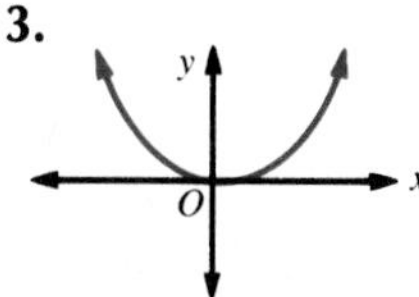

3.3 **Domain = $\{-2, -1, 1, 2\}$. For each function, draw an arrow diagram, list the range, and then graph the function.**

14. $r: a \to a^3$ 15. $S(u) = -u^2$ 16. $F(x) = \frac{1}{x - 4}$

If $g(x) = 3x^2 + 4x + 1$, find the missing coordinates so that the point is on the graph of g.

17. (0, ?) 18. (1, ?) 19. $\left(\frac{1}{3}, ?\right)$ 20. $\left(\frac{1}{2}, ?\right)$

Determine the domain of each function so that the range is in $\mathcal{R}$.

21. $f(x) = \sqrt{6 - x}$ 22. $f(x) = \frac{2}{x^2 - 4}$ 23. $f(x) = \sqrt{x^2 + 2}$ 24. $f(x) = \frac{\sqrt{3x - 9}}{5 - x}$

3.4 25. If f is a linear function and $f(2) = -5$ and $f(-2) = -8$, find $f(x)$.

26. Sketch the graph of $g(x) = x^2 + 4x + 1$.

27. Sketch the graph of $k(x)$ if $k(x) = -1$ if $x < 0$, $k(x) = 3$ if $x > 0$, and $k(x) = 0$ if $x = 0$.

28. In an electrical transformer, voltage is directly proportional to the number of turns on the coil. If 110 volts comes from 55 turns, what is the voltage produced by 66 turns?

3.5 **Domain = $\{-3, -2, -1, 1, 2, 3\}$. Write the inverse of each function as a set of ordered pairs. Tell if the inverse is also a function.**

29. $\{(x, y) : y = 2x\}$ 30. $E(x) = 2^x$ 31. $G(x) = |3 - x|$

Find $f^{-1}(x)$ for each of the following.

32. $f(x) = 3x$ 33. $f(x) = 2x - 5$ 34. $f(x) = \frac{3 - x}{7}$

3.6 **Identify the graph of each conic section. Then sketch the graph of the relation.**

35. $x^2 + y^2 = 25$ 36. $xy = 8$ 37. $y^2 + 4x^2 = 16$

Find the x- and y-intercepts, if they exist, of the graphs of each equation.

38. $y^2 - 5x^2 = 25$ 39. $4x^2 + 9y^2 = 36$ 40. $x^2 - 4x - 12y = 32$

3.7 **For the following exercises, $f(x) = x^2$, $g(x) = 3x$, and $h(x) = x - 1$. Find each composite.**

41. $g(f(-2))$ 42. $f(h(-3))$ 43. $g(h(-2))$ 44. $h(g(-2))$

45. $f(h(x))$ 46. $g(f(x))$ 47. $h(f(x))$ 48. $f(g(x))$

Chapter Test

If the domain is {−3, −2, −1, 0, 1, 2, 3}, graph each relation and state the range.

1. $\{(x, y): y = x^3\}$

2. $\{(x, y): y = \sqrt{x + 3}\}$

3. $\{(x, y): y = |x| - 2\}$

State the values of *x*, if any, for which each expression is undefined or is imaginary.

4. $\sqrt{x}$

5. $\frac{5x}{x^2 + 5x - 6}$

6. $\frac{3}{x^2 + 1}$

7. $\frac{1}{\sqrt{x + 1}}$

Find *f*(−2) for each of the following.

8. $f(x) = \frac{3}{x - 4}$

9. $f: x \to x + 1$

10. $f(x) = 5$

11. $f(x) = 8 - 2x^2$

If $f(x) = 4x^3 - 3$, find each of the following.

12. $f(0)$

13. $f(5)$

14. $f(-2)$

15. $f\left(-\frac{1}{2}\right)$

Determine the domain of each function so that the range is in $\mathcal{R}$.

16. $f(x) = \frac{x + 4}{2x - 1}$

17. $f(x) = \frac{1}{2x^2 - 2}$

18. $T(x) = \frac{x - 8}{x^2 + 5x + 1}$

Graph each of the following functions.

19. $g(x) = x^2 - x$

20. $H(x) = \begin{cases} 1 \text{ if } x \geq 0 \\ x \text{ if } x < 0 \end{cases}$

21. $p(x) = [2x]$

Which of the following are functions?

22. $\{(x, y): y = 5\}$

23. $\{(x, y): x = |y|\}$

24. $\{(x, y): x + y = 8\}$

Write the inverse of each function as a set of ordered pairs. Tell if the inverse is also a function.

25. $\{(1, 5), (-2, 7)\}$

26. $\{(0, 2), (2, 3)\}$

27. $\{(-5, 5), (5, -4), (-4, 5)\}$

Find $f^{-1}(x)$ for each of the following.

28. $f(x) = 2x$

29. $f(x) = 3x + 4$

30. $f(x) = \frac{1}{2}x - 6$

Identify the graph of each conic section. Then sketch the graph of the relation.

31. $x^2 - 9 = y$

32. $4x^2 - 25y^2 = 100$

33. $x^2 + 4y^2 = 36$

34. If $f(x) = x^2 - 5$ and $g(x) = x + 2$, find $f(g(-2))$.

35. The number of gallons of gasoline used by a moped varies directly as the number of miles it travels. If a moped uses 2 gallons of gasoline to travel 115 miles, how many gallons will it need to travel 264.5 miles?

36. The volume of any gas varies inversely as the pressure if the temperature remains constant. The volume of a certain gas is 120 cubic feet under 6 pounds of pressure. What is the volume at the same temperature if the pressure is 8 pounds?

An Introduction to Transformation Geometry

Chapter

4

Crystals often occur in nature. Others are man-made. They are important in the development of transistors and solid state electronics. Many of the ideas from transformation geometry have been helpful to scientists in understanding the properties and chemical behavior of crystals.

4.1 Line Reflections

A function, or mapping, pairs each member of one set, called the domain, with exactly one member of another set, called the range. Many of the mappings studied so far have had domains and ranges consisting of real numbers. In this chapter, mappings will be considered in which the domain and range are the set of all points in a plane.

Definition of a Transformation

A **transformation of the plane** is a one-to-one mapping whose domain and range are the set of all points in the plane.

The scene shown at the right suggests one type of transformation.

When the water is perfectly calm, there is a mirror-like reflection of the things on the shore.

Think of the water line along the shore as line ℓ. Then point A, for example, is flipped or reflected over line ℓ onto point A', *read A prime*. B' is called the **image** of point B, and C' is the image of C. Notice that point E, which is on line ℓ, is its own image.

The figure at the right shows the same situation as the one above it. This time, however, line segments connect points and their images. Notice that

1. ℓ is perpendicular to each segment, and
2. ℓ bisects each segment.

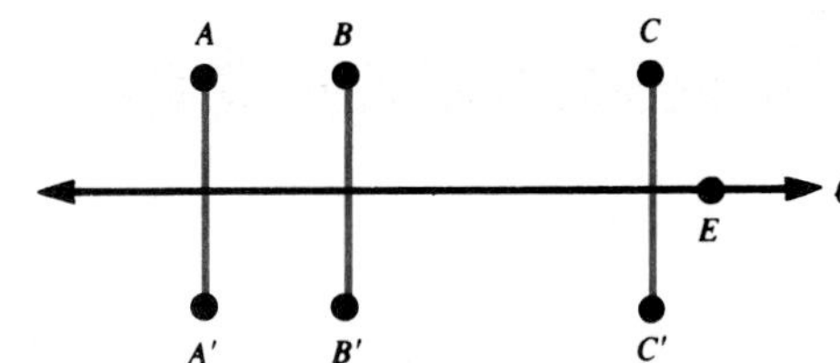

ℓ is the perpendicular bisector of the line segment joining a point and its image.

If this page is considered a portion of a plane, then the situation shown above is an example of a *line reflection*. In this case, it is a line reflection over line ℓ. Note that a line reflection is a one-to-one mapping since different points map to different images. Therefore, a line reflection is a transformation of the plane.

Definition of Line Reflection

A reflection over line ℓ is a transformation of the plane that maps each point P of the plane onto a point P' as follows.

1. If P is on line ℓ, then the image of P is P.
2. If P is not on ℓ, then the image of P is the point P' such that ℓ is the perpendicular bisector of $\overline{PP'}$.

The symbol $A \xrightarrow{R_\ell} A'$ means that A maps to A' under the line reflection R_ℓ. Recall that A' is the *image* of A. Similarly, A is the **pre-image** of A'.

Using function notation, we write $R_\ell(A) = A'$, which means R_ℓ maps A to A'.

Examples

1 **If $\triangle ABC$ is flipped, or reflected, over line n, its image is $\triangle A'B'C'$.**

That is, $\triangle ABC \xrightarrow{R_n} \triangle A'B'C'$.

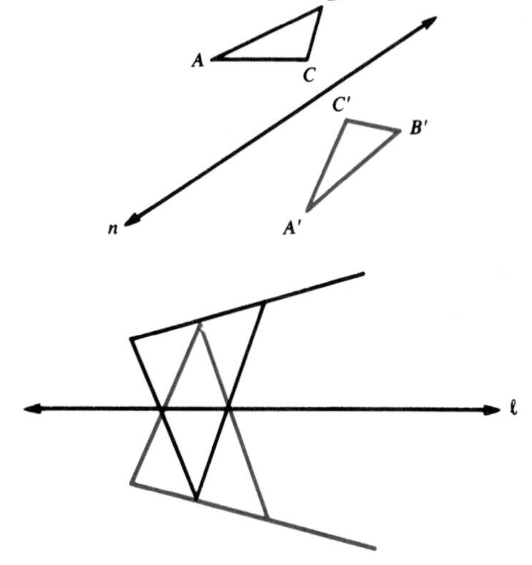

2 **The flag figure in red is the image of the one in black when you reflect it over line ℓ.**

You can use a ruler or compass to check that each point satisfies the requirements of a reflection.

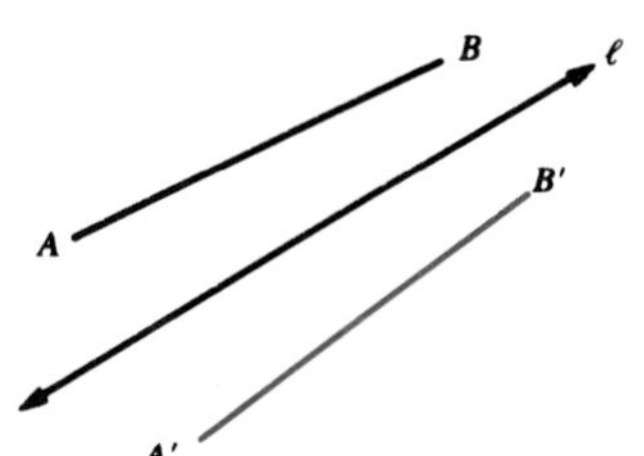

Look carefully at the line reflection over line ℓ shown at the left. It appears that segment AB and its image $\overline{A'B'}$ have the same length. That is, the distance from A to B is the same as the distance from A' to B'. This observation can be summarized as follows.

Line reflections preserve distance.

Can you think of any other property of a figure that seems to be preserved by a line reflection? Are there any properties that are not preserved? You will be asked to think about these questions in the exercises that follow.

Exercises

Exploratory **Justify your answer to each of the following.**

1. Can a transformation map two different points onto the same point?
2. Can a transformation map every point onto itself?
3. Can a transformation assign two different points to a point P?

Name the images of each of the following under line reflection R_ℓ.

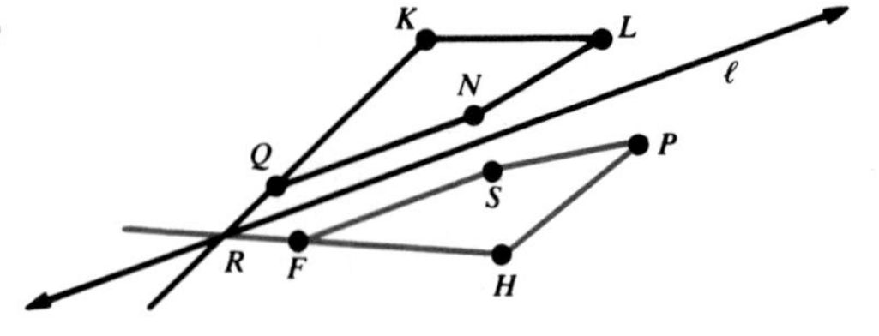

4. Point R
5. Point L
6. Point P
7. $\overline{RH}$
8. $\overline{SF}$
9. $\angle PSF$

Use the figure above to complete each of the following.

10. $R_\ell(F) = ?$
11. $R_\ell(\overline{PH}) = ?$
12. $R_\ell(\angle KLN) = ?$
13. $H \xrightarrow{R_\ell} ?$
14. $R \xrightarrow{R_\ell} ?$
15. $? \xrightarrow{R_\ell} F$
16. Explain how to find the image of a point under a line reflection using compass and straight edge only. Justify the construction.

Written **Copy each figure onto your paper. Then sketch as carefully as you can the image of each figure under R_m. Label image points using prime notation, that is A, A', B, B', and so on.**

1.

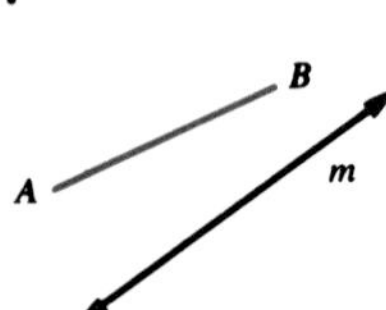

2.

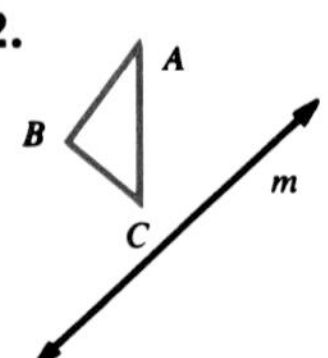

3.

4.

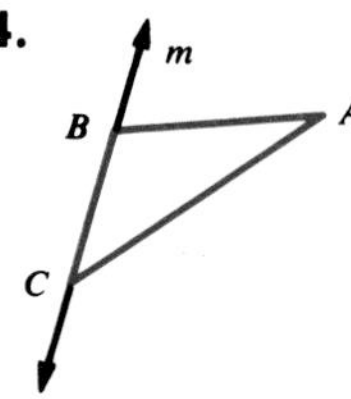

5.

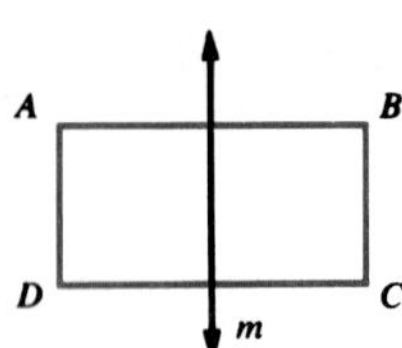

6.

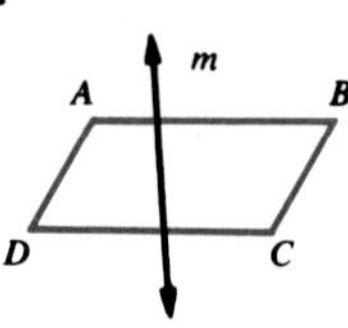

7.

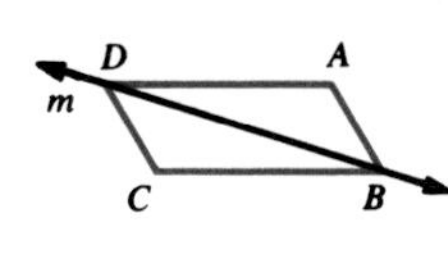

8.

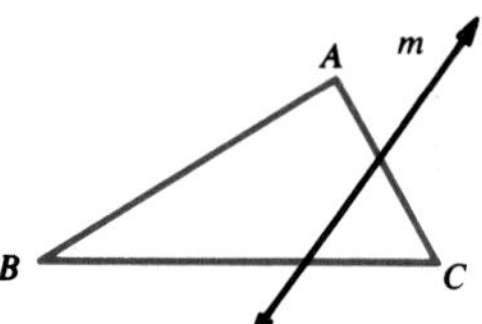

9. Find the image of points A, B, C, and D under R_ℓ.

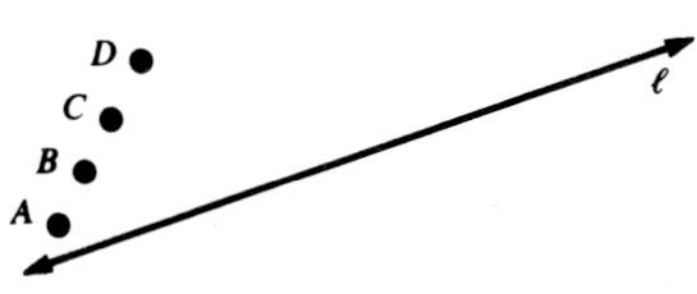

Explain how your results in exercise 9 suggest that the following statements in exercises 10 and 11 are true.

10. Line reflections preserve collinearity.
11. Line reflections preserve betweenness.
12. Do you think line reflections preserve angle measure? Explain.
13. Are a triangle and its image congruent under a line reflection?
14. Try to find a property of geometric figures not preserved by line reflections. *Hint: Think of a line reflection of a clock.*

4.2 Line Symmetry

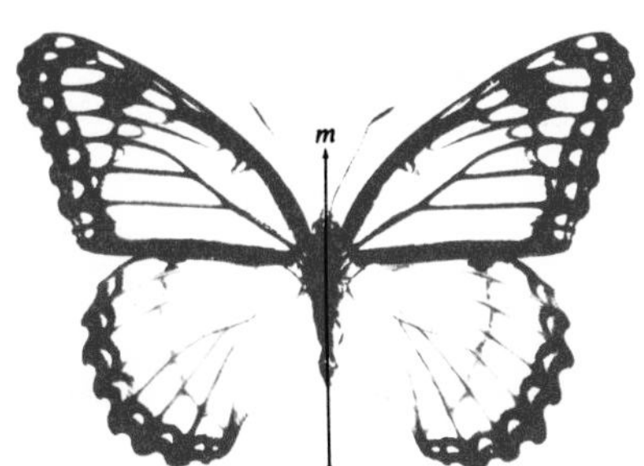

Nature and art provide many interesting examples of shapes that are their own images under a line reflection. For example, if you reflect the butterfly over line *m*, each point on the butterfly has an image which is also on the butterfly. That is, the butterfly is its own image when it is reflected over line *m*. Such a figure has line symmetry over line *m* or is symmetrical with respect to line *m*.

Definition of Line Symmetry

If a figure is its own image under a line reflection over line *m*, then the figure is said to have line symmetry over line *m*, or to be **symmetrical with respect to line *m*.** Line *m* is called a **line of symmetry.**

Examples

1 **Isosceles trapezoid *ABCD* is symmetrical with respect to line ℓ. It is not, however, symmetrical about line *k*.**

For example, if you reflect point *D* over line *k*, its image is clearly not on the trapezoid.

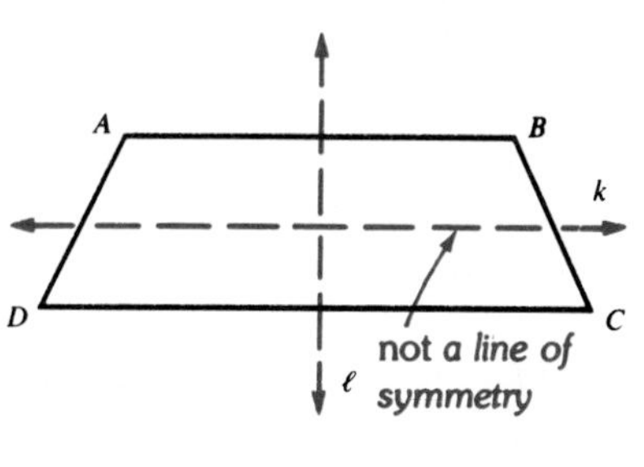

2 **The word MOM has line symmetry with respect to line *m*. The word DAD does not have line symmetry with respect to either line *n* or line *p*.**

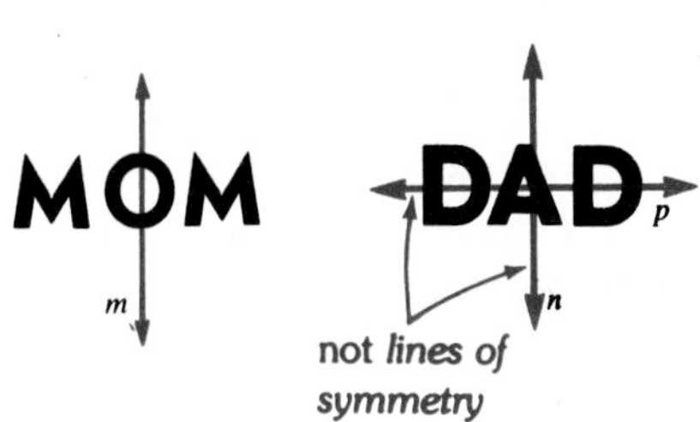

3 **Can a figure have more than one line of symmetry?**

Consider the figures below.

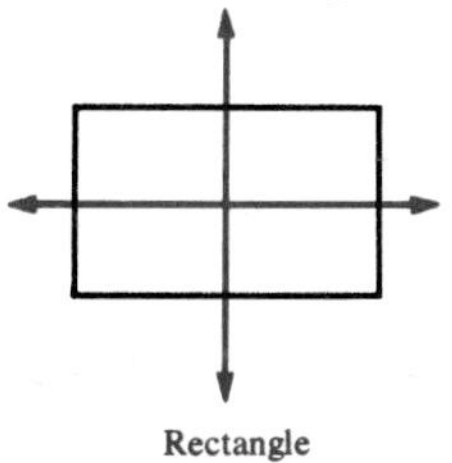
Rectangle

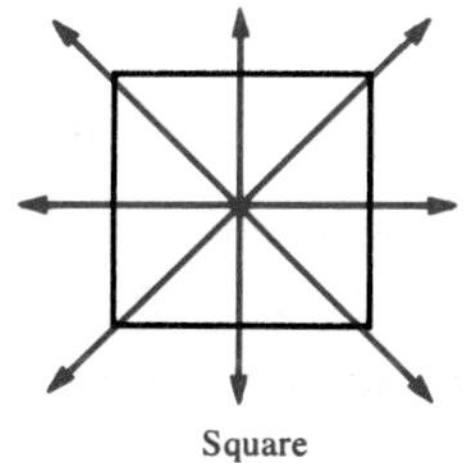
Square

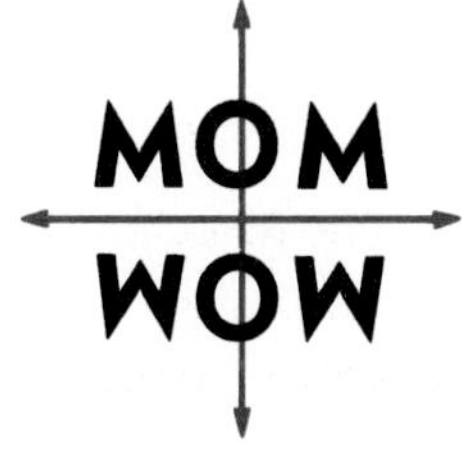

The rectangle has two lines of symmetry. The square has four lines of symmetry. The figure MOM-WOW has two lines of symmetry.

Exercises

Exploratory **Answer each of the following.**

1. Explain what is meant by line symmetry.
2. Can a figure have no lines of symmetry? Explain.

Consider the letters of the alphabet shown below. Then answer the following questions.

A B C D E F G H I J K L M

N O P Q R S T U V W X Y Z

3. What letters have a vertical line of symmetry?
4. What letters have a horizontal line of symmetry?
5. Do any letters have lines of symmetry other than vertical or horizontal? If so, what are they?

Sketch a figure for each of the following. Then tell how many, if any, lines of symmetry exist.

6. rectangle
7. square
8. rhombus
9. line
10. line segment
11. angle

Written **Copy each figure in exercises 1–12, and sketch all lines of symmetry. If there are none, write *none*.**

1. 

2.

3. 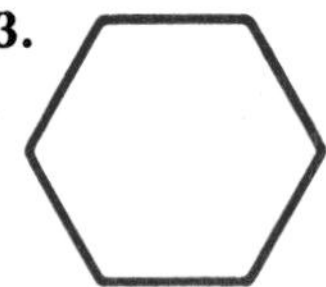

4. 

5.

6. 

7. 

8. 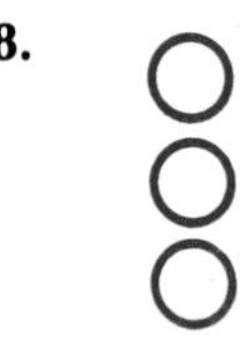

9. 

10. 

11. 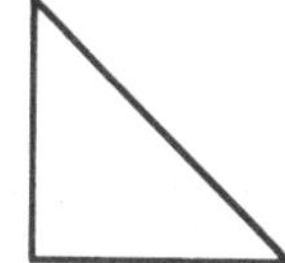

12. 

13. Draw a triangle that has exactly one line of symmetry.
14. Draw a triangle that has exactly three lines of symmetry.
15. Can a triangle have exactly two lines of symmetry? Explain.
16. Some words have horizontal lines of symmetry as in the word, BOOK. Find five other words that have horizontal line symmetry.
17. Some words have vertical lines of symmetry as in the example shown at the right. Find five other words that have vertical line symmetry.

Challenge **How would you define plane symmetry in 3-space? Decide if each of the following has planes of symmetry.**

1. 
sphere

2. 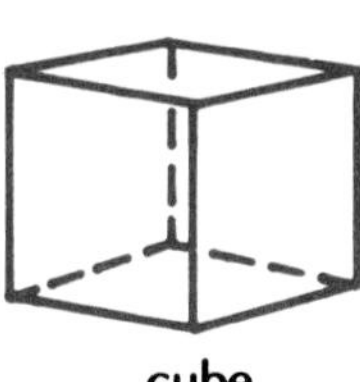
cube

3. 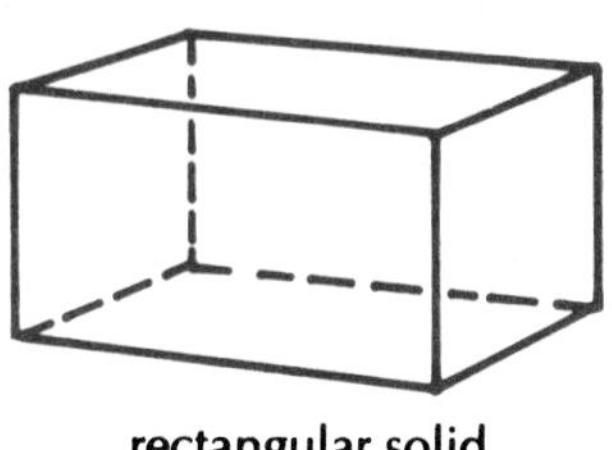
rectangular solid

Mixed Review

Study the first figure and the four patterns that follow. Then answer each question.

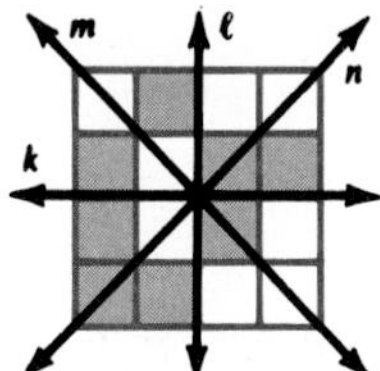

a. 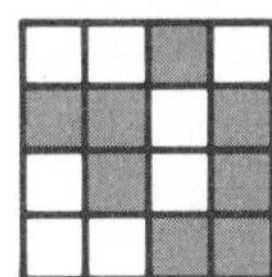

b. 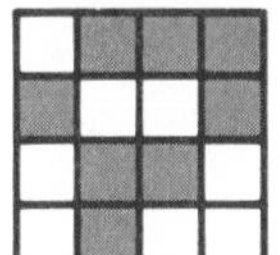

c. 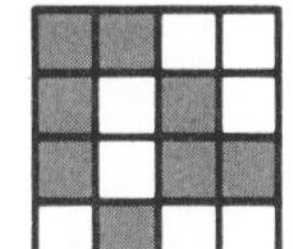

d. 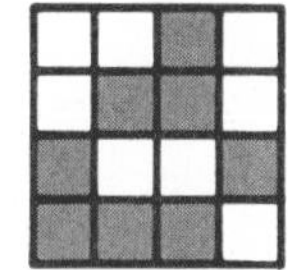

1. Which pattern is a reflection of the first figure over line k?
2. Which pattern is a reflection of the first figure over line ℓ?
3. Which pattern is a reflection of the first figure over line m?
4. Which pattern is a reflection of the first figure over line n?
5. Determine a domain of $f(x) = \sqrt{4 - x^2}$ so that f^{-1} is also a function. Then write an equation for $f^{-1}(x)$.

4.3 Line Reflections in the Coordinate Plane

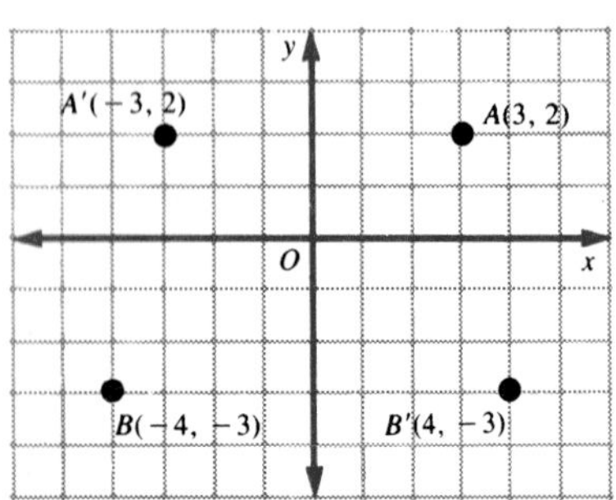

Suppose you have a point in the coordinate plane and want to reflect it over the *y*-axis. For example, if we reflect point $A(3, 2)$ over the *y*-axis, its image is $A'(-3, 2)$.

What do you think is the image of $(-4, -3)$ under a reflection over the *y*-axis? If you said $(4, -3)$ you are correct. In each case, it appears that the *y*-coordinate of the original point and its image are the same. The *x*-coordinate of the image is the additive inverse of the original *x*-coordinate.

This rule can be expressed in the following way.

Reflection over the *y*-axis

> **If point $P(x, y)$ is reflected over the *y*-axis, its image is $P'(-x, y)$.**
>
> $$(x, y) \xrightarrow{R_{y\text{-axis}}} (-x, y)$$

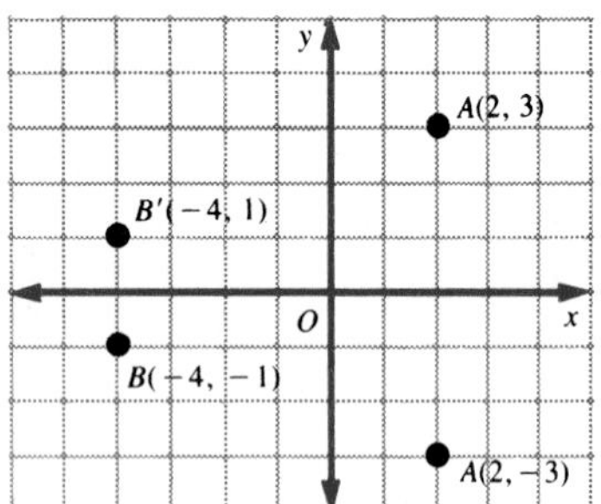

In the figure at the left, A and B are reflected over the *x*-axis. Notice the following results.

$$A(2, 3) \xrightarrow{R_{x\text{-axis}}} A'(2, -3)$$

$$B(-4, -1) \xrightarrow{R_{x\text{-axis}}} B'(-4, 1)$$

What is the pattern? How is this similar to a reflection over the *y*-axis?

Reflection over the *x*-axis

> **If point $P(x, y)$ is reflected over the *x*-axis, its image is $P'(x, -y)$.**
>
> $$P(x, y) \xrightarrow{R_{x\text{-axis}}} P'(x, -y)$$

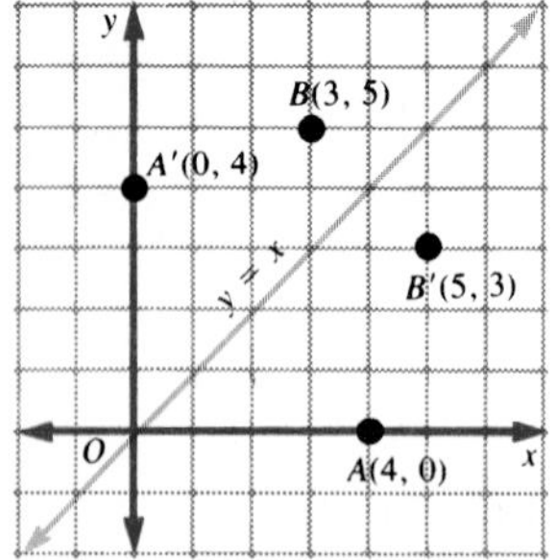

Now consider a reflection over the line whose equation is $y = x$. In the figure at the left, notice the following results.

$$A(0, 4) \xrightarrow{R_{y=x}} A'(4, 0)$$

$$B(3, 5) \xrightarrow{R_{y=x}} B'(5, 3)$$

What is the pattern of *x*-coordinates and *y*-coordinates? These and many other similar examples suggest the following rule.

Reflection over the line $y = x$

If point $P(x, y)$ is reflected over the line whose equation is $y = x$, its image is $P'(y, x)$.

$$P(x, y) \xrightarrow{R_{y=x}} P'(y, x)$$

Examples

1 **Graph $\triangle ABC$ with vertices $A(-1, 2)$, $B(-4, 1)$, and $C(-3, -2)$. Then graph their images and $\triangle A'B'C'$ under $R_{y\text{-axis}}$.**

$$A(-1, 2) \xrightarrow{R_{y\text{-axis}}} A'(1, 2)$$

$$B(-4, 1) \xrightarrow{R_{y\text{-axis}}} B'(4, 1)$$

$$C(-3, -2) \xrightarrow{R_{y\text{-axis}}} C'(3, -2)$$

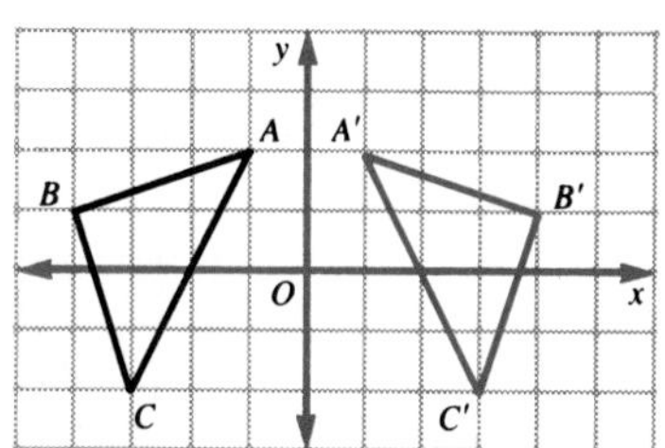

2 **Using points A and B of example 1, show that $AB = A'B'$.**

Recall the distance formula, $d = \sqrt{(x_2 - x_1)^2 + (y_2 - y_1)^2}$.

$$AB = \sqrt{(-4 + 1)^2 + (1 - 2)^2} = \sqrt{9 + 1} \text{ or } \sqrt{10}$$

$$A'B' = \sqrt{(4 - 1)^2 + (1 - 2)^2} = \sqrt{9 + 1} \text{ or } \sqrt{10}$$

Therefore, $AB = A'B'$.

Exercises

Exploratory **Complete the following rules.**

1. $(x, y) \xrightarrow{R_{y\text{-axis}}} ?$

2. $(x, y) \xrightarrow{R_{x\text{-axis}}} ?$

3. $(x, y) \xrightarrow{R_{y=x}} ?$

Find the image of each of the following under a reflection over the y-axis.

4. $(3, 2)$

5. $(6, -4)$

6. $\left(2\frac{1}{2}, 7\right)$

7. $(0, 3)$

8. $(-4, -4)$

9. $\left(-23, 3\frac{3}{4}\right)$

10. $(3, -8)$

11. $(-6, 0)$

Find the image of each of the following under a reflection over the x-axis.

12. (3, 1) **13.** (0, 5) **14.** (−4, 3) **15.** (3, −5)

16. $\left(3\frac{1}{2}, 2\frac{1}{2}\right)$ **17.** (−8, 0) **18.** (0, −5) **19.** (0, 0)

Find the image of each of the following under a reflection over the line for $y = x$.

20. (2, 1) **21.** (3, 0) **22.** (−5, 1) **23.** (3, −8)

24. (−5, −1) **25.** (7, 2) **26.** (0, 6) **27.** (0, 0)

State which of the following curves is symmetric with respect to the y-axis, the x-axis, both, or neither.

28.

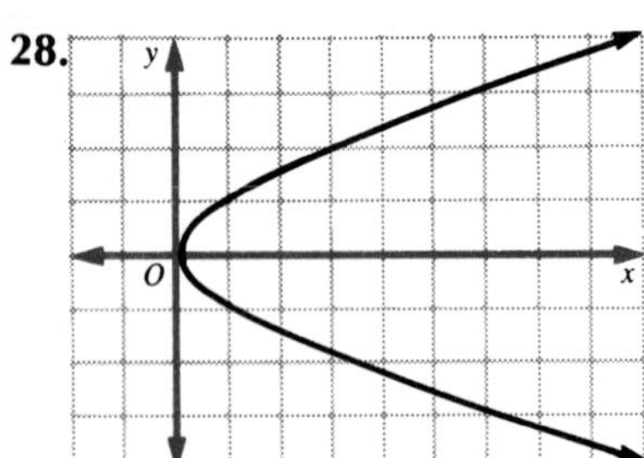

29.

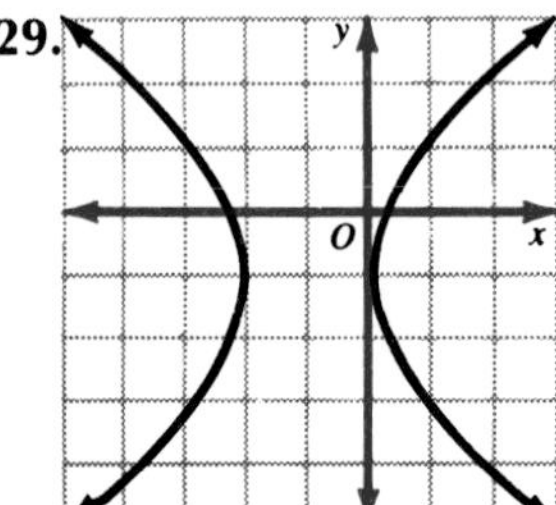

30.

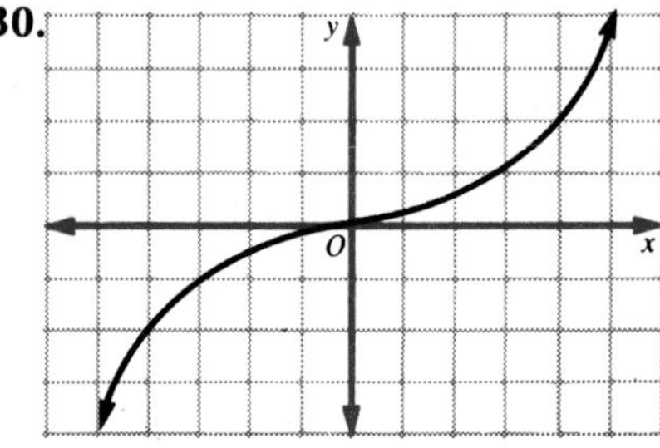

31.

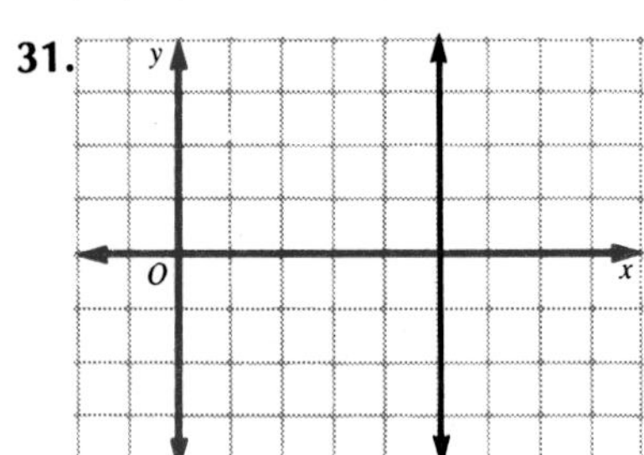

32.

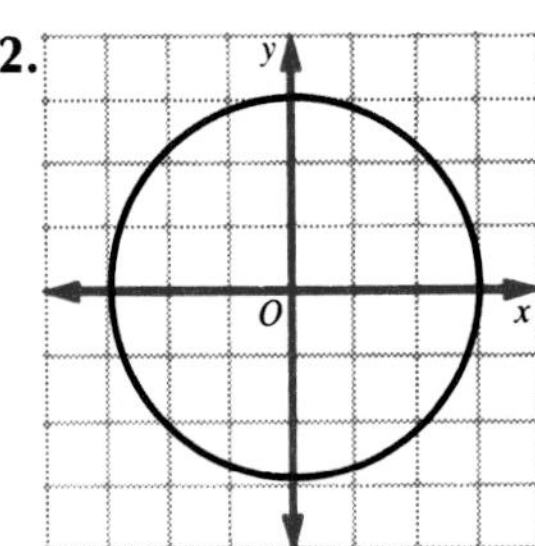

33.

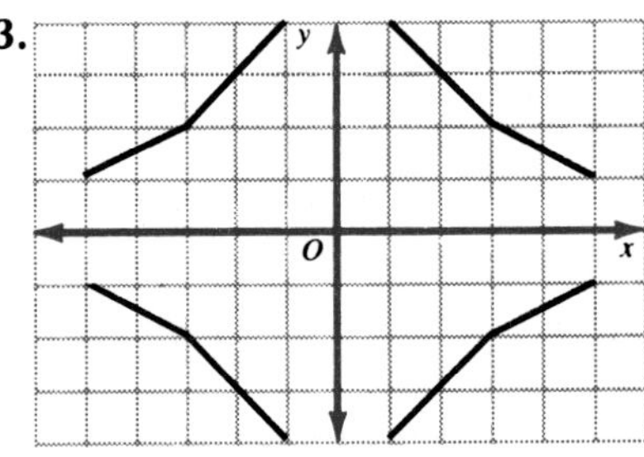

Written **For each pair of points *A* and *B*, find their images under $R_{y\text{-axis}}$. Then use the distance formula to compute *AB* and *A′B′*.**

1. $A(4, 2), B(1, 5)$ **2.** $A(0, 3), B(3, -1)$ **3.** $A(1, 4), B(6, 2)$

4. $A(-3, 4), B(5, 6)$ **5.** $A(-2, -3), B(5, 1)$ **6.** $A(7, 0), B(0, 7)$

For each pair of points *A* and *B*, find their images under $R_{x\text{-axis}}$. Then use the distance formula to compute *AB* and *A′B′*.

7. $A(5, 6), B(4, 2)$ **8.** $A(0, 0), B(7, 6)$ **9.** $A(-1, 3), B(2, -1)$

10. $A(4, 2), B(5, 1)$ **11.** $A(1, 1), B(-1, -6)$ **12.** $A(1, 2), B(2, 6)$

For each pair of points *A* and *B*, find their images under $R_{y=x}$. Then use the distance formula to compute *AB* and *A′B′*.

13. $A(4, 2), B(-3, -1)$ **14.** $A(0, 3), B(-3, 4)$ **15.** $A(-2, -3), B(0, -5)$

16. $A(0, 5), B(5, 0)$ **17.** $A(4, -1), B(3, -5)$ **18.** $A(1, -5), B(-3, -2)$

Find the coordinates of the images of $A(5, -2)$ and $B(-2, 3)$ under a reflection over the line represented by each of the following equations.

19. $x = -1$ **20.** $y = -3$ **21.** $x = -2$ **22.** $y = 8$

Use graph paper to find the image of each point under a reflection over the line $y = -x$.

23. (3, 0) **24.** (0, 5) **25.** (2, −1) **26.** (−4, 3)

27. Use your answers to exercises 23–26 to help you complete this rule.

$$(x, y) \xrightarrow{R_{y=-x}} \underline{\quad ? \quad}$$

Graph the quadrilateral with the vertices $A(2, 1)$, $B(5, -2)$, $C(9, 4)$, and $D(6, 7)$. Then answer the following.

28. Find the images of A, B, C, and D under $R_{y\text{-axis}}$.

29. Find the slopes of $\overleftrightarrow{AB}$ and $\overleftrightarrow{CD}$.

30. Find the slopes of $\overleftrightarrow{A'B'}$ and $\overleftrightarrow{C'D'}$.

31. Explain how the results of exercises 29 and 30 suggest that line reflections preserve parallelism.

Challenge

For exercises 1–9, refer to the figure at the right.

In the figure, $A \xrightarrow{R_{y\text{-axis}}} B$, $B \xrightarrow{R_{x\text{-axis}}} C$, $C \xrightarrow{R_{y\text{-axis}}} D$, and $D \xrightarrow{R_{x\text{-axis}}} A$.

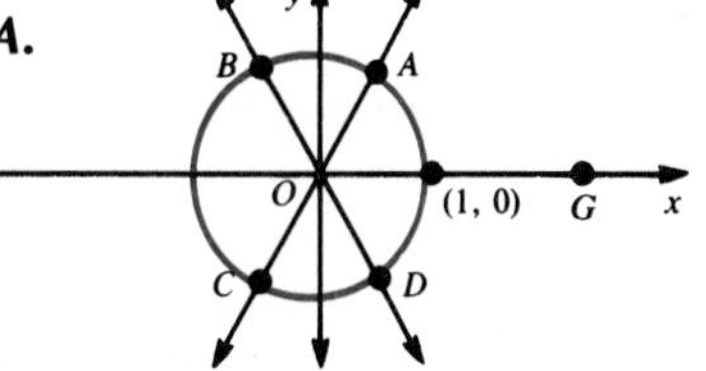

For each $m\angle GOA$ given below, find $m\angle GOB$.

1. 30 **2.** 60 **3.** 50

4. 83 **5.** 58 **6.** 72

If A has the following x-coordinate, find the coordinates of B, C, and D.

7. $\frac{\sqrt{2}}{2}$ **8.** $\frac{\sqrt{3}}{2}$ **9.** $\frac{2}{3}$

Find an equation for each line of symmetry for the graphs of each of the following equations.

10. $y = x^2 - 5$ **11.** $x = y^2$ **12.** $y = \frac{1}{x}$ **13.** $|x| + |y| = 6$

Mathematical Excursions

A Figure to Figure

Given $\triangle ABC$ is an equilateral triangle and $BCDEF$ and $AGHIC$ are regular pentagons, find the measure of $\angle GCE$.

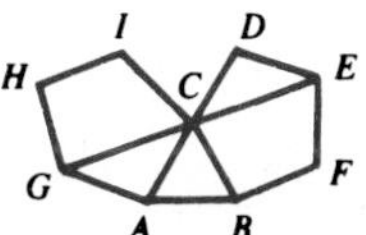

4.4 Translations

A line reflection is but one type of transformation. There are other types.

Consider "shifting" or "sliding" $\triangle ABC$ two inches in the direction shown.

In this case, A maps to A', B maps to B', and C maps to C'. Note that the segments connecting each point and its image are parallel and congruent.

Definition of Translation

A **translation** is a transformation of the plane that "moves" every point in the plane the same distance in the same direction.

Examples

1 **Translate each figure 3 centimeters in the direction shown.**

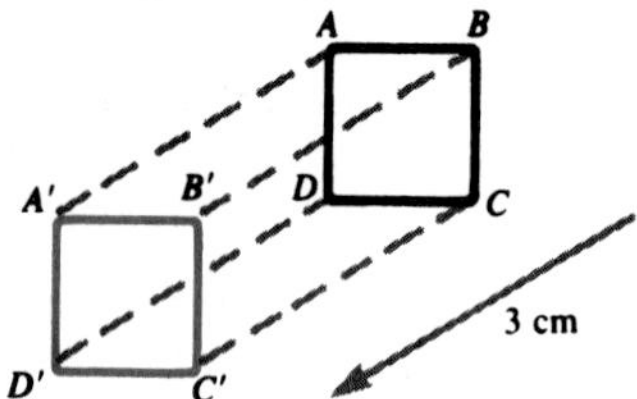

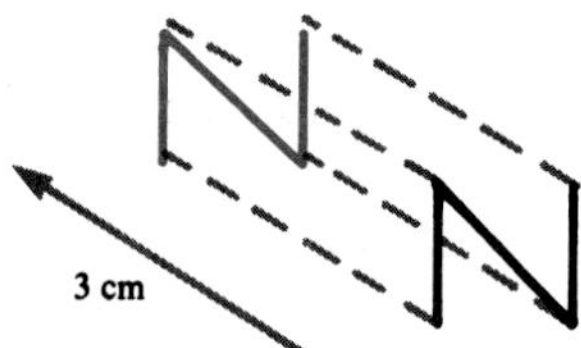

2 **For each figure below describe a transformation under which one triangle appears to be the image of the other.**

In figure 1, $\triangle ABD$ and $\triangle ACD$ appear to be images under a line reflection over $\overleftrightarrow{AD}$.

In figure 2, it appears that $\triangle ABC$ has been "moved" or translated to form its image $\triangle DEF$.

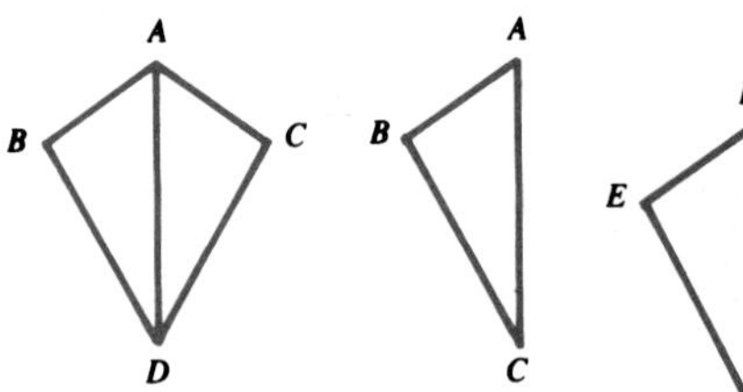

Figure 1 Figure 2

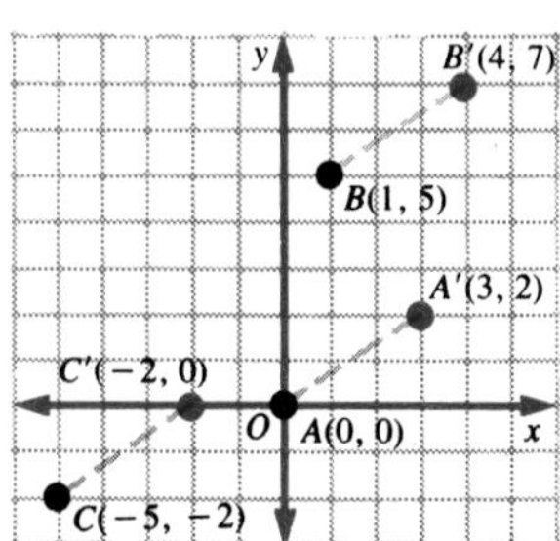

It is also possible to consider translations in the coordinate plane. In the figure, every point is moved three units to the right and then two units up. Here are some points and their images.

$$A(0, 0) \longrightarrow A'(3, 2)$$
$$B(1, 5) \longrightarrow B'(4, 7)$$
$$C(-5, -2) \longrightarrow C'(-2, 0)$$

The rule of this translation can be expressed in the following way.

$$(x, y) \xrightarrow{T_{3,\,2}} (x + 3, y + 2)$$

Rule for Translation

If point $P(x, y)$ is translated a units horizontally and b units vertically, its image is $P'(x + a, y + b)$.

$$(x, y) \xrightarrow{T_{a,\,b}} (x + a, y + b)$$

Examples

3 **Given points $A(3, 2)$, $B(-1, 4)$, and $C(1, -2)$, find the image of $\triangle ABC$ under $T_{-6,\,-2}$.**

The rule for $T_{-6,\,-2}$ is

$$(x, y) \xrightarrow{T_{-6,\,-2}} (x - 6, y - 2).$$

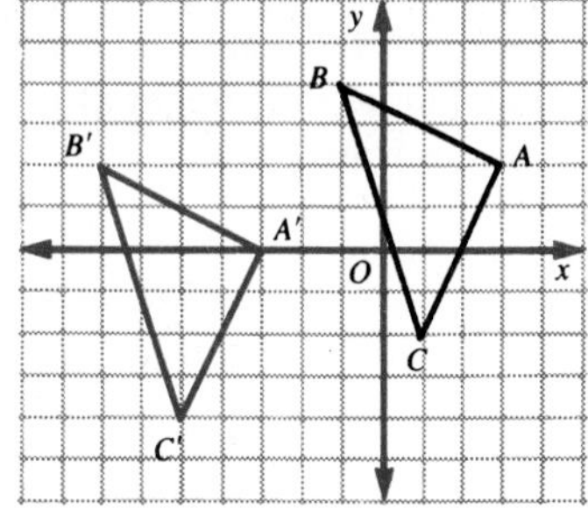

Therefore, $A(3, 2) \longrightarrow A'(-3, 0)$

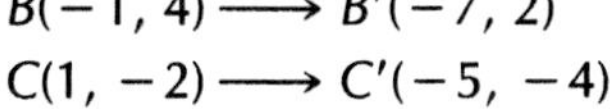

$$B(-1, 4) \longrightarrow B'(-7, 2)$$
$$C(1, -2) \longrightarrow C'(-5, -4)$$

4 **Graph $\triangle A''B''C''$, the image of $\triangle A'B'C'$ in example 3, under $T_{-1,\,7}$. What single transformation maps $\triangle ABC$ to $\triangle A''B''C''$?**

$$A'(-3, 0) \longrightarrow A''(-4, 7)$$
$$B'(-7, 2) \longrightarrow B''(-8, 9)$$
$$C'(-5, -4) \longrightarrow C''(-6, 3)$$

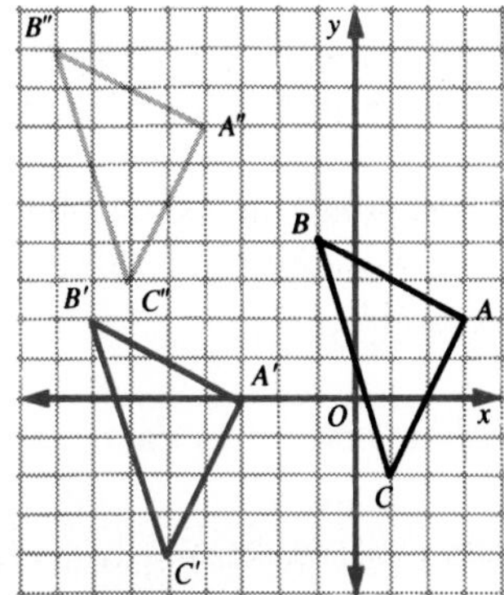

To find the transformation, compare $\triangle ABC$ and $\triangle A''B''C''$.

$$A(3, 2) \longrightarrow A''(-4, 7)$$

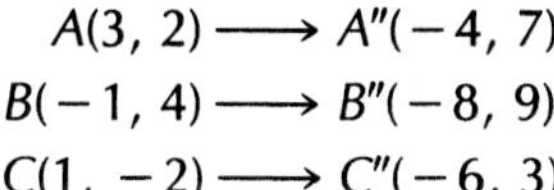

$$B(-1, 4) \longrightarrow B''(-8, 9)$$
$$C(1, -2) \longrightarrow C''(-6, 3)$$

This transformation is $T_{-7,\,5}$.

5 Given points $A(-5, 2)$ and $B(3, -1)$ and translation $T_{1,4}$, show that $AB = A'B'$.

The rule for $T_{1,4}$ is $(x, y) \xrightarrow{T_{1,4}} (x + 1, y + 4)$

Therefore, $A(-5, 2) \xrightarrow{T_{1,4}} A'(-4, 6)$, and

$B(3, -1) \xrightarrow{T_{1,4}} B'(4, 3)$.

Using the distance formula gives us these results.

$$AB = \sqrt{(-5 - 3)^2 + [2 - (-1)]^2} = \sqrt{64 + 9} = \sqrt{73}$$

$$A'B' = \sqrt{(-4 - 4)^2 + (6 - 3)^2} = \sqrt{64 + 9} = \sqrt{73}$$

We see that $AB = A'B'$.

Exercises

Exploratory **In the figure, D is the image of E under a translation. Name the images of each of the following under the same translation.**

1. M
2. H
3. $\overline{GP}$
4. $\angle JEF$
5. $\angle INM$
6. $\overrightarrow{IN}$

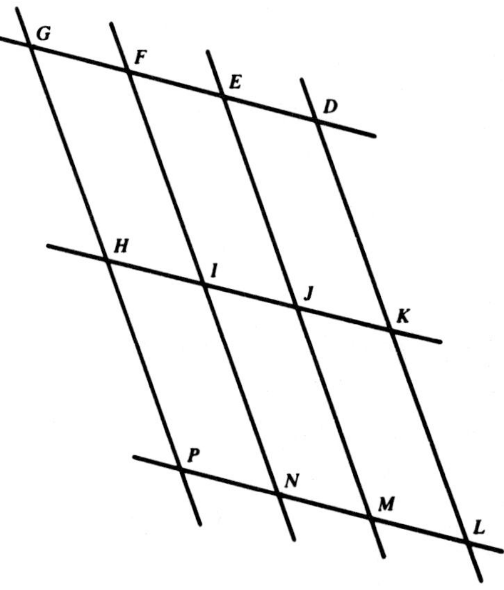

In the figure for exercises 1–6, suppose H is the image of F under a translation. Name the images of each of the following under the same translation.

7. E
8. K
9. I
10. $\overline{DE}$
11. $\overline{DK}$
12. $\angle JIF$
13. $\angle JKD$
14. $\overrightarrow{EJ}$

Find the coordinate rule of a translation under which B is the image of A.

15. $A(1, 1)$, $B(4, 2)$
16. $A(3, 5)$, $B(-1, 7)$
17. $A(1, -3)$, $B(-3, -1)$
18. $A(12, -5)$, $B(0, 0)$
19. $A(-1, 5)$, $B(-5, -14)$
20. $A(4\sqrt{2}, 1)$, $B(\sqrt{2}, -2)$

Explain what is meant by each statement.

21. Under a transformation of the plane, every point in the plane has a unique image.
22. Under a transformation of the plane, every point in the plane has a unique pre-image.

Written **Find the image of $(-4, 9)$ under each transformation.**

1. $T_{4, 3}$
2. $T_{-5, 6}$
3. $T_{-8, -4}$
4. $T_{-3, -2}$
5. $R_{y\text{-axis}}$
6. $R_{x\text{-axis}}$
7. $R_{y = x}$
8. $R_{y = -x}$

In each of the following, B' is the image of B under a translation. Find the rule for each translation. Find the image of $(-5, 6)$ under the same translation.

9. $B(1, 5)$, $B'(-3, 8)$
10. $B(-4, 6)$, $B'(-2, -8)$
11. $B(0, -1)$, $B'(-3, 4)$
12. $B(-10, -4)$, $B'(9, -2)$
13. $B(-6, 10)$, $B'(9, -4)$
14. $B(4\sqrt{3}, -5\sqrt{2})$, $B'(\sqrt{3}, 2\sqrt{2})$

Under a translation, the image of $(4, -1)$ is $(7, 5)$. Find the preimage of each of the following under the same translation.

15. $(5, 8)$
16. $(9, 1)$
17. $(-4, 5)$
18. $(0, -1)$
19. $(-6, -9)$
20. $\left(3\frac{1}{2}, 4\frac{1}{4}\right)$
21. $(\sqrt{3}, \sqrt{5})$
22. $(\sqrt{3} + 1, \sqrt{5} - 2)$

Find the images of the points $A(1, 3)$ and $B(5, 1)$ under $T_{4, -1}$. Then justify each of the following.

23. $AB = A'B'$
24. $\overleftrightarrow{AB} \parallel \overleftrightarrow{A'B'}$
25. $AA' = BB'$
26. $\overleftrightarrow{AA'} \parallel \overleftrightarrow{BB'}$
27. $ABB'A'$ is a parallelogram.
28. $\overline{AB'}$ and $\overline{A'B}$ bisect each other.

Plot the points $A(-3, -1)$, $B(0, 5)$, $C(4, 2)$ and $D(1, -3)$. Then find and graph their images under the following rules.

29. $(x, y) \rightarrow (x, -y)$
30. $(x, y) \rightarrow (x^2, y)$
31. $(x, y) \rightarrow (x - 7, y)$
32. $(x, y) \rightarrow (x, y + 6)$
33. $(x, y) \rightarrow (y, x)$
34. $(x, y) \rightarrow (|x|, |y|)$

Which of the rules in the following exercises are transformations of the coordinate plane? Explain.

35. exercise 29
36. exercise 30
37. exercise 31
38. exercise 32
39. exercise 33
40. exercise 34

Graph the triangle with vertices $A(-5, 3)$, $B(-4, -2)$, and $C(-1, 1)$.

41. Find the image of $\triangle ABC$ under reflection over the y-axis. Label it $A'B'C'$.
42. Reflect $\triangle A'B'C'$ over the line $x = 7$. Label the image $A''B''C''$.
43. Compare $\triangle ABC$ and $\triangle A''B''C''$. Is there a single transformation that takes $\triangle ABC$ to $\triangle A''B''C''$? Explain.

Graph the triangle with vertices $A(-2, 0)$, $B(1, 5)$, and $C(3, -1)$.

44. Graph the image of $\triangle ABC$ under $T_{-2, 3}$. Label the image $A'B'C'$.
45. Graph the image of $\triangle A'B'C'$ under $T_{-3, 4}$. Label the image $\triangle A''B''C''$.
46. What single transformation maps $\triangle ABC$ to $\triangle A''B''C''$?

4.5 Rotations

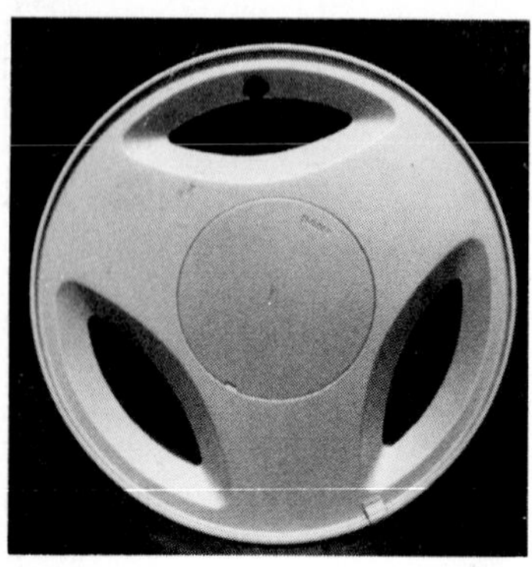

The hubcaps pictured above suggest another type of transformation called a *rotation*.

Definition of Rotation

A **rotation** is a transformation that maps every point to an image point in the plane by rotating the plane around a fixed point. This fixed point is called the **center of the rotation** and is its own image.

Example

1 With center at Q, rotate the plane 110° counterclockwise. Show the image of $\triangle ABC$.

$\triangle A'B'C'$ is the image of $\triangle ABC$.

The symbol $\text{Rot}_{Q,\ 110^\circ}$ stands for a rotation with center Q through an angle of 110° counterclockwise. Thus, from the figure in example 1,

$$\triangle ABC \xrightarrow{\text{Rot}_{Q,\ 110^\circ}} \triangle A'B'C'.$$

For a rotation with center Q through an angle of 110° clockwise, the figure in example 1 shows

$$\triangle A'B'C' \xrightarrow{\text{Rot}_{Q,\ -110^\circ}} \triangle ABC.$$

Examples

2 **Find the image of $\overline{AB}$ under a rotation of 75° clockwise with center O.**

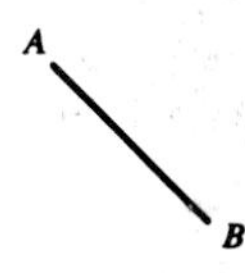

First we need to find the image of A.

Step 1 Draw $\overline{OA}$.

Step 2 Using your protractor, draw a 75° angle with vertex O and sides $\overline{OA}$ and $\overline{OX}$.

Step 3 Using your compass, locate A' on $\overrightarrow{OX}$ such that $OA' = OA$. A' is the image of A.

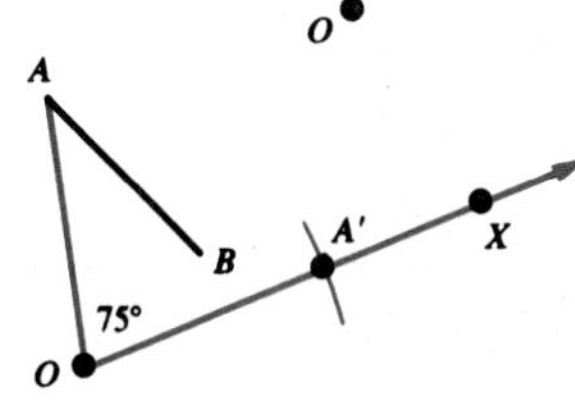

Now repeat steps 1–3 to find the image of B.

Thus, $\overline{AB} \xrightarrow{\text{Rot}_{O,\,-75°}} \overline{A'B'}$

3 **In the coordinate plane, find the images of $A(-3, 0)$, $B(2, -1)$, and $C(1, 4)$ under a counterclockwise rotation of 90° about the origin.**

From the figure we see

$$A(-3, 0) \xrightarrow{\text{Rot}_{O,\,90°}} A'(0, -3)$$

$$B(2, -1) \xrightarrow{\text{Rot}_{O,\,90°}} B'(1, 2)$$

$$C(1, 4) \xrightarrow{\text{Rot}_{O,\,90°}} C'(-4, 1).$$

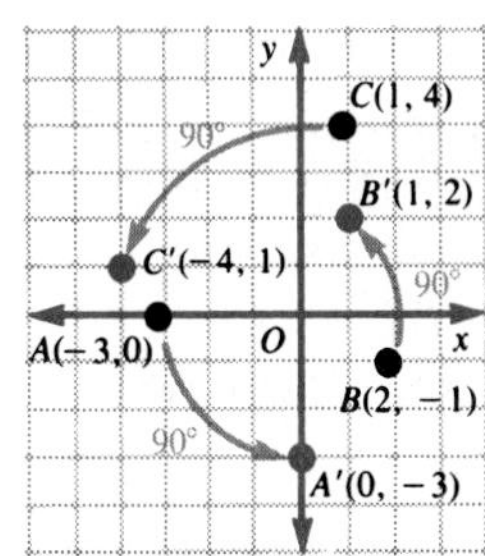

The results of example 3 suggest the following rule.

Rule for $\text{Rot}_{O,\,90°}$

If point $P(x, y)$ is rotated 90° counterclockwise about the origin, its image is $P'(-y, x)$.

$$P(x, y) \xrightarrow{\text{Rot}_{O,\,90°}} P'(-y, x)$$

Just as some figures have line symmetry, so others have **rotational symmetry.**

Definition of Rotational Symmetry

A figure has rotational symmetry if it is the image of itself under some rotation. The angle of rotation must be greater than 0° and less than 360°.

Example

4 The figure at the right has rotational symmetry about point *O*.

If you rotate the figure about *O* through 120° clockwise, or counterclockwise, it is its own image. The figure may also be rotated through an angle of 240° either clockwise or counterclockwise and again produce its own image.

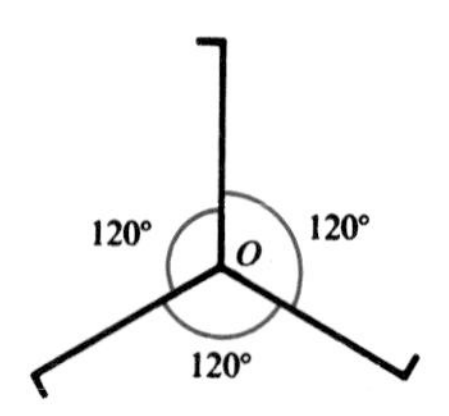

Exercises

Exploratory **The figure at the right shows a rotation of $\triangle KTP$.**

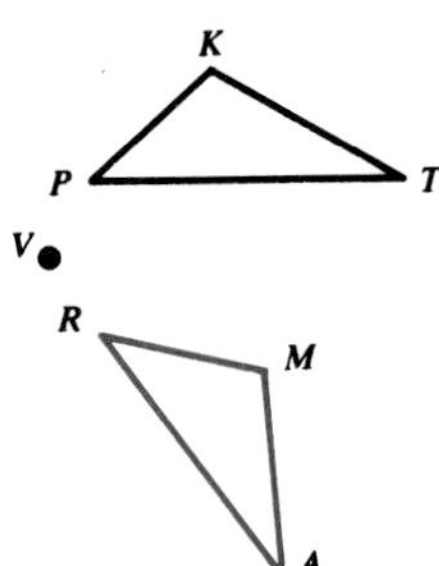

1. Name the center of rotation.
2. Name the image of *K*.
3. Estimate the angle of rotation.
4. Use a protractor to measure the angle of rotation. How close was your estimate in exercise 3?

In the figure, points are equally spaced on circle *O*. Find the image of each point below under the given rotation.

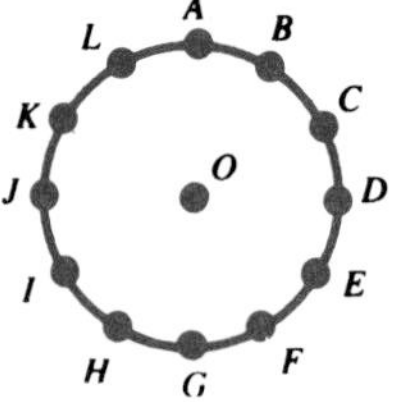

5. A, $\text{Rot}_{O,\ 90°}$
6. D, $\text{Rot}_{O,\ 90°}$
7. G, $\text{Rot}_{O,\ 180°}$
8. K, $\text{Rot}_{O,\ 180°}$
9. J, $\text{Rot}_{O,\ 60°}$
10. L, $\text{Rot}_{O,\ -60°}$
11. F, $\text{Rot}_{O,\ 120°}$
12. E, $\text{Rot}_{O,\ 270°}$
13. I, $\text{Rot}_{O,\ -240°}$
14. B, $\text{Rot}_{O,\ 300°}$
15. H, $\text{Rot}_{O,\ 180°}$
16. B, $\text{Rot}_{O,\ 180°}$

Which of the following figures has rotational symmetry? Explain.

17. A
18. D
19. E
20. H
21. M
22. N
23. O
24. S
25. Z
26.
27.
28.
29.
30.

Written **Copy and enlarge the figure. Draw the image of $\overline{AB}$ under the following.**

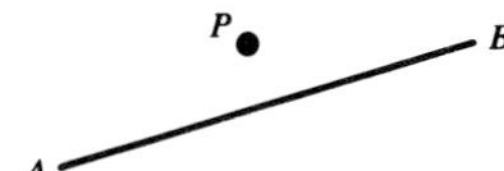

1. $Rot_{P, 90°}$
2. $Rot_{P, -90°}$
3. $Rot_{P, 30°}$
4. $Rot_{P, -210°}$

Find the coordinates of the image of each of the following under $Rot_{O, 90°}$.

5. (4, 2)
6. (−4, 5)
7. (−6, 9)
8. (−3, −7)
9. (6, −5)
10. (8, 0)
11. (−8, −8)
12. (10, −7)

Copy and enlarge the figure. Draw the image under each of the following.

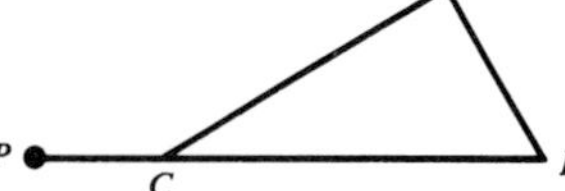

13. $Rot_{P, 90°}$
14. $Rot_{P, -125°}$
15. $R_{\overline{BP}}$
16. $R_{\overline{AC}}$

△*ABC* has the following vertices. Graph △*ABC* and its image △*A′B′C′* under $Rot_{O, 90°}$.

17. *A*(1, 1), *B*(1, 6), *C*(5, 6)
18. *A*(2, 1), *B*(2, 5), *C*(4, 1)
19. *A*(−4, 3), *B*(−1, 4), *C*(−1, −1)
20. *A*(1, 4), *B*(2, −3), *C*(3, −3)
21. *A*(−3, −1), *B*(3, −1), *C*(3, 1)
22. *A*(−2, −3), *B*(0, 0), *C*(2, −4)

Rectangle *EFGH* has the following vertices. Graph rectangle *EFGH* and its image under $Rot_{O, 270°}$.

23. *E*(1, 4), *F*(9, 4), *G*(9, 8), *H*(1, 8)
24. *E*(−2, 2), *F*(0, 0), *G*(5, 5), *H*(3, 7)
25. *E*(−1, 0), *F*(2, −3), *G*(5, 0), *H*(2, 3)
26. *E*(1, 2), *F*(1, 6), *G*(3, 6), *H*(3, 2)

27. Complete the following rule. $(x, y) \xrightarrow{Rot_{O, -90°}} \underline{\quad ? \quad}$

Challenge **Write an equation for the image of each of the following under $Rot_{O, 90°}$.**

1. $x = y^2$
2. $y = x^2$
3. $y = |x|$
4. $y = \frac{1}{x}$

Mixed Review

Find the image of (3, −1) under each transformation.

1. $R_{y\text{-axis}}$
2. $R_{y = x}$
3. $T_{-3, 3}$
4. $Rot_{O, 90°}$

Find the preimage of (2, 5) under each transformation.

5. $R_{x\text{-axis}}$
6. $R_{x = -1}$
7. $T_{4, 1}$
8. $Rot_{O, 90°}$

9. If $f(x) = 1 - 2x$ and $g(x) = x^2 - 4x + 3$, find $(f \circ g)(-2)$ and $(g \circ f)(x)$.
10. Solve $|x^2 - 5x| \leq 6$. *Hint: You will need to solve two quadratic inequalities.*
11. Bonnie's Print Shop charges $8.00 to copy a 100-page document. The charge is $12.50 to copy a 175-page document. Write an equation to describe the relationship between the number of pages copied and the total charge, assuming the relationship is linear.

4.6 Half-Turns

Rotations of 180° about a point occur so frequently that they are given a special name.

Definition of Half-Turn

A rotation of 180° about a point P is called a **half-turn** with center P. A half-turn is also called a *point reflection*. The symbol $\text{Rot}_{P,\,180°}$ represents a half-turn about point P.

Example

1 Find the image of $\overline{AB}$ under a half-turn with center P.

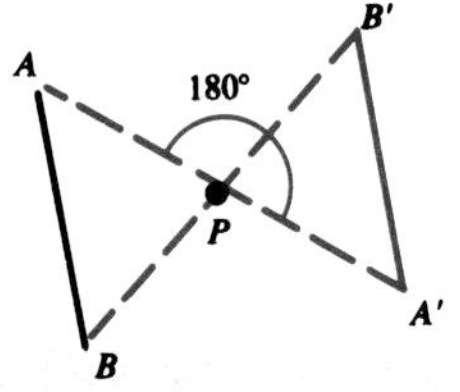

First find the image of A under a 180° rotation about P. Then find the image of B. Segment $A'B'$ is the image of $\overline{AB}$ under a half-turn with center P.
In symbols, this can be expressed as

$$\overline{AB} \xrightarrow{\text{Rot}_{P,\,180°}} \overline{A'B'}$$

Look carefully at how A' and B' were obtained in example 1. Note that point P is the midpoint of $\overline{AA'}$ and $\overline{BB'}$. A half-turn maps every point A to a point A' such that P is the midpoint of AA'.

Examples

2 Find the image of $\triangle ABC$ under a half-turn with center P.

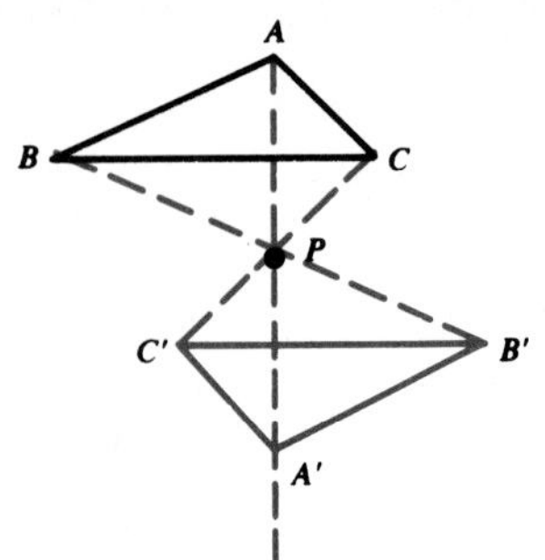

First draw $\overrightarrow{AP}$. Using a compass, locate A' such that P is the midpoint of $\overline{AA'}$. In the same way, locate points B' and C'. Therefore, $\triangle A'B'C'$ is the image of $\triangle ABC$ under a half-turn with center P.

3 **Graph $\triangle ABC$ with vertices $A(2, 3)$, $B(-3, 5)$, and $C(1, 6)$. Then graph the image of $\triangle ABC$ under a half-turn about the origin.**

$\triangle ABC$ is graphed at the right. Locate A', B', and C' such that O is the midpoint of $\overline{AA'}$, $\overline{BB'}$, and $\overline{CC'}$.

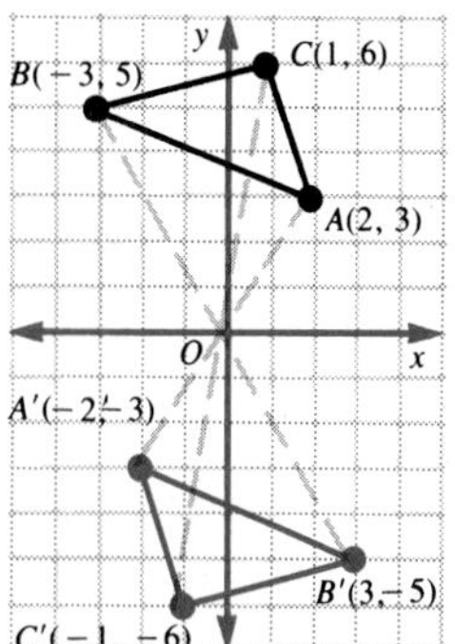

$$A(2, 3) \longrightarrow A'(-2, -3)$$
$$B(-3, 5) \longrightarrow B'(3, -5)$$
$$C(1, 6) \longrightarrow C'(-1, -6)$$

Example 3 suggests the following rule.

Rule for a Half-Turn about the Origin

If point $P(x, y)$ is rotated 180° either clockwise or counterclockwise about the origin, its image is $P'(-x, -y)$.

$$P(x, y) \xrightarrow{Rot_{O,\ 180°}} P°(-x, -y)$$

If a figure is its own image under some half-turn with center P, then it is said to have **point symmetry** with respect to P.

Example

4 **Each of the figures shown below has point symmetry with respect to point P.**

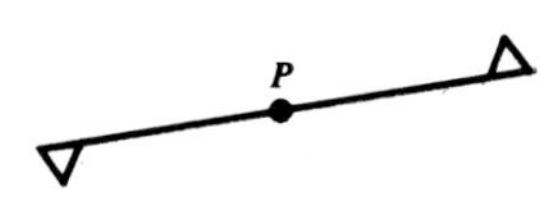

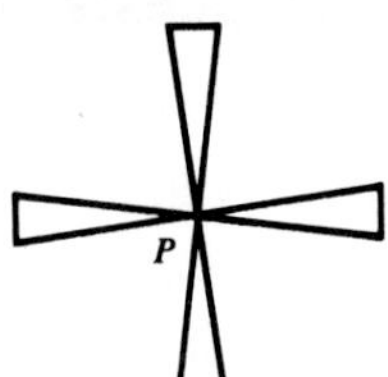

Exercises

Exploratory **Find the image of each of the following points under a half-turn whose center is the origin of the coordinate plane.**

1. $(2, 3)$
2. $(-4, 5)$
3. $(-5, 0)$
4. $(\frac{1}{2}, \frac{1}{4})$
5. $(3, -5)$
6. $(-8, -4)$
7. $(-6, 2)$
8. $(2a, -3b)$

Find the image of $(-4, 5)$ under each of the following.

9. $\text{Rot}_{O,\ 180^\circ}$ **10.** $\text{Rot}_{O,\ 90^\circ}$ **11.** $T_{-5,\ 6}$

12. $R_{x\text{-axis}}$ **13.** $R_{y\text{-axis}}$ **14.** $R_{y=x}$

Written **Copy and enlarge the figure. Then use compass and straightedge to find the image of $\triangle PQR$ under a half-turn when the center is each of the following points.**

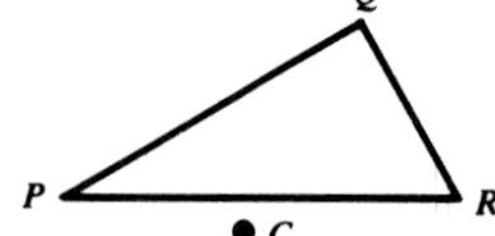

1. P **2.** Q **3.** R **4.** C

Find the image of (3, 1) under a half-turn whose center is each of the following points. Use graph paper.

5. $(0, -2)$ **6.** $(-2, 0)$ **7.** $(-2, 5)$ **8.** $(3, 1)$

For each of the following, if A' is the image of A under $\text{Rot}_{P,\ 180^\circ}$, find the coordinates of P.

9. $A(6, 3), A'(-2, 3)$ **10.** $A(4, 5), A'(-2, -1)$ **11.** $A(-2, -3), A'(-2, 0)$

12. $A(8, -2), A'(-3, 6)$ **13.** $A(-10, 4), A'(3, -1)$ **14.** $A(-1, 5), A'(2, -4)$

Find the images of $A(-1, 3)$ and $B(-7, -4)$ under a half-turn whose center is the origin. Then justify each of the following.

15. $AB = A'B'$ **16.** $\overleftrightarrow{AB} \parallel \overleftrightarrow{A'B'}$ **17.** $\overleftrightarrow{AB'} \parallel \overleftrightarrow{A'B}$

Find the images of $A(-3, 2)$ and $B(6, -4)$ under $\text{Rot}_{O,\ 90^\circ}$. Then justify each of the following.

18. $\overleftrightarrow{AB} \perp \overleftrightarrow{A'B'}$ **19.** $\overleftrightarrow{AA'} \parallel \overleftrightarrow{BB'}$ **20.** $AB = A'B'$

Use $\triangle ABC$ with vertices $A(1, 1)$, $B(6, 1)$ and $C(6, 4)$ to do the following.

21. Graph $\triangle ABC$ and its image $\triangle A'B'C'$ under $\text{Rot}_{O,\ 180^\circ}$.

22. On the same axes graph $\triangle A''B''C''$, the image of $\triangle A'B'C'$ under $R_{y\text{-axis}}$.

23. Name a single transformation that maps $\triangle ABC$ to $\triangle A''B''C''$.

24. On a different set of axes graph $\triangle ABC$ and its image $\triangle A'B'C'$ under $R_{y\text{-axis}}$.

25. On the same axes graph $\triangle A''B''C''$, the image of $\triangle A'B'C'$ under $R_{x\text{-axis}}$.

26. Name a single transformation that maps $\triangle ABC$ to $\triangle A''B''C''$.

For each figure in exercises 27–32, describe a transformation under which one triangle is the image of the other.

27.

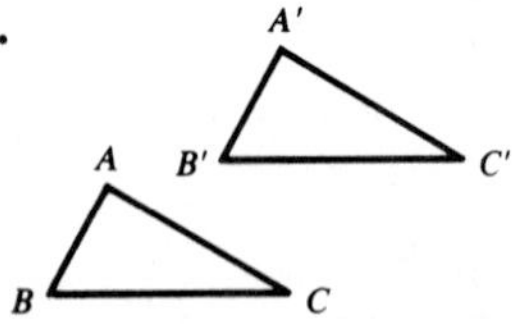

28.

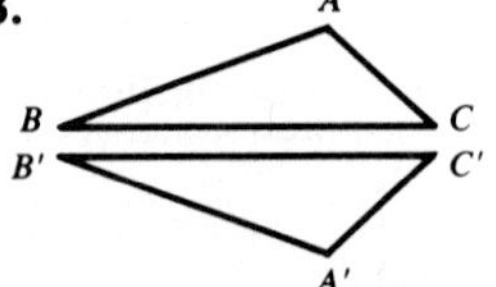

29.

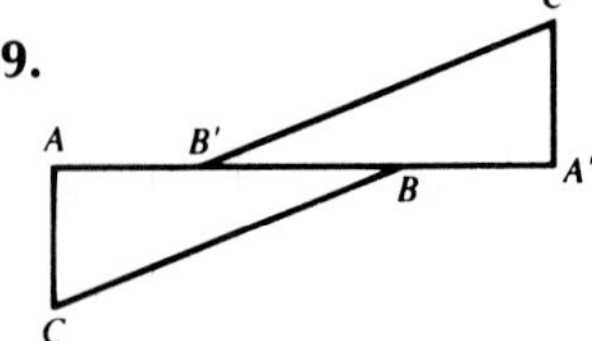

30.

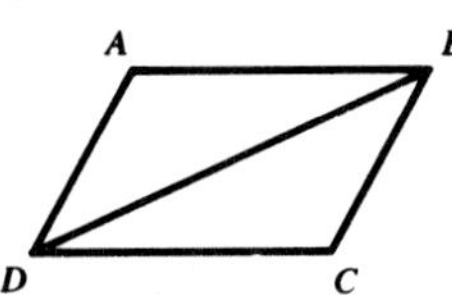

31.

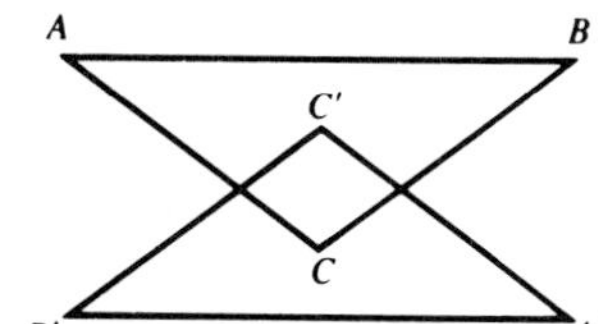

32.

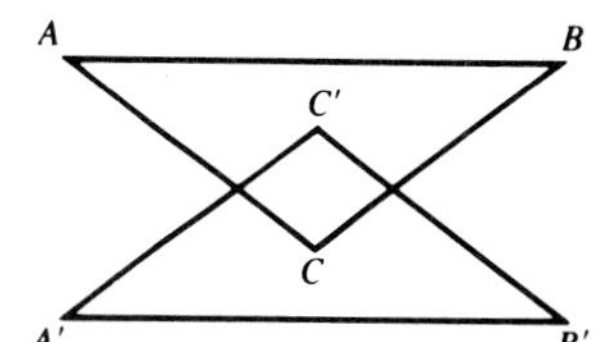

Which of the following figures has point symmetry?

33. rectangle **34.** rhombus **35.** isosceles triangle

36. **C** **37.** **N** **38.** 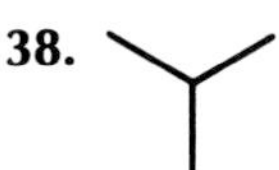 **39.** ⬭ **40.** 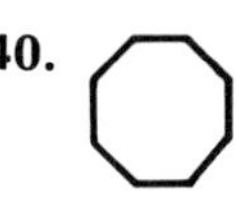 **41.** 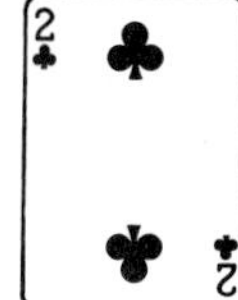

42. Sketch a figure that has point symmetry but not line symmetry.

Mathematical Excursions

Crystals

Crystals are classified according to symmetry, a balanced arrangement of faces. There are three basic aspects of crystal symmetry—plane of symmetry, axis of symmetry, and center of symmetry. Crystals can be grouped into 32 possible combinations of symmetry. These combinations can then be classified into the seven general crystal systems shown below.

System Name	Description	Example
Isometric	cube	pyrite
Tetragonal	prism with rectangular sides and square bottom and top	rutile
Hexagonal	prism with six rectangular faces	apatite
Rhombohedral	six rhomboidal faces	quartz
Orthorhombic	prism with three sets of unequal rectangular faces that meet at right angles	barite
Monoclinic	two rhomboidal faces and four rectangular ones, with top and bottom surfaces inclined	gypsum
Triclinic	different shaped faces that do not meet at right angles	plagioclase feldspar

4.7 Graphing Functions and Their Inverses

A knowledge of transformations is helpful when investigating the relationship between the graph of a function and the graph of its inverse.

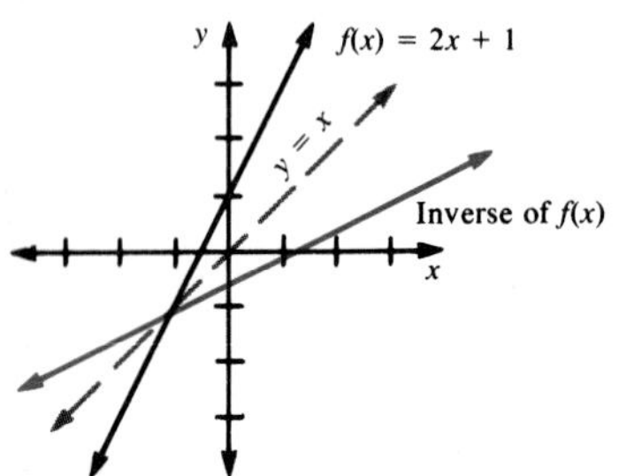

Consider a function f such that $f(x) = 2x + 1$. The inverse of f is obtained by interchanging the x and y coordinates of each ordered pair of the function. Therefore, whenever the graph of f contains the point (a, b), the graph of its inverse f^{-1} must contain the point (b, a). The points (a, b) and (b, a) are symmetric with respect to the line $y = x$. This means that the graph of the inverse of f can be obtained by reflecting the graph of f over the line $y = x$. This situation is shown in the figure at the left.

To simplify terminology, we will use "line $y = x$" to refer to the line whose equation is $y = x$.

Examples

1 Sketch the graph of the inverse of $f(x) = x^2$. Then find an equation for the graph of the inverse.

First graph $y = x^2$. Then draw the line $y = x$. The reflection of $f(x) = x^2$ over this line is shown in the figure at the right.

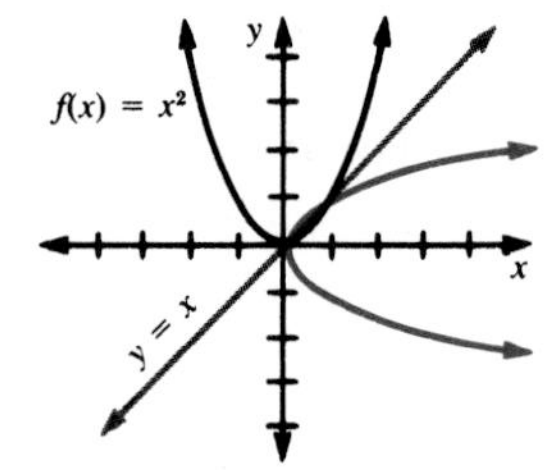

The equation for the graph of f is $y = x^2$. To obtain an equation for the inverse, interchange x and y in the equation and solve for y. Thus, an equation for the inverse of f is $x = y^2$ or $y = \pm\sqrt{x}$.

2 The red curve in the figure at the right is a graph of a one-to-one function since it passes the horizontal line test. Its inverse, shown in black, is also a function.

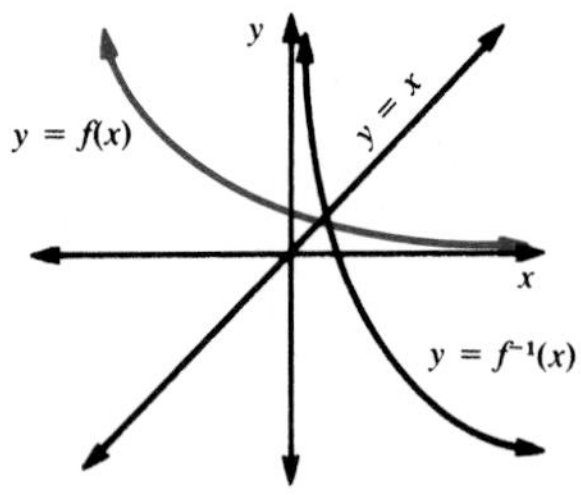

Note that the domain of f is the same as the range of f^{-1} and the range of f is the same as the domain of f^{-1}.

The next example shows that a knowledge of other symmetries is helpful in obtaining other graphs from the graph of $y = f(x)$.

Example

3 **Given the graph of $y = f(x)$, sketch the graphs of each of the following functions.**

a. $y = -f(x)$
b. $y = f(-x)$
c. $y = |f(x)|$

a. Recall that $(x, y) \xrightarrow{R_{x\text{-axis}}} (x, -y)$. Therefore, the graph of $y = -f(x)$ is obtained by reflecting the graph of $y = f(x)$ over the x-axis.

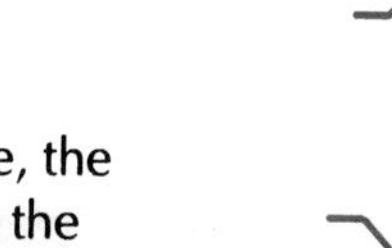

b. Recall that $(x, y) \xrightarrow{R_{y\text{-axis}}} (-x, y)$. Therefore, the graph of $y = f(-x)$ is obtained by reflecting the graph of $y = f(x)$ over the y-axis.

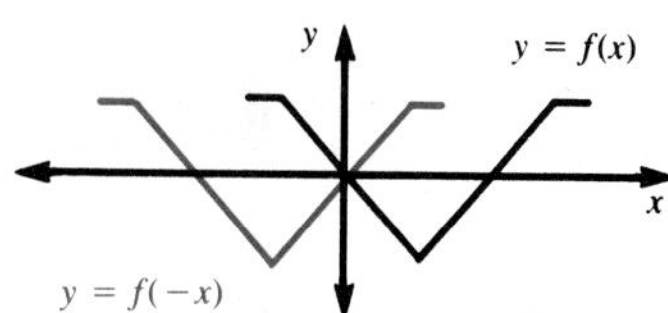

c. For any x, if $f(x) \geq 0$, then $f(x) = |f(x)|$. This means that the graphs of $y = f(x)$ and $y = |f(x)|$ are the same for points on or above the x-axis. For any x, if $f(x) < 0$, then $-f(x) = |f(x)|$. Therefore, reflect over the x-axis that portion of the graph of $y = f(x)$ that lies below the x-axis.

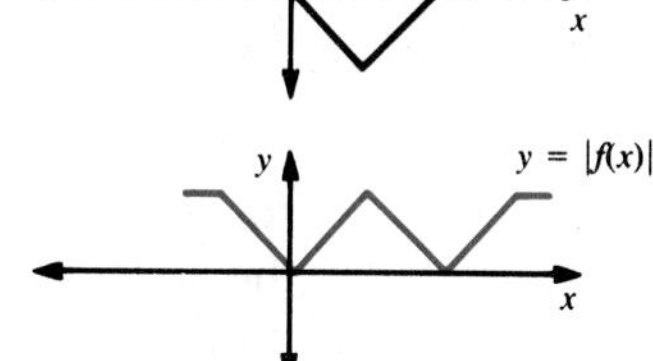

Exercises

Exploratory **Graph each function and its inverse on the same set of axes. Tell if the inverse is also a function.**

1. $\{(2, 3), (4, 6)\}$
2. $\{(1, 2), (2, 4), (3, 4)\}$
3. $\{(-1, 6), (0, 1), (1, 0), (6, -1)\}$
4. $\{(-2, -2), (-1, 5), (2, 1)\}$

Write an equation for the inverse of each function f.

5. $f(x) = 3x$
6. $f(x) = 10x$
7. $f(x) = x + 3$
8. $f(x) = x - 7$
9. $f(x) = 2x - 1$
10. $f(x) = 3x + 7$
11. $f(x) = 8 - 5x$
12. $f(x) = x^2$
13. $f(x) = \sqrt{4 - x^2}$

Written **Copy and enlarge each graph. Then sketch the inverse. Tell if the inverse is a function.**

1. 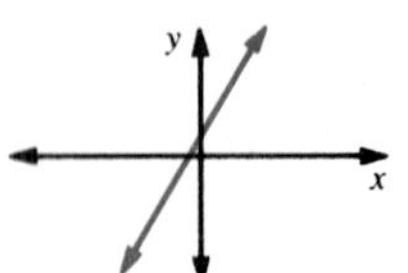
2. 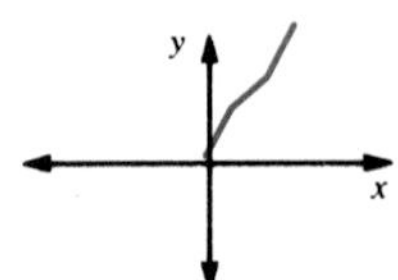
3. 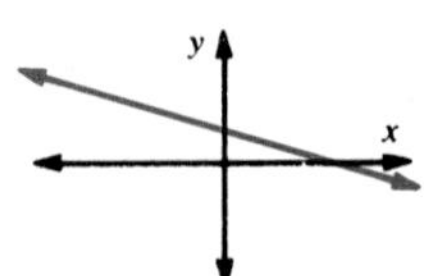
4. 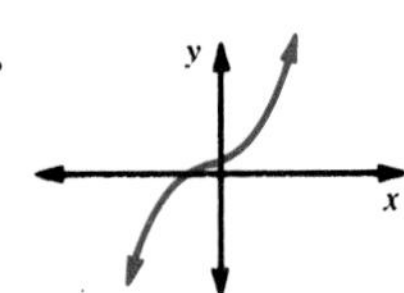
5. 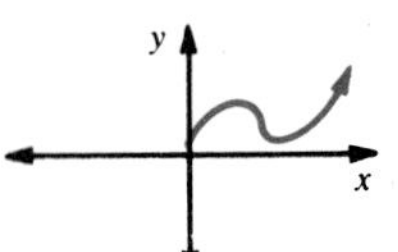
6. 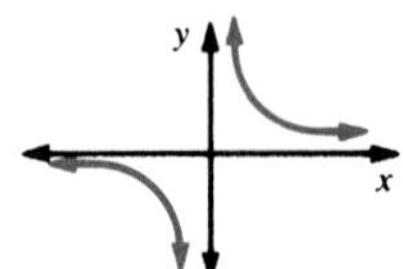
7. 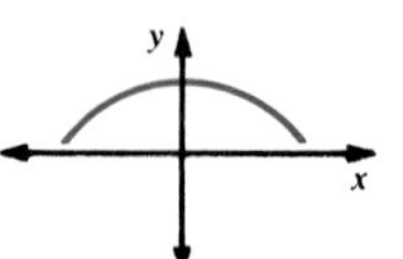
8. 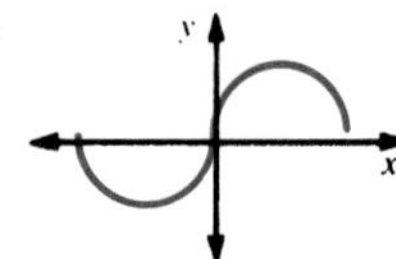
9. 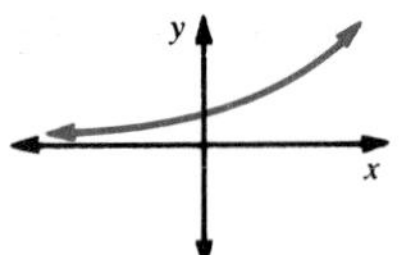
10. 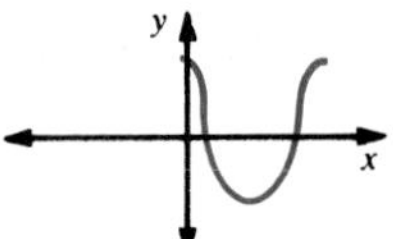
11. 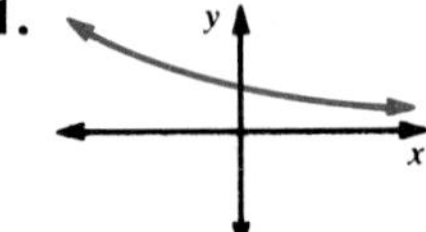
12. 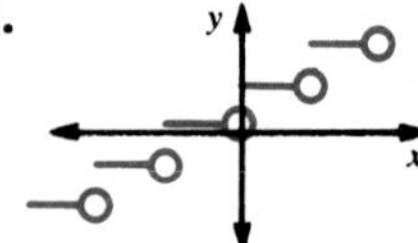

For each of the following draw the graph of f. On the same axes sketch the graph of f^{-1}.

13. $f(x) = 2x$
14. $f(x) = 3$
15. $f(x) = -4x$
16. $f(x) = -\frac{2}{3}x$
17. $f(x) = \frac{1}{5}x$
18. $f(x) = x + 1$
19. $f(x) = 2x + 3$
20. $f(x) = 5 - 2x$
21. $f(x) = -3x - 4$
22. $f(x) = -x^2$
23. $f(x) = -x^2 + 2$
24. $f(x) = x^2 + 3$
25. $f(x) = -x^3$
26. $f(x) = |x|$
27. $f(x) = |x| + 2$

Copy and enlarge the indicated graph from exercises 1–9. Use the graph to draw the graphs of each of the following functions.

a. $y = -f(x)$ **b.** $y = f(-x)$ **c.** $y = |f(x)|$

28. exercise 1
29. exercise 2
30. exercise 3
31. exercise 4
32. exercise 5
33. exercise 6
34. exercise 7
35. exercise 8
36. exercise 9

Challenge **Devise a procedure for testing an equation to see if its graph has symmetry with respect to each of the following.**

1. the x-axis
2. the y-axis
3. the line $y = x$
4. the origin

Test each equation for each of the symmetries in exercises 1–4. Then use this information to graph the equation.

5. $|x| = |y|$
6. $|x| + |y| = 6$
7. $|x| + |y| = 4$
8. $|x + y| = 6$
9. $|x + y| = 4$
10. $x^2 + y^2 = 16$
11. $x^2 + 4y^2 = 16$
12. $x^2 - y^2 = 16$
13. $4y^2 - x^2 = 16$

4.8 Dilations

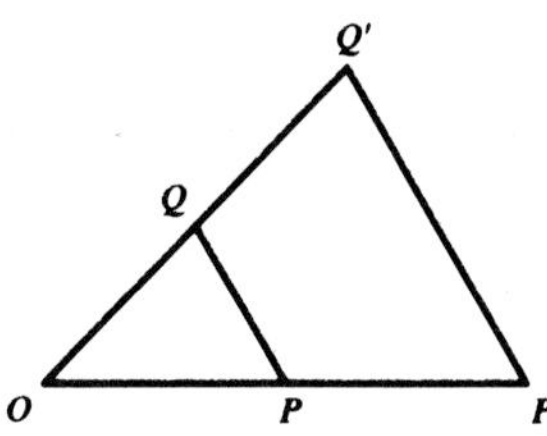

The transformations presented thus far have all preserved distance. Suppose a plane could be stretched or shrunk at will. In the figure at the left, a stretching of the plane, with point O held fixed, might produce $\triangle OP'Q'$ from $\triangle OPQ$.

In the figure, P' is assigned to P, Q' is assigned to Q, and O is assigned to itself. Note also that P' is on $\overrightarrow{OP}$ and Q' is on $\overrightarrow{OQ}$. In this particular case, $OQ' = 2 \cdot OQ$ and $OP' = 2 \cdot OP$. This transformation is an example of a **dilation** with center O and scale factor 2.

Example

1 Find the image of $\triangle ABC$ under a dilation with center O and scale factor 2.

First, draw $\overrightarrow{OA}$, $\overrightarrow{OB}$, and $\overrightarrow{OC}$.
Then, locate A', B', and C' such that

$OA' = 2 \cdot OA$,
$OB' = 2 \cdot OB$, and
$OC' = 2 \cdot OC$.

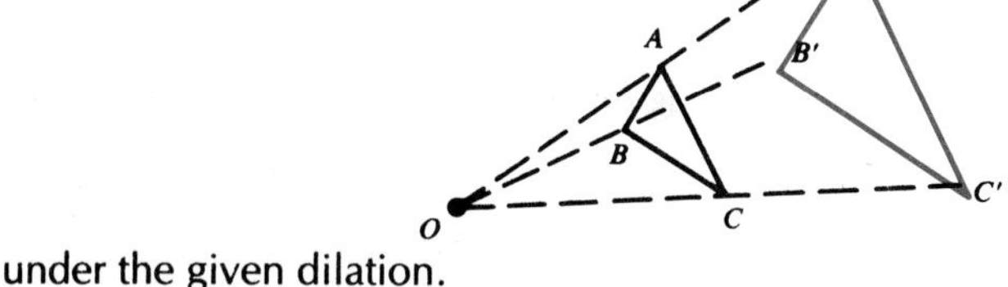

$\triangle A'B'C'$ is the image of $\triangle ABC$ under the given dilation.

The scale factor of a dilation with center O can be negative. When this is the case, P' is located on the ray opposite to $\overrightarrow{OP}$, as shown in the following example.

Example

2 Find the image of $\triangle ABC$ under a dilation with center O and scale factor -2.

Locate A', B', and C' on the rays opposite $\overline{OA}$, $\overline{OB}$, and $\overline{OC}$ such that $OA' = 2 \cdot OA$, $OB' = 2 \cdot OB$, and $OC' = 2 \cdot OC$.

$\triangle A'B'C'$ is the image of $\triangle ABC$ under the given dilation.

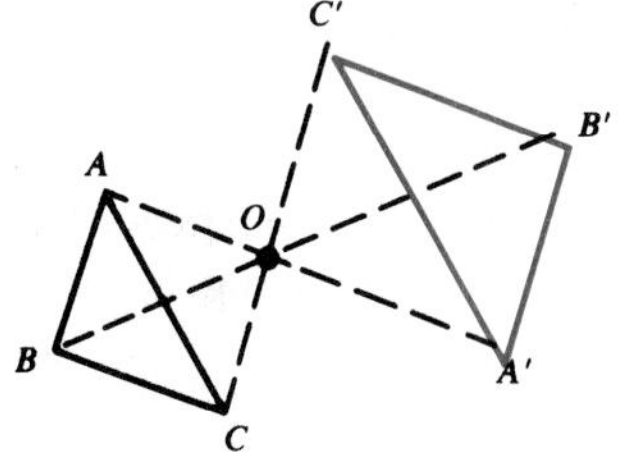

Examples 1 and 2 suggest that the images of figures under a dilation retain some characteristics of the original figure. But one important feature is not retained—size! Thus, dilations do not preserve distance.

As with other transformations, we can also perform dilations in the coordinate plane. The symbol $D_{O,\,k}$ represents the dilation with center at the origin and scale factor k.

Example

3 **Graph $A(-1, 3)$ and $B(3, 2)$. Then find the images of A and B under $D_{O,\,2}$.**

First, draw $\overrightarrow{OA}$ and $\overrightarrow{OB}$. Then locate A' and B' such that $OA' = 2 \cdot OA$ and $OB' = 2 \cdot OB$.

$$A(-1, 3) \xrightarrow{D_{O,\,2}} A'(-2, 6)$$

$$B(3, 2) \xrightarrow{D_{O,\,2}} B'(6, 4)$$

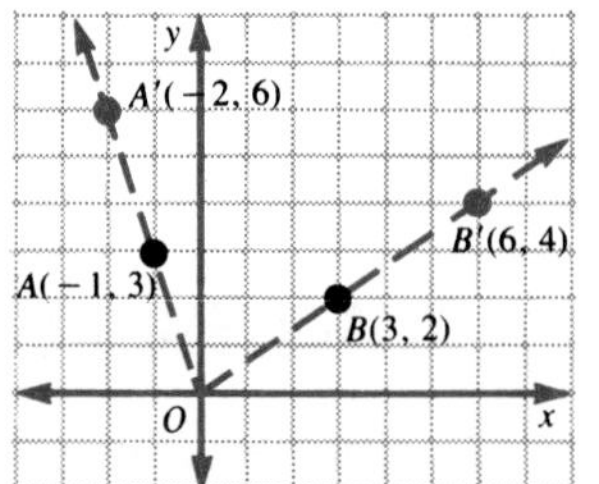

The results of example 3 suggest the following rule.

Rule for $D_{O,\,k}$

If point $P(x, y)$ is dilated with center at the origin and scale factor k, its image is $P'(kx, ky)$.

$$P(x, y) \xrightarrow{D_{O,\,k}} P'(kx, ky)$$

Example

4 **Find the image of the triangle with vertices $A(1, 1)$, $B(1, 4)$, and $C(5, 1)$ under $D_{O,\,3}$.**

Using the rule for dilation we obtain

$$(x, y) \xrightarrow{D_{O,\,3}} (3x, 3y).$$

Therefore, $A(1, 1) \longrightarrow A'(3, 3)$
$B(1, 4) \longrightarrow B'(3, 12)$
$C(5, 1) \longrightarrow C'(15, 3)$.

$\triangle A'B'C'$ is the image of $\triangle ABC$ under the given dilation.

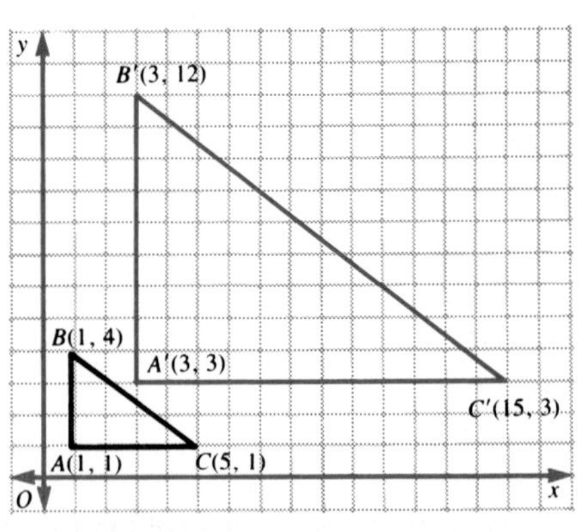

Look carefully at example 4. Although distance is not preserved by dilations, another important property is: *ratio of distance*.

In example 4, $A'B' = 3 \cdot AB$, $A'C' = 3 \cdot AC$, and $B'C' = 3 \cdot BC$. Therefore, $\frac{AB}{A'B'} = \frac{AC}{A'C'} = \frac{BC}{B'C'}$. This means that $\triangle ABC \sim \triangle A'B'C'$.

Exercises

Exploratory **Find the image of each of the following points under $D_{O,\,5}$.**

1. $(2, 4)$
2. $(-3, 1)$
3. $(4, -6)$
4. $(-2, -3)$
5. $\left(1\frac{1}{2}, 2\frac{1}{5}\right)$
6. $\left(-\frac{3}{4}, \frac{5}{2}\right)$
7. $(2\sqrt{2}, -3\sqrt{2})$
8. $(-3\sqrt{5}, 2\sqrt{5})$

In each of the following, A' is the image of A under $D_{O,\,k}$. Find the value of k for each.

9. $A(3, 2)$, $A'(12, 8)$
10. $A(0, 5)$, $A'(0, 10)$
11. $A(4, 8)$, $A'(2, 4)$
12. $A(-6, 4)$, $A'(3, -2)$
13. $A(-8, 9)$, $A'\left(4, -4\frac{1}{2}\right)$
14. $A(6, 10)$, $A'(-9, -15)$

Determine whether the following mappings could be the result of a dilation with center O.

15. $A(2, 3) \rightarrow A'(-6, -9)$
16. $A(2, -5) \rightarrow A'\left(1, \frac{5}{2}\right)$
17. $A(-9, -6) \rightarrow A'(3, 2)$
18. $A(3, 2) \rightarrow A(-9, -6)$
19. $A(7, 3) \rightarrow A'(2.1, 1.2)$
20. $A(8, -5) \rightarrow A'\left(-5\frac{1}{3}, 3\frac{1}{3}\right)$

Written **Copy the figure at the right. Then use a ruler to draw the image of $\triangle ABC$ under each dilation.**

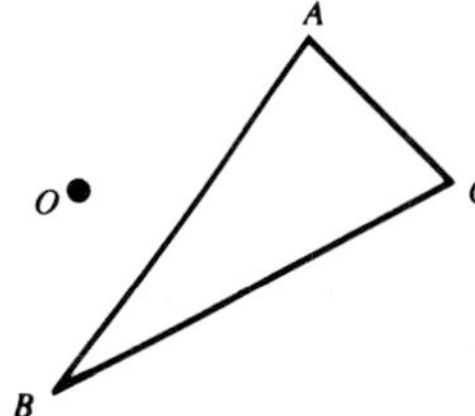

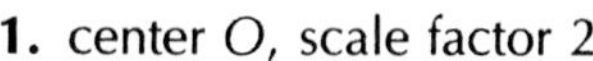

1. center O, scale factor 2
2. center O, scale factor 3
3. center O, scale factor $\frac{1}{2}$
4. center A, scale factor -1
5. center B, scale factor $\frac{2}{3}$
6. center C, scale factor 2

A triangle has vertices $A(0, 6)$, $B(2, 2)$, and $C(4, 2)$. Graph $\triangle ABC$. Then find its image under each of the following dilations.

7. $D_{O,\,2}$
8. $D_{O,\,-1}$
9. $D_{O,\,-3}$
10. $D_{O,\,\frac{1}{2}}$

Under a dilation with center P and scale factor k, A' is the image of A, and B' is the image of B. Find k under the given conditions.

11. $PA = 4$ and $PA' = 12$
12. $PB = 16$ and $PB' = 8$
13. $PA' = 4$ and $AA' = 6$
14. $AB = 4$ and $A'B' = 24$

Find the value of each ? in the following rules.

15. $\left(\frac{1}{2}, \frac{\sqrt{3}}{2}\right) \xrightarrow{D_{O,\,2}} (?, ?)$
16. $(?, ?) \xrightarrow{D_{O,\,4}} (2\sqrt{2}, 2\sqrt{2})$
17. $\left(\frac{1}{2}, -\frac{\sqrt{3}}{2}\right) \xrightarrow{D_{O,\,?}} \left(-\frac{11}{4}, \frac{11\sqrt{3}}{4}\right)$
18. $(a, b) \xrightarrow{D_{O,\,r}} (?, ?)$

Which of the following properties are preserved by dilations? Justify your answer and include a sketch.

19. collinearity
20. angle measure
21. distance
22. ratio of distances
23. parallelism of lines
24. perpendicularity of lines

Let p be "Transformation t is a line reflection"; let r be "Transformation t is a dilation with scale factor $k = 1$"; let s be "Transformation t does not preserve distance." Which of the following are true? Justify your answers.

25. $p \rightarrow s$
26. $r \rightarrow s$
27. $\sim s \rightarrow r$
28. $\sim r \rightarrow \sim s$
29. $p \rightarrow \sim s$

30. Show that a dilation with scale factor -1 is a half-turn.

Challenge **Find an equation of the image of the graph of $y = x^2$ under each of the following.**

1. $D_{O,\,2}$
2. $D_{O,\,\frac{1}{3}}$
3. $R_{x\text{-axis}}$
4. $T_{-1,\,3}$
5. $Rot_{O,\,90°}$
6. $Rot_{O,\,180°}$
7. $R_{y\text{-axis}}$
8. $T_{5,\,-3}$

Mathematical Excursions

Tessellations

A tessellation is a design in which the plane is completely covered by congruent figures without any gaps or overlapping of figures. Often they are composed of only one kind of polygon. However, when a few different types of polygons are used there is a repeated pattern creating an interesting effect.

Tessellations can have line, point, rotational, or translational symmetry. Often a series of transformations can be developed to describe how a certain tessellation is formed from a single polygon.

M. C. Escher, a Dutch artist, is famous for his varied designs that are tessellations of patterns he creates. An example of his work can be seen on the opening page of Chapter 10. You too can create some special effects using color with line drawings of tessellations. Use the ones shown below or create some of your own to create different visual perceptions of the same tessellation.

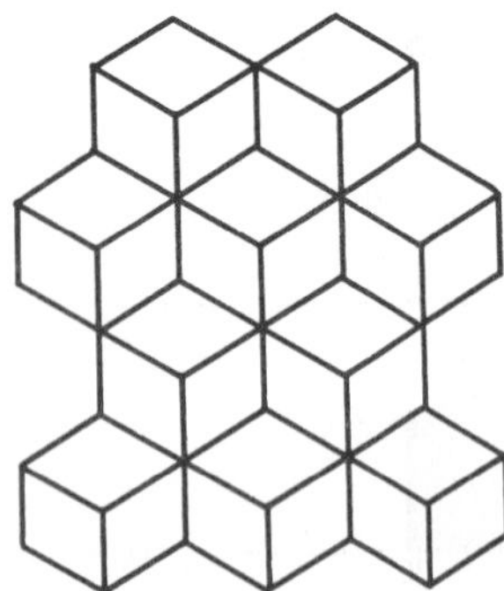
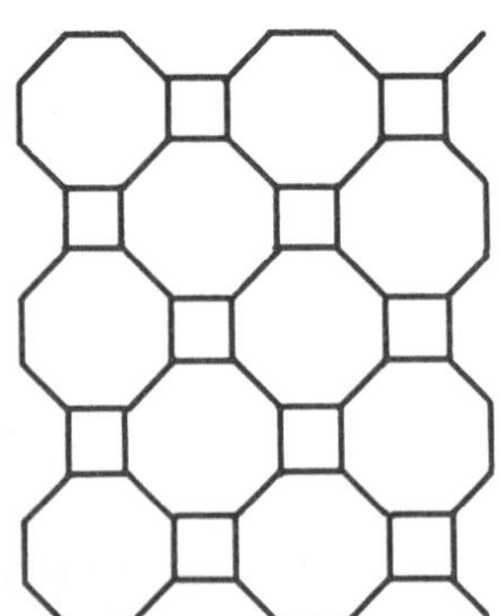
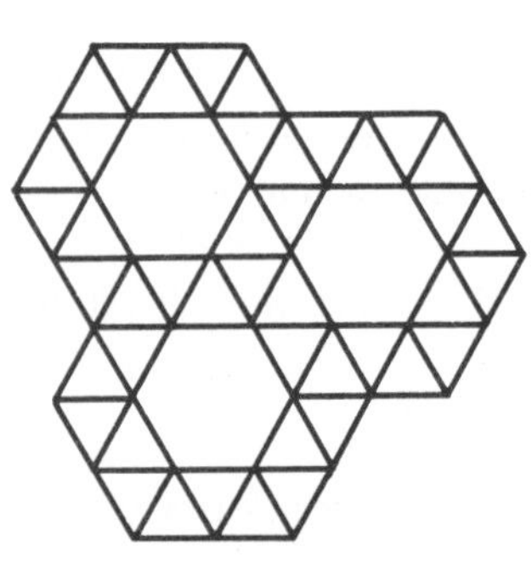

Problem Solving Application: Linear Programming

Certain real-world problems are solved by determining minimum or maximum values subject to certain conditions, called *constraints*. Often, the value to be minimized or maximized can be represented by an expression in two variables, and the constraints can be expressed as linear inequalities. Then, a procedure called linear programming can be used to solve the problem.

The following procedure can be used to solve linear programming problems.

Linear Programming Procedure

1. Define variables
2. Write a system of inequalities for the constraints.
3. Graph the system.
4. Determine the vertices of the polygon formed. The maximum or minimum value always occurs at a vertex of the polygon.
5. Write an expression to be maximized or minimized.
6. Substitute values from the vertices into the expression.
7. Select the greatest or least result. Answer the problem.

Example

1 **A dressmaking shop makes dresses and suits. The equipment in the shop allows for making at most 30 dresses and 20 suits in a week. It takes 10 worker-hours to make a dress and 20 worker-hours to make a suit, and there are 500 worker-hours available per week in the shop. If the profit on a dress is $30 and the profit on a suit is $45, how many of each should be made each week to maximize the shop's profit?**

Define variables. Let d = the number of dresses made per week.
Let p = the number of suits made per week.

Write inequalities.

$0 \le d \le 30$ At most 30 dresses can be made.

$0 \le p \le 20$ At most 20 suits can be made.

$10d + 20p \le 500$ At most 500 worker-hours are available.

Graph the system and determine the vertices of the polygon.

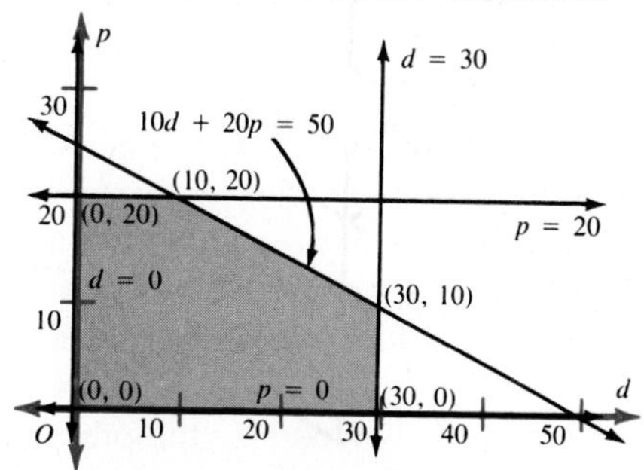

Any point in the shaded region or its boundaries will satisfy the conditions of this problem.

The vertices of the polygon are (0, 0), (0, 20), (10, 20), (30, 10), and (30, 0).

Write expression to be maximized. profit = profit on dresses + profit on suits

$P = 30d + 45p$

Recall that the maximum or minimum value always occurs at a vertex.

Substitute values from the vertices into the expression.

(d, p)	(0, 0)	(0, 20)	(10, 20)	(30, 10)	(30, 0)
30d + 45p	\$0	\$900	\$1200	\$1350	\$900

Answer the problem. The dressmaking shop will have the greatest profit, \$1350, by making 30 dresses and 10 suits per week.

Exercises

Written Solve the problem in example 1 if the following changes are made while all other things remain the same.

1. The profit on suits is increased to \$65.
2. The number of worker-hours available per week is increased to 750.
3. A painter has exactly 32 units of yellow dye and 54 units of green dye. He plans to mix the dyes to make as many gallons of color A and color B as possible. Each gallon of color A requires 4 units of yellow dye and 1 unit of green dye. Each gallon of color B requires 1 unit of yellow dye and 6 units of green dye.
   a. Let A represent the number of gallons of color A and B represent the number of gallons of color B. Write a system of inequalities to represent the constraints.
   b. Graph the system of inequalities, and determine the vertices of the polygon formed.
   c. Find the values of A and B that produce the maximum number of gallons, $A + B$, possible.
4. Jerome Beck works a maximum of 20 hours per week while in school. He receives \$7 per hour for mowing lawns and \$10 per hour for tutoring math students. He wants to spend at least 3 hours, but not more than 8 hours, tutoring math students.
   a. Let m represent the number of hours spent mowing lawns and t represent the number of hours tutoring math students. Write a system of inequalities to represent the constraints.
   b. Graph the system of inequalities, and determine the vertices of the polygon formed.
   c. Find the values of m and t that produce the maximum weekly earnings, $7m + 10t$, possible.
5. A farmer has 20 days in which to plant corn and beans. Corn can be planted at a rate of 10 acres per day, and beans can be planted at a rate of 15 acres per day. The farmer has 250 acres available for planting.
   a. Let c represent the number of acres of corn and b represent the number of acres of beans. Write a system of inequalities to represent the constraints.
   b. Graph the system of inequalities, and determine the vertices of the polygon formed.
   c. If the profit on corn is \$30 per acre and the profit on beans is \$25 per acre, find the values of c and b that will maximize the profit.
   d. If the profit on corn is \$29 per acre and the profit on beans is \$30 per acre, find the values of c and b that will maximize the profit.

Vocabulary

transformation
line reflection
image
pre-image
line symmetry
line of symmetry
reflection over the y-axis
reflection over the x-axis
reflection over the line $y = x$
translation
rotation
rotational symmetry
half-turn
point symmetry
dilation
scale factor

Chapter Summary

1. Under a line reflection, any point on the line of reflection is its own image. For any point P not on the line and its image P', the line of reflection is the perpendicular bisector of $\overline{PP'}$.
2. Line reflections preserve distance.
3. If point $P(x, y)$ is reflected over the y-axis, its image is $P'(-x, y)$.

$$(x, y) \xrightarrow{R_{y\text{-axis}}} (-x, y)$$

4. If point $P(x, y)$ is reflected over the x-axis, its image is $P'(x, -y)$.

$$(x, y) \xrightarrow{R_{x\text{-axis}}} (x, -y)$$

5. If point $P(x, y)$ is reflected over the line whose equation is $y = x$, its image is $P'(y, x)$.

$$(x, y) \xrightarrow{R_{y = x}} (y, x)$$

6. If point $P(x, y)$ is translated a units horizontally and b units vertically, its image is $P'(x + a, y + b)$.

$$(x, y) \xrightarrow{T_{a, b}} (x + a, y + b)$$

7. Translations preserve distance.
8. If point $P(x, y)$ is rotated 90° counterclockwise about the origin, its image is $P'(-y, x)$.

$$(x, y) \xrightarrow{Rot_{O, 90^\circ}} (-y, x)$$

9. If a figure has rotational symmetry, the angle of rotation must be greater than 0° but less than 360°.
10. A half-turn is also called a point reflection.
11. If point $P(x, y)$ is rotated a half-turn about the origin, its image is $P'(-x, -y)$.

$$(x, y) \xrightarrow{Rot_{O,\ 180°}} (-x, -y)$$

12. If point $P(x, y)$ is dilated with center at the origin and scale factor k, its image is $P'(kx, ky)$.

$$(x, y) \xrightarrow{D_{O,\ k}} (kx, ky)$$

13. Dilations, unlike translations, reflections, and rotations, do not preserve distance.
14. When a polygon is dilated, the result is two similar polygons.

Chapter Review

4.1 **The two figures at the right are images under R_m. Name the image of each of the following.**

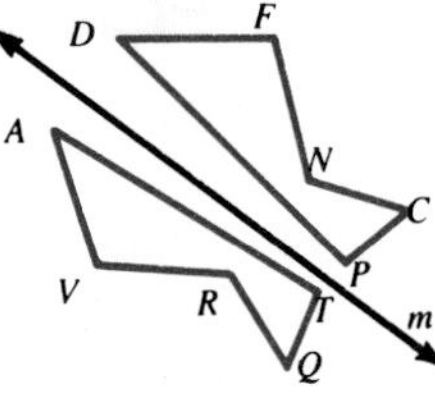

1. A
2. $\overline{TQ}$
3. $\angle NFD$
4. F
5. $\overline{NF}$
6. $\angle ATQ$

4.2 **Find all the lines of symmetry in those figures that have line symmetry.**

7.

8.

9. HAH

10. 

11. 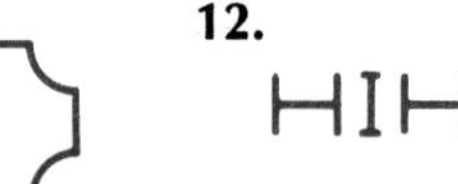

12.

4.3 **Find the image of $(-6, 4)$ under each of the following.**

13. $R_{y\text{-axis}}$
14. $R_{x\text{-axis}}$
15. $R_{y=x}$
16. $R_{y=-x}$
17. $R_{x=4}$
18. $R_{y=-3}$

4.4 **Find the coordinate rule of a translation in which *B* is the image of *A*.**

19. $A(7, 3)$, $B(10, -1)$
20. $A(2, -4)$, $B(1, -5)$
21. $A(-4, 6)$, $B(7, 1)$

Use the same set of axes for exercises 22–24.

22. Graph $\triangle ABC$ with vertices $A(-2, -2)$, $B(-1, 2)$ and $C(2, 1)$.
23. Graph $\triangle A'B'C'$, the image of $\triangle ABC$, under $T_{-1,\ 4}$.

24. Graph $\triangle A''B''C''$, the image of $\triangle A'B'C'$, under $T_{5,\,-2}$.

25. What single transformation maps $\triangle ABC$ to $\triangle A''B''C''$?

4.5 Find the image of $(-4, 6)$ under each of the following.

26. $Rot_{O,\,90°}$

27. $Rot_{O,\,180°}$

4.6 Which of the figures below have rotational symmetry? point symmetry?

28.

29.

30.

31.

32.

33.

34.

35.

36.

37.

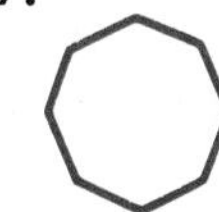

38.

39.

40. Given $A(-2, 3)$ and $B(5, -2)$ under a half-turn about the origin, show that $AB = A'B'$.

4.7 For each of the following functions, sketch the graph of f and then, on the same set of axes, sketch the graph of f^{-1}.

41. $f(x) = 2x + 3$

42. $f(x) = |x|$

43. $f(x) = x^2 - 3$

4.8 Let RST be the triangle with vertices $R(-1, 0)$, $S(1, -3)$ and $T(4, 2)$. Let $\triangle R'S'T'$ be the image of $\triangle RST$ under $D_{O,\,3}$.

44. Graph $\triangle RST$ and $\triangle R'S'T'$ on the same set of axes.

45. Show that $\overline{ST} \parallel \overline{S'T'}$.

46. Show that $OT' = 3\ OT$.

For each set of figures, describe a transformation under which one triangle is the image of the other.

47.

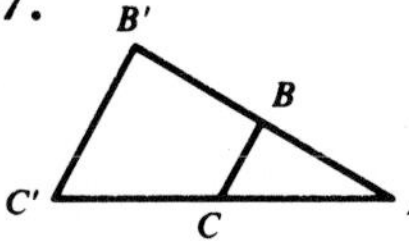

48.

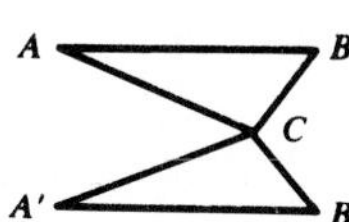

49.

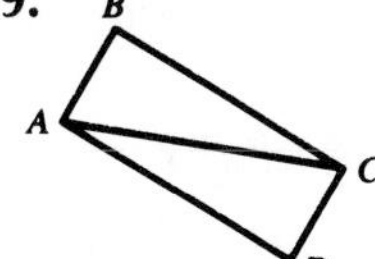

50.

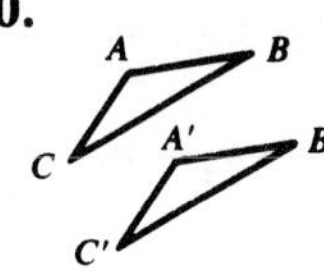

51.

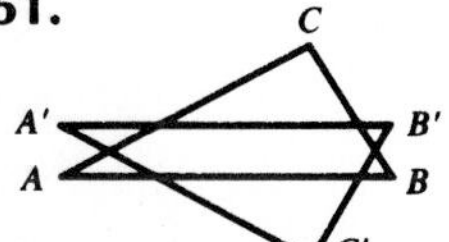

52.

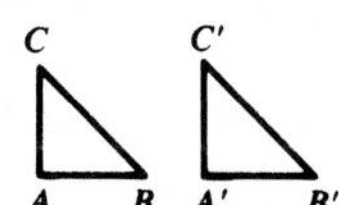

53.

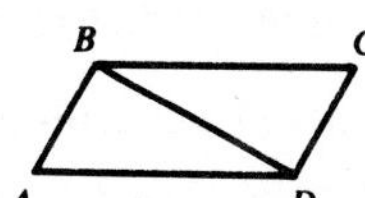

54.

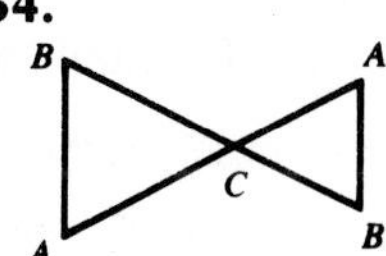

Chapter Test

State whether each statement below is true or false.

1. All transformations preserve distance.
2. Line reflections preserve collinearity.
3. Every figure has at least one line of symmetry.
4. In a transformation each point must have exactly one pre-image.
5. Every figure with rotational symmetry also has line symmetry.
6. Every figure with point symmetry also has rotational symmetry.
7. Ratio of distance is preserved by dilations.

Choose the correct answer.

8. The letter "N" has
   a. point symmetry only. **b** line symmetry only
   c. both point and line symmetry **d.** neither point nor line symmetry
9. The graphs of a function and its inverse are symmetrical to the
   a. x-axis **b.** y-axis **c.** origin **d.** line $y = x$.
10. How many lines of symmetry does a rectangle have?
    a. 1 **b.** 2 **c.** 3 **d.** 4
11. Assume that the polygons named below are regular. Which has 60° rotational symmetry?
    a. pentagon **b.** hexagon **c.** square **d.** triangle
12. Which property is not preserved under a line reflection?
    a. distance **b.** parallelism **c.** betweenness **d.** orientation

Find the image of (4, −2) under each of the following transformations.

13. $R_{x\text{-axis}}$ 14. $R_{y=x}$ 15. $R_{x=2}$ 16. $T_{2,\,4}$
17. $T_{5,\,-8}$ 18. $T_{-5,\,6}$ 19. $Rot_{O,\,90°}$ 20. $Rot_{O,\,180°}$
21. $D_{O,\,3}$ 22. $D_{O,\,\frac{1}{2}}$ 23. $D_{O,\,-3}$ 24. $Rot_{O,\,270°}$

25. Label three non-collinear points A, B, and C. Then sketch the image of $\overline{AB}$ under a half-turn with center C.
26. Sketch the graph of $y = x^3$. Then sketch its inverse.

Graph each of the following on the same set of axes.

27. $\triangle ABC$ with vertices $A(2, 1)$, $B(6, 4)$, and $C(8, 1)$
28. $\triangle A'B'C'$, the image of $\triangle ABC$ under $R_{x\text{-axis}}$
29. $\triangle A''B''C''$, the image of $\triangle A'B'C'$ under $T_{-6,\,-2}$
30. $\triangle A'''B'''C'''$, the image of $\triangle A''B''C''$ under $Rot_{O,\,90°}$

Exponential and Logarithmic Functions

Chapter 5

Some of the many applications of the ideas you will encounter in this chapter are found in the study of population growth and decline, radioactive decay, compound interest, and inflation. Pictured below is a frog egg during cell division. Each step results in the number of cells being doubled. The number of cells at any given time can be determined by an exponential function.

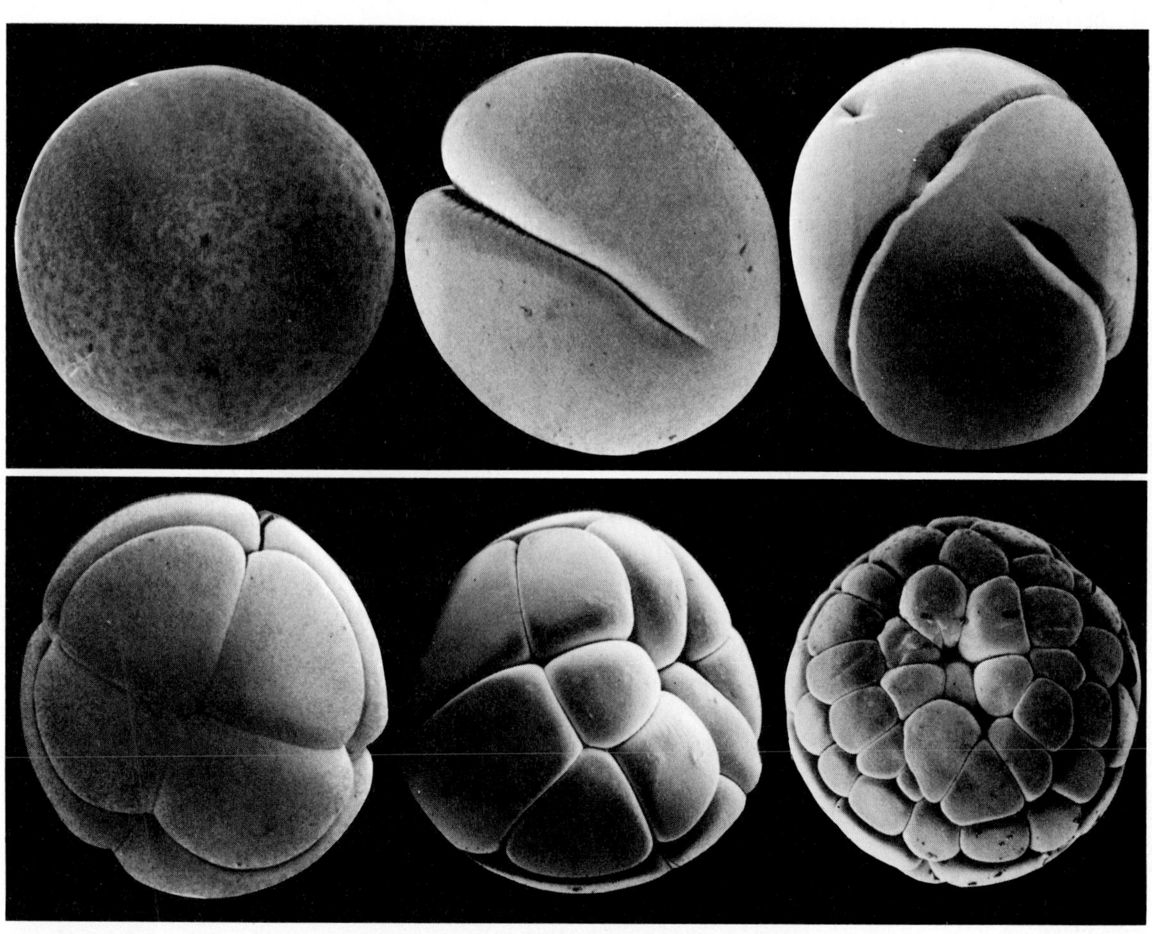

5.1 Rational Exponents

Before considering expressions such as $5^{\frac{2}{3}}$, it is helpful to review the following definitions.

Definition of Zero and Negative Exponents

For any real number r, except $r = 0$, and for any positive integer n,

$$r^0 = 1 \text{ and } r^{-n} = \frac{1}{r^n}.$$

Examples

Evaluate each of the following sums.

1 $-4^2 + 8^0$

$$\begin{aligned} -4^2 + 8^0 &= -16 + 1 \\ &= -15 \end{aligned}$$

2 $4^{-2} + \left(\frac{1}{2}\right)^{-1}$

$$\begin{aligned} 4^{-2} + \left(\frac{1}{2}\right)^{-1} &= \frac{1}{4^2} + \frac{1}{\frac{1}{2}} \\ &= \frac{1}{16} + 2 \\ &= 2\frac{1}{16} \end{aligned}$$

Some familiar rules for operating with expressions containing exponents are listed below. In each case, m and n represent integers, and x and y are real numbers. Since 0^0 is not defined and since division by 0 is not possible, we agree that all variables in exponential expressions are appropriately restricted so that these situations do not occur.

Laws of Exponents

Law 1	$x^m \cdot x^n = x^{m+n}$
Law 2	$(x^m)^n = x^{mn}$
Law 3	$(xy)^m = x^m y^m$
Law 4	$\frac{x^m}{x^n} = x^{m-n}$
Law 5	$\left(\frac{x}{y}\right)^m = \frac{x^m}{y^m}$

Examples

Simplify and express each with positive exponents.

3 $(x^{-2})^4$

$$(x^{-2})^4 = x^{-8} = \frac{1}{x^8}$$

4 $\frac{x^{-2}}{x^4}$

$$\frac{x^{-2}}{x^4} = x^{-2-4} = x^{-6} = \frac{1}{x^6}$$

5 $\frac{a^2(r^{-3}s)^{-2}}{a^5r^3s^3}$

$$\frac{a^2(r^{-3}s)^{-2}}{a^5r^3s^3} = \frac{a^2r^6s^{-2}}{a^5r^3s^3} = a^{2-5}r^{6-3}s^{-2-3} = a^{-3}r^3s^{-5} = \frac{r^3}{a^3s^5}$$

What meaning can be given to expressions such as $3^{\frac{1}{2}}$? If Law 2 is applied, then the following must be true.

$$(3^{\frac{1}{2}})^2 = 3^{\frac{1}{2}\cdot 2} = 3^1 = 3$$

This statement says that the square of $3^{\frac{1}{2}}$ is 3. Therefore $3^{\frac{1}{2}}$ must be the square root of 3. That is, $3^{\frac{1}{2}} = \sqrt{3}$. This example suggests the following definition.

Definition of $b^{\frac{1}{n}}$

If b is any real number, n is any integer, and $n > 1$, then

$$b^{\frac{1}{n}} = \sqrt[n]{b},$$

except when b is negative and n is even.

Note that b cannot be negative when n is even because numbers like $\sqrt{-4}$ would be imaginary.

If Law 2 is used again, then the following statements about $5^{\frac{2}{3}}$ must also be true.

$$5^{\frac{2}{3}} = (5^{\frac{1}{3}})^2 = (\sqrt[3]{5})^2 \qquad 5^{\frac{2}{3}} = (5^2)^{\frac{1}{3}} = \sqrt[3]{5^2}$$

Therefore, $5^{\frac{2}{3}}$ must be equal to $(\sqrt[3]{5})^2$ or $\sqrt[3]{5^2}$. This example suggests the following definition.

Definition of $b^{\frac{p}{q}}$

If b is any real number, p and q are integers, and $q > 1$, then

$$b^{\frac{p}{q}} = (\sqrt[q]{b})^p = \sqrt[q]{b^p},$$

except when b is negative and q is even.

All the laws of exponents stated for integral exponents apply to rational exponents as well.

Examples

Evaluate each expression.

6 $27^{\frac{1}{3}}$

$27^{\frac{1}{3}} = \sqrt[3]{27} = 3$

7 $16^{\frac{1}{4}}$

$16^{\frac{1}{4}} = \sqrt[4]{16} = 2$

8 $16^{\frac{3}{4}}$

$$\begin{aligned} 16^{\frac{3}{4}} &= (\sqrt[4]{16})^3 \\ &= 2^3 \\ &= 8 \end{aligned}$$

9 $(-8)^{-\frac{5}{3}}$

$$\begin{aligned} (-8)^{-\frac{5}{3}} &= (\sqrt[3]{-8})^{-5} \\ &= (-2)^{-5} \\ &= \frac{1}{(-2)^5} \\ &= \frac{1}{-32} \text{ or } -\frac{1}{32} \end{aligned}$$

10 **Evaluate $x^{\frac{3}{2}} - 5x^0$ if $x = 4$.**

$$\begin{aligned} x^{\frac{3}{2}} - 5x^0 &= 4^{\frac{3}{2}} - 5(4^0) && \textit{Substitution} \\ &= (\sqrt{4})^3 - 5(1) && \textit{Definition of } b^{\frac{p}{q}} \textit{ and zero power.} \\ &= (2)^3 - 5 && \textit{Simplify.} \\ &= 8 - 5 \text{ or } 3 \end{aligned}$$

11 **Simplify $(8x^2y^3)^{\frac{2}{3}}$.**

$$\begin{aligned} (8x^2y^3)^{\frac{2}{3}} &= 8^{\frac{2}{3}} \cdot (x^2)^{\frac{2}{3}} \cdot (y^3)^{\frac{2}{3}} && \textit{Law 3} \\ &= (\sqrt[3]{8})^2 \cdot x^{\frac{4}{3}} \cdot y^2 && \textit{Definition of } b^{\frac{p}{q}} \textit{ and Law 2} \\ &= 4x^{\frac{4}{3}}y^2 && (\sqrt[3]{8})^2 = 2^2 \text{ or } 4 \end{aligned}$$

Exercises

Exploratory **Evaluate.**

1. 6^0 **2.** 8^{-1} **3.** 2^{-3} **4.** 3^{-3} **5.** 9^{-2} **6.** $\left(\frac{1}{4}\right)^2$ **7.** $\left(\frac{1}{4}\right)^{-1}$

8. $\left(\frac{3}{4}\right)^{-1}$ **9.** $\left(-\frac{4}{5}\right)^{-1}$ **10.** $\left(-\frac{5}{8}\right)^{-1}$ **11.** $\left(\frac{1}{2}\right)^{-3}$ **12.** $\left(\frac{2}{3}\right)^{-2}$ **13.** $\left(-\frac{1}{4}\right)^{-2}$ **14.** $\left(-\frac{3}{5}\right)^{-2}$

15. $\left(-\frac{1}{2}\right)^{-4}$ **16.** $\left(-\frac{3}{2}\right)^{-3}$ **17.** $8^{\frac{1}{3}}$ **18.** $9^{\frac{1}{2}}$ **19.** $16^{\frac{1}{4}}$ **20.** $125^{\frac{1}{3}}$ **21.** $8^{\frac{2}{3}}$

22. $100^{\frac{3}{2}}$ **23.** $81^{\frac{3}{4}}$ **24.** $64^{\frac{2}{3}}$ **25.** $(-27)^{\frac{2}{3}}$ **26.** $8^{-\frac{1}{3}}$ **27.** $16^{-\frac{1}{2}}$ **28.** $100^{-\frac{1}{2}}$

29. $125^{-\frac{1}{3}}$ **30.** $16^{-\frac{3}{2}}$ **31.** $\left(\frac{1}{4}\right)^{\frac{1}{2}}$ **32.** $\left(-\frac{8}{27}\right)^{-\frac{1}{3}}$ **33.** $\left(\frac{16}{81}\right)^{\frac{1}{4}}$ **34.** $\left(\frac{27}{64}\right)^{-\frac{1}{3}}$ **35.** $\left(\frac{1}{32}\right)^{-\frac{3}{5}}$

Write each of the following as an equivalent expression with positive exponents.

36. y^{-3} 37. a^{-4} 38. $2m^{-3}$ 39. $5x^{-7}$ 40. $\frac{4}{y^{-1}}$ 41. $\frac{9}{x^{-2}}$

42. $\left(\frac{x}{y}\right)^{-1}$ 43. $\frac{6}{t^{-5}}$ 44. $\sqrt{3y}$ 45. $\sqrt[3]{5y}$ 46. $a\sqrt[3]{b}$ 47. $\sqrt[3]{rp^3}$

Written Evaluate.

1. 4^{-3} 2. 2^{-5} 3. 5^{-3} 4. 6^{-3} 5. $\left(\frac{2}{3}\right)^{-3}$ 6. $\left(\frac{2}{5}\right)^{-3}$

7. $9^{\frac{5}{2}}$ 8. $4^{\frac{3}{2}}$ 9. $9^{-\frac{3}{2}}$ 10. $16^{-\frac{3}{4}}$ 11. $(-32)^{-\frac{1}{5}}$ 12. $(-64)^{-\frac{2}{3}}$

13. $(5^0)^{\frac{2}{3}}$ 14. $(7^{\frac{1}{3}})^0$ 15. $64^{-\frac{5}{6}}$ 16. $81^{-\frac{3}{4}}$ 17. $\left(\frac{1}{27}\right)^{\frac{1}{3}}$ 18. $\left(\frac{4}{9}\right)^{\frac{1}{2}}$

19. $\left(\frac{8}{27}\right)^{\frac{1}{3}}$ 20. $\left(\frac{9}{16}\right)^{-\frac{1}{2}}$ 21. $\left(\frac{4}{25}\right)^{-\frac{1}{2}}$ 22. $\left(-\frac{1}{125}\right)^{-\frac{2}{3}}$ 23. $(16)^{\frac{3}{4}}$ 24. $(121^0)^{\frac{1}{2}}$

25. $5^{-1} \cdot 5^{-3}$ 26. $5^{-2} \cdot 6^2$ 27. $4^{-3} + 8^{-2}$ 28. $6^0 + 6^{-1}$

29. $(3^0 + 3)^{\frac{1}{2}}$ 30. $-9^0 + 9^{\frac{1}{2}}$ 31. $(\sqrt[3]{216})^2$ 32. $81^{\frac{1}{2}} - 81^{-\frac{1}{2}}$

33. $(8^{243} + 121^{11})^0$ 34. $(15^0 + 15)^{\frac{1}{2}}$ 35. $(3^{-1} + 3^{-2})^{-1}$ 36. $(16^{\frac{1}{2}} + 16^{\frac{1}{4}})^2$

37. $\frac{1}{4^{-2}} + \frac{1}{4^{-1}}$ 38. $\frac{64}{64^{\frac{2}{3}}}$ 39. $\frac{9^{\frac{1}{2}}}{27^{-\frac{1}{3}}}$ 40. $\frac{25^{\frac{3}{4}}}{25^{\frac{1}{4}}}$

Evaluate each expression if $x = 9$.

41. $x^0 + x^{-1}$ 42. $4x^0 + (2x)^0$ 43. $x^{-1} + x^{\frac{1}{2}}$ 44. $3x^{-2} + (3x)^0$

45. $-x^0 - x^{\frac{1}{2}}$ 46. $-x^{-\frac{1}{2}} + 2x^{-1}$ 47. $x^{\frac{3}{2}} + (x^{\frac{3}{5}})^0$ 48. $(3x)^{\frac{4}{3}} - 6x^0$

Simplify and express each with positive exponents.

49. $\frac{5}{y^{-3}}$ 50. $\frac{x^{-8}}{x^{-2}}$ 51. $2x^{-1}y^{-2}y$ 52. $(3x^{-2}y)(4x^5y^{-4})$

53. $(2x^3y)^{-2}$ 54. $(3a^{-1}b)^{-3}$ 55. $(16x^{-2})^{\frac{1}{2}}$ 56. $(8y^6x^{-3})^{\frac{1}{3}}$

57. $(4x^{-1})(2x)^{-2}$ 58. $\frac{x^2}{x^{-2}y^{-1}}$ 59. $\left(\frac{2x}{3y}\right)^{-2}$ 60. $\frac{18x^{-2}y}{9xy^{-3}}$

61. $(27x^{-3}y^{-9})^{-\frac{1}{3}}$ 62. $\sqrt[4]{r^8s^{12}}$ 63. $\sqrt[3]{27a^{-3}b}$ 64. $\sqrt[7]{-x^{14}y^{28}}$

Evaluate each of the following if $f(x) = x^{-\frac{2}{3}} + x^{-3}$.

65. $f(1)$ 66. $f(-8)$ 67. $f(27)$ 68. $f(1000)$ 69. $f(0.001)$

Challenge

1. Ted claims that, for a given number b such that $b > 0$ and $b \neq 1$, b^x increases as x increases. Do you agree? Explain.
2. Solve for x if $x^{\frac{1}{2}} - 10x^{\frac{1}{4}} + 24 = 0$.

Mathematical Excursions

Scientific Notation

Scientific notation is often used for very large or very small numbers. Many of these numbers can be found in science. For example, the mass of the earth can be written as 6×10^{24} kg. The mass of an electron is written as 9.1083×10^{-31} kg. The distance from the sun to earth is 9.30×10^{7} miles. The following gives a definition for writing great or small numbers in scientific notation.

Definition of Scientific Notation

A number is expressed in scientific notation when it is in the following form.
$a \times 10^n$ where $1 \leq a < 10$ and n is an integer

Example **Express 72,500 in scientific notation.**

$$72500 = 7.25 \times 10^4$$

4 3 2 1

Example **Express 0.00236 in scientific notation.**

$$0.00236 = 2.36 \times 10^{-3}$$

1 2 3

Exercises **Express each of the following in scientific notation.**

1. 68,430
2. 143,000
3. 628
4. 0.00235
5. 0.0482
6. 206.5
7. 384.29
8. 4,000,000
9. 0.00000003
10. 4,030,000
11. 0.000000408
12. 0.0000638

Express each of the following in decimal notation.

13. 2.86×10^4
14. 3.82×10^5
15. 9.8×10^6
16. 3.21×10^{-2}
17. 5.92×10^{-4}
18. 1.06×10^{-8}

Write each expression first in simplest scientific notation and then in decimal notation.

19. $(1.3 \times 10^3)(2 \times 10^2)$
20. $(2.1 \times 10^4)(1.3 \times 10^{-2})$
21. $\dfrac{8 \times 10^{-1}}{16 \times 10^{-2}}$
22. $\dfrac{4.8 \times 10^5}{6 \times 10^2}$
23. $\dfrac{(8 \times 10^{-5})^2}{4 \times 10^{-6}}$
24. $\dfrac{(1.5 \times 10^{-4})^2}{3 \times 10^3}$

5.2 Exponential Functions

Number of hours x	Number of organisms $f(x)$
0	1
1	2
2	4
3	8
4	16
5	32
.	.
.	.
.	.
x	f(x)

A biologist prepares a culture containing one single-cell organism. At the end of one hour, the organism has divided into two organisms. At the end of the second hour, each of these organisms has divided, making a total of four organisms. At the end of the third hour, each of these divides, and so on. These results are shown in the chart.

Notice that this situation defines a function f, where $f(x)$ is the number of organisms at the end of x hours. The rule for the function f can be written in the following ways.

$$f{:}x \rightarrow 2^x \quad \text{or} \quad f(x) = 2^x \quad \text{or} \quad y = 2^x$$

The function f is an example of an exponential function.

Definition of an Exponential Function

A function f, where $f(x) = b^x$ and b is a positive real number other than 1, is called an **exponential function with base b.**

To obtain an accurate graph of $y = 2^x$, the table of values in the chart above must be extended. Then several of the ordered pairs are plotted. A partial graph of the function appears below.

The graph of <u>*any*</u> *exponential function passes through the point (0, 1) since $b^0 = 1$ for any positive real number b.*

x	$y = 2^x$
−3	$\frac{1}{8}$ or 0.125
−2	$\frac{1}{4}$ or 0.25
−1.5	0.4
−1	0.5
0	1
0.5	1.4
1	2
1.5	2.8
2	4
3	8

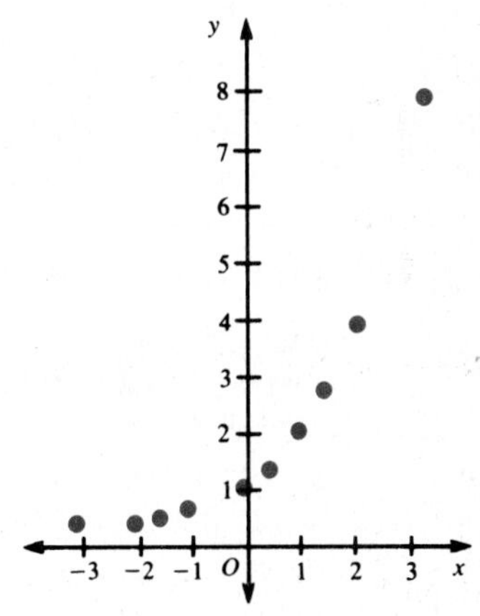

The graph suggests that the points should be connected to show the complete graph of $y = 2^x$. The smooth curve that is formed is known as an **exponential curve.**

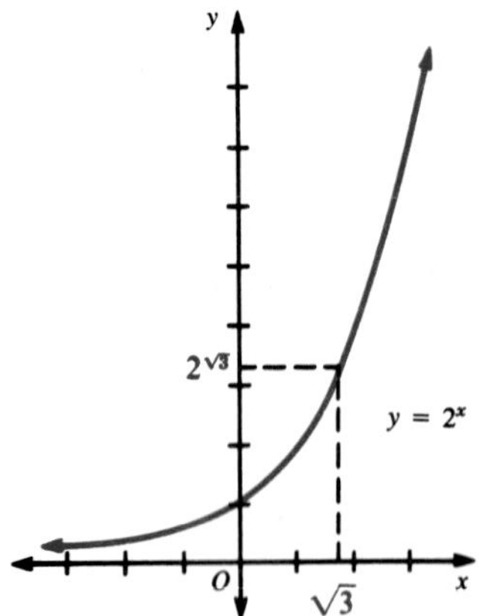

By connecting these points, we assume that numbers such as $2^{\sqrt{3}}$ and $2^{\sqrt{7}}$ exist. In fact, mathematicians have defined such numbers so that they can be found on the graph at the points anticipated. The graph shows the point $(\sqrt{3}, 2^{\sqrt{3}})$.

Note that although the graph of $y = 2^x$ gets closer and closer to the x-axis, as x becomes increasingly negative, it will never intersect the x-axis. Why?

A table of values for $y = 8^x$ is shown. Notice that the graph of the function has the same basic shape as the graph of $y = 2^x$. In fact, the graph of any exponential function defined by $y = b^x$, where $b > 1$, has this shape.

If $b = 1$, the graph of $y = b^x$ is the horizontal line $y = 1$, which is *not* an exponential curve.

What is the shape of the graph of $y = b^x$, when $0 < b < 1$? Example 1 provides an answer.

x	y
-1	$\frac{1}{8}$
$-\frac{2}{3}$	$\frac{1}{4}$
$-\frac{1}{3}$	$\frac{1}{2}$
0	1
$\frac{1}{3}$	2
$\frac{2}{3}$	4
1	8

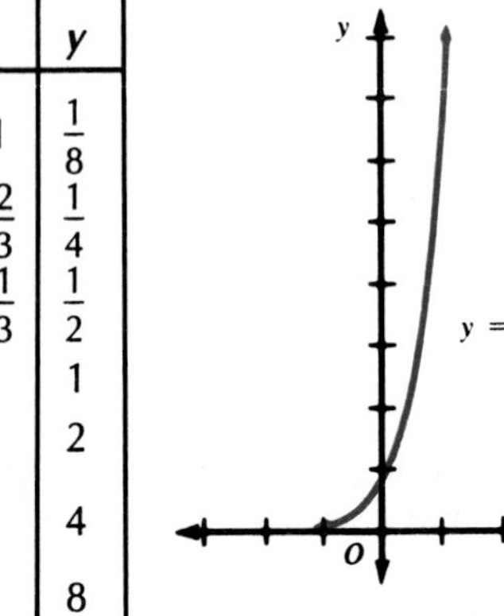

Examples

1 Sketch the graph of $y = \left(\frac{1}{2}\right)^x$.

First observe that

$$y = \left(\frac{1}{2}\right)^x = (2^{-1})^x = 2^{-x}$$

Now, compare $y = 2^x$ and $y = 2^{-x}$. For any ordered pair (x, y) satisfying the equation $y = 2^x$, the ordered pair $(-x, y)$ satisfies the equation $y = 2^{-x}$. This means the graph of $y = 2^{-x}$ is the image of the graph of $y = 2^x$ under a line reflection in the y-axis.

The graphs are symmetric about the y-axis.

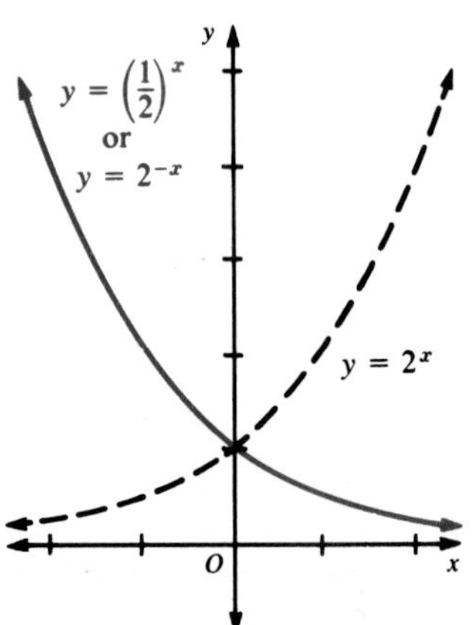

The graph can also be drawn by using ordered pairs to plot points to be connected by a smooth unbroken curve.

x	-3	-2	-1	0	1	2	3
y	8	4	2	1	$\frac{1}{2}$	$\frac{1}{4}$	$\frac{1}{8}$

2 State the domain and range of the function graphed in Example 1.

The domain is the set of real numbers.

The range is the set of positive real numbers.

The preceding examples suggest the following conclusions.

1. The graphs of $y = b^x$ and $y = b^{-x}$, where $b > 0$ and $b \neq 1$, are symmetrical with respect to the y-axis.
2. For any exponential function, the domain is the set of real numbers and the range is the set of positive real numbers.

Exercises

Exploratory **In exercises 1–4, state which of the following defines an exponential function.**

1. $f:x \rightarrow x^2$ 2. $g:x \rightarrow \left(\frac{1}{3}\right)^x$ 3. $y = 5^x$ 4. $y = x^{-1}$
5. Explain the relationship between the graph of $y = 2^x$ and $y = \left(\frac{1}{2}\right)^x$.
6. How can the graph of $y = 4^{-x}$ be obtained from the graph of $y = 4^x$?
7. In what quadrants does the graph of an exponential function lie?
8. State the domain and range of an exponential function.
9. Are exponential functions one-to-one? Explain.
10. At what point do the graphs of $y = 6^x$ and $y = 6^{-x}$ intersect?
11. How can the graph of $y = -2^x$ be obtained from the graph of $y = 2^x$?
12. Explain how to use the graph of $y = 3^x$ to approximate the value of $3^{\sqrt{2}}$.

Written **Evaluate each of the following if $g(x) = 16^x$.**

1. $g(-1)$ 2. $g(0.5)$ 3. $g(0.75)$ 4. $g(1.5)$ 5. $g(-1.25)$

Sketch the graph of each of the following functions.

6. $f(x) = 3^x$ 7. $g(x) = 4^x$ 8. $h(x) = 10^x$ 9. $H(x) = \left(\frac{1}{3}\right)^x$
10. $F(x) = \left(\frac{1}{4}\right)^x$ 11. $G(x) = \left(\frac{1}{10}\right)^x$ 12. $t(x) = \left(\frac{3}{2}\right)^x$ 13. $m(x) = -3^x$
14. $k(x) = 5^{-x}$ 15. $M(x) = 6^{-x}$ 16. $s(x) = \left(\frac{5}{2}\right)^x$ 17. $q(x) = 3^x + 1$

Sketch the general shape of the graph of $y = b^x$ for each condition given in exercises 18–20.

18. $0 < b < 1$ **19.** $b > 1$ **20.** $b = 1$

21. Try to graph $y = (-4)^x$. Explain any difficulties you encounter.

22. For what values of b is $f(x) = b^x$ an exponential function?

Challenge **Sketch the graphs of f and g on the same axes. Then describe the relationship between the two graphs.**

1. $f(x) = 2^x,\ g(x) = -(2^{-x})$ **2.** $f(x) = 2^x,\ g(x) = 2^x + 3$

3. $f(x) = 2^{-x},\ g(x) = 2^{-x} - 2$ **4.** $f(x) = 3(2^x),\ g(x) = 3(2^{-x})$

Use the graph of $y = 2^x$ to evaluate x or y to the nearest tenth. Then use a calculator to check your results.

5. $y = 2^{4.5}$ **6.** $y = 2^{3.8}$ **7.** $y = 2^{-2.6}$

8. $3.7 = 2^x$ **9.** $4.1 = 2^x$ **10.** $\sqrt{7} = 2^x$

11. $12 = 2^x$ **12.** $y = 2^{2.9}$ **13.** $48 = 2^x$

An amoeba divides once every hour. Beginning with a single amoeba, how many amoebae will there be after the given amount of time?

14. 3 hours **15.** 7 hours **16.** 10 hours **17.** 1 day

Mathematical Excursions

Irrational Exponents

Previously, it has been stated that the five Exponential Laws on page 168 hold true for integral and rational values of n and m. These laws also apply to irrational exponents. For example

$$x^{\sqrt{2}} \cdot x^{\sqrt{3}} = x^{\sqrt{2}+\sqrt{3}}.$$

Exercises **Simplify each expression using the Exponential Laws.**

1. $3^{\sqrt{2}} \cdot 3^{\sqrt{8}}$ **2.** $2^{\sqrt{2}} \cdot 2^{3\sqrt{2}}$ **3.** $9^{\sqrt{2}} \cdot 3^{\sqrt{2}}$ **4.** $4^{\pi} \cdot 2^{\frac{\sqrt{5}}{\pi}}$

5. $(5^{1+\sqrt{2}})^{1-\sqrt{2}}$ **6.** $\dfrac{4^{1+\sqrt{2}}}{2^{2\sqrt{2}}}$ **7.** $\left(\dfrac{2^{1+\sqrt{3}}}{2^{1-\sqrt{3}}}\right)^{\sqrt{3}}$

8. $(m^{\sqrt{2}} + n^{\sqrt{2}})^2$ **9.** $(x^{\sqrt{3}} + y^{\sqrt{2}})^2$ **10.** $(x^{\sqrt{2}} - y^{\sqrt{2}})(x^{\sqrt{2}} + y^{\sqrt{2}})$

5.3 Equations and Exponents

Some equations contain rational exponents. To solve such equations the method used is similar to the one used for solving radical equations in Chapter 2.

Examples

1 **Solve $3x^{-\frac{1}{2}} = 6$.**

$3x^{-\frac{1}{2}} = 6$

$x^{-\frac{1}{2}} = 2$ *Divide by 3.*

$(x^{-\frac{1}{2}})^{-2} = (2)^{-2}$ *Raise both sides to the same power.*

$x^1 = \frac{1}{2^2}$ *Law 2, Definition of negative exponents*

$x = \frac{1}{4}$

Check: $3x^{-\frac{1}{2}} = 6$

$3\left(\frac{1}{4}\right)^{-\frac{1}{2}} \stackrel{?}{=} 6$

$3(2) \stackrel{?}{=} 6$

$6 = 6$

2 **Solve $x^{\frac{3}{2}} - 1 = 7$.**

$x^{\frac{3}{2}} - 1 = 7$

$x^{\frac{3}{2}} = 8$ *Additive Property*

$(x^{\frac{3}{2}})^{\frac{2}{3}} = (8)^{\frac{2}{3}}$ *Raise both sides to same power.*

$x = (\sqrt[3]{8})^2$ *Definition of rational exponent.*

$x = 4$

The check is left for you.

In the preceding section, we saw that exponential functions are one-to-one. This fact can be restated as the Property of Equality.

Property of Equality for Exponential Functions

If $b > 0$ and $b \neq 1$, then $b^{x_1} = b^{x_2}$ if and only if $x_1 = x_2$.

This property can be used to solve certain *exponential equations*. In exponential equations the variable appears in the exponent. The following examples illustrate a method for solving some of these exponential equations. In a later section, another more general method will be presented.

Examples

Find x in the following equations.

3

$$3^{x-2} = 27$$
$$3^{x-2} = 27$$
$$3^{x-2} = 3^3$$ ⟵ *Express each side as a power of the same base.*
$$x - 2 = 3$$ ⟵ *Property of Equality*
$$x = 5$$

4

$$8^{2x} = 32^{x-4}$$
$$8^{2x} = 32^{x-4}$$
$$(2^3)^{2x} = (2^5)^{x-4}$$ ⟵ *Express each side as a power of the same base.*
$$2^{6x} = 2^{5x-20}$$
$$6x = 5x - 20$$ ⟵ *Property of Equality*
$$x = -20$$

The checks are left for you.

Exercises

Exploratory **For what value of m is each expression equal to x?**

1. $(x^{\frac{1}{2}})^m$
2. $(x^{\frac{2}{5}})^m$
3. $(x^{-3})^m$
4. $(x^{-\frac{3}{4}})^m$

Express each of the numbers as a power with the least possible positive integral base.

5. 49
6. 36
7. 27
8. 32
9. 81
10. 10,000
11. 0.01
12. $\frac{1}{25}$
13. $\frac{1}{8}$
14. $\frac{1}{64}$

Written **Solve and check.**

1. $x^{\frac{1}{3}} = 2$
2. $x^{\frac{1}{2}} = 7$
3. $x^{\frac{3}{2}} = 8$
4. $y^{-2} = 16$
5. $t^{-\frac{1}{2}} = 4$
6. $y^{\frac{3}{2}} + 1 = 9$
7. $x^{\frac{4}{3}} + 2 = 18$
8. $y^{\frac{4}{3}} - 5 = 11$
9. $x^{-\frac{3}{4}} - 1 = 7$
10. $4x^{\frac{4}{3}} = 324$
11. $2x^{-\frac{1}{4}} + 5 = 6$
12. $(x + 1)^{\frac{2}{3}} = 25$
13. $2^{x+3} = 2^4$
14. $5^{2x+1} = 5^{x+4}$
15. $3^{2x} = 3$
16. $7^{3x+6} = 7^{x+2}$
17. $2^x = 2^{3x-2}$
18. $3^{2x} = 3^{x-5}$
19. $5^{x^2} = 5^{15-2x}$
20. $5^{x+1} = 25$
21. $36^x = 6^{x+1}$
22. $2^{3x+2} = 8^{2x}$
23. $9^{2x} = 3^{2x-2}$
24. $5^{4x} = 25^{x+1}$
25. $8^{x-3} = 2^{x-3}$
26. $81^x = 3^{x+1}$
27. $25^{x^2+2} = 125^{x+1}$
28. $16^x = 2^{3x-1}$
29. $32^x = 4^{x+4}$
30. $27^x = 9^{x-1}$
31. $9^{2x} = 27^{3x-4}$
32. $8^{x+4} = 16^{3x}$
33. $\left(\frac{1}{2}\right)^{x-2} = 4^{5x}$
34. $\left(\frac{1}{3}\right)^{4x-1} = 3^{x-1}$
35. $8^{2x-2} = 16^{6x}$
36. $\left(\frac{1}{3}\right)^{2-x} = 27^{3+x}$

Challenge **Solve.** *Hint: Try to express each equation in quadratic form.*

1. $x^{\frac{1}{2}} - 6x^{\frac{1}{4}} + 8 = 0$
2. $4x^{\frac{4}{3}} - 25x^{\frac{2}{3}} + 36 = 0$

Mixed Review

Exercises **Choose the best answer.**

1. Which of the following is a rational number?
   a. $\sqrt{\frac{1}{3}}$ **b.** π **c.** $\sqrt{\frac{16}{25}}$ **d.** $\sqrt{8}$
2. Which is the best description of the graph of the solution set of $-5 < x < 3$?
   a. numbers to the right of -5 not including -5
   b. numbers between -5 and 3 including 3
   c. numbers to the right of 3 not including 3 and to the left of and including -5
   d. none of these
3. Which number is not in the range of the relation $\{(1, 4), (1, 2), (2, 5)\}$?
   a. 1 **b.** 4 **c.** 2 **d.** 5
4. The multiplicative inverse of $1 + i$ is which of the following?
   a. $-1 - i$ **b.** 1 **c.** 0 **d.** $\frac{1}{2} - \frac{1}{2}i$
5. The conjugate of $-4 + 5i$ is which of the following?
   a. $4 + 5i$ **b.** $4 - 5i$ **c.** $-4 - 5i$ **d.** none of these
6. What is the image of $(-4, 2)$ under $T_{-1,2}$?
   a. $(-5, -4)$ **b.** $(5, -4)$ **c.** $(-5, 4)$ **d.** $(4, 4)$
7. Which word has line symmetry?
   a. ODD **b.** EVEN **c.** COMPOSITE **d.** PRIME
8. Which of the following does not represent or define a function?
   a. $y = 3x$ **b.** $y = |x|$ **c.** $\{(1, 1), (2, 1)\}$ **d.** $\{(1, 1), (1, 2)\}$
9. If P' is the image of P under a reflection over line ℓ then which of the following is true?
   a. ℓ is the perpendicular bisector of $\overline{PP'}$ **b.** $\overline{PP'}$ is parallel to ℓ.
   c. P and P' must both be on line ℓ **d.** none of these
10. If $a = -2$ and $b = -3$ the value of $a^2 - b$ is which of the following?
    a. 1 **b.** -1 **c.** -7 **d.** 7
11. The relation whose graph is a parabola symmetric to the y-axis is most likely which of the following?
    a. a one-to-one function **b.** not a function
    c. a quadratic function **d.** none of these
12. Which number *cannot* be in the domain of the relation $\left\{(x, y): y = \frac{3}{x - 5}\right\}$?
    a. 0 **b.** 1 **c.** 5 **d.** π
13. Describe the roots of $x^2 + x - 1 = 0$.
    a. real, rational, and unequal **c.** real, irrational, and unequal
    b. real, rational, and equal **d.** imaginary
14. Which regular polygon has 60° rotational symmetry?
    a. regular pentagon **b.** regular hexagon **c.** square **d.** equilateral triangle
15. If $f(x) = (-x)^{\frac{3}{2}} - (x + 1)^{\frac{2}{3}}$, what is the value of $f(-9)$?
    a. 31 **b.** -31 **c.** 23 **d.** none of these

5.4 Inverses of Exponential Functions

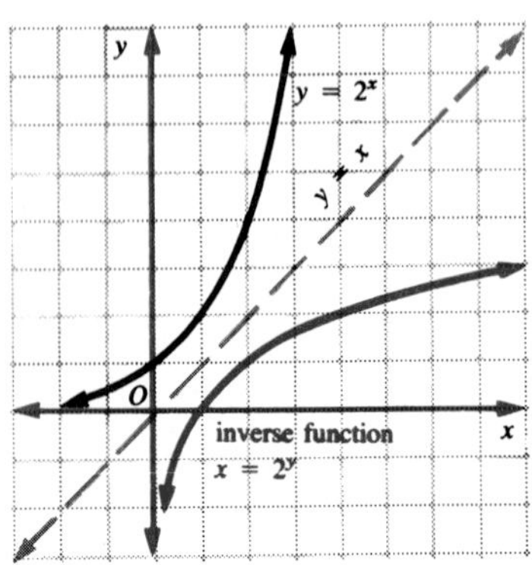

Because exponential functions are one-to-one, their inverses are also functions. In fact, the graph of the inverse function is obtained by a reflection over the line $y = x$.

Shown at the left is the graph of $y = 2^x$ and its inverse function. To obtain an equation for this inverse, the x and y are interchanged in the equation $y = 2^x$. The result is $x = 2^y$. Before this equation can be solved for y, some new terminology must be introduced.

The equation $x = 2^y$ can be stated in words as follows.

y is the exponent of base 2 that produces x.

Remember: A logarithm is an exponent.

A common way of expressing this relationship is

y is the logarithm of x to the base 2.

More compactly, this is written in symbolic form as

$$y = \log_2 x$$

and is read "y equals log x to the base 2."

Therefore, the equations $x = 2^y$ and $y = \log_2 x$ are equivalent.

Definition of $y = \log_b x$

If $b > 0$ and $b \neq 1$, then $y = \log_b x$ if and only if $x = b^y$.

The graph of any logarithmic function passes through the point (1, 0) since a logarithmic function is the inverse of an exponential function.

The logarithmic function defined by $y = \log_b x$ is the inverse function of the exponential function defined by $y = b^x$, where $b > 0$ and $b \neq 1$. The domain of $y = \log_b x$ is the set of positive real numbers. The range is the set of real numbers.

The following chart shows some equivalent exponential and logarithmic equations.

Exponential Form	Logarithmic Form
$x = 3^y$	$y = \log_3 x$
$x = 10^y$	$y = \log_{10} x$
$x = \left(\frac{1}{2}\right)^y$	$y = \log_{\frac{1}{2}} x$

Examples

Evaluate the following logarithms.

1 $\log_3 9$

$$\begin{aligned} \text{Let } \log_3 9 &= y \\ \text{Then } 3^y &= 9 \\ 3^y &= 3^2 \\ y &= 2 \\ \text{Thus, } \log_3 9 &= 2 \end{aligned}$$

2 $\log_4 8$

$$\begin{aligned} \text{Let } \log_4 8 &= y \\ \text{Then } 4^y &= 8 \\ 2^{2y} &= 2^3 \\ 2y &= 3 \\ y &= \frac{3}{2} \\ \text{Thus, } \log_4 8 &= \frac{3}{2} \end{aligned}$$

3 **Find x if $\log_9 (x^2 - 1) = \frac{1}{2}$.**

$$\begin{aligned} \log_9 (x^2 - 1) &= \frac{1}{2} \\ x^2 - 1 &= 9^{\frac{1}{2}} && \textit{Convert to exponential form.} \\ x^2 - 1 &= 3 \\ x^2 - 4 &= 0 && \textit{Write in standard form.} \\ (x - 2)(x + 2) &= 0 && \textit{Factor.} \\ x = 2 \text{ or } x &= -2 && \textit{The check is left for you.} \end{aligned}$$

4 **Sketch the graph of $y = \log_{10} x$**

The logarithmic function defined by $y = \log_{10} x$ is the inverse of the exponential function defined by $y = 10^x$. First, graph $y = 10^x$. Then reflect in the line $y = x$ to obtain the graph of $y = \log_{10} x$.

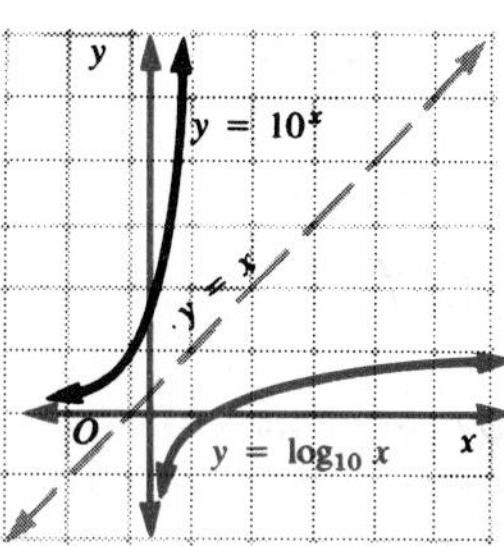

Like exponential functions, logarithmic functions are one-to-one functions. Thus, $\log_b x_1 = \log_b x_2$ if and only if $x_1 = x_2$. This property is used in the next example.

Example

5 **Solve $\log_2 (x^2 - 3) = \log_2 (3x - 5)$. Check your solution.**

$$\log_2 (x^2 - 3) = \log_2 (3x - 5)$$
$$x^2 - 3 = 3x - 5$$
$$x^2 - 3x + 2 = 0$$
$$(x - 2)(x - 1) = 0$$
$$x = 2 \quad \text{or} \quad x = 1$$

Check for x = 1:

$$\log_2 (1^2 - 3) \stackrel{?}{=} \log_2 (3 \cdot 1 - 5)$$
$$\log_2 (-2) \stackrel{?}{=} \log_2 (-2)$$

Since $\log_2 (-2)$ does not exist, we reject $x = 1$.

Check for x = 2:

$$\log_2 (2^2 - 3) \stackrel{?}{=} \log_2 (3 \cdot 2 - 5)$$
$$\log_2 1 \stackrel{?}{=} \log_2 1$$
$$0 = 0$$

From the checks you can see the only solution is $x = 2$.

Exercises

Exploratory **Express each equation in logarithmic form.**

1. $2^3 = 8$ **2.** $5^2 = 25$ **3.** $3^3 = 27$ **4.** $7^2 = 49$

5. $4^{-\frac{1}{2}} = \frac{1}{2}$ **6.** $10^{-2} = \frac{1}{100}$ **7.** $17^0 = 1$ **8.** $100^{-\frac{1}{2}} = \frac{1}{10}$

9. $(\sqrt{5})^2 = 5$ **10.** $8^{-\frac{1}{3}} = \frac{1}{2}$ **11.** $9^{\frac{3}{2}} = 27$ **12.** $\left(\frac{1}{81}\right)^{-\frac{1}{2}} = 9$

Express each equation in exponential form.

13. $\log_3 9 = 2$ **14.** $\log_2 \frac{1}{32} = -5$ **15.** $\log_{27} 9 = \frac{2}{3}$ **16.** $\log_3 243 = 5$

17. $\log_{\frac{1}{2}} 16 = -4$ **18.** $\log_{\frac{1}{9}} 81 = -2$ **19.** $\log_3 \frac{1}{9} = -2$ **20.** $\log_{\sqrt{2}} 2 = 2$

21. $\log_{10} \frac{1}{10} = -1$ **22.** $\log_8 4 = \frac{2}{3}$ **23.** $\log_5 \frac{1}{25} = -2$ **24.** $\log_{\frac{1}{3}} 81 = -4$

Name the quadrants in which the graph of each equation lies.

25. $y = 4^x$ **26.** $y = \log_4 x$ **27.** $y = \log_{\frac{1}{2}} x$ **28.** $y = -2^x$

State the domain and range of the function defined by each equation in exercises 29–32.

29. $y = 3^x$ **30.** $y = \log_{10} x$ **31.** $y = \log_{\frac{1}{4}} x$ **32.** $y = \left(\frac{1}{2}\right)^x$

33. Are the graphs of $x = 11^y$ and $y = \log_{11} x$ the same? Explain.

34. How are the graphs of $y = 4^x$ and $y = \log_4 x$ related? Why?

Written **Evaluate.**

1. $\log_{10} 1000$ **2.** $\log_7 49$ **3.** $\log_{10} 1$ **4.** $\log_9 3$ **5.** $\log_2 4$

6. $\log_4 64$ **7.** $\log_{10} 0.01$ **8.** $\log_8 2$ **9.** $\log_{32} 2$ **10.** $\log_4 2$

11. $\log_{\frac{1}{2}} 32$ **12.** $\log_9 27$ **13.** $\log_{\frac{1}{2}} 8$ **14.** $\log_7 \frac{1}{343}$ **15.** $\log_8 16$

Tell whether each statement is *true* or *false*. Justify your answer.

16. $\log_2 4 + \log_4 16 = 4$

17. $\log_5 125 + \log_5 1 = \log_5 125$

18. $\log_2 8 + \log_2 4 = \log_2 12$

19. $\log_2 8 + \log_2 4 = \log_2 32$

20. $\log_2 (4 + 4) = \log_2 4 + \log_2 4$

21. $\log_3 27 = 3 \log_3 3$

22. $\log_3 27 + \log_3 3 = \log_3 81$

23. $\log_5 25 \cdot \log_{25} 5 = 1$

Sketch the graph of each function and its inverse on the same set of axes. Then use log notation to write an equation for the inverse of the function.

24. $f(x) = 2^x$ **25.** $f(x) = 3^x$ **26.** $f(x) = 4^x$ **27.** $f(x) = 10^x$

28. $f(x) = \left(\frac{1}{2}\right)^x$ **29.** $f(x) = \left(\frac{1}{8}\right)^x$ **30.** $f(x) = \left(\frac{1}{4}\right)^x$ **31.** $f(x) = \left(\frac{3}{2}\right)^x$

Solve each equation.

32. $\log_2 x = 4$ **33.** $\log_6 x = 2$ **34.** $\log_4 x = -2$ **35.** $\log_9 x = 0$

36. $\log_{\frac{1}{2}} x = 3$ **37.** $\log_3 x = -3$ **38.** $\log_4 64 = y$ **39.** $\log_{10} 100 = y$

40. $\log_{\frac{1}{2}} 16 = y$ **41.** $\log_b 36 = 2$ **42.** $\log_b 49 = 2$ **43.** $\log_b 125 = 3$

44. $\log_b 81 = 4$ **45.** $\log_b 18 = 1$ **46.** $\log_b 0.01 = 2$ **47.** $\log_b 36 = -2$

48. $\log_b 0.1 = -\frac{1}{2}$ **49.** $\log_b \frac{1}{27} = -3$ **50.** $\log_4 x = \frac{3}{2}$ **51.** $\log_{\frac{1}{3}} 27 = y$

52. $\log_4 x = -\frac{1}{2}$ **53.** $\log_{\sqrt{2}} x = 6$ **54.** $\log_b \sqrt{5} = \frac{1}{4}$ **55.** $\log_b \sqrt[3]{7} = \frac{1}{3}$

56. $\log_2 (2x - 2) = 3$

57. $\log_3 (5x + 2) = 3$

58. $\log_{10} (2x - 8) = 1$

59. $\log_4 (x + 3) = \frac{1}{2}$

Determine whether each statement is *always, sometimes,* or *never* true.

60. $\log_{10} x$ is positive.

61. $\log_2 x$ is negative.

62. $\log_2 x < \log_3 x$

63. $\log_2 x < 2^x$

Solve and check.

64. $\log_3 (2x + 1) = \log_3 (3x - 6)$

65. $\log_{10} (4 + y) = \log_{10} 2y$

66. $\log_{10} (2m + 1) = \log_{10} (m + 5)$

67. $\log_{10} (x^2 + 36) = \log_{10} 100$

68. $\log_{10} (x^2 + 3x) = 1$

69. $\log_6 (x^2 + x) = 1$

70. $\log_{10}(x^2 + 9x) = 1$

71. $\log_2(x^2 - 6x) = 4$

Challenge **Evaluate.**

1. $5^{\log_5 3}$

2. $9^{\log_9 2}$

3. $12^{\log_{12} 7}$

4. $b^{\log_b 10}$

Solve and check.

5. $\log_{10}[\log_3(\log_5 x)] = 0$

6. $7^{\log_7 x^2} = x + 20$

7. $\log_4[\log_2(\log_3 x)] = \frac{1}{2}$

8. $6^{\log_6 x^2} = x + 30$

Mathematical Excursions

Richter Scale

A logarithmic scale called the **Richter scale** is used to measure the strength of an earthquake. Each increase of one on the Richter scale corresponds to a ten-times increase in intensity. In other words, an earthquake that registers 8 on the Richter scale is ten times as intense as an earthquake that registers 7. An earthquake that registers 9 is ten times as intense as the one registering 8, and one hundred times as intense as the one registering 7.

The table below gives the effects of earthquakes of various intensities.

Richter Number	Intensity	Effect
1	10^1	only detectable by seismograph
2	10^2	hanging lamps sway
3	10^3	can be felt
4	10^4	glass breaks, buildings shake
5	10^5	furniture collapses
6	10^6	wooden houses damaged
7	10^7	buildings collapse
8	10^8	catastrophic damage

Exercises

1. It is believed that the 1906 San Francisco earthquake would have registered 8.3 on the Richter scale. How much stronger was this earthquake than one registering 8.0?

2. On October 17,1989, an earthquake registering 7.1 on the Richter scale hit San Francisco. How much weaker was this earthquake than the 1906 earthquake?

5.5 Working with Logarithms

Since logarithms are exponents, there are three laws of logarithms which are based on the laws of exponents. The first law is illustrated below.

$$\begin{aligned} \log_2 (8 \cdot 4) &= \log_2 (2^3 \cdot 2^2) && \textit{8 = 2^3, 4 = 2^2} \\ &= \log_2 (2^{3+2}) && \textit{Exponential Law 1} \\ &= 3 + 2 && \textit{Definition of log} \\ &= \log_2 8 + \log_2 4 && \textit{Log}_2\ 8 = 3,\ \textit{Log}_2\ 4 = 2 \end{aligned}$$

Laws of Logarithms

Assume m and *n* are positive, $b > 0$ and $b \neq 1$.

Law 1 $\log_b (m \cdot n) = \log_b m + \log_b n$

Law 2 $\log_b \frac{m}{n} = \log_b m - \log_b n$

Law 3 $\log_b m^p = p \cdot \log_b m$

The proof of Law 1 is given below. You will be asked to prove the other laws in the exercises.

Proof of Law 1

Let $\log_b m = u$ and $\log_b n = v$.
Then, $b^u = m$ and $b^v = n$.
Observe that $mn = b^u \cdot b^v$ or b^{u+v}.
The logarithmic form of $mn = b^{u+v}$ is

$$\log_b mn = u + v.$$

Substituting the values of u and v, we have

$$\log_b mn = \log_b m + \log_b n.$$

Examples

1 **Evaluate $\log_6 9 + \log_6 4$.**

$$\begin{aligned} \log_6 9 + \log_6 4 &= \log_6 (9 \cdot 4) && \textit{Law 1} \\ &= \log_6 (36) \\ &= \log_6 (6^2) \\ &= 2 \end{aligned}$$

If $\log_2 7 = 2.807$ and $\log_2 5 = 2.322$, evaluate the following logs.

2 $\log_2 1.4$

$$\begin{aligned}\log_2 1.4 &= \log_2 \frac{7}{5}\\ &= \log_2 7 - \log_2 5 \quad \textit{Law 2}\\ &= 2.807 - 2.322\\ &= 0.485\end{aligned}$$

3 $\log_2 \sqrt{5}$

$$\begin{aligned}\log_2 \sqrt{5} &= \log_2 5^{\frac{1}{2}}\\ &= \frac{1}{2}\log_2 5 \quad \textit{Law 3}\\ &= \frac{1}{2}(2.322)\\ &= 1.161\end{aligned}$$

4 **Solve $\log_5 x - \frac{1}{2}\log_5 16 = \log_5 25$.**

$$\begin{aligned}\log_5 x - \frac{1}{2}\log_5 16 &= \log_5 25\\ \log_5 x - \log_5 16^{\frac{1}{2}} &= \log_5 25 \quad \textit{Law 3}\\ \log_5 x - \log_5 4 &= \log_5 25\\ \log_5 \frac{x}{4} &= \log_5 25 \quad \textit{Law 2}\\ \frac{x}{4} &= 25 \quad \textit{Property of equality}\\ x &= 100\end{aligned}$$

5 **Express $\log_b \frac{x^3}{y}$ in terms of $\log_b x$ and $\log_b y$.**

$$\begin{aligned}\log_b \frac{x^3}{y} &= \log_b x^3 - \log_b y \quad \textit{Law 2}\\ &= 3\log_b x - \log_b y \quad \textit{Law 3}\end{aligned}$$

6 **Solve $\log_6 (x + 1) + \log_6 (x - 4) = 1$.**

$$\begin{aligned}\log_6 (x + 1) + \log_6 (x - 4) &= 1\\ \log_6 (x + 1)(x - 4) &= 1\\ (x + 1)(x - 4) &= 6^1\\ x^2 - 3x - 4 &= 6\\ x^2 - 3x - 10 &= 0\\ (x - 5)(x + 2) &= 0\\ x = 5 \text{ or } x &= -2\end{aligned}$$

Check for $x = 5$:

$$\begin{aligned}\log_6 (5 + 1) + \log_6 (5 - 4) &\stackrel{?}{=} 1\\ \log_6 6 + \log_6 1 &\stackrel{?}{=} 1\\ 1 + 0 &\stackrel{?}{=} 1\\ 1 &= 1\end{aligned}$$

Check for $x = -2$:

$$\begin{aligned}\log_6 (-2 + 1) + \log_6 (-2 - 4) &\stackrel{?}{=} 1\\ \log_6 (-1) + \log_6 (-6) &\stackrel{?}{=} 1\end{aligned}$$

Since the left side is undefined, we reject $x = -2$.

From the checks you can see the only solution is $x = 5$.

Exercises

Exploratory **Use the laws of logarithms to expand each expression.**

1. $\log_5 xy$ **2.** $\log_4 \frac{a}{b}$ **3.** $\log_3 x^2$ **4.** $\log_2 x^{\frac{1}{3}}$

5. $\log_5 abc$ **6.** $\log_7 m^5$ **7.** $\log_5 \sqrt[3]{x}$ **8.** $\log_3 m^2n$

9. $\log_b \frac{r^2}{s}$ **10.** $\log_{10} \sqrt[5]{x^2y}$ **11.** $\log_{10} \sqrt[4]{xy}$ **12.** $\log_{10} \frac{a}{bc}$

Use the laws of logarithms to rewrite each expression as a single logarithm.

13. $\log_4 a + \log_4 b$ **14.** $2 \log_5 x$ **15.** $\log_4 x - \log_4 y$

16. $\log_3 m - \log_3 n$ **17.** $3 \log_5 2x$ **18.** $2 \log_7 m + \log_7 n$

19. $\log_3 r + \log_3 s + \log_3 t$ **20.** $\log_5 a + \log_5 b - \log_5 c$

21. $3 \log_2 r - 2 \log_2 s$ **22.** $\log_2 a - \log_2 b - \log_2 c$

23. $\frac{1}{2}(\log_{10} x - \log_{10} y)$ **24.** $\frac{1}{3}(\log_{10} x + \log_{10} y - \log_{10} z)$

Written **If $\log_3 6 = 1.631$ and $\log_3 4 = 1.262$, evaluate each expression.**

1. $\log_3 36$ **2.** $\log_3 16$ **3.** $\log_3 24$

4. $\log_3 1.5$ **5.** $\log_3 \sqrt[4]{6}$ **6.** $\log_3 144$

Solve each equation.

7. $\log_2 3 + \log_2 7 = \log_2 x$ **8.** $\log_5 4 + \log_5 x = \log_5 36$

9. $\log_4 18 - \log_4 x = \log_4 6$ **10.** $\log_3 56 - \log_3 8 = \log_3 x$

11. $\log_7 x = 3 \log_7 2$ **12.** $\log_5 x = 2 \log_5 9$

13. $\log_2 r = \frac{1}{2} \log_2 81$ **14.** $\log_3 5 + \log_3 4 = \log_3 x$

15. $\log_2 7 + \log_2 9 = \log_2 x$ **16.** $\log_5 t - \log_5 4 = \log_5 3$

17. $\log_4 24 - \log_4 3 = \log_4 m$ **18.** $\log_7 4 + \log_7 x = \log_7 32$

19. $\log_5 2m + \log_5 3 = \log_5 30$ **20.** $5 \log_7 x - \log_7 8 = \log_7 4$

21. $\log_7 x = \frac{1}{2} \log_7 144 - \frac{1}{3} \log_7 8$ **22.** $\log_3 t = \frac{1}{3} \log_3 64 + \frac{1}{2} \log_3 121$

23. $4 \log_5 x - \log_5 4 = \log_5 4$ **24.** $2 \log_3 x + \log_3 \frac{1}{10} = \log_3 5 + \log_3 2$

25. $\log_5 (x + 3) - \log_5 x = \log_5 4$ **26.** $\frac{1}{2} \log_7 x = 4 \log_7 2 - \log_7 4$

27. $\log_4 (t + 3) + \log_4 (t - 3) = 2$ **28.** $\log_4 (m - 1) + \log_4 (m - 1) = 2$

29. $\log_3 (x - 4) + \log_3 (x + 4) = 2$ **30.** $\log_7 (2x - 3) - \log_7 (x + 1) = \log_7 2$

31. $\log_{10} (x + 21) + \log_{10} x = 2$ **32.** $\log_2 (9y + 5) - \log_2 (y^2 - 1) = 2$

Tell whether each statement is *true* or *false*. Justify your answer.

33. If $\log_{10} 123 = x$, then $1 < x < 2$.

34. $\log_b 35 + \log_b 4 = \log_b 420 - \log_b 3$

35. $\log_b (x + y) = \log_b x + \log_b y$

36. $2 \log_b 8 = 6 \log_b 2$

37. $\log_b \frac{m}{n} = \frac{\log_b m}{\log_b n}$

38. $\log_b x - \log_b y = \frac{\log_b x}{\log_b y}$

39. $\log_b x^{n-2} = n \log_b x - 2 \log_b x$

40. If $x_1 > x_2$, then $\log_b x_1 > \log_b x_2$.

Prove each of the following laws of logarithms.

41. Law 2

42. Law 3

Challenge

1. Give the domain and range of $f(x) = \log_2 (x - 1)$.
2. Graph $f(x) = \log_2 (x - 1)$ and $g(x) = \log_2 x$ on the same set of axes. What transformation maps the graph of g to the graph of f? Explain.
3. Graph $g(x) = \log_2 x$ and $h(x) = \log_2 x - 1$ on the same set of axes. Describe the relationship of the two graphs.
4. On the same set of axes, sketch graphs of $y = 2^x$, $y = 2^{x+3}$, and $y = 2^x + 3$.
5. Show that $\log_B M = \frac{\log_b M}{\log_b B}$.

Mathematical Excursions

The Chambered Nautilus

The chambered nautilus (NAHT ul us) is a member of the mollusk family. It moves about by using its tentacles to maneuver in the water. The shell of the nautilus takes on a spiral shape because of its growth pattern and the formation of small chambers of increasing size.

For each growth stage the nautilus forms a new chamber, leaving behind a smaller one. The nautilus lives in the newest formed chamber. The size of each chamber is 6% greater than the one before it. This can be expressed as a logarithmic function.

If s represents the size of the original chamber, then the size of any chamber formed thereafter can be calculated by the following formula.

$$S = s(1.06)^n$$

In this formula, S represents the size of the chamber in question and n represents the number of growth stages after the original chamber was formed.

5.6 Common Logarithms

Since our number system is a base 10 place-value system, logarithms to base 10 are often used in mathematics. Logarithms to base 10 are called **common logarithms.** To simplify notation, $\log_{10} x$ is written as $\log x$.

To find common logarithms of numbers between 1 and 10, the table of common logarithms on pages 616 and 617 is used. Part of the table is shown below.

The entries in the table are four-place decimal approximations. The decimal points are omitted.

N	0	1	2	3	4	5	6	7	8	9
10	0000	0043	0086	0128	0170	0212	0253	0294	0334	0374
11	0414	0453	0492	0531	0569	0607	0645	0682	0719	0755
12	0792	0828	0864	0899	0934	0969	1004	1038	1072	1106
13	1139	1173	1206	1239	1271	1303	1335	1367	1399	1430
14	1461	1492	1523	1553	1584	1614	1644	1673	1703	1732
15	1761	1790	1818	1847	1875	1903	1931	1959	1987	2014
16	2041	2068	2095	2122	2148	2175	2201	2227	2253	2279

To find log 1.38, read across the row labeled 13 and down the column labeled 8.

$\log 1.38 = 0.1399$ *Although log 1.38 is approximately 0.1399, it is customary to use an equal sign.*

To find the common logarithm of a positive number less than 1 or greater than 10, express the number in scientific notation. Then, use Law 1 and the table of common logarithms.

Example

1 Find log 14,300.

$\log 14{,}300 = \log(1.43 \times 10^4)$ *$14{,}300 = 1.43 \times 10^4$*

$= \log 1.43 + \log 10^4$ *Law 1*

$= \log 1.43 + 4$ *$\log_{10} 10^4 = 4$*

$= 0.1553 + 4$ or 4.1553 *From the table, $\log 1.43 = 0.1553$*

Every common logarithm has two parts. The integral part is called the **characteristic.** The positive decimal part is called the **mantissa.**

$$\log 14{,}300 = \underset{\text{characteristic}}{4}.\underset{\text{mantissa}}{\underbrace{1533}}$$

Example

2 Find log 0.0143.

$\log 0.0143 = \log(1.43 \times 10^{-2})$		*$0.0143 = 1.43 \times 10^{-2}$*
$= \log 1.43 + \log 10^{-2}$		*Law 1*
$= 0.1553 + (-2)$		*From the table and $\log_{10} 10^{-2} = -2$*
$= 0.1553 - 2$		

The logarithm table includes only positive mantissas. Therefore, in example 2 it was not necessary to simplify $0.1553 - 2$. Other ways of writing this logarithm are shown below.

$$\log 0.0143 = 8.1553 - 10$$
$$\log 0.0143 = 28.1553 - 30$$

If $\log N = y$, then N is the number whose common logarithm is y. The number N is called the **antilogarithm** of y. Thus, if $\log N = y$, then $N =$ antilog y.

Examples

3 Find N if $\log N = 3.1931$.

Since the characteristic is 3, we know N is of the following form.

$$N = (\text{number between 1 and 10}) \times 10^3$$

To find the missing number, locate 0.1931 in the table of mantissas. It appears in the row labeled 15 and column headed 6.

$$\text{Therefore, } N = 1.56 \times 10^3 \quad \text{or} \quad 1560$$

4 Find antilog (0.5142 − 4) correct to three decimal places.

This means the same as find N if $\log N = 0.5142 - 4$.

$$N = (\text{some number between 1 and 10}) \times 10^{-4}$$

The mantissa 0.5142 lies between 0.5132 and 0.5145. It is closer to 0.5145, which corresponds to 327 in the table.

$$N = 3.27 \times 10^{-4} \text{ or } 0.000327$$
$$\text{antilog}(0.5142 - 4) = 0.000327$$

-1.5229

Exercises

Exploratory **State the characteristic of the common logarithm of each number.**

1. 348
2. 54.2
3. 8.13
4. 0.912
5. 0.00315
6. 36,400

For each of the following common logarithms, state the characteristic and mantissa.

7. 2.3819
8. 0.4615 − 2
9. 28.8152 − 30
10. 3.7998
11. 0.4193 + 3

Find the common logarithm of each number without using a table.

12. 1
13. 10
14. 10,000
15. 0.1
16. 0.001
17. 100
18. 100,000
19. 0.00001
20. $10^{0.237}$
21. $10^2 \times 10^{0.81}$

Written **Find the common logarithm of each number. Use the table on pages 616–617.**

1. 4.78
2. 47.8
3. 53.7
4. 816
5. 147
6. 9320
7. 0.0352
8. 0.007
9. 2.09
10. 0.000624
11. 62,700
12. 7

Find N correct to three decimal places.

13. $\log N = 0.4099$
14. $\log N = 0.6435$
15. $\log N = 0.8531$
16. $\log N = 1.7619$
17. $\log N = 3.9877$
18. $\log N = 0.5514 - 1$
19. $\log N = 0.9562 - 2$
20. $\log N = 9.8069 - 10$
21. $\log N = 7.7664 - 10$
22. $\log N = 4.1038$
23. $\log N = 0.5287$
24. $\log N = 0.7115 - 3$

Find each of the following.

25. antilog 0.3345
26. antilog 1.5922
27. antilog 3.9090
28. antilog (0.6513 − 2)
29. antilog (9.8287 − 10)
30. antilog (0.7846 − 3)
31. $10^{2.8488}$
32. $10^{0.6405-1}$
33. $10^{0.4456} \cdot 10^{1.1812}$

If $10^{0.4771} = 3$, evaluate each of the following.

34. $10^{1.4771}$
35. $10^{4.4771}$
36. $10^{0.4771-1}$
37. $10^{8.4771-10}$

Find each of the following if $\log m = 0.4132$ and $\log x = 3.4832$.

38. $\log 10m$
39. $\log 100m$
40. $\log \frac{m}{1000}$
41. $\log 10m^2$
42. $\log x^2$
43. $\log \sqrt{x}$
44. $\log \sqrt[4]{x}$
45. $\log \frac{100}{x^3}$

Challenge **Solve for x.**

1. $(x - 3) \log x > 0$
2. $(x - 3) \log (x - 1) \leq 0$
3. $(x + 1) \log (x + 3) > 0$
4. $(x + 2) \log_r r^x = x$

Logarithms and Calculators

The calculator is a useful tool for computing logarithms and antilogarithms. On most calculators, the [log] key is used to find common logarithms.

Example: **Find log 76500.**

The [log] key represents $\log_{10}$.

ENTER: 76500 [log]

DISPLAY: 76500 4.88366144

The value of log 76500 is approximately 4.8837.

The mantissa is 0.8837 and the characteristic is 4.

Example: **Find log 0.00431.**

ENTER: 0.00431 [log]

DISPLAY: 0.00431 −2.36552273

The value of log 0.00431 is approximately −2.3655.

Notice that this value has a *negative* mantissa. You can rewrite −2.3655 as a logarithm with a positive mantissa by adding and then subtracting the same number. While any number can be used, the number most often used is 10.

$$-2.3655 = -2.3655 + (10 - 10) \quad \text{or} \quad (-2.3655 + 10) - 10$$

ENTER: 2.3655 [+/−] [+] 10 [=]

DISPLAY: 2.3655 −2.3655 10 7.6345

Thus, log 0.00431 = 7.6345 − 10.

The mantissa is 0.6345 and the characteristic is 7 − 10 or −3.

From the definition of logarithm, you know that $x = 10^a$ if and only if $a = \log x$. Since $x = \text{antilog } a$ if $a = \log x$, we can conclude that antilog $a = 10^a$. Thus, the [10^x] key can be used to find antilogarithms.

Example: **Find antilog 1.0899.**

ENTER: 1.0899 [10^x]

DISPLAY: 1.0899 12.2998552

The [y^x] key can also be used to compute antilogarithms.

The value of antilog 1.0889 is approximately 12.3.

Exercises **Use a calculator to find the logarithm of each number.**

1. 5.273 **2.** 7.184 **3.** 27.53 **4.** 604.7

5. 0.1952 **6.** 0.07635 **7.** 0.0001139 **8.** 863,100

Use a calculator to find the antilogarithm of each logarithm.

9. 0.8506 **10.** 0.4861 **11.** 3.6748 **12.** 2.7792

13. 0.3915 − 3 **14.** 28.2651 − 30 **15.** 6.5688 − 10 **16.** 5.0039

5.7 Calculating With Logarithms

After logarithms were developed in the late sixteenth century, they were used chiefly as a tool for evaluating complicated expressions. Later, logarithms were used to construct slide rules. Today, logarithmic functions and equations are used in many different fields. The increasing availability of calculators has reduced the need for using logarithmic tables to evaluate expressions.

Examples

1 **Use logarithms to calculate $(38.7)^8$.**

$$\begin{aligned} \text{Let } N &= (38.7)^8 \\ \text{Then, } \log N &= \log (38.7)^8 \\ &= 8 \log 38.7 \\ &= 8\,(1.5877) \\ &= 12.7016 \\ \text{Therefore, } N &= \text{antilog } 12.7016 \\ &= 5.03 \times 10^{12} \quad \text{or} \quad 5{,}030{,}000{,}000{,}000 \end{aligned}$$

The value of $(38.7)^8$ is approximately 5,030,000,000,000.

2 **Use logarithms to calculate $\sqrt[3]{0.878}$.**

$$\begin{aligned} \text{Let } N &= \sqrt[3]{0.878} = (0.878)^{\frac{1}{3}} \\ \text{Then, } \log N &= \log (0.878)^{\frac{1}{3}} \\ &= \frac{1}{3} \log 0.878 \\ &= \frac{1}{3}(0.9435 - 1) \end{aligned}$$

Since you must multiply by $\frac{1}{3}$, it is convenient to write $0.9435 - 1$ so that its characteristic is a multiple of 3. This is accomplished by rewriting the expression $0.9435 - 1$ as $2.9435 - 3$. You could also use $29.9435 - 30$.

$$\begin{aligned} \log N &= \frac{1}{3}(2.9435 - 3) \\ &= 0.9812 - 1 \\ N &= \text{antilog } (0.9812 - 1) \\ N &= 9.58 \times 10^{-1} \quad \text{or} \quad 0.958 \end{aligned}$$

The answer is accurate to three decimal places.

The following example illustrates the procedure for computing products and quotients. This method was widely used by mathematicians and scientists before modern computing devices were developed.

Example

3 Use logarithms to calculate $\frac{24.6}{147 \times 7.09}$.

Let $N = \frac{24.6}{147 \times 7.09}$

Then, $\log N = \log\left(\frac{24.6}{147 \times 7.09}\right)$

$= \log 24.6 - (\log 147 + \log 7.09)$ *$\log 24.6 = 1.3909$*

$= 1.3909 - 3.0179$ *$\log 147 + \log 7.09 = 2.1673 + 0.8506$ or 3.0179*

Performing the last subtraction would result in a negative mantissa. Since the table of logarithms contains only positive mantissas, rewrite 1.3909 as $4.3909 - 3$.

$\log N = (4.3909 - 3) - 3.0179$

$4.3903 - 3$
-3.0179
$1.3730 - 3$

$= 1.3730 - 3$ or $0.3730 - 2$

$\log N = 0.3730 - 2$

$N = \text{antilog}\,(0.3730 - 2)$

$N = 2.36 \times 10^{-2}$ or 0.0236

To get better approximations of logarithms, a method called **interpolation** is used. The next example demonstrates this process.

Examples

4 Find log 1.342.

a. Log 1.342 is between log 1.340 and log 1.350. Use the table of common logarithms to find the mantissas. Arrange this information in a chart as shown with the greatest number at the top.

Number	Mantissa
1350	1303
1342	?
1340	1271

b. Find the differences between the numbers and the corresponding mantissas, as indicated at the right.

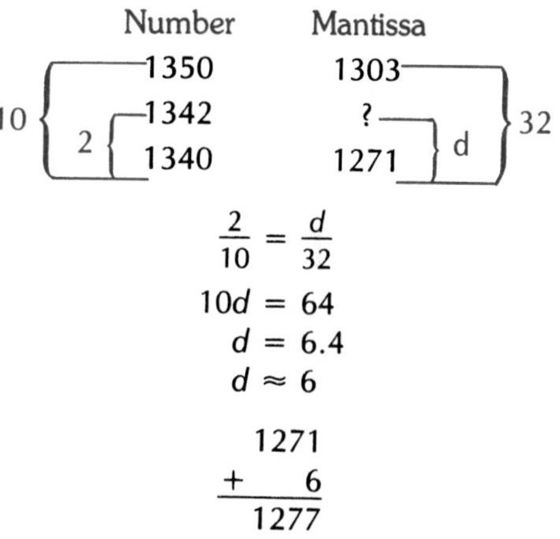

c. Write and solve the proportion formed from the ratios of these differences. Round the value of d to the nearest integer.

$$\frac{2}{10} = \frac{d}{32}$$
$$10d = 64$$
$$d = 6.4$$
$$d \approx 6$$

d. Since d is the difference between the unknown mantissa and 1271, add the value of d to 1271 to find the required mantissa.

$$\begin{array}{r} 1271 \\ +\quad 6 \\ \hline 1277 \end{array}$$

The value of log 1.342 is 0.1277. *Use a calculator to verify this result.*

5 Find log 134.2.

$$\begin{aligned} \log 134.2 &= \log(1.342 \times 10^2) \\ &= \log 1.342 + \log 10^2 \\ &= \log 1.342 + 2 \\ &= 0.1277 + 2 \\ &= 2.1277 \end{aligned}$$

Use the results of example 4 to find log 1.342.

6 Find antilog 2.5343.

antilog 2.5343 = (antilog 0.5343) × 10^2 *Use interpolation similar to example 4.*

a. In the table, 0.5343 lies between 0.5353 and 0.5340. Complete a chart and calculate the differences in the mantissas and their respective numbers.

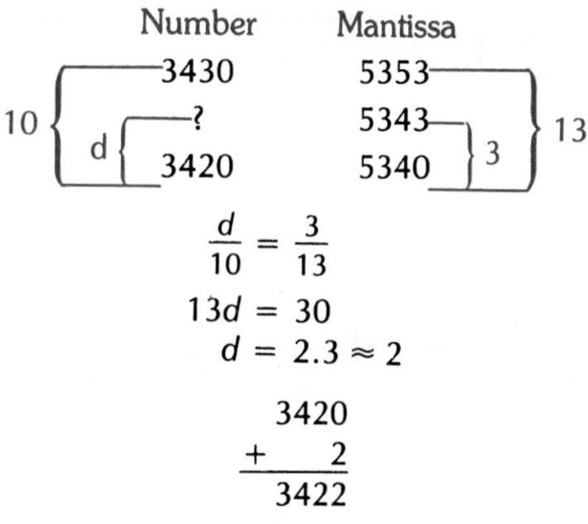

b. Write and solve the proportion based on the differences in the chart.

$$\frac{d}{10} = \frac{3}{13}$$
$$13d = 30$$
$$d = 2.3 \approx 2$$

c. Add the value of d to 3420 to find the required number.

$$\begin{array}{r} 3420 \\ +\quad 2 \\ \hline 3422 \end{array}$$

d. Find the required antilog.

Use a calculator to verify this result.

$$\begin{aligned} \text{antilog } 2.5343 &= (\text{antilog } 0.5343) \times 10^2 \\ &= 3.422 \times 10^2 \\ &= 342.2 \end{aligned}$$

Exercises

Exploratory **Evaluate each expression using logarithms.**

1. $(3.04)^4$
2. $(63.9)^3$
3. $(282)^5$
4. $(0.08)^6$
5. $\sqrt{248}$
6. $\sqrt[3]{2.08}$
7. $\sqrt[3]{0.650}$
8. $\sqrt[5]{0.482}$
9. $\sqrt[3]{0.0782}$
10. $(384)^{\frac{3}{5}}$
11. $\dfrac{14.8}{0.00821}$
12. $\dfrac{0.0428}{0.0789}$

Solve each equation for *n*.

13. $\log n = 2 \log 7$
14. $\log n = \frac{1}{2} \log 16$
15. $\log_5 n = 3 \log_5 4$
16. $\log 2n = 3 \log 2$
17. $\log n = \log 5 + \log 7$
18. $\log_4 n = \log_4 5 - \log_4 10$

Written **Evaluate each expression using logarithms.**

1. $(8.26)(43.4)$
2. $4570 \div 246$
3. $(0.00721)^{\frac{3}{5}}$
4. $46\sqrt{0.831}$
5. $(8.43)(5.67)^3$
6. $15\sqrt[3]{0.204}$
7. $\dfrac{\sqrt{286}}{8.42}$
8. $\sqrt[3]{\dfrac{8640}{3.14}}$
9. $\sqrt{\dfrac{3(4.23)}{63.5}}$
10. $863(2.72)^{12}$
11. $3.14(178)^5$
12. $\sqrt[3]{94.6(0.019)}$
13. $\dfrac{62.8(3.84)}{4.07}$
14. $\dfrac{82.1\sqrt{314}}{(0.252)^2}$
15. $\sqrt[5]{\dfrac{7.21(52)^2}{863}}$
16. $\sqrt[3]{0.471}$

Use interpolation to find the logarithm of each number.

17. 5.273
18. 7.184
19. 4.234
20. 27.53
21. 604.7
22. 0.1952
23. 0.07635
24. 8631
25. 4167
26. 0.04729
27. 0.001979
28. 0.005364

Use interpolation to find the antilog of each logarithm.

29. 0.5506
30. 0.4861
31. 3.5748
32. 2.7792
33. $0.3353 - 2$
34. $0.6173 - 3$
35. $0.6409 - 2$
36. 3.4193
37. $0.5915 - 3$
38. 5.5958
39. $28.2651 - 30$
40. $6.5688 - 10$

Evaluate each expression using logarithms.

41. $\sqrt[3]{1827}$
42. $\sqrt[5]{0.2815}$
43. $(2.038)^{10}$
44. $\sqrt[3]{14.23}$

Challenge **Tell whether the following statements are *true* or *false*. Justify your answer.**

1. $\log \dfrac{x}{y^2} = \log x + 2 \log y$
2. $\log \sqrt{\dfrac{x}{y}} = \dfrac{1}{2} \log x - \dfrac{1}{2} \log y$
3. $2 < \log 248 < 3$
4. $\log 100x^2 = 2(1 + \log x)$
5. $\log a + \log b - \log c = \log \dfrac{a + b}{c}$
6. $\log x^{n+1} = n \log x + x$

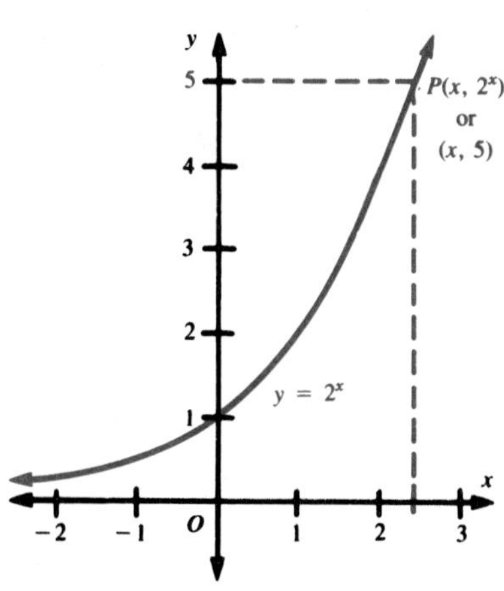

5.8 Equations and Exponents

The graph of $y = 2^x$ is shown at the left. Suppose we wish to find the coordinates of point P. Clearly, the y-coordinate is 5. To find the x-coordinate, solve the following exponential equation.

$$\begin{aligned} 2^x &= 5 \\ \log 2^x &= \log 5 \\ x \log 2 &= \log 5 \\ x &= \frac{\log 5}{\log 2} \\ x &= \frac{0.6990}{0.3010} \text{ or } 2.322 \end{aligned}$$

Therefore, the x-coordinate of point P is approximately 2.322.

Example

1 **Express $\log_3 42$ in terms of common logarithms and then find its value.**

$$\begin{aligned} \text{Let } x &= \log_3 42. \\ \text{Then, } 3^x &= 42. \\ \log 3^x &= \log 42 \\ x \log 3 &= \log 42 \\ x &= \frac{\log 42}{\log 3} \\ \text{Therefore, } \log_3 42 &= \frac{\log 42}{\log 3} \end{aligned}$$

Now find the value of $\log_3 42$.

$$\log_3 42 = \frac{\log 42}{\log 3} = \frac{1.6232}{0.4771} \approx 3.4022$$

Note that since $3^3 = 27$ and $3^4 = 81$, it is reasonable that $3^{3.4022} \approx 42$.

Example 1 is a specific case of the following formula.

Change of Base Formula

If n, a, and b are positive, $a \neq 1$ and $b \neq 1$, then

$$\log_b n = \frac{\log_a n}{\log_a b}.$$

If $a = 10$, then $\log_b n = \frac{\log n}{\log b}$.

Examples

2 **Find $\log_3 35$.**

$$\begin{aligned} \log_3 35 &= \frac{\log 35}{\log 3} && \textit{Change of Base Formula for } a = 10. \\ &= \frac{1.5441}{0.4771} \\ &\approx 3.2364 \end{aligned}$$

3 **Solve $x^{\frac{2}{3}} = 24$.**

$$x^{\frac{2}{3}} = 24$$
$$(x^{\frac{2}{3}})^{\frac{3}{2}} = (24)^{\frac{3}{2}}$$ *Raise both sides to a power to result in x^1.*
$$x = 24^{\frac{3}{2}}$$
$$\log x = \frac{3}{2}\log 24$$
$$= \frac{3}{2}(1.3802)$$
$$= 2.0703$$
$$x = \text{antilog } 2.0703$$
$$= (\text{antilog } 0.0703) \times 10^2$$
$$= 1.176 \times 10^2$$
$$= 117.6$$

Exercises

Exploratory **Use logarithms to state an equation equivalent to each of the following.**

1. $2^x = 3$
2. $4.2^x = 3$
3. $5^x = 61$
4. $7^{2x} = 42$
5. $10^{3x} = 191$
6. $3.7^x = 52.8$
7. $x^5 = 28$
8. $x^{\frac{2}{3}} = 51$

Express each logarithm in terms of common logarithms.

9. $\log_3 8$
10. $\log_5 10$
11. $\log_5 3$
12. $\log_{11} 7$
13. $\log_{20} 60$
14. $\log_{37} 52$
15. $\log_\pi \sqrt{2}$
16. $\log_{10} 2.718$

Written **Solve for x.**

1. $2^x = 53$
2. $3^x = 32$
3. $6^x = 18$
4. $4^x = 12$
5. $5^x = 42.7$
6. $14^x = 283$
7. $2.7^x = 52.3$
8. $4.3^x = 78.5$
9. $x = \log_4 51.6$
10. $x = \log_3 19.8$
11. $x = \log_6 144$
12. $7^{2x} = 83$
13. $10^{4x} = 186$
14. $2^{-x} = 5$
15. $3^x = \sqrt{13}$
16. $2^x = 3\sqrt{2}$
17. $2.1^{x-4} = 8.6$
18. $x^{\frac{2}{3}} = 18$
19. $x^{\frac{3}{5}} = 15$
20. $x^{\frac{2}{3}} = 17$
21. $3^{3x} = 2^{2x+3}$
22. $32^{2x} = 5^{4x+1}$
23. $2^{5x-1} = 3^{2x+1}$

Approximate each logarithm to three decimal places.

24. $\log_3 7$
25. $\log_7 12$
26. $\log_4 22$
27. $\log_6 11$
28. $\log_4 24$
29. $\log_6 72$
30. $\log_5 104$
31. $\log_{3.21} 10$

Solve for x without using a table of common logs.

32. $2^{2x+4} = 8^{2x}$
33. $9^{x-1} = 27^x$
34. $\log_{\sqrt{5}} 25 = x$
35. $5^x = 5^{2x^2-1}$
36. $3^{\frac{1}{x-4}} = 3^{\frac{x+2}{9}}$
37. $\log_3 (x^2 - 1) = \log_3 15$

A piece of equipment either appreciates (increases in value) or depreciates (decreases in value) with time. If the equipment has an initial value of P, then its value, V_n, at the end of n years is given by the formula $V_n = P(1 + r)^n$. The constant r is positive when it represents appreciation and negative when it represents depreciation. Use this formula to answer each question.

38. A piece of machinery valued at \$50,000 depreciates 10% per year. After how many years will its value have depreciated to \$25,000?

39. Electronic equipment valued at \$150,000 depreciates 20% per year. After how many years will its value have depreciated to \$15,000?

40. Lois purchased a new car 10 years ago for \$9000. The car is now worth \$900. Assuming a steady depreciation, what was the yearly rate of depreciation?

41. Louis bought a new house 15 years ago for \$87,000. The house is now worth \$217,500. Assuming a steady appreciation, what was the yearly rate of appreciation?

42. Computer components valued at \$200,000 appreciated 15% per year for 3 years and then depreciated 10% per year after that. After how many years will their value have depreciated to \$125,000?

Challenge

1. If $h(x) = \log (x^2 + 6)$, find functions f and g such that $h(x) = (f \circ g)(x)$.

2. If $f(x) = 10^{2x+5}$, find functions g and h such that $f(x) = (g \circ h)(x)$.

Mixed Review

For each exercise, copy the pattern at the right. Then shade squares so that the resulting pattern has the given symmetry or symmetries.

1. horizontal line of symmetry

2. point symmetry

3. exactly one diagonal line of symmetry

4. horizontal *and* vertical lines of symmetry

Solve each equation.

5. $2^{x^2 - 1} = 8^{3x - 5}$

6. $\log_x 9 = -4$

7. $\left(\frac{1}{27}\right)^{3-2x} = 81^{4x}$

8. $3^x = 2\sqrt{27}$

9. $\log_8 (x^2 - x) = \frac{1}{3}$

10. $\log_5 32 - 4 \log_5 x = 3 \log_5 \frac{1}{2}$

11. $\log_3 (y + 4) + \log_3 (y - 2) = 3$

12. $\log_6 (x - 3) - \log_6 (2x + 1) = 2$

13. If the solution set of $x^2 + px + q$ is $\{1 + 2i, 1 - 2i\}$, find the value of p.

14. Determine a domain of $g(x) = x^2 + 4$ so that $g^{-1}(x)$ is also a function. Then write an equation for $g^{-1}(x)$.

15. The frequency of a vibrating string varies inversely with the length of the string. A 36-inch piano string vibrates at a specific frequency. Find the length of the piano string that vibrates at three times this frequency.

5.9 Applications

Many formulas used in fields such as biology, business, and social science involve exponents and logarithms. For example, almost everyone at some time needs to understand the idea of compound interest.

Example

1 **Jill deposits \$1000 in the Evergreen Savings Bank, which pays 5% interest compounded quarterly. This means that every three months (one-fourth of a year), the bank pays 1.25% interest (one-fourth of 5%). Each three-month interval is called an *interest period.***

Derive a formula for the amount *(A)* after *t* years.

At the end of one interest period, Jill has \$1000 + \$1000 (0.0125) or \$1000 (1.0125). At the end of the next interest period Jill has \$1000 (1.0125) plus 1.25% of that amount. The process continues for each interest period so that the new balance equals the old balance plus 1.25% of the old balance.

The chart below demonstrates these calculations.

Interest Period	Previous Balance	Interest	New Balance
0	1000		
1	1000	1000 (0.0125)	1000 (1.0125)
2	1000 (1.0125)	[1000 (1.0125)] (0.0125)	$1000\,(1.0125)^2$
3	$1000\,(1.0125)^2$	$[1000\,(1.0125)^2]\,(0.0125)$	$1000\,(1.0125)^3$
4	$1000\,(1.0125)^3$	$[1000\,(1.0125)^3]\,(0.0125)$	$1000\,(1.0125)^4$

From the chart it appears that at the end of the n interest periods, Jill has \$1000 $(1.0125)^n$.

Since there are four interest periods per year, the amount *(A)* after t years can be expressed as

$$A = 1000\,(1.0125)^{4t}$$

The formula derived in Example 1 is a special case of the **compound interest formula.**

Compound Interest Formula

If A is the amount accumulated, P is the number of dollars invested (or principal), r is the interest rate per year, n is the number of interest periods per year, and t is the number of years,

$$A = P\left(1 + \frac{r}{n}\right)^{nt}.$$

Example

2 **If \$4000 is invested in a bank paying 9% interest compounded monthly, how much money is accumulated after five years?**

$A = P\left(1 + \frac{r}{n}\right)^{nt}$ *Compound Interest Formula*

$A = 4000\left(1 + \frac{0.09}{12}\right)^{12 \cdot 5}$ *$p = \$4000$, $r = 0.09$, $n = 12$, $t = 5$*

$A = 4000\left(1 + \frac{0.09}{12}\right)^{60}$

$A = \$6263$ *Use a calculator or logarithms.*

One formula for population growth is similar to the compound interest formula.

Population Growth Formula

If y is the new population, y_0 is the original population, r is the rate of growth per year, and n is the number of years,

$$y = y_0\,(1 + r)^n.$$

Example

3 **In 1989 the population of Pine City was 43,600. If the population increases at the rate of 2% per year, what will be the population in 2000?**

$y = y_0\,(1 + r)^n$

$y = 43{,}600\,(1 + 0.02)^{11}$ *Substitute into the population growth formula.*

$y = 54{,}211$ *Use a calculator or logarithm.*

The population will be about 54,211.

Of special importance in mathematics is the irrational number e, which is approximately 2.7182818. When bank interest is compounded continuously, this number is used in the following formula.

Continuously Compounded Interest Formula

If the amount A is accumulated after t years at an annual interest rate r for the original principal P,

$$A = Pe^{rt}.$$

Example

4 **If $1000 is invested at 6% a year compounded continuously for ten years, how much money is accumulated after that time?**

$A = Pe^{rt}$

$A = 1000\,(e)^{0.06(10)}$

$A = 1000\,(e)^{0.6}$

$A \approx 1000\,(2.72)^{0.6}$ — *$e \approx 2.72$*

$A \approx \$1823$ — *Use a calculator or logarithms.*

Exercises

Exploratory **Suppose $100 is deposited in a bank paying 8% annual interest. Find the amount in three years if the interest is compounded as follows.**

1. annually
2. semi-annually
3. quarterly
4. monthly

Written **A bank pays 8% interest per year compounded quarterly. How much will a $2500 investment be worth in the following number of years?**

1. 3
2. 5
3. 10
4. 25

Solve.

5. A bank pays 10% interest per year compounded quarterly. How much should be invested now in order to have $8000 in ten years?

6. In 1626, Native Americans were paid $24 for Manhattan Island. If the $24 had been invested at 6% per year compounded annually, what would have been the amount in 1980?

7. Which yields more, \$275 invested at 10% compounded annually over 15 years, or \$275 invested at 9.5% compounded quarterly over 13 years?

8. Which yields more, \$700 invested at 9% compounded semi-annually over 20 years, or \$570 invested at $8\frac{1}{2}$% compounded quarterly over 25 years?

The population of Pineville was 3,410 in 1970. What will the population be in the year 2000 if the population increases each year at the following rates?

9. 2%

10. 1%

11. 4%

12. $5\frac{1}{4}$%

Solve.

13. If \$100 is invested at 8% annual interest compounded continuously, what will be the amount in three years?

14. If \$100 is invested at 6% annual interest compounded continuously, when will the money be double the original amount?

Radioactive substances decay with time. Beginning with N grams, the number of grams, y, after t years is given by $y = Ne^{kt}$, where k is a negative constant. Use this formula to compute the results for exercises 15 and 16.

15. For a certain radioactive substance, the constant is -0.08042. How long will it take 250 grams of the substance to reduce to 50 grams?

16. In 10 years, 200 grams of a radioactive substance is reduced to 100 grams. Find the constant k for this substance.

In an experiment it is found that the number y of Glorfiums after t days is given by $y = 200(2^{\frac{t}{5}})$.

17. How many Glorfiums were there when $t = 0$?

18. How many Glorfiums can be expected in four days?

19. How long does it take for the number of Glorfiums to double?

20. In how many days can you expect to have 500 Glorfiums?

The following chart shows the population growth of North Pine City at various times.

Year	1950	1960	1970	1980	1990	2000
Population, y	120	180	274	415	620	?

21. By what percent does it appear that the population is increasing every ten years?
22. Is the population growing exponentially? Explain.
23. What would you predict the population to be in 2000? in 2010?
24. Which of the following is the best formula for predicting the population, y, after t years?
    a. $y = 120^t$
    b. $y = 120(0.5)^t$
    c. $y = 120(1.5)^{\frac{t}{10}}$
    d. $y = 120(1.5)^t$
25. Predict the population in 2005.
26. Most likely, what was the population in 1930? in 1940?

Problem Solving Application: Using Tables and Charts

A useful problem solving strategy is to organize relevant information by using a table or chart. In some cases, the solution to a problem can be taken directly from the table or chart. In other cases, the table or chart can help you find a pattern that is useful in solving the problem.

Examples

1 **The half-life of a radioactive element is the amount of time it takes for half of the element to decay. A certain 10-gram sample has a half-life of 13 years. When will less than 1 gram of the sample remain?**

Explore The sample has an initial weight of 10 grams. After every 13 years (the half-life), half of the sample will have decayed.

Plan Construct a chart showing the beginning and ending weight for each consecutive half-life that occurs.

Solve

Time (years)	0	13	26	39	52
Initial Weight (g)	10	10	5	2.5	1.25
Weight after Half-life (g)		5	2.5	1.25	0.625

According to the chart, there will be less than 1 gram left after 52 years. *Examine the solution.*

2 **Assume the number of cells in a fertilized egg doubles every hour, and there are 100 cells present initially. Determine when the number of cells will total 204,800.**

Explore There are 100 cells initially present. The number of cells doubles each hour. You want to know the number of cells after each hour.

Plan Make a chart that lists the number of cells after each time interval. Look for a pattern.

Solve

Number of Hours	Number of Cells
0	$100 = 2^0 (100)$
1	$200 = 2^1 (100)$
2	$400 = 2^2 (100)$
3	$800 = 2^3 (100)$
4	$1600 = 2^4 (100)$

From the pattern in the chart, you can conclude that after n hours, there will be 2^n (100) cells in the egg. Using this pattern, you can solve the problem.

$$\begin{aligned} 2^n (100) &= 204{,}800 \\ 2^n &= 2048 \\ \log 2^n &= \log 2048 \\ n \log 2 &= \log 2048 \\ n &= \frac{\log 2048}{\log 2} \\ n &= 11 \end{aligned}$$

The number of cells in a fertilized egg will total 204,800 in 11 hours.

Examine By extending the chart, you can verify the solution.

Exercises

Written Use a table or chart to solve each problem.

1. Assume that the number of bacteria in an experiment doubles every four hours. There are 2500 bacteria initially present. How many bacteria are present after 12 hours? After how many hours are 160,000 bacteria present?
2. A 50-gram sample of a certain radioactive substance has a half-life of 21 days. How long will it take until there are only $3\frac{1}{8}$ grams left?
3. Sheila deposits \$500 in her savings account. The annual interest rate is 8%. Sheila does not make any additional deposits or withdrawals. How much will she have at the end of 5 years if the interest is compounded semiannually?
4. Marcus despoits \$500 into his savings account. He does not make any additional deposits or withdrawals. The bank will pay an annual interest rate of 6% compounded quarterly. How much will Marcus have in 2 years?
5. It costs 25¢ for a stamp and 18¢ for a postcard. What combinations of each can you buy if you spend between \$1.90 and \$2.20 and you buy at least six stamps?
6. Find the sum of the first 3 consecutive even integers and of the first 4 consecutive even numbers. Then find the sum of the first 50 consecutive even integers.
7. Connie tore a sheet of paper into 3 pieces. Then she tore each of the resulting pieces into 3 pieces. If she continues this process 10 more times, how many pieces of paper will she have at the end?
8. The measure of the perimeter of a rectangle is twice the measure of its area. What are the possible dimensions of the rectangle if the measures of the length and width must be natural numbers?
9. Scientists introduce 2 different types of bacteria into an experiment. Bacteria A doubles in number every 2 hours. Bacteria B triples in number every 2 hours. Assuming the bacteria have no effect on each other and there are 1000 of each bacteria present initially, how long will the scientists have to wait for the experiment to have a total of 35,000 bacteria?

Vocabulary

zero exponent
negative exponent
exponential function
exponential curve
property of equality for exponential functions
logarithm
common logarithm
antilogarithm
characteristic
mantissa
interpolation
change of base formula
compound interest formula
population growth formula
continuously compounded interest formula

Chapter Summary

1. The five exponential laws govern the multiplication and division of monomial terms.
2. If b is any real number, n, p, and q are integers, $n > 1$ and $q > 1$, then $b^{\frac{1}{n}} = \sqrt[n]{b}$ and $b^{\frac{p}{q}} = (\sqrt[q]{b})^p$ or $\sqrt[q]{b^p}$.
3. The graphs of $y = b^x$ and $y = b^{-x}$, where $b > 0$ and $b \neq 1$, are symmetrical with respect to the y-axis.
4. For any exponential function, the domain is the set of real numbers and the range is the set of positive real numbers.
5. The three logarithmic laws, which are derived from the exponential laws, enable logarithms of products and quotients to be written in simpler forms.
6. Interpolation can frequently be used to find numbers or mantissas not readily available in the table of common logarithms.
7. Logarithmic functions have uses in practical applications such as computing compound interest, population growth, and radioactive decay.

Chapter Review

5.1 Evaluate.

1. 12^0 **2.** 6^{-1} **3.** 5^{-3} **4.** -6^{-2} **5.** $(-3)^{-3}$ **6.** $\left(\frac{1}{2}\right)^{-3}$

7. $16^{\frac{1}{2}}$ **8.** $81^{-\frac{1}{3}}$ **9.** $\left(\frac{2}{3}\right)^{-2}$ **10.** $(-64)^{\frac{2}{3}}$ **11.** $8^{-\frac{2}{3}}$ **12.** $16^{-\frac{3}{4}}$

Evaluate each expression if $a = 27$ and $b = 125$.

13. $4a^0 + b^{\frac{1}{3}}$ **14.** $3a^{-\frac{1}{3}} - 2b^{\frac{2}{3}}$ **15.** $a^{-\frac{2}{3}} + (b + 1)^0$

Simplify and express the result with positive exponents only.

16. $\frac{x^2y^{-1}}{xy^{-3}}$ **17.** $(3x^{\frac{1}{2}}y)(2x^{\frac{3}{2}}y^{-1})$ **18.** $\frac{(4x^2y^4)^2}{8x^3y^{-5}}$

5.2 **Sketch the graph of each function. State the domain and range.**

19. $y = 5^x$ **20.** $y = \left(\frac{1}{3}\right)^x$ **21.** $y = 4^{-x}$

5.3 **Solve each equation.**

22. $x^{\frac{5}{2}} = 32$ **23.** $x^{-2} - 4 = 77$ **24.** $2x^{\frac{3}{4}} - 28 = 100$

25. $3^{x+1} = 3^5$ **26.** $3^{2x-4} = \left(\frac{1}{9}\right)^{x-3}$ **27.** $8^{2x} = 4^{5x+2}$

5.4 **Evaluate.**

28. $\log_3 27$ **29.** $\log_{16} 4$ **30.** $\log_{10} 0.001$ **31.** $\log_4 \sqrt{2}$

Sketch the graph of each function. State the domain and range.

32. $f(x) = \log_2 x$ **33.** $f(x) = \log_{\frac{1}{3}} x$

Write an equation for the inverse of each function.

34. $y = 4^x$ **35.** $y = \log_{10} x$ **36.** $y = 3^{-x}$

Solve each equation.

37. $\log_5 x = 2$ **38.** $\log_9 x = \frac{1}{2}$ **39.** $\log_6 12 = \log_6 (5x - 3)$

5.5 **40.** $\log_3 x - \log_3 4 = \log_3 12$ **41.** $\log_5 x + \log_5 3 = \log_5 15$

42. $\log_5 7 + \frac{1}{2}\log_5 4 = \log_5 x$ **43.** $\log_3 (x - 4) + \log_3 (x + 2) = 3$

5.6 **Find each of the following.**

44. log 6.87 **45.** log 4390 **46.** log 0.000108

47. antilog 1.5502 **48.** antilog 0.8645 − 1 **49.** antilog 6.7619 − 10

5.7 **Use logarithms to evaluate each expression.**

50. $\sqrt[5]{2.61}$ **51.** $(49.8)^8$ **52.** $\sqrt[3]{0.198}$

Use interpolation to find each of the following.

53. log 463.2 **54.** log 0.02832 **55.** antilog 8.8490 − 10

5.8 **Solve each equation.**

56. $2^x = 53$ **57.** $\log_4 11.2 = x$ **58.** $x^{\frac{3}{4}} = 24$

Evaluate each of the following.

59. $\log_2 7$ **60.** $\log_5 2$ **61.** $\log_4 3$

5.9 **62.** If \$200 is invested at 12% per year compounded quarterly, when will the investment be worth \$300?

63. If \$1200 is invested at 8% per year compounded continuously, how much is accumulated after 15 years?

Chapter Test

Evaluate.

1. 14^0
2. 12^{-1}
3. $\left(\frac{3}{4}\right)^{-2}$
4. $\left(\frac{4}{25}\right)^{\frac{1}{2}}$
5. $4^{-\frac{3}{2}}$
6. $\left(\frac{1}{8}\right)^{-\frac{1}{3}}$

Evaluate each of the following if $a = 8$ and $b = 16$.

7. $a^{\frac{2}{3}} + b^{\frac{1}{4}}$
8. $2a^0 + 3b^{\frac{3}{4}}$
9. $(4a^{\frac{1}{3}} - 2b^{\frac{1}{4}})^{-2}$

Simplify and express with positive exponents only.

10. $\dfrac{a^3y^{-2}}{a^2y^{-4}}$
11. $2(a^2b^{-3})(ab^{-2})^{-3}$
12. $\dfrac{(3a^4b^2)^3}{9a^{-2}b^{-6}}$

Sketch the graph of each function. State the domain and range.

13. $y = 3^x$
14. $y = \log_3 x$

Solve each equation.

15. $4^{x-2} = 16$
16. $8^{2x-2} = 16^x$
17. $\log_5 25 = x$
18. $\log 0.0001 = x$
19. $\log_{16} 4 = x$
20. $\log_7 x = 2$
21. $\log_x \frac{1}{4} = -1$
22. $x^{\frac{2}{3}} = 16$

Express $\log N$ in terms of $\log a$ and $\log b$.

23. $N = ab$
24. $N = \dfrac{a^2}{b}$
25. $N = a\sqrt{b}$
26. $N = a^2\sqrt[3]{b}$

Solve each equation.

27. $\log_2 x = \frac{1}{3} \log_2 27$
28. $\log_3 x - \log_3 4 = \log_3 12$
29. $\log_5 7 + \frac{1}{2} \log_5 4 = \log_5 x$
30. $\log_6 (x - 3) + \log_6 (x + 2) = 1$

Find each of the following. Use interpolation if necessary.

31. log 589
32. log 1832
33. antilog 0.6839 − 3
34. antilog 9.8346 − 10
35. Use logarithms to calculate $\sqrt[5]{1790}$.
36. Solve for x if $3^x = 35$.
37. Suppose $2500 is invested at 7% interest compounded quarterly. How long will it take for this amount to triple?
38. The population of Pineburg was 45,000 in 1989. What will be the population in 1999 if the annual rate of growth is 2.5%?
39. How much money should have been deposited in 1980 in a bank paying 8% per year compounded continuously, in order to have $8000 in the year 2000?

Tell if each statement is true or false. Justify your answer.

40. If $\log x = m$, then $\log x^2 = 2m$.
41. If $A = \pi r^2$, then $\log A = 2\pi \log r$.
42. If $\log_b 4 = 1$, then $\log_2 b = 2$.
43. If $\log N = 2.7409$, then $\log 100N = 274.09$.
44. If $0 < x < 1$, then $\log x$ is negative.
45. If $0 < \log x < 1$, then $1 < x < 10$.

Chapter

Circles 6

Of all the familiar geometric shapes, the circle is probably the one that has most captivated man's imagination. Circles are seen in all forms of art and architecture as well as in scientific and industrial designs. The circle also serves as a symbol for the Heavens, the Earth, the Sun, and other celestial bodies. In this chapter, some of the mathematical concepts associated with the circle will be examined.

6.1 Parts of a Circle

Before beginning a study of the circle and its many important properties, it may be helpful to review some of the ideas you have already encountered. Unless otherwise stated, all geometric figures in this chapter will be assumed to lie in one plane.

Definition of Circle

A **circle** is the set of all points that are equally distant from a given fixed point, called the **center of the circle.** The distance from the center to any point on the circle, as well as any actual segment from the center to a point on the circle, is called a **radius** (plural: radii) of the circle.

It follows from this definition that *all radii of a circle are equal.*

A circle is named by its center if there is no possibility of confusion. For example, a circle with center O is symbolized $\odot O$.

Definitions of Equal Circles and Concentric Circles

Equal circles are circles with equal radii.

Concentric circles are circles with the same center, though not necessarily the same radius.

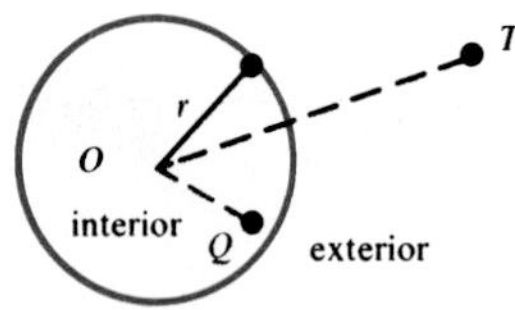

A circle separates the plane into three regions: the points *on the circle itself;* the points in the *interior of the circle;* and the points in the *exterior of the circle*. It follows that if Q is a point in the interior of $\odot O$ with radius r, then $OQ < r$, and conversely. If T is a point in the exterior of the same circle, then $OT > r$, and conversely.

Definition of Chord

A **chord** is a segment whose endpoints are points on a circle.

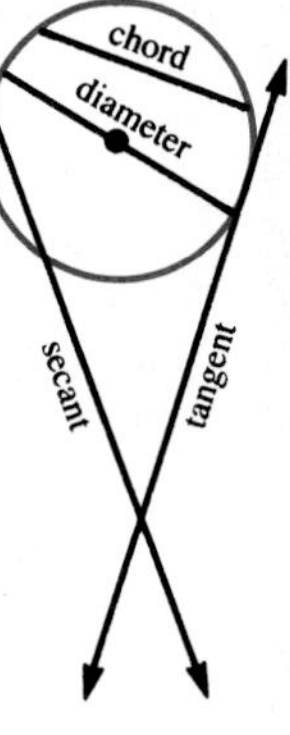

Definition of Diameter

A **diameter** of a circle is a chord passing through the center of the circle. It follows that a diameter is equal in length to two radii, or $d = 2r$.

Definition of Secant

A **secant** is a line that intersects the circle in two distinct points.

Definition of Tangent

A **tangent** is a line which intersects the circle in exactly one point.

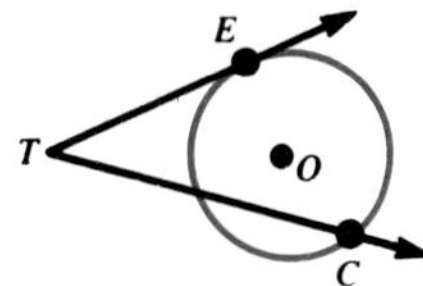

Segment *TC* and ray *TC* are *secants drawn to the circle from an outside point,* while $\overline{TE}$ and $\overrightarrow{TE}$ are *tangents drawn to the circle from an outside point. E* is the *point of tangency.*

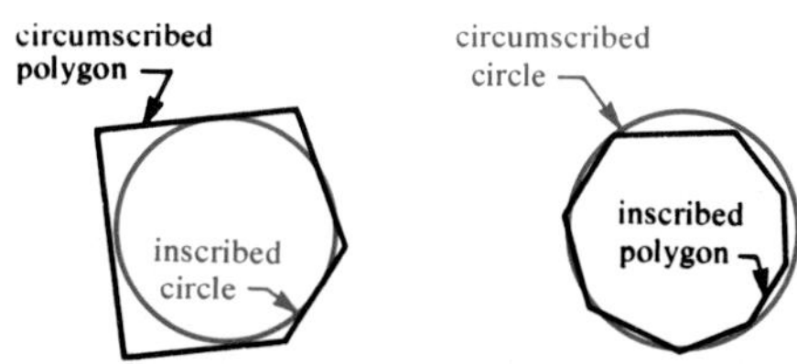

A polygon all of whose sides are tangent to circle *O* is said to be **circumscribed** about circle *O*. The circle is then **inscribed** in the polygon. A polygon whose vertices are points of a circle is **inscribed** in a circle, and the circle is **circumscribed** about the polygon.

Definition of a Central Angle

A **central angle** of a circle is an angle whose vertex is the center of the circle. Every central angle cuts off or *intercepts* an *arc* of the circle.

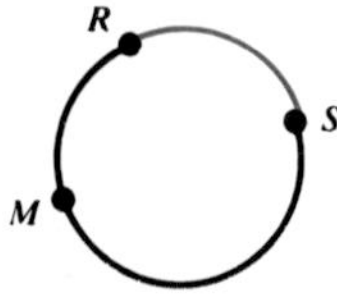

An **arc** is part of a circle. There are three kinds of arcs: a **semicircle,** which is "half a circle"; a **minor arc,** which is less than a semicircle; and a **major arc,** which is greater than a semicircle. The symbol for arc *AB* is $\widehat{AB}$. Two points are used to name a minor arc. A major arc is named by three points. In the figure at the left, $\widehat{RS}$ is in red and $\widehat{RMS}$ is in black. $\widehat{RS}$ is a minor arc. $\widehat{RMS}$ is a major arc.

The measure of an arc, in degrees, is given by the measure of the central angle which intercepts it.

The degree measure of $\widehat{RS}$ *is denoted* $m\widehat{RS}$.

Example

If ⊙*O* in the figure at the right has radius 5 units and $m\angle AOB = 90$, find each of the following.

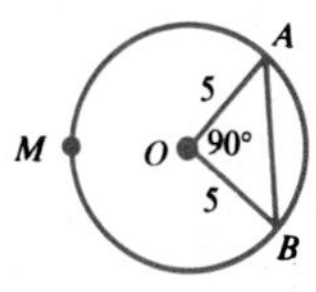

1 Measure of chord $\overline{AB}$

Since $\triangle AOB$ is a right triangle, we can use the Pythagorean Theorem. We know $OB = OA = 5$, so

$$(AB)^2 = 5^2 + 5^2$$
$$= 50$$
$$AB = 5\sqrt{2}$$

2 $m\widehat{AB}$ and $m\widehat{AMB}$

$\widehat{AB}$ is measured by central angle *AOB*.

$$m\widehat{AB} = 90$$
$$m\widehat{AMB} = 360 - m\widehat{AB}$$
$$= 360 - 90 \text{ or } 270$$

Exercises

Exploratory **Explain the meaning of each of the following. Sketch each situation.**

1. A polygon is inscribed in a circle.
2. A circle is inscribed in a polygon.
3. A polygon is circumscribed about a circle.
4. A circle is circumscribed about a polygon.

For exercises 5–21, refer to ⊙*O* at the right.

5. Name two chords in the diagram.
6. Name two central angles that are supplementary.
7. Name two diameters.
8. If segment *DE* were drawn, it would be a ___?___.
9. Name a secant line if there is one.
10. Name a tangent line if there is one.
11. Name two major arcs.
12. Name two minor arcs.
13. Name the angle whose measure determines that of $\widehat{AC}$.
14. How many radii are shown in the diagram?
15. Is $\overline{AE}$ necessarily parallel to $\overline{CD}$? Explain.
16. If $\overline{AE}$ were parallel to $\overline{CD}$, what two angles would necessarily be equal? Explain.
17. Is $\triangle OBD$ necessarily isosceles? Explain.
18. What can you say about $\angle B$ and $\angle D$?
19. If $m\angle BOD = 34$, then $m\angle D =$ ___?___
20. Is $m\widehat{ACD}$ necessarily equal to $m\widehat{CAB}$? Explain.
21. Is $m\widehat{CE}$ necessarily equal to $m\widehat{AD}$? Explain.
22. Describe the regions into which a circle separates a plane.

Written **Find each of the following, if $\overline{TW}$ and $\overline{RS}$ are diameters of ⊙*C*, $m\angle TCM = 80$, and $m\widehat{MS} = 30$.**

1. $m\widehat{TM}$	2. $m\angle MCS$	3. $m\widehat{TS}$	4. $m\angle SCW$
5. $m\angle TCR$	6. $m\angle RCW$	7. $m\widehat{RW}$	8. $m\widehat{RWM}$
9. $m\widehat{WRT}$	10. $m\widehat{WRM}$	11. $m\widehat{MW}$	12. $m\widehat{RM}$
13. $m\widehat{MTS}$	14. $m\widehat{RMW}$	15. $m\widehat{SWT}$	16. $m\angle MCW$

Sketch each of the following.

17. ⊙O and ⊙Q with ⊙O and ⊙Q concentric.
18. ⊙O with chords $\overline{RM}$ and $\overline{MW}$ such that $\overline{RM} \perp \overline{MW}$.
19. ⊙O concentric to ⊙C with chord $\overline{AB}$ in ⊙O tangent to ⊙C at *H*.
20. ⊙C inscribed in rhombus *RHBS*.
21. ⊙O circumscribed about rectangle *RECT*.

6.2 Circumference, Area, and Pi

Before continuing our study of circles, we must first review the formulas dealing with the circumference and area of a circle.

The circumference, C, of a circle of radius r units and diameter d units is $2\pi r$ units or πd units where π is an irrational number approximately equal to 3.14.

The area, A, of a circle of radius r units is πr^2 square units.

Examples

1 **Find the circumference and area of a circle of radius 3 cm.**

$C = 2\pi r = 2\pi \cdot 3$ or 6π cm $\quad A = \pi r^2 = \pi(3)^2$ or 9π cm^2

2 **The area of a circle is 1000 square units. Find an approximation of the radius to the nearest hundredth.**

$$\pi r^2 = A$$
$$\pi r^2 = 1000$$
$$r^2 = \frac{1000}{\pi}$$
$$r = \sqrt{\frac{1000}{\pi}} \approx \sqrt{\frac{1000}{3.14}} \text{ or } 17.85 \text{ units}$$

3 **Find the area of the shaded region of the figure at the right if the diameter of the circle is 12 cm.**

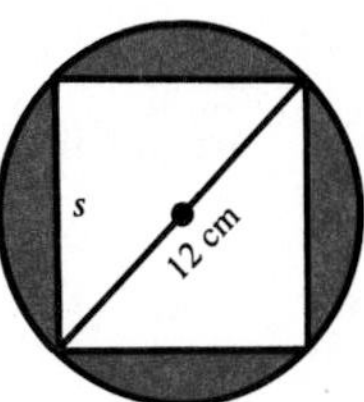

Area of shaded region =
(Area of circle) − (Area of square)

Since the diameter of the circle is 12 cm, the radius is 6 cm.

$$\text{Area of circle} = \pi r^2 = \pi(6)^2 \text{ or } 36\pi \text{ cm}^2$$

The area of the square is s^2. Use the Pythagorean Theorem to find a value for s.

$$s^2 + s^2 = 12^2$$
$$2s^2 = 144$$
$$s^2 = 72$$

Since s^2 is the area of the square, we can stop at this point.

The area of the shaded region is $(36\pi - 72)$ cm^2.

Definition of Sector of a Circle

A **sector** of a circle is the region bounded by two radii and an arc of the circle.

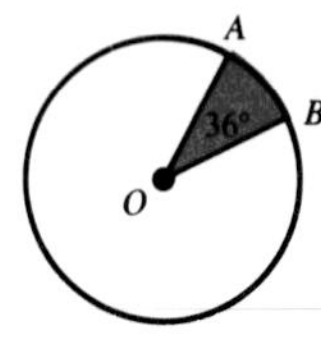

Sector AOB is shaded in the figure at the left. The unshaded region of the circle is also considered a sector.

Suppose $\odot O$ has a radius of 4 units. To find the area of sector AOB and the length of arc AB, we can proceed as follows.

Since 36° is $\frac{1}{10}$ of 360°, the area of the sector is $\frac{1}{10}$ the area of the whole circle. Since the area of the circle is $\pi(4)^2$ or 16π, it follows that the area of sector AOB is $\frac{1}{10} \cdot 16\pi$ or $\frac{8}{5}\pi$. Similarly, the length of $\widehat{AB}$ is $\frac{1}{10}$ of the circumference or $\frac{1}{10}$ of 8π, or $\frac{4}{5}\pi$.

For $\odot O$ of radius r units with an arc AB of degree measure n,

$$\text{Area of sector } AOB = \frac{n}{360} \cdot \pi r^2 \text{ square units, and}$$

$$\text{Length of arc } AB = \frac{n}{360} \cdot 2\pi r \text{ units.}$$

Example

4 **Find the area of the shaded region in the figure.**

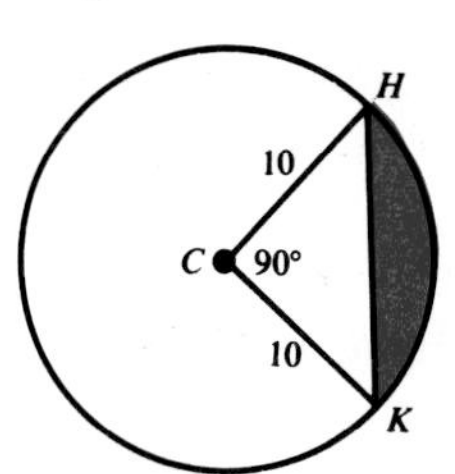

Area of the shaded region =
(Area of sector HCK) − (Area of $\triangle HCK$)

$$\text{Area of sector } HCK = \frac{90}{360} \cdot \pi(10)^2$$

$$= \frac{1}{4} \cdot 100\pi \text{ or } 25\pi \text{ square units}$$

Since $\triangle HCK$ is a right triangle,

$$\text{Area of } \triangle HCK = \frac{1}{2} \cdot 10 \cdot 10 \text{ or } 50 \text{ square units.}$$

Therefore, the area of the shaded region is $(25\pi - 50)$ square units.

The shaded region in example 4 is an example of a *segment* of a circle. A segment of a circle is a region bounded by a chord and the minor arc it intercepts.

Exercises

Exploratory **Find the *exact* circumference of circles with the following radii.**

1. 5
2. 10
3. $\frac{1}{2}$
4. 2
5. $3\sqrt{5}$
6. 0.013

Find the *exact* circumference of circles with the following diameters.

7. 0.12
8. $\sqrt{5}$
9. 3.016
10. $\sqrt[3]{2}$
11. π

Find rational approximations for the circumferences of the circles with these radii. Use $\pi = 3.14$ or $\pi = \frac{22}{7}$.

12. 6
13. $\frac{1}{4}$
14. $\frac{7}{15}$
15. $3\sqrt{2}$
16. $5\sqrt{3}$
17. Describe a method, using tape-measure and assorted tin cans, you could use to convince a younger student that the circumference of a circle is a little more than three times its diameter.

Written **Find the *exact* area of circles with the following radii.**

1. 6
2. 0.06
3. $\frac{1}{2}$
4. $\frac{3}{2}$
5. $3\sqrt{2}$
6. $\frac{11}{3}$

Find the *exact* radius of circles with the following areas.

7. 25π
8. 0.04π
9. 10π
10. 0.0001π
11. π^2
12. 20

Explain how to find each of the following.

13. The length of an arc cut off by a central angle of c° in a circle with radius r units.
14. The area of a sector of a circle of radius r units determined by a central angle of c°.

In each of the following, the radius of $\odot O$ and the measure of central angle O are given. Find the exact area of the sector determined by the central angle and the exact length of the intercepted arc.

15. $r = 6, m\angle O = 60$
16. $r = 6, m\angle O = 30$
17. $r = 1, m\angle O = 30$
18. $r = 12, m\angle O = 120$
19. $r = 10, m\angle O = 72$
20. $r = 18, m\angle O = 108$
21. $r = 5, m\angle O = 150$
22. $r = 3\sqrt{2}, m\angle O = 90$
23. $r = 0.4, m\angle O = k$
24. How does the area of a circle with radius 10 units compare with that of a circle of radius 20 units?
25. What happens to the area of a circle if its radius is doubled? is tripled? is multiplied by k^2? Verify your answers.
26. The radius of $\odot C$ is $t + 1$ units and the radius of $\odot O$ is $t - 1$ units. Find the difference in their areas.

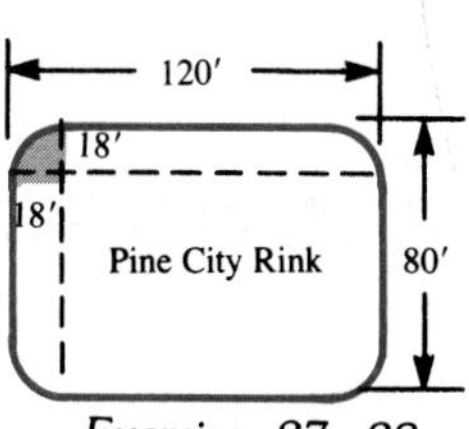

Exercises 27–28

The shaded region is a quarter circle.

27. The Pine City Skating Rink is rectangular, with circular corners, and the dimensions shown. Find the exact area of the rink. Then, find an approximation of the area to the nearest hundredth.
28. What is the exact perimeter of the Pine City Skating Rink?

Find the exact area of the shaded portions of the figures.

29.

30.

31.

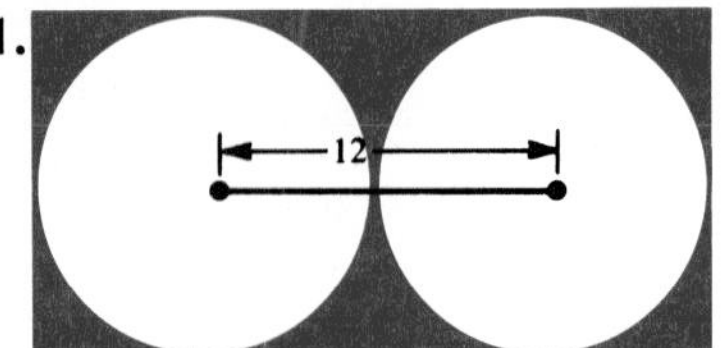

32.

33.

34.

35.

36.

Find the total length of wire required to produce the following patterns. In each case, give the exact length and a rational approximation to the nearest tenth.

37.

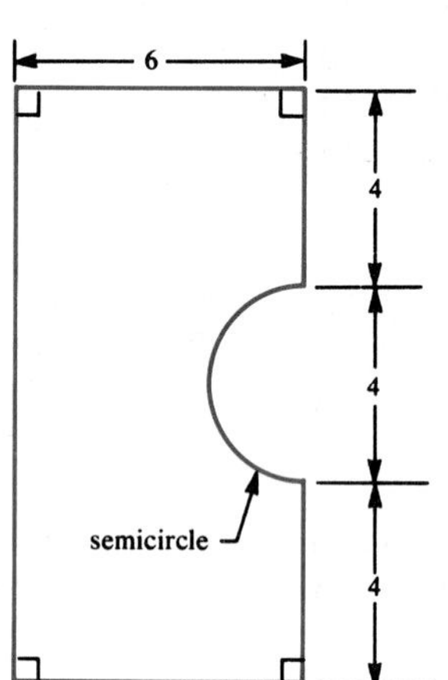

38.

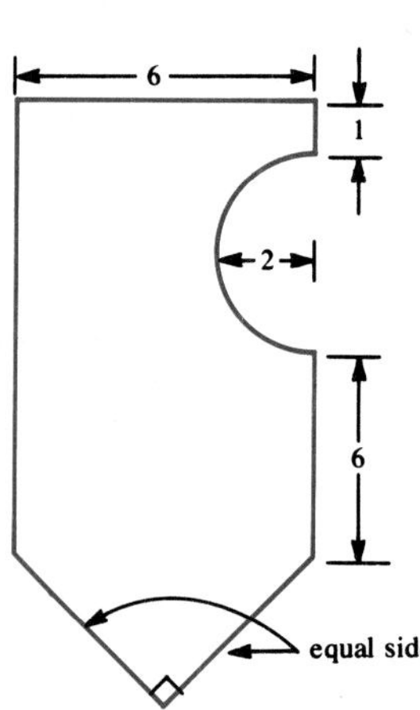

39.

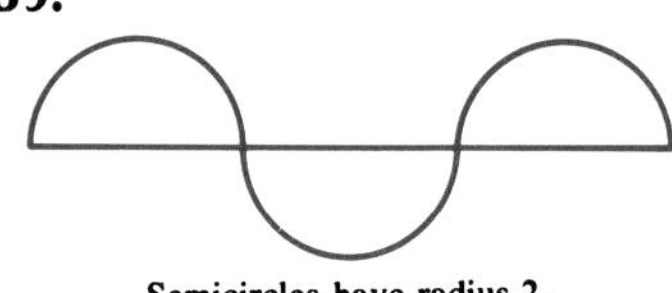

Semicircles have radius 2.

40.

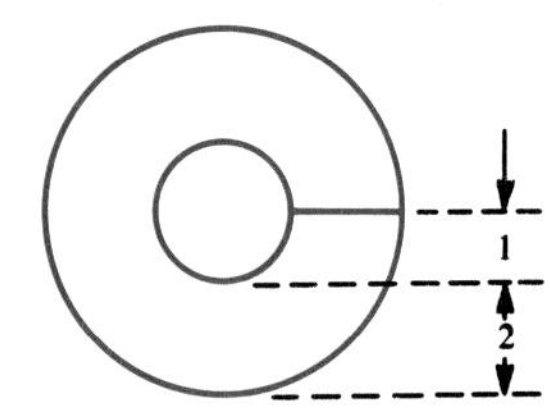

Challenge

1. The **apothem** of a regular polygon is the length of a radius of its inscribed circle. If a is the apothem of a regular polygon and p is its perimeter, show that the area of the polygon is equal to $\frac{1}{2}ap$. *(Hint: Divide the polygon into congruent triangles)*
2. Express the area of pentagon SNAKE if $CQ = 5$ and $NA = k$.
3. Professor Smythe refers to the circumference of a circle as "the limiting value of the perimeters of regular polygons inscribed in the circle as the number of sides increases without bound." Explain what he has in mind, and include a few illustrations.
4. Formulate a method for finding the area of a segment of a circle.

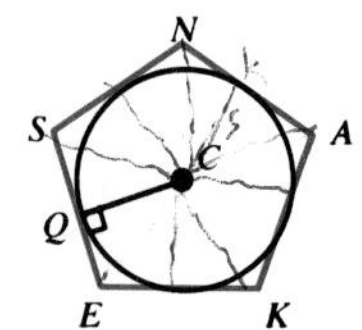

6.3 Tangents

One of the most important theorems of circle geometry concerns the relationship between a tangent and the radius drawn to the point of tangency. Before investigating this relationship, we must first consider a few preliminary ideas.

First, recall that through a point not on a given line, there exists one and only one line perpendicular to that given line. We also need the following theorem.

Shortest Distance Theorem

The perpendicular segment from a point to a line is the shortest segment from the point to the line.

This fact appears fairly obvious but is tricky to prove. A plan for the proof of this theorem can be found in the Challenge Exercises.

Now, to discover the idea of our main theorem on your own, draw a circle, and a line tangent to it at any point. Then, draw the radius to the tangent at the point of tangency. The radius should appear to be perpendicular to the tangent. This is stated in Theorem 6–1.

Theorem 6–1

A line tangent to a circle is perpendicular to the radius drawn to the point of tangency.

To prove this theorem, we will use an *indirect method*. The plan will be to *assume the theorem is false,* and then to show that *this assumption is false*. Symbolically, assume $\sim p$ and show that $\sim p$ leads to a contradiction. Therefore, $\sim p$ must be false. We then conclude that p must be true.

Proof of Theorem 6–1

Given: $\odot O$, with t tangent to $\odot O$ at P, radius $\overline{OP}$
Prove: $\overline{OP} \perp t$

Assume that $\overline{OP}$ is not perpendicular to t. Draw the unique line through O perpendicular to t. Call the point of intersection of t with this new line Q. Then, $\overline{OQ} \perp t$. Since P and Q both lie on the same line and since the Shortest Distance Theorem states that the perpendicular segment from a point to a line is the shortest segment from the point to the line, we have $OQ < OP$. However, t is tangent to circle O at P. Therefore, Q must be exterior to $\odot O$ since all points on a tangent other than the point of tangency lie outside the circle. This means the length of $\overline{OQ}$ is greater than the length of radius $\overline{OP}$ which forces $OP < OQ$.

Thus, the assumption that $\overline{OP}$ is not perpendicular to t has led to a contradiction. Therefore, the assumption is false and $\overline{OP} \perp t$.

The converse of Theorem 6–1 is also true.

Theorem 6–2

If a line is perpendicular to a radius at a point P on the circle, then the line is tangent to the circle at P.

Theorem 6–1 has some important consequences. One of them is the next theorem.

Theorem 6–3

Two tangent segments to a circle from the same exterior point have equal lengths.

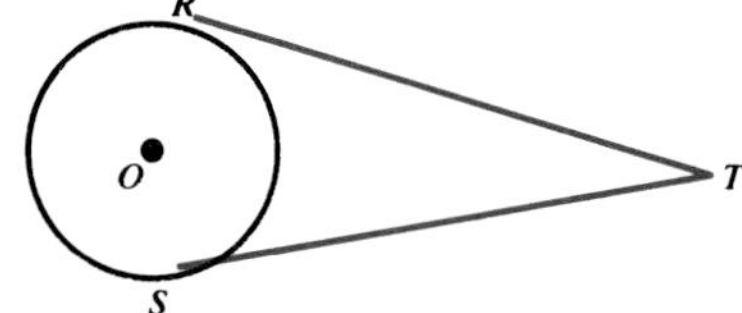

Theorem 6–3 asserts that $TR = TS$.

The proofs of Theorems 6–2 and 6–3 are left for the exercises.

Example

1 **Circle P is inscribed in $\triangle ABC$. Find CD if $AC = 17$, $AB = 32$, and $BE = 20$.**

Since $\odot P$ is inscribed in $\triangle ABC$, $\overline{AB}$, $\overline{AC}$, and $\overline{BC}$ are all tangent to $\odot P$. By Theorem 6–3, $AD = AF$ and $BE = BF$.

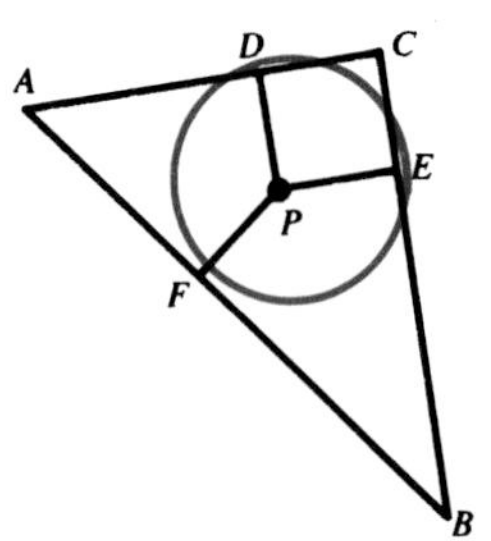

$CD = AC - AD$	*Segment Addition Theorem*
$= AC - AF$	*Substitution*
$= AC - (AB - BF)$	*Segment Addition Theorem*
$= AC - (AB - BE)$	*Substitution*
$= 17 - (32 - 20)$	
$= 5$	

A line tangent to each of two circles is a **common tangent.** If the tangent intersects the line segment joining the centers of the two circles, it is a **common internal tangent.** If it does not, the tangent is a **common external tangent.** The **line of centers** is the line containing the centers of the two circles.

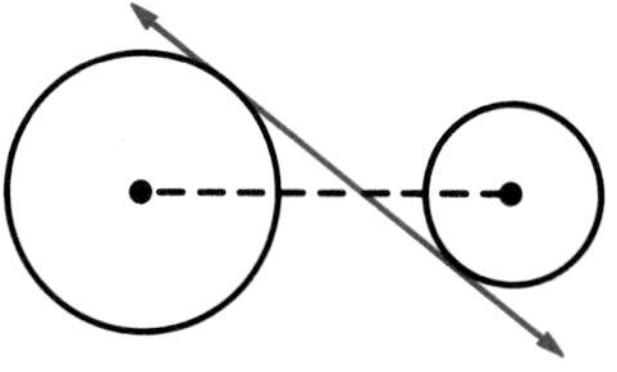
common internal tangent

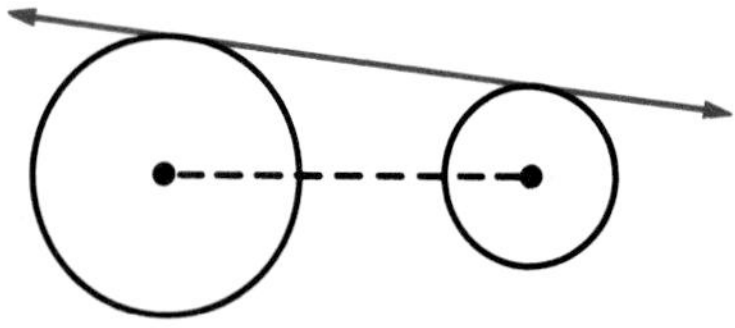
common external tangent

Two circles can be tangent to each other when they are tangent to the same line at the same point. The circles can be either **externally tangent** or **internally tangent.**

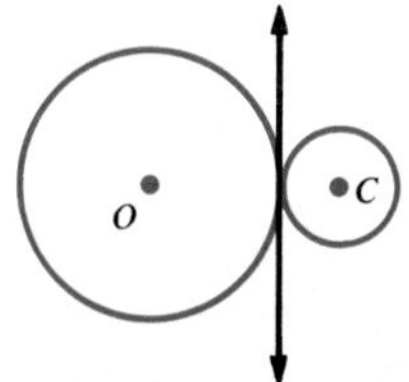

$\odot O$ and $\odot C$ are externally tangent.

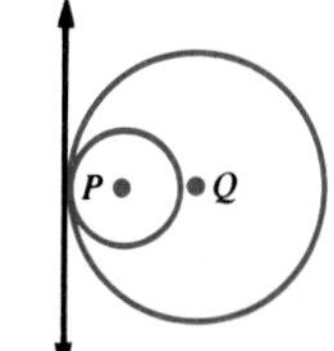

$\odot P$ and $\odot Q$ are internally tangent.

Exercises

Exploratory In exercises 1–3, refer to $\odot O$ at the right to name each of the following.

1. two different tangent lines.
2. two different tangent segments
3. two different points of tangency

Tell if the following statements are *always, sometimes,* or *never* true.

4. From point B not in line ℓ there exist two lines perpendicular to ℓ.
5. The segment RM from point R to point $M \in$ line ℓ is the shortest segment from R to ℓ.
6. If line t is tangent to $\odot C$ at point P, then $\overline{CP} \perp t$.
7. If line s is tangent to $\odot C$, s intersects $\odot C$ at P and Q, with $P \neq Q$.
8. If $\overline{BD}$ is a diameter of $\odot O$ and t is tangent to $\odot O$ at D, then $\overline{BD} \perp t$.
9. A quadrilateral $ABCD$ may be inscribed in $\odot O$.
10. If $\overline{TL}$ is tangent to $\odot C$ at L and $\overline{TM}$ tangent to $\odot C$ at M, then $TL = TM$.
11. If line m is tangent to $\odot C$ at B and line n is tangent to $\odot C$ at D, then $\overline{BD}$ is a diameter of $\odot C$.
12. Two circles tangent to each other are internally tangent.
13. Three circles tangent to each other (each one tangent to each of the other two) are externally tangent.
14. $\odot C$ and $\odot O$ are tangent to line k. Then $\overline{OC} \perp k$.

Written **Exercises 1–9 refer to ⊙O. $\overleftrightarrow{RS}$ is tangent to ⊙O at L.**

1. What is the measure of $\angle SLO$?
2. If $OW = 3$, find OL and LW.
3. If $OW = 3$, $\overline{OR}$ is drawn and $OR = 5$, find LR.
4. If $OW = 3$ and $SL = 3$, find OS.
5. If $OW = 3$ and $SL = 3$, find SW.

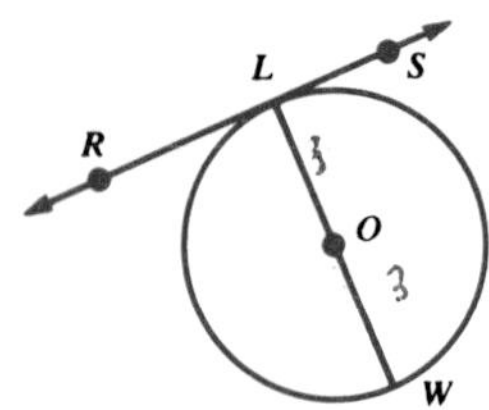

Find the length of $\overline{RW}$ (if drawn) for each of the following situations.

6. $RL = 8$ and $LW = 15$
7. $RL = 11$ and $LO = 3$
8. $RL = 3\sqrt{2}$ and $LW = 5\sqrt{2}$
9. $RL = 4\sqrt{3}$ and $OW = 6$

If $\overline{KL}$ is a diameter of ⊙C and $\overrightarrow{KP}$ is tangent to ⊙O at K, find the exact measure indicated by each ?.

	KC	KL	KP	PL
10.	3	?	8	?
11.	?	8	6	?
12.	?	12	$2\sqrt{3}$	?
13.	$\sqrt{2}$	?	?	10
14.	?	?	6	7

	KC	KL	KP	PL
15.	6	?	9	?
16.	?	8	?	17
17.	?	?	7	$7\sqrt{2}$
18.	?	$5\sqrt{3}$	?	9
19.	6	?	?	17

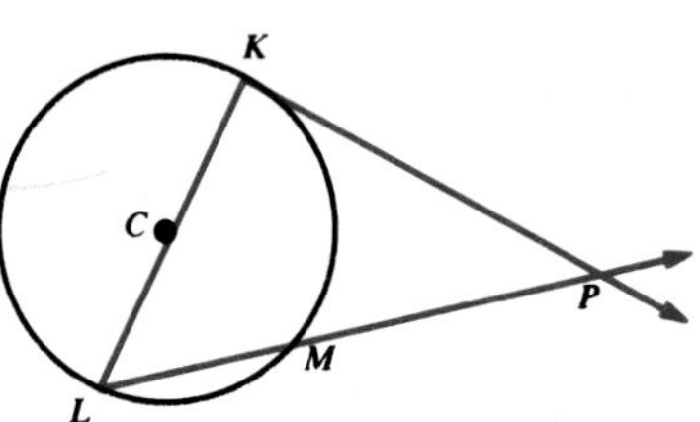

Find the length marked x in each of the following.

20.

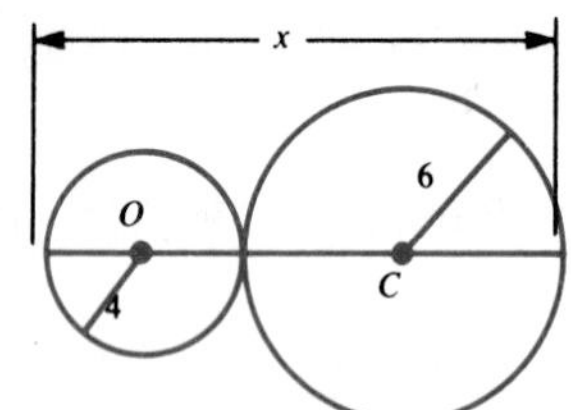

21.

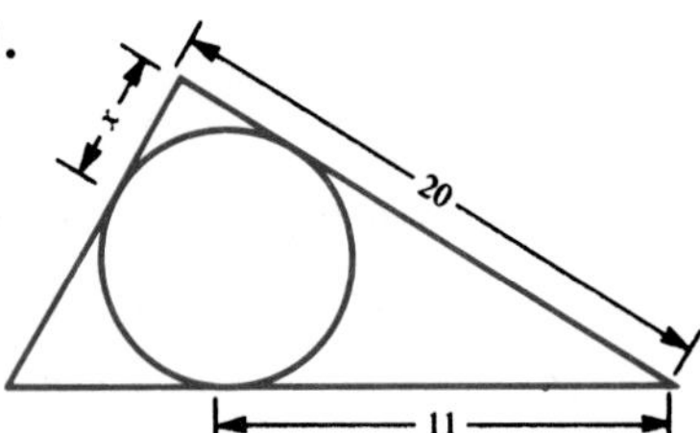

Exercises 22–32 refer to ⊙O and ⊙C which are tangent at W. $\overleftrightarrow{KM}$ is tangent to ⊙O and ⊙C.

22. What is true about $\overline{OK}$ and $\overline{KL}$? $\overline{CL}$ and $\overline{KL}$? Why?
23. What is true about $\overline{OK}$ and $\overline{CL}$? Why?
24. What is true about $\triangle OKM$ and $\triangle CLM$? Why?

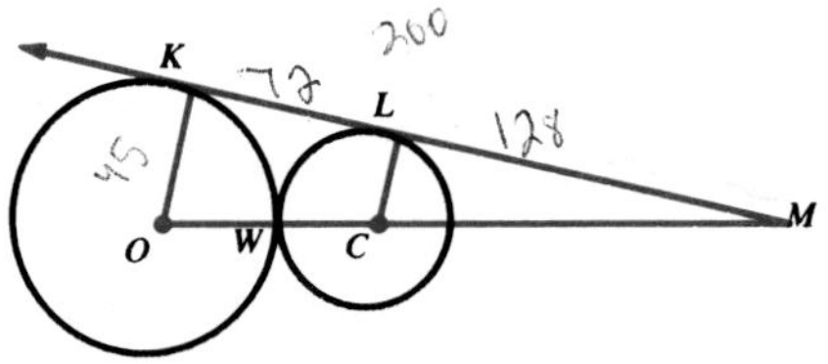

Find the measure indicated by ? if possible. If not possible, explain why.

25. $OK = 9$	$KM = 40$	$OM = ?$	
26. $OK = 45$	$KM = 200$	$KL = 72$	$LC = ?$
27. $OK = 6$	$KM = 8$	$LM = 2$	$LC = ?$
28. $OK = 6$	$KM = 8$	$LM = 2$	$OC = ?$
29. $OW = 45$	$WC = 20$	$KM = 108$	$LM = ?$
30. $KL = 12$	$OK = 9$	$OC = 13$	$LM = ?$

31. $LM = 14$	$CM = 15$	$WC = ?$
32. $LM = 12$	$CM = 14$	$OK = ?$

Copy these figures onto your paper. Then sketch in all common internal and external tangents.

33. 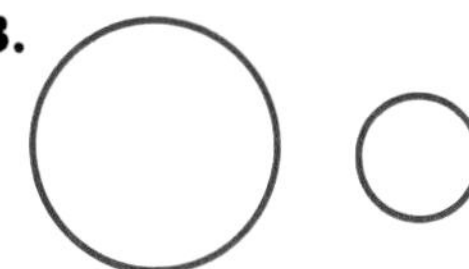

34. 

35. 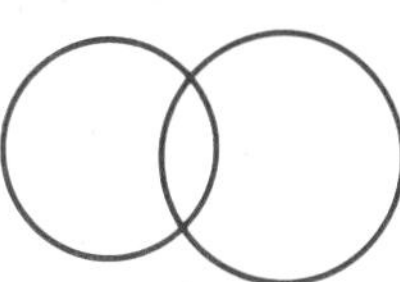

36. Draw a sketch of two circles having two common external tangents and one common internal tangent.

37. Justify this statement, and include a sketch. If circle O and circle C are externally tangent at point T, and the common tangent line is t, then $\overleftrightarrow{OC} \perp t$.

For exercises 38–39, $\overleftrightarrow{AB}$ and $\overleftrightarrow{RS}$ are tangent to both $\odot O$ and $\odot C$. Show that $AB = RS$.

38. 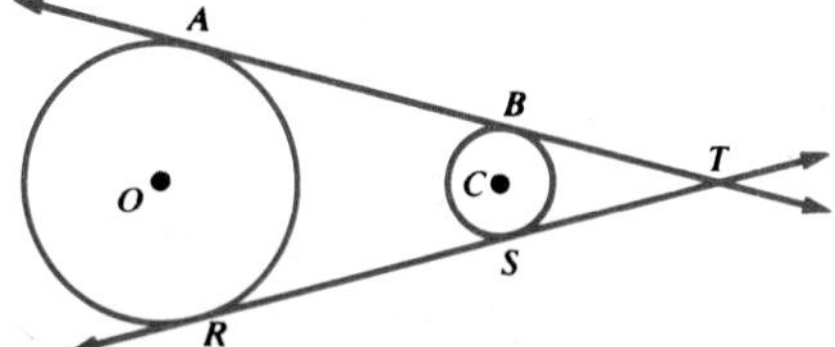

39. 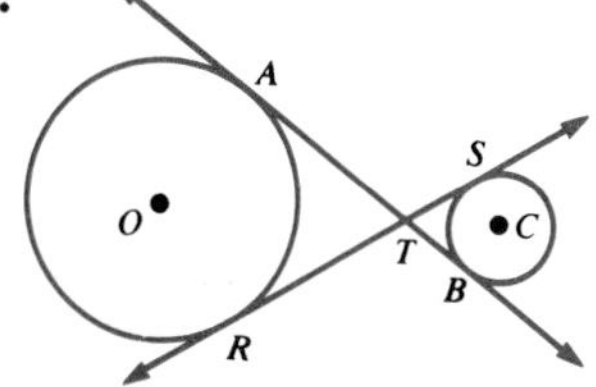

40. Prove Theorem 6–3. Include a figure.

Challenge

1. Prove Theorem 6–2. (Hint: Use indirect proof. Let Q be a point $\neq P$ on the "alleged" tangent. Show $OQ < OP$. Why is this a contradiction?)
2. Prove: If $\overline{AB}$ is a diameter of $\odot O$, and t is tangent to $\odot O$ at A, and r is tangent to $\odot O$ at B, then $t \parallel r$.
3. Prove the Shortest Distance Theorem. Use the following plan.

   Given: Line ℓ; $\overline{PD} \perp \ell$, E on ℓ, $E \neq D$

   Prove: $PD < PE$

   Plan: Use the facts you learned about inequalities in angle measure of triangles, and the sides opposite them. Show $PD < PE$ because, in $\triangle PED$, $m\angle E < m\angle D$.

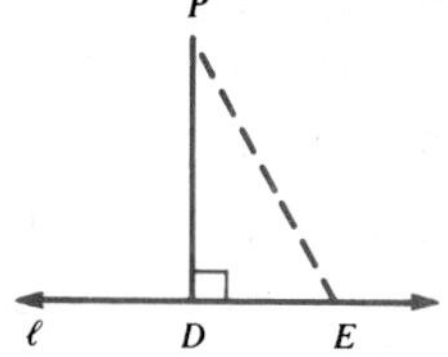

Mathematical Excursions

A Problem of Tangent Circles

Two congruent circles are externally tangent. A line segment is drawn from the center of one circle tangent to the other circle. The length of the tangent segment is 24. Find the length of a radius of one of these congruent circles.

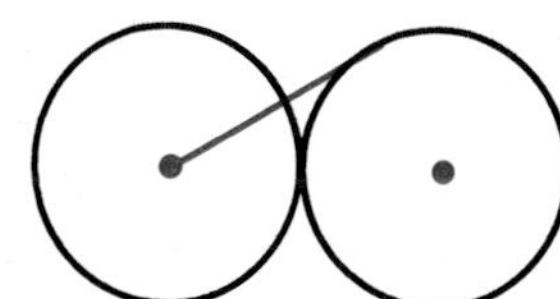

6.4 Chords

This section is devoted to investigating several important facts about chords of a circle. Below are some key ideas with which you should be familiar.

Reflexive Property of Equality

Every number, or quantity, is equal to itself.

Substitution Property of Equality

If two elements are equal, then one may be substituted for the other in any expression.

Addition Property of Equality

For real numbers a, b, c, if $a = b$, then $a + c = b + c$.

Multiplication Property of Equality

For real numbers a, b, c, if $a = b$, then $ac = bc$.

Some Basic Geometric Axioms and Theorems

1. For any two points, there exists one and only one line containing them.
2. If parallel lines are cut by a transversal, then the alternate interior angles formed have equal measure.
3. The measure of the exterior angle of a triangle is equal to the sum of the measures of the remote interior angles.
4. If two sides of a triangle are congruent, then the angles opposite them are congruent. (The converse of this statement is also true.)
5. Two triangles are congruent if the following pairs of correponding parts are congruent.
   a. Three sides (SSS Axiom)
   b. Two sides and the included angle (SAS Axiom)
   c. Two angles and the included side (ASA Axiom)
   d. Two angles and a non-included side (AAS Theorem)
   e. In a right triangle, the hypotenuse and one leg (HL Theorem)
6. Two triangles are similar if two pairs of corresponding angles are congruent (AA Axiom).

Now, let us consider some theorems dealing with the relationship between the chords and arcs of circles.

Theorem 6–4

In the same circle, or in equal circles, equal chords have equal arcs.

To simplify terminology, we will use the term equal *to indicate equal measures that result directly from the understood congruences.*

Proof of Theorem 6–4

Given: Equal circles O and C, $AB = RS$
Prove: $m\widehat{AB} = m\widehat{RS}$

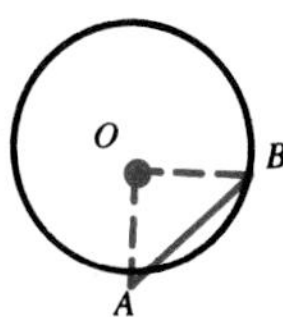

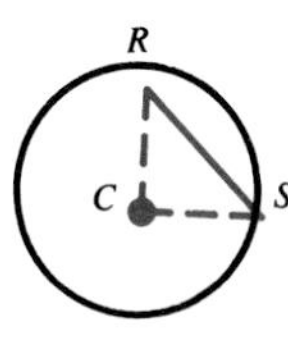

STATEMENTS	REASONS
1. Draw $\overline{OA}$, $\overline{OB}$, $\overline{CR}$, $\overline{CS}$.	1. For any two points, there exists one and only one line containing them.
2. $OA = CR$; $OB = CS$	2. Radii of equal circles are equal.
3. $AB = RS$	3. Given
4. $\triangle OAB \cong \triangle CRS$	4. SSS Axiom
5. $m\angle O = m\angle C$	5. CPCTC (Corresponding Parts of Congruent Triangles are Congruent.)
6. $m\widehat{AB} = m\angle O$; $m\widehat{RS} = m\angle C$	6. The measure of an arc is determined by the measure of its central angle.
7. $m\widehat{AB} = m\widehat{RS}$	7. Substitution Property of Equality

Theorem 6–5

In the same circle, or in equal circles, equal arcs have equal chords.

Theorem 6–6

A line passing through the center of a circle and perpendicular to a chord bisects the chord and its arcs.

Proof of Theorem 6–6

Given: $\odot O$, chord $\overline{AB}$ intersecting diameter $\overline{CD}$ at E, $\overline{CD} \perp \overline{AB}$
Prove: $AE = EB$, $m\widehat{AD} = m\widehat{DB}$, $m\widehat{AC} = m\widehat{CB}$

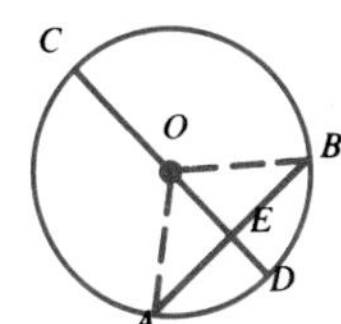

STATEMENTS	REASONS
1. Draw $\overline{OA}$ and $\overline{OB}$.	1. For any two points, there exists one and only one line containing them.
2. $OA = OB$	2. Radii of the same circle are equal.
3. $OE = OE$	3. Reflexive Property of Equality
4. $\overline{CD} \perp \overline{AB}$	4. Given
5. $\triangle OEA$ and $\triangle OEB$ are right triangles.	5. Definition of right triangle
6. $\triangle OEA \cong \triangle OEB$	6. HL Theorem
7. $AE = EB$; $m\angle AOD = m\angle BOD$	7. CPCTC.

8. $m\angle AOC = m\angle BOC$	**8.** If two angles are supplementary to congruent angles, then they are congruent to each other.
9. $m\widehat{AD} = m\angle AOD$, $m\widehat{DB} = m\angle BOD$, $m\widehat{AC} = m\angle AOC$, $m\widehat{CB} = m\angle BOC$	**9.** The measure of an arc is determined by the measure of its central angle.
10. $m\widehat{AD} = m\widehat{DB}$, $m\widehat{AC} = m\widehat{CB}$	**10.** Substitution Property of Equality

Theorem 6–7 In the same circle, or in equal circles, equal chords are equally distant from the center.

Theorem 6–8 In the same or in equal circles, chords equally distant from the center are equal.

You will be asked to prove Theorems 6–5, 6–7, and 6–8 in the exercises.

Exercises

Exploratory **Exercises 1–14 refer to $\odot O$. If $\overline{AB} \perp \overline{CD}$, find each measure indicated by ?.**

1. $CE = 3, OE = 4, OD = ?$
2. $OC = 17, OE = 8, CD = ?$
3. $CD = 24, OE = 5, OB = ?$
4. $OD = 10, CD = 16, OE = ?$
5. $OB = 17, CD = 30, OE = ?$
6. $OE = 48, CD = 28, OC = ?$
7. $OC = 30, OE = 24, CD = ?$
8. $m\angle COD = 90, OE = 7, CD = ?$
9. $m\angle COD = 60, OD = 16, OE = ?$
10. $m\angle COD = 90, OD = \sqrt{18}, OE = ?$
11. $OA = 17, AE = 2, CD = ?$
12. $OA = 41, AE = 1, CD = ?$
13. $CD = 24, AE = 8, OD = ?$
14. $CD = 16, AE = 2, OB = ?$

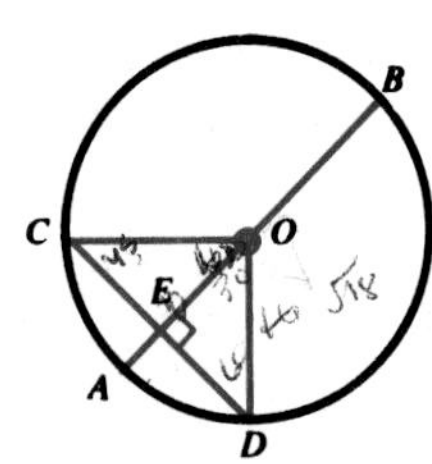

Written **Find the measure marked *k* in each case.**

1.
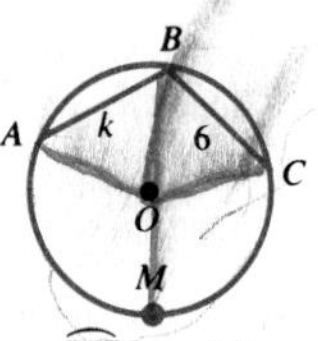

$m\widehat{BC} = 75$
$m\widehat{AMC} = 210$

2.
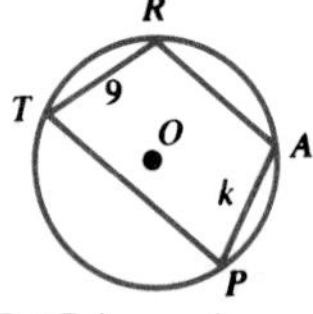

TRAP is an isosceles trapezoid with $m\widehat{TR} = 58$.

3.
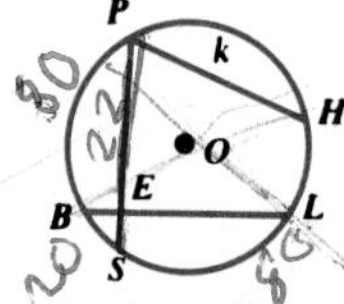

$PS = 22, m\widehat{PB} = 80$
$m\widehat{BS} = 20, m\widehat{SH} = 160$

4.
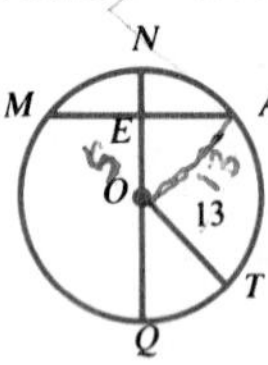

$OE = 5, MA = k, \overline{MA} \perp \overline{NQ}$

5.
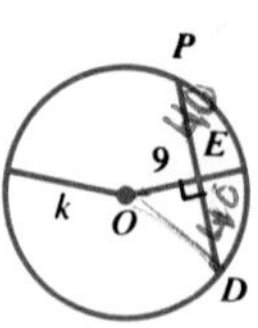

$OE = 9, PD = 80$

6.
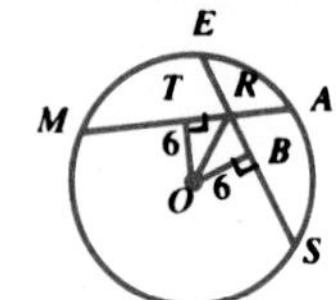

$MA = 13, ES = k$

7. What is the logical relationship between Theorems 6–5 and 6–4?
8. Re-prove Theorem 6–4 for the case of equal chords in the same circle.
9. Prove Theorem 6–5.
   Given: Equal circles O and C, $m\widehat{AB} = m\widehat{RS}$
   Prove: $AB = RS$
10. Prove Theorem 6–7.
    Given: Equal circles O and C, $AB = RS$, $\overline{OE} \perp \overline{AB}$, $\overline{CM} \perp \overline{RS}$
    Prove: $OE = CM$
11. Prove Theorem 6–8.
    Given: Equal circles O and C, $OE = CM$, $\overline{OE} \perp \overline{AB}$, $\overline{CM} \perp \overline{RS}$
    Prove: $AB = RS$

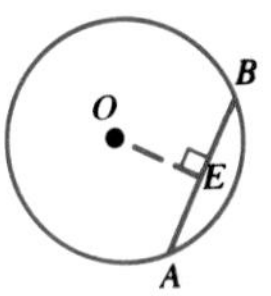

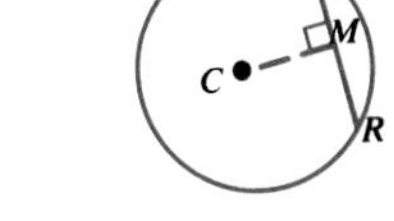

Exercises 9–11

12. How is this statement related to Theorem 6–6? "A line passing through the center of a circle and perpendicular to a chord bisects the chord."
13. Write the converse of the statement in exercise 12. Begin, "If a line bisects a chord, . . ."
14. Prove or disprove the statement you formed in exercise 13.
15. What is the converse of Theorem 6–6? Begin, "If a line bisects a chord and its arc, . . ."
16. **Prove:** A line which bisects a chord and its arc is perpendicular to the chord.
17. Give a convincing argument for this statement: "The perpendicular bisector of a chord of a circle passes through the center of the circle." (You will have to use the definition of *center*.)

Prove each of the following.

18. **Given:** $AE = EB$
    Prove: $m\widehat{AM} = m\widehat{MB}$

19. **Given:** $m\angle A = m\angle B$
    Prove: $m\widehat{BC} = m\widehat{AC}$

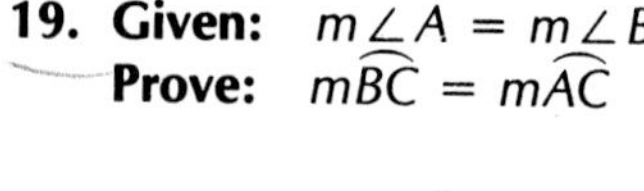

20. **Given:** $\overline{CB}$ is a diameter of $\odot O$, $\overline{AB} \parallel \overline{CD}$
    Prove: $AB = CD$

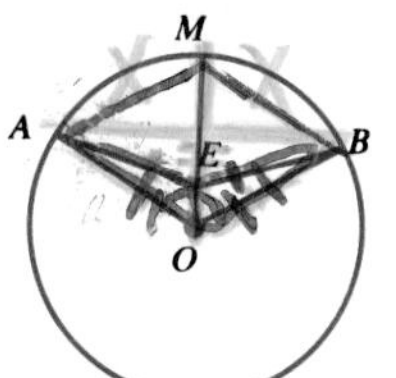

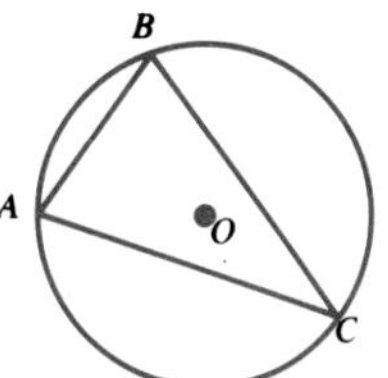

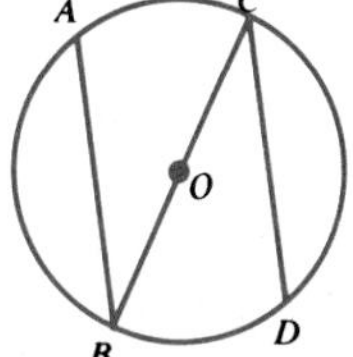

21. **Given:** $\odot O$, $\overline{AB} \parallel \overline{CD}$, O is the midpoint of $\overline{KL}$
    Prove: $AB = CD$

22. **Given:** $m\widehat{AB} = m\widehat{BC}$, $\overline{OX} \perp \overline{BC}$, $\overline{OY} \perp \overline{AB}$
    Prove: $\triangle OXB \cong \triangle OYB$

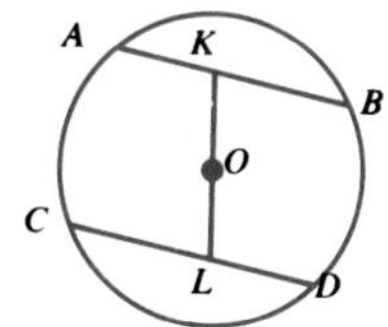

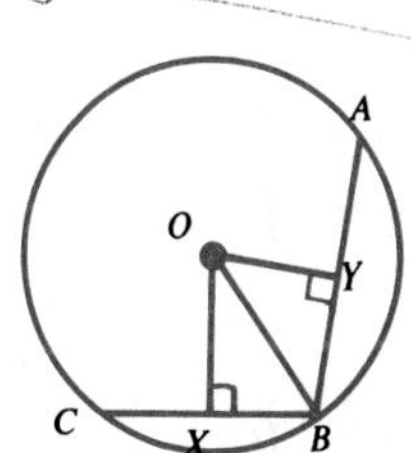

6.5 Inscribed Angles

We know that a central angle has its vertex at the center of a circle, and that its measure is equal to that of the arc it intercepts. In this section, we will look at the relationship between an *inscribed* angle and the arc it intercepts.

Definition of Inscribed Angle

An **inscribed angle** is an angle whose vertex is on a circle and whose sides contain chords of the circle.

Theorem 6–9

An inscribed angle is measured by one-half its intercepted arc. That is, the measure of the arc is twice the measure of the inscribed angle.

There are three cases in the proof of Theorem 6–9. Each of these is illustrated below.

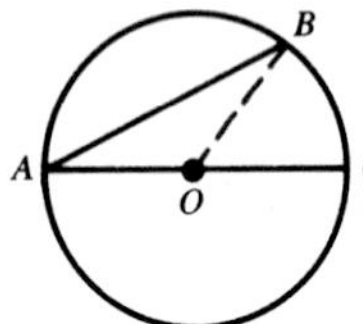

Case I
One side of $\angle BAC$ is a diameter of $\odot O$.

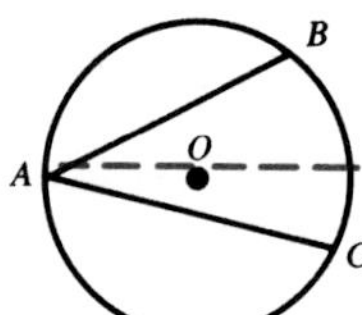

Case II
Sides of $\angle BAC$ are on opposite sides of diameter of $\odot O$.

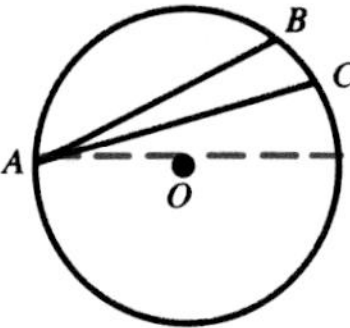

Case III
Both sides of $\angle BAC$ are on same side of diameter of $\odot O$.

Proof of Case I of Theorem 6–9

Given: $\odot O$ with diameter $\overline{AC}$ **Prove:** $m\angle BAC = \frac{1}{2} m\widehat{BC}$

STATEMENTS	REASONS
1. Draw radius $\overline{OB}$	1. For any two points, there exists one and only one line containing them.
2. $OB = OA$	2. Radii of the same circle are equal.
3. $m\angle BAC = m\angle ABO$	3. If two sides of a triangle are congruent, then the angles opposite them are congruent.
4. $m\angle BOC = m\angle BAC + m\angle ABO$	4. The measure of an exterior angle of a triangle is equal to the sum of the measures of the two remote interior angles.
5. $m\angle BOC = 2m\angle BAC$	5. Substitution Property of Equality
6. $m\angle BOC = m\widehat{BC}$	6. The measure of an arc is determined by the measure of its central angle.
7. $2m\angle BAC = m\widehat{BC}$	7. Substitution Property of Equality
8. $m\angle BAC = \frac{1}{2} m\widehat{BC}$	8. Multiplication Property of Equality

A *corollary to a theorem* is itself a theorem, but one which follows directly from the first theorem. Here are three corollaries to Theorem 6–9.

Corollaries to Theorem 6–9

1. If two inscribed angles intercept the same arc, they have the same measure.
2. An angle inscribed in a semicircle is a right angle.
3. If a quadrilateral is inscribed in a circle, then its opposite angles are supplementary.

The proofs of Cases II and III of Theorem 6–9 and these corollaries are left for the exercises.

Theorem 6–10 gives us an important fact about chords which can be proven with the aid of Theorem 6–9. You will be asked to do the proof in the exercises.

Theorem 6–10

If two chords intersect within a circle, then the product of the lengths of the segments of one chord is equal to the product of the lengths of the segments of the other.

Examples

1 **Find $m\angle DFC$ and $m\widehat{AB}$ if $m\widehat{AD} = 100$, $m\widehat{BC} = 60$, and $m\angle DAC = 75$.**

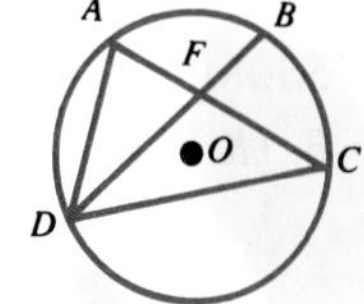

Using Theorem 6–9, we have $m\angle FDC = 30$ and $m\angle FCD = 50$. Since the sum of the measures of the angles of a triangle is 180,

$$m\angle DFC = 180 - m\angle FCD - m\angle FDC$$
$$= 180 - 50 - 30 \text{ or } 100.$$

Again, by Theorem 6–9, $m\widehat{DC} = 2(m\angle DAC)$ or 150.

$$m\widehat{AB} = 360 - m\widehat{AD} - m\widehat{DC} - m\widehat{BC}$$
$$= 360 - 100 - 150 - 60 \text{ or } 50$$

2 **Find the lengths of all four chord segments shown.**

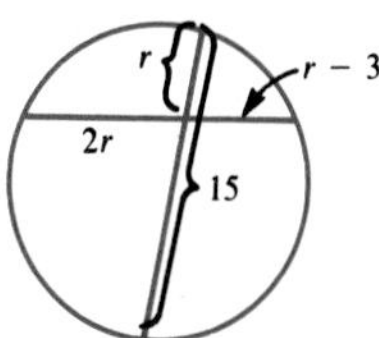

Using Theorem 6–10, we have

$$2r(r - 3) = r(15 - r).$$
$$2r^2 - 6r = 15r - r^2$$
$$3r^2 - 21r = 0$$
$$3r(r - 7) = 0$$
$$3r = 0 \text{ or } r - 7 = 0$$
$$r = 0 \text{ or } \quad r = 7$$

Since $r = 0$ gives us a negative length, we reject that value. Thus the lengths of the four segments are 4 units, 7 units, 8 units, and 14 units.

Exercises

Exploratory

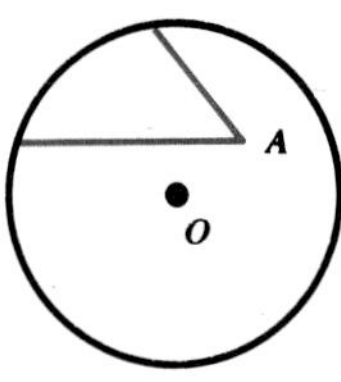

Exercise 2

1. What is the difference between a central angle and an inscribed angle?
2. Is angle A inscribed, central, or neither? Can angle A be interpreted as one of the types of angles of this section?
3. In $\odot D$, G is an inscribed angle, and H is a central angle intercepting the same arc. What is the relationship between $m\angle G$ and $m\angle H$?

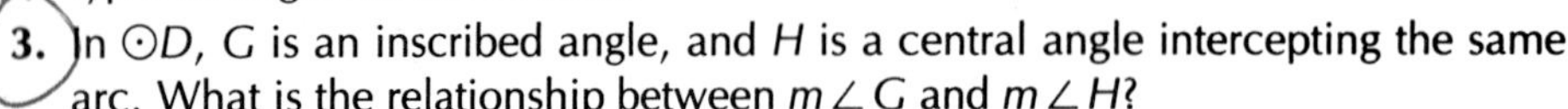

For exercises 4–21, refer to $\odot O$ given below. If $m\widehat{MR} = 70$, $m\widehat{ST} = 54$, and $m\angle VRO = 18$, find the measure of each arc or angle indicated.

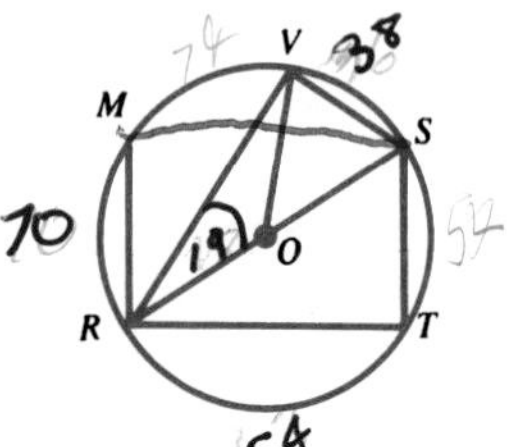

4. $\widehat{VS}$
5. $\widehat{VT}$
6. $\widehat{RT}$
7. $\widehat{MV}$
8. $\widehat{MSR}$
9. $\angle MRT$
10. $\angle VOS$
11. $\angle RVO$
12. $\widehat{RMS}$
13. $\angle MRV$
14. $\angle MRS$
15. $\angle ROV$
16. $\angle RTS$
17. $\angle RST$
18. $\angle RSV$
19. $\angle TSV$
20. Does $RV = RT$? Why or why not?
21. If $\overline{MS}$ were drawn, what kind of triangle would RMS be? Why?

Exercises 22–26 refer to $\odot Q$ at the right. Find each measure indicated by ?.

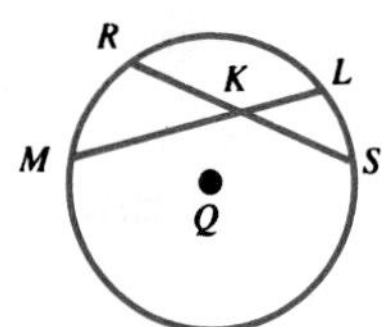

22. $KS = 5$, $KL = 3$, $KM = 10$, $RK = ?$
23. $KM = 7$, $RK = 14$, $KS = 2$, $KL = ?$
24. $RK = 30$, $KL = 20$, $ML = 80$, $RS = ?$
25. $RS = 55$, $MK = 30$, $KS = 10$, $KL = ?$
26. $KS = 8$, $LK = 4$, $KM = 10$, $RS = ?$

Written

Exercises 1–15 refer to $\odot O$ at the right. Find all possible values of r for each situation.

1. $AE = 6$, $EB = 4$, $ED = 8$, $CE = r$
2. $AB = 10$, $AE = 7$, $ED = 6$, $CE = r$
3. $AE = 9$, $EB = 6$, $CE = 6$, $CD = r$
4. $AE = 8$, $EB = 5$, $m\widehat{AB} = m\widehat{CD}$, $CD = r$
5. $CE = r$, $ED = 2r$, $AE = EB = 4$
6. $ED = r$, $CE = r + 5$, $AE = 2r$, $EB = r + 1$
7. $EB = r$, $AB = 8$, $CE = r + 1$, $ED = r + 2$
8. $AB = 30$, $CD = 33$, $CE = 8$, $EB = r$
9. $CE = ED = r + 6$, $EB = r$, $AE = 6r + 1$
10. $EB = r$, $AE = 8r$, $CE = 4r$, $ED = r + 1$
11. $CE = 6$, $ED = 4$, $AE:EB = 6:1$, $AE = r$
12. $AE = 16$, $EB = 5$, $CE:ED = 4:5$, $CE = r$
13. $AE = r$, $ED = r - 3$, $EB = r - 1$, $CE = 3r - 5$
14. $CE = r$, $EB = r - 5$, $AE = 2r$, $ED = r - 3$
15. $EB = r + 2$, $AE = 2r + 2$, $CE = 3r + 1$, $ED = r + 1$
16. Complete this statement, and explain: "If $\triangle ABC$ is inscribed in a circle in a way such that $\overline{AB}$ is a diameter, then $\triangle ABC$ is ________."
17. In $\odot O$, $\overline{AB}$ is a diameter. Point C on $\odot O$ is selected, and $\overline{AC}$, $\overline{CB}$, and $\overline{OC}$ are drawn. What kind of triangle is $\triangle ABC$? What is true about $\overline{OC}$, $\overline{AO}$, and $\overline{OB}$? Explain.

18. Prove Theorem 6–10: If two chords intersect within a circle, then the product of the lengths of the segments of one chord is equal to the product of the lengths of the segments of the other.

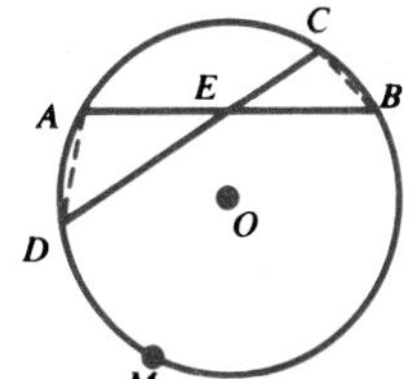

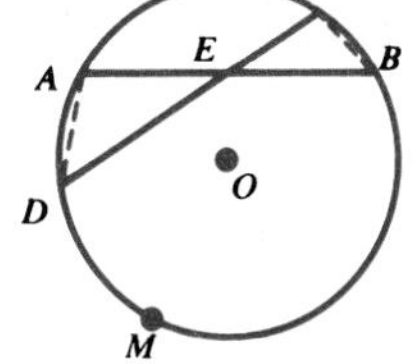

Given: $\odot O$, chords $\overline{AB}$ and $\overline{CD}$ intersecting at E
Prove: $(AE)(EB) = (DE)(EC)$ *Hint: Draw $\overline{AD}$ and $\overline{BC}$.*

Prove each of the following.

19. Case II of Theorem 6–9

20. Case III of Theorem 6–9

21. Corollary 1 of Theorem 6–9

22. Corollary 2 of Theorem 6–9

23. Corollary 3 of Theorem 6–9

24. The length of the median drawn to the hypotenuse of a right triangle is equal to one-half the length of the hypotenuse.

25. Given: S bisects $\widehat{RW}$
Prove: $m\angle A = m\angle B$

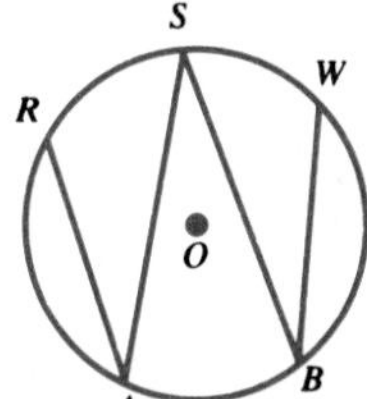

26. Given: $BC = AD$
Prove: $\triangle ADE \cong \triangle BCE$

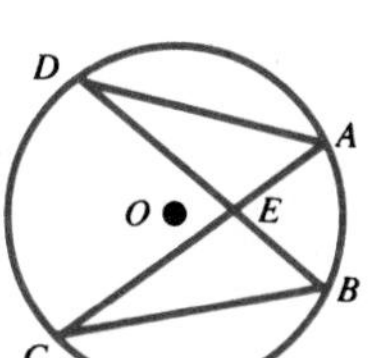

27. Given: $AE = BE$
Prove: $m\widehat{AC} = m\widehat{BD}$

Challenge

1. On $\odot O$, points R, S, T are chosen at random, and $\overline{RS}$, $\overline{ST}$, and $\overline{RT}$ are drawn. Determine the sum of the measures of $\angle R$, $\angle S$, and $\angle T$, and relate it to the sum of the intercepted arcs. Should this result surprise you? Explain fully.

Mixed Review

1. If $A'(3, -5)$ is the image of $A(-2, -3)$ under a translation, find the rule for the translation.

2. If $B'(7, 4)$ is the image of $B(7, -2)$ under a line reflection, find the equation of the line of reflection.

3. Under a dilation with center P and scale factor k, C' is the image of C. Find the value of k if $PC' = 6$ and $CC' = 10$.

4. If \$750 is invested at 7.5% annual interest compounded monthly, what will be the amount in 5 years?

5. If $N = b^3a^{-2}$, express $\log_b N$ in terms of $\log a$ and $\log b$.

6. If $f(x) = 2^x$, write an equation for $f^{-1}(x)$.

7. Three circles are externally tangent to each other. If R, S, and T are the points of tangency, find $m\widehat{RS} + m\widehat{ST} + m\widehat{TR}$.

8. In circle P, chord $\overline{XY}$ is 15 cm from P. If the length of $\overline{XY}$ is 16 cm, what is the radius of circle P?

6.6 Angles Formed by Chords, Tangents, Secants

After studying inscribed angles, a question to be considered is "What other angles can be formed using chords, secants, and tangents?" If we investigate this question, we find that there are five other angles that can be formed. These angles are formed in the following ways.

1. Two chords intersecting within a circle
2. A chord and a tangent intersecting on a circle
3. Two secants intersecting outside a circle
4. A secant and a tangent intersecting outside a circle
5. Two tangents intersecting outside a circle

Theorem 6–11

An angle formed by two chords intersecting within a circle is measured by one-half the sum of the arcs intercepted by the angle and its vertical angle.

Proof of Theorem 6–11

Given: $\odot O$ with chords $\overline{AB}$ and $\overline{CD}$ intersecting at E

Prove: $m\angle x = \frac{1}{2}(m\widehat{AC} + m\widehat{DB})$

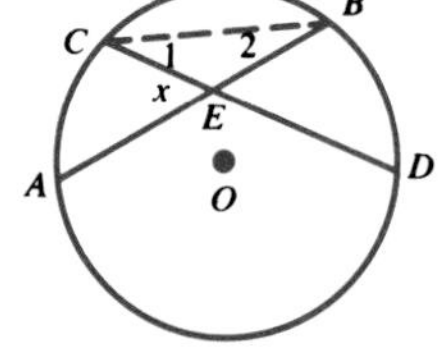

STATEMENTS	REASONS
1. Draw $\overline{CB}$.	1. For any two points, there exists one and only one line containing them.
2. $m\angle x = m\angle 1 + m\angle 2$	2. The measure of an exterior angle of a triangle is equal to the sum of the measures of the two remote interior angles.
3. $m\angle 1 = \frac{1}{2}m\widehat{DB}, m\angle 2 = \frac{1}{2}m\widehat{AC}$	3. An inscribed angle is measured by one-half its intercepted arc.
4. $m\angle x = \frac{1}{2}m\widehat{DB} + \frac{1}{2}m\widehat{AC}$	4. Substitution Property of Equality
5. $m\angle x = \frac{1}{2}(m\widehat{AC} + m\widehat{DB})$	5. Distributive Property

Theorem 6–12

An angle formed by a chord and a tangent is measured by one-half the intercepted arc.

There are three cases in the proof of Theorem 6–12. Each of these is illustrated at the top of the next page.

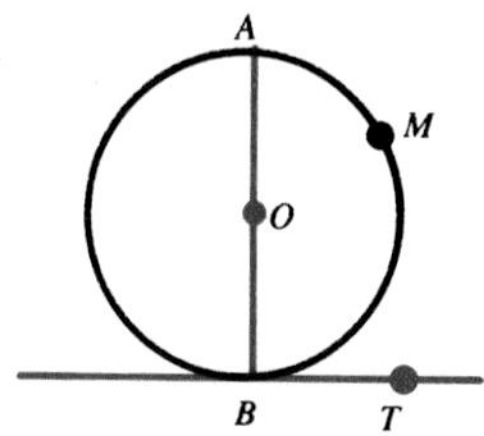

Case I
The chord is a diameter.

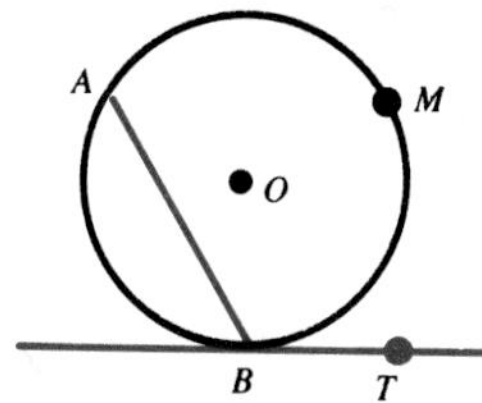

Case II
The center of the circle is in the interior of $\angle ABT$.

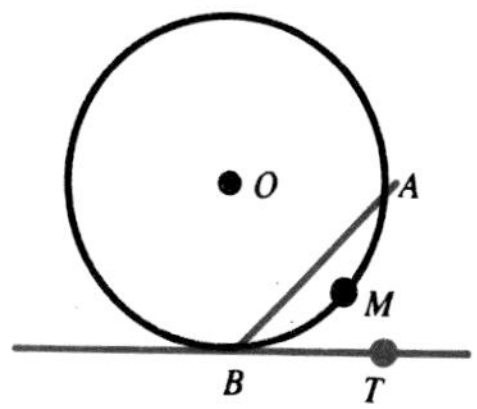

Case III
The center of the circle is in the exterior of $\angle ABT$.

Proof of Case I of Theorem 6–12

Given: $\odot O$ with $\overleftrightarrow{BT}$ tangent to $\odot O$ at B, diameter $\overline{AB}$

Prove: $m\angle ABT = \frac{1}{2}m\widehat{AMB}$

STATEMENTS	REASONS
1. $\overline{AB} \perp \overline{BT}$	1. A line tangent to a circle is perpendicular to the radius drawn to the point of tangency.
2. $m\angle ABT = 90 = \frac{1}{2}(180)$	2. Definition of perpendicular lines
3. $\widehat{AMB}$ is a semicircle.	3. A diameter separates a circle into two semicircles.
4. $m\widehat{AMB} = 180$	4. Measure of a semicircle is 180.
5. $m\angle ABT = \frac{1}{2}m\widehat{AMB}$	5. Substitution Property of Equality

Theorem 6–13

An angle formed by two secants intersecting outside a circle is measured by one-half the difference of the intercepted arcs.

Theorem 6–14

An angle formed by a secant and a tangent intersecting outside a circle is measured by one-half the difference of the intercepted arcs.

Theorem 6–15

An angle formed by two intersecting tangents is measured by one-half the difference of the intercepted arcs.

The proofs of Cases II and III of Theorem 6–12, Theorem 6–13, Theorem 6–14, and Theorem 6–15 are left for the exercises.

Examples

1 **Find $m\angle KAL$ and $m\angle LPT$ if $\overleftrightarrow{PT}$ is tangent to $\odot O$, $m\widehat{KL} = 100$, $m\widehat{LM} = 60$ and, $m\widehat{MP} = 70$.**

By Theorem 6–11, $m\angle KAL = \frac{1}{2}(m\widehat{KL} + m\widehat{MP})$
$= \frac{1}{2}(100 + 70)$ or 85.

By Theorem 6–12, $m\angle LPT = \frac{1}{2}m\widehat{LP}$
$= \frac{1}{2}(m\widehat{LM} + m\widehat{MP})$
$= \frac{1}{2}(60 + 70)$ or 65.

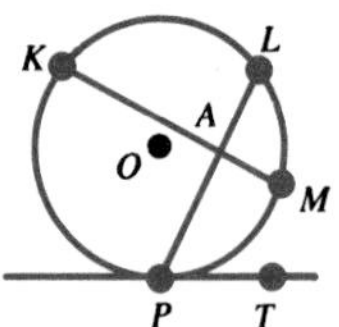

2 **Find $m\angle BRC$, $m\angle CRD$, and $m\angle ARD$, if $\overrightarrow{RA}$ and $\overrightarrow{RD}$ are tangents to $\odot O$, $m\widehat{ED} = 40$, $m\widehat{EF} = 32$, and $m\widehat{AB} = m\widehat{BC} = m\widehat{CD} = 82$.**

By Theorem 6–13, $m\angle BRC = \frac{1}{2}(m\widehat{BC} - m\widehat{EF})$
$= \frac{1}{2}(82 - 32)$ or 25.

By Theorem 6–14, $m\angle CRD = \frac{1}{2}(m\widehat{CD} - m\widehat{DE})$
$= \frac{1}{2}(82 - 40)$ or 21.

By Theorem 6–15, $m\angle ARD = \frac{1}{2}(m\widehat{ABD} - m\widehat{AD})$
$= \frac{1}{2}(246 - 114)$ or 66.

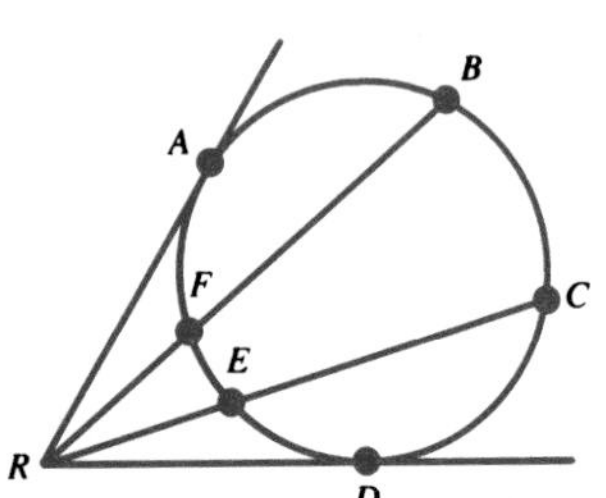

Exercises

Exploratory **Exercises 1–9 refer to $\odot O$ at the right. If $m\widehat{SR} = 80$ and $m\widehat{RT} = 70$, then find the measure indicated.**

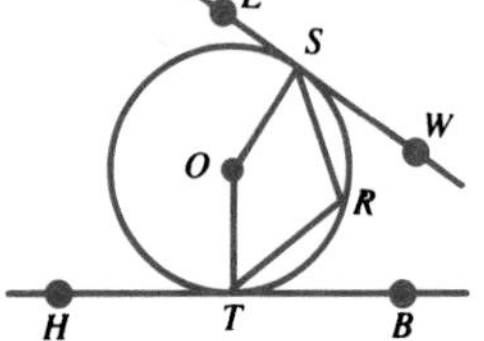

1. $m\angle RSW$
2. $m\angle RTB$
3. $m\angle TRS$
4. $m\angle LSR$
5. $m\angle HTR$
6. $m\angle OTH$
7. $m\angle OTR$
8. $m\angle TOS$
9. $m\angle OSR$
10. Do the theorems in this section cover the case of two secants intersecting within a circle? two tangents intersecting within a circle? Explain.

Written **In exercises 1–12, find the measure of each arc or angle indicated by x.**

1. 40°, x°, O, 80°

2.

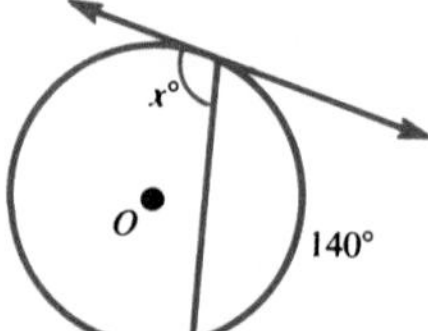

3.

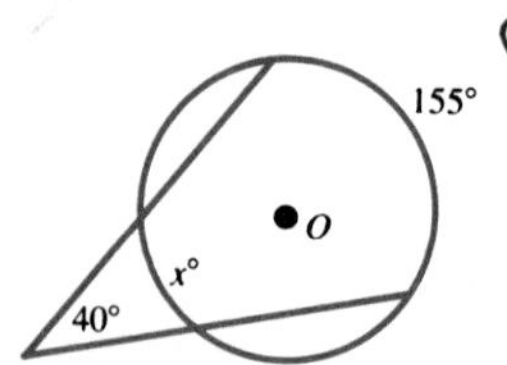

4.

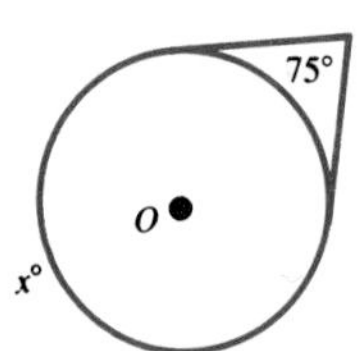

5.

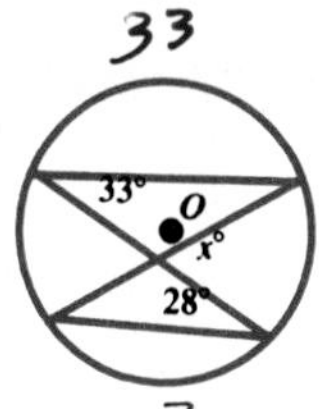

6.

150°
x°
25°
O

7.

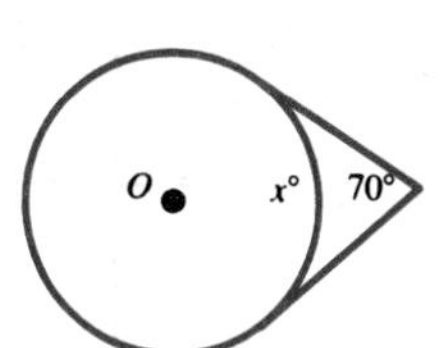

8.

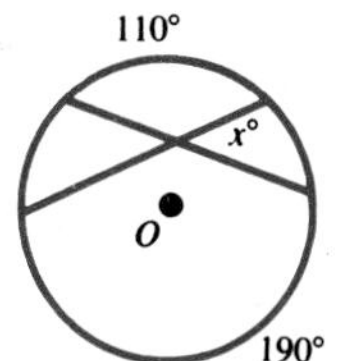

9.

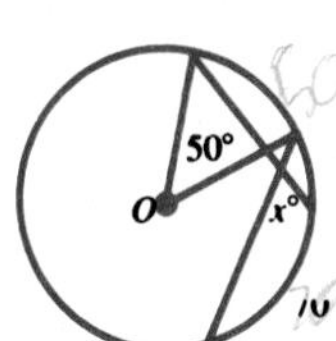

10.

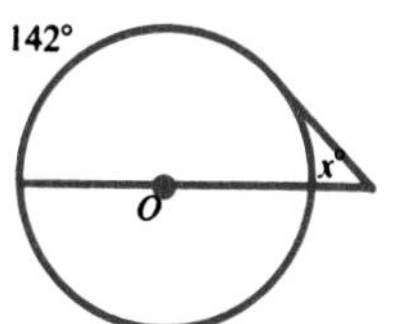

11.

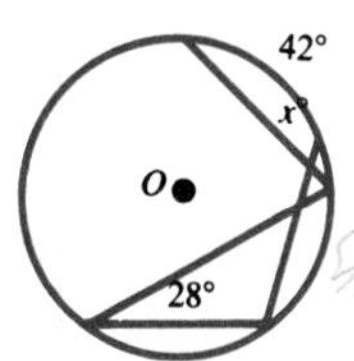

12.

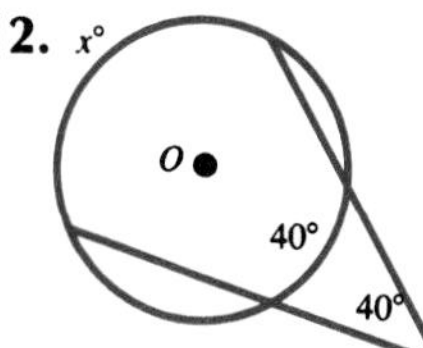

Exercise 13–30 refer to ⊙O at the right. If $AB = CD$, $m\angle FGE = 60$, $m\widehat{AC} = 104$, and $m\widehat{AB} : m\widehat{BC} = 25:1$, find each of the following.

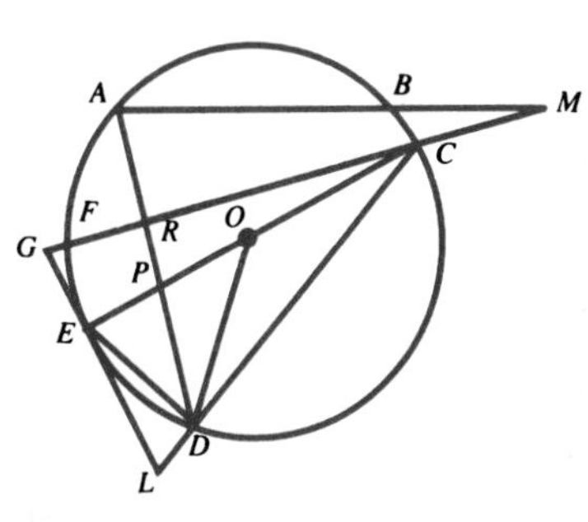

13. $m\widehat{AB}$
14. $m\widehat{BC}$
15. $m\widehat{CD}$
16. $m\widehat{ED}$
17. $m\angle DEL$
18. $m\angle COD$
19. $m\angle ECD$
20. $m\angle EPD$
21. $m\angle ODC$
22. $m\widehat{EF}$
23. $m\widehat{AF}$
24. $m\angle FCE$
25. $m\angle ARC$
26. $m\angle CPD$
27. $m\angle AMF$
28. $m\angle ELD$
29. $m\angle GEC$
30. $m\angle PDO$

Copy and complete the following proof.

31. Theorem 6–13: An angle formed by two secants intersecting outside a circle is measured by one-half the difference of the intercepted arcs.

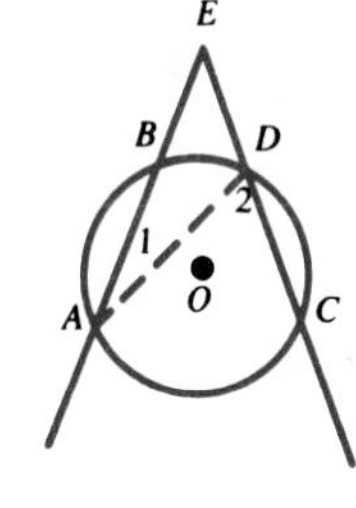

Given: ⊙O with secants $\overleftrightarrow{AB}$ and $\overleftrightarrow{CD}$ intersecting at E

Prove: $m\angle E = \frac{1}{2}(m\widehat{AC} - m\widehat{BD})$

Proof:

STATEMENTS	REASONS
1. Draw $\overline{AD}$	1. ________
2. $m\angle 2 = m\angle 1 + m\angle E$	2. ________
3. $m\angle E = m\angle 2 - m\angle 1$	3. ________
4. $m\angle 2 = \frac{1}{2}m\widehat{AC}$, $m\angle 1 = \frac{1}{2}m\widehat{BD}$	4. ________
5. $m\angle E = \frac{1}{2}m\widehat{AC} - \frac{1}{2}m\widehat{BD}$	5. ________
6. $m\angle E = \frac{1}{2}(m\widehat{AC} - m\widehat{BD})$	6. ________

Prove each of the following.

32. Case II of Theorem 6–12
33. Case III of Theorem 6–12
34. Theorem 6–14
35. Theorem 6–15

6.7 More About Secants and Tangents

In section 6.6, we learned about the angles formed when two secants or a secant and a tangent intersect outside a circle. Now, we will examine the *segments* formed under either of these situations.

Theorem 6–16

Let two secants be drawn to a circle from an outside point. Then the product of the lengths of one secant and its external segment equals the product of the lengths of the other secant and its external segment.

Proof of Theorem 6–16

Given: $\odot O$, secants $\overleftrightarrow{AB}$ and $\overleftrightarrow{CD}$ intersecting at P

Prove: $(PA)(PB) = (PC)(PD)$

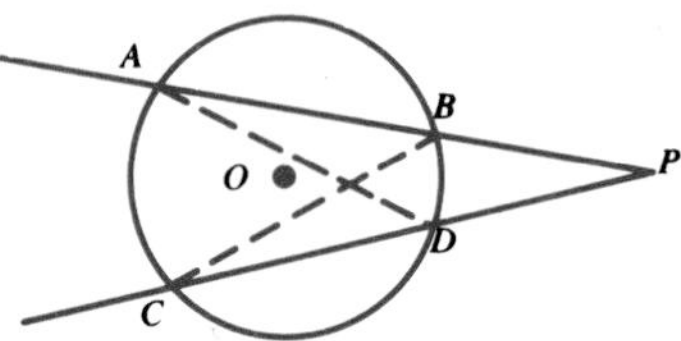

STATEMENTS	REASONS
1. Draw $\overline{AD}$ and $\overline{BC}$.	1. For any two points, there exists one and only one line containing them.
2. $m\angle P = m\angle P$	2. Reflexive Property of Equality
3. $m\angle BAD = m\angle BCD$	3. Corollary 1 of Theorem 6–9
4. $\triangle ADP \sim \triangle CBP$	4. AA Axiom
5. $\frac{PA}{PC} = \frac{PD}{PB}$	5. Definition of similar triangles
6. $(PA)(PB) = (PC)(PD)$	6. Means-Extremes Property

Theorem 6–17

Let a tangent and a secant be drawn to a circle from an outside point. Then the square of the length of the tangent segment is equal to the product of the lengths of the secant and its external segment.

Proof of Theorem 6–17

Given: $\odot O$, $\overleftrightarrow{AP}$ tangent to $\odot O$ at A, secant $\overleftrightarrow{CP}$ intersecting $\overleftrightarrow{AP}$ at P

Prove: $(AP)^2 = (PC)(PB)$

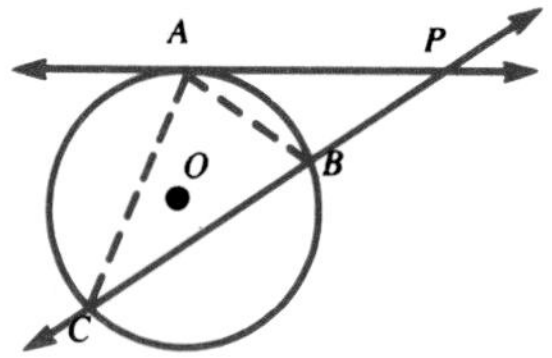

STATEMENTS	REASONS
1. Draw $\overline{AC}$ and $\overline{AB}$.	1. For any two points, there exists one and only one line containing them.
2. $m\angle P = m\angle P$	2. Reflexive Property of Equality

3. $m\angle ACP = \frac{1}{2}m\widehat{AB}$	3. Theorem 6–9
4. $m\angle BAP = \frac{1}{2}m\widehat{AB}$	4. Theorem 6–12
5. $m\angle ACP = m\angle BAP$	5. Substitution Property of Equality
6. $\triangle PAC \sim \triangle PBA$	6. AA Axiom
7. $\frac{PA}{PC} = \frac{PB}{PA}$	7. Definition of similar triangles
8. $(AP)^2 = (PC)(PB)$	8. Means-Extremes Property

Examples

1 **Find the measure indicated by *x*.**

By Theorem 6–16, $(7 + 5)5 = (4 + x)x.$

$$60 = 4x + x^2$$
$$0 = x^2 + 4x - 60$$
$$0 = (x + 10)(x - 6)$$
$$x + 10 = 0 \text{ or } x - 6 = 0$$
$$x = -10 \text{ or } x = 6$$

Since $x = -10$ gives us a negative length, we reject that value. Thus, $x = 6$.

2 **Find the measure indicated by *z*.**

By Theorem 6–17, $6^2 = (2 + z)2.$

$$36 = 4 + 2z$$
$$32 = 2z$$
$$16 = z$$

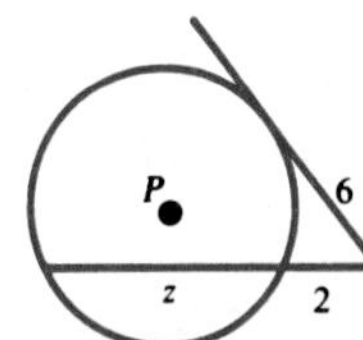

Exercises

Exploratory **Exercises 1–12 refer to ⊙O and ⊙P at the right.**

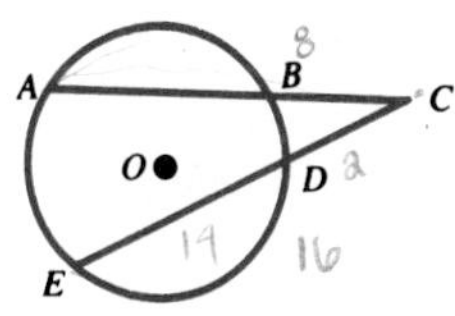

1. Name any secants or tangents, if they exist, in ⊙O and state what equalities must hold according to either Theorem 6–16 or Theorem 6–17.
2. Repeat exercise 1 for ⊙P.

Find each measure indicated by ?.

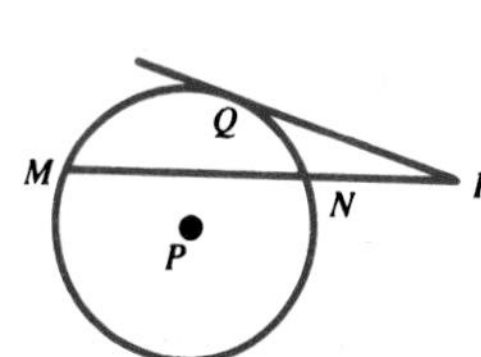

3. $AC = 8$, $EC = 16$, $DC = 2$, $BC = ?$
4. $BC = 6$, $EC = 18$, $DC = 3$, $AC = ?$
5. $BC = 5$, $AB = 15$, $EC = 25$, $DC = ?$
6. $DC = 4$, $AB = 1$, $BC = 7$, $ED = ?$
7. $AC = 15$, $BC = 4$, $EC = 12$, $ED = ?$
8. $NR = 4$, $QR = 10$, $MR = ?$
9. $MR = 27$, $NR = 3$, $QR = ?$
10. $MN = 9$, $NR = 3$, $QR = ?$
11. $MN:NR = 3:1$, $QR = 8$, $MN = ?$
12. $MN = 15$, $QR = 10$, $NR = ?$

Written **In each of the following, find the measure indicated by y or z.**

1.

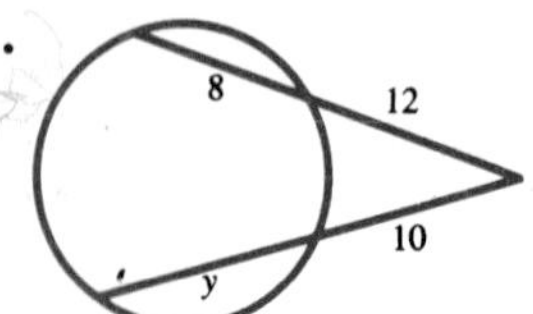

2.

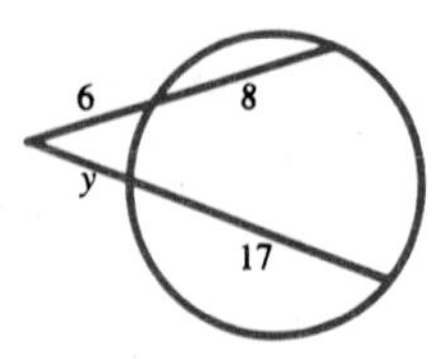

3.

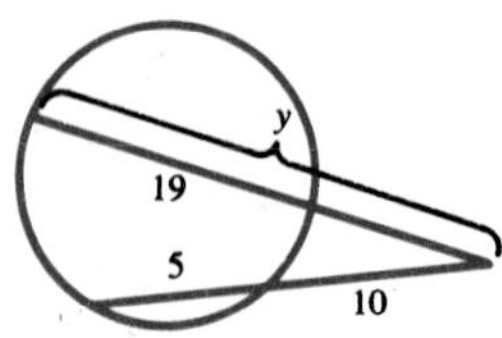

4.

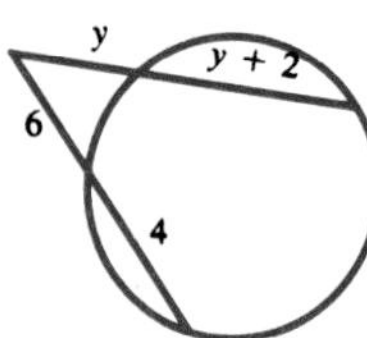

5.

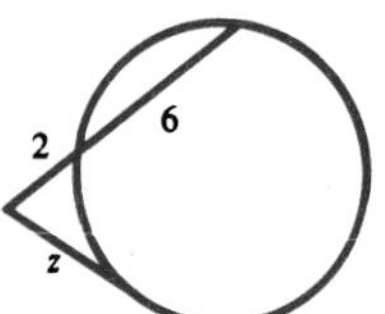

6.

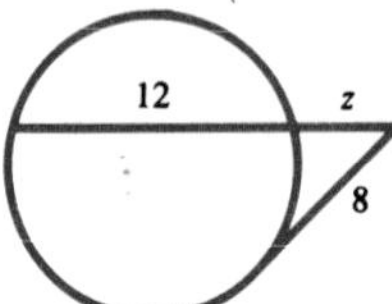

7.

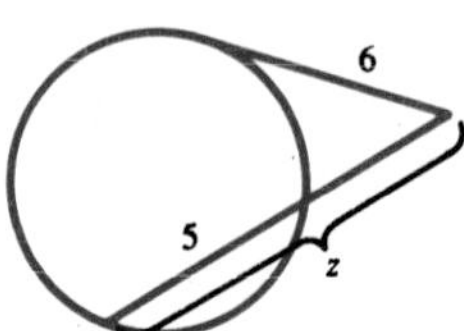

8.

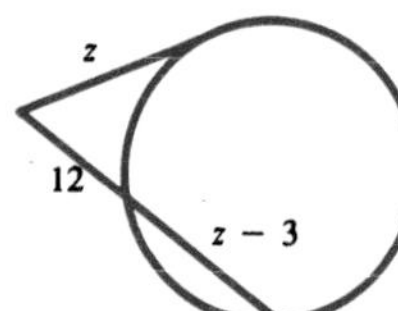

For exercises 9–14, refer to $\odot O$ at the right. If $\overrightarrow{MB}$ is tangent to $\odot O$ at B, find each measure indicated by x and y.

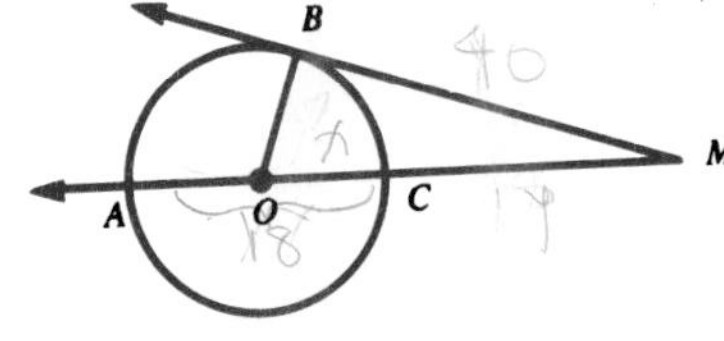

9. $MB = 15$, $MC = 9$, $AC = x$, $OB = y$
10. $AC = 20$, $MC = 16$, $OC = x$, $MB = y$
11. $OB = 7$, $MC = 18$, $AC = x$, $MB = y$
12. $OB = 5$, $MB = 12$, $AC = x$, $MC = y$
13. $AC = 18$, $MB = 40$, $OC = x$, $MC = y$
14. Is there another way that you could solve exercises 9–12? Explain.

For exercises 15–26 refer to $\odot O$ at the right. If $AP = 8$, $BH = 30$, $CH = 32$, $DG = 18$, $QG = 7$, $RE = 12$, $JK = 27$, and $CK = 7$, find each of the following.

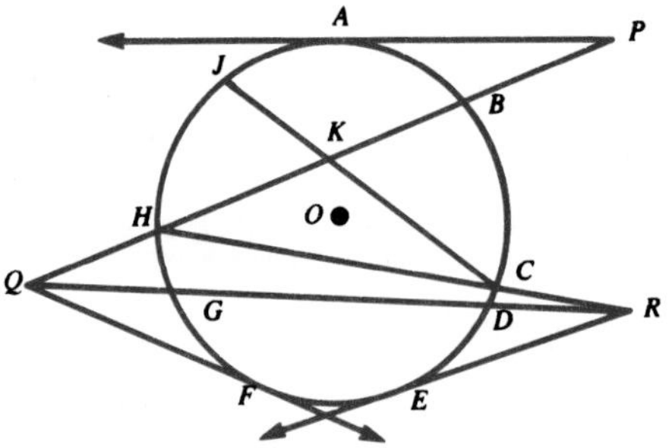

15. PB	16. HQ	17. QF
18. RD	19. RC	20. KB

If $AP = 16$, $PB = 8$, $HQ = 6$, $QG = 5$, $RC = 8$, $RD = 9$, $KB = 10$, and $JC = 33$, find each of the following.

21. BH	22. QF	23. GD
24. CH	25. RE	26. JK

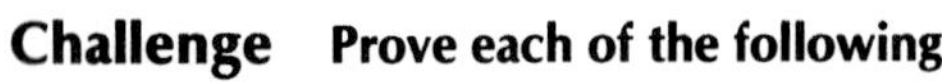

Challenge **Prove each of the following.**

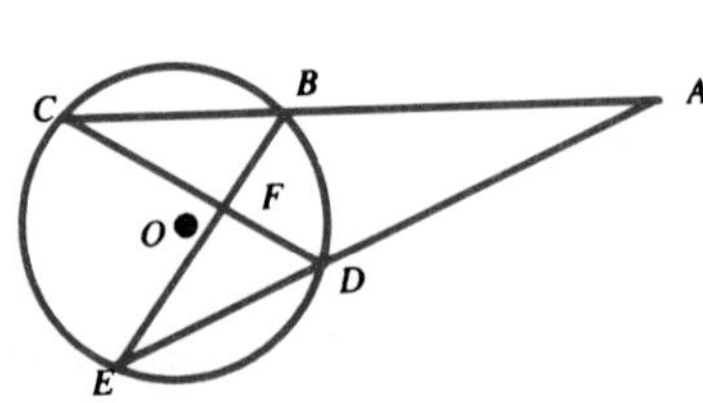

1. **Given:** $AB = AD$
   Prove: $\triangle BCF \cong \triangle DEF$
2. **Given:** $BC = DE$
   Prove: $\triangle ABE \cong \triangle ADC$

6.8 Some Constructions

In previous courses, you studied the basic principles of geometric constructions. Recall that the compass and straightedge were the only implements allowed. Now, we will study some constructions having circles as their major feature. First, we need to review some constructions you have already studied. You should practice each construction a few times before continuing.

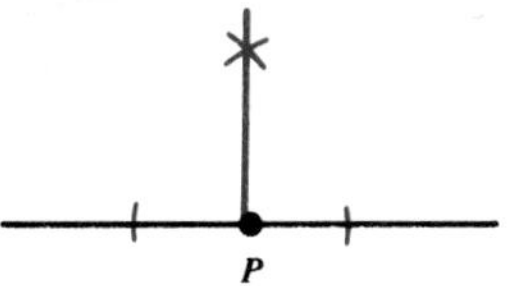

1. To construct a perpendicular to a given line at a given point on the line

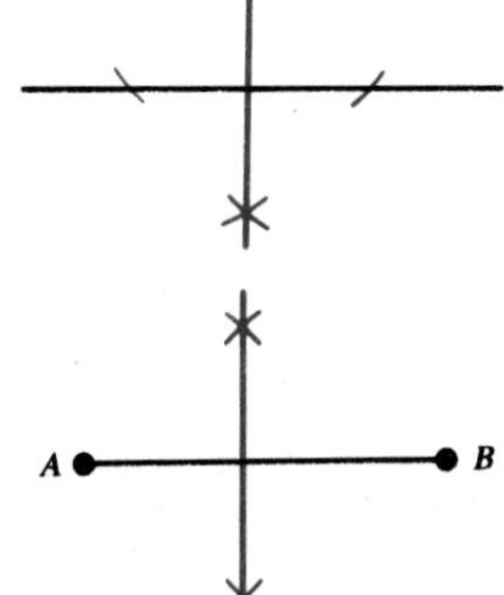

2. To construct a perpendicular to a given line from a given point not on the line

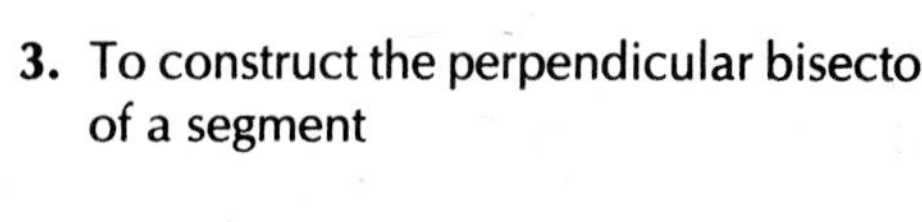

3. To construct the perpendicular bisector of a segment

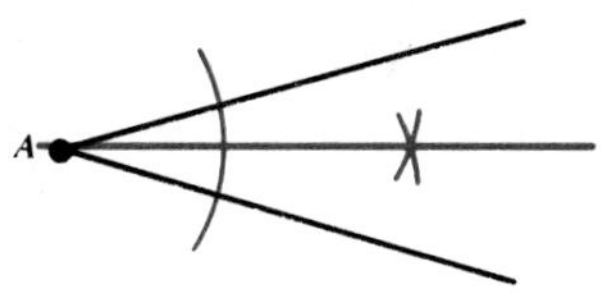

4. To construct the bisector of a given angle

CONSTRUCTION 1: Construct a tangent to a given circle at a given point on the circle.

Given: $\odot O$, point B on $\odot O$
Method:

1. Draw $\overrightarrow{OB}$.

2. Construct a perpendicular to $\overleftrightarrow{OB}$ at B. This line is tangent to $\odot O$ at B.

Justification of this construction is left for the exercises.

CONSTRUCTION 2: Construct a tangent to a circle from an outside point.

Given: $\odot O$, point B outside $\odot O$.
Method:

1. Draw $\overline{OB}$.
2. Find the midpoint of $\overline{OB}$ by constructing its perpendicular bisector. Call the midpoint M.
3. Using M as a center and MO as a radius, draw a circle. $\odot M$ will intersect $\odot O$ at some point A.
4. Draw $\overleftrightarrow{BA}$. This line is tangent to $\odot O$ at A.

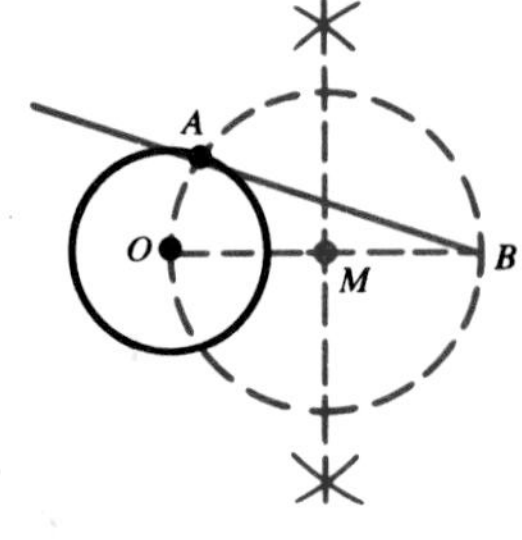

Justification: If $\overline{OA}$ is drawn, then $\angle OAB$ is a right angle, by Corollary 2 of Theorem 6–9. Then, $\overleftrightarrow{BA}$ is perpendicular to radius $\overline{OA}$ at A. Therefore, by Theorem 6–2, $\overleftrightarrow{BA}$ is tangent to $\odot O$ at A.

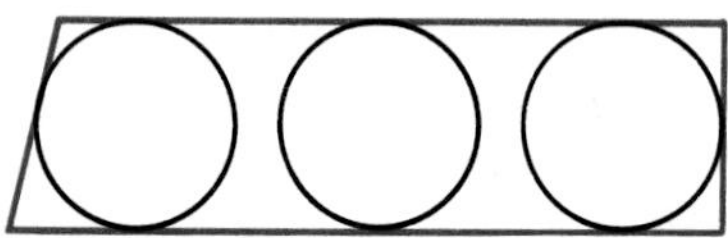

Figure 1

Figure 2

In our discussion of circles, we have assumed that a circle can be inscribed in, or circumscribed about, any regular polygon. But what about any polygon—say any quadrilateral? For a quick answer to that one, look at Figures 1 and 2 at the left. Figure 1 seems to suggest that there exists at least one quadrilateral in which no circle can be inscribed. As for circumscribing a circle around any quadrilateral, try doing that with the quadrilateral in Figure 2!

With triangles, however, things are different. As we will see in Constructions 3 and 4, we can always inscribe a circle in, or circumscribe a circle about, any triangle.

Before examining these constructions, let us recall a few more necessary theorems. These may be familiar to you from previous studies.

Theorem 6–18

If a point lies on the bisector of an angle, then it is equidistant from the sides of the angle.

Theorem 6–19

If a point is equidistant from the sides of an angle, then it lies on the bisector of the angle.

Theorem 6–20

If a point lies on the perpendicular bisector of a segment, then it is equidistant from the ends of the segment.

Theorem 6–21 If a point is equidistant from the ends of a segment, then it lies on the perpendicular bisector of the segment.

Theorem 6–22 The bisectors of the angles of a triangle intersect in a point equidistant from the three sides of the triangle.

Proof of Theorem 6–22

Given: $\triangle ABC$ with bisectors of $\angle A$, $\angle B$, $\angle C$
Prove: Angle bisectors intersect at a unique point equidistant from sides $\overline{AB}$, $\overline{AC}$, $\overline{BC}$.

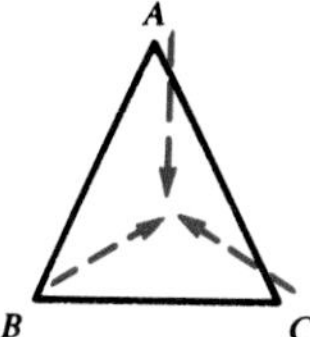

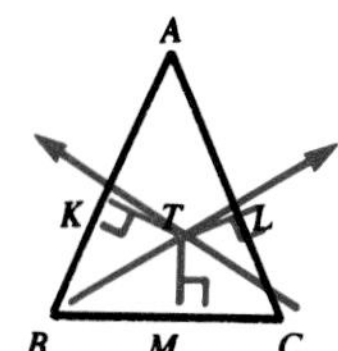

STATEMENTS	REASONS
1. The bisectors of $\angle B$ and $\angle C$ meet at some unique point; call it T.	1. Two nonparallel lines intersect at a unique point.
2. Draw perpendiculars $\overline{TL}$, $\overline{TK}$, and $\overline{TM}$.	2. There exists one and only one perpendicular to a line from an outside point.
3. $TK = TM$; $TM = TL$	3. Theorem 6–18
4. $TK = TL$	4. Substitution Property of Equality
5. T must lie on the bisector of $\angle A$	5. Theorem 6–19
6. Angle bisectors intersect at T and T is equidistant from $\overline{AB}$, $\overline{AC}$, and $\overline{BC}$.	6. Definition of distance from a point to a line and definition of equidistant.

Theorem 6–23 The perpendicular bisectors of the sides of a triangle intersect in a point equidistant from the three vertices of the triangle.

CONSTRUCTION 3: Circumscribe a circle about a given triangle.

Given: $\triangle ABC$
Method:

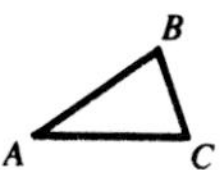

1. Construct the perpendicular bisectors of any two sides of $\triangle ABC$. Call their point of intersection O.

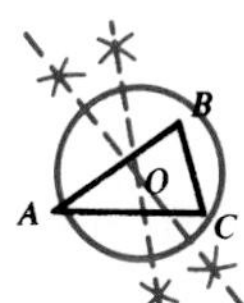

2. Using O as a center and OA as a radius, draw circle O. $\odot O$ is the required circumscribed circle.

Justification of this construction is left for the exercises.

CONSTRUCTION 4: Inscribe a circle in a given triangle.

Given: $\triangle ABC$
Method:

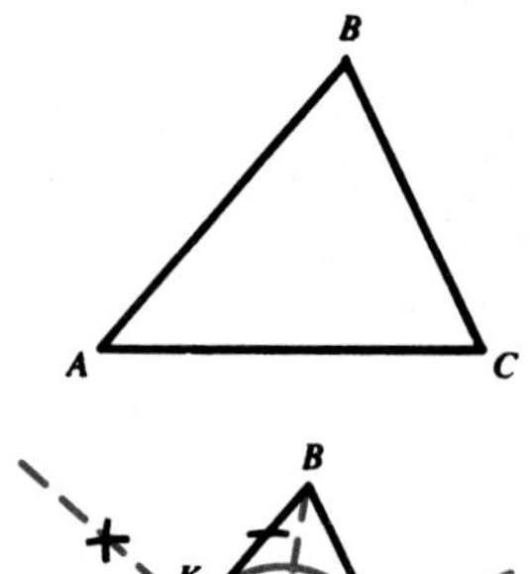

1. Construct the angle bisectors of any two angles, say A and B. Call their point of intersection O.
2. Construct a perpendicular from O to $\overline{AB}$. Call the point where it intersects $\overline{AB}$, K.
3. Using O as a center and OK as a radius, draw $\odot O$. $\odot O$ is the required inscribed circle.

You will be asked to do the justification in the exercises.

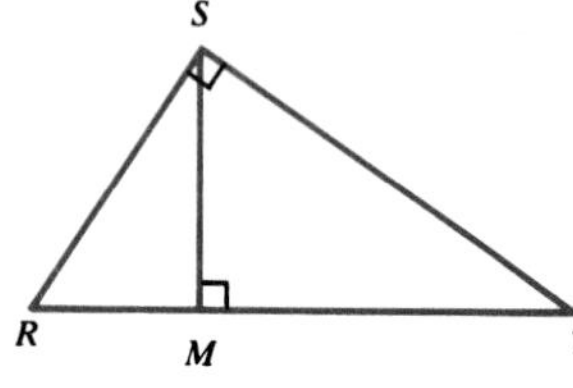

Before attempting Construction 5, we need to review the idea of a *mean proportional* or *geometric mean*. The mean proportional of two numbers x and y is $\sqrt{xy}$. In the figure at the left, $\triangle RMS \sim \triangle SMT$. By definition of similar triangles, we have $\frac{RM}{SM} = \frac{SM}{MT}$ or $(SM)^2 = (RM)(MT)$. Therefore, SM is the mean proportional of RM and MT. We will use this fact in Construction 5.

CONSTRUCTION 5: Construct the geometric mean of two given segments.

Given: Segments of length k and ℓ
Method:

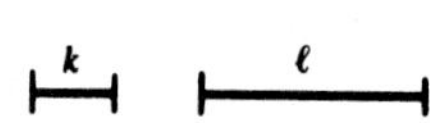

1. Draw a line. Mark off $\overline{RT}$ such that $RM = k$ and $MT = \ell$.
2. Construct the perpendicular bisector of $\overline{RT}$ to find the midpoint of $\overline{RT}$. Call the midpoint O.
3. Using O as a center and OR as a radius, construct a bit more than a semicircle.
4. Construct a perpendicular to $\overleftrightarrow{RT}$ at M. This line will intersect the semicircle at some point. Call that point H.
5. The length of $\overline{HM}$ is the required geometric mean.

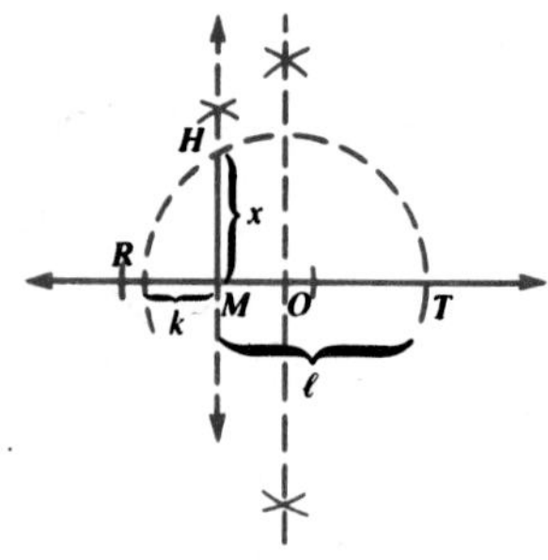

Justification of this construction is left for the exercises.

CONSTRUCTION 6: Construct a regular hexagon with side *k*.

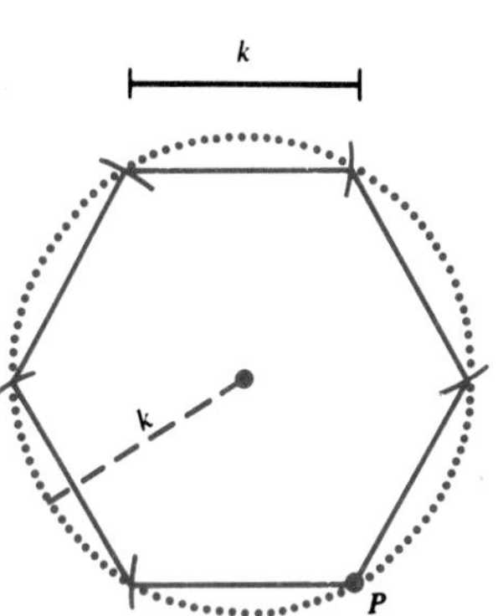

Given: Segment of length k
Method:

1. Using length k as a radius, draw a circle.
2. Choose any point P on the circle. Using the same radius as before, mark off consecutive points on the circle until you get back to the point P.
3. Using a straightedge, connect the consecutive points produced in Step 2. The resulting polygon will be the required regular hexagon.

Justification for this construction is left for the exercises.

Exercises

Exploratory **Copy the figure at the right. Then, perform the indicated construction.**

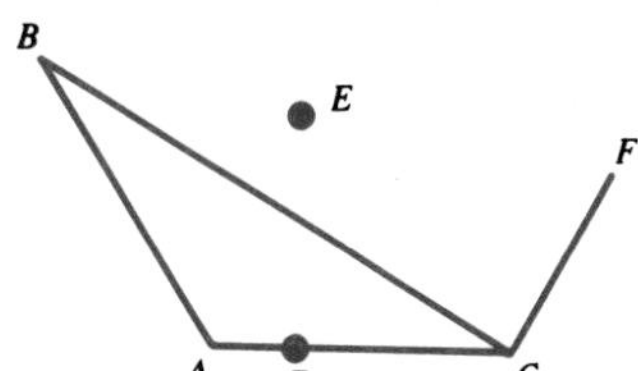

1. Construct a line perpendicular to $\overline{AC}$ at D.
2. Construct a line perpendicular to $\overline{BC}$ containing E.
3. Find the midpoint of $\overline{CF}$.
4. Construct the perpendicular bisector of $\overline{AB}$.
5. Construct the bisector of $\angle B$.
6. Construct the bisector of $\angle A$.
7. How are the constructions in exercises 3 and 4 different?

Written **Copy the figure at the right. Then, perform the indicated construction.**

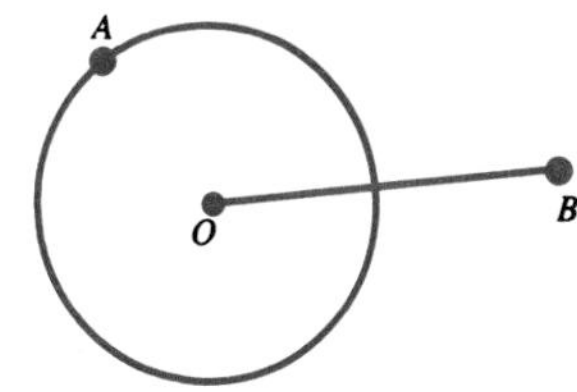

1. Construct a tangent to $\odot O$ at A.
2. Construct a tangent to $\odot O$ containing B.
3. Construct a tangent to $\odot O$ containing B different from the tangent constructed in exercise 2.
4. Is it possible to construct a perpendicular to $\overline{OB}$ tangent to $\odot O$? If so, do the construction.

For exercises 5–8, inscribe a circle in the figure, then circumscribe a circle about the figure.

5. an acute triangle
6. another acute triangle
7. an obtuse triangle
8. another obtuse triangle

Construct a triangle *KLM* with sides k, ℓ, and m.

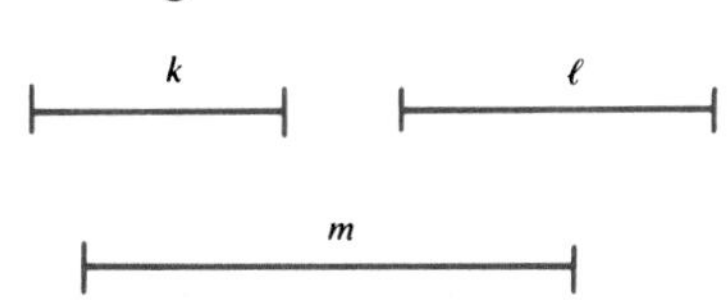

9. Justify your construction of $\triangle KLM$.
10. Circumscribe a circle about $\triangle KLM$.
11. Inscribe a circle in $\triangle KLM$.

Use Construction 5 to construct a segment with the measure indicated.

12. $\sqrt{2}$ **13.** $\sqrt{3}$ **14.** $\sqrt{5}$ **15.** $\sqrt{6}$ **16.** $\sqrt{7}$ **17.** $\sqrt{11}$

18. How else could we construct the segments in exercises 12–17?

Use the procedure and justification for Construction 2 to answer exercises 19–21.

19. Why can we draw $\overline{OB}$?

20. Why is $\angle OAB$ a right angle?

21. Why is $\overleftrightarrow{BA}$ necessarily tangent to circle O at A?

Write a justification for each construction.

22. Construction 1

23. Construction 3

24. Construction 4

25. Construction 5

26. Construction 6

Give a method for constructing each of the following. *Hint: Make a change in Construction 6*

27. regular dodecagon

28. equilateral triangle

29. regular 24-gon

30. regular octagon

Examine the proof of Theorem 6–22 to answer exercises 31–34.

31. Why are the three dashed segments in the first figure not extended until they meet at point T?

32. In Step 1, why can the bisectors of $\angle B$ and $\angle C$ be so extended?

33. Why are $\overline{TL}$, $\overline{TK}$, and $\overline{TM}$ drawn perpendicular to the sides of $\triangle ABC$?

34. What do perpendicular segments have to do with the statement of the theorem?

Prove each of the following.

35. Theorem 6–18

Given: $\overline{AT}$ is the bisector of $\angle BAC$.

Prove: $TB = TC$

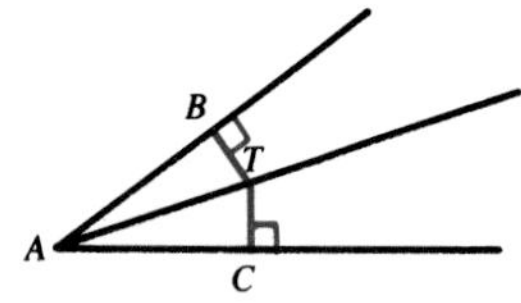

36. Theorem 6–19

Given: $MB = MC$, $\overline{MB} \perp \overline{AB}$, $\overline{MC} \perp \overline{AC}$

Prove: M is on the bisector of $\angle BAC$

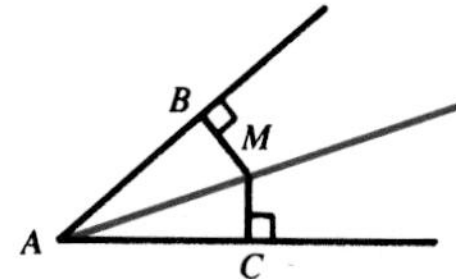

37. Theorem 6–20

Given: T is on the perpendicular bisector of $\overline{BC}$.

Prove: $TB = TC$

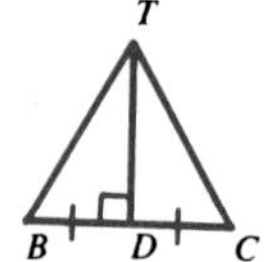

38. Theorem 6–21

Given: $TB = TC$

Prove: T is on the perpendicular bisector of $\overline{BC}$.

39. Theorem 6–23

Given: Perpendicular bisectors of $\overline{AC}$, $\overline{AB}$, and $\overline{BC}$

Prove: Perpendicular bisectors meet at some unique point M and $MA = MB = MC$.

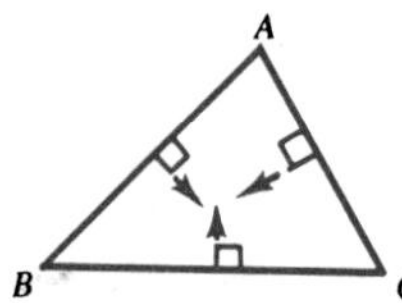

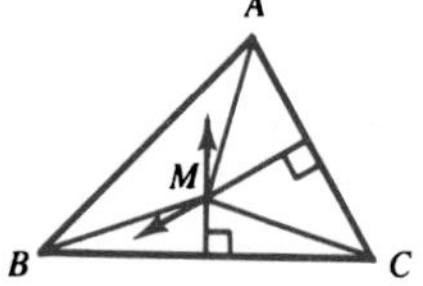

Problem Solving Application: **Using a Diagram**

Many problems are more easily solved when a picture or diagram is drawn to represent the situation.

Examples

1 **Jennifer is camping at Sequoia Campgrounds, 140 meters from the bank of Beech Creek. She needs to go to the creek, get a pail of water, and take it to Kyle who is camping on the same side of Beech Creek at Evergreen Campgrounds. Kyle's campsite is 60 meters from the bank of the creek and 170 meters from Jennifer's campsite. To what point along the bank should she go so that the distance she travels from her campsite to the creek to Kyle's campsite will be a minimum?**

Explore Draw a diagram to represent the situation. We must find point Q on line n so that the path from J to Q to K ($JQ + QK$) is the shortest path.

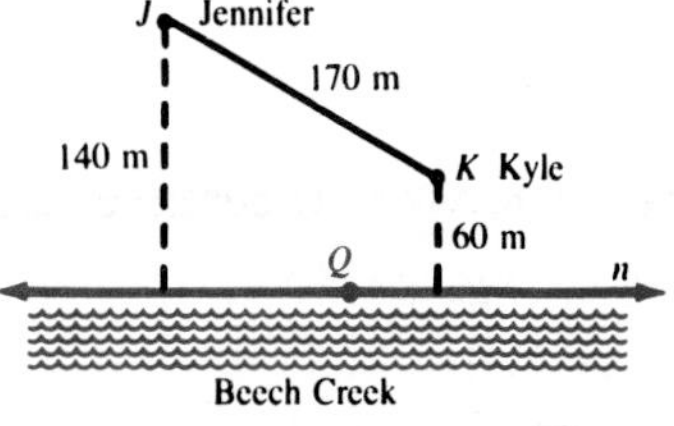

Plan If Kyle's campsite was on the *opposite* side of line n, then the shortest path would be the segment joining the campsites. Thus, if K' is the image of K under R_n, then $\overline{JK'}$ is the shortest path, and Q is the intersection of $\overline{JK'}$ and line n.

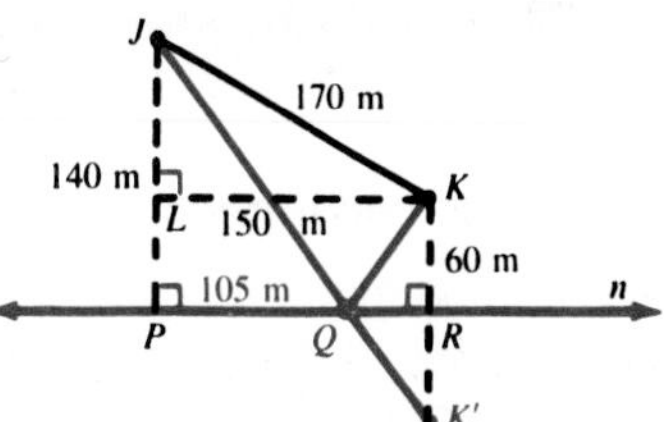

Solve Since $\overline{JP} \perp n$ and $\overline{KK'} \perp n$ and since line reflections preserve angle measure, it can be proved that $\triangle JQP \sim \triangle KQR$. Thus, we can find PQ by solving the following proportion.

$$\frac{PQ}{QR} = \frac{JP}{KR} \quad \text{or} \quad \frac{PQ}{PR - PQ} = \frac{140}{60}$$

Since $JPRK$ is a rectangle, $LK = PR$. Thus, by the Pythagorean Theorem, $PR = \sqrt{170^2 - 80^2} = \sqrt{22500}$ or 150.

$$\frac{PQ}{150 - PQ} = \frac{140}{60}$$

$$60(PQ) = 21000 - 140(PQ)$$

$$200(PQ) = 21000$$

$$(PQ) = 105$$

Thus, point Q is located 105 meters along the bank from point P.

Examine Since $\overline{JK'}$ is the shortest path from J to K' and $QK' = QK$ (line reflections preserve distance), $JQ + QK$ is the shortest distance. Thus, the choice of point Q produces the shortest path.

2 Square *ABCD* is inscribed in circle *O* and circumscribed about circle *P*. If the radius of circle *P* is 8 cm, find the exact area of circle *O*.

Explore Draw a diagram to represent the situation. Notice that circles *O* and *P* must be concentric. Thus, *O* and *P* are the same point.

Plan The radius, $\overline{PX}$, of circle *P* is one-half the length of the side of square *ABCD*. The radius, $\overline{OB}$, of circle *O* is one-half the length of the diagonal of square *ABCD*.

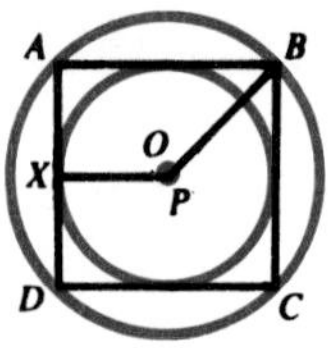

Solve

$$\begin{aligned} \text{area of circle } O &= \pi(OB)^2 \\ &= \pi[(PX)^2 + (XB)^2] \\ &= \pi[(PX)^2 + (PX)^2] \quad XB = \tfrac{1}{2}(AB) = PX \\ &= \pi(8^2 + 8^2) \\ &= 128\pi \end{aligned}$$

The area of circle *O* is 128π cm^2. *Examine this solution.*

Exercises

Written **Solve each problem.**

1. For the problem in example 1, determine the total distance, to the nearest tenth of a meter, that Jennifer must travel to bring the water to Kyle's campsite.
2. Solve the problem in example 1 if Kyle's campsite is on the same side of Beech Creek, but now is 50 meters from the bank and 150 meters from Jennifer's campsite.
3. Circle *Q* is inscribed in square *DEFG* and circumscribed about square *KLMN*. If the radius of circle *Q* is 12 cm, find the area of each square.
4. Equilateral triangle *ABC* is inscribed in circle *P* and circumscribed about circle *R*. If the side of $\triangle ABC$ is 8 cm, find the area of each circle.
5. How many games will occur in a 8-team soccer tournament if the teams play each other exactly once?
6. You can cut a pizza into 7 pieces with only 3 straight cuts. What is the greatest number of pieces you can make with 6 straight cuts?
7. Pine City is 6 km from Susanton, 9 km from Blockburg, and 13 km from Beckville. Also, Susanton, Blockburg, and Beckville do not lie in a straight line. Bonnie begins in Blockburg and drives directly to Susanton, then to Beckville, and then back to Blockburg. What is the maximum distance she could have traveled on this trip? What is the minimum distance she could have traveled?
8. A survey of 100 students who came into Kampus Kove showed that 18 students bought a snack, a soda, and played a video game, 26 bought a snack and a soda, 30 bought a snack and played a video game, and 24 bought a soda and played a video game. Also, 6 only bought a snack, 5 only bought a soda, and 7 only played a video game. How many of the 100 students did not buy a snack or a soda, or play a video game?

Vocabulary

circle
center of the circle
radius
equal circles
concentric circles
chord
diameter
secant
tangent
circumscribed
inscribed
central angle
arc
semicircle
minor arc
major arc
sector
common tangent
common internal tangent
common external tangent
line of centers
externally tangent
internally tangent
inscribed angle

Chapter Summary

1. A circle with radius r units has circumference $2\pi r$ units and area πr^2 square units.
2. For a circle O with radius r units and arc AB of degree measure n, the area of sector AOB is $\frac{n}{360} \cdot \pi r^2$ square units and the length of arc AB is $\frac{n}{360} \cdot 2\pi r$ units.
3. Shortest Distance Theorem: The perpendicular segment from a point to a line is the shortest segment from the point to the line.
4. Theorem 6–1: A line tangent to a circle is perpendicular to the radius drawn to the point of tangency.
5. Theorem 6–2: If a line is perpendicular to a radius at a point P on the circle, then the line is tangent to the circle at P.
6. Theorem 6–3: Tangents to a circle from an outside point are equal.
7. Theorem 6–4: In the same circle, or in equal circles, equal chords have equal arcs.
8. Theorem 6–5: In the same circle, or in equal circles, equal arcs have equal chords.
9. Theorem 6–6: A line passing through the center of a circle and perpendicular to a chord bisects the chord and its arcs.
10. Theorem 6–7: In the same circles, or in equal circles, equal chords are equally distant from the center.
11. Theorem 6–8: In the same circle, or in equal circles, chords equally distant from the center are equal.
12. Theorem 6–9: An inscribed angle is measured by one-half its intercepted arc.

13. Theorem 6–10: If two chords intersect within a circle, then the product of the lengths of the segments of one chord is equal to the product of the lengths of the segments of the other.
14. Theorem 6–11: An angle formed by two chords intersecting within a circle is measured by one-half the sum of the arcs intercepted by the angle and its vertical angle.
15. Theorem 6–12: An angle formed by a chord and a tangent is measured by one-half the intercepted arc.
16. Theorem 6–13: An angle formed by two secants intersecting outside a circle is measured by one-half the difference of the intercepted arcs.
17. Theorem 6–14: An angle formed by a secant and a tangent intersecting outside a circle is measured by one-half the difference of the intercepted arcs.
18. Theorem 6–15: An angle formed by two intersecting tangents is measured by one-half the difference of the intercepted arcs.
19. Theorem 6–16: Let two secants be drawn to a circle from an outside point. Then the product of the lengths of one secant and its external segment equals the product of the lengths of the other secant and its external segment.
20. Theorem 6–17: Let a tangent and a secant be drawn to a circle from an outside point. Then the square of the length of the tangent segment is equal to the product of the lengths of the secant and its external segment.
21. Theorem 6–18: If a point lies on the bisector of an angle, then it is equidistant from the sides of the angle.
22. Theorem 6–19: If a point is equidistant from the sides of an angle, then it lies on the bisector of the angle.
23. Theorem 6–20: If a point lies on the perpendicular bisector of a segment, then it is equidistant from the ends of the segment.
24. Theorem 6–21: If a point is equidistant from the ends of a segment, then it lies on the perpendicular bisector of the segment.
25. Theorem 6–22: The bisectors of the angles of a triangle intersect in a point equidistant from the three sides of the triangle.
26. Theorem 6–23: The perpendicular bisectors of the sides of a triangle intersect in a point equidistant from the three vertices of the triangle.

Chapter Review

6.1 Sketch each of the following.

1. A hexagon inscribed in a circle.
2. A triangle circumscribed about a circle.
3. $\odot O$ with chords $\overline{RS}$ and $\overline{UV}$ such that $\overline{RS} \parallel \overline{UV}$.
4. $\odot O$ concentric to $\odot Q$; chord $\overline{AB}$ in $\odot O$ is tangent to $\odot Q$ at F.

6.2 **In each of the following, you are given the radius of circle O and the measure of central angle O. Find the exact area of the sector determined by the central angle and the exact length of the intercepted arc.**

5. $r = 10$; $m\angle O = 110$ **6.** $r = 10$; $m\angle O = 90$ **7.** $r = 6$; $m\angle O = 25$

8. $r = 3\sqrt{2}$; $m\angle O = 90$ **9.** $r = 4$; $m\angle O = 150$ **10.** $r = 0.5$; $m\angle O = k$

6.3 **Exercises 11–14 refer to $\odot Q$ at the right.**

11. Name a tangent, a tangent segment, and the point of tangency.

12. What is $m\angle TWQ$?

13. If $QR = 2$, $\overline{QS}$ is drawn, and $QS = 4$, find SW.

14. If $QR = 4$ and $WT = 6$, find TR.

6.4 **15.** Find the radius of a circle if a chord of length 12 cm is 10 cm from the center of the circle.

Prove each of the following.

16. Given: $m\angle A = m\angle C$
Prove: $m\widehat{AB} = m\widehat{BC}$

17. Given: $AC = CB$
Prove: $m\widehat{AM} = m\widehat{MB}$

18. Given: $m\widehat{ST} = m\widehat{TU}$, $m\widehat{SW} = m\widehat{UV}$
Prove: $\triangle STW \cong \triangle UTV$

6.5
6.6
and
6.7 **In the figure at the right, $\overrightarrow{PA}$ and $\overrightarrow{PB}$ are tangent to $\odot Q$. If $m\widehat{CB} = 90$ and $m\angle APB = 70$, find each of the following.**

19. $m\widehat{AB}$ **20.** $m\widehat{AD}$

21. $m\angle DEC$ **22.** $m\angle PAC$

23. $m\angle ABP$ **24.** $m\angle DBC$

In the figure at the right, $\overline{PD}$ is the tangent to $\odot O$. Find the measure indicated by ?.

25. $AB = 8$, $CE = 16$, $PE = 5$, $PB = ?$

26. $PD = 10$, $PB = 4$, $AB = ?$

27. $CE = 30$, $AF = 25$, $FD = 8$, $CF = ?$

28. $PD = 8$, $AB = 12$, $PA = ?$

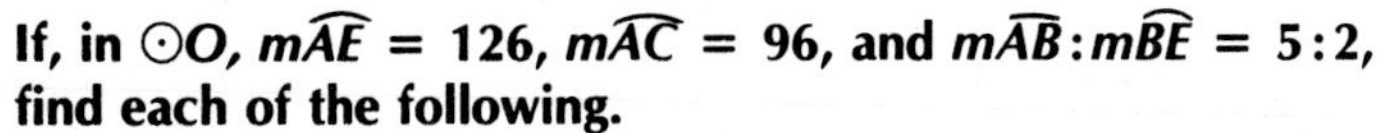

If, in $\odot O$, $m\widehat{AE} = 126$, $m\widehat{AC} = 96$, and $m\widehat{AB} : m\widehat{BE} = 5:2$, find each of the following.

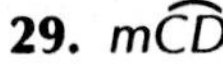

29. $m\widehat{CD}$ **30.** $m\widehat{BE}$ **31.** $m\angle APC$

32. $m\angle CFD$ **33.** $m\angle BAD$ **34.** $m\angle CPD$

6.8 **Inscribe a circle in and circumscribe a circle about the following figures.**

35. an isosceles triangle **36.** an equilateral triangle **37.** a right triangle

Chapter Test

1. Find the circumference of a circle whose area is 49π in^2.
2. In circle O, $\angle AOB$ is a central angle with measure 135. If the radius is 8 cm, find the area of sector AOB.
3. A chord is 5 cm from the center of a circle with radius 13 cm. Find the length of the chord.
4. Find the area of the shaded portion of the figure at the right.
5. Point P is 25 cm from the center of a circle of radius 7 cm. If $\overline{PA}$ is tangent to the circle at A, find PA.

Find the measure indicated by x in each of the following.

6. 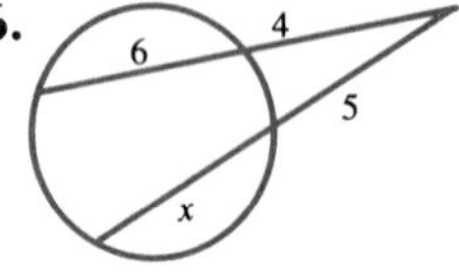

7. 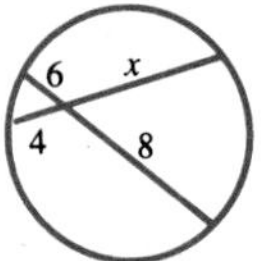

8. 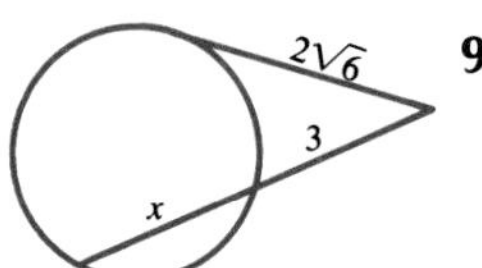

9. 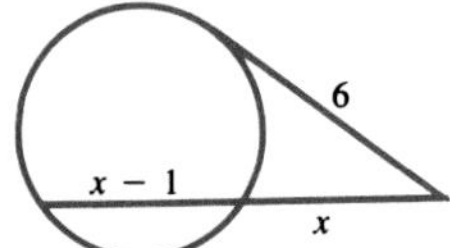

In the figure at the right, $\overrightarrow{PA}$ is tangent to $\odot O$, $m\angle P = 42$, and $m\widehat{AC}:m\widehat{AB} = 5:2$. Find each of the following.

10. $m\widehat{AC}$
11. $m\widehat{BC}$
12. $m\angle ACP$
13. $m\angle ABP$

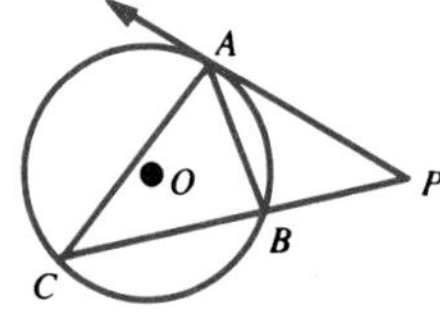

In the figure at the right, $m\widehat{AE}:m\widehat{ED} = 4:2$, $m\widehat{ED}:m\widehat{DC} = 2:1$, and $m\widehat{DC}:m\widehat{BC} = 1:2$. Find each of the following.

14. $m\widehat{AE}$
15. $m\widehat{BC}$
16. $m\widehat{DC}$
17. $m\angle EFA$
18. $m\angle DAB$
19. $m\angle AGC$
20. Draw a large acute triangle. Construct the inscribed circle of the triangle.

21. Draw another large acute triangle. Construct the circumscribed circle of the triangle.

Prove each of the following.

22. **Given:** $AC = BC$
    Prove: $\triangle CDE$ is isosceles

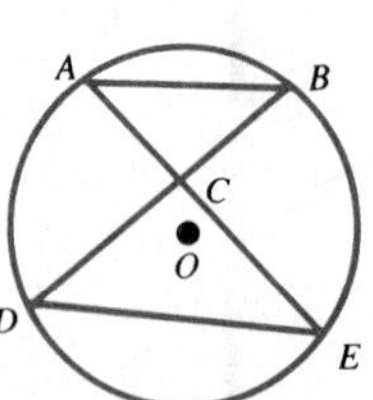

23. **Given:** $m\widehat{AC} = m\widehat{BC}$
    Prove: $\triangle DAC \cong \triangle DBC$

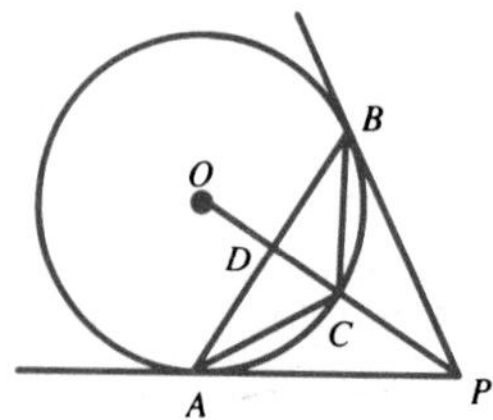

An Introduction to Circular Functions

Chapter **7**

We can approach the study of circular functions by using our knowledge of the circle and other aspects of geometry.

One of the functions we will study is the sine function. This function is helpful in describing phenomena like musical vibrations and light waves.

An oscilloscope may be used to analyze the fluctuations of energy produced by such phenomena. For example, sound waves, when they are converted to electrical impulses, cause curves to be displayed on the screen of an oscilloscope. A pure musical tone produces a curve that is the graph of the sine function.

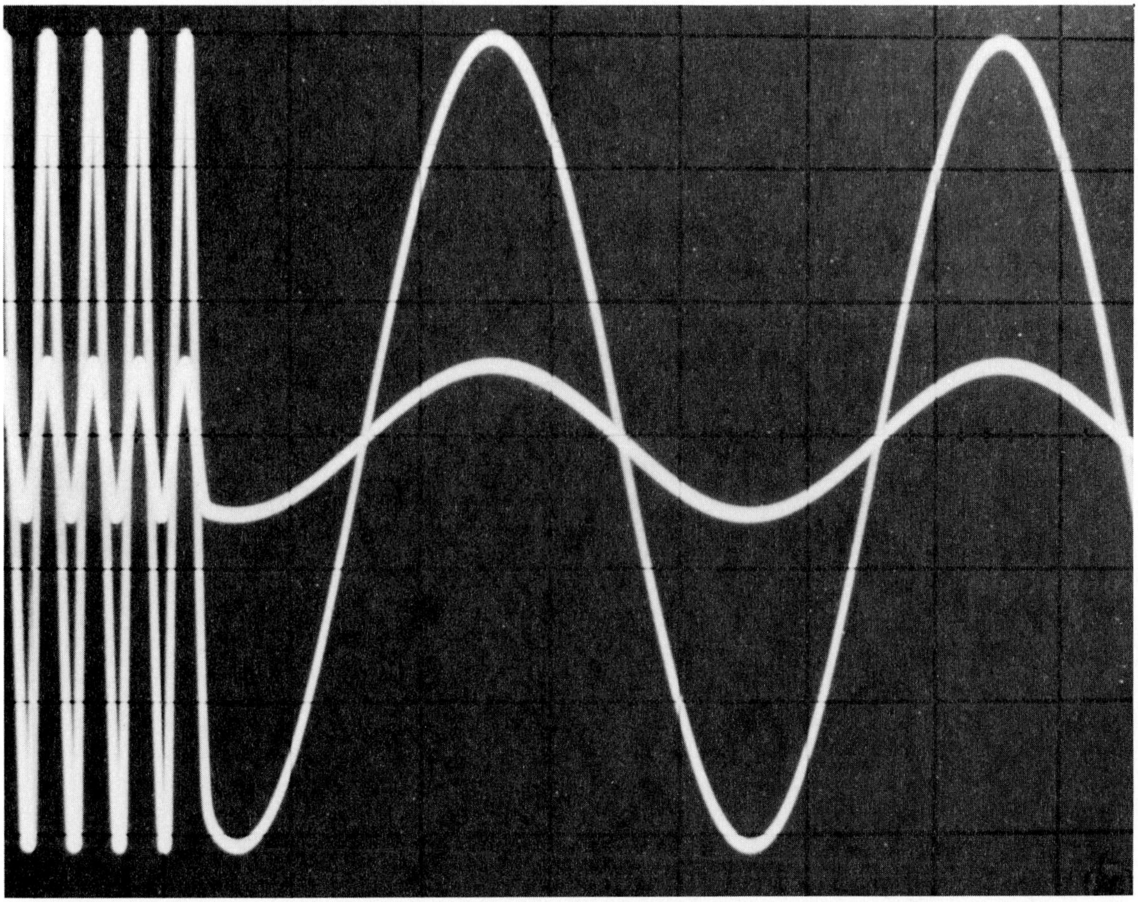

7.1 Working With Circles and Angles

To begin our study of trigonometry take a circle and center it on a pair of coordinate axes. Imagine two rays with a common endpoint at the center of the circle. One of the rays is fixed along the positive x-axis. The other is allowed to rotate around the circle. These rays will form angles of different measures.

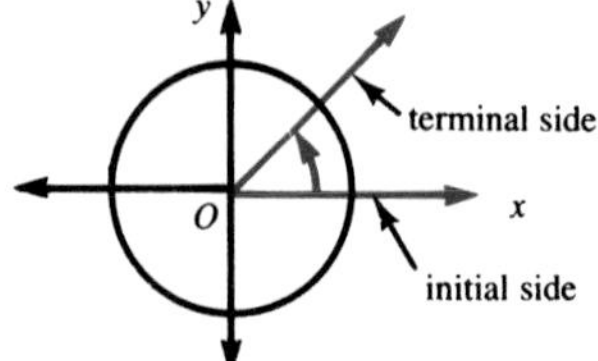

The fixed ray along the x-axis is called the **initial side of the angle.** The "moving" ray is the **terminal side of the angle.** An angle with its vertex at the origin and its initial side along the positive x-axis is said to be in **standard position.**

The amount of rotation of the terminal side can be measured in degrees. Let us agree that the measurement of an angle that is equivalent to one complete counterclockwise rotation is 360 degrees. Then, one half of a complete counterclockwise rotation is 180 degrees. One quarter of a counterclockwise rotation is 90 degrees.

If the terminal side of an angle moves *counterclockwise,* the measure of the angle is *positive.* If the terminal side moves *clockwise,* the measure of the angle is *negative.*

Recall that an (x, y)-coordinate system separates the plane into four quadrants as shown at the right.

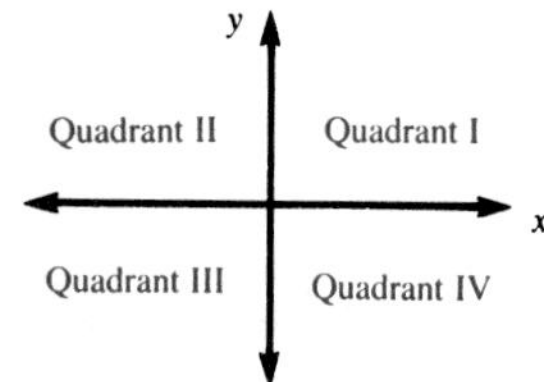

Angles in standard position are classified according to the quadrant in which the terminal side lies.

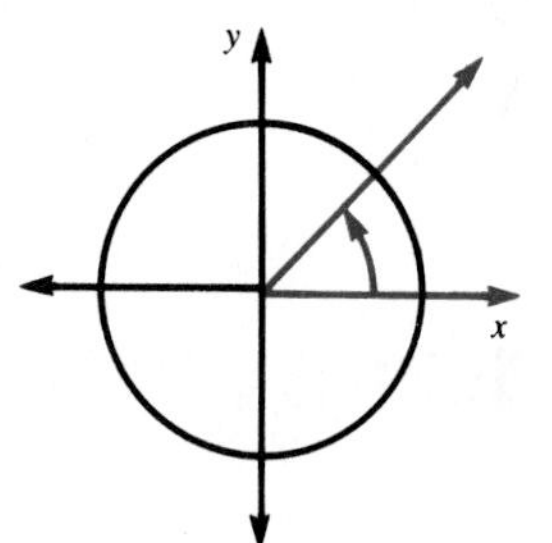

Angle in the First Quadrant

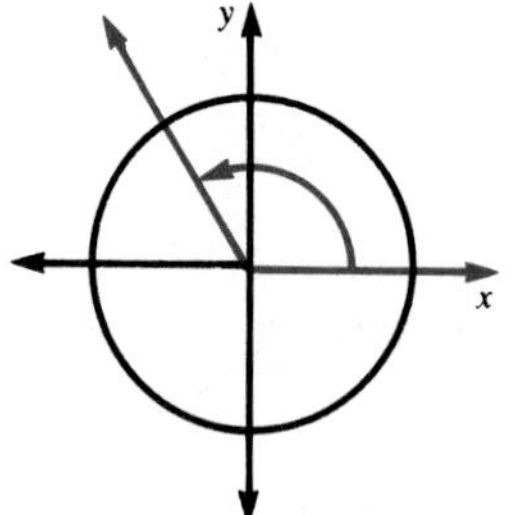

Angle in the Second Quadrant

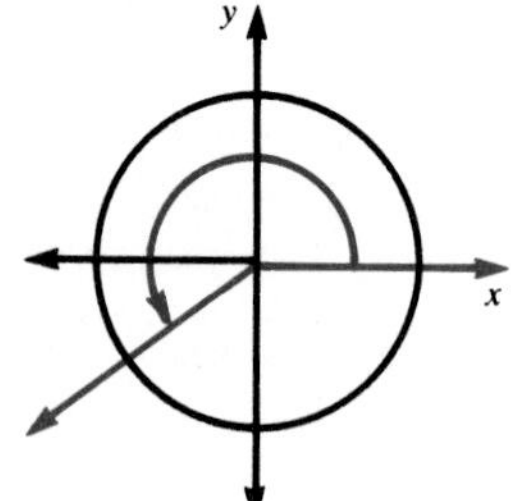

Angle in the Third Quadrant

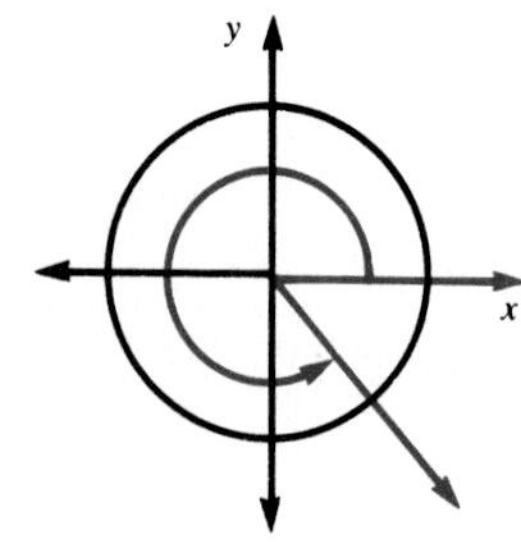

Angle in the Fourth Quadrant

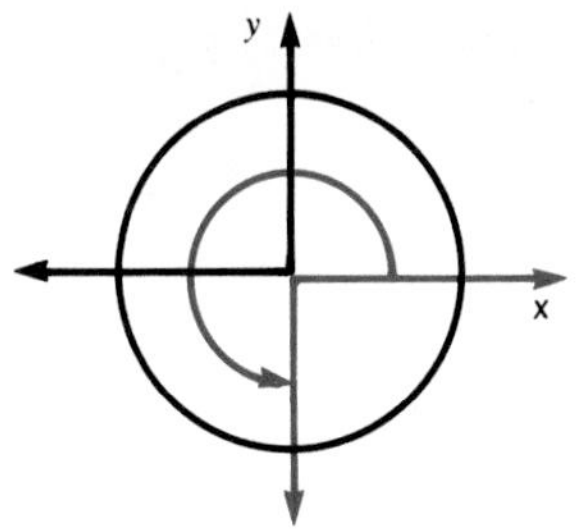

An angle in standard position that has its terminal side on one of the coordinate axes is called a **quadrantal angle.**

An angle of 270°, as shown in the figure, is a quadrantal angle.

The terminal side may rotate more than one rotation around the circle. For example, if it rotates $2\frac{1}{4}$ rotations counterclockwise, then its measurement is 360° + 360° + 90° or 810°. Some other examples appear in the figures below.

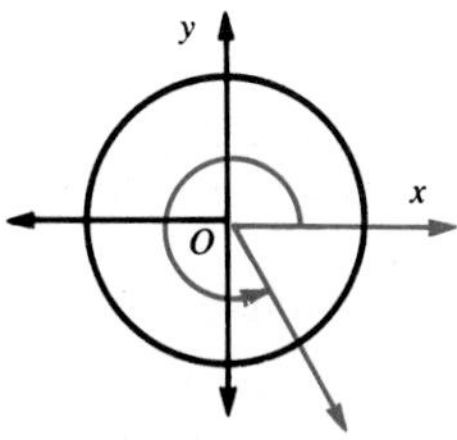

Angle measures 300°
Equivalent to
$\frac{5}{6}$ rotation counterclockwise

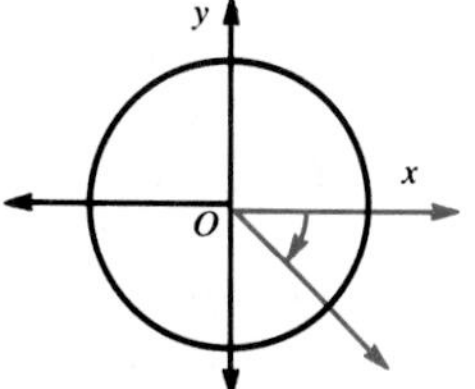

Angle measures −45°
Equivalent to
$\frac{1}{8}$ rotation clockwise

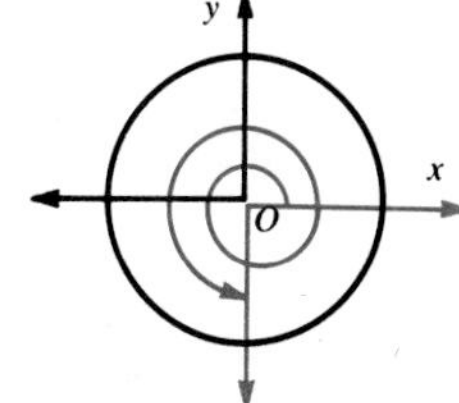

Angle measures 630°
Equivalent to
$1\frac{3}{4}$ rotations counterclockwise

Example

1 An angle measures 1100°. How many complete rotations does this represent, and in which direction?

The rotation is counterclockwise since the measure is positive. To find the number of rotations, divide 360 into 1100. An angle of 1100° is equivalent to 20° more than 3 complete rotations.

$$\begin{array}{r} 3 \\ 360\overline{)1100} \\ \underline{1080} \\ 20 \end{array}$$

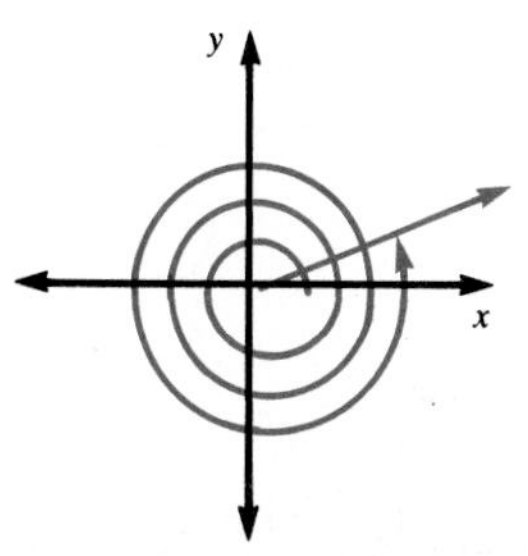

An angle of 1100° and an angle of 20° have the "same" terminal side. These angles differ by complete rotations, that is, by multiples of 360°. Such angles are said to be **coterminal angles.**

Note that angles of 0° and 360° are coterminal.

Example

2 **Which of the following pairs of angles are coterminal?**

a. 140° and −140° **b. 30° and −330°** **c. 40° and 320°**

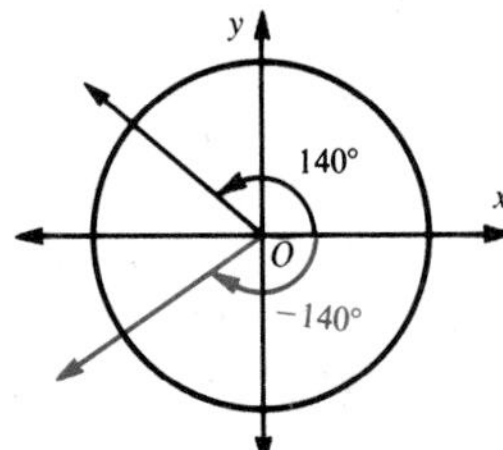

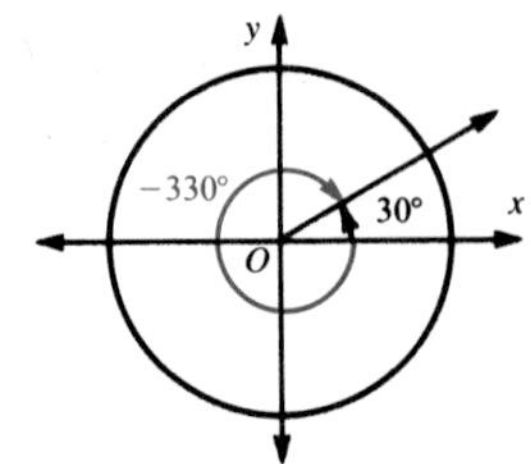

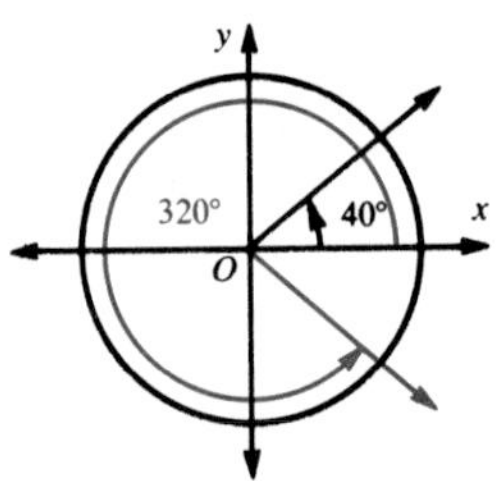

The illustrations above show that only the pair in part **b.** are coterminal.

Exercises

Exploratory **Draw an angle in standard position for each of the following. Identify the initial and terminal sides.**

1. 90°	**2.** 180°	**3.** 60°	**4.** 210°	**5.** 0°
6. 270°	**7.** 240°	**8.** −45°	**9.** 330°	**10.** −120°
11. −315°	**12.** −400°	**13.** 1080°	**14.** −540°	**15.** −300°

Written **Name the quadrant in which an angle in standard position with the following measurement lies.**

1. 48°	**2.** 130°	**3.** 89°	**4.** −40°	**5.** 328°
6. −91°	**7.** −240°	**8.** 210°	**9.** 920°	**10.** −198°
11. −356°	**12.** 480°	**13.** −780°	**14.** −315°	**15.** −181°

In each column below select the angles, if any, which are coterminal with the angle described at the head of that column.

16. 30°	**17.** 100°	**18.** −45°	**19.** 210°	**20.** 720°
60°	300°	315°	150°	0°
330°	−260°	675°	−150°	−720°
−330°	460°	−405°	−210°	10800°

Find the least positive degree measure of an angle that is coterminal with an angle of the following measurement.

21. 380°	**22.** 390°	**23.** 570°	**24.** −10°	**25.** −50°
26. −330°	**27.** −120°	**28.** 790°	**29.** −420°	**30.** −800°

Challenge **Suppose the two given angles are in standard position. Name the line, in each of the following, in which the terminal sides of the angles are each other's reflection.**

1. 30°, 60°	**2.** 30°, 150°	**3.** 240°, 300°	**4.** 120°, 150°

7.2 Radian Measure

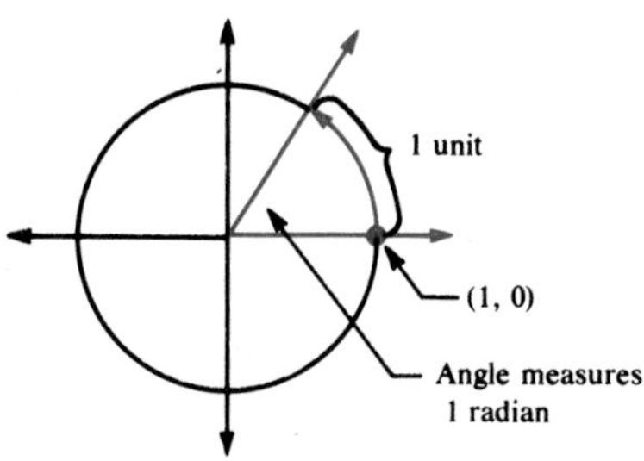

Until now, the *degree* has been used as a unit for measuring angles. Another important unit of angle measure is the **radian.** It is used much more extensively than the degree in advanced work, especially in *calculus.*

To see how radian measure works, think of a circle with center at the origin and radius of one unit. Such a circle is called a **unit circle.** Suppose an angle in standard position intercepts an arc whose length is the same as the radius of this unit circle. That is, the arc length is one unit. This angle measures **one radian.**

Definition of Radian Measure

If an angle in standard position on the unit circle intercepts an arc of length s units, then the angle measures s radians.

Examples

1 **Draw an angle that measures approximately 2 radians.**

The angle must intercept an arc whose length is about 2 units on the unit circle. The angle is pictured in the figure at the right.

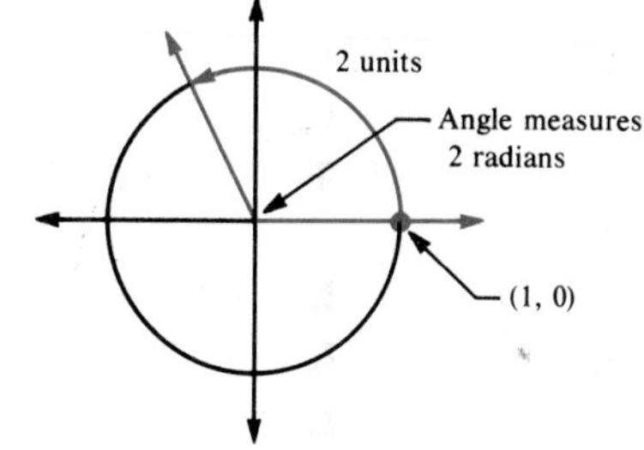

2 **What is the radian measure of the angle shown at the right?**

You must find the length of the intercepted arc. Here, the length of the arc is one-half that of the circumference of the unit circle. The circumference is $2\pi \cdot 1$ or 2π units. Therefore, the angle pictured has a radian measure $\frac{1}{2} \cdot 2\pi$, or π radians.

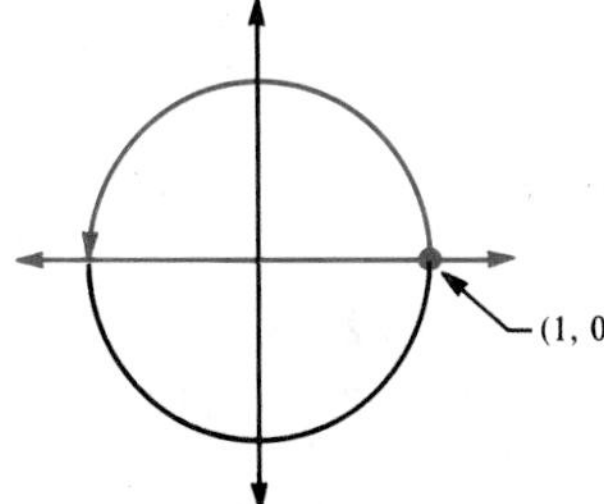

The angle shown in Example 2 measures 180° or π radians. Thus,

$$180° = \pi \text{ radians.}$$

Divide both sides by 180.

$$1° = \frac{\pi}{180} \text{ radians}$$

Divide both sides by π.

$$\frac{180°}{\pi} = 1 \text{ radian}$$

Remember that radian measure is more widely used in advanced work than degree measure. In fact, when no unit of measure is given, radian measure is generally understood. For example, if we say that the measure of angle B is $\frac{\pi}{6}$, it should be understood that the measure is $\frac{\pi}{6}$ radians.

Examples

3 Convert 45° to radian measure.

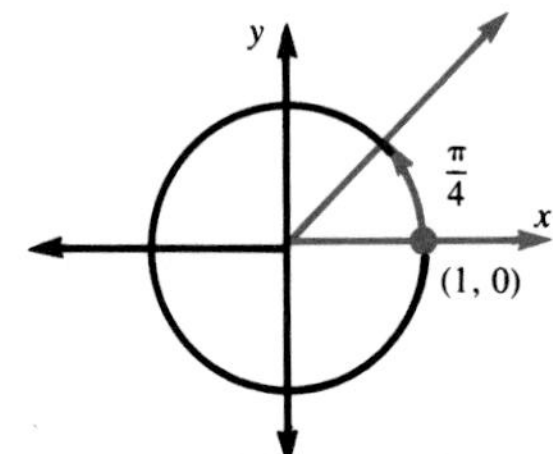

Recall that $1° = \frac{\pi}{180}$ radians. Therefore, the radian measure equivalent of 45° is $45 \cdot \frac{\pi}{180}$ or $\frac{\pi}{4}$ radians.

4 Convert $\frac{5\pi}{6}$ radians to degree measure.

Recall that 1 radian $= \frac{180°}{\pi}$. Therefore the degree measure equivalent of $\frac{5\pi}{6}$ radians is

$$\frac{5\pi}{6} \cdot \frac{180°}{\pi} \quad \text{or} \quad 150°.$$

5 Suppose $\angle AOC$ has a measure of 3 radians. Find the length of the arc intercepted by $\angle AOC$ on a circle with center at the origin and radius of 5 units.

Let s represent the length of arc AC. For the same angle, the ratio of arc length to radius length is constant for circles of differing radii.

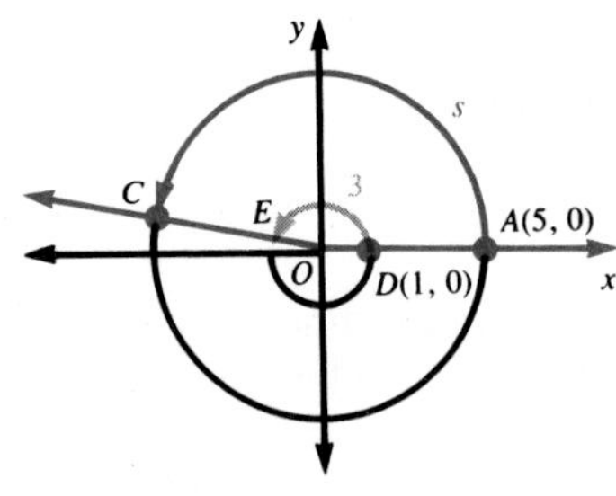

Therefore, $\frac{\text{length of } \widehat{DE}}{OD} = \frac{\text{length of } \widehat{AC}}{OA}$.

$$\frac{3}{1} = \frac{s}{5}$$

$$s = 15$$

Thus, the length of $\widehat{AC}$ is 15 units.

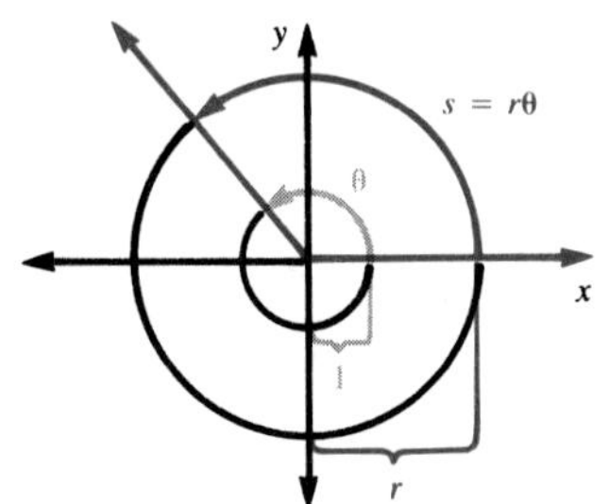

In general, if θ is the radian measure of a central angle, and r is the radius, then s, the length of the intercepted arc, is given by

$$s = r\,\theta.$$

θ is a Greek letter pronounced "thay-ta"

Example

6 In a circle, a central angle of 3 radians intercepts an arc of 18 cm. Find the radius of the circle.

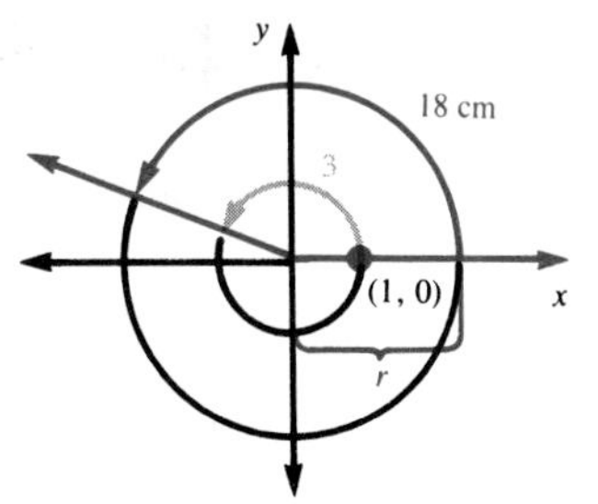

$$s = r\,\theta$$
$$18 = r \cdot 3$$
$$6 = r$$

The radius is 6 cm.

Exercises

Exploratory **Draw a unit circle centered at the origin. Draw angles that have approximately the following measurements.**

1. 1 radian **2.** 2 radians **3.** 3 radians

4. $\frac{3}{4}$ radian **5.** $1\frac{1}{2}$ radians **6.** $-1\frac{1}{2}$ radians

7. 3.14 radians **8.** π radians **9.** 6.28 radians

Written **Convert each of the following to radian measure.**

1. 90° **2.** 30° **3.** 60° **4.** 120°

5. −90° **6.** 270° **7.** 150° **8.** 300°

9. 45° **10.** 315° **11.** −135° **12.** −210°

13. 765° **14.** 450° **15.** 420° **16.** −240°

Convert each of the following to degree measure. *For exercises 31–36, remember to use $\pi \approx 3.14$.*

17. π **18.** $\frac{\pi}{6}$ **19.** $\frac{\pi}{4}$ **20.** $\frac{2\pi}{3}$

21. $-\frac{\pi}{3}$ **22.** 2π **23.** $\frac{3\pi}{2}$ **24.** $-\frac{5\pi}{4}$

25. $\frac{11\pi}{6}$ **26.** $-\frac{8\pi}{3}$ **27.** $-\frac{7\pi}{4}$ **28.** $\frac{3\pi}{4}$

29. $-\frac{5\pi}{6}$ **30.** $5\frac{1}{2}\pi$ **31.** $6\frac{1}{2}$ **32.** 2

33. 0 **34.** 1.571 **35.** −1.571 **36.** 3.14

Let θ be the radian measure of a central angle that intercepts an arc of *s* units on a circle with a radius of *r* units.

37. Find s if $r = 2$ cm and $\theta = 4$.
38. Find s if $r = 1.5$ cm and $\theta = 5$.
39. Find r if $\theta = 2$ and $s = 10$ cm.
40. Find r if $\theta = 3$ and $s = 15$ cm.
41. Find r if $\theta = 5$ and $s = 20$ cm.
42. Find θ if $s = 18$ cm and $r = 6$ cm.
43. Find θ if $s = 14$ cm and $r = 7$ cm.
44. Find θ if $s = 10$ cm and $r = 2.5$ cm.

Find the radian measure of the angle determined by the following counterclockwise rotations.

45. $\frac{1}{4}$ of a rotation **46.** $\frac{2}{3}$ of a rotation **47.** $\frac{3}{4}$ of a rotation **48.** $\frac{1}{2}$ of a rotation

49. $41\frac{1}{2}$ rotations **50.** $\frac{3}{8}$ of a rotation **51.** $\frac{5}{6}$ of a rotation **52.** $\frac{3}{10}$ of a rotation

53. Sketch a central angle that intercepts an arc of length 5 cm on a circle with a radius of 1 cm.

54. Sketch a central angle that intercepts an arc of length 5 cm on a circle with a radius of 2 cm.

55. Approximate to the nearest tenth the number of degrees in 1 radian.

Challenge

1. In circle P, a central angle of 30° intercepts an arc of 5π inches. Find the length of the radius.
2. A wheel of radius 18 inches rotates at the rate of 50 revolutions per minute. Find the rate, in inches per second, of a point on the rim of the wheel.
3. The length of the minute hand of a clock is 8 cm. What distance, in centimeters, does the tip of the hand travel in 40 minutes?

Points $A(3, 0)$, $C(-3, 0)$, and $E(0, -3)$ all lie on the circle of radius 3 cm centered at the origin, O. If $\overline{BD}$ is a diameter of circle O, $\angle AOB$ is in standard position, and $m\angle AOB = 60$, find the length of each arc.

4. $\widehat{AB}$ **5.** $\widehat{BC}$ **6.** $\widehat{CD}$ **7.** $\widehat{DE}$ **8.** $\widehat{CA}$ **9.** $\widehat{DA}$

Mixed Review

In the figure at the right, $AE = 3r - 5$, $BE = r + 11$, $CE = r + 1$, $DE = 2r - 5$, $AB = 10$, $CD = 5$, $PC = 16$, $m\angle APC = 25$, and $m\widehat{AC} = 80$. Find each of the following.

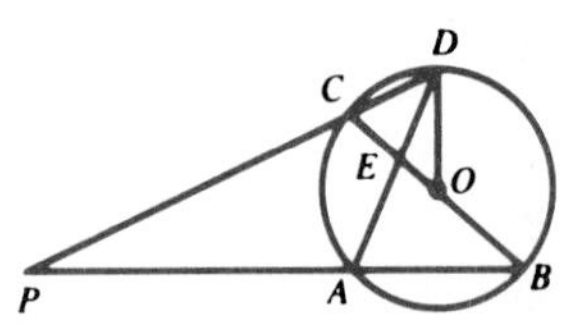

1. EC **2.** BC **3.** AD **4.** PB

5. $m\angle ABC$ **6.** $m\widehat{BD}$ **7.** $m\angle BOD$ **8.** $m\angle CED$

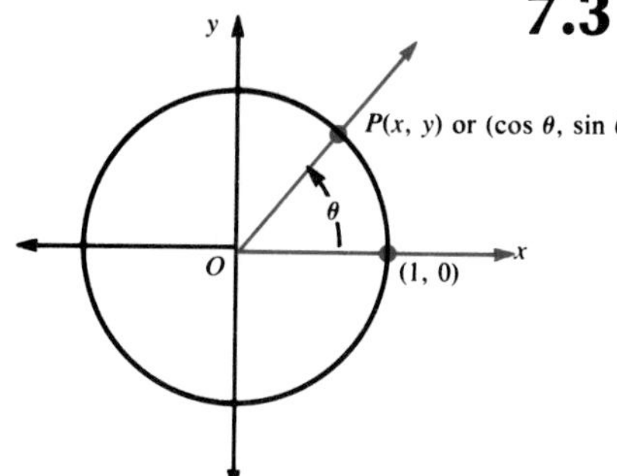

7.3 Two Circular Functions

Suppose we have an angle in standard position on a unit circle. The situation is pictured at the left.

Let θ represent the measure of the angle. The terminal side of the angle intersects the unit circle at point P with coordinates (x, y). We call the x-coordinate the **cosine of θ** and the y-coordinate the **sine of θ.**

Definition of Sine and Cosine Functions

Let θ be the measure of an angle in standard position. Let $P(x, y)$ be the point of intersection of the terminal side of the angle with the unit circle. Then the function **sine** pairs θ with the y-coordinate of P and the function **cosine** pairs θ with the x-coordinate of P.

Using function notation, we write sine $\theta = y$ and cosine $\theta = x$. The expression sine θ is abbreviated **sin θ** and cosine θ as **cos θ.**

Examples

1 Find cos 0° and sin 0°.

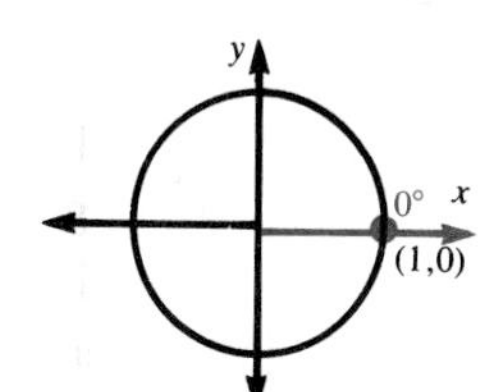

Draw an angle of 0° in standard position.

The terminal side intersects the unit circle at (1, 0). Therefore,

$$\cos 0° = 1 \quad \text{and} \quad \sin 0° = 0.$$

2 Find cos 180° and sin 180°.

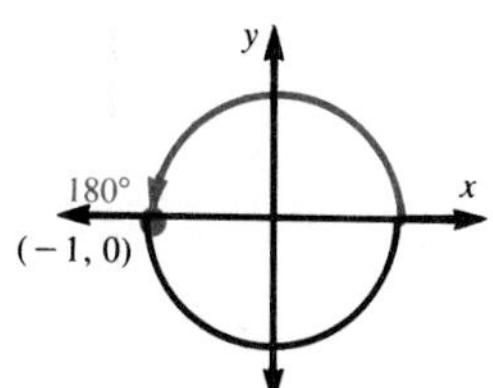

Draw an angle of 180° in standard position.

The terminal side intersects the unit circle at (−1, 0). Therefore,

$$\cos 180° = -1 \quad \text{and} \sin 180° = 0.$$

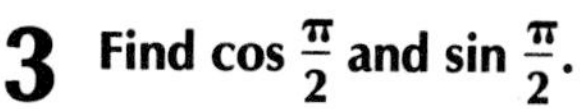

3 Find cos $\frac{\pi}{2}$ and sin $\frac{\pi}{2}$.

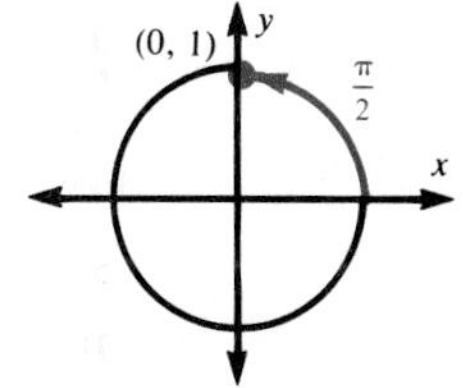

The figure at the right shows that

$$\cos \frac{\pi}{2} = 0 \quad \text{and} \quad \sin \frac{\pi}{2} = 1.$$

4 Find $\cos \frac{3\pi}{2}$ and $\sin \frac{3\pi}{2}$.

Draw an angle of $\frac{3\pi}{2}$ radians in standard position.

$$\cos \frac{3\pi}{2} = 0 \quad \text{and} \quad \sin \frac{3\pi}{2} = -1$$

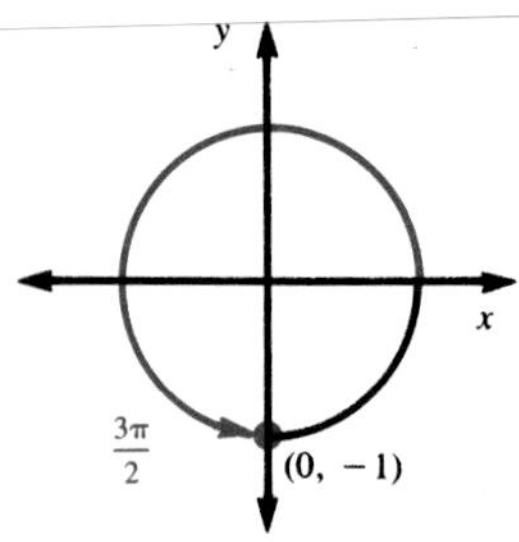

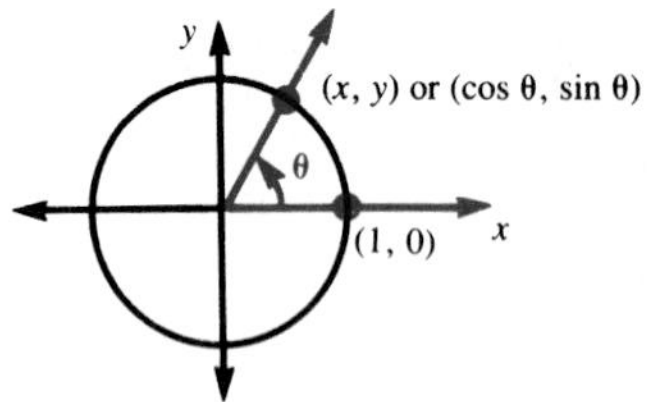

Previously you learned that every point (x, y) on a unit circle satisfies the equation

$$x^2 + y^2 = 1.$$

Since $\cos \theta = x$ and $\sin \theta = y$, we have

$$(\cos \theta)^2 + (\sin \theta)^2 = 1.$$

The expression $(\cos \theta)^2$ is written as $\cos^2 \theta$. Similarly, the expression $(\sin \theta)^2$ is written as $\sin^2 \theta$. We have, then, the following trigonometric identity. *Recall that an identity is an equation that is true for all values of the variable.*

Pythagorean Identity

Let θ be the measure of an angle in standard position. Then

$$\sin^2 \theta + \cos^2 \theta = 1.$$

If B is an angle with measure θ, we often write $\sin B$ to mean $\sin \theta$. For example, if $m\angle A = 90$ or $\frac{\pi}{2}$ radians, $\sin A$ means $\sin 90°$ or $\sin \frac{\pi}{2}$. Also, an expression such as "θ is in quadrant II" means "θ is the measure of an angle that lies in quadrant II."

Example

5 Find $\cos \theta$ if $\sin \theta = \frac{3}{5}$ and θ is in quadrant II.

Pictured at the right is an angle in quadrant II with measure θ and $\sin \theta = \frac{3}{5}$.

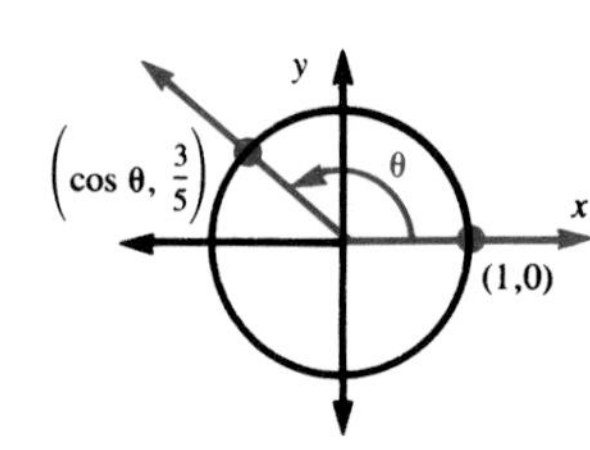

We use the Pythagorean Identity and solve for $\cos \theta$.

$$\sin^2 \theta + \cos^2 \theta = 1$$

Now substitute $\frac{3}{5}$ for $\sin \theta$.

The Pythagorean Identity becomes

$$\left(\frac{3}{5}\right)^2 + \cos^2\theta = 1.$$

$$\frac{9}{25} + \cos^2\theta = 1$$

$$\cos^2\theta = \frac{16}{25}$$

$$\cos\theta = \pm\frac{4}{5}$$

We reject $\frac{4}{5}$ since the x-coordinate of a point in quadrant II must be negative.

Therefore, $\cos\theta = -\frac{4}{5}$.

Example 5 illustrates that for an angle in quadrant II, the sine is positive and the cosine is negative. The figures shown below illustrate all possible cases.

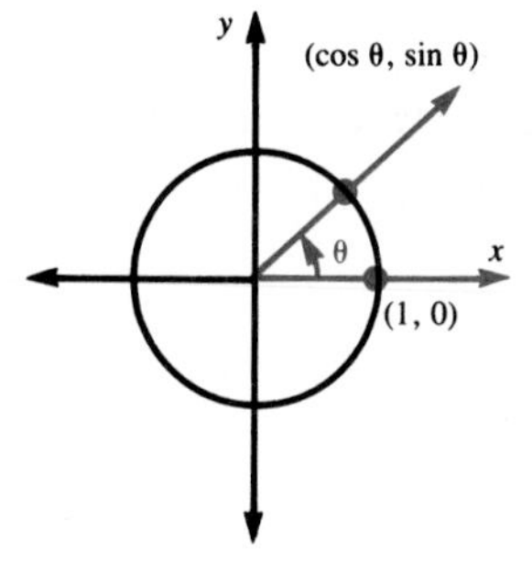

Quadrant I
sin θ and cos θ both positive

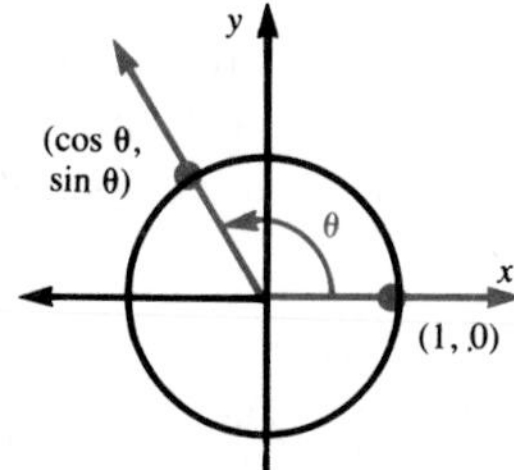

Quadrant II
sin θ positive
cos θ negative

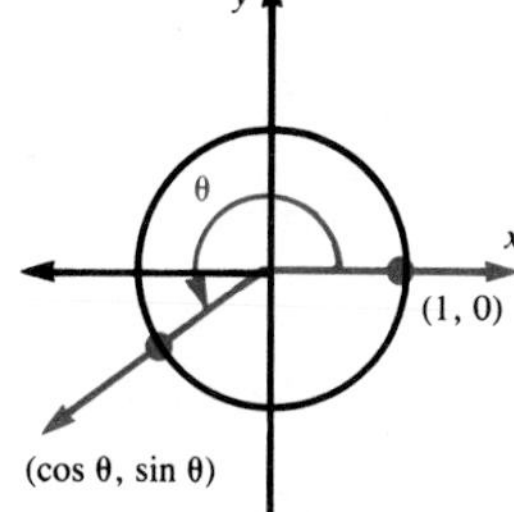

Quadrant III
sin θ and cos θ both negative

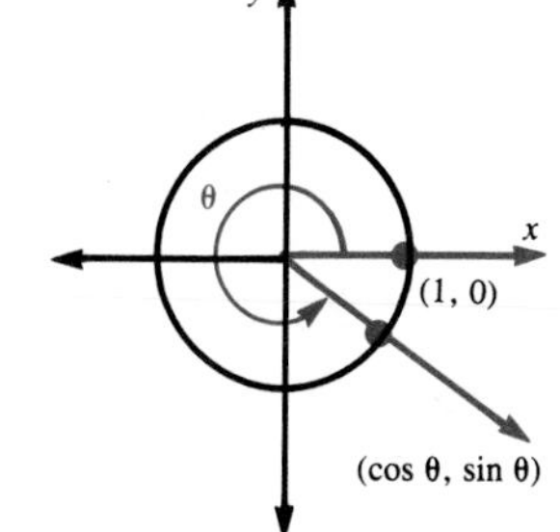

Quadrant IV
sin θ negative
cos θ positive

Exercises

Exploratory **Find each of the following.**

1. $\sin 90°$ **2.** $\cos 0°$ **3.** $\sin 270°$ **4.** $\cos(-180°)$

5. $\cos\frac{\pi}{2}$ **6.** $\cos\left(-\frac{\pi}{2}\right)$ **7.** $\sin 180°$ **8.** $\sin\left(-\frac{3\pi}{2}\right)$

9. $\sin 720°$ **10.** $\cos 540°$ **11.** $\sin(-450°)$ **12.** $\cos(-990°)$

13. $-\sin 90°$ **14.** $-\cos 180°$ **15.** $-\sin 270°$ **16.** $-\cos 0°$

Tell whether each of the following is a positive or negative number. Justify your answer.

17. $\sin 30°$ **18.** $\cos 45°$ **19.** $\sin 125°$ **20.** $\cos 230°$

21. $\sin 240°$ **22.** $\sin(-52°)$ **23.** $\cos(-343°)$ **24.** $\cos 315°$

25. $\sin\frac{5\pi}{6}$ **26.** $\cos\left(-\frac{5\pi}{4}\right)$ **27.** $\cos\left(-\frac{7\pi}{4}\right)$ **28.** $\sin\left(-\frac{7\pi}{3}\right)$

Written **Determine the quadrant in which θ lies when the following conditions are satisfied.**

1. $\sin\theta > 0$ and $\cos\theta > 0$.
2. $\sin\theta < 0$ and $\cos\theta < 0$.
3. $\sin\theta > 0$ and $\cos\theta < 0$.
4. $\sin\theta < 0$ and $\cos\theta > 0$.

If θ lies in the given quadrant and sin θ has the given value, find cos θ.

5. $\sin\theta = \frac{4}{5}$; quadrant I
6. $\sin\theta = \frac{5}{13}$; quadrant II
7. $\sin\theta = -\frac{3}{5}$; quadrant IV
8. $\sin\theta = -\frac{1}{\sqrt{2}}$; quadrant III

If θ lies in the given quadrant and cos θ has the given value, find sin θ.

9. $\cos\theta = -0.6$; quadrant II
10. $\cos\theta = \frac{1}{4}$; quadrant IV
11. $\cos\theta = -\frac{5}{13}$; quadrant III
12. $\cos\theta = -\frac{\sqrt{5}}{3}$; quadrant II

Evaluate.

13. $(\sin 90°)(\cos 180°)$
14. $\sin 180° + \cos 0°$
15. $\sin^2 \frac{3\pi}{2}$
16. $3\cos 0° - 5\sin 270°$
17. $\cos(70° + 20°)$
18. $\sin^2 42° + \cos^2 42°$
19. $\cos\pi - \cos 180°$
20. $3\sin\frac{\pi}{2}$
21. $5\sin\left[6\left(\frac{\pi}{4}\right)\right]$

Tell whether each statement is true or false. Justify your answer.

22. $\sin 180° = 2\sin 90°$
23. $\cos 40° = \cos(-40°)$
24. $\sin 30° = \sin 150°$
25. $\cos 210° = -\cos 30°$
26. $\cos^2\theta = \cos\theta^2$
27. $\sin\theta + \cos\theta = 1$
28. The sine of an angle is never greater than 1.
29. The cosine of an angle can be equal to $-\frac{3}{2}$.
30. If $\sin\theta\cos\theta > 0$ then θ is in quadrant I.

Challenge **For exercises 1–9, refer to the figure in which** $B \xrightarrow{R_{y\text{-axis}}} C$, $C \xrightarrow{R_{x\text{-axis}}} D$, **and** $D \xrightarrow{R_{y\text{-axis}}} E$. **The angles *AOB*, *AOC*, *AOD*, and *AOE* are in standard position with the following measures.**

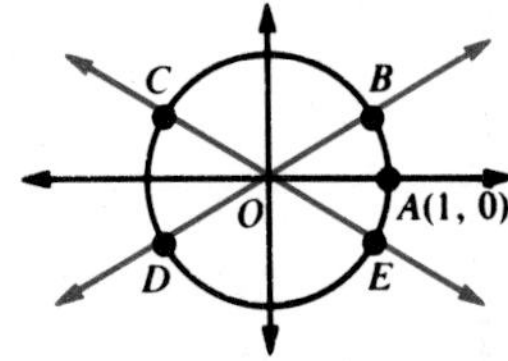

$m\angle AOB = \theta_1$ $m\angle AOC = \theta_2$ $m\angle AOD = \theta_3$ $m\angle AOE = \theta_4$

Tell whether the following are true or false. Justify your answer.

1. $\cos\theta_1 = \cos\theta_4$
2. $\sin\theta_1 = \sin\theta_2$
3. $\sin\theta_1 = \sin\theta_3$
4. $\cos\theta_2 = \cos\theta_3$
5. $\sin\theta_2 = -\sin\theta_3$
6. $\cos\theta_3 = \cos\theta_4$
7. $\cos\theta_1 = \cos\theta_3$
8. $\cos\theta_2 = -\cos\theta_4$
9. $\sin\theta_1 = \sin\theta_4$
10. What are the domain and range of the sine function? of the cosine function?

7.4 Special Angles

To find the sine and cosine of angle measurements such as 30°, 45°, or 60°, it is helpful to recall the following relationships in certain right triangles.

45° – 45° – 90° Triangle

In a 45° – 45° – 90° triangle, the lengths of the legs are the same. Each leg is one-half the length of the hypotenuse times $\sqrt{2}$.

30° – 60° – 90° Triangle

In a 30°–60°–90° triangle, the length of the side opposite the 30° angle is one-half the length of the hypotenuse. The length of the side opposite the 60° angle is one-half the length of the hypotenuse times $\sqrt{3}$.

Example

1 **Find cos 45° and sin 45°.**

The figure shows an angle of 45° in standard position. The right triangle is a 45° – 45° right triangle. Since the hypotenuse has a length of 1 unit, each leg has a length of $\frac{1}{2}\sqrt{2}$ or $\frac{\sqrt{2}}{2}$ units. Therefore, the coordinates of P are $\left(\frac{\sqrt{2}}{2}, \frac{\sqrt{2}}{2}\right)$. We conclude that $\cos 45° = \frac{\sqrt{2}}{2}$ and $\sin 45° = \frac{\sqrt{2}}{2}$.

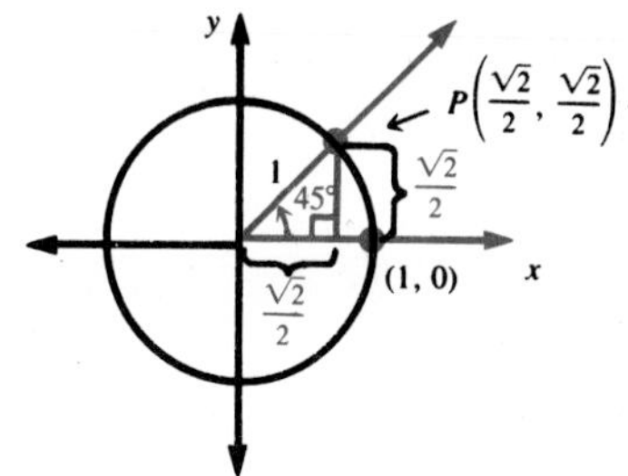

2 **Find cos 30° and sin 30°.**

The figure shows an angle of 30° in standard position. The right triangle is a 30° – 60° right triangle. Since the hypotenuse has a length of 1 unit, the side opposite the 30° angle has a length of $\frac{1}{2}$ unit. The other side, opposite the 60° angle, has a length of $\frac{\sqrt{3}}{2}$ units. Therefore, the coordinates of P are $\left(\frac{\sqrt{3}}{2}, \frac{1}{2}\right)$. We conclude that $\cos 30° = \frac{\sqrt{3}}{2}$ and $\sin 30° = \frac{1}{2}$.

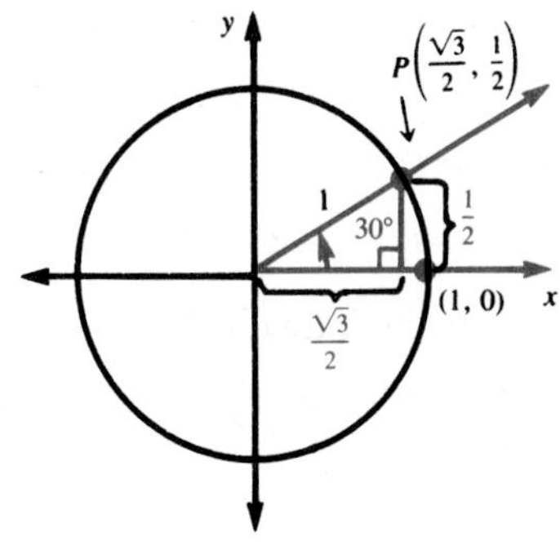

3 **Find cos 60° and sin 60°.**

This time an angle of 60° is in standard position. Using the 30°–60° right triangle relationship, we find that the coordinates of P are $\left(\frac{1}{2}, \frac{\sqrt{3}}{2}\right)$. Thus, $\cos 60° = \frac{1}{2}$ and $\sin 60° = \frac{\sqrt{3}}{2}$.

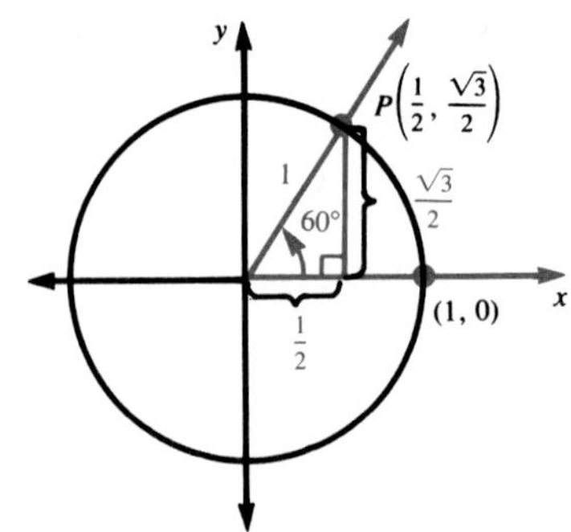

4 **Find cos 150° and sin 150°.**

The terminal side of a 150° angle is a reflection of the terminal side of a 30° angle over the y-axis.

Recall that $(x, y) \xrightarrow{R_{y\text{-axis}}} (-x, y)$.

Thus, $\left(\frac{\sqrt{3}}{2}, \frac{1}{2}\right) \xrightarrow{R_{y\text{-axis}}} \left(-\frac{\sqrt{3}}{2}, \frac{1}{2}\right)$.

Hence, $\cos 150° = -\frac{\sqrt{3}}{2}$ and $\sin 150° = \frac{1}{2}$.

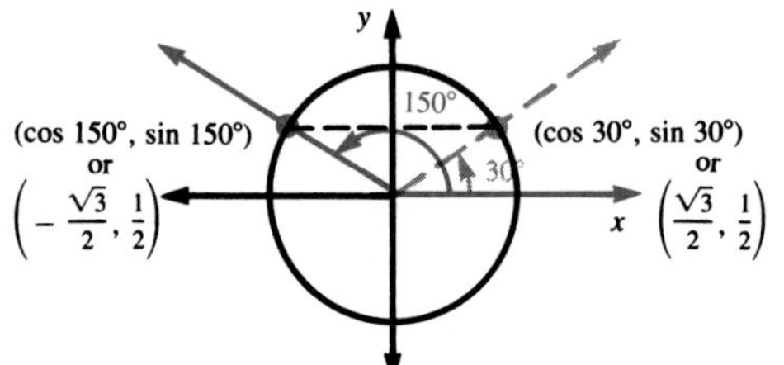

Example 4 is a special case of a general method for expressing the sine and cosine of angles in quadrants II, III, and IV in terms of the sine and cosine of angles in quadrant I.

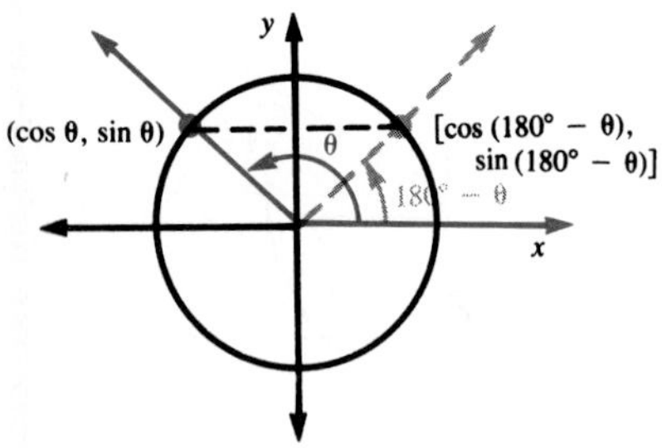

The angle with measure 180° − θ is the reference angle.

Consider an angle in quadrant II with measure θ. Under a reflection over the y-axis, the image of the terminal side is in quadrant I. The angle formed in quadrant I, called the **reference angle,** measures $180° - \theta$. Using the fact that $(x, y) \xrightarrow{R_{y\text{-axis}}} (-x, y)$, we have the following rules.

If θ is the measure of an angle in quadrant II,

then $\cos \theta = -\cos (180° - \theta)$

and $\sin \theta = \sin (180° - \theta)$.

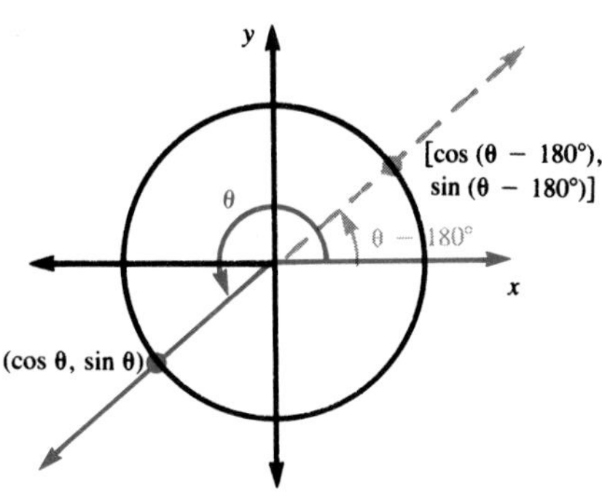

Pictured at the left is an angle in quadrant III. Under a point reflection in the origin, the image of the terminal side is in quadrant I. The reference angle formed has a degree measure of $\theta - 180°$.

Using the fact that $(x, y) \xrightarrow{Rot_{O, 180°}} (-x, -y)$, we have the following rule.

If θ is the measure of an angle in quadrant III,

then $\cos \theta = -\cos (\theta - 180°)$

and $\sin \theta = -\sin (\theta - 180°)$.

The situation of a quadrant IV angle with degree measure θ is pictured at the right. Under a reflection over the x-axis, the image of the terminal side is in quadrant I. The reference angle has measure $360° - \theta$.

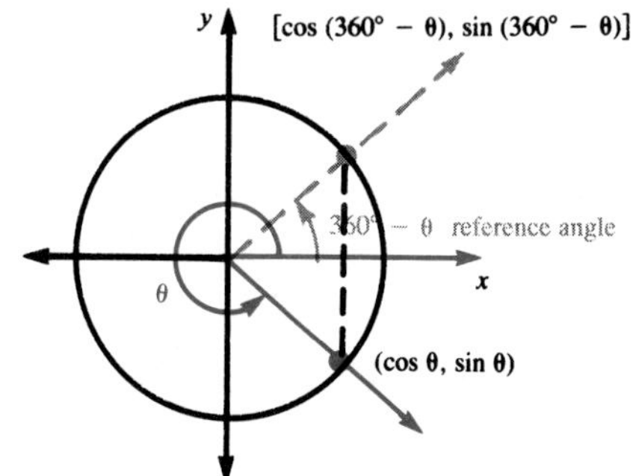

Using the fact that $(x, y) \xrightarrow{R_{x\text{-axis}}} (x, -y)$, we have the following rule.

If θ is the measure of an angle in quadrant IV,

then $\cos \theta = \cos (360° - \theta)$

and $\sin \theta = -\sin (360° - \theta)$.

Examples

5 Find cos 240° and sin 240°.

Sketch an angle of 240° in standard position. The angle lies in quadrant III. The reference angle in quadrant I is $240° - 180°$ or $60°$. In quadrant III, both the sine and cosine have negative values. Thus,

$$\cos 240° = -\cos 60° = -\frac{1}{2}$$

and $$\sin 240° = -\sin 60° = -\frac{\sqrt{3}}{2}.$$

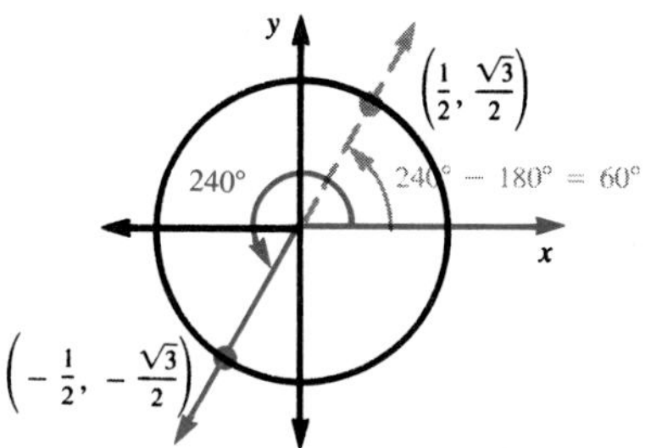

6 Express cos 280° as a function of a positive acute angle.

Since 280° is in quadrant IV, we have

$$\cos 280° = \cos (360° - 280°)$$
$$= \cos 80°.$$

The reference angle is 80°. The cosine of a quadrant IV angle is positive.

Exercises

Exploratory **Find each of the following.**

1. $\sin 30°$
2. $\cos 45°$
3. $\cos 30°$
4. $\sin 45°$
5. $\cos 60°$
6. $\sin 60°$
7. $\sin 45° + \cos 45°$
8. $\sin 30° \cos 60°$
9. $\sin^2 60° + \cos^2 60°$
10. $\cos 180° + 2 \cos 60°$

Written **Find each of the following. Include a sketch.**

1. $\cos 120°$
2. $\sin 120°$
3. $\sin 135°$
4. $\cos 135°$
5. $\cos 150°$
6. $\sin 315°$
7. $\cos 315°$
8. $\sin 240°$
9. $\sin 225°$
10. $\cos 225°$
11. $\cos 240°$
12. $\sin 330°$
13. $\cos 330°$
14. $\sin 210°$
15. $\cos 300°$
16. $\cos 210°$
17. $\sin \frac{\pi}{6}$
18. $\cos \frac{2\pi}{3}$
19. $\sin (-30°)$
20. $\cos \frac{7\pi}{4}$
21. $\sin \frac{4\pi}{3}$
22. $\cos (-60°)$
23. $\cos \left(-\frac{3\pi}{4}\right)$
24. $\cos \frac{11\pi}{4}$
25. $\sin \left(-\frac{\pi}{6}\right)$
26. $\cos (-390°)$
27. $\sin 660°$
28. $\cos (-1560°)$

Express as a function of a positive acute angle.

29. $\sin 130°$
30. $\cos 140°$
31. $\sin 310°$
32. $\sin 190°$
33. $\cos 350°$
34. $\cos 165°$
35. $\sin 156°$
36. $\cos 138°$
37. $\sin 284°$
38. $\cos 310°$
39. $\sin 215°$
40. $\cos 200°$
41. $\sin 400°$
42. $\cos 670°$
43. $\sin (-40°)$
44. $\cos (-50°)$
45. $\sin (-160°)$
46. $\sin (-350°)$
47. $-\cos (-250°)$
48. $-\cos (-200°)$
49. $\cos \frac{8\pi}{7}$
50. $-\sin \frac{11\pi}{12}$
51. $-\sin \left(-\frac{\pi}{4}\right)$
52. $\cos \frac{7\pi}{4}$

Evaluate.

53. $\sin 30° + \cos 60°$
54. $4(\sin 30°)(\cos 60°)$
55. $\sin 300° + \sin 240°$
56. $\sin 240° \cos 120°$
57. $\cos \frac{\pi}{2} + 3 \sin \frac{5\pi}{4}$
58. $\cos \left(-\frac{3\pi}{4}\right) - 3 \sin \frac{7\pi}{4}$
59. $\sin \frac{4\pi}{3} \sin \frac{\pi}{3}$
60. $2 \sin \left(-\frac{5\pi}{6}\right) - 4 \cos \frac{2\pi}{3}$
61. $8 \sin 120° \cos 120°$

Challenge **If $0° \le \theta < 360°$, find all values of θ that satisfy the equation.**

1. $\cos \theta = \frac{\sqrt{3}}{2}$
2. $\sin \theta = \frac{\sqrt{2}}{2}$
3. $\cos \theta = -\frac{1}{2}$
4. $\sin \theta = -\frac{\sqrt{3}}{2}$
5. $\sin \theta = -\frac{1}{2}$
6. $\cos \theta = -\frac{\sqrt{2}}{2}$

7.5 The Sine and Cosine Graphs

Recall that a function pairs each element of one set, called the *domain* of the function, with one and only one element of another set, called the *range*. For both the sine and cosine functions, the domain is the set of real numbers in the form of radian measures or the set of all degree measures. The range of the sine and cosine functions is the set of real numbers between -1 and 1, that is, $\{y: -1 \le y \le 1\}$.

The first step in drawing the graph of $y = \sin \theta$ is to prepare a table of values as shown below.

θ	*degrees*	0	30	45	60	90	120	135	150	180	210	225	240	270	300	315	330	360
	radians	0	$\frac{\pi}{6}$	$\frac{\pi}{4}$	$\frac{\pi}{3}$	$\frac{\pi}{2}$	$\frac{2\pi}{3}$	$\frac{3\pi}{4}$	$\frac{5\pi}{6}$	π	$\frac{7\pi}{6}$	$\frac{5\pi}{4}$	$\frac{4\pi}{3}$	$\frac{3\pi}{2}$	$\frac{5\pi}{3}$	$\frac{7\pi}{4}$	$\frac{11\pi}{6}$	2π
sin θ	exact value	0	$\frac{1}{2}$	$\frac{\sqrt{2}}{2}$	$\frac{\sqrt{3}}{2}$	1	$\frac{\sqrt{3}}{2}$	$\frac{\sqrt{2}}{2}$	$\frac{1}{2}$	0	$-\frac{1}{2}$	$-\frac{\sqrt{2}}{2}$	$-\frac{\sqrt{3}}{2}$	-1	$-\frac{\sqrt{3}}{2}$	$-\frac{\sqrt{2}}{2}$	$-\frac{1}{2}$	0
	rational approx.	0	0.5	0.71	0.87	1	0.87	0.71	0.5	0	-0.5	-0.71	-0.87	-1	-0.87	-0.71	-0.5	0

Next, choose a pair of coordinate axes. The horizontal axis is for the values of θ in either degrees or radians. The vertical axis is for the values of $\sin \theta$. To minimize distortion, we choose the same unit of length on each axis. The distance from 0 to 1 on the vertical axis should be about the same as the distance from 0 to $\frac{\pi}{3}$ on the horizontal axis, since $\frac{\pi}{3}$ is approximately $\frac{3.14}{3}$ or 1.05.

After plotting the points from the table of values, draw a smooth, continuous curve.

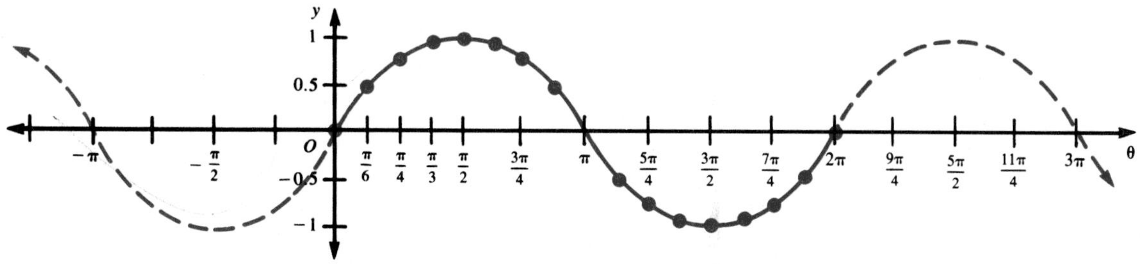

The graph is the *basic sine curve* or *wave*. The oscilloscope shown at the opening of this chapter is one of the many places sine waves occur in scientific work.

If the table of values is extended to include values of θ less than 0 or greater than 2π, then the values of $\sin\theta$ repeat, as shown in the dashed part of the graph. One full cycle of the sine curve occurs in every interval of 360° or 2π radians. In symbols, we have

$$\left.\begin{aligned} \sin\theta &= \sin(\theta + 360^\circ n) \\ \text{or} \quad \sin\theta &= \sin(\theta + 2\pi n) \end{aligned}\right\} \text{ where } n \text{ is any integer.}$$

The sine function is an example of a **periodic function.** A function f is called **periodic** if there exists a nonzero number a such that for all x in the domain of f, $f(x) = f(x + a)$. The least positive value of a is called the **period** of f.

The period of the function defined by $y = \sin\theta$ is 360° or 2π radians.

A table of values for $y = \cos\theta$ and the graph appear below.

θ	0	$\frac{\pi}{6}$	$\frac{\pi}{3}$	$\frac{\pi}{2}$	$\frac{2\pi}{3}$	$\frac{5\pi}{6}$	π	$\frac{7\pi}{6}$	$\frac{4\pi}{3}$	$\frac{3\pi}{2}$	$\frac{5\pi}{3}$	$\frac{11\pi}{6}$	2π
$\cos\theta$	1	0.87	0.5	0	−0.5	−0.87	−1	−0.87	−0.5	0	0.5	0.87	1

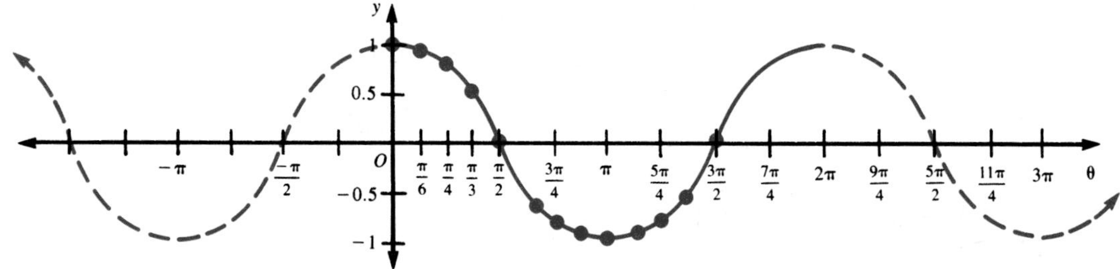

The cosine function is also periodic. Its period is 360° or 2π radians.

Examples

1 **On the same axes, draw the graphs of $y = 2\sin x$ and $y = \frac{1}{2}\sin x$ for x in the interval from 0 to 2π.**

First, prepare a table of values as shown below.

x	0	$\frac{\pi}{6}$	$\frac{\pi}{3}$	$\frac{\pi}{2}$	$\frac{2\pi}{3}$	$\frac{5\pi}{6}$	π	$\frac{7\pi}{6}$	$\frac{4\pi}{3}$	$\frac{3\pi}{2}$	$\frac{5\pi}{3}$	$\frac{11\pi}{6}$	2π
$\sin x$	0	$\frac{1}{2}$	$\frac{\sqrt{3}}{2}$	1	$\frac{\sqrt{3}}{2}$	$\frac{1}{2}$	0	$-\frac{1}{2}$	$-\frac{\sqrt{3}}{2}$	−1	$-\frac{\sqrt{3}}{2}$	$-\frac{1}{2}$	0
$2\sin x$	0	1	$\sqrt{3}$ or 1.7	2	$\sqrt{3}$ or 1.7	1	0	−1	$-\sqrt{3}$ or −1.7	−2	$-\sqrt{3}$ or −1.7	−1	0
$\frac{1}{2}\sin x$	0	$\frac{1}{4}$	$\frac{\sqrt{3}}{4}$ or 0.4	$\frac{1}{2}$	$\frac{\sqrt{3}}{4}$ or 0.4	$\frac{1}{4}$	0	$-\frac{1}{4}$	$-\frac{\sqrt{3}}{4}$ or −0.4	$-\frac{1}{2}$	$-\frac{\sqrt{3}}{4}$ or −0.4	$-\frac{1}{4}$	0

Then sketch the graph of each function.

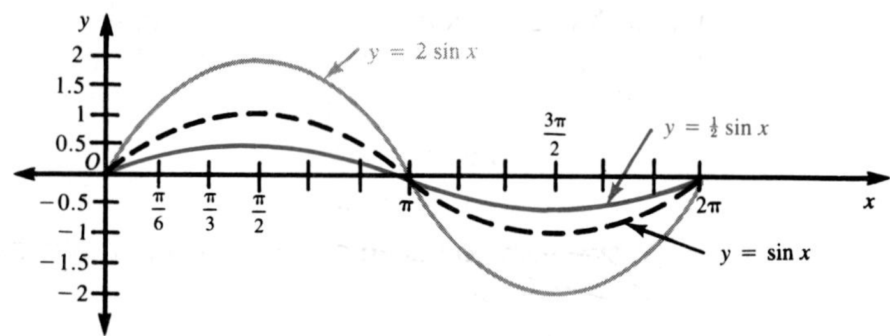

2 **Draw the graph of $y = \cos 2x$ as x varies from 0 to 2π. State the period of the graph.**
Careful! This is not the same as $y = 2 \cos x$.

x	0	$\frac{\pi}{12}$	$\frac{\pi}{6}$	$\frac{\pi}{4}$	$\frac{\pi}{3}$	$\frac{5\pi}{12}$	$\frac{\pi}{2}$
2x	0	$\frac{\pi}{6}$	$\frac{\pi}{3}$	$\frac{\pi}{2}$	$\frac{2\pi}{3}$	$\frac{5\pi}{6}$	π
cos 2x	1	0.9	0.5	0	−0.5	−0.9	−1

x	$\frac{7\pi}{12}$	$\frac{2\pi}{3}$	$\frac{3\pi}{4}$	$\frac{5\pi}{6}$	$\frac{11\pi}{12}$	π
2x	$\frac{7\pi}{6}$	$\frac{4\pi}{3}$	$\frac{3\pi}{2}$	$\frac{5\pi}{3}$	$\frac{11\pi}{6}$	2π
cos 2x	−0.9	−0.5	0	0.5	0.9	1

Note that one complete cycle of the curve spans 180°, or π radians.

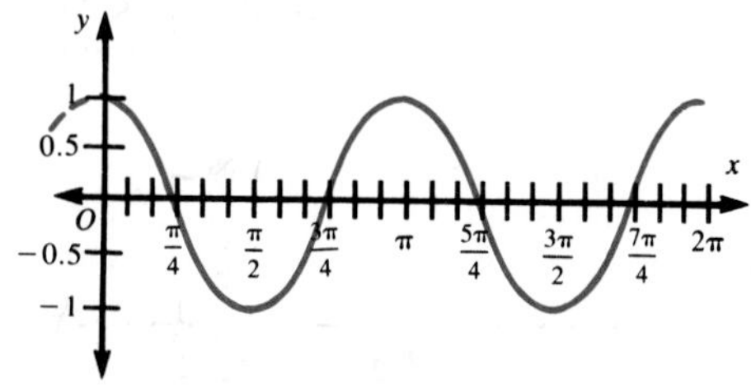

Since the function repeats after 180° or π radians, the period is 180° or π radians.

In Example 2, observe that there are two full cycles of the cosine curve in the interval $0 \leq x \leq 2\pi$. We say that the **frequency** of the curve $y = \cos 2x$ is 2. In general, the frequency is the number of cycles that occur within an interval of 2π radians or 360°.

For the function $y = \sin x$, the greatest or maximum y-value is 1. We say that the **amplitude** of $y = \sin x$ is 1. In Example 1, notice that the greatest y-value of $y = 2 \sin x$ is 2. Thus, the amplitude of $y = 2 \sin x$ is 2. The minimum value of $y = \sin x$ is -1 and the minimum value of $y = 2 \sin x$ is -2.

Amplitude, Period, and Frequency

For functions of the form $y = a \sin bx$ and $y = a \cos bx$, the **amplitude** is $|a|$, the **period** is $\frac{2\pi}{|b|}$ or $\frac{360°}{|b|}$, and the **frequency** is $|b|$.

Examples

3 **Find the amplitude, period, and frequency of $y = 3 \cos \frac{1}{2}x$.**

The amplitude is $|3|$ or 3.

The period is $\frac{2\pi}{\left|\frac{1}{2}\right|}$ or 4π radians. *In degrees, the period is 720°.*

The frequency is $\left|\frac{1}{2}\right|$ or $\frac{1}{2}$. *Within 2π radians, one -half of a cosine curve appears.*

4 **Write an equation for the graph shown at the right.**

The maximum value of y is 3, so the amplitude of the function is 3. The graph completes one-half of a cycle in the interval $0 \leq x \leq 2\pi$, so the frequency is $\frac{1}{2}$. The equation is either $y = \pm 3 \cos \frac{1}{2}x$ or $y = \pm 3 \sin \frac{1}{2}x$. Since the values of y for $x = 0$ to 2π are positive we can rule out $y = -3 \cos \frac{1}{2}x$ and $y = -3 \sin \frac{1}{2}x$.

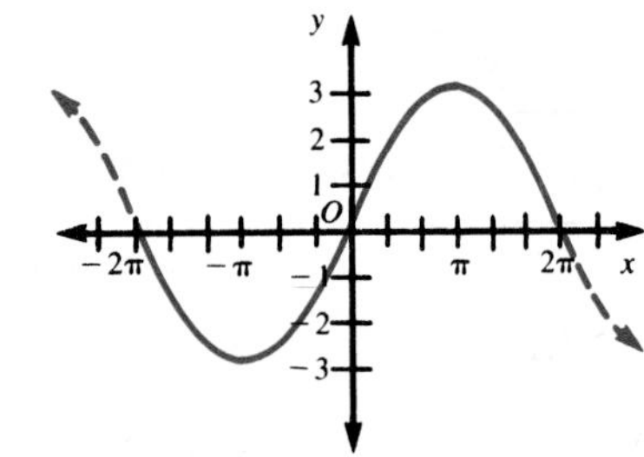

The maximum y-value of the graph occurs at $x = \pi$ and since the maximum value of $y = \sin x$ occurs at $x = \frac{\pi}{2}$ (and the maximum value of $y = \cos x$ occurs at $x = 0$), this is the graph of $y = 3 \sin \frac{1}{2}x$.

Exercises

Exploratory **State the amplitude, period, and frequency of each function in exercises 1–16.**

1. $y = \sin \theta$

2. $y = \frac{1}{2} \cos \theta$

3. $y = 2 \sin x$

4. $y = 3 \cos \theta$

5. $y = \sin 2x$

6. $y = \cos \frac{1}{2} \theta$

7. $y = \sin 4x$

8. $y = \cos \frac{2}{3} \theta$

9. $y = 3 \cos 2x$ **10.** $y = 2 \sin 3x$ **11.** $y = \frac{1}{2} \sin 2x$ **12.** $y = 13 \sin 4x$

13. $y = 5 \sin \frac{1}{3}x$ **14.** $y = \frac{3}{5} \cos \frac{1}{2}x$ **15.** $y = -3 \sin x$ **16.** $y = 2 \sin \left(-\frac{1}{4}x\right)$

Written **Find the range of each function.**

1. $y = \sin x$ **2.** $y = \cos x$ **3.** $y = 3 \cos x$ **4.** $y = \sin 2x$

5. $y = 5 \sin 2x$ **6.** $y = -\cos x$ **7.** $y = 3 \sin \frac{1}{2}x$ **8.** $y = -2 \sin 3x$

Find the values of x in the interval $0 \leq x \leq 2\pi$ for which each of the following has a maximum value and a minimum value.

9. $y = \sin x$ **10.** $y = \cos x$ **11.** $y = 2 \sin x$ **12.** $y = 3 \cos x$

13. $y = \sin 2x$ **14.** $y = \cos 2x$ **15.** $y = 3 \sin \frac{1}{2}x$ **16.** $y = 2 \cos \frac{1}{2}x$

Write an equation for each graph.

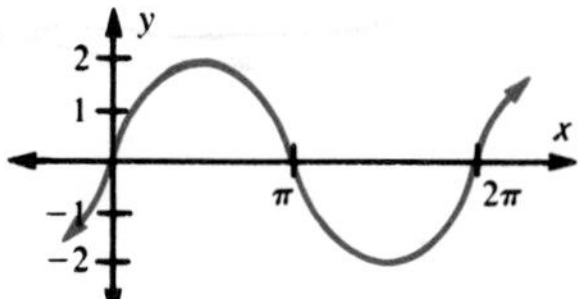

18.

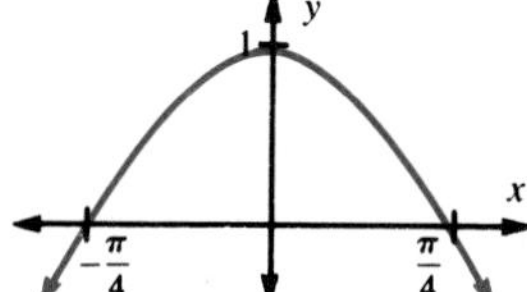

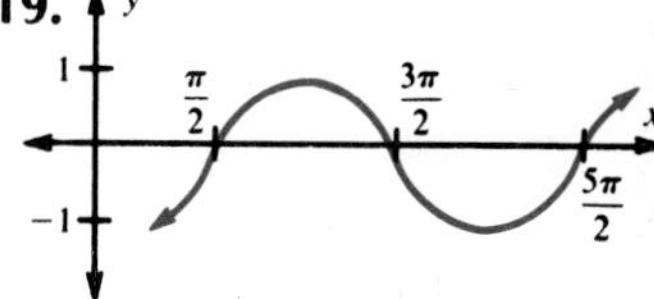

Draw the graph of each of the following as x varies from 0 to 2π.

20. $y = 2 \cos x$ **21.** $y = \sin 2x$ **22.** $y = \frac{1}{2} \cos x$ **23.** $y = 3 \sin x$

24. $y = 3 \cos 2x$ **25.** $y = -\sin x$ **26.** $y = -\cos x$ **27.** $y = |\sin x|$

28. Draw a graph of $y = \sin x$ in the interval $-2\pi \leq x \leq 2\pi$. What points or lines of symmetry does the graph have?

29. Repeat Exercise 28 for the graph of $y = \cos x$.

30. If the graphs of $y = \sin x$ and $y = \cos x$ are drawn on the same set of axes, what transformation maps one to the other?

Mathematical Excursions

A Problem for the Scholar

You may need to review parts of chapter 6 before answering this question.

What is the shortest distance from the center of the circle to the longest chord if three parallel chords of lengths 4, 8, and 10 lie in a circle on the same side of its center and the distance between them is the same?

7.6 Sketching $y = a \sin bx$ and $y = a \cos bx$

A knowledge of the amplitude and period help us to sketch the graph of a function of the form $y = a \sin bx$ or $y = a \cos bx$.

Example

1 **Sketch the graph of $y = 3 \cos 2x$ as x varies from 0 to 2π inclusive.**

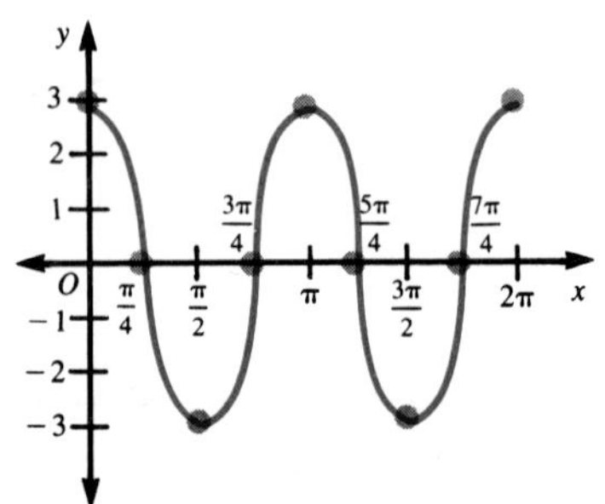

The amplitude is $|3|$ or 3.

The period is $\frac{2\pi}{2}$ or π.

Plot the "key" points whose x-coordinates divide the interval from 0 to π into quarters, and the interval from π to 2π into quarters. Connect the points with a smooth cosine curve.

From previous work you know that a system of two equations can be solved graphically. The following example demonstrates how graphs may be used to determine the number of solutions of certain trigonometric equations.

Example

2 **On the same set of axes, sketch the graphs of $y = \cos 2x$ and $y = \sin \frac{1}{2}x$ as x varies from 0 to 2π. Find the number of values of x between 0 and 2π that satisfy the equation $\cos 2x = \sin \frac{1}{2}x$.**

Use the procedure of Example 1 to locate "key" points for each graph.

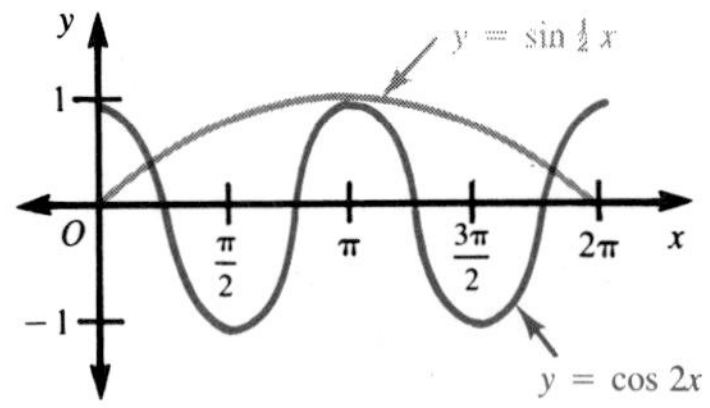

$y = \cos 2x$	$y = \sin \frac{1}{2}x$
amplitude = 1	amplitude = 1
period = π	period = 4π

The graphs intersect at three points. The x-coordinates of each of these points are the values of x for which $\cos 2x = \sin \frac{1}{2}x$. Therefore, there are three values of x that satisfy the equation.

Exercises

Exploratory **State the amplitude, period, and equation of each graph.**

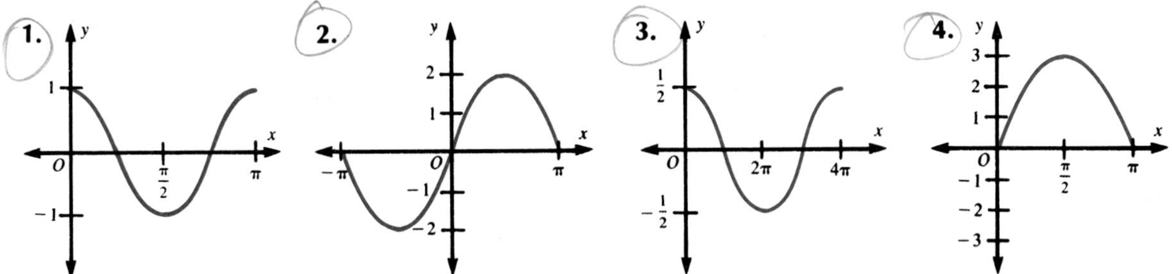

Written **Sketch the graph of each function in the indicated interval.**

1. $y = \sin 2x;\ 0 \le x \le 2\pi$
2. $y = 2 \cos x;\ 0 \le x \le 2\pi$
3. $y = 3 \sin x;\ -\pi \le x \le \pi$
4. $y = \sin \frac{1}{2}x;\ 0 \le x \le \pi$
5. $y = 3 \cos \frac{1}{2}x;\ 0 \le x \le 2\pi$
6. $y = 3 \cos \frac{1}{2}x;\ -\pi \le x \le \pi$
7. $y = \sin \frac{1}{3}x;\ 0 \le x \le 2\pi$
8. $y = \cos \frac{1}{3}x;\ 0 \le x \le 2\pi$
9. $y = 3 \sin 2x;\ -2\pi \le x \le 2\pi$
10. $y = \frac{1}{2} \cos 3x;\ -\frac{\pi}{2} \le x \le \frac{\pi}{2}$
11. $y = -2 \cos \frac{1}{2}x;\ -\pi \le x \le \pi$
12. $y = -\frac{1}{2} \sin 2x;\ -\pi \le x \le \pi$

On the same set of axes, sketch the graphs of *f* and *g* in the interval $0 \le x \le 2\pi$. Then, determine the number of values of *x* in the interval $0 \le x \le 2\pi$ that satisfy the equation $f(x) = g(x)$.

13. $f(x) = 2 \cos x$
    $g(x) = \sin \frac{1}{2}x$
14. $f(x) = 3 \cos x$
    $g(x) = \sin 2x$
15. $f(x) = 2 \cos 3x$
    $g(x) = \sin x$
16. $f(x) = 2 \sin x$
    $g(x) = 2 \cos \frac{1}{2}x$
17. $f(x) = \sin \frac{1}{2}x$
    $g(x) = 2 \sin 2x$
18. $f(x) = -\sin x$
    $g(x) = 2 \cos x$

Challenge **Draw the graph of each function.**

1. $y = \cos\left(x - \frac{\pi}{3}\right)$
2. $y = \sin (2x - \pi)$
3. $y = -2 \cos (2x + \pi)$
4. $y = \sin 2x - 1$
5. $y = 4 \sin (2x - \pi)$
6. $y = 3 \sin (2x - \pi) + 1$
7. $y = \sin x + \cos x$
8. $y = |\sin 2x|$
9. $y = -2 \sin (2x - \pi)$

7.7 The Tangent Function

Another important circular function is the *tangent* function.

Definition of Tangent Function

Let θ be the measure of an angle in standard position. The tangent function pairs θ with the fraction $\frac{\sin \theta}{\cos \theta}$ when $\cos \theta \neq 0$.

The value of the tangent function at θ is symbolized by **tan θ** which is read "tangent of θ." Thus, $\tan \theta = \frac{\sin \theta}{\cos \theta}$ when $\cos \theta \neq 0$.

A geometric interpretation of tan θ is shown at the left. Notice that tan θ is equal to the length of $\overline{PR}$ divided by the length of $\overline{OR}$. That is, $\tan \theta = \frac{\sin \theta}{\cos \theta} = \frac{y}{x}$, $x \neq 0$.

Examples

1 **Find tan 45°.**

$$\tan 45° = \frac{\sin 45°}{\cos 45°}$$

$$= \frac{\frac{\sqrt{2}}{2}}{\frac{\sqrt{2}}{2}} \text{ or } 1$$

2 **Find tan 30°.**

$$\tan 30° = \frac{\sin 30°}{\cos 30°}$$

$$= \frac{\frac{1}{2}}{\frac{\sqrt{3}}{2}} = \frac{1}{\sqrt{3}} \text{ or } \frac{\sqrt{3}}{3}$$

3 **Find tan 120°.**

$$\tan 120° = \frac{\sin 120°}{\cos 120°}$$

Recall that $\sin 120° = \sin(180° - 120°)$

$$= \sin 60° = \frac{\sqrt{3}}{2}$$

and $\cos 120° = -\cos(180° - 120°)$

$$= -\cos 60° = -\frac{1}{2}.$$

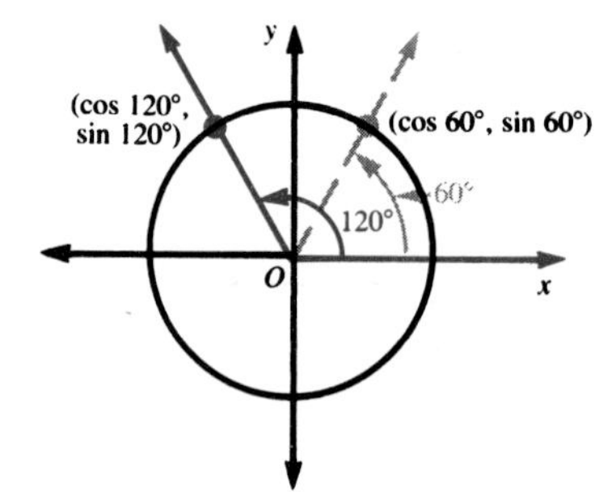

Thus, $\tan 120° = \frac{\frac{\sqrt{3}}{2}}{-\frac{1}{2}}$ or $-\sqrt{3}$.

4 **Which of the following is not in the domain of the tangent function?**

a. 0 **b.** $\frac{\pi}{2}$ **c.** $-\pi$ **d.** $-\frac{\pi}{4}$

$\tan\theta = \frac{\sin\theta}{\cos\theta}$ when $\cos\theta \neq 0$.

$$\tan\frac{\pi}{2} = \frac{\sin\frac{\pi}{2}}{\cos\frac{\pi}{2}}$$

But $\cos\frac{\pi}{2} = 0$. Thus, $\frac{\pi}{2}$ is not in the domain of the tangent function.

The answer is b, $\frac{\pi}{2}$.

You should verify that $\tan 0 = 0$, $\tan(-\pi) = 0$, *and* $\tan\left(-\frac{\pi}{4}\right) = -1$.

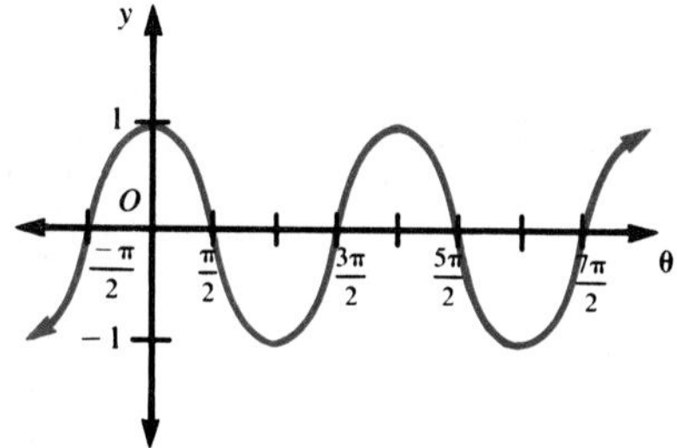

In Example 4, we see that $\frac{\pi}{2}$ is not in the domain of the tangent function, since $\cos\frac{\pi}{2} = 0$. Are there other values of θ for which $\tan\theta$ is undefined? Look carefully at the sketch of $y = \cos\theta$. Notice that $\cos\theta$ is 0 when θ is $-\frac{\pi}{2}, \frac{\pi}{2}, \frac{3\pi}{2}, \frac{5\pi}{2}, \frac{7\pi}{2}$, or any odd multiple of $\frac{\pi}{2}$. Thus, the domain of the tangent function is the set of all real numbers except the odd multiples of $\frac{\pi}{2}$.

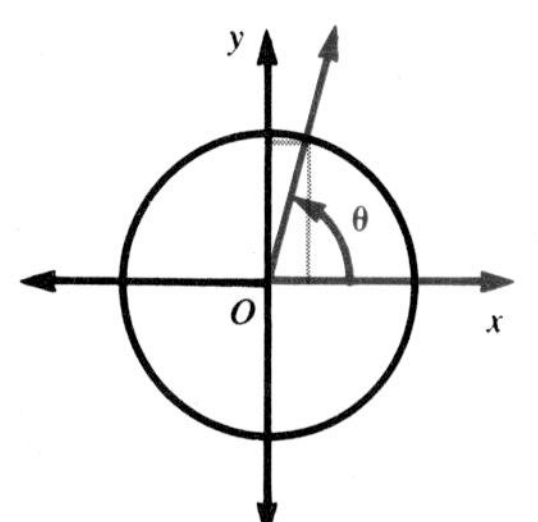

In the figure at the left, notice that the value of $\frac{\sin\theta}{\cos\theta}$ becomes greater as θ approaches $\frac{\pi}{2}$. Thus, the tangent function is not bounded by -1 and 1 as the sine and cosine functions are. Therefore, the range of the tangent function is the set of real numbers.

Now we can prepare a table of values for $y = \tan\theta$ and draw its graph for values of x from 0 to 2π.

θ	0	$\frac{\pi}{6}$	$\frac{\pi}{4}$	$\frac{\pi}{3}$	$\frac{\pi}{2}$	$\frac{2\pi}{3}$	$\frac{3\pi}{4}$	$\frac{5\pi}{6}$	π	$\frac{7\pi}{6}$	$\frac{5\pi}{4}$	$\frac{4\pi}{3}$	$\frac{3\pi}{2}$	$\frac{5\pi}{3}$	$\frac{7\pi}{4}$	$\frac{11\pi}{6}$	2π
tan θ	0	$\frac{\sqrt{3}}{3}$	1	$\sqrt{3}$	not defined	$-\sqrt{3}$	-1	$-\frac{\sqrt{3}}{3}$	0	$\frac{\sqrt{3}}{3}$	1	$\sqrt{3}$	not defined	$-\sqrt{3}$	-1	$-\frac{\sqrt{3}}{3}$	0

Note that the period of the tangent function is 180° or π radians.

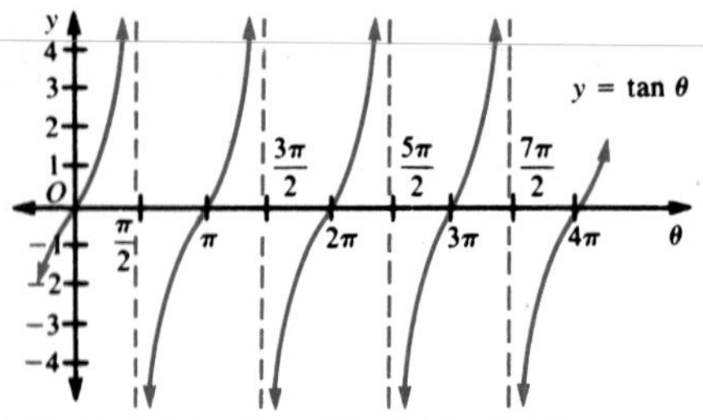

The line for $\theta = \frac{\pi}{2}$ is an example of an **asymptote.** In general, if a graph can be made as close as you wish to a line, the line is called an asymptote of the graph. The graph is said to be asymptotic to the line.

The tangent function does not have an amplitude.

The dashed lines are asymptotes.

Function	Quadrant I	Quadrant II	Quadrant III	Quadrant IV
sin	pos	pos	neg	neg
cos	pos	neg	neg	pos
tan	pos	neg	pos	neg

The sketches of $y = \sin x$, $y = \cos x$, and $y = \tan x$ help us to summarize the signs of these functions in the different quadrants. These results are shown in the table at the left.

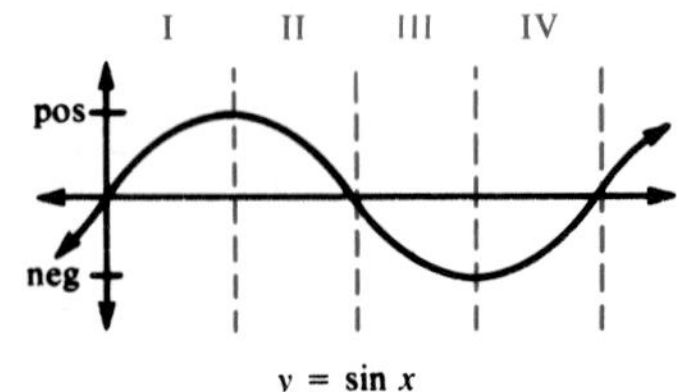

$y = \sin x$

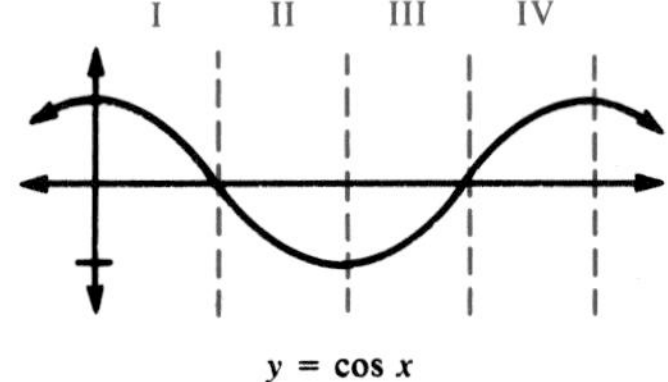

$y = \cos x$

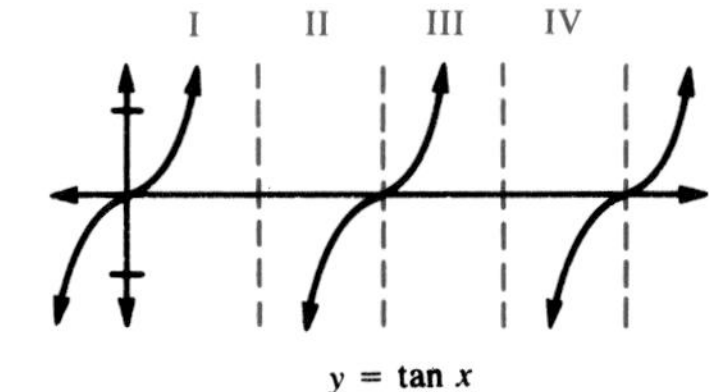

$y = \tan x$

Example

5 If $\sin\theta = \frac{1}{2}$ and θ is in quadrant II, find $\tan\theta$.

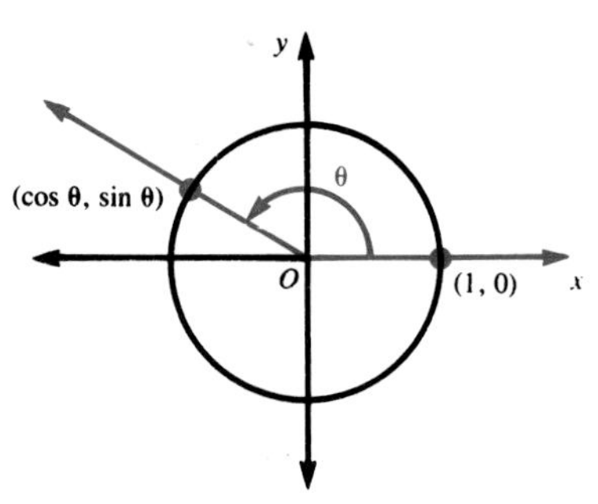

$\sin^2\theta + \cos^2\theta = 1$ *Pythagorean Identity*

$$\left(\frac{1}{2}\right)^2 + \cos^2\theta = 1$$

$$\cos^2\theta = 1 - \frac{1}{4}$$

$$\cos^2\theta = \frac{3}{4}$$

$$\cos\theta = \pm\frac{\sqrt{3}}{2}$$

Since θ is in quadrant II, $\cos\theta$ is negative. Therefore, $\cos\theta = -\frac{\sqrt{3}}{2}$.

$$\text{Thus, } \tan\theta = \frac{\sin\theta}{\cos\theta} = \frac{\frac{1}{2}}{-\frac{\sqrt{3}}{2}}$$

$$= -\frac{1}{\sqrt{3}} \text{ or } -\frac{\sqrt{3}}{3}.$$

Exercises

Exploratory **Determine the quadrant in which θ lies when the following conditions are satisfied.**

1. $\sin \theta > 0$ and $\cos \theta > 0$
2. $\sin \theta > 0$ and $\tan \theta < 0$
3. $\tan \theta > 0$ and $\cos \theta < 0$
4. $\cos \theta < 0$ and $\tan \theta < 0$
5. $\sin \theta < 0$ and $\cos \theta < 0$
6. $\tan \theta < 0$ and $\sin \theta < 0$
7. State the period of the tangent function.
8. If $0° \leq \theta \leq 360°$, state the values of θ for which $\tan \theta$ is undefined.

Written **Evaluate.**

1. $\tan 30°$
2. $\tan 45°$
3. $\tan 60°$
4. $\tan 240°$
5. $\tan(-45°)$
6. $\tan 180°$
7. $\tan 150°$
8. $\tan 330°$
9. $\tan 135°$
10. $\tan 210°$
11. $\tan 225°$
12. $\tan 420°$
13. $\tan(-\pi)$
14. $\tan \frac{5\pi}{6}$
15. $\tan\left(-\frac{\pi}{6}\right)$
16. $\tan\left(-\frac{2\pi}{3}\right)$
17. $\tan 135° + \tan(-315°)$
18. $\cos 90° + \tan 0°$
19. $\cos 0° \tan 300°$

Sketch the graph of each function in the indicated interval.

20. $y = \tan x;\ 0 \leq x \leq 2\pi$
21. $y = \tan x;\ -2\pi \leq x \leq 2\pi$
22. $y = \frac{1}{2} \tan x;\ 0 \leq x \leq 2\pi$
23. $y = 2 \tan x;\ -\pi \leq x \leq \pi$

On the same set of axes, sketch the graphs of f and g in the interval $-\pi \leq x \leq \pi$. Then determine the number of values of x in the interval $-\pi \leq x \leq \pi$ that satisfy the equation $f(x) = g(x)$.

24. $f(x) = \tan x$
    $g(x) = \cos x$
25. $f(x) = \tan x$
    $g(x) = 2 \sin x$
26. $f(x) = \tan x$
    $g(x) = \cos 2x$

Express as a function of a positive acute angle.

27. $\tan 142°$
28. $\tan 320°$
29. $\tan 200°$
30. $\tan(-80°)$
31. $\tan(-112°)$
32. $\tan 260°$
33. $\tan \frac{11\pi}{12}$
34. $\tan\left(-\frac{\pi}{6}\right)$

Given the following information, find $\tan \theta$.

35. $\sin \theta = \frac{3}{5}$ and θ in quadrant I
36. $\cos \theta = \frac{3}{5}$ and θ in quadrant I
37. $\sin \theta = \frac{3}{5}$ and θ in quadrant II
38. $\cos \theta = \frac{12}{13}$ and θ in quadrant IV
39. $\cos \theta = -\frac{4}{5}$ and θ in quadrant III
40. $\sin \theta = -\frac{1}{2}$ and θ in quadrant IV
41. $\sin \theta = \frac{1}{\sqrt{2}}$ and $\cos \theta < 0$
42. $\cos \theta = -\frac{4}{5}$ and $\sin \theta > 0$

If $f(x) = \sin x + 2 \cos x$, find each of the following.

43. $f(\pi)$
44. $f\left(-\frac{\pi}{2}\right)$
45. $f\left(\frac{\pi}{4}\right)$
46. $f\left(\frac{3\pi}{2}\right)$
47. $f(-2\pi)$
48. $f\left(\frac{\pi}{3}\right)$
49. $f\left(\frac{3\pi}{4}\right)$
50. $f\left(-\frac{7\pi}{4}\right)$

7.8 The Reciprocal Circular Functions

Three other functions related to sine, cosine, and tangent are **secant, cosecant,** and **cotangent.** These functions are abbreviated **sec, csc,** and **cot,** respectively.

Definition of Secant, Cosecant, and Cotangent Functions

(The Reciprocal Circular Functions)

Let θ be the measure of an angle in standard position.

$$\sec\theta = \frac{1}{\cos\theta} \text{ when } \cos\theta \neq 0$$

$$\csc\theta = \frac{1}{\sin\theta} \text{ when } \sin\theta \neq 0$$

$$\cot\theta = \frac{\cos\theta}{\sin\theta} \text{ when } \sin\theta \neq 0$$

If θ is not the measure of a quadrantal angle, then

$$\cot\theta\tan\theta = \frac{\cos\theta}{\sin\theta}\cdot\frac{\sin\theta}{\cos\theta} = 1.$$

Thus, $\tan\theta = \dfrac{1}{\cot\theta}$.

Examples

1 Evaluate cot 60°.

$$\cot 60° = \frac{\cos 60°}{\sin 60°} = \frac{\frac{1}{2}}{\frac{\sqrt{3}}{2}} = \frac{1}{\sqrt{3}} \text{ or } \frac{\sqrt{3}}{3}$$

2 Evaluate sec²150°.

$$\sec^2 150° = \left(\frac{1}{\cos 150°}\right)^2 = \left(\frac{1}{-\frac{\sqrt{3}}{2}}\right)^2 = \frac{4}{3}$$

3 Find csc π.

$$\csc\pi = \frac{1}{\sin\pi}$$

But $\sin\pi = 0$.

Hence csc π is not defined.

4 Express csc θ tan θ in terms of cos θ.

$$\csc\theta\tan\theta = \frac{1}{\sin\theta}\cdot\frac{\sin\theta}{\cos\theta}$$ *Definition of csc θ and tan θ*

$$= \frac{1}{\cos\theta}$$ *Divide numerator and denominator by sin θ.*

The graphs of the secant, cosecant, and cotangent functions appear below. The graphs of the cosine, sine, and tangent functions appear for reference as the dashed curves.

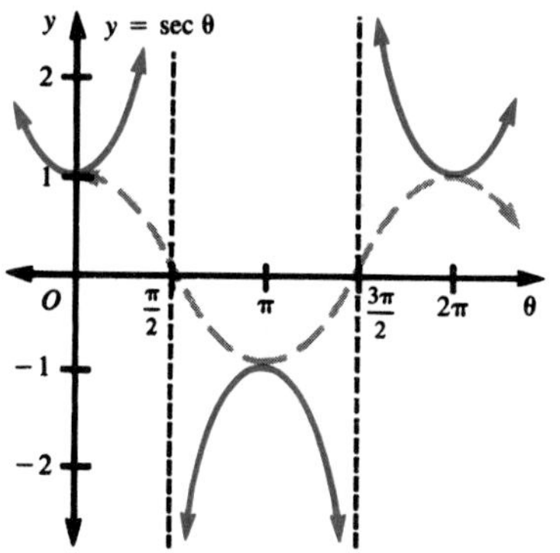

The period of $y = \sec\theta$ is 2π.

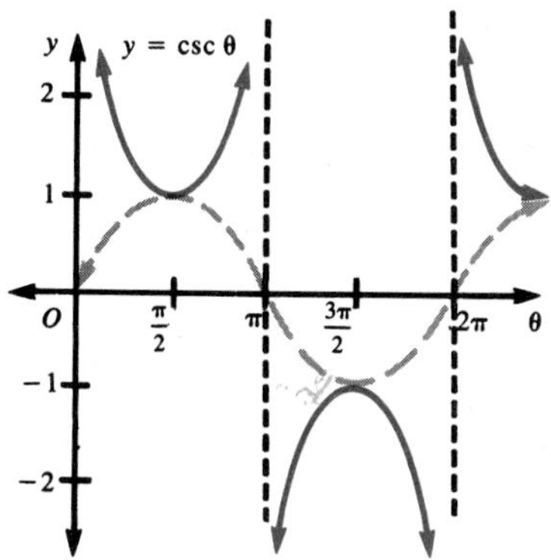

The period of $y = \csc\theta$ is 2π.

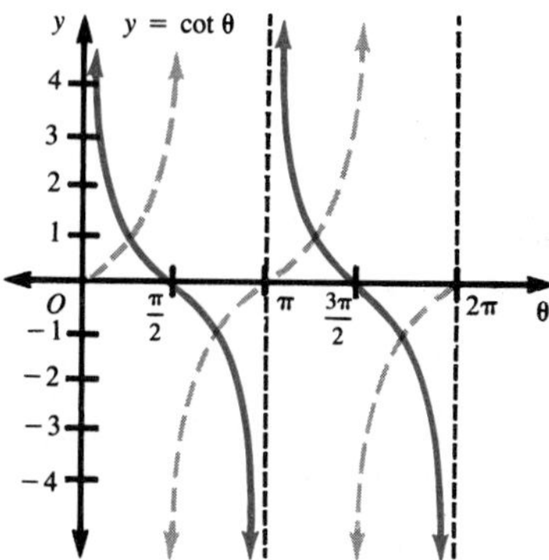

The period of $y = \cot\theta$ is π.

Example

5 State the domain and range of the secant function.

$\sec\theta = \frac{1}{\cos\theta}$ when $\cos\theta \neq 0$. Therefore, all values of θ for which $\cos\theta = 0$ must be excluded from the domain. As can be seen from the graph of $y = \sec\theta$ above, θ cannot be $\frac{\pi}{2}, \frac{3\pi}{2}$, or in fact, any odd multiple of $\frac{\pi}{2}$. Therefore, the domain of the secant function is the set of real numbers except the odd multiples of $\frac{\pi}{2}$.

The graph of $y = \sec\theta$ does not contain any point with y-coordinate between -1 and 1. We conclude that the range of the secant function is $\{y \in \mathcal{R}: y \leq -1 \text{ or } y \geq 1\}$. *The range can also be written as* $\{y \in \mathcal{R}: |y| \geq 1\}$.

Exercises

Exploratory **Express each of the following in terms of sin θ or cos θ and then simplify if possible.**

1. $\sec\theta$
2. $\csc\theta$
3. $\cot\theta$
4. $\sec\theta\cot\theta$
5. $\frac{\sin\theta}{\csc\theta}$
6. $\frac{\cos\theta}{\sec\theta}$
7. $\frac{\cot\theta}{\csc\theta}$
8. $\frac{\tan\theta}{\csc\theta}$

Determine the quadrant in which θ lies if the following conditions are satisfied.

9. $\tan\theta < 0$ and $\sec\theta < 0$
10. $\tan\theta > 0$ and $\sec\theta < 0$
11. $\sin\theta > 0$ and $\cot\theta < 0$
12. $\csc\theta > 0$ and $\cos\theta > 0$
13. $\csc\theta > 0$ and $\sec\theta < 0$
14. $\sec\theta > 0$ and $\cot\theta < 0$

If $-2\pi \le \theta \le 2\pi$, state the values of θ for which each is undefined.

15. $\tan\theta$
16. $\cot\theta$
17. $\sec\theta$
18. $\csc\theta$

State the period of each function.

19. $y = \sec\theta$
20. $y = \csc\theta$
21. $y = \cot\theta$
22. $y = \csc 2\theta$
23. $y = 3\cot\theta$
24. $y = \sec\frac{1}{2}\theta$
25. $y = 3\csc\frac{1}{2}\theta$
26. $y = \tan\frac{1}{2}\theta$

Written **Copy and complete the following table.**

1.

θ	$\sin\theta$	$\cos\theta$	$\tan\theta$	$\cot\theta$	$\sec\theta$	$\csc\theta$
0°						
30°						
45°						
60°						
90°						

Evaluate.

2. cot 135°
3. csc 240°
4. sec 180°
5. sec 150°
6. csc 120°
7. cot 210°
8. sec (−120°)
9. cot (−60°)
10. $\cot\frac{7\pi}{4}$
11. $\csc\frac{3\pi}{2}$
12. $\sec\frac{5\pi}{4}$
13. $\csc\left(-\frac{7\pi}{6}\right)$
14. $\csc^2 150°$
15. $\tan^2 45° + \sec^2 30°$
16. $\tan^2 120° \csc^2 240°$
17. $\csc\frac{3\pi}{2} - \cot\frac{3\pi}{4}$
18. $\tan\left(-\frac{5\pi}{6}\right) + \csc\frac{4\pi}{3}$
19. cot 270° sec 390°
20. $\cot^2\left(-\frac{\pi}{6}\right) + \csc^2\left(-\frac{\pi}{6}\right)$
21. $\cot^2\frac{9\pi}{4} + \csc^2\frac{\pi}{2}$
22. $\cot^2(-600°)\sec^2(-30°)$

Given the following information, find the values of the remaining circular functions.

23. $\cos\theta = \frac{1}{2}$ and θ in quadrant I
24. $\csc\theta = \frac{5}{3}$ and $\cos\theta < 0$
25. $\sec\theta = -\frac{15}{13}$ and $\cot\theta > 0$
26. $\sin\theta = -\frac{1}{4}$ and $\cot\theta < 0$
27. $\sec\theta = 3$ and $\csc\theta < 0$
28. $\csc\theta = \frac{41}{9}$ and $\sec\theta < 0$

State the domain and range of each function.

29. $y = \sec\theta$
30. $y = \csc\theta$
31. $y = \cot\theta$
32. $y = 3\cot\theta$
33. $y = \frac{1}{2}\csc\theta$
34. $y = 3\sec\theta$
35. $y = \frac{3}{4}\csc 2\theta$
36. $y = -\sec\frac{1}{2}\theta$

State whether the values of the given function are increasing or decreasing in each of the four quadrants.

37. $y = \sin\theta$
38. $y = \cos\theta$
39. $y = \tan\theta$
40. $y = \cot\theta$
41. $y = \sec\theta$
42. $y = \csc\theta$

Graph each function in the interval $-2\pi \le x \le 2\pi$. What points or lines of symmetry does each graph have?

43. $y = \tan x$
44. $y = \cot x$
45. $y = \sec x$
46. $y = \csc x$

Challenge

1. Is $\frac{1}{\tan\theta} = \frac{\cos\theta}{\sin\theta}$ for all values of θ? Explain.

Solve each equation in the interval $-2\pi \le \theta \le 2\pi$.

2. $\tan\theta = \cot\theta$
3. $\sec\theta = \cos\theta$
4. $\csc\theta = \sin\theta$

Mixed Review

1. Find the solution set of $3^{2x-1} = 3^{x^2}$.
2. If $\log 2 = A$ and $\log 3 = B$, express $\log 6$ in terms of A and B.
3. Solve for x: $\frac{1}{x} + 1 = \frac{3}{2x}$.
4. If $\overline{QR}$ is tangent to circle O at R and $m\angle OQR = 20$, find $m\angle ROQ$.
5. Express $(3 - 2i)^2$ in simplest $a + bi$ form.
6. Find the value of $16^{\frac{3}{4}}$.
7. Solve for x: $\sqrt{2x + 3} - 5 = 0$.
8. What is the image of the point $(5, -2)$ under the translation $T_{2,1}$?
9. What types of symmetry does an equilateral triangle have?
10. Write the inverse of the function $\{(-1, 2), (0, 5), (3, -2)\}$.
11. If $f(x) = 2x - 7$, find $f^{-1}(x)$.
12. In what quadrants does the graph of $y = \left(\frac{1}{2}\right)^x$ lie?
13. Describe the nature of the roots of $x^2 + x + 1 = 0$.
14. For what value of k is the graph of $y = x^2 + 4x + k$ tangent to the x-axis?
15. Solve for x in terms of a and b: $\frac{1}{a} - \frac{1}{b} = \frac{1}{x}$.
16. Simplify $\left(\frac{a}{b} - 1\right) \div \left(b - \frac{a^2}{b}\right)$.
17. Points A, B, and C are on circle O such that $\overline{PC}$ is tangent to circle O at C and $\overline{PAB}$ is a secant. If $PA = 3$ and $AB = 9$, find PC.
18. If the domain of $f(x) = 2x - 1$ is $-3 \le x \le 3$, find the greatest value in the range of f.
19. Express in simplest $a + bi$ form the multiplicative inverse of $2 + i$.
20. If $a = 8$, find the value of $(a^{-1})^{\frac{2}{3}}$.

7.9 Finding Trigonometric Values

So far we have found values such as sin 30°, cos 0°, tan $\frac{\pi}{4}$, and so forth. But what about values like sin 14° or cos 201°? Decimal approximations for these values can be found by using a scientific calculator or by using a trigonometric table. Such a table, giving values accurate to four decimal places, begins on page 618 of this book.

The "trig table" gives values of the six trigonometric functions for angles of 0° to 90° at *ten-minute* intervals. A **minute** is one sixtieth of a degree. The symbol for minute is ′. For example, $2\frac{1}{2}$ degrees can be written as 2°30′ which is read as "two degrees, thirty minutes." A minute is divided into **seconds.** One second, written 1″, is equal to $\frac{1}{60}$ minute.

To use the table for angles between 0° and 45°, read down on the left side and use the column headings at the top of the page. For angles from 45° to 90°, read up on the right side and use the headings on the bottom of the page.

Examples

1 **Find sin 18°20′ and tan 71°50′ using the trig table on page 620.**

Part of the table is shown below.

Since 0° < 18°20′ < 45°, we read down the column labeled Sin. Thus, sin 18°20′ = 0.3145.

Since 45° < 71°50′ < 90°, we read up the column labeled Tan. Thus, tan 71°50′ = 3.047.

We write = instead of ≈ even though the values in the table are rational approximations of the true values.

Angle	Sin	Cos	Tan	Cot	Sec	Csc	
18°00′	0.3090	0.9511	0.3249	3.078	1.051	3.236	72°00′
10′	0.3118	0.9502	0.3281	3.047	1.052	3.207	50′
20′	0.3145	0.9492	0.3314	3.018	1.053	3.179	40′
30′	0.3173	0.9483	0.3346	2.989	1.054	3.152	30′
40′	0.3201	0.9474	0.3378	2.960	1.056	3.124	20′
50′	0.3228	0.9465	0.3411	2.932	1.057	3.098	10′
27°00′	0.4540	0.8910	0.5095	1.963	1.122	2.203	63°00′
	Cos	Sin	Cot	Tan	Csc	Sec	Angle

2 **Find cos 58°20′ using a calculator.** *Be sure the calculator is in degree mode.*

Change the angle measure to decimal form by dividing the minutes portion by 60.

ENTER:	58	+	20	÷	60	=	cos
DISPLAY:	58		20		60	58.3333333	0.52497658

Thus, cos 58°20′ = 0.5250.

3 **Find cos 126° and sin 126°.**

First express cos 126° and sin 126° as functions of positive acute angles. Then use the trig tables.

Since 126° is in quadrant II, the reference angle is 180° − 126° = 54°. In quadrant II, the sine is positive and the cosine is negative. Thus,

$$\cos 126° = -\cos 54° = -0.5878$$

$$\text{and } \sin 126° = \sin 54° = 0.8090.$$

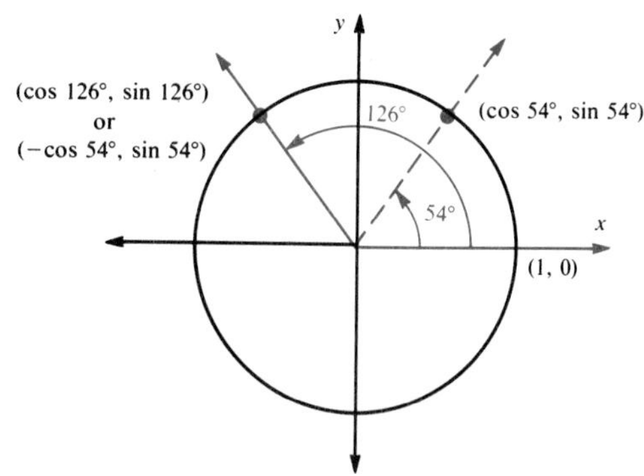

4 **If 0° ≤ θ ≤ 360° and tan θ = 0.8796, find θ.**

The tangent is positive in quadrants I and III. In the trig table, we find 0.8769 in the column with "Tan" at the top. The angle in the left column is 41°20′.

In quadrant III, the angle with reference angle 41°20′ is 180° + 41°20′ or 221°20′. We conclude that θ = 41°20′ or 221°20′.

Sometimes we want approximations of trigonometric values more accurate than the ones given in the trig table. To do this, we use the method of *interpolation,* which is illustrated in the following examples.

Examples

5 **Find the value of sin 28°23′ to four decimal places.**

Use the trig table to find the sines of the angles closest to 28°23′. Set up a chart as shown at the right. Use the chart to write the following proportion.

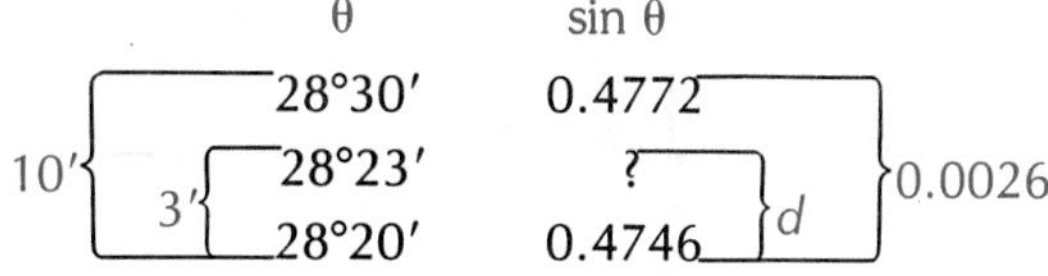

$$\frac{3}{10} = \frac{d}{0.0026}$$

$$10d = 3(0.0026)$$

$$10d = 0.0078$$

$$d = 0.00078 \text{ or } 0.0008$$

Round d to four decimal places since the information in the table is accurate only to four decimal places.

Add 0.0008 to the value of sin 28°20′.

sin 28°23′ = 0.4746 + 0.0008 or 0.4754

Thus, sin 28°23′ = 0.4754.

6 **If $\cos \theta = 0.9121$, find θ to the nearest minute if θ is a positive acute angle.**

	θ	$\cos \theta$		
10	24°10′	0.9124	0.0012	*Use the table to find the cosine values closest to 0.9121.*
d	?	0.9121	0.0009	
	24°20′	0.9112		

$$\frac{d}{10} = \frac{0.0009}{0.0012}$$

$$\frac{d}{10} = \frac{3}{4}$$

$$4d = 30$$

$$d = 7.5 \text{ or } 8$$ *Round to the nearest integer.*

Thus, $\theta = 24°20' - 8'$ or $24°12'$.

7 **If $\sin \theta = 0.3546$ and $0° \leq \theta < 360°$, use a calculator to find θ to the nearest minute.** *Be sure the calculator is in degree mode.*

The calculator will give the value of θ in decimal form. To find the value to the nearest minute, multiply the decimal portion of the angle measure by 60.

ENTER:	0.3546	INV	sin	−	20	=	×	60	=	
DISPLAY:	0.3546		20.7689313		20	0.7689313		60	46.1358781	

Thus, one value of θ is 20°46′.

The sine is positive in quadrants I and II. In quadrant II, the angle with reference angle 20°46′ is 180° − 20°46′ or 159°14′.

Therefore, $\theta = 20°46'$ or $159°14'$.

Exercises

Exploratory **Find each value.**

1. sin 14°
2. sin 29°20′
3. cos 34°40′
4. tan 18°50′
5. sin 10°20′
6. cos 44°40′
7. tan 29°50′
8. tan 47°10′
9. cos 51°20′
10. sin 63°10′
11. tan 72°40′
12. cos 88°10′

Find the value of x in degrees and minutes if $0° < x < 90°$.

13. $\tan x = 0.1554$
14. $\cos x = 0.9877$
15. $\sin x = 0.8910$
16. $\sin x = 0.4950$
17. $\cot x = 0.6412$
18. $\tan x = 2.414$
19. $\cos x = 0.7050$
20. $\tan x = 1.199$
21. $\sin x = 0.9989$

Written **First express each function as a function of a positive acute angle. Then use the trig table to find the value.**

1. sin 100°	**2.** cos 170°	**3.** tan 310°	**4.** cos 260°
5. tan 140°	**6.** tan 195°	**7.** sin 352°	**8.** sin 430°
9. tan 160°	**10.** tan 350°	**11.** cos 236°	**12.** sin (−70°)
13. cos (−132°)	**14.** cos (−283°)	**15.** tan (−390°)	**16.** sin (−127°)
17. sin 153°10′	**18.** tan 91°10′	**19.** cos 143°50′	**20.** sin 201°30′

Find each value to four decimal places.

21. sin 32°15′	**22.** sin 78°52′	**23.** tan 26°14′	**24.** cos 38°43′
25. cos 14°58′	**26.** tan 71°19′	**27.** tan 18°28′	**28.** cos 20°45′
29. cos 63°47′	**30.** sin 7°32′	**31.** tan 82°12′	**32.** cos 11°4′
33. cos 29°18′	**34.** sin 41°3′	**35.** cos 62°38′	**36.** tan 7°58′

If $0° < x < 90°$, find the value of x to the nearest minute.

37. $\cos x = 0.5132$	**38.** $\sin x = 0.3291$	**39.** $\tan x = 0.3147$
40. $\tan x = 1.7050$	**41.** $\cos x = 0.3333$	**42.** $\sin x = 0.8081$
43. $\sin x = \frac{3}{4}$	**44.** $\tan x = \frac{4}{3}$	**45.** $\sec x = \frac{7}{5}$

If $0° \le x \le 360°$, find the values of x to the nearest degree.

46. $\cos x = 0.2843$	**47.** $\sin x = 0.9005$	**48.** $\tan x = 0.9438$
49. $\sin x = 0.8671$	**50.** $\tan x = 6.8273$	**51.** $\cos x = 0.2451$

Use a sketch of the graph of $y = \sin x$, $y = \cos x$, or $y = \tan x$ and your knowledge of line and point symmetry to justify each of the following.

52. $\sin(90° + x) = \sin(90° - x)$	**53.** $\sin(180° - x) = -\sin(180° + x)$
54. $\tan x = \tan(\pi + x)$	**55.** $\cos(\pi - x) = \cos(\pi + x)$
56. $\cos x = \cos(2\pi - x)$	**57.** $\tan\left(\frac{\pi}{2} - x\right) = -\tan\left(\frac{\pi}{2} + x\right)$
58. $\sin x = -\sin(-x)$	**59.** $\cos(-x) = \cos(360° + x)$
60. $2\cos(-x) = 2\cos x$	**61.** $\tan x = -\tan(2\pi - x)$
62. $\sin x = -\sin(2\pi - x)$	**63.** $\sin(\pi + x) = \sin(2\pi - x)$

Challenge

1. 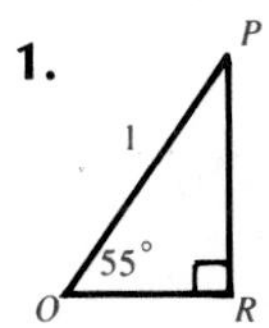

Find to the nearest tenth the length of $\overline{OR}$ and the length of $\overline{PR}$.

2. A 7-meter ladder leans against a building. The ladder forms an angle with the ground measuring 74°. How far is the foot of the ladder from the base of the building? Round the answer to the nearest tenth.

Problem Solving Application: Networks

Diagrams called **networks** are often useful in solving problems.

Examples

1 **For the annual spring goofy games, students are allowed to register for any 7 games. Twelve students have registered for the games indicated on the chart below. In order to complete the games in the shortest amount of time, some of the games will be played at the same time. What is the minimum number of time periods needed so that each student can play all of their choices?**

Students

Games	1	2	3	4	5	6	7	8	9	10	11	12
A	√			√			√		√			√
B	√	√			√							
C		√					√	√				
D			√	√	√					√		
E				√				√			√	
F	√		√			√						
G		√					√				√	

Explore You must find the least number of time periods needed so that there are no conflicts.

Plan To solve this problem, let each game be represented by a dot. All of the dots representing games being played by an individual student should be connected by line segments. For example, since student number 1 is playing games A, B, and F, the dots representing A, B, and F should be connected. These games cannot be played at the same time. Continue to connect the games played by each of the other students. After the network is completed, color or label the dots so that no two dots connected by a line segment have the same color or symbol. Try to use the least number of colors or symbols.

Solve

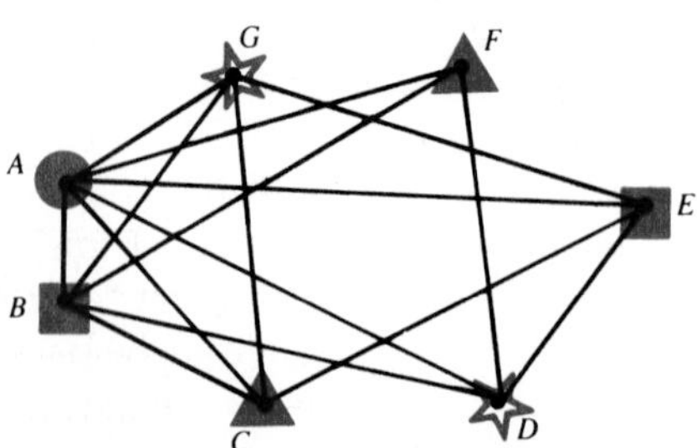

Since there are four distinguishing symbols, four time periods are needed.

Why aren't C and D represented by the same symbol?

Examine Carefully check your work. Are all the dots representing conflicting games connected? Do any of the connected dots have the same symbol? Is there any way to use only 3 symbols?

Exercises

Written **Solve each problem.**

1. In order for a small company to maintain its service agreements, certain employees cannot take vacations at the same time. In the network at the right, each dot represents an employee. The line segments connect the dots of employees who cannot take vacations at the same time. What is the least number of vacation times needed so everyone gets a vacation, and all service agreements are maintained?

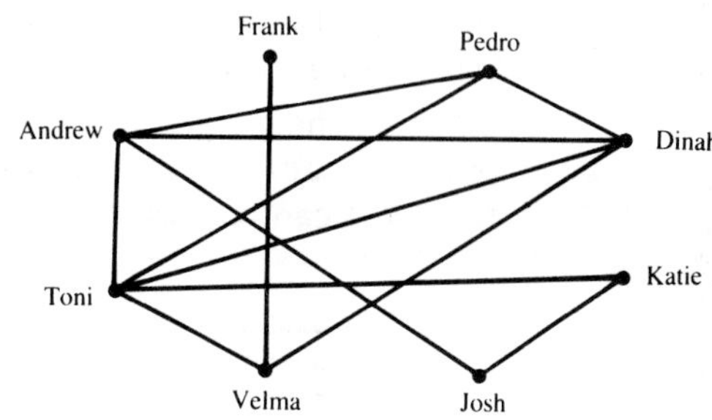

2. The chart below indicates the subjects taken by 10 students. If final exams are given by subject, what is the least number of exam periods needed so that there are no conflicts for these students?

Students

Sub.	1	2	3	4	5	6	7	8	9	10
A	√		√							√
B		√		√				√		
C						√	√	√		
D	√		√			√			√	
E	√				√				√	
F		√			√				√	
G			√	√			√			
H		√			√		√			√

3. Ten teachers attending a conference have indicated which sessions they would like to attend. Each session will only be presented once. What is the least number of meeting times needed so that each teacher can attend all of the sessions that were chosen?

Teachers

Ses.	1	2	3	4	5	6	7	8	9	10
A		√				√			√	
B	√	√			√			√	√	√
C				√			√			
D			√		√		√	√		
E	√			√					√	
F	√			√						

4. Suppose you are planning a math contest. Seven teams have entered the various events as shown at the right. Each event is to be scheduled for 15 minutes followed by a 5-minute break. You want to schedule the contest, beginning at 10 AM, so that the total time is minimal. When will the contest end?

Teams

Events	1	2	3	4	5	6	7
A	√			√			
B		√	√				
C	√		√			√	
D		√				√	
E					√	√	
F							√

Vocabulary

initial side	sine	tangent
terminal side	cosine	asymptote
standard position	reference angle	secant
quadrantal angle	periodic function	cosecant
coterminal	period	cotangent
radian	frequency	minute
unit circle	amplitude	second

Chapter Summary

1. One complete counterclockwise rotation around a circle measures an angle of 360° or 2π radians. If the rotation is clockwise, the measurement of the angle is negative.
2. $180° = \pi$ radians. $1° = \frac{\pi}{180}$ radians. $\frac{180°}{\pi} = 1$ radian.
3. If θ is the radian measure of a central angle and r is the radius, then s, the length of the intercepted arc, is given by $s = r\,\theta$.
4. The Pythagorean Identity: Let θ be the measure of an angle in standard position. Then $\sin^2\theta + \cos^2\theta = 1$.
5. If θ is the measure of an angle in quadrant II, then $\cos\theta = -\cos(180° - \theta)$ and $\sin\theta = \sin(180° - \theta)$.
6. If θ is the measure of an angle in quadrant III, then $\cos\theta = -\cos(\theta - 180°)$ and $\sin\theta = -\sin(\theta - 180°)$.
7. If θ is the measure of an angle in quadrant IV, then $\cos\theta = \cos(360° - \theta)$ and $\sin\theta = -\sin(360° - \theta)$.
8. For both the sine and cosine functions, the domain is the set of real numbers and the range is the set of real numbers between -1 and 1, inclusive.
9. For functions of the form $y = a \sin bx$ and $y = a \cos bx$, the amplitude is $|a|$, the period is $\frac{2\pi}{|b|}$ or $\frac{360°}{|b|}$, and the frequency is $|b|$.
10. The domain of the tangent function is the set of all real numbers except odd multiples of $\frac{\pi}{2}$. The range is the set of real numbers. The period of the tangent function is 180° or π radians and it has no amplitude.

Chapter Review

7.1 **Find the least positive degree measure of an angle that is coterminal with an angle of the following measurement.**

1. 420° **2.** $-40°$ **3.** 755° **4.** $-462°$

7.2 **Convert each of the following to radian measure.**

5. 100° **6.** 40° **7.** 450° **8.** −220°

Convert each of the following to degree measure.

9. $\frac{2\pi}{5}$ **10.** $\frac{7\pi}{4}$ **11.** $-\frac{3\pi}{2}$ **12.** $2\frac{5}{6}\pi$

In exercises 13–14, a central angle of radian measure θ intercepts an arc of length *s* on a circle with radius *r*.

13. Find s if $\theta = 3$ and $r = 2$. **14.** Find θ if $s = 18$ and $r = 3$.

7.3 **15.** If θ lies in quadrant II and $\sin\theta = \frac{12}{13}$, find $\cos\theta$.

16. If θ lies in quadrant III and $\cos\theta = -0.28$, find $\sin\theta$.

Evaluate.

17. sin 180° cos 90° **18.** 4 cos 360° − 7 sin 90° **19.** $\sin^2 74° + \cos^2 74°$

7.4 **20.** sin 120° **21.** cos (−330°) **22.** sin 585°

23. $2\cos\pi + 3\sin\frac{\pi}{2}$ **24.** $4\sin\frac{5\pi}{4}\cos\frac{7\pi}{4}$ **25.** $\sin^2\frac{\pi}{4} + 2\cos^2\frac{3\pi}{4}$

26. Express cos 192° as a function of a positive acute angle.

7.5 **State the amplitude, period, and range of each function.**

27. $y = 3\sin 2x$ **28.** $y = 2\cos\frac{1}{2}x$ **29.** $y = -\frac{1}{2}\sin 3x$

7.6 **30.** On the same set of axes, sketch the graphs of $y = \sin 2x$ and $y = 2\cos x$ in the interval $0 \le x \le \pi$. Then state the number of values of x in the interval $0 \le x \le \pi$ that satisfy the equation $\sin 2x = 2\cos x$.

7.7 **31.** If $\sin\theta = \frac{1}{3}$ and $\cos\theta < 0$, find $\tan\theta$.

32. On the same set of axes, sketch the graphs of $y = \cos\frac{x}{2}$ and $y = \tan x$ in the interval $-\pi \le x \le \pi$. Then state the number of values of x in the interval $-\pi \le x \le \pi$ that satisfy the equation $\cos\frac{x}{2} = \tan x$.

Evaluate.

33. tan 135° **34.** tan 300° **35.** tan (−120°)

7.8 **36.** $\cot\frac{3\pi}{2}$ **37.** sec 120° **38.** $\csc\left(-\frac{5\pi}{6}\right)$

39. Express csc 340° as a function of a positive acute angle.

If $f(x) = 3\sec x$, find each of the following.

40. $f\left(\frac{\pi}{4}\right)$ **41.** $f(-\pi)$ **42.** $f\left(\frac{2\pi}{3}\right)$ **43.** $f\left(-\frac{5\pi}{6}\right)$

7.9 **Find each value correct to four decimal places.**

44. sin 32°20′ **45.** cos 86°50′ **46.** tan 100°

If θ is the measure of an acute angle, find the measure of θ to the nearest minute.

47. $\sin\theta = 0.8162$ **48.** $\cos\theta = 0.5669$ **49.** $\tan\theta = 0.2971$

If $0° \le \theta \le 360°$, find the measure of θ to the nearest degree.

50. $\sin\theta = 0.5000$ **51.** $\tan\theta = 1$ **52.** $\cos\theta = -0.5000$

Chapter Test

1. Convert $\frac{7\pi}{3}$ radians to degree measure.
2. Convert 140° to radian measure.
3. State the amplitude of the function $y = 3 \sin \frac{1}{3}x$.
4. In a circle, a central angle of 2 radians intercepts an arc of 10 cm. Find the length, in centimeters, of the radius of the circle.
5. If $\sin \theta = 0.5319$ and θ is in quadrant I, find the value of θ to the nearest minute.
6. Express $\cos(-230°)$ as a function of a positive acute angle.
7. If $\tan \theta < 0$ and $\csc \theta < 0$, in which quadrant does θ lie?
8. State the period of the function $y = 2 \cos \frac{1}{2}x$.
9. If $f(x) = 2 \cos 2x$, find the value of $f\left(\frac{\pi}{3}\right)$.
10. If $\cos x = -\frac{3}{4}$ and $\sin x < 0$, find $\tan x$.

Evaluate.

11. $\sin 120°$
12. $\cos 315°$
13. $\cot 210°$
14. $\cos 600°$
15. $2 \sin 90° - 3 \cos 180°$
16. $4 \cos 0° + \csc 270° - 3 \tan 225°$
17. On the same set of axes, sketch the graphs of $y = 3 \cos x$ and $y = \sin \frac{x}{2}$ in the interval $0 \leq x \leq 2\pi$. Then determine the number of values of x in the interval $0 \leq x \leq 2\pi$ that satisfy the equation $3 \cos x = \sin \frac{x}{2}$.

Tell whether each statement is true or false.

18. The number $\frac{\pi}{2}$ is in the domain of the function $y = \tan x$.
19. The number $\frac{\pi}{2}$ is in the domain of the function $y = \cot x$.
20. If $\sin \theta \cos \theta < 0$ then θ is in either quadrant I or III.
21. The angles with radian measure $\frac{\pi}{6}$ and $-\frac{5\pi}{6}$ are coterminal.
22. The expression $\cos x \sec x$ is equivalent to 1.
23. The value of $\sin^2 37° + \cos^2 37°$ is 1.
24. The maximum value of the function $y = 2 \cos 3x$ is 3.
25. As x increases from $\frac{\pi}{2}$ to π, the value of $\sin x$ decreases.
26. The number -3 is in the range of the function $y = 5 \sin x$.
27. If $\csc x < 0$ then $\tan x < 0$.
28. If x and y are measures of acute angles then $\sin (x + y) = \sin x + \sin y$.
29. The graph of $y = \sin x$ is symmetric with respect to the origin.
30. If θ is the measure of an angle in the third quadrant, then $\sec \theta = -\sec (\theta - 180°)$.

Applications of Circular Functions

Chapter 8

The knowledge of circular functions can be applied to the study of sciences such as aerodynamics, the study of the forces acting on objects moving in space. Scientists have studied the aerodynamics of such popular items as the Frisbee.™ An understanding of the forces acting on the spinning disc can help players in analyzing the moves required for special stunts with the Frisbee™.

8.1 Trigonometric Coordinates

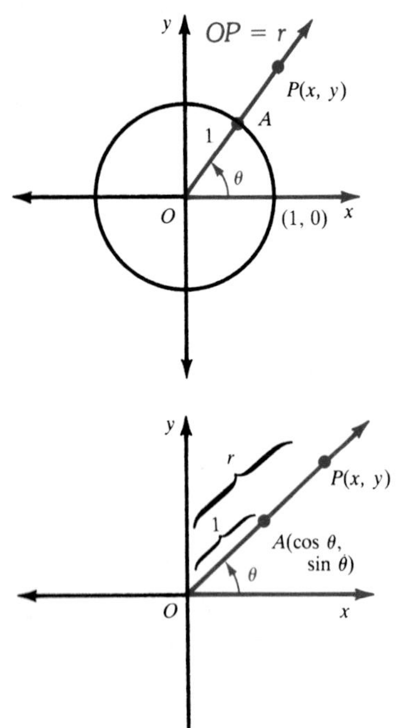

In the figure at the left, point A is the intersection of the unit circle and the terminal side of θ. The expression "the terminal side of θ" means "the terminal side of an angle in standard position that has measure θ."

Suppose we wish to express the coordinates of P in terms of r and θ. To do this, note that under the dilation $D_{O,\,r}$, point P is the image of point A. That is,

$$A \xrightarrow{D_{O,\,r}} P.$$

Since A is on the unit circle, A has coordinates $(\cos\theta, \sin\theta)$. Using the rule for a dilation, we have

$$A \xrightarrow{D_{O,\,r}} P$$

$$\text{or } (\cos\theta, \sin\theta) \xrightarrow{D_{O,\,r}} (r\cos\theta, r\sin\theta).$$

Since P has coordinates (x, y), it follows that

$$x = r\cos\theta \text{ and } y = r\sin\theta.$$

We see that the coordinates of $P(x, y)$ can be expressed as

$$(r\cos\theta, r\sin\theta).$$

Coordinates in Trigonometric Form

If P is a point located r units from the origin on the terminal side of an angle in standard position that has measure θ, then the coordinates of P expressed in **trigonometric form** are

$$(r\cos\theta, r\sin\theta).$$

The coordinates (x, y) of P are often called the **rectangular coordinates** *of P.*

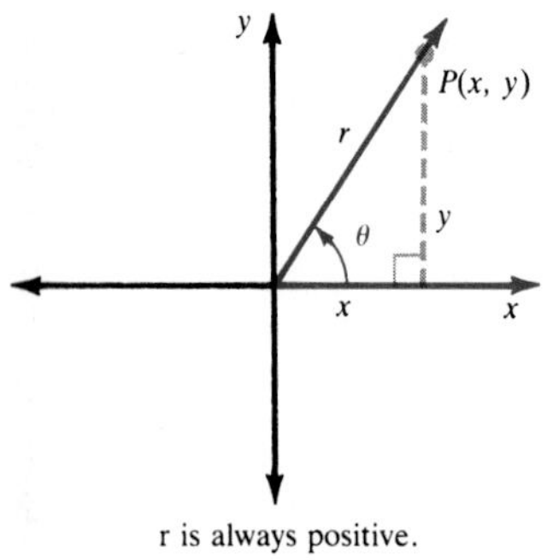

r is always positive.

Since $x = r\cos\theta$ and $y = r\sin\theta$, it follows that

$$\cos\theta = \frac{x}{r} \text{ and } \sin\theta = \frac{y}{r}.$$

Since $\tan\theta = \frac{\sin\theta}{\cos\theta}$, we then have

$$\tan\theta = \frac{\frac{y}{r}}{\frac{x}{r}} \text{ or } \frac{y}{x}.$$

Examples

1 **Find the rectangular coordinates of point P if $OP = 6$ and $\theta = 240°$.**

$$x = r \cos \theta \qquad y = r \sin \theta$$
$$= 6 \cos 240° \qquad = 6 \sin 240°$$
$$= 6\left(-\frac{1}{2}\right) \qquad = 6\left(-\frac{\sqrt{3}}{2}\right)$$
$$x = -3 \qquad y = -3\sqrt{3}$$

The rectangular coordinates of P are $(-3, -3\sqrt{3})$.

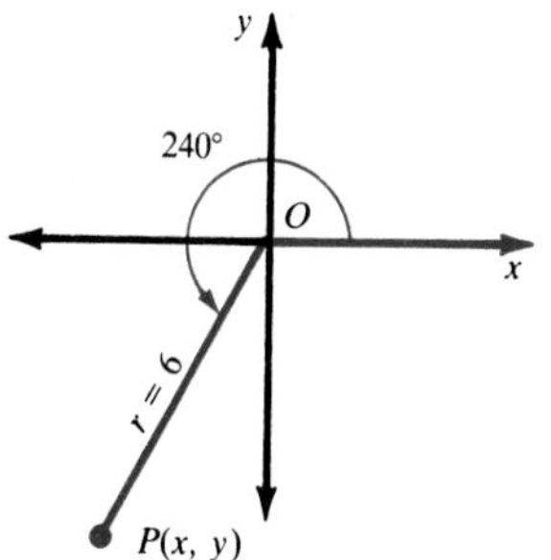

2 **Find $\sin \theta$ and $\cos \theta$ if the terminal side of θ contains the point $(1, -3)$.**

First draw a sketch.

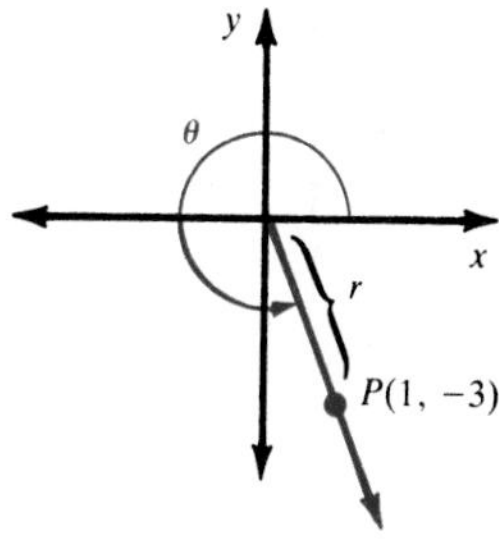

$$\cos \theta = \frac{x}{r} \quad \text{and} \quad \sin \theta = \frac{y}{r}$$
$$= \frac{1}{r} \qquad = \frac{-3}{r}$$

Substitute $x = 1$ and $y = -3$.

To find r, use the distance formula.

$$r = \sqrt{(x - 0)^2 + (y - 0)^2}$$
$$= \sqrt{1^2 + (-3)^2}$$
$$r = \sqrt{10}$$

$-\sqrt{10}$ is rejected because distance is positive.

Therefore, $\cos \theta = \frac{1}{\sqrt{10}}$ or $\frac{\sqrt{10}}{10}$ and $\sin \theta = \frac{-3}{\sqrt{10}}$ or $-\frac{3\sqrt{10}}{10}$.

3 **Find the trigonometric coordinates of $P(-2\sqrt{3}, 2)$.**

First draw a sketch.

The trigonometric coordinates of P are $(r \cos \theta, r \sin \theta)$.

To find r, use the distance formula.

$$r = \sqrt{(x - 0)^2 + (y - 0)^2}$$
$$= \sqrt{(-2\sqrt{3})^2 + 2^2}$$
$$r = 4$$

To find θ, note that $\tan \theta = \frac{y}{x}$

$$= \frac{2}{-2\sqrt{3}} = -\frac{\sqrt{3}}{3}$$

Recall that $\tan 30° = \frac{\sqrt{3}}{3}$.

Since θ is in quadrant II, $\theta = 180° - 30°$ or $150°$.

Thus, the trigonometric coordinates of P are $(4 \cos 150°, 4 \sin 150°)$.

4 **If $\tan \theta = -\frac{\sqrt{5}}{2}$ and $\sin \theta > 0$, find $\cos \theta$.**

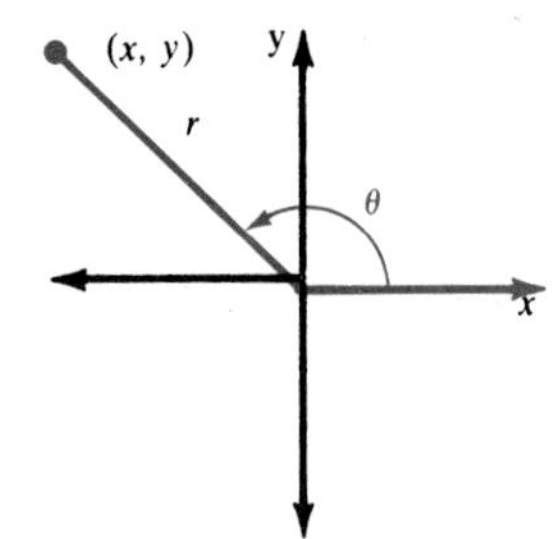

Since $\tan \theta$ is negative and $\sin \theta$ is positive, θ is in quadrant II. Since $\cos \theta = \frac{x}{r}$, we must find x and r.

$$\tan \theta = \frac{y}{x}$$

$$= -\frac{\sqrt{5}}{2} \text{ or } \frac{\sqrt{5}}{-2}$$

In quadrant II, y is positive and x is negative.

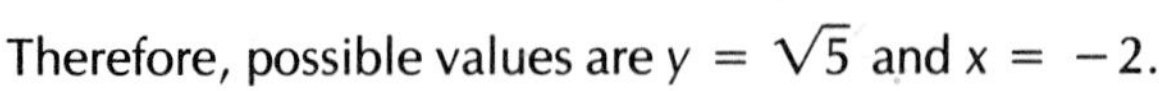

Therefore, possible values are $y = \sqrt{5}$ and $x = -2$.

$$r = \sqrt{(x - 0)^2 + (y - 0)^2}$$ *Distance Formula*

$$= \sqrt{(-2)^2 + (\sqrt{5})^2}$$

$$r = 3$$

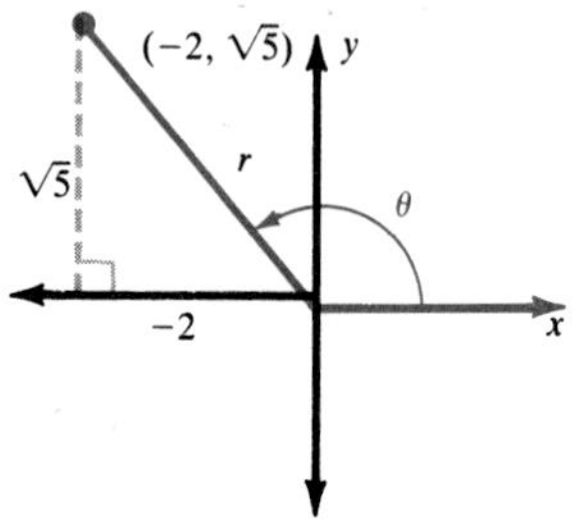

We conclude that $\cos \theta = \frac{x}{r}$

$$= \frac{-2}{3} \text{ or } -\frac{2}{3}.$$

Exercises

Exploratory **Plot each of the following points.**

1. $(3 \cos 90°, 3 \sin 90°)$
2. $(2 \cos 270°, 2 \sin 270°)$
3. $(6 \cos 45°, 6 \sin 45°)$
4. $(\cos 180°, \sin 180°)$
5. $(8 \cos 315°, 8 \sin 315°)$
6. $(4 \cos 135°, 4 \sin 135°)$

Find the rectangular coordinates (x, y) of a point located r units from the origin and on the terminal side of an angle in standard position that has measure θ.

7. $r = 4, \theta = 90°$
8. $r = 5, \theta = 60°$
9. $r = 10, \theta = 30°$
10. $r = 1, \theta = 45°$
11. $r = 7, \theta = 240°$
12. $r = 3, \theta = 180°$
13. $r = 2, \theta = 150°$
14. $r = 9, \theta = 330°$
15. $r = 6, \theta = 210°$
16. $r = 14, \theta = \frac{2\pi}{3}$
17. $r = \sqrt{2}, \theta = 315°$
18. $r = \sqrt{3}, \theta = 300°$

Written **Find the rectangular coordinates (x, y) of each point and plot the point on graph paper.**

1. $(4 \cos 180°, 4 \sin 180°)$
2. $(6 \cos 60°, 6 \sin 60°)$
3. $(5 \cos 270°, 5 \sin 270°)$
4. $(\sqrt{2} \cos 225°, \sqrt{2} \sin 225°)$
5. $(8 \cos 210°, 8 \sin 210°)$
6. $(10 \cos 26°, 10 \sin 26°)$

Express the coordinates of each point in trigonometric form.

7. $(0, 5)$ **8.** $(6, 0)$ **9.** $(-7, 0)$ **10.** $(0, -3)$
11. $(2, 2\sqrt{3})$ **12.** $(2\sqrt{3}, 2)$ **13.** $(-2, 2\sqrt{3})$ **14.** $(\sqrt{2}, \sqrt{2})$
15. $(3\sqrt{2}, 3\sqrt{2})$ **16.** $(-3\sqrt{3}, 3)$ **17.** $(-4, 4)$ **18.** $(5, -5)$
19. $(-4\sqrt{3}, -4)$ **20.** $(3\sqrt{2}, -3\sqrt{2})$ **21.** $(-5\sqrt{3}, 5)$ **22.** $(-7, -7\sqrt{3})$

Find sin θ, cos θ, and tan θ if the terminal side of θ contains the following point.

23. $(4, 3)$ **24.** $(5, 12)$ **25.** $(3, -4)$ **26.** $(-6, 8)$
27. $(-12, 5)$ **28.** $(2\sqrt{2}, 1)$ **29.** $(-5, -5)$ **30.** $(-2\sqrt{3}, 2)$
31. $(-1, \sqrt{15})$ **32.** $(8, -15)$ **33.** $(1, -\sqrt{3})$ **34.** $(-6, 3)$

Given the following information, find the values of all the trigonometric functions of θ.

35. $\tan\theta = -\frac{5}{12}$ and $\sin\theta > 0$

36. $\tan\theta = \sqrt{15}$ and $\sin\theta < 0$

37. $\cot\theta = -\frac{3}{4}$ and $\cos\theta > 0$

38. $\cot\theta = \frac{12}{5}$ and $\sin\theta < 0$

39. $\sin\theta = -\frac{1}{3}$ and $\cos\theta > 0$

40. $\csc\theta = -\frac{13}{5}$ and $\tan\theta > 0$

41. $\sec\theta = -\frac{4}{3}$ and $\cot\theta > 0$

42. $\tan\theta = \frac{2}{5}$ and $\csc\theta < 0$

Challenge

1. Find the length of $\overline{AB}$.

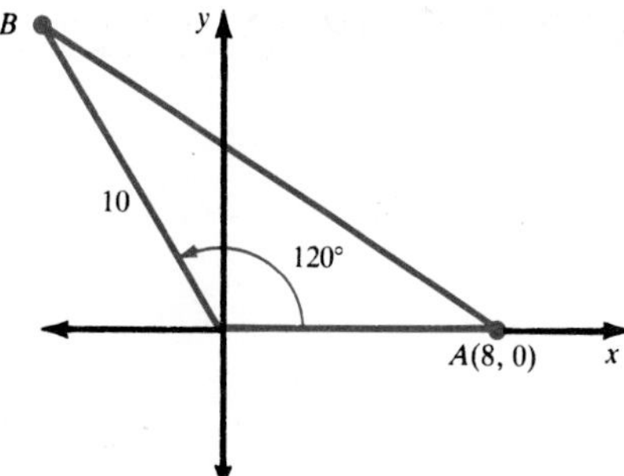

2. Find the measure of $\angle AOB$ to the nearest degree.

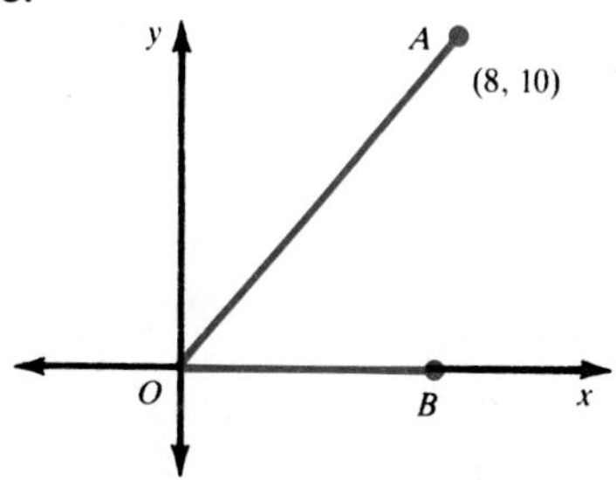

8.2 The Law of Cosines

Consider any triangle ABC. Place the triangle on a pair of coordinate axes, as shown at the right. In this case, $\angle C$ is obtuse.

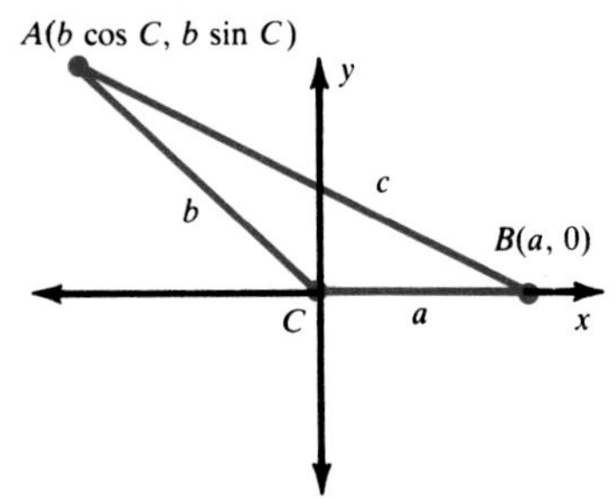

The side opposite vertex A has length a units.
The side opposite B has length b units.
The side opposite C has length c units.

The coordinates of B are $(a, 0)$. Since $\angle C$ is in standard position and A is b units from the origin, the coordinates of A in trigonometric form are $(b \cos C, b \sin C)$.

c is the length of $\overline{AB}$.
A is (x_2, y_2).
B is (x_1, y_1).

Now the distance formula, $d^2 = (x_2 - x_1)^2 + (y_2 - y_1)^2$, can be used to compute c, the length of $\overline{AB}$. Let A represent (x_2, y_2) and B represent (x_1, y_1). By substituting, we obtain

$$
\begin{aligned}
c^2 &= (b \cos C - a)^2 + (b \sin C - 0)^2 \\
c^2 &= b^2 \cos^2 C - 2ab \cos C + a^2 + b^2 \sin^2 C \\
c^2 &= b^2 \cos^2 C + b^2 \sin^2 C + a^2 - 2ab \cos C \\
c^2 &= b^2 (\cos^2 C + \sin^2 C) + a^2 - 2ab \cos C \\
c^2 &= a^2 + b^2 - 2ab \cos C
\end{aligned}
$$

Since the triangle could have been positioned with any of its sides along the x-axis, there are two other similar formulas. All three formulas, known as the *Law of Cosines,* are given below.

Law of Cosines

$$
\begin{aligned}
a^2 &= b^2 + c^2 - 2bc \cos A \\
b^2 &= a^2 + c^2 - 2ac \cos B \\
c^2 &= a^2 + b^2 - 2ab \cos C
\end{aligned}
$$

The Law of Cosines can be used in the following cases.

1. To find the length of the third side of a triangle when given the lengths of the other two sides and the measure of their included angle.
2. To find the measure of an angle of a triangle if the lengths of the three sides are given. In such cases, it is useful to rewrite the Law of Cosines in one of these forms.

$$\cos A = \frac{b^2 + c^2 - a^2}{2bc} \qquad \cos B = \frac{a^2 + c^2 - b^2}{2ac} \qquad \cos C = \frac{a^2 + b^2 - c^2}{2ab}$$

Examples

1 **In $\triangle ABC$, $b = 7$, $c = 5$, and $m\angle A = 45$. Find a to the nearest tenth.**

Use the following form of the Law of Cosines.

$$\begin{aligned} a^2 &= b^2 + c^2 - 2bc \cos A \\ &= 7^2 + 5^2 - 2 \cdot 7 \cdot 5 \cdot \cos 45^\circ \\ &= 49 + 25 - 70\,(0.7071) \\ &= 74 - 49.497 \\ &= 24.503 \\ a &= \sqrt{24.503} \approx 4.95 \end{aligned}$$

Note that $\cos 45^\circ = \frac{\sqrt{2}}{2} \approx 0.7071.$

$a = 5.0$ to the nearest tenth.

2 **The lengths of two adjacent sides of a parallelogram are 7 and 12 units and an angle measures 120°. Find the length of the longer diagonal to the nearest tenth.**

First sketch the parallelogram as accurately as possible. The longer diagonal is opposite the angle measuring 120°.

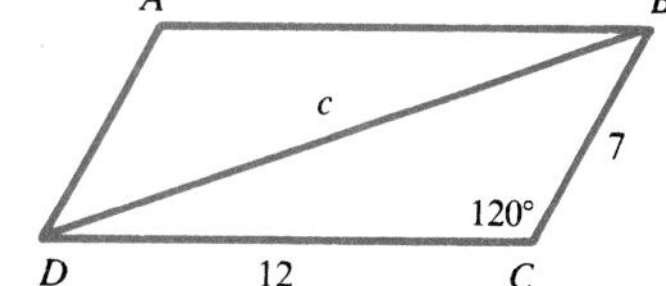

In $\triangle BCD$, use the following form of the Law of Cosines.

$$\begin{aligned} c^2 &= b^2 + a^2 - 2ba \cos C \\ &= 12^2 + 7^2 - 2(12)(7)(\cos 120^\circ) \\ &= 144 + 49 - 2(12)(7)\left(-\frac{1}{2}\right) \\ &= 277 \\ c &= \sqrt{277} \approx 16.64 \end{aligned}$$

Note that $\cos 120^\circ = -\frac{1}{2}.$

The length of the longer diagonal is 16.6 units to the nearest tenth.

3 **In $\triangle ABC$, $a = 9$, $b = 10$, and $c = 3$. Find, to the nearest degree, the measure of the greatest angle.**

First sketch the triangle as accurately as possible. We must find $m\angle B$ since $\angle B$ is opposite the longest side.

$$\begin{aligned} \cos B &= \frac{a^2 + c^2 - b^2}{2ac} \\ &= \frac{9^2 + 3^2 - 10^2}{2 \cdot 9 \cdot 3} \\ &= \frac{90 - 100}{54} \\ &\approx -0.1852 \end{aligned}$$

Using a calculator: **ENTER:** −0.1852 [INV] [cos]
DISPLAY: −0.1852 100.672793 ≈ 101

Since cos B is negative, $\angle B$ is a quadrant II angle. This means $\angle B$ is obtuse. The table gives $0.1851 = \cos 79^\circ 20'$. To the nearest degree, this is 79. Since $\angle B$ is a quadrant II angle, $m\angle B = 180 - 79$ or $m\angle B = 101$ to the nearest degree. *Note that the answer conforms to the sketch.*

Exercises

Exploratory **In $\triangle ABC$, the lengths of two sides and the cosine of the included angle are given. Find the length of the third side. Leave answers in radical form or compute to the nearest tenth as directed by your teacher.**

1. $b = 5, c = 4, \cos A = \frac{1}{8}$

2. $b = 5, c = 4, \cos A = \frac{4}{5}$

3. $a = 5, b = 6, \cos C = \frac{1}{5}$

4. $a = 7, b = 5, \cos C = \frac{1}{7}$

5. $a = 6, c = 4, \cos B = -\frac{1}{8}$

6. $a = 7, b = 5, \cos C = -\frac{1}{10}$

7. $a = 8, c = 4, \cos B = -\frac{1}{4}$

8. $a = 12, c = 10, \cos B = -\frac{1}{3}$

The lengths of the three sides of $\triangle RST$ are given. Find the cosine of the greatest angle.

9. $r = 4, s = 5, t = 6$

10. $r = 5, s = 7, t = 8$

11. $r = 3, s = 4, t = 2$

12. $r = 6, s = 3, t = 4$

13. $r = 8, s = 2, t = 7$

14. $r = 10, s = 13, t = 5$

Written **Given the following measures in $\triangle DEF$, find the length of the third side to the nearest tenth.**

1. $e = 7, f = 8, m\angle D = 30$

2. $d = 10, e = 8, m\angle F = 60$

3. $e = 3, f = 5, m\angle D = 58$

4. $d = 3, f = 10, m\angle E = 120$

5. $d = 2, f = \sqrt{2}, m\angle E = 45$

6. $d = 4, e = 9, m\angle F = 110$

7. $d = 4, e = 8, m\angle F = 120$

8. $e = 2, f = 6\sqrt{2}, m\angle D = 135$

9. $d = 4, f = 2\sqrt{2}, m\angle E = 135$

10. $e = 5\sqrt{2}, f = 4\sqrt{3}, m\angle D = 140$

Given the following lengths of the sides of $\triangle PQR$, find the measure of the indicated angle to the nearest degree.

11. $p = 7, q = 5, r = 8; m\angle P$

12. $p = 7, q = 5, r = 8; m\angle Q$

13. $p = 6, q = 7, r = 8; m\angle P$

14. $p = 10, q = 7, r = 8; m\angle R$

15. $p = 1, q = 2, r = \sqrt{3}; m\angle R$

16. $p = 5, q = 7, r = 10; m\angle R$

17. $p = 3, q = 4, r = 6; m\angle Q$

18. $p = 12, q = 2, r = 11; m\angle P$

The lengths of two adjacent sides and the measurement of an angle of a parallelogram are given. Find the length of the longer diagonal to the nearest centimeter.

19. 6 cm, 8 cm, 120°

20. 8 cm, 10 cm, 120°

21. 4 cm, 8 cm, 60°

22. 12 cm, 24 cm, 60°

23. 21 cm, 18 cm, 40°

24. 20 cm, 14 cm, 38°10′

25. 12 cm, 15 cm, 56°40′

26. 100 cm, 90 cm, 82°30′

27. The lengths of the diagonals of a parallelogram are 8 and 10 units and they intersect at an angle of 60°. Find, to the nearest tenth, the length of each side of the parallelogram.

Answer exercise 27 if the diagonals and angle of intersection have the following measures.

28. 10, 4, 60°

29. 8, 12, 120°

30. 26, 32, 48°

31. 28, 40, 150°

32. 16, 9, 110°

33. 13, 7, 137°

34. The distance across the Pine County Swamp cannot be measured directly. If surveyors have the information in the sketch, find d to the nearest tenth.

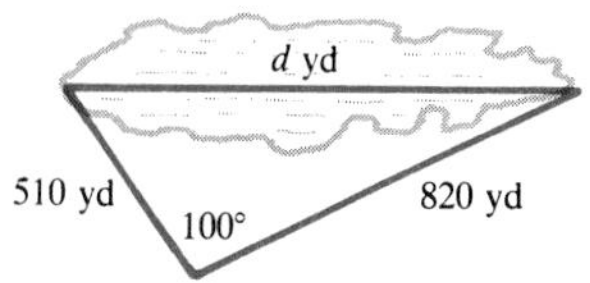

35. Find, to the nearest degree, the measure of a base angle of an isosceles triangle whose sides measure 14, 14, and 8 units.

36. Find, to the nearest degree, the measure of the acute angle formed by the two diagonals of a parallelogram if the diagonals measure 86 and 68 units and the shorter side of the parallelogram measures 18 units.

37. At 2 PM Tom leaves the Brayford ranch and travels at 30 mph. At the same time, Otto leaves from the same spot as Tom, but on a different road, and travels at 35 mph. If the two roads intersect at an angle of 54°, find to the nearest mile the distance between Tom and Otto at 4 PM.

Repeat exercise 37 if the following information is changed.

38. Otto travels at 44 mph.

39. Otto travels at 55 mph.

40. Tom travels at 40 mph, Otto at 50 mph, and the roads meet at 110°.

41. Answer exercise 37 for the distance between Tom and Otto at 5 PM.

Find to the nearest degree the measures of the angles of a parallelogram if the sides, s_1 and s_2, and a diagonal d have the following measures.

42. $s_1 = 10, s_2 = 6, d = 13$

43. $s_1 = 11, s_2 = 7, d = 15$

44. $s_1 = 5, s_2 = 7, d = 3$

45. $s_1 = 9, s_2 = 4, d = 6$

46. $s_1 = 50, s_2 = 60, d = 90$

47. $s_1 = 75, s_2 = 6\sqrt{2}, d = 80$

48. In golf, a *slice* is a shot to the right of its intended path (for a right-handed player) and a *hook* is off to the left. Mr. Kelly, a math teacher, estimates that he has hooked his last drive 18° and that the ball is now 200 yards from the tee. If the hole for which he was aiming is 180 yards from the tee, how far, to the nearest yard, is the ball from the hole?

49. Answer exercise 48 if the ball is hooked 23° and lands 250 yards from the tee.

50. Answer exercise 48 if the ball is sliced 12° and lands 150 yards from the tee.

51. Apply the Law of Cosines and Pythagorean Theorem to $\triangle ABC$, with right angle at C, in order to find the values for $\cos A$ and $\cos B$. What do you notice?

52. Zeb claims that the Law of Cosines can be used to prove the Pythagorean Theorem. Comment on Zeb's assertion.

Find, to the nearest degree, the measures of the angles of the triangles with the following vertices.

53. (0, 0), (1, 4), (5, −1)

54. (2, 3), (4, 0), (−1, −1)

55. (−1, 1), (3, 7), (4, −1)

56. (2, 2), (3, 3), (4, −2)

57. (1, 2), (7, 6), (3, −2)

58. (4, 4), (−3, −3), (7, 1)

Challenge

In exercises 1 and 2, find the measure of x in the figure at the right if a, b, c, and d have the following measures.

1. $a = 16, b = 12, c = 10, d = 22$

2. $a = 18, b = 9, c = 11, d = 26$

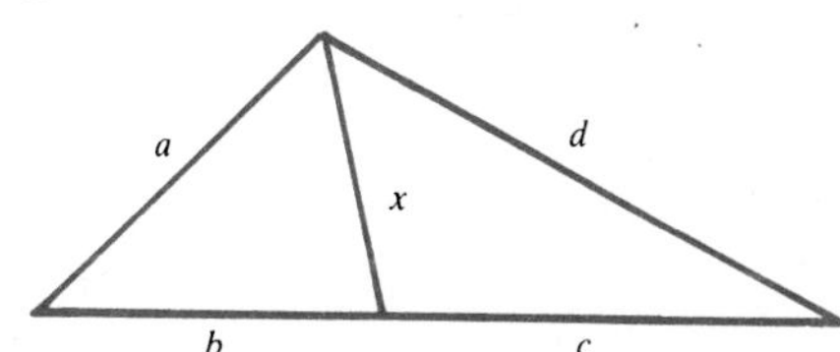

3. If $\triangle RST$ is isosceles with $r = t$, show that $s^2 = 2r^2(1 - \cos S)$.

8.3 Area of a Triangle

The principles of trigonometry can be used to derive a formula for the area of a triangle. Any triangle *ABC* can be set up on the coordinate plane in the three ways shown below. Using the given coordinates, we can find the area, symbolized by **A.** Recall for any triangle, **A** $= \frac{1}{2}bh$.

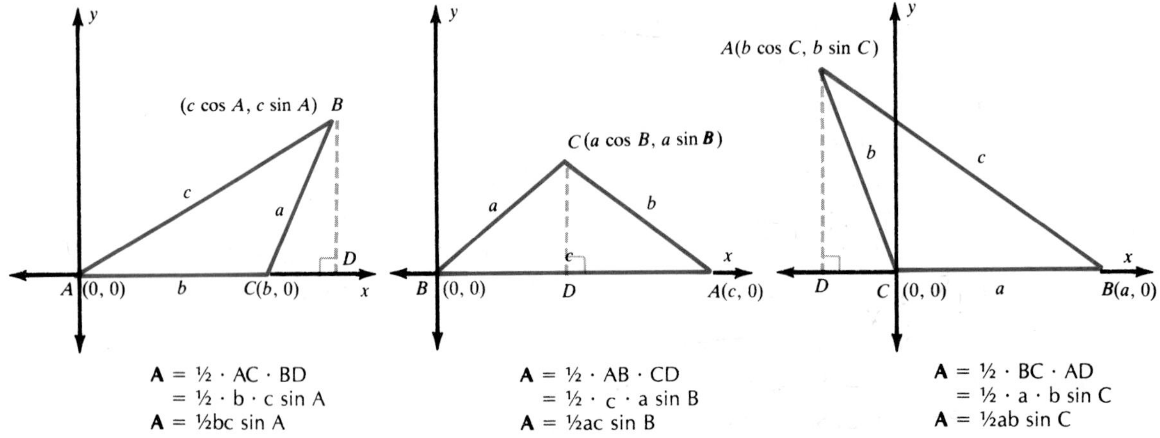

A $= \frac{1}{2} \cdot AC \cdot BD$
$= \frac{1}{2} \cdot b \cdot c \sin A$
A $= \frac{1}{2}bc \sin A$

A $= \frac{1}{2} \cdot AB \cdot CD$
$= \frac{1}{2} \cdot c \cdot a \sin B$
A $= \frac{1}{2}ac \sin B$

A $= \frac{1}{2} \cdot BC \cdot AD$
$= \frac{1}{2} \cdot a \cdot b \sin C$
A $= \frac{1}{2}ab \sin C$

The result of this investigation can be summarized as follows.

Area of a Triangle

The area, **A**, of any triangle is equal to one half the product of the lengths of any two sides and the sine of the included angle.

Examples

1 **Find the area of $\triangle ABC$ if $a = 12$, $b = 8$, and $m\angle C = 30$.**

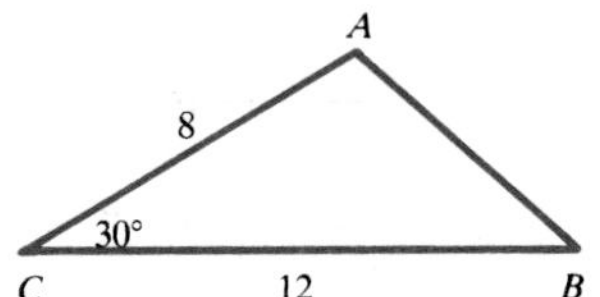

$\mathbf{A} = \frac{1}{2}ab \sin C = \frac{1}{2}(12)(8)(\sin 30°)$
$= \frac{1}{2}(12)(8)(\frac{1}{2})$

The area of $\triangle ABC$ is 24 square units.

2 **Find the exact area of an equilateral triangle whose sides measure 10 units.** *Recall that each angle measures 60°.*

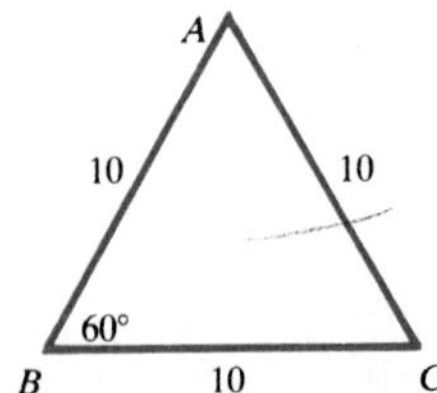

$\mathbf{A} = \frac{1}{2}ac \sin B = \frac{1}{2}(10)(10)(\sin 60°)$
$= \frac{1}{2}(10)(10)\frac{\sqrt{3}}{2}$

The exact area is $25\sqrt{3}$ square units.
Can you think of another method of solving this problem?

Exercises

Exploratory **Write an equation that can be used to find the area of $\triangle ABC$.**

1. $a = 10, b = 17, m\angle C = 46$
2. $b = 15, c = 20, m\angle A = 63$
3. $a = 15, b = 30, m\angle C = 90$
4. $a = 6, c = 4, m\angle B = 52$
5. $a = 18, c = 26, m\angle B = 142$
6. $b = 24, c = 42, m\angle A = 162$

Written **Find the exact area of $\triangle ABC$.**

1. $a = 8, b = 9, \sin C = \frac{1}{4}$
2. $b = 14, c = 6, \sin A = \frac{1}{7}$
3. $a = 11, c = 24, \sin B = \frac{5}{6}$
4. $b = 17, c = 28, \sin A = \frac{3}{4}$
5. $a = 6, b = 10, m\angle C = 90$
6. $b = 2\sqrt{3}, c = 6, m\angle A = 60$
7. $b = 16, c = 18, m\angle A = 150$
8. $a = 12, c = 3\sqrt{2}, m\angle B = 135$
9. $b = 5\sqrt{3}, c = 4, m\angle A = 120$
10. $a = b = 14, m\angle B = 75$
11. $a = 10, m\angle B = m\angle C = 60$
12. $a = 15, b = 6\sqrt{2}, m\angle C = 135$

Find, to the nearest tenth, the area of each triangle.

13. In $\triangle DEF$, $d = 12$, $e = 12$, and $m\angle F = 50$.
14. In $\triangle RST$, $\angle R$ measures $50°6'$, $s = 11$, and $t = 5$.
15. In $\triangle PQR$, $q = 4$, $r = 19$, and $\angle P$ measures $73°24'$.
16. In $\triangle TUV$, $t = 9.4$, $v = 13.5$, and $m\angle U = 95$.

Find, to the nearest tenth, the area of $\square ABCD$ if the sides, s_1 and s_2, and angle A have the following measures.

17. $s_1 = 18, s_2 = 14, m\angle A = 60$
18. $s_1 = 4, s_2 = 18, m\angle A = 150$
19. $s_1 = 19, s_2 = 23, m\angle A = 120$
20. $s_1 = 23, s_2 = 31, m\angle A = 42$

21. If the area of $\triangle ABC$ is 21 square units, find $\sin A$ if $b = 14$ and $c = 9$.
22. If the area of $\triangle ABC$ is 90 square units, find a if $b = 20$ and $m\angle C = 150$.
23. If the area of $\triangle RST$ is 13.5 square units, find r if $s = 18$ and $m\angle T = 120$.
24. If the area of $\triangle PQR$ is 66 square units, find $\sin P$ if $q = 22$ and $r = 21$.

25. The area of a triangle is 108 cm^2 and the lengths of two sides are 18 and 24 cm. Find, to the nearest degree, the measure of the included angle if it is acute.
26. In $\triangle DEF$, $\angle D$ is obtuse, $e = 16$, and $f = 5.5$. If the area of $\triangle DEF$ is $10\sqrt{3}$ yd^2, find $m\angle D$ to the nearest degree.

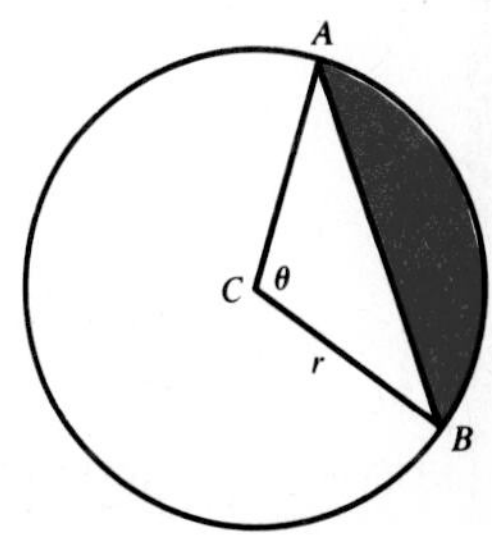

Challenge Recall that in the interior of a circle, the set of points bounded by a chord and its arc is called a **segment** of the circle. Show that with $m\angle C = \theta$ measured in radians and r the radius of the circle.

$$\textit{Area of a segment} = \tfrac{1}{2}r^2(\theta - \sin\theta).$$

8.4 The Law of Sines

In the preceding section we saw that there are three equal expressions for the area of $\triangle ABC$.

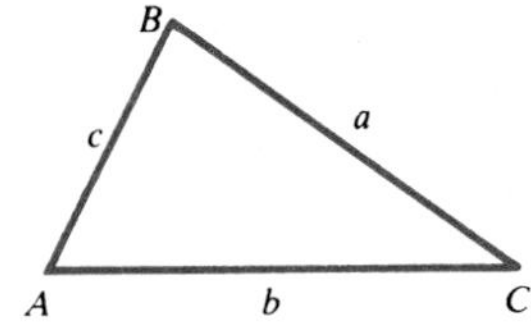

$$\frac{1}{2} bc \sin A = \frac{1}{2} ac \sin B = \frac{1}{2} ab \sin C$$

If we divide each of these three equal expressions by $\frac{1}{2} abc$, we obtain

$$\frac{\sin A}{a} = \frac{\sin B}{b} = \frac{\sin C}{c}.$$

This equality is known as the *Law of Sines.*

Law of Sines

In any triangle, the sines of the angles are proportional, respectively, to the lengths of the sides opposite the angles.

In $\triangle ABC$, $\dfrac{\sin A}{a} = \dfrac{\sin B}{b} = \dfrac{\sin C}{c}$.

You should note that the proportions established in the Law of Sines can also be written in other ways. For example,

$\dfrac{\sin A}{a} = \dfrac{\sin B}{b}$ is equivalent to

$$\frac{\sin A}{\sin B} = \frac{a}{b} \quad \text{and} \quad \frac{a}{\sin A} = \frac{b}{\sin B}.$$

Examples

1 In $\triangle ABC$, $a = 8$, $m\angle A = 40$, and $m\angle B = 62$. Find b to the nearest tenth.

$$\frac{b}{\sin B} = \frac{a}{\sin A}$$

$$\frac{b}{\sin 62^\circ} = \frac{8}{\sin 40^\circ}$$

$$b = \frac{8 \sin 62^\circ}{\sin 40^\circ}$$

Use the trig tables or a calculator to find $\sin 62^\circ$ and $\sin 40^\circ$.

$$b = \frac{8(0.8829)}{0.6428}$$

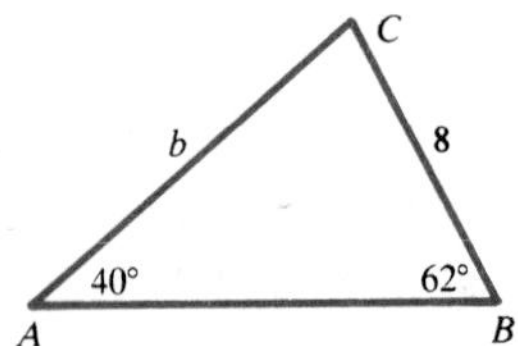

$$b \approx 10.98$$

$b = 11.0$ to the nearest tenth.

2 **In $\triangle KLM$, $m\angle K = 45$, $k = 20$, $m\angle L = 105$. Find m.**

First, sketch the triangle.

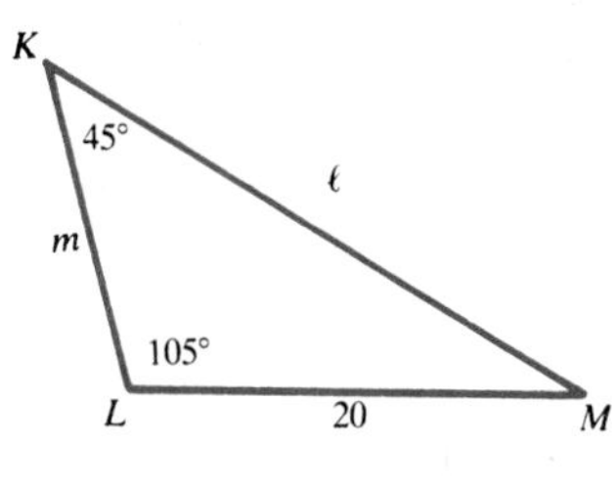

$$\frac{m}{k} = \frac{\sin M}{\sin K}$$

When using the Law of Sines, it is a good idea to set up your proportion so that the unknown part is the numerator of the first fraction.

$$\frac{m}{20} = \frac{\sin 30°}{\sin 45°}$$

Note that $m\angle M = 180 - (45 + 105)$ $= 30$

$$\frac{m}{20} = \frac{\frac{1}{2}}{\frac{\sqrt{2}}{2}}$$

$$\frac{m}{20} = \frac{\sqrt{2}}{2}$$

$$m = \frac{20\sqrt{2}}{2}$$

$$= 10\sqrt{2} \text{ units}$$

Exercises

Exploratory **Complete the following statements of the Law of Sines.**

1. $\frac{a}{\sin A} = \frac{b}{?}$
2. $\sin B = \frac{b \sin C}{?}$
3. $\frac{\sin A}{\sin C} = \frac{?}{?}$
4. $b = \frac{?}{\sin A}$
5. $c \sin A = ?$
6. $\frac{b \sin C}{\sin B} = ?$

Given the following information about $\triangle ABC$, find the measure indicated.

7. $a = 16$, $\sin A = \frac{1}{2}$, $\sin B = \frac{1}{4}$, $b = ?$
8. $b = 20$, $\sin B = \frac{2}{3}$, $\sin C = \frac{1}{2}$, $c = ?$
9. $c = 24$, $\sin B = \frac{1}{2}$, $\sin C = \frac{2}{5}$, $b = ?$
10. $b = 12$, $\sin A = \frac{3}{4}$, $\sin B = \frac{3}{5}$, $a = ?$
11. $a = 20$, $\sin A = 0.4$, $\sin C = 0.6$, $c = ?$
12. $b = 25$, $\sin A = 0.6$, $\sin B = 0.75$, $a = ?$

Written **Find the indicated measures for $\triangle ABC$. Use exact values when possible. If a table or calculator is used, round lengths to the nearest tenth and angles to the nearest degree.**

1. $m\angle A = 45$, $m\angle B = 60$, $a = 2\sqrt{6}$, $b = ?$
2. $m\angle B = 30$, $b = c = 14$, $a = ?$
3. $m\angle C = 105$, $m\angle B = 45$, $b = 8\sqrt{2}$, $a = ?$
4. $m\angle A = 45$, $a = 16$, $b = 12$, $m\angle B = ?$
5. $m\angle C = 75$, $m\angle A = 45$, $b = 5\sqrt{6}$, $a = ?$
6. $m\angle A = 30$, $m\angle B = 105$, $a = 10\sqrt{2}$, $c = ?$

7. $m\angle B = 45$, $m\angle C = 60$, $c = 12$, $b = ?$
8. $m\angle C = 140$, $b = 8$, $c = 16$, $m\angle B = ?$
9. $m\angle A = 30$, $m\angle C = 120$, $a = 3\sqrt{3}$, $c = ?$
10. $m\angle B = m\angle C = 70$, $a = 8$, $b = ?$
11. $m\angle B = 135$, $m\angle C = 15$, $a = 18$, $b = ?$
12. $m\angle A = 100$, $m\angle C = 30$, $c = 15$, $a = ?$
13. $m\angle C = 68$, $m\angle B = 82$, $a = 12$, $c = ?$
14. $m\angle B = 60$, $m\angle C = 90$, $b = 6\sqrt{3}$, $a = ?$
15. $b = 13$, $a = 8$, $m\angle B = 22$, $m\angle A = ?$
16. $m\angle C = 95$, $m\angle A = 70$, $b = 3$, $a = ?$

Solve each problem. Round answers to the nearest tenth.

17. Two angles of a triangle measure 40° and 56°. If the longest side is 28 cm, find the length of the shortest side.
18. Two angles of a triangle measure 34° and 71°. If the shortest side is 83 cm, find the length of the longest side.
19. A base angle of an isosceles triangle measures 64°20′. If the base is 18.4 cm, find the lengths of the congruent sides.
20. In $\triangle RST$, $m\angle R = 118$, $r = 10$ and $s = 3$. Find $m\angle S$ to the nearest degree.
21. Find the perimeter of an isosceles triangle that has a base of 22 cm and a vertex angle of 36°.
22. In $\triangle ABC$, the measurements of $\angle A$ and $\angle B$ respectively are 33°40′ and 53°20′ and $BC = 17.4$ cm. Find AC.

A surveyor measures a fence from *A* to *B*. Then she takes the bearings of a landmark, *C*, from *A* and *B*. Given the following information, find the length of $\overline{AC}$, rounded to two decimal places.

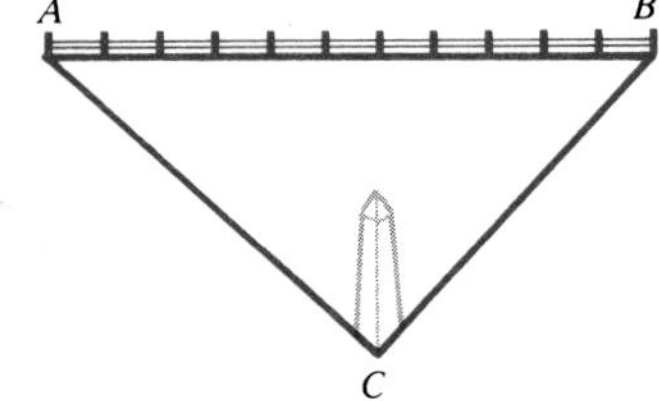

23. $\overline{AB}$ is 440 meters long, $\angle A$ measures 48°, and $\angle B$ measures 75°.
24. $\overline{AB}$ is 326 meters long, $\angle A$ measures 32°40′, and $\angle B$ measures 65°50′.
25. $\overline{AB}$ is 158 meters long, $\angle A$ measures 29°50′, and $\angle B$ measures 58°10′.
26. $\overline{AB}$ is 18.2 meters long, $\angle A$ measures 61°20′, and $\angle B$ measures 19°30′.

The Dolden Metalworks Company manufactures zinc alloy frames in the shape of isosceles triangles as shown. Find the length of the equal sides if the other specifications are as follows.

27. vertex angle: 30°; base: 8 dm
28. vertex angle: 24°; base: 7.8 dm
29. vertex angle: 32°; base: 8.5 dm
30. vertex angle: 22°; base: 6 dm

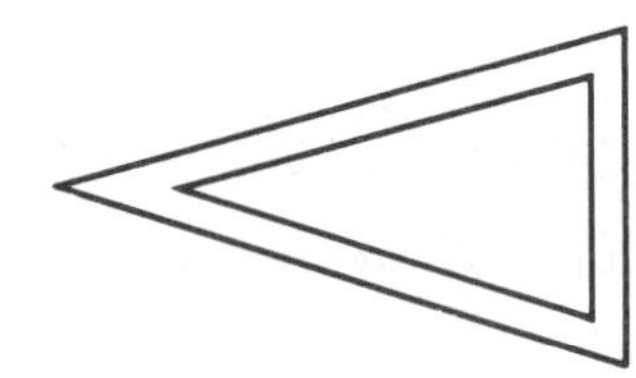

Challenge **Prove each of the following.**

1. In $\triangle ABC$, if $a \cos A = b \cos B$ then $\triangle ABC$ is either isosceles or right.
2. $\dfrac{b}{a + b} = \dfrac{\sin B}{\sin A + \sin B}$
3. $\dfrac{a - b}{b} = \dfrac{\sin A - \sin B}{\sin B}$
4. $\dfrac{a + b}{a - b} = \dfrac{\sin A + \sin B}{\sin A - \sin B}$

8.5 The Ambiguous Case

When the measures of two sides and an angle not included between them are given, there may be one triangle, two triangles, or no triangle at all! For this reason, a side, side, angle triangle situation (SSA) is sometimes called the *ambiguous case*. A description of the conditions under which these circumstances occur is given below.

Case I. Suppose $\angle A$ is acute. *The red dashed perpendicular has length* b *sin* A.

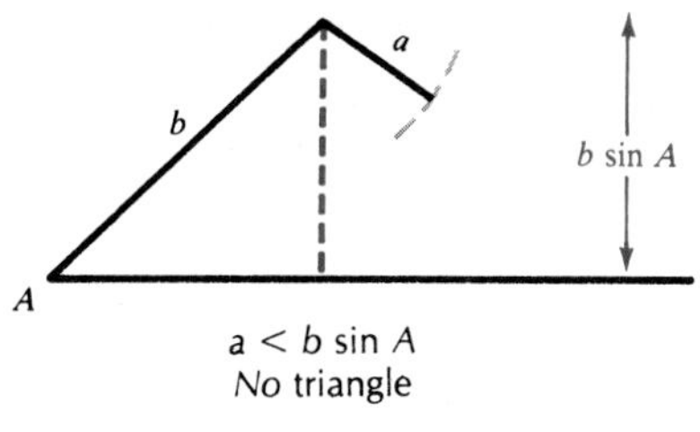

$a < b \sin A$
No triangle

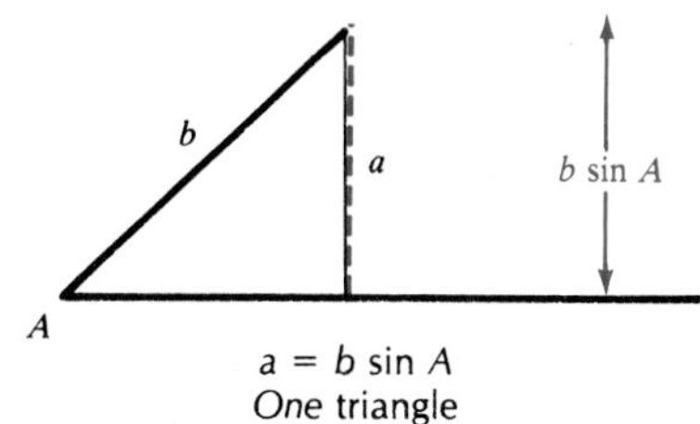

$a = b \sin A$
One triangle

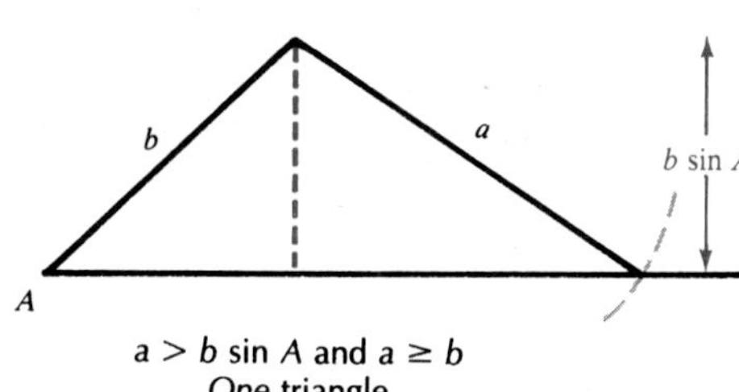

$a > b \sin A$ and $a \geq b$
One triangle

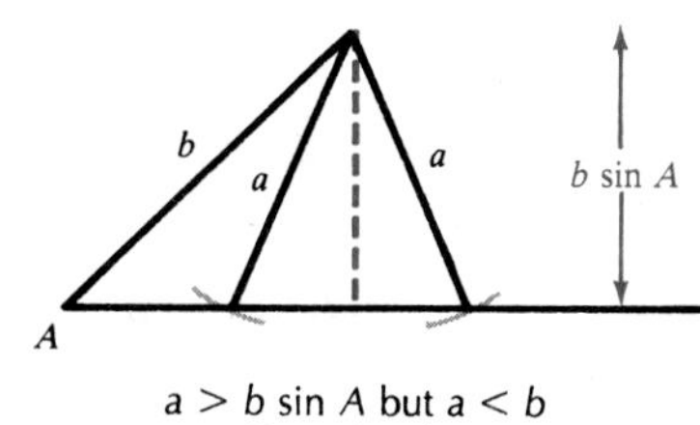

$a > b \sin A$ but $a < b$
Two triangles

Case II. Suppose $\angle A$ is obtuse.

In the examples, trig tables or a calculator can be used to find values for sin A or angle measures.

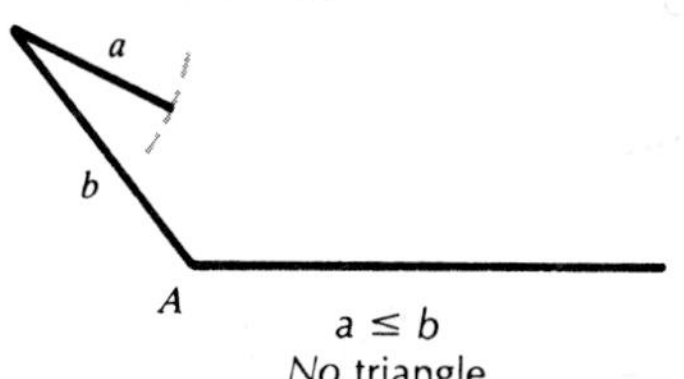

$a \leq b$
No triangle

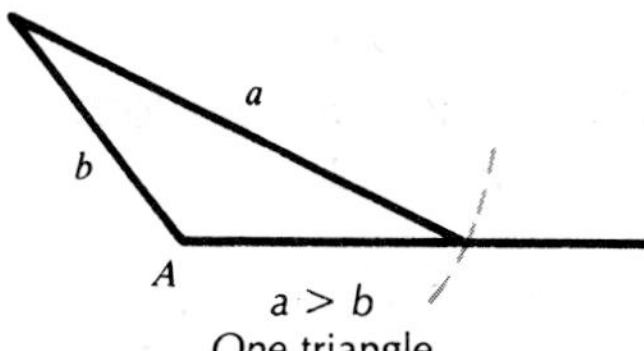

$a > b$
One triangle

Examples

1 Determine the number of triangles that have $m\angle A = 40$, $a = 3$, and $b = 10$.

First, try to sketch the triangle.

$b \sin A = 10 \sin 40°$
$= 10(0.6428)$
$= 6.428$

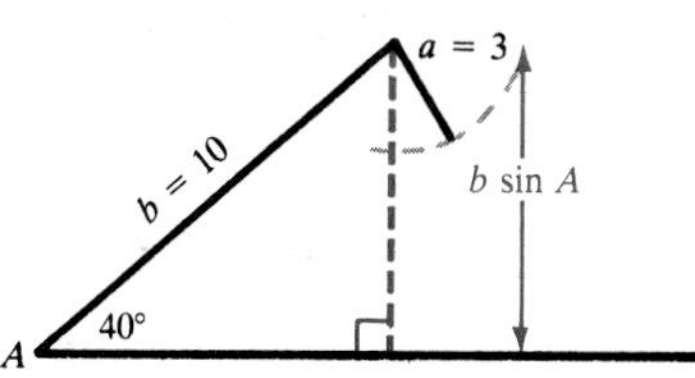

Since $3 < 6.428$, no triangle exists with the given measures.

2 **Determine the number of triangles that have $m\angle A = 40$, $a = 7.5$, and $b = 10$.**

Try to sketch a triangle.

$$b \sin A = 10 \sin 40°$$
$$= 10(0.6428)$$
$$= 6.428$$

Since $7.5 > 6.428$ and $7.5 < 10$, this indicates that there are two triangles.

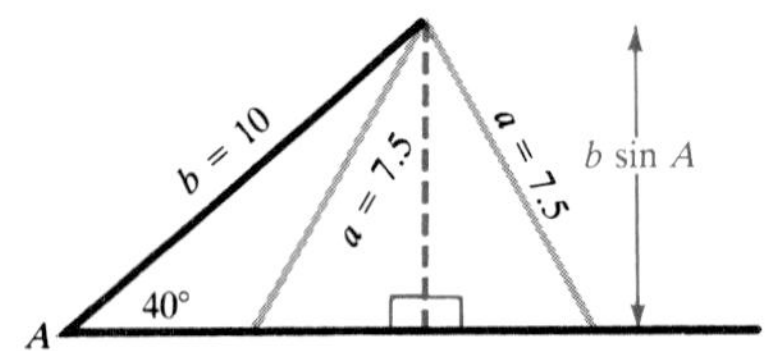

3 **In $\triangle ABC$, $m\angle A = 25$, $a = 8$, and $b = 10$. Find $m\angle B$ to the nearest ten minutes.**

This is an SSA situation where $\angle A$ is acute.

$$b \sin A = 10 \sin 25°$$
$$= 10(0.4226)$$
$$= 4.226$$

Since $8 > 4.226$ and $8 < 10$, there are two triangles. They are sketched below.

To find $m\angle B$, use the Law of Sines.

$$\frac{\sin B}{b} = \frac{\sin A}{a}$$

$$\sin B = \frac{b \sin A}{a}$$

$$= \frac{10 \sin 25°}{8}$$

$$\sin B = 0.5283$$

$\angle B$ measures 31°50′, to the nearest ten minutes, if B is acute.

$\angle B$ measures 180° − 31°50′ or 148°10′, to the nearest ten minutes, if B is obtuse.

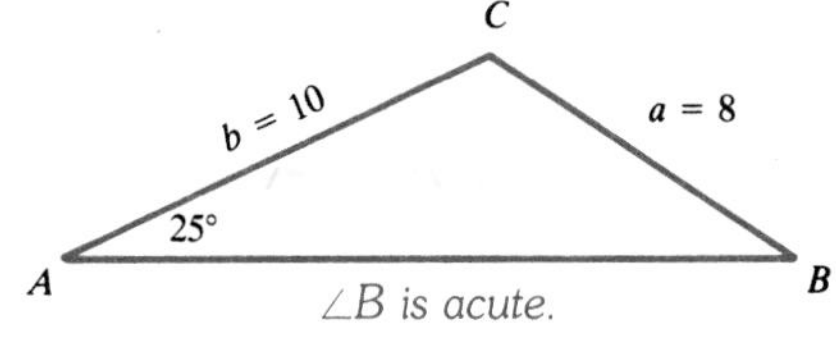

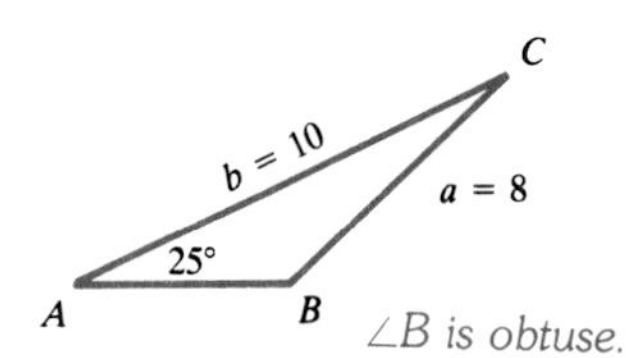

Exercises

Exploratory **Determine the number of triangles that can be constructed with the given measures. Include a sketch.**

1. $m\angle A = 140$, $b = 10$, $a = 3$

2. $m\angle A = 118$, $b = 11$, $a = 17$

3. $m\angle A = 30$, $a = 4$, $b = 8$

4. $m\angle A = 43$, $b = 20$, $a = 11$

5. $m\angle A = 50$, $a = 13$, $b = 16$

6. $m\angle A = 38$, $b = 10$, $a = 8$

7. $m\angle A = 135$, $b = 15$, $a = 13$

8. $m\angle A = 58$, $b = 13$, $a = 17$

9. $m\angle C = 45$, $c = 8$, $b = 8\sqrt{2}$

10. $m\angle B = 60$, $b = 10$, $c = 6\sqrt{3}$

11. $m\angle A = 125$, $a = 18$, $b = 26$

12. $m\angle C = 60$, $b = 18\sqrt{3}$, $c = 29$

13. $m\angle B = 45$, $a = 20\sqrt{2}$, $b = 18$

14. $\angle A$ measures 18°40′, $b = 35$, $a = 20$

Written **Classify $\triangle ABC$ as acute, obtuse, or right.**

1. $m\angle A = 30$, $a = 5$, $b = 10$
2. $m\angle A = 30$, $a = 5$, $b = 8$
3. $m\angle A = 60$, $a = 10$, $b = 10$
4. $m\angle A = 45$, $a = 6\sqrt{2}$, $b = 12$

For each of the following, first determine the number of triangles that can be constructed with the given measures. Then find, if possible, the indicated angle measures to the nearest degree.

5. $m\angle A = 30$, $a = 7$, $b = 8$; $m\angle B = ?$
6. $m\angle A = 40$, $a = 20$, $c = 22$; $m\angle C = ?$
7. $m\angle A = 50$, $a = 18$, $c = 24$; $m\angle C = ?$
8. $m\angle A = 128$, $a = 19$, $b = 20$; $m\angle B = ?$
9. $m\angle B = 78$, $a = 26$, $b = 27$; $m\angle A = ?$
10. $m\angle B = 48$, $b = 40$, $c = 60$; $m\angle C = ?$
11. $m\angle A = 114$, $a = 20$, $c = 18$; $m\angle B = ?$
12. $m\angle B = 18$, $a = 40$, $b = 50$; $m\angle C = ?$
13. $\angle C$ measures $57°20'$, $c = 95$, $b = 110$; $m\angle B = ?$
14. $\angle C$ measures $82°40'$, $c = 119.5$, $b = 120$; $m\angle B = ?$
15. An engineering student was given an assignment to construct a triangular model with three steel girders. Two of the girders measured 20 cm and 15 cm, and the angle opposite the 15 cm girder had to be 61°. Could she construct the triangular model? If so, what should be the length of the third girder?

Mixed Review

Choose the best answer.

1. Which function is the inverse of $y = 2^x$?
   a. $y = \log x$ b. $y = x^2$ c. $y = \log_2 x$ d. $x = \log_2 y$
2. If $\overline{PX}$ is tangent to circle C at X and secant $\overline{PYZ}$ intersects circle C at Y and Z, then which of the following is true?
   a. $(PX)^2 = (PY)(YZ)$ b. $PX = PY$
   c. $(PX)^2 = (PY)(PZ)$ d. $(PZ)^2 = (PX)(PY)$
3. What is the period of the graph of $y = 3 \cos 4x$?
   a. 3 b. 4 c. $\frac{\pi}{4}$ d. $\frac{\pi}{2}$
4. If $\sin \theta = \frac{8}{17}$ and $\cot \theta > 0$, what is the value of $\sec \theta$?
   a. $\frac{17}{15}$ b. $\frac{17}{8}$ c. $\frac{15}{8}$ d. $\frac{15}{17}$
5. Under which transformation is the graph of $y = \cot x$ *not* its own image?
   a. $Rot_{O,180°}$ b. $T_{2\pi,0}$ c. $R_{y\text{-axis}}$ d. $T_{\pi,0}$
6. Points X, Y, and Z are on circle P such that $XY = 13$, $YZ = 16$, and $XZ = 10$. If $m\widehat{XZ} = 130$, which expression represents the area of $\triangle XYZ$?
   a. 5(16) sin 65° b. 5(16) sin 130° c. 8(13) sin 65° d. 10(16) sin 65°
7. In circle Q, $\angle AQB$ is a central angle. If $m\angle AQB = 72$ and the diameter of circle Q is 10 cm, what is the length of $\widehat{AB}$?
   a. 4π cm b. 360 cm c. 2π cm d. π cm

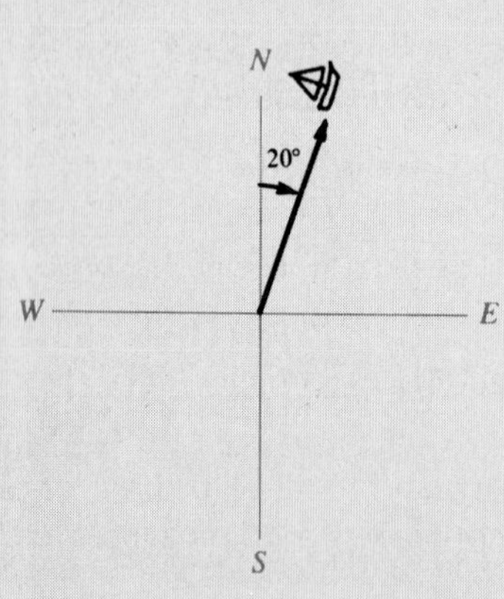

There are different ways that navigators, and others concerned with expressing directions at sea or in the air, express *the direction*. Often, the *heading*, or *bearing*, of a ship or plane is given as a number of degrees *clockwise from north*. When this method of giving direction is used, no additional compass direction is given.

In addition, distances at sea are measured in *nautical miles*. A nautical mile is about 1.15 land miles. Speed at sea is in *knots:* one knot is one nautical mile per hour. Note that we do not say "knots per hour."

Furthermore, we must make an important assumption: Angles, triangles, areas, and so on, that we deal with are in a *plane*. Of course, they are *actually* on a curved surface of a sphere since the world is, after all, round. Now, the assumption that the world is flat may seem to be a step backward to you, but for relatively small regions of the globe, we can simplify our computations greatly and get good approximate answers by making the assumption. For larger areas, other methods must be used.

Exercises **Answer each question.**

1. The sloop MIM-III left its home port and sailed at 12 knots on a heading of 105°. After three hours, it changed its speed to 14 knots and its course to 140°. After two hours at the new heading, how far was the sloop from its home port?
2. Suppose that the sloop in exercise 1 changed course again (after the five hours indicated in the problem) to a heading of 210°. After three hours, how far would it then be from its home port?
3. A cruiser is 60 miles east of a fleet of ships moving at 18 knots with a bearing of 45°. What should be the heading of the cruiser if the aim of its captain is to catch up with the fleet as quickly as possible and if the cruiser can attain a speed of 27 knots?

The chart below gives today's information about four ships of the New Hanseatic Ship Company. The home port is the same for all ships. What is the distance between each pair of ships at 5 PM?

4. Hansa II, Lubeck
5. Hansa II, Wisby
6. Hansa II, Bremen
7. Lubeck, Bremen
8. Lubeck, Wisby
9. Wisby, Bremen

Ship	Time left home port	Heading	Speed
Hansa II	Noon	86°10′	20 knots
Lubeck	10 AM	91°40′	18 knots
Wisby	1 PM	166°00′	24 knots
Bremen	3 PM	112°10′	20 knots

10. How many different exercises like exercises 4–9, for two ships, could be asked?
11. At what time will the Hansa II and the Lubeck be 130 miles apart?
12. When will the Hansa II and the Bremen be 200 miles apart?

8.6 Solving Triangles

In the examples, trig tables or a calculator can be used to find values of trigonometric functions or angle measures.

Triangles, as you know, have six parts: three angles and three sides. In preceding sections, you used the Law of Cosines and the Law of Sines to find one part of a triangle, given some of the others. The process of finding all the remaining parts of a triangle when we are given some of the parts is called **solving the triangle.**

Examples

1 **Solve $\triangle ABC$ if $b = 8$, $c = 7$, and $m\angle A = 28$.** *Unless otherwise specified, find lengths to the nearest tenth and angle measures to the nearest degree.*

First sketch the triangle.

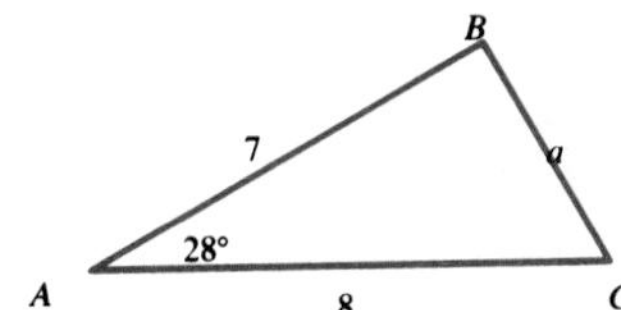

We have two sides and the included angle (SAS). In this case the Law of Sines cannot be used because the three given parts do not give us two Sine Law ratios. We must use the Law of Cosines to find a.

$$\begin{aligned} a^2 &= b^2 + c^2 - 2bc \cos A \\ &= 64 + 49 - 2(8)(7) \cos 28° \\ &= 113 - 112\,(0.8829) \\ &= 14.1152 \\ a &\approx 3.76 \\ a &= 3.8 \text{ to the nearest tenth} \end{aligned}$$

Now find the two remaining angles. Although $\angle B$ is the greatest angle, we cannot be sure from our sketch if $\angle B$ is acute or obtuse. The least angle, $\angle C$, must be acute because there can be at most one obtuse angle in a triangle. Therefore, we find $m\angle C$ first.

$$\begin{aligned} \frac{\sin C}{c} &= \frac{\sin A}{a} \\ \sin C &= \frac{c \sin A}{a} \\ \sin C &\approx \frac{7 \sin 28°}{3.76} \\ \sin C &\approx 0.874 \\ m\angle C &\approx 61 \end{aligned}$$

The third angle, $\angle B$, is found using the fact that the sum of the measures of three angles of a triangle is 180°. We have

$$\begin{aligned} m\angle B &\approx 180 - (61 + 28) \\ &\approx 91. \end{aligned}$$

Thus, the two remaining angles measure 61 and 91 to the nearest degree.

2 **Solve $\triangle ABC$ if $a = 10$, $b = 16$, and $c = 19$.**

First do a sketch.

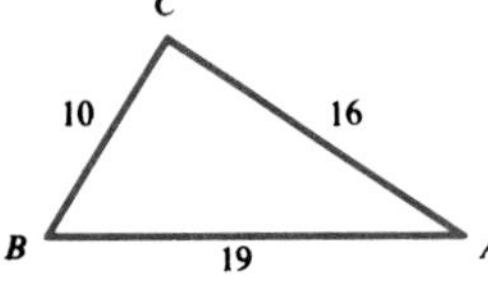

In this case we have SSS, the three sides. Therefore, we must use the Law of Cosines to find one of the angles. Usually it is a good idea to find the greatest first. (This angle will be the one opposite the longest side.) Then we know the other two must be acute.

$$\cos C = \frac{a^2 + b^2 - c^2}{2ab}$$

$$\cos C = \frac{100 + 256 - 361}{2 \cdot 10 \cdot 16}$$

$$\cos C = \frac{-5}{320} \approx -0.0156$$

The cosine is negative, so we know $\angle C$ is in quadrant II. Therefore, $\angle C$ is obtuse. The measurement of $\angle C$ is $90°54'$.

Use the Law of Sines to find one of the other angles, say A.

The third angle is $180° - (90°54' + m\angle A)$.

We will leave it to you to complete the solution of this triangle in the exercises.

3 **Solve $\triangle ABC$ if $m\angle A = 40$, $a = 7.5$, and $b = 10$.** *This is an SSA situation.*

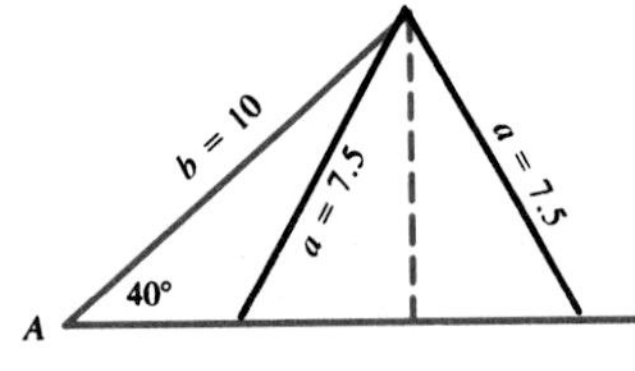

$$b \sin A = 10 \sin 40°$$
$$= 10(0.6428) \text{ or } 6.428$$

Since $a > b \sin A$ and $a < b$, two solutions exist.

$$\frac{\sin B}{b} = \frac{\sin A}{a}$$

$$\sin B = \frac{b \sin A}{a}$$

$$\sin B = \frac{10 \sin 40°}{7.5}$$

$$\sin B = \frac{10(0.6428)}{7.5}$$

$$\sin B = 0.8571$$

$m\angle B = 59$ or $m\angle B = 121$

$m\angle C = 180 - (40 + m\angle B)$

$m\angle C = 81$ or $m\angle C = 19$

We now have all three angles and sides a and b. It remains to find the two possibilities for side c, and this we leave to you in the exercises.

Summary Of Methods For Solving Triangles	
Step 1 Sketch the triangle! Determine what is given, and what must be found.	
Step 2 Then use the appropriate procedure.	
Given	**Procedure**
Two sides, included angle (SAS)	Use the Law of Cosines for the third side. Then use the Law of Sines for the other angles. Find the lesser one first.
Three sides (SSS)	Use the Law of Cosines to find the greatest angle first. Then use the Law of Sines.
Two angles, one side (SAA or ASA)	Find the third angle (180° less the sum of the other two). Use the Law of Sines.
Two sides, angle not included (SSA)	Carefully examine the situation for the number of solutions. Use the Law of Sines.

Exercises

Exploratory **Answer each of the following.**

1. What does it mean to "solve a triangle"?
2. Lyle says to Joni, "I'll take you to the Pine Burger if you can find the greatest angle in a triangle whose sides are 2, 3, and 8." Could Joni go to the Pine Burger with Lyle if she wished?
3. Describe the things you have to be careful about when using the Law of Sines to solve a triangle.
4. Given the three sides of a triangle, how can you be sure which is the greatest angle in the triangle?
5. Why is it a good idea to find the greatest angle first when solving a triangle given the three sides?
6. Complete example 2.
7. Complete example 3.

Written **For the written exercises, find all lengths to the nearest tenth, all angle measures to the nearest degree. In exercises 1–16, solve $\triangle ABC$, if possible.**

1. $b = 10, m\angle A = 60, c = 14$
2. $a = 8, b = 12, c = 15$
3. $m\angle C = 45, a = 12, b = 18$
4. $b = 11, m\angle A = 30, c = 8$
5. $a = 8.5, m\angle B = 20, m\angle C = 70$
6. $a = 8.5, b = 12.5, c = 10$
7. $a = 3.6, c = 5.4, m\angle B = 100$
8. $m\angle B = 115, a = 2, c = 5$
9. $a = 2.12, b = 1.9, c = 1.7$
10. $a = 0.12, b = 1.9, c = 1.7$
11. $m\angle A = 146, a = 11, c = 12.5$
12. $a = 14, b = 1, c = 14$
13. $m\angle A = 40, m\angle B = 45, c = 4$
14. $a = 31, b = 14, c = 12$
15. $m\angle A = 30, m\angle B = 45, c = 14$
16. $m\angle A = 26, b = 10, c = 12$

In exercises 17–26, solve $\triangle ABC$ using these steps: 1. Draw a sketch. 2. Decide on the number of solutions: 0, 1, or 2. Find all existing solutions.

17. $m\angle A = 42$, $b = 12$, $c = 15$

18. $m\angle A = 120$, $b = 16$, $a = 30$

19. $m\angle A = 45$, $b = 18$, $a = 9\sqrt{2}$

20. $m\angle C = 10$, $a = 6$, $c = 1.2$

21. $m\angle C = 56$, $a = 4.4$, $c = 6.2$

22. $m\angle A = 88$, $a = 10$, $b = 9$

23. $m\angle B = 148$, $c = 14$, $b = 8.1$

24. $m\angle B = 72$, $c = 15$, $b = 8$

25. $m\angle C = 25$, $a = 40$, $c = 20$

26. $m\angle B = 38$, $a = 22$, $b = 14$

27. A triangular plot of land has sides with lengths 50 m, 70 m, and 85 m. Find the measure of the angle opposite the shortest side.

28. A triangular plot of land has two sides with lengths of 400 ft and 600 ft which intersect at an angle of 46°20′. Find the perimeter and area of the triangle.

29. The sides of a triangle are 6.8 cm, 8.4 cm, and 4.9 cm. Find the measure of the least angle.

30. A ship at sea is 70 miles from one radio transmitter and 130 miles from another. The angle between the signals measures 130. How far apart are the transmitters?

31. The sides of a parallelogram are 20 cm and 32 cm. If the longer diagonal measures 40 cm, find the measures of the angles of the parallelogram.

32. The sides of a parallelogram are 55 cm and 71 cm. Find the length of each diagonal if the greater angle is 106°.

Challenge **Interpreting blueprints requires the ability to select and use trigonometric functions and geometric properties. Assume that lines that appear parallel or perpendicular are. Select the relevant information and appropriate functions to find the unknown values as indicated for each problem.**

1. Find x and y.

2. Find $m\angle C$ and $m\angle D$.

3. Find $m\angle A$, $m\angle B$, r, and t.

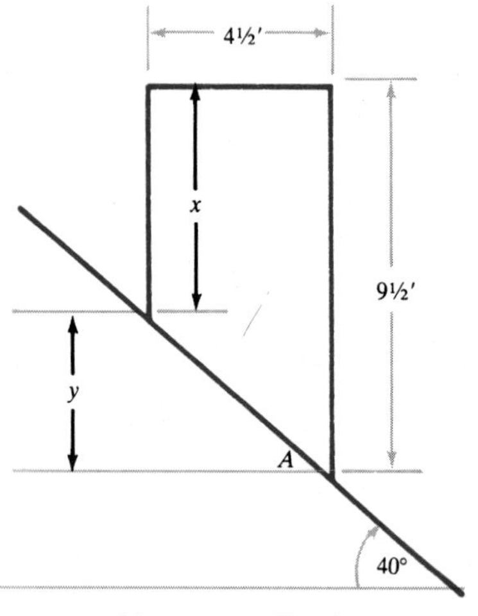

Chimney on Roof

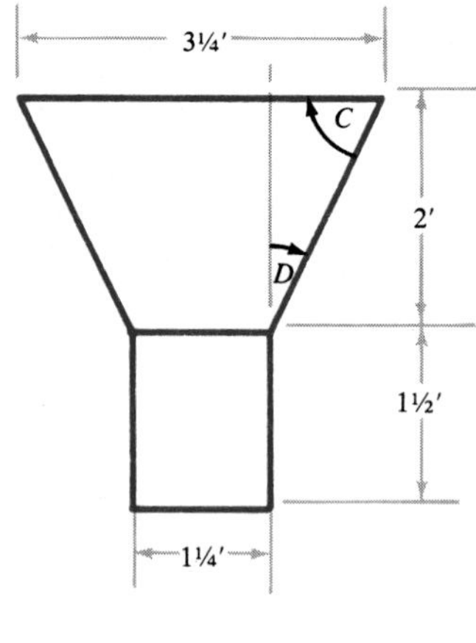

Air Vent

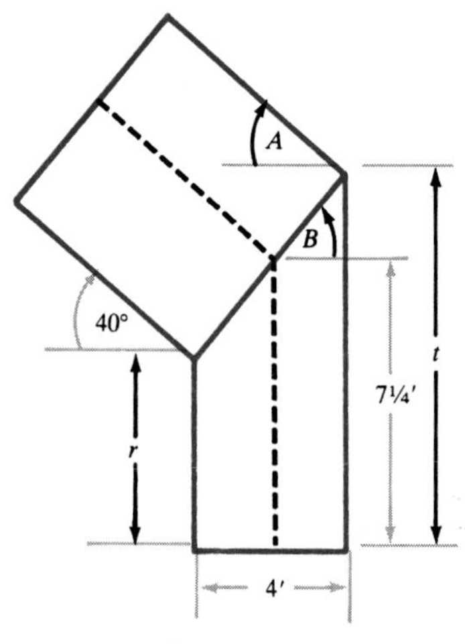

Elbow Joint

8.7 Trigonometry of the Right Triangle

The trigonometry of the right triangle has always had a special place in mathematical studies.

Consider right triangle ABC shown at the right. Using the Law of Sines, we have

$$\frac{\sin A}{\sin C} = \frac{a}{c}.$$

Since $m\angle C = 90$, $\sin C = 1$.

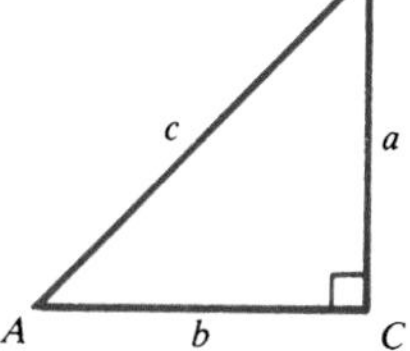

$\overline{AC}$ and $\overline{BC}$ are legs and $\overline{AB}$ is the hypotenuse of right $\triangle ABC$.

Therefore, $\sin A = \frac{a}{c}$. *a is the side opposite $\angle A$. c is the hypotenuse.*

Similarly, $\frac{\sin B}{\sin C} = \frac{b}{c}$ *b is the side opposite $\angle B$. c is the hypotenuse.*

or $\sin B = \frac{b}{c}$.

These results can be summarized as follows.

The sine of an acute angle in a right triangle equals the length of the side opposite the angle divided by the length of the hypotenuse. This is abbreviated as

$$\sin A = \frac{\text{side opp } A}{\text{hypotenuse}}$$

$$\text{or } \sin A = \frac{\text{opp}}{\text{hyp}}.$$

Using the Pythagorean Identity, we have

$$\sin^2 A + \cos^2 A = 1.$$

$$\left(\frac{a}{c}\right)^2 + \cos^2 A = 1$$

$$\cos^2 A = 1 - \frac{a^2}{c^2}$$

$$= \frac{c^2 - a^2}{c^2}$$

$$= \frac{b^2}{c^2}$$

In right $\triangle ABC$, $a^2 + b^2 = c^2$ or $b^2 = c^2 - a^2$.

$$\cos A = \frac{b}{c}$$

Similarly, $\cos B = \frac{a}{c}$.

This result can be summarized as follows.

The cosine of an acute angle in a right triangle equals the length of the side adjacent to the angle divided by the length of the hypotenuse. This is abbreviated by

$$\cos A = \frac{\text{side adj } A}{\text{hypotenuse}} \text{ or } \cos A = \frac{\text{adj}}{\text{hyp}}.$$

Since $\tan A = \frac{\sin A}{\cos A}$, we have

$$\tan A = \frac{\frac{a}{c}}{\frac{b}{c}} \text{ or } \frac{a}{b}. \qquad \textit{Similarly, } \tan B = \frac{b}{a}.$$

$$\text{In a right triangle, } \tan A = \frac{\text{side opp } A}{\text{side adj } A} \text{ or } \tan A = \frac{\text{opp}}{\text{adj}}.$$

A helpful mnemonic device for remembering these special ratios for right triangles is

S O H – C A H – T O A.

$$\text{sine} = \frac{\text{opp}}{\text{hyp}} \qquad \cos = \frac{\text{adj}}{\text{hyp}} \qquad \tan = \frac{\text{opp}}{\text{adj}}$$

Examples

1 What size angle does the diagonal of a 5′ by 8′ rectangle make with the longer side?

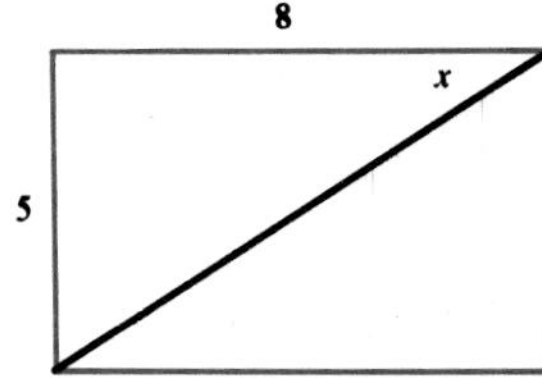

We are looking for angle x. The side opposite angle x has length 5′, and the side adjacent to angle x has length 8′. We use the tangent ratio.

$$\tan x = \frac{5}{8} \text{ or } 0.625$$

$$x \approx 32°$$

The diagonal makes an angle of about 32° with the longer side.

2 It has been suggested that a ladder is in the safest position when the angle it makes with the ground is 75°. How far from the foot of the wall should a 14′ ladder be placed?

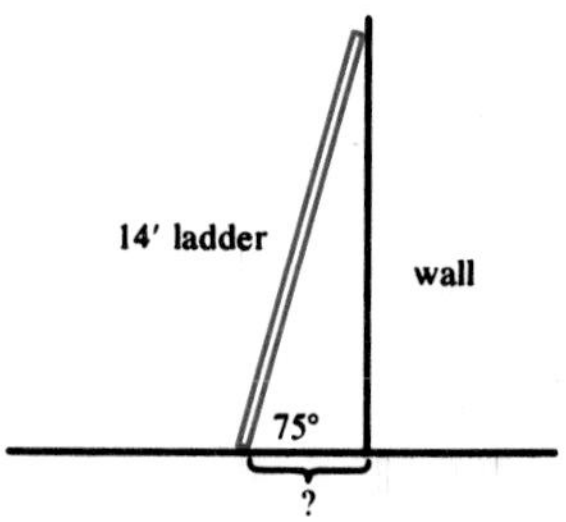

$$\cos 75° = \frac{x}{14}$$

$$x \approx 14(0.2588) \text{ or } 3.62$$

Place the ladder a bit more than $3\frac{1}{2}$ feet from the wall.

We often encounter situations where distant objects are sighted at a given angle from the line of sight of the observer.

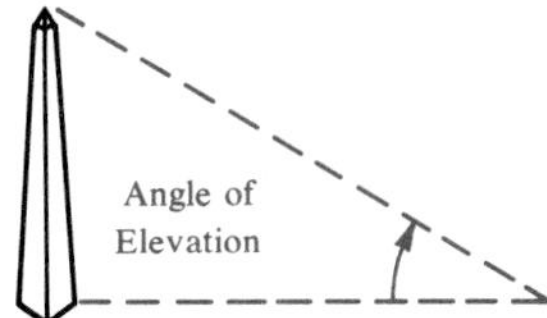

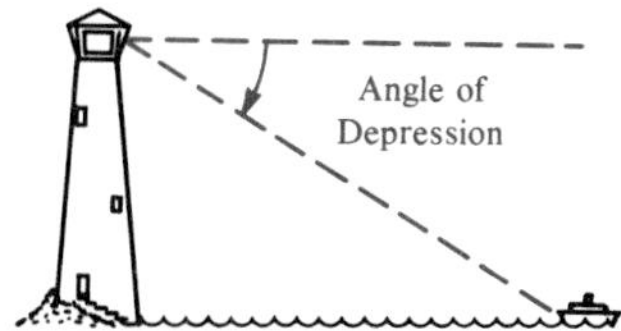

An **angle of elevation** is the angle between the horizontal and the line of sight when an object *elevated* from the viewer is sighted.

An **angle of depression** is the angle between the horizontal and the line of sight when an object *beneath* the viewer is sighted.

Example

3 **Clem is standing on top of a cliff 200 feet above a lake. The measurement of the angle of depression to a boat on the lake is 21°. How far is the boat from the base of the cliff?**

Computation is easier if the cotangent function is used.

$$\cot 21° = \frac{x}{200}$$

$$2.6051 = \frac{x}{200}$$

$$x = 521 \text{ feet}$$

(to the nearest foot)

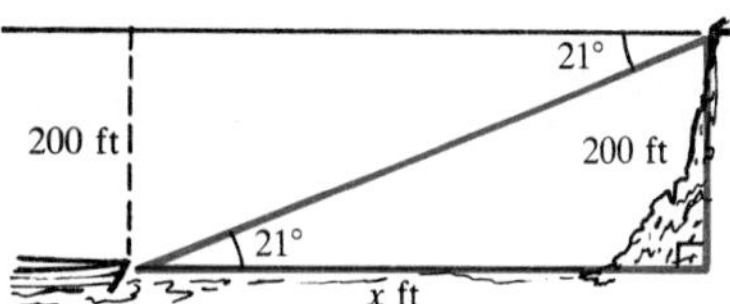

Exercises

Exploratory **Find the indicated measures in right triangle ABC with right angle at C. Round lengths to the nearest tenth and angle measures to the nearest 10 minutes. Let A = the degree measure of $\angle A$.**

1. $A = 28°, b = 7, c = ?$
2. $A = 15°, c = 37, a = ?$
3. $A = 76°, a = 13, b = ?$
4. $B = 16°, a = 13, c = ?$
5. $a = 7, b = 12, A = ?$
6. $b = 11, a = 18, B = ?$
7. $A = 68°, c = 15, b = ?$
8. $b = 101, c = 158, A = ?$
9. $b = 6, c = 13, A = ?$
10. $A = 78°, c = 82, a = ?$
11. $B = 8°20', a = 12, b = ?$
12. $a = 120, c = 140, B = ?$
13. $a = b\sqrt{3}, c = ?, A = ?, B = ?$
14. $a = \sqrt{21}, b = \sqrt{7}, B = ?$

For each of the following dimensions of a rectangle, find to the nearest degree the measure of the angle that a diagonal of that rectangle makes with the shorter side.

15. 5 ft by 7 ft
16. 9 ft by 14 ft
17. 1 in. by 3 in.
18. 3 mm by 0.5 mm

Assume the same safety advice given in example 2. Find to the nearest foot the distance from the wall that ladders with the following lengths should be placed.

19. 8 ft
20. 16 ft
21. 18 ft
22. 20 ft
23. 5 m

Written **Solve each problem. Round lengths to the nearest tenth and angle measures to the nearest degree.**

1. A lighthouse keeper views a small boat in the water beneath him. How far from the foot of the lighthouse is the boat if the lighthouse is thirty feet high and the angle of depression of the boat is 36°?

Answer exercise 1 for the following changes in information.

2. Lighthouse is 40 ft high.
3. Angle of depression is 18°.
4. Lighthouse is 28 ft high, angle of depression is 8°.
5. A reconnaissance plane is 1000 feet over its carrier. The pilot sights a submarine in the water at an angle of depression of 12°. How far is the sub from the carrier?

In exercises 6–7, answer exercise 5 for these circumstances.

6. The plane is 750 ft in the air.
7. The plane is 1250 ft in the air and the angle of depression is 24°.
8. From a point 1000 feet away, along level ground, the angle of elevation of the Pine City Obelisk is 10°. How tall is the obelisk?
9. The Hornetville Obelisk is 155 feet high. How far away is a viewer if the angle of elevation from his vantage point is 8°? (Neglect the viewer's height.)
10. Answer exercise 9 if you take into account the fact that the viewer's eyes are six feet from the ground.

Answer exercise 8 for the obelisks in exercises 11–13.

11. South Beeville Obelisk; from 750 ft, the angle of elevation is 13°
12. West Beeville Obelisk; from 900 ft, the angle of elevation is 11°
13. The famous Mata-lisk; angle of elevation from 2500 ft, 8°
14. After it takes off, a plane travels a distance of 3000 feet to rise to a height of 400 ft. At what angle from the ground did it take off?

Answer exercise 14 for these distances and heights.

15. 4500 ft, 200 ft
16. 1500 m, 150 m
17. 1280 m, 180 m
18. A plane takes off at an angle of 16° to the horizontal. How high does it rise after it travels a distance of 3800 feet?

Answer exercise 18 for these angles and distances traveled.

19. 11°50′, 1000 m
20. 17°50′, 1280 m
21. 25°, 1600 m

22. An aircraft ascends at an angle of 12°30′ with the horizontal. How far has it traveled when it is 230 meters above the ground?

Answer exercise 22 for these angles and rises.

23. 11°, 300 m
24. 8°, 100 m
25. 9°30′, 150 m

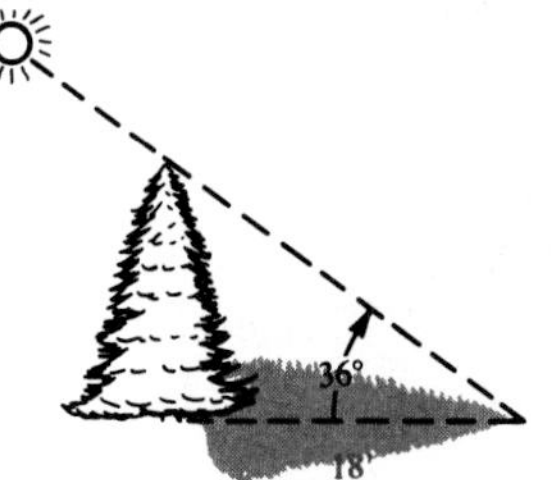

26. How tall is a tree which casts an 18-foot shadow when the angle of elevation, as shown, is 36°?

Answer exercise 26 for these shadow lengths and angles of elevation.

27. 16 m, 20°
28. 10 ft, 78°
29. 12.5 ft, 46°
30. 40 m, 33°

31. The angle of elevation to the top of a building from a point on the ground is 38°20′. From a point 50 feet closer to the building, the angle of elevation is 45°. What is the height of the building?

32. Two observers 200 feet apart are in line with the base of a flagpole due east of the observers. The angles of elevation to the top of the flagpole are 30° and 60°. How far is the flagpole from each observer?

33. The base of a television antenna is in line with two points on the ground due west of the antenna. The two points are 100 feet apart. From the two points, the angles of elevation to the top of the antenna are 30° and 20°. Find the height of the antenna.

34. A television antenna sits atop a building. From a point 200 feet from the base of the building, the angle of elevation of the top of the antenna is 80°. The angle of elevation of the bottom of the antenna from the same point is 75°. How tall is the antenna?

35. To find the height of a mountain peak, two points, A and B, were located on a plain in line with the peak. The angles of elevation were measured from each point. The angle at A was 36°40′ and the angle at B was 21°10′. The distance from A to B was 720 feet. How high is the peak above the level of the plain?

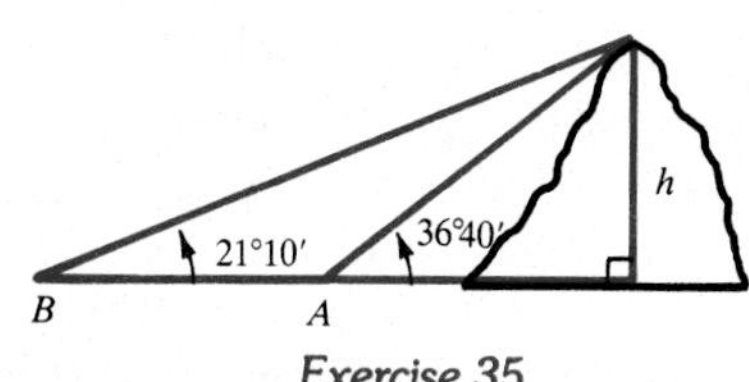

Exercise 35

36. A pendulum 50 centimeters long is moved 40° from the vertical. How far did the tip of the pendulum rise?

37. A building 60 feet tall is on top of a hill. A surveyor is at a point on the hill and observes that the angle of elevation to the top of the building is 42° and to the bottom of the building is 18°. How far is the surveyor from the building?

38. A 40 foot television antenna stands on top of a building. From a point on the ground, the angles of elevation of the top and bottom of the antenna, respectively have measures of 56 and 42. How tall is the building?

39. In $\triangle ABC$, $\angle B$ is obtuse and $m\angle A = 15$. The length of altitude $\overline{CX}$, which is drawn to the line that contains side $\overline{AB}$, is 16 cm. If the length of $\overline{BX}$ is 12 cm, find the perimeter of $\triangle ABC$.

8.8 Trigonometry and Physics

Although the ideas of trigonometry are important in many fields of study, they are particularly useful in understanding physical ideas.

Consider the situation of two children arguing over possession of a toy, as sketched at the right. Our sense of this situation tells us that the forces acting on the toy would tend to propel it toward neither of the children. It would be propelled instead in a direction between them as shown by the red arrow.

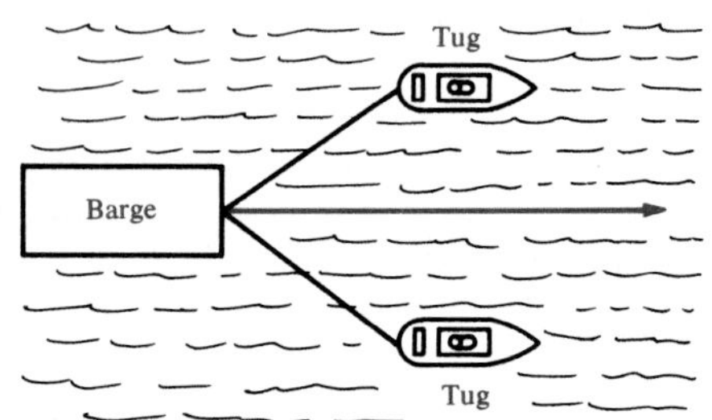

In another case, consider the action of two tugboats pulling a barge down a river, as shown at the left. The barge, of course, does not move toward either of the tugs; rather, it moves in a course down the river.

In both of these cases, the two forces acting on an object result in the *same effect that would be produced by a single force* pulling in the direction of the red arrow. This single force is called the **resultant** of the other two forces.

The effect of a force upon a body depends on two things: its strength, or **magnitude,** and its **direction.** A force is an example of a **vector.** In general, a vector describes a quantity that has both magnitude and direction.

A vector is represented graphically by an **arrow diagram.** The length of the segment, related to a given scale, represents the magnitude of the vector. The direction of the arrow corresponds to the direction of the vector.

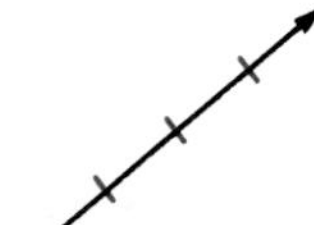

The magnitude of this vector is 400 lb. The direction corresponds to the direction of the arrow.

When two forces act in the same or in exactly opposite directions, it is easy to find the magnitude of the resultant. For example, if the engine of a plane propels it 150 mph due west and a tailwind of 50 mph is blowing due west, then the resultant force is 200 mph.

The examples below consider two forces acting at an angle to each other.

Examples

1 **A plane's engine without a wind would take it 150 mph due east. A wind is blowing southeast at 50 mph. The direction of the wind forms an angle of 45° with the direction of the plane. What is the actual speed and actual direction of the plane?**

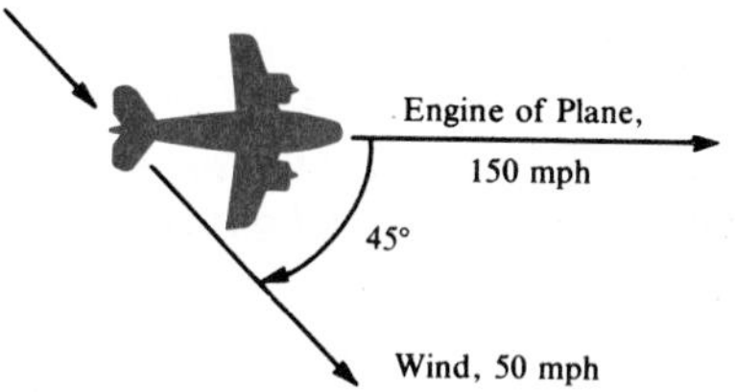

The situation is pictured at the right.

Complete a **parallelogram of forces** as shown at the right.

The gray dashed vector shows the actual direction of the plane.

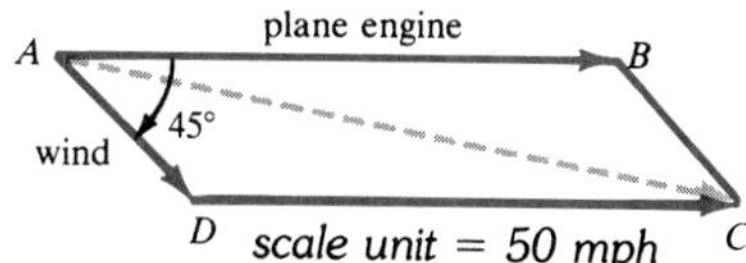

scale unit = *50 mph*

To find the actual speed of the plane, we use the Law of Cosines in $\triangle ABC$. Note that in the parallelogram, $BC = AD$ because opposite sides have the same length. Since the consecutive angles of a parallelogram are supplementary, $m\angle B = 135$.

$$b^2 = a^2 + c^2 - 2ac\cos 135°$$
$$= (50)^2 + (150)^2 - 15000\left(-\frac{\sqrt{2}}{2}\right)$$
$$\approx 2500 + 22500 - 15000(-0.707)$$
$$\approx 25000 + 10605$$
$$\approx 35605$$
$$b \approx 188.69 \text{ (choosing the positive value)}$$

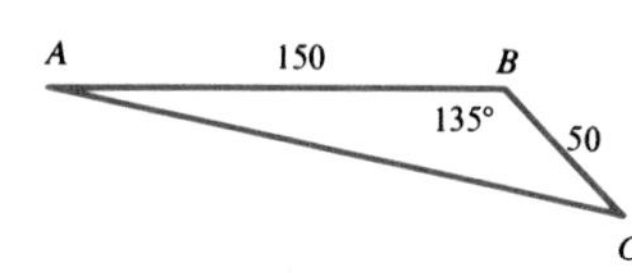

The actual speed of the plane is about 188.7 miles per hour (to the nearest tenth).

2 **In example 1, find $m\angle BAC$ to the nearest ten minutes.**

Either the Law of Sines or the Law of Cosines may be used. In $\triangle ABC$, use the following form of the Law of Sines.

$$\frac{\sin A}{a} = \frac{\sin B}{b}$$
$$\frac{\sin A}{50} = \frac{0.707}{188.69}$$
$$\sin A = \frac{50(0.707)}{188.69}$$
$$\approx 0.1873$$

$\angle BAC$ measures $10°50'$ to the nearest ten minutes.

Exercises

Exploratory

1. A plane flying at 300 mph meets a 50 mph wind in the opposite direction. What are its speed and direction for the same energy output?
2. Answer exercise 1 if the wind is in the same direction as the plane.

Find the resultant of the two forces or velocities in each exercise. The scale is one unit to 50 lb, where applicable. Include a careful diagram. Use protractors and rulers to estimate your answers.

3. **4.** **5.** **6.**

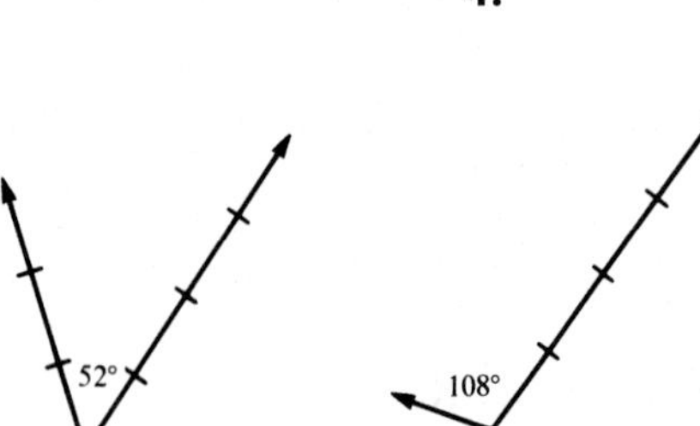

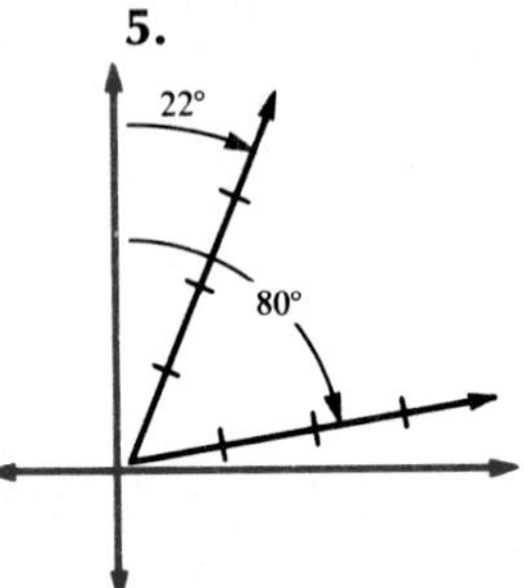

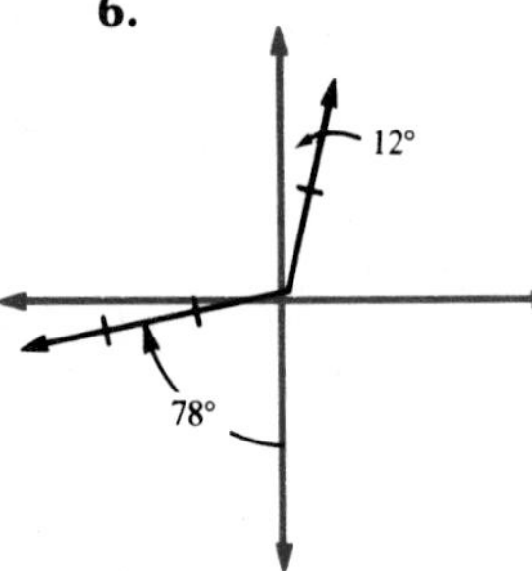

Written

In each exercise, find the magnitude of the resultant force of the two given forces acting on the same point at the given angle. Use the methods of trigonometry and express answers to the nearest unit.

1. **2.** **3.**

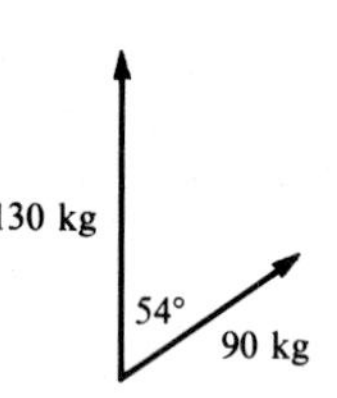

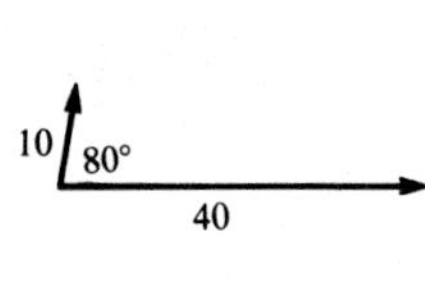

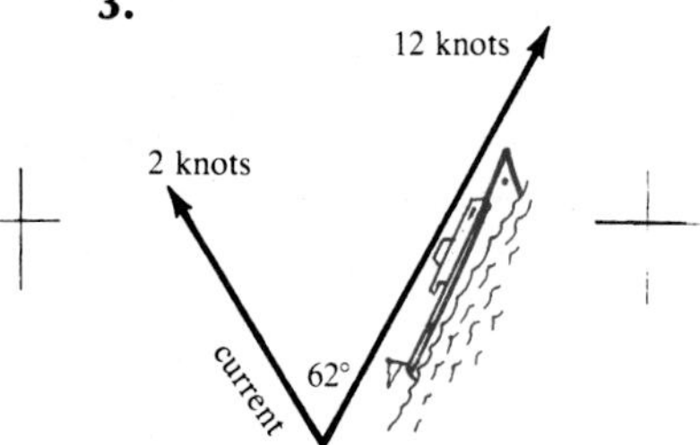

4. **5.** **6.**

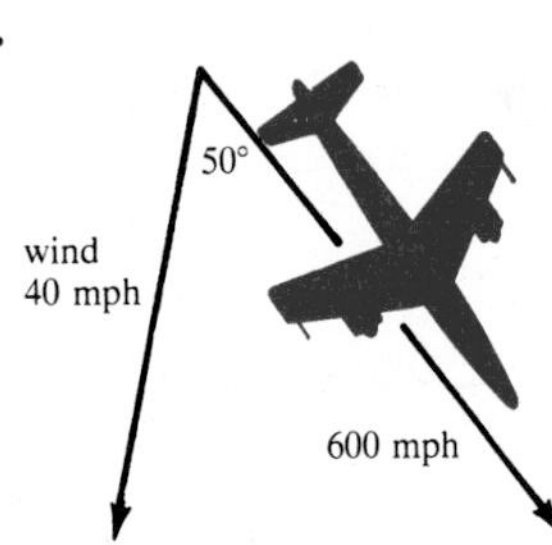

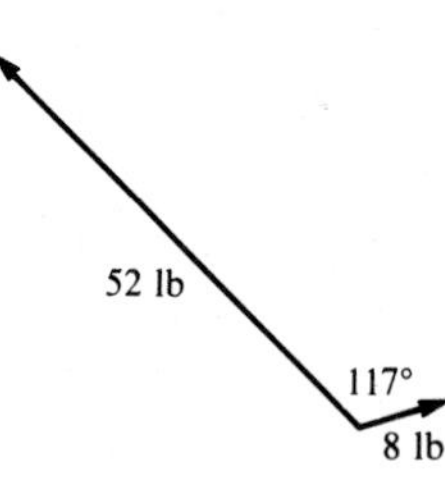

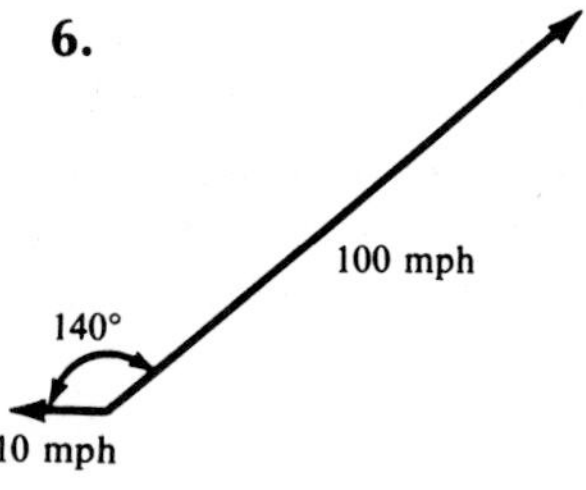

7. **8.** **9.**

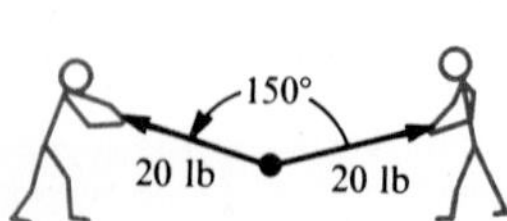

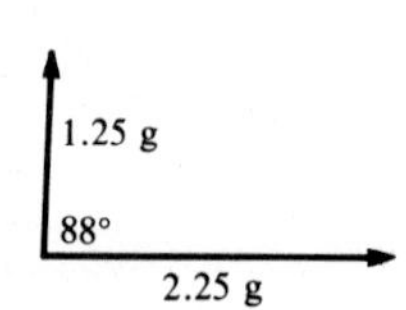

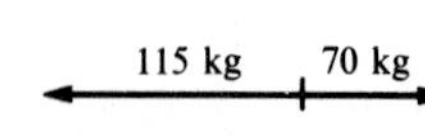

Solve. Find lengths to the nearest tenth and angle measures to the nearest ten minutes.

10. Two forces of 18 lb and 20 lb act on a body at an angle of 120°. Find the magnitude of the resultant force.
11. The resultant of two forces acting on a body has a magnitude of 80 lb. The angles between the resultant and the forces are 20° and 52°. Find the magnitude of the larger force.

Answer exercise 11 if the resultant force and angles have the following measures.

12. 32 lb; 72°, 46°
13. 56 lb; 34°, 63°
14. The resultant of two forces has a magnitude of 60 lb. The angle between the 28-lb force and the resultant measures 50°. Find the magnitude of the other force.

Two forces of the given magnitude act on the same point and are equivalent to a resultant force, as given. Find the angle between the given forces F_1 and F_2 and the angle between the resultant and the larger force.

15. $F_1 = 8$, $F_2 = 10$, $R = 11$
16. $F_1 = 5$, $F_2 = 8.2$, $R = 6.8$
17. $F_1 = 40$, $F_2 = 50$, $R = 60$
18. $F_1 = F_2 = 9\sqrt{2}$, $R = 18$
19. $F_1 = 30{,}000$, $F_2 = 40{,}000$, $R = 50{,}000$
20. $F_1 = 1.12$, $F_2 = 0.86$, $R = 0.90$
21. $F_1 = 8\sqrt{2}$, $F_2 = 7\sqrt{3}$, $R = 12$
22. $F_1 = 120$, $F_2 = 76$, $R = 100$
23. Every year the Pine County Fair opens with the releasing of a hot-air balloon. Last year the balloon rose vertically at a rate of eight feet per second but at the same time was being blown horizontally by a wind of ten feet per second. Find the angle (with the horizontal) at which the balloon rose.
24. What was the distance of the balloon in exercise 23 from the starting point after 90 seconds?
25. The annual Grande Foire de la Ville du Pin begins in the same way. (See exercise 23.) If the balloon rose vertically at 5 meters per second and was blown horizontally with a force of 10 meters per second, at what angle did it rise?

Challenge **Alternate proofs of the Law of Sines and the Law of Cosines appear below. In each case, give reasons for the steps taken, and complete the proof.**

Proof of the Law of Cosines

Case I. $c^2 = h^2 + (a - x)^2$
$= h^2 + a^2 - 2ax + x^2$

Also, $b^2 = h^2 + x^2$.

Hence $c^2 - b^2 = a^2 - 2ax$, or
$c^2 = a^2 + b^2 - 2ax$.

But $x = b \cos C$.

Finally, $c^2 = a^2 + b^2 - 2ab \cos C$.

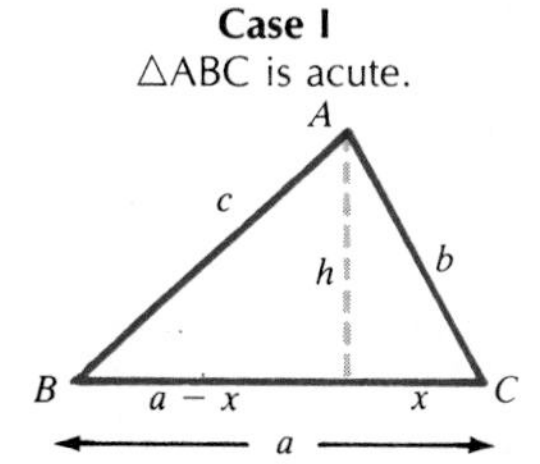

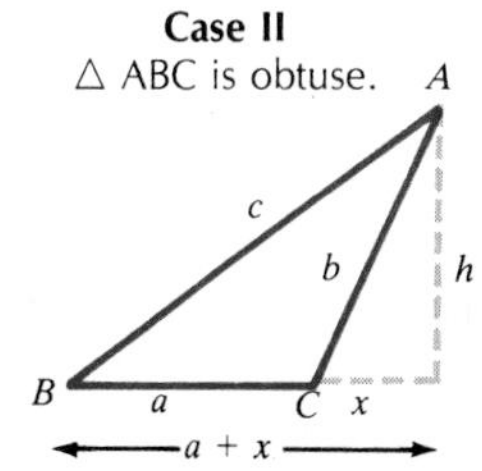

Case II is for you!

Proof of the Law of Sines

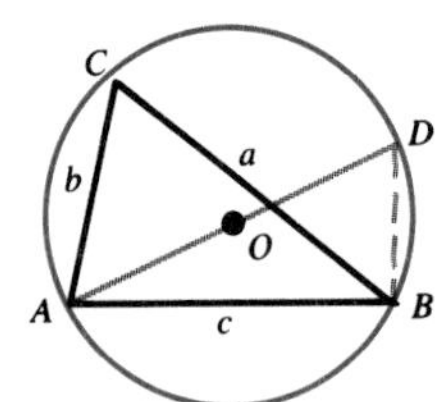

1. Circumscribe circle O about $\triangle ABC$.
2. Draw diameter through A and O intersecting $\odot O$ at D. Let $2r$ be length of diameter.

3. Draw $\overline{DB}$. Then $m\angle C = m\angle D$.
4. Now, $m\angle ABD = 90$. Hence $\sin C = \sin D = \frac{c}{2r}$.
5. Then, $\frac{c}{\sin C} = 2r$.

Complete the proof by showing that $\frac{b}{\sin B} = \frac{a}{\sin A} = 2r$.

Mathematical Excursions

Polar Form of a Complex Number

The coordinates (x, y) of a point on the terminal side of an angle θ in standard position can be written $(r\cos\theta, r\sin\theta)$. Thus,

$$x = r\cos\theta \text{ and } y = r\sin\theta.$$

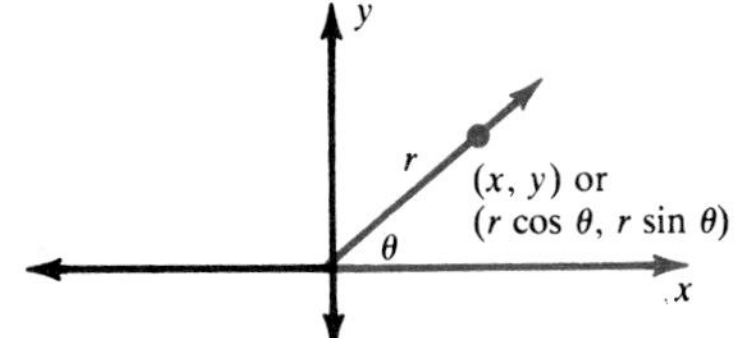

Therefore, the complex number $x + yi$ can be written as follows.

$$\begin{aligned} x + yi &= r\cos\theta + (r\sin\theta)i \\ &= r\cos\theta + i(r\sin\theta) \\ &= r(\cos\theta + i\sin\theta) \end{aligned}$$

Polar Form of a Complex Number

The polar form of the complex number $x + yi$ is $r(\cos\theta + i\sin\theta)$, where r is called the **modulus** and θ is called the **amplitude.**

Polar form and trigonometric form mean the same thing.

Example **Write 3(cos 60° + i sin 60°) in $x + yi$ form.**

$$\begin{aligned} 3(\cos 60^\circ + i\sin 60^\circ) &= 3\left(\frac{1}{2} + i\cdot\frac{\sqrt{3}}{2}\right) \\ &= \frac{3}{2} + \frac{3}{2}i\sqrt{3} \end{aligned}$$

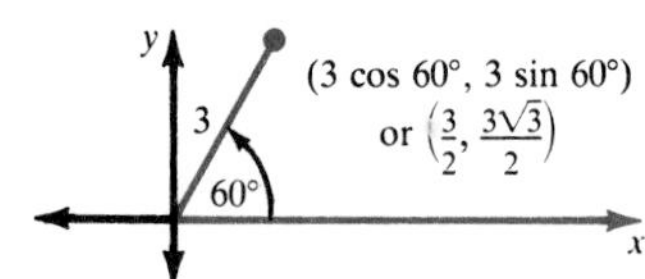

Exercises **Express each complex number in $x + yi$ form.**

1. $2\cos 60^\circ + 2i\sin 60^\circ$
2. $2\left(\cos\frac{3\pi}{2} + i\sin\frac{3\pi}{2}\right)$

Express each complex number in polar form.

3. $1 + i$
4. $2 - 2i$
5. $-2\sqrt{3} - 2i$
6. $8\pi + 15\pi i$

8.9 Additional Applications to Physics

Just as two forces acting on a body concurrently can be replaced by a single force, so can a single force be replaced by two. The two "replacement" forces are called **components** and the process is called **resolution of forces.**

Investigation 1 While Mr. Brullman prepares dinner, Mrs. Brullman is mowing the lawn. Suppose that the handle of the lawn mower makes an angle of 30° with the ground, and that Mrs. Brullman exerts a force of 40 pounds in the direction of the handle. Clearly, though, the lawn mower *does not move in the direction of her force.* It moves horizontally along the ground.

We can think of the force of 40 pounds along the handle as having two components: the forward component and the downward component. The *useful* part is the forward component.

To find the forward and downward components we use a parallelogram of forces and the principles of trigonometry.

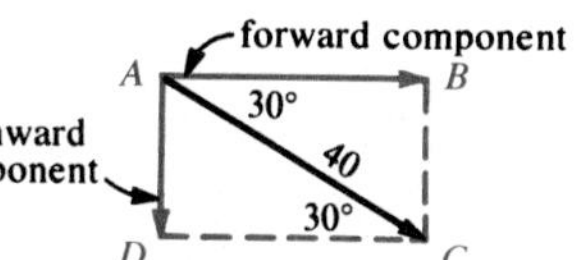

Since $\angle BAC$ and $\angle DCA$ are alternate interior angles of parallel lines and $m\angle DCA = 30$, then $m\angle BAC = 30$.

$$\cos 30° = \frac{AB}{40} \qquad \cos = \frac{adj}{hyp} \text{ in right } \triangle ABC$$

$$\frac{\sqrt{3}}{2} = \frac{AB}{40}$$

$$AB = \frac{40\sqrt{3}}{2} \text{ or } 20\sqrt{3} \approx 34.64$$

The forward component is about 34.64 lb.

For AD, the downward component, note that $AD = BC$.

$$\sin 30° = \frac{BC}{40} \qquad \sin = \frac{opp}{hyp} \text{ in right } \triangle ABC$$

$$\frac{1}{2} = \frac{BC}{40}$$

$$BC = 20$$

The downward component is 20 lb.

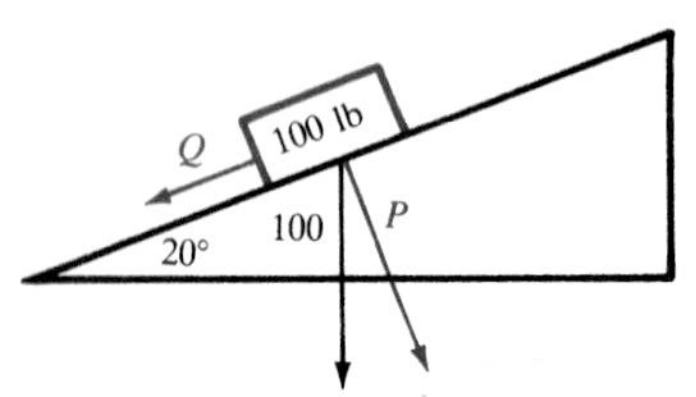

Investigation 2 Suppose that an object weighing 100 pounds is resting on an inclined plane which slopes up at 20° to the horizontal, as shown at the left. The force of 100 pounds acts straight down because of gravity. Obviously, however, the object cannot move *through* the plane. Its hundred-pound force can be considered to be resolved into two perpendicular forces, one. P, acting directly on (at right angles to) the plane, and the other, Q, tending to cause the object to slide down the incline.

To find the forces P and Q, proceed as follows.

Consider the parallelogram of forces, shown at the left, with the 100-pound force to be resolved into forces P and Q.

Now, $m\angle CAD$ must be 20. (Why?) Therefore,

$$P = 100 \cos 20° \quad \text{and} \quad Q = 100 \sin 20°.$$

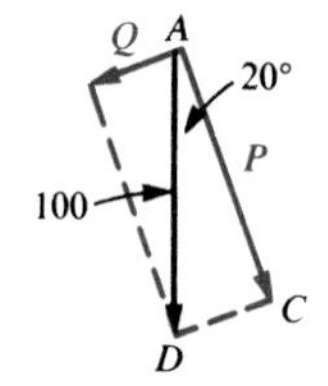

The force P acting at right angles to the plane is $100 \cos 20° = 100(0.9397)$ or 93.97 pounds. The force Q tending to cause the object to slide down the plane is $100 \sin 20° = 100(0.3420)$ or 34.20 pounds.

If the object does not slide down the plane, then a force *equal in magnitude*, but *opposite in direction*, to the force pulling it down the plane must be acting on the body. This force which keeps the object from falling is called a **friction force.** Often we encounter problems where the friction along the slope or other surface can be disregarded. In such a case we say that we have a *smooth* slope.

Exercises

Exploratory **In Investigation 1, find the forward and downward components for each change in the data.**

1. Lawn mower handle makes angle 40° with ground
2. Lawn mower handle makes angle 24° with ground
3. Force on handle is 42 lb

In exercises 4–7, resolve each of the forces into horizontal and vertical components.

4.

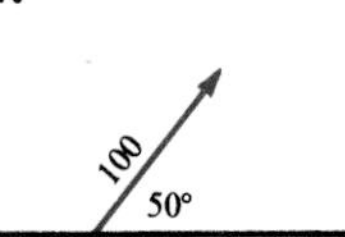

5.

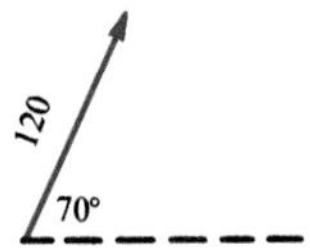

6.

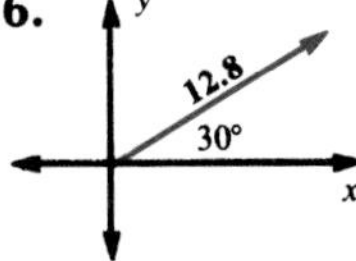

7.

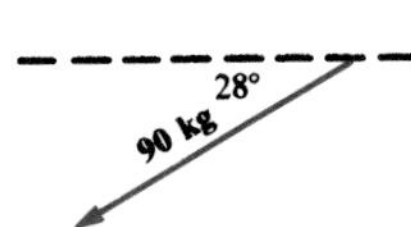

8. Clearly explain why $m\angle CAD = 20$ in Investigation 2.

Written **In exercises 1–6, the slope of an inclined plane and the weight of an object on the inclined plane are given. Find, to the nearest unit, the force acting at a right angle to the inclined plane and the force required to keep the object from sliding down the plane.**

1. 16°, 120 lb
2. 8°, 1000 kg
3. 22°, 40 kg
4. 14°, 3.2 tons
5. 25°30′, 32 kg
6. 32°10′, 480 lb

7. A disabled snowmobile can be pulled by a force of 120 lb parallel to the ground. What force must be applied at an angle of 25° to the ground?

8. A block of ice weighing 300 lb is held on the ice slide at the Pine City Ski Resort by a rope parallel to the slide. If the slide slopes upward at an angle of 22°, find the pull on the rope and the force on the ice slide.

9. What is the minimum force required to pull a block of ice weighing 525 lb on the Pine City ice slide (exercise 8)?

10. Consider the barge pulled by two tugs, as shown. Suppose that it requires 6000 lb of force to pull the barge in the indicated direction and that the angles made by the ropes attaching the tugs to the barge are θ_1 and θ_2. Find F_1 and F_2 if $\theta_1 = 40°$ and $\theta_2 = 42°$.

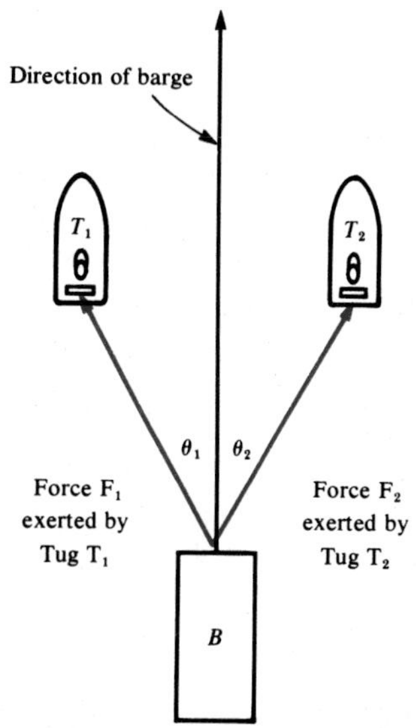

Answer exercise 10 for the following information.

	Force required to pull barge in correct direction	θ_1	θ_2
11.	6000 lb	20°	22°
12.	8000 lb	20°	30°
13.	7500 lb	30°	42°
14.	9800 lb	28°	32°

Mathematical Excursions

Using Trigonometric Coordinates

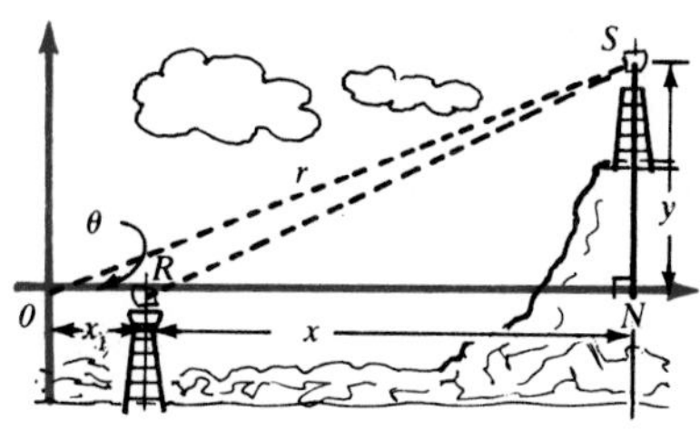

The figure at the left shows the location of a transmitting tower and a receiving antenna. Point S represents the point from which the signal is emitted. Point R represents the location for the receiving antenna. Point O is a fixed reference point, r units from S.

Finding the distance from S to R helps determine the strength of the signal between S and R. As this distance increases, the strength of the signal diminishes.

Exercises

1. Use the Pythagorean Theorem and trigonometric coordinates to derive a formula for the distance from S to R. *Hint: Trigonometric coordinates of S are $(r \cos \theta, r \sin \theta)$.*

Use the formula derived in exercise 1 to solve each problem.

2. Find the distance from R to S if $\theta = 60°$, $\overline{OS}$ is 28 km long, and $\overline{OR}$ is 7 km long.

3. Find θ if $\overline{OS}$ is 30 km long, $\overline{OR}$ is 11 km long, and $\overline{RS}$ is 25 km long.

Problem Solving Application: Using Trigonometry

The Law of Sines, Law of Cosines, or right triangle trigonometry can be applied to solve various types of problems involving triangles.

Example

1 **A pilot, flying from Olympia, Washington to Albany, New York, encounters a large area of severe thunderstorm activity. To avoid this area, she changed course by turning 40° counterclockwise. She flies on this course for 75 miles, and then turns 60° clockwise. She then flies on this course until she can turn back onto her original course. How many additional miles were added to the flight?**

Explore Draw a diagram to represent this situation.

Since m ∠ A = 40 and m ∠ B = 120, m ∠ C = 20.

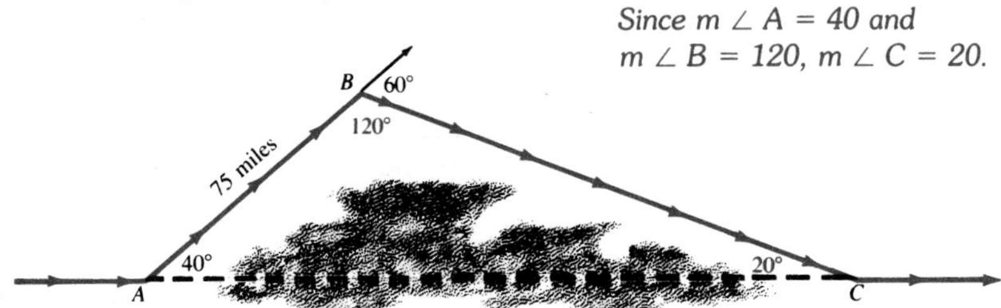

Plan In order to determine the number of additional miles flown, we need to determine the lengths of *AC* and *BC*. Since the measure of all three angles and one side are known, we can apply the Law of Sines.

Solve

$$\frac{AC}{\sin B} = \frac{AB}{\sin C} \qquad \frac{BC}{\sin A} = \frac{AB}{\sin C}$$

$$AC = \frac{AB \sin B}{\sin C} \qquad BC = \frac{AB \sin A}{\sin C}$$

$$AC = \frac{75 \sin 120°}{\sin 20°} \qquad BC = \frac{75 \sin 40°}{\sin 20°}$$

$$AC = \frac{75(0.8660)}{0.3420} \qquad BC = \frac{75(0.6428)}{0.3420}$$

$$AC \approx 190 \qquad BC \approx 141$$

Use trig tables or a calculator to find sin A, sin B, and sin C.

Thus, the pilot flew approximately (141 + 75) − 190 or 26 additional miles in order to avoid the area of thunderstorms.

Examine Since the sides and angles of the triangle have the proper relationships (smallest side opposite smallest angle, largest side opposite largest angle, and so forth), the answer seems reasonable.

Exercises

Written **Solve each problem. Round all lengths to the nearest unit and all angle measures to the nearest degree.**

1. Solve the problem in example 1 if the pilot turns 30° counterclockwise, flies for 100 miles, and then turns 50° clockwise and flies until she can turn back onto her original course.

2. A tree is broken by the wind. The top of the tree touches the ground 15 meters from its base and makes an angle with the ground measuring 32°. How tall was the tree before it was broken?

3. A regular pentagon is inscribed in a circle with radius 15 cm. What is the perimeter of this pentagon? What is the area of the region inside the circle but outside the pentagon?

4. A pilot is flying from Buffalo to Miami, a distance of 1400 miles. He starts his flight 12° off course and flies for 250 miles. How far is he from Miami and by how much should he correct his error?

5. A baseball diamond is a square with sides that are 90 feet long. The pitcher's mound is 60.5 feet from home plate on the diagonal connecting second base and home plate. How far is the pitcher's mound from second base and from first base?

6. The Leaning Tower of Pisa makes an angle with the ground measuring 82°. The angle of elevation of the top of the tower to a point on the ground 28 meters from the base of the tower is 42°. What is the height of the Leaning Tower of Pisa?

7. A ship sailed for 75 kilometers. It then turned 30° clockwise and sailed for 60 kilometers. Finally, it turned 20° clockwise and sailed 50 kilometers. How far is the ship from its starting point?

8. From the roof of Leshlock Tower, the angle of elevation to the top of Beck Tower is 51°, and the angle of depression to the base of Beck Tower is 47°. If the height of Leshlock Tower is 20 meters, find the height of Beck Tower.

9. A wire is strung from the base of one antenna to the top of a shorter antenna and then back to the top of the first antenna. The angle of elevation from the top of the shorter antenna to the top of the other antenna is 38°. How long is the wire if the shorter antenna is 6 meters high and the antennas are 10 meters apart?

10. A ladder is braced by a 12-foot rod placed at a right angle to the ladder. One end of the rod is 5 feet up from the base of the ladder. The other end is 7 feet from the base of the wall on which the ladder is resting. How far is the top of the ladder from the base of the wall? What is the measure of the angle formed by the ladder and the wall?

11. In order to find the height of chimney CT, the angle of elevation of the top T is measured by means of a transit from point A, whose distance from the chimney is not known. Then the transit is turned through a horizontal angle of 90° and point B is located. At B, the angle of elevation of the top T is measured again. Find the height of the chimney if $m\angle CAT = 37$, $m\angle CBT = 24$, and $\overline{AB}$ is 75 meters long.

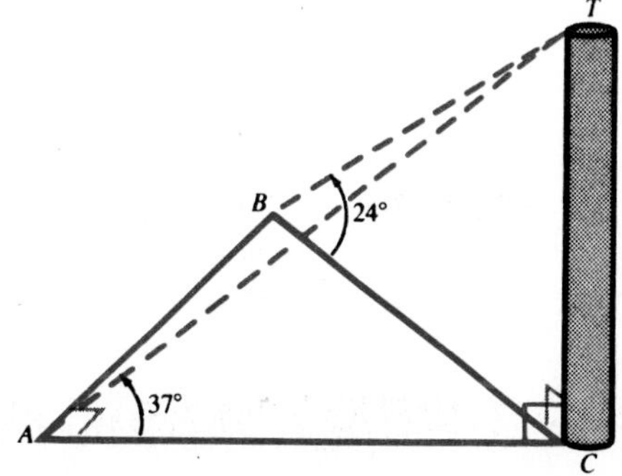

Vocabulary

trigonometric form	magnitude	arrow diagram
segment of the circle	direction	components
solving the triangle	resultant	resolution of forces
angle of elevation	vector	friction force
angle of depression		

Chapter Summary

1. If P is a point r units from the origin and located on the terminal side of an angle θ in standard position, then the coordinates of P expressed in trigonometric form are $(r \cos \theta, r \sin \theta)$.
2. The Law of Cosines can be expressed in the following three forms.
$$a^2 = b^2 + c^2 - 2bc \cos A$$
$$b^2 = a^2 + c^2 - 2ac \cos B$$
$$c^2 = a^2 + b^2 - 2ab \cos C$$
3. The area of a triangle is equal to one half the product of the lengths of two sides and the sine of the included angle. It is symbolized by the letter **A.** An example is $\mathbf{A} = \frac{1}{2}bc \sin A$.
4. The Law of Sines states that in any triangle, the sines of the angles are proportional, respectively, to the lengths of the sides opposite the angles.
For example, in $\triangle ABC, \frac{\sin A}{a} = \frac{\sin B}{b} = \frac{\sin C}{c}$.
5. An ambiguous case occurs when the measures of two sides and an angle not included between them are given (*SSA*).
6. If A is an acute angle in a right triangle, the following trigonometric ratios hold.
$$\sin A = \frac{\text{side opposite}}{\text{hypotenuse}}$$
$$\cos A = \frac{\text{side adjacent}}{\text{hypotenuse}}$$
$$\tan A = \frac{\text{side opposite}}{\text{side adjacent}}$$
7. A vector is represented graphically by an arrow diagram. The length of the segment represents the magnitude of the vector. The direction of the arrow corresponds to the direction of the vector.
8. Two forces acting at an angle to each other result in a single force called the resultant.
9. One force can be resolved into two component forces.

Chapter Review

8.1 **Find $\sin\theta$, $\cos\theta$, and $\tan\theta$ if the terminal side of θ contains the following points.**

1. $(0, 4)$ **2.** $(5, -12)$ **3.** $(-3, -4)$ **4.** $(-1, \sqrt{3})$

5. If $\sin\theta = -\frac{2}{3}$ and $\cos\theta > 0$, find $\cot\theta$.

8.2 **6.** In $\triangle RST$, $m\angle R = 110$, $t = 18$ and $s = 20$, find r to the nearest tenth.

7. The diagonals of a parallelogram intersect at an angle of 150°. If they measure 12 cm and 18 cm, find to the nearest tenth the length of the longer side of the parallelogram.

8. Find, to the nearest degree, the measure of a base angle of an isosceles triangle whose sides measure 20, 20, and 14 units.

8.3 **Find the exact area of $\triangle ABC$.**

9. $\sin C = 0.75$, $a = 8$, $b = 6$ **10.** $m\angle B = 60$, $a = 8$, $c = 10$

11. $m\angle A = 120$, $b = 8\sqrt{3}$, $c = 6$

8.4 **12.** Two angles of a triangle measure 42° and 64°. If the shortest side is 68 cm, find to the nearest tenth the length of the longest side.

8.5 **Determine the number of triangles that can be constructed with the given measures.**

13. $m\angle A = 30$, $a = 5$, $b = 6$ **14.** $m\angle A = 45$, $a = 5\sqrt{2}$, $b = 8$

15. $m\angle R = 40$, $t = 12$, $r = 7$

8.6 **Solve each triangle. Find lengths to the nearest tenth and angle measures to the nearest degree.**

16. $m\angle A = 34$, $b = 12$, $c = 10$ **17.** $a = 20$, $b = 15$, $c = 30$

18. $m\angle A = 52$, $a = 31$, $c = 38$

8.7 **19.** From a point on the ground 50 meters from the base of a flagpole, the angle of elevation of the top measures 48°. Find the height of the pole to the nearest meter.

20. A pilot 3000 feet above the ocean notes that the angle of depression of a ship is 42°. Find, to the nearest foot, the distance of the plane to the ship.

21. The base of a monument and two points all lie in a horizontal line. The two points, both due north of the monument, are 50 meters apart. The angles of elevation to the top of the monument are 45° and 25°. Find, to the nearest meter, the height of the monument.

8.8 **22.** Two forces act on a body at an angle of 82°. If the forces measure 26 lb and 32 lb, find to the nearest pound the magnitude of the resultant.

23. Find, to the nearest degree, the measure of the angle between a force of 5 lb and a force of 12 lb if the resultant has magnitude 10 lb.

8.9 **24.** A wagon can be pulled by a force of 85 lb parallel to the ground. What force must be applied at an angle of 20° to the ground?

Chapter Test

1. In $\triangle ABC$, if $a = 5$, $\sin A = \frac{1}{8}$, and $\sin B = \frac{1}{4}$, find the value of b.
2. In $\triangle ABC$, if $a = 3$, $b = 4$, and $m\angle C = 60$, find the value of c.
3. In $\triangle ABC$, if $a = 5$, $b = 7$, and $c = 6$, find $\cos B$.
4. In $\triangle RST$, if $m\angle R = 30$, $s = 12$, and $r = 10$, find $\sin S$.
5. If a tree 35 feet high casts a shadow 45 feet long, find to the nearest degree the angle of elevation of the sun.
6. In $\triangle ABC$, $a = 6$, $b = 5$, and $\cos C = \frac{3}{4}$. Find the exact value of c.
7. If $\tan \theta = -\frac{\sqrt{6}}{3}$ and $\sin \theta < 0$, find $\cos \theta$ in simplest radical form.
8. If the terminal side of θ contains the point $(3, -5)$, find $\sin \theta$.
9. Find the area of $\triangle PQR$ if $m\angle P = 150$, $q = 24$, and $r = 10$.
10. If the area of $\triangle ABC$ is 80 square units $b = 40$, and $m\angle C = 30$, find a.

Determine the number of possible solutions. If a solution exists, solve the triangle. Find lengths to the nearest tenth and angle measures to the nearest degree.

11. $m\angle A = 65$, $b = 21$, $a = 6$
12. $a = 13$, $b = 11$, $c = 17$
13. $m\angle A = 46$, $m\angle B = 77$, $a = 6$
14. $m\angle A = 44$, $a = 12$, $b = 14$
15. A 32-foot ladder leans against a building. The top touches the building 26 feet above the ground. Find, to the nearest degree, the measure of the angle formed by the ladder with the ground.

The lengths of two sides of a triangle are 10 cm and 24 cm and the included angle is 32°20′.

16. Find, to the nearest centimeter, the length of the third side.
17. Find, to the nearest square centimeter, the area of the triangle.

Two adjacent sides of a parallelogram measure 8 cm and 10 cm. The length of the longer diagonal is 14 cm.

18. Find, to the nearest degree, the measure of the largest angle of the parallelogram.
19. Find, to the nearest square centimeter, the area of the parallelogram.

In trapezoid $ABCD$, $\overline{AB}$ and $\overline{DC}$ are parallel bases and $\overline{AD} \perp \overline{DC}$, $m\angle C = 128$, $m\angle DAC = 42$, and $AD = 26$.

20. Find AB to the nearest whole number.

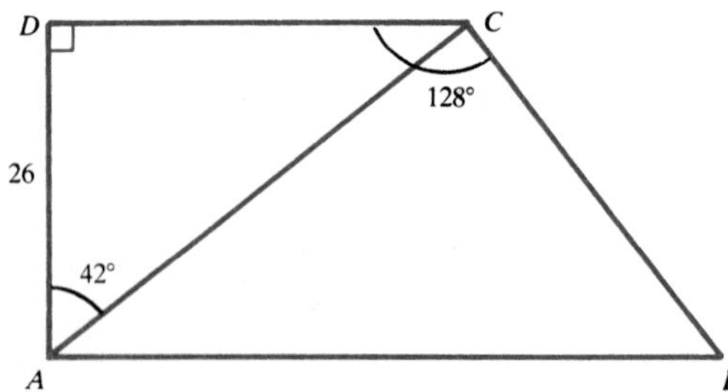

Trigonometric Identities and Equations

Chapter 9

It has been found that musical sounds are made of precise patterns of waves. These vibrational waves can be described by trigonometric functions. These functions are used in physics to study items such as a tuning fork. The equations used in these studies are often simplified by using trigonometric identities.

9.1 Trigonometric Identities

Frequently, you have come across open sentences that are true for any replacement of the variable. Some examples are

$$2x + x = 3x \quad \text{and} \quad x^2 - 1 = (x + 1)(x - 1).$$

Such open sentences are called **identities.**

The following equations are identities since they are true for all replacements of the variable for which the expression is defined.

Basic Trigonometric Identities

$$\tan x = \frac{\sin x}{\cos x} \qquad \sec x = \frac{1}{\cos x}$$

$$\cot x = \frac{\cos x}{\sin x} \qquad \csc x = \frac{1}{\sin x}$$

To this list, add the following all-important identity.

$$\sin^2 x + \cos^2 x = 1$$

This identity follows immediately from the fact that on the unit circle, the coordinates of *every* point satisfy $x^2 + y^2 = 1$.

Two other important identities are

$$\tan^2 x + 1 = \sec^2 x \quad \text{and} \quad \cot^2 x + 1 = \csc^2 x.$$

A proof of the first identity is shown below. You will be asked to prove the second identity in the exercises.

Proof of $\tan^2 x + 1 = \sec^2 x$

$$\tan^2 x + 1 = \left(\frac{\sin x}{\cos x}\right)^2 + 1 \qquad \textit{Recall } \tan x = \frac{\sin x}{\cos x}.$$

$$= \frac{\sin^2 x}{\cos^2 x} + 1$$

$$= \frac{\sin^2 x}{\cos^2 x} + \frac{\cos^2 x}{\cos^2 x} \qquad \textit{Use } \cos^2 x \textit{ as a common denominator.}$$

$$= \frac{\sin^2 x + \cos^2 x}{\cos^2 x}$$

$$= \frac{1}{\cos^2 x} \qquad \textit{Use } \sin^2 x + \cos^2 x = 1.$$

$$= \sec^2 x \qquad \textit{Use } \frac{1}{\cos x} = \sec x.$$

The identities listed above are called the *basic trigonometric identities.* The last three are called *Pythagorean identities.* Trigonometric expressions can often be simplified by using the basic identities. These basic identities can also be used to prove more complex identities.

Examples

1 **Express $\cos x \csc x \tan x$ in simplest form containing one circular function or a constant.**

Look at each factor of the expression in relation to the basic identities. Choose substitutions that will best simplify the expression.

$$\cos x \csc x \tan x = \cos x \cdot \frac{1}{\cancel{\sin x}} \cdot \frac{\cancel{\sin x}}{\cos x}$$

$$= \cancel{\cos x} \cdot \frac{1}{\cancel{\cos x}}$$

$$= 1$$

Remember that this is an identity only for those values of x for which the expression is defined.

2 **Prove that $\tan x + \cot x = \csc x \sec x$.**

$$\tan x + \cot x = \frac{\sin x}{\cos x} + \frac{\cos x}{\sin x}$$

$$= \frac{\sin^2 x}{\sin x \cos x} + \frac{\cos^2 x}{\sin x \cos x}$$ *Rewrite with a common denominator.*

$$= \frac{\sin^2 x + \cos^2 x}{\sin x \cos x}$$

$$= \frac{1}{\sin x \cos x}$$ *$\sin^2 x + \cos^2 x = 1$*

$$= \frac{1}{\sin x} \cdot \frac{1}{\cos x} \quad \text{or} \quad \csc x \sec x$$

3 **Prove that $\frac{1 - \cos \alpha}{\sin \alpha} = \frac{\sin \alpha}{1 + \cos \alpha}$.**

Begin by simplifying the right side.

$$\frac{1 - \cos \alpha}{\sin \alpha} \stackrel{?}{=} \frac{\sin \alpha}{1 + \cos \alpha}$$

$$\stackrel{?}{=} \frac{\sin \alpha (1 - \cos \alpha)}{(1 + \cos \alpha)(1 - \cos \alpha)}$$ *Multiply numerator and denominator by $1 - \cos \alpha$.*

$$\stackrel{?}{=} \frac{\sin \alpha (1 - \cos \alpha)}{1 - \cos^2 \alpha}$$ *Since $\sin^2 \alpha + \cos^2 \alpha = 1$, $\sin^2 \alpha = 1 - \cos^2 \alpha$.*

$$\stackrel{?}{=} \frac{\sin \alpha (1 - \cos \alpha)}{\sin^2 \alpha}$$

$$\frac{1 - \cos \alpha}{\sin \alpha} = \frac{1 - \cos \alpha}{\sin \alpha}$$

Note that we could also have begun by simplifying the left side.

Exercises

Exploratory **Answer each of the following.**

1. What is an identity?
2. State the three Pythagorean Identities.
3. $1 - \cos^2 x = ?$
4. $1 - \sin^2 x = ?$
5. $\csc^2 x - 1 = ?$
6. $\sec^2 x - 1 = ?$
7. $\tan^2 x - \sec^2 x = ?$
8. $\cot^2 x - \csc^2 x = ?$
9. Show that $\sin x + \cos x = 1$ is not an identity.
10. Show that $\cot^2 x + 1 = \csc^2 x$ is an identity.

Written **Use the basic identities to find each value. Express your answers in simplest form.**

1. $\cos x$, if $\sin x = \frac{1}{4}$
2. $\cos x$, if $\sin x = \frac{2}{3}$
3. $\sec x$, if $\tan^2 x = \frac{1}{2}$
4. $\csc^2 x$, if $\cot^2 x = 0.36$
5. $\tan x$, if $\sec x = 2.5$
6. $\csc x$, if $\cot x = 2$

Express each of the following as a single term containing one circular function or constant.

7. $\sin^2 x - 1$
8. $\sin x \cot x$
9. $\sin \theta \csc \theta \cot \theta$
10. $\cos x \tan x \csc x$
11. $2 - 2\cos^2 \theta$
12. $\sec \theta \cot \theta$
13. $\sin \theta \csc \theta \sec \theta$
14. $5 \tan x \cos x$
15. $\sin x \sec x \cot x$
16. $\cos x\,(1 + \tan^2 x)$
17. $\cos \theta \sec \theta - \cos^2 \theta$
18. $\cos^2 \theta \csc \theta \sec \theta$
19. $\cos^2 \theta\,(\cot^2 \theta + 1)$
20. $\sin^2 x\,(1 + \cot^2 x)$
21. $(\sec x - \tan x)(1 + \sin x)$
22. $\dfrac{\cos \theta}{\sec \theta}$
23. $\dfrac{\sec \theta}{\tan \theta}$
24. $\dfrac{\csc \theta}{\cot \theta}$
25. $\dfrac{\cos \theta \tan \theta}{\sin \theta \sec \theta}$
26. $\dfrac{\cos^2 \theta - 1}{\sin \theta}$
27. $\dfrac{\sin^2 \theta - 1}{\cos \theta}$
28. $\dfrac{5}{\cos^2 \theta} - 5$
29. $\dfrac{\csc \theta - \sin \theta}{\cot \theta}$
30. $\dfrac{\sec \theta - \cos \theta}{\tan \theta}$
31. $\dfrac{\sin^2 \theta - \cos^2 \theta + 1}{2 \sin \theta \cos \theta}$
32. $\dfrac{\tan^2 x - \sin^2 x}{\sin^2 x}$
33. $\dfrac{\cot^2 x - \cos^2 x}{\cos^2 x}$

Prove that each of the following equations is an identity.

34. $\tan x \cos x = \sin x$
35. $\tan x \csc x = \sec x$
36. $\cot x \sin x = \cos x$
37. $\cos x \csc x = \cot x$
38. $\dfrac{\sin^2 x + \cos^2 x}{\cos x} = \sec x$
39. $\dfrac{\sin y + 1}{\sin y} = \csc y + 1$
40. $\cot \theta = \dfrac{\csc \theta}{\sec \theta}$
41. $\dfrac{\tan \beta}{\sec \beta} = \dfrac{\cos \beta}{\cot \beta}$
42. $\sin \theta + \cot \theta \cos \theta = \csc \theta$
43. $\dfrac{\sin y - 1}{\cos y} = \tan y - \sec y$
44. $\dfrac{\csc^2 x - 1}{\cos x} = \csc x \cot x$
45. $1 - \cos^2 x = \sin x \cos x \tan x$
46. $(\sin x - \cos x)^2 = 1 - 2 \sin x \cos x$
47. $\cos^2 t + \sin^2 t = \tan t \cot t$
48. $\cos^2 x = \sec^2 x - \tan^2 x - \sin^2 x$
49. $\cos^2 t(\sec^2 t - 1) = \sin^2 t$
50. $(\cot x + \tan x)^2 = \csc^2 x + \sec^2 x$

51. $\sec \beta \csc \beta = \dfrac{\sec^2 \beta}{\tan \beta}$

52. $\dfrac{\tan x}{1 - \cos^2 x} = \csc x \sec x$

53. $\sec x + 1 = \dfrac{\tan^2 x}{\sec x - 1}$

54. $\dfrac{\cot \theta + \tan \theta}{\csc^2 \theta} = \tan \theta$

55. $\dfrac{1 - \sin \theta}{\cos \theta} = \dfrac{\cos \theta}{1 + \sin \theta}$

56. $\dfrac{\tan \theta - \sin \theta \cos \theta}{\sin^2 \theta} = \tan \theta$

57. $1 - \cos C = 1 - \sin C \cot C$

58. $\cot^2 \theta + \sin^2 \theta = \csc^2 \theta - \cos^2 \theta$

59. $\dfrac{\sec \theta}{\sin \theta} - \dfrac{\sin \theta}{\cos \theta} = \cos \theta \csc \theta$

60. $\csc \theta - \dfrac{\sin \theta}{1 + \cos \theta} = \cot \theta$

61. $\sec t + \tan t = \dfrac{\cos t}{1 - \sin t}$

62. $\dfrac{\sin t}{1 + \cos t} = \csc t - \cot t$

63. $\dfrac{(\cos x + \sin x)^2}{1 + 2 \sin x \cos x} = \cos x \tan x \csc x$

64. $\dfrac{\tan \theta - \sin \theta}{\sec \theta} = \dfrac{\sin^3 \theta}{1 + \cos \theta}$

65. $\dfrac{\cot^2 \beta}{1 + \cot^2 \beta} = 1 - \sin^2 \beta$

66. $\cot x \sec x = \dfrac{1}{2}\left(\dfrac{\sin x}{1 + \cos x} + \dfrac{1 + \cos x}{\sin x}\right)$

67. $\dfrac{1 - \sin \beta}{1 + \sin \beta} = 2 \tan^2 \beta - 2 \sec \beta \tan \beta + 1$

68. $4 \tan x \sec x = \dfrac{1 + \sin x}{1 - \sin x} - \dfrac{1 - \sin x}{1 + \sin x}$

Mixed Review

For exercises 1–13, refer to circle O at the right. If $m\widehat{AB} = 58$, $m\angle YZX = 45$, $m\angle ARF = 10$, $m\angle CPD = 40$, and $m\widehat{CD} = 56$, find each measure.

1. $m\widehat{DE}$
2. $m\angle EFC$
3. $m\widehat{AF}$
4. $m\widehat{BC}$
5. $m\angle SCF$
6. $m\angle EQF$

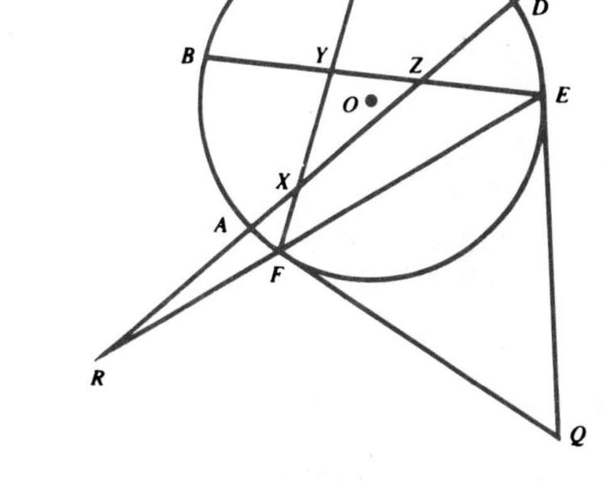

If $CP = 18$, $DP = 9$, $DZ = 10$, $RF = 13$, $EF = AR + 11$, $BY = EZ$, $YZ = \frac{1}{2}BY - 2$, and $EZ:QF = 2:3$, find each measure.

7. AD
8. AR
9. EF
10. EZ
11. BZ
12. QE

13. In circle O, the length of $\overline{CF}$ is 24 cm. If $\overline{CF}$ is 3.5 cm from O, what is the radius of circle O?

14. Two airplanes, flying to different cities, leave an airport at the same time. One flies at 110 miles per hour and the other at 125 miles per hour. If their flight paths form an 80° angle, how far apart are the planes after 3 hours?

15. A surveyor is 370 meters from the base of a mountain peak. The height of this peak is 3920 meters. If the surveyor's instrument is 1.7 meters above the ground, find to the nearest degree the angle of elevation to the top of the peak.

16. Towns X and Y are located on opposite sides of a valley. Tower T is 60 kilometers from town X. What is the width of the valley if $m\angle YXT = 108$ and $m\angle YTX = 35$?

9.2 Solving Trigonometric Equations

You have had a good deal of experience in solving "regular" algebraic equations like $2x = 1$, which is a linear equation, or $x^2 - 3x - 2 = 0$, which is a quadratic equation. In this section, we will consider linear and quadratic equations where the *variable is a trigonometric expression*. These trigonometric equations may be solved by the same methods used to solve other algebraic equations.

Examples

1 Find all values of θ in the interval $0° \leq \theta < 360°$ that satisfy the equation $2 \cos \theta = 1$.

This equation is similar to the equation $2x = 1$. Begin by dividing both sides of the equation by 2. This produces $\cos \theta = \frac{1}{2}$. Consulting the graph below, we see that there are two values in the given interval that satisfy $\cos \theta = \frac{1}{2}$. These are $\theta = 60°$ and $\theta = 300°$.

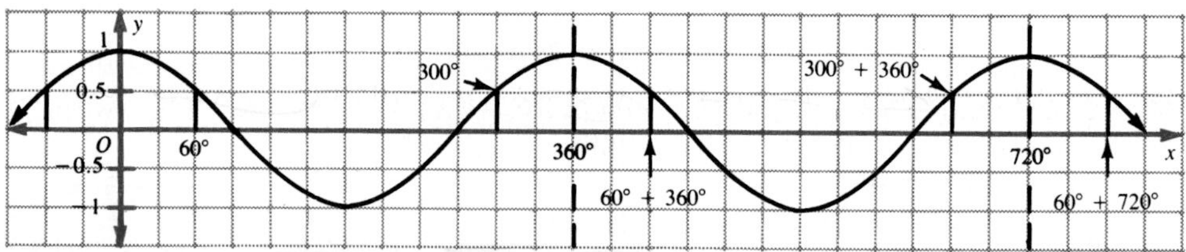

2 Find *all* values of θ satisfying $2 \cos \theta = 1$.

The results from example 1 indicate that $\cos \theta = \frac{1}{2}$ when $\theta = 60°$ or when $\theta = 300°$. Since $\cos \theta$ is a cyclic function, the value of $\frac{1}{2}$ occurs every 360°. Therefore, $\cos \theta = \frac{1}{2}$ at 60°, 60° ± 360°, 60° ± 720°, and so on. Likewise, $\cos \theta = \frac{1}{2}$ at 300°, 300° ± 360°, 300° ± 720°, and so on. All of these solutions can be expressed as $\theta = 60° + 360°n$ or $\theta = 300° + 360°n$ for $n \in \mathbb{Z}$. In radian measure, these can be expressed as $\theta = \frac{\pi}{3} + 2\pi n$ and $\theta = \frac{5\pi}{3} + 2\pi n$, for $n \in \mathbb{Z}$.

3 Find all values of θ (to the nearest degree) in the interval $0° \leq \theta < 360°$ that satisfy the equation $3 \cos \theta = 1$.

$$3 \cos \theta = 1$$
$$\cos \theta = \tfrac{1}{3} \approx 0.3333$$

Using the table on page 618 and the sketch of the cosine curve, we have $\theta \approx 71°$ or $\theta \approx 360° - 71°$ or 289°.

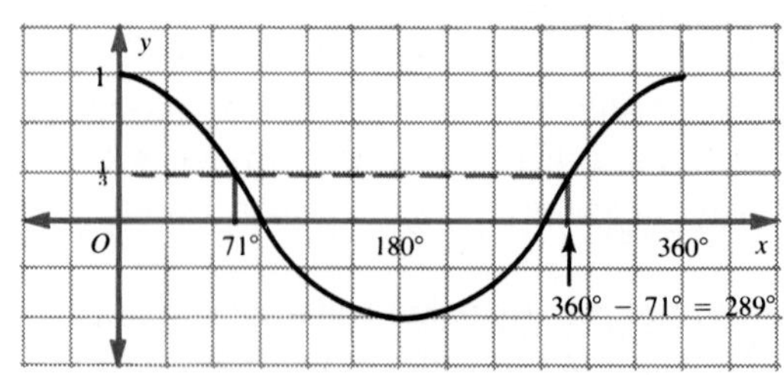

4 **Find *all* values of x satisfying $2 \sin x + 5 = 10 - \sin x$.**

$$2 \sin x + 5 = 10 - \sin x$$
$$3 \sin x = 5$$
$$\sin x = \frac{5}{3}$$

Since sin x can never be greater than 1, there are *no* values of x satisfying the original equation.

5 **Find all values of x in the interval $0° \leq x < 360°$ that satisfy $\sin x + 2 \sin x \cos x = 0$.**

$$\sin x + 2 \sin x \cos x = 0$$
$$\sin x (1 + 2 \cos x) = 0 \quad \textit{Factor out common factor, sin x.}$$

$$\sin x = 0 \quad \vee \quad 1 + 2 \cos x = 0$$
$$x = 0°, 180° \qquad 2 \cos x = -1$$
$$\cos x = -\tfrac{1}{2}$$
$$x = 120°, 240°$$

The solutions are 0°, 120°, 180°, and 240°.

Exercises

Exploratory **Solve each equation for all values of θ in the interval $0° \leq \theta < 360°$.**

1. $2 \sin \theta = 1$
2. $2 \cos \theta = \sqrt{3}$
3. $\tan \theta - 1 = 0$
4. $3 \tan \theta = \sqrt{3}$
5. $2 \sin \theta + 1 = 0$
6. $7 \sin \theta = -7$
7. $3 \sin \theta + 2 = 2$
8. $5 \tan \theta = 5$
9. $-2 \sin \theta - \sqrt{2} = 0$
10. $3 \cos \theta + 5 = 0$
11. $2 \tan \theta + 2\sqrt{3} = 0$
12. $2 - 3 \cos \theta = 5 + 3 \cos \theta$

Written **Solve each equation for all values of θ in the interval $0° \leq \theta < 360°$.**

1. $5 \tan \theta = 5\sqrt{3}$
2. $-\sin \theta = \frac{\sqrt{3}}{2}$
3. $-2 \sin \theta = \sqrt{2}$
4. $2 \tan \theta + 2 = 0$
5. $3(\sin \theta - 1) = 0$
6. $3 \cos \theta = \cos \theta - 1$
7. $4(\sin \theta + 1) = 3 + 2 \sin \theta$
8. $\cos \theta + 3 = 2(\cos \theta + 1)$
9. $\cot \theta + 2 = 2 \cot \theta + 3$
10. $4 \csc \theta + 5 = 4 + 2 \csc \theta$
11. $3 \cos \theta + 1 = \frac{9}{2}$
12. $\cos \theta + 1 = \frac{\sqrt{2} + 2}{2}$
13. $\sin \theta - 3 = \frac{6 - \sqrt{3}}{2}$
14. $\sqrt{2} + \sin \theta = 3 \sin \theta - \sqrt{5}$
15. $-12 \sin \theta = \sqrt{108}$

Solve the equation in each indicated exercise for all values of θ. Use radian measure.

16. exercise 1 **17.** exercise 2 **18.** exercise 3 **19.** exercise 4
20. exercise 6 **21.** exercise 7 **22.** exercise 8 **23.** exercise 9

Solve each equation for all values of x if $0° \leq x < 360°$. Give answers to the nearest degree.

24. $3 \sin x = 1$
25. $4 \cos x = -1$
26. $\tan x + 5 = 8$
27. $5 \cos x + 1 = 0$
28. $3 \tan x = \sqrt{18}$
29. $\tan x + 2 = 2.5$
30. $2 \cos x = \sqrt{5}$
31. $2 \sin x = 1.2861$
32. $2 \sin x + 3 = 4.0088$
33. $-\cos x + 1 = 0.0662$
34. $\frac{1}{3} \csc x + 1 = 2$
35. $9 - \sec x = 11.5$
36. $2 \cos x \sin x - \cos x = 0$
37. $\sin x + 2 \sin x \cos x = 0$
38. $2 \sin x \cos x - \sin x = 0$
39. $2 \cos x \sin x + \sqrt{2} \sin x = 0$
40. $\cot x + 2 \cot x \sin x = 0$
41. $\tan x \sec x - 2 \tan x = 0$
42. $\sin x + 3 \sin x \cos x = 0$
43. $2 \sin x \tan x = \tan x$
44. $\tan^2 x = \tan x$
45. $\sin x \cos x - \cos x = 2 \cos x$
46. $\sec x \csc x = \csc x$
47. $\sqrt{8} \sin x = \sqrt{18} \cos x \sin x$

Challenge Solve $\sin^2 \theta = \cos^2 \theta \sin^2 \theta$ for all values of θ.

Mathematical Excursions

Trigonometry Underwater

The design of underwater diving equipment uses information about human breathing patterns. One complete cycle of a breathing pattern, inhaling and exhaling, can be represented by the sine curve.

The vital capacity is the maximum volume of air inhaled and exhaled at each breath for a given size lung. Note that the amplitude of the breathing cycle at rest is much smaller than that of the vital capacity. The lung capacity of a body at rest is about 20% of its vital capacity.

In a laboratory, scientists can determine how long an air tank will last for a diver by using the changes in the sine curve for various activities and the knowledge of the effects of water pressure on the air in the tank.

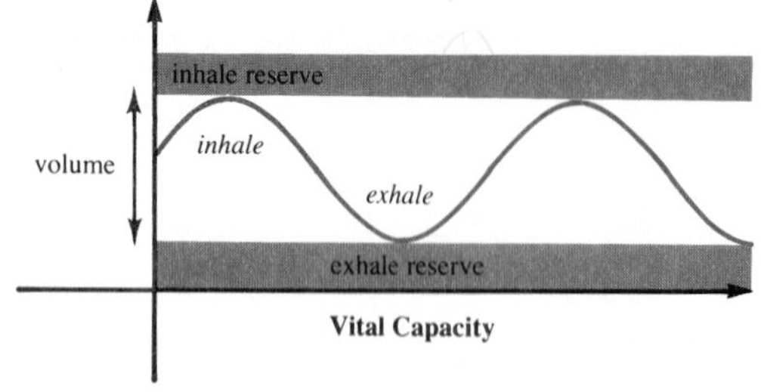

Vital Capacity

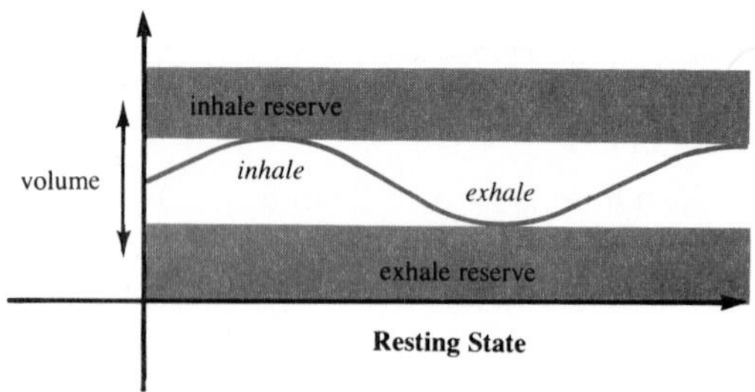

Resting State

9.3 More Trigonometric Equations

Quadratic equations such as $x^2 - 3x + 2 = 0$ may be solved by factoring or by using the quadratic formula. Quadratic trigonometric equations may also be solved by using these methods.

Examples

1 **Solve $\sin^2 \theta = 1$ for the values of θ if $0° \leq \theta < 360°$.**

Treat the expression $\sin \theta$ as the variable. Just as $s^2 - 1$ factors into $(s + 1)(s - 1)$, so $\sin^2 \theta - 1$ factors into $(\sin \theta + 1)(\sin \theta - 1)$.

$$\sin^2 \theta - 1 = 0$$
$$(\sin \theta + 1)(\sin \theta - 1) = 0$$
$$\sin \theta = -1 \quad \vee \quad \sin \theta = 1$$
$$\theta = 270° \qquad\qquad \theta = 90°$$

Check $\theta = 270°$.

$$\sin^2 \theta = 1$$
$$\sin^2 270° \stackrel{?}{=} 1$$
$$(-1)^2 \stackrel{?}{=} 1$$
$$1 = 1$$

Check $\theta = 90°$.

$$\sin^2 \theta = 1$$
$$\sin^2 90° \stackrel{?}{=} 1$$
$$(1)^2 \stackrel{?}{=} 1$$
$$1 = 1$$

For the interval $0° \leq \theta < 360°$, the roots are 90°, 270°.

2 **Find all values of θ in the interval $0 \leq \theta < 2\pi$ that satisfy $2 \sin^2 \theta - 7 \sin \theta = -3$.**

$$2 \sin^2 \theta - 7 \sin \theta = -3$$
$$2 \sin^2 \theta - 7 \sin \theta + 3 = 0 \qquad \textit{Rewrite in standard form.}$$
$$(\sin \theta - 3)(2 \sin \theta - 1) = 0 \qquad \textit{Factor.}$$

$\sin \theta = 3 \quad \vee \quad 2 \sin \theta - 1 = 0$

Since $\sin \theta$ cannot exceed 1, $\sin \theta = 3$ has no solution.

$$2 \sin \theta = 1$$
$$\sin \theta = \frac{1}{2}$$
$$\theta = \frac{\pi}{6}, \frac{5\pi}{6}$$

For the interval $0 \leq \theta < 2\pi$, the roots are $\frac{\pi}{6}, \frac{5\pi}{6}$.

Very often trigonometric equations do not work out as nicely as those in the examples. The solutions are not special angles like 0° or 120°. In such cases, the trigonometric tables or a calculator can be used to find an approximate value of θ.

Examples

3 **Solve $\sin^2\theta + 5\sin\theta + 3 = 0$ for θ in the interval $0° \le \theta < 360°$. Round your answer to the nearest degree.**

This is a quadratic equation with variable $\sin\theta$. Since it is not factorable, we must use the quadratic formula.

In this case, $x = \sin\theta$, $a = 1$, $b = 5$, and $c = 3$.

Recall that if $ax^2 + bx + c = 0$, $x = \dfrac{-b \pm \sqrt{b^2 - 4ac}}{2a}$

$$\sin\theta = \frac{-5 \pm \sqrt{25 - 4(1)(3)}}{2(1)} = \frac{-5 \pm \sqrt{13}}{2}$$

$$\sin\theta = -0.697 \quad \text{or} \quad \sin\theta = -4.303$$

$\sqrt{13} \approx 3.606$

Since $\sin\theta$ can never be less than -1, $\sin\theta = -4.303$ has no solutions.

For $\sin\theta = -0.697$, θ must be in the third or fourth quadrant. Use trig tables or a calculator to find an approximation for $\sin\theta = 0.697$. We find that $\sin 44° = 0.6947$.

In quadrant III, $\theta = 44° + 180°$ or $224°$.
In quadrant IV, $\theta = 360° - 44°$ or $316°$.

The roots are 224° and 316° to the nearest degree.

4 **Solve $\sec^2\theta - \tan\theta - 3 = 0$ for θ in the interval $0° \le \theta < 360°$.**

Use the basic identities to rewrite the equation in terms of one circular function.

$$\sec^2\theta - \tan\theta - 3 = 0$$
$$\tan^2\theta + 1 - \tan\theta - 3 = 0 \qquad \textit{\sec^2\theta = \tan^2\theta + 1}$$
$$\tan^2\theta - \tan\theta - 2 = 0$$

You will be asked to complete the example in the exercises.

Exercises

Exploratory **Solve each equation for θ in the interval $0° \le \theta < 360°$.**

1. $\cos^2\theta - 1 = 0$ **2.** $\tan^2\theta - 1 = 0$ **3.** $2\sin^2\theta = 2$

4. $\tan^2\theta = 3$ **5.** $\sec^2\theta + 2 = 0$ **6.** $\cos^2\theta = \cos\theta$

7. $4\sin^2\theta = 2$ **8.** $\cos\theta = \dfrac{1}{\cos\theta}$ **9.** $\sin\theta = \sqrt{3}\cos\theta$

10. Find the least measure of θ in the interval $0° \leq \theta < 360°$ that satisfies $2\cos^2\theta - 3\cos\theta = -1$.

11. Complete the solution in example 4 on page 338.

Written **Solve each equation for θ in the interval $0° \leq \theta < 360°$. It may be necessary to approximate answers to the nearest degree.**

1. $(\sin\theta + 3)(\sin\theta - 1) = 0$
2. $(\cos\theta + 2)(\tan\theta - 1) = 0$
3. $(2\sin\theta + 1)(\tan\theta - 1) = 0$
4. $(2\sin\theta - 3)(\cos\theta - 1) = 0$
5. $(2\sin\theta - \sqrt{3})(\sin\theta + 4) = 0$
6. $\cos^2\theta - \cos\theta = 0$
7. $\cos^2\theta - 8\cos\theta = 0$
8. $\sin^2\theta + 3\sin\theta + 2 = 0$
9. $\sin^2\theta + 4\sin\theta + 3 = 0$
10. $\sin^2\theta - \sin\theta - 2 = 0$
11. $\cos^2\theta - 2\cos\theta = 3$
12. $2\cos^2\theta = \cos\theta$
13. $2\sin^2\theta + \sin\theta - 1 = 0$
14. $\cos^2\theta - \cos\theta + 3 = 0$
15. $3\sin^2\theta + 2\sin\theta - 1 = 0$
16. $5\cos^2\theta - 6\cos\theta + 1 = 0$
17. $\cos^2\theta + 2\cos\theta = 5$
18. $\tan^2\theta + \tan\theta - 2 = 0$
19. $2\tan^2\theta + 5\tan\theta = 3$
20. $\tan^2\theta - 3\tan\theta = 0$
21. $3\csc^2\theta + 2\csc\theta = 1$
22. $\sec^2\theta = \sec\theta + 2$
23. $5\tan^2\theta - 11\tan\theta + 2 = 0$
24. $36\cos^2\theta - 1 = 0$
25. $8\cos^2\theta + 3\cos\theta = 0$
26. $\cot^2\theta + 4\cot\theta + 3 = 0$
27. $16\sin^2\theta - 1 = 0$
28. $2\sin\theta + \dfrac{2}{\sin\theta} = 5$
29. $3\sin\theta + \dfrac{1}{\sin\theta} = 4$
30. $2\tan\theta + \dfrac{6}{\tan\theta} = 7$
31. $3\cos\theta + \dfrac{4}{\cos\theta} = 8$
32. $25\sin\theta = \dfrac{4}{\sin\theta}$
33. $\sin^2\theta + 5\sin\theta - 1 = 0$
34. $\cos^2\theta = 3\cos\theta - 1$
35. $2\sin^2\theta + 2\sin\theta - 1 = 0$
36. $\cos^2\theta - 4\cos\theta = 1$
37. $3\cos^2\theta = 2 + 2\cos\theta$
38. $\cos^2\theta = 2\sin\theta$
39. $\sin^2\theta = 1 + \cos^2\theta$
40. $\sin\theta - 2\csc\theta + 1 = 0$
41. $3\sin^2\theta = 1 + \cos^2\theta$
42. $3\sin\theta - 2\cos^2\theta = -3$
43. $\cos\theta - \sin^2\theta = 1$
44. $4\cos^2\theta + 4\sin\theta = 5$
45. $2\cos^2\theta + 3\sin\theta = 0$
46. $6 - \tan\theta = 2\sec^2\theta$
47. $\sec^2\theta = \tan\theta + 3$
48. $8\sin\theta - \cos^2\theta = 8$
49. $\cot^2\theta = \csc\theta + 5$
50. $\sin^2\theta + 4\cos\theta = 3$
51. $\sin\theta - 3\cos^2\theta = -2$
52. $2\sec^2\theta - 5\tan\theta = 0$
53. $\cos\theta = \cot\theta$
54. $\tan\theta = \dfrac{1}{\csc\theta}$
55. $\cot\theta = 2\sin\theta\cos\theta$
56. $3(1 - 2\sin^2\theta) + 2 = -5\cos\theta$

9.4 Cos (α − β) and Related Expressions

The exact value of the sine, cosine, and tangent of special angles such as 0°, 30°, and 45° can be computed. Trigonometric tables can be used for approximations of other values. But suppose you want to find the *exact* value of cos 15°.

It is tempting to try to find the value by noting that cos 15° equals cos (45° − 30°). But does cos (45° − 30°) = cos 45° − cos 30°? *In general, does* $\cos(\alpha - \beta) = \cos\alpha - \cos\beta$?

It is often easy to prove a generalization false by counterexample. For example, cos (60° − 60°) is surely not equal to cos 60° − cos 60° since cos (60° − 60°) = cos 0° or 1 and, obviously, cos 60° − cos 60° = 0.

We will prove the following identity.

cos (α − β)

$$\cos(\alpha - \beta) = \cos\alpha\cos\beta + \sin\alpha\sin\beta$$

Carefully follow the demonstration given below. This approach uses many familiar properties of algebra and geometry such as the distance formula.

Recall that the distance between $P_1(x_1, y_1)$ *and* $P_2(x_2, y_2)$ *is given by the equation* $d = \sqrt{(x_2 - x_1)^2 + (y_2 - y_1)^2}$.

Proof of $\cos(\alpha - \beta) = \cos\alpha\cos\beta + \sin\alpha\sin\beta$

Step 1 Let $\alpha = m\angle AOC$, and $\beta = m\angle AOB$.

Then, by definition, point C has coordinates $(\cos\alpha, \sin\alpha)$ and point B has coordinates $(\cos\beta, \sin\beta)$. The distance d between the two points is given by the following equation.

$$\begin{aligned}
d &= \sqrt{(\cos\alpha - \cos\beta)^2 + (\sin\alpha - \sin\beta)^2} \\
d^2 &= (\cos\alpha - \cos\beta)^2 + (\sin\alpha - \sin\beta)^2 \\
&= (\cos^2\alpha - 2\cos\alpha\cos\beta + \cos^2\beta) + (\sin^2\alpha - 2\sin\alpha\sin\beta + \sin^2\beta) \\
&= \cos^2\alpha + \sin^2\alpha + \cos^2\beta + \sin^2\beta - 2\cos\alpha\cos\beta - 2\sin\alpha\sin\beta \\
&= 1 + 1 - 2\cos\alpha\cos\beta - 2\sin\alpha\sin\beta \\
&= 2 - 2\cos\alpha\cos\beta - 2\sin\alpha\sin\beta
\end{aligned}$$

Step 2 Now, reposition $\angle BOC$ in standard position.

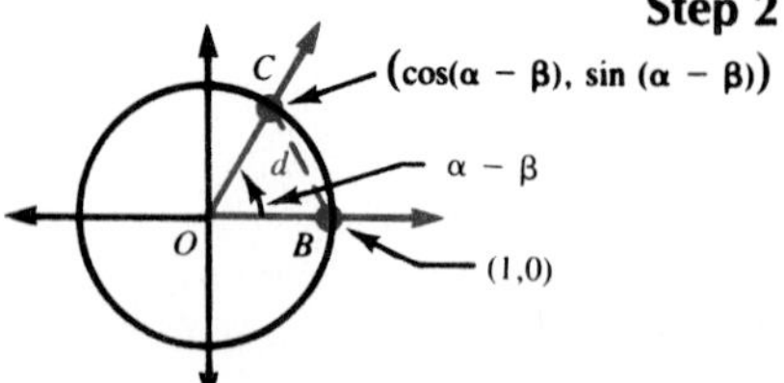

Remember that $m\angle BOC = (\alpha - \beta)$. The coordinates of the intersection of the terminal ray with the unit circle are $(\cos(\alpha - \beta), \sin(\alpha - \beta))$. Recompute the distance d. Remember that this is the same distance as in step 1.

$$
\begin{aligned}
d &= \sqrt{[\cos(\alpha - \beta) - 1]^2 + [\sin(\alpha - \beta) - 0]^2} \\
d^2 &= [\cos(\alpha - \beta) - 1]^2 + [\sin(\alpha - \beta) - 0]^2 \\
&= \cos^2(\alpha - \beta) - 2\cos(\alpha - \beta) + 1 + \sin^2(\alpha - \beta) \\
&= \cos^2(\alpha - \beta) + \sin^2(\alpha - \beta) - 2\cos(\alpha - \beta) + 1 \\
&= 1 - 2\cos(\alpha - \beta) + 1 \\
&= 2 - 2\cos(\alpha - \beta)
\end{aligned}
$$

Step 3 Set the results of steps 1 and 2 equal to each other. Simplify the equation.

$$
\begin{aligned}
2 - 2\cos(\alpha - \beta) &= 2 - 2\cos\alpha\cos\beta - 2\sin\alpha\sin\beta \\
-1 + \cos(\alpha - \beta) &= -1 + \cos\alpha\cos\beta + \sin\alpha\sin\beta \\
\cos(\alpha - \beta) &= \cos\alpha\cos\beta + \sin\alpha\sin\beta
\end{aligned}
$$

Examples

1 Find the exact value of cos 15°.

Rename 15° as a difference. Then use the $\cos(\alpha - \beta)$ identity.

$$
\begin{aligned}
\cos 15° = \cos(60° - 45°) &= \cos 60° \cos 45° + \sin 60° \sin 45° \\
&= \frac{1}{2} \cdot \frac{\sqrt{2}}{2} + \frac{\sqrt{3}}{2} \cdot \frac{\sqrt{2}}{2} \\
&= \frac{\sqrt{2}}{4} + \frac{\sqrt{6}}{4} \quad \text{or} \quad \frac{\sqrt{2} + \sqrt{6}}{4}
\end{aligned}
$$

2 Show that $\sin\theta = \cos(90° - \theta)$, for all values of θ.

$$
\begin{aligned}
\cos(90° - \theta) &= \cos 90° \cos\theta + \sin 90° \sin\theta \\
&= 0 \cdot \cos\theta + 1 \cdot \sin\theta \\
&= \sin\theta
\end{aligned}
$$

Similarly, $\sin(90° - \theta) = \cos\theta$. *You will be asked to prove this in the exercises.*

The identity $\cos(\alpha - \beta) = \cos\alpha\cos\beta + \sin\alpha\sin\beta$ can be used to prove other important identities. Consider the proof of $\cos(-\beta) = \cos\beta$.

$$
\begin{aligned}
\cos(-\beta) &= \cos(0° - \beta) \\
&= \cos 0° \cos\beta + \sin 0° \sin\beta \\
&= 1 \cdot \cos\beta + 0 \cdot \sin\beta \\
&= \cos\beta
\end{aligned}
$$

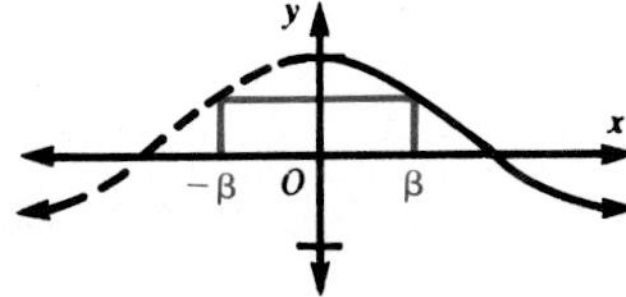

This identity can also be shown using transformation geometry. The cosine curve is symmetric about the line $x = 0°$. From the graph we can see that the $\cos(-\beta) = \cos\beta$.

Likewise, it can be shown that $\sin(-\beta) = -\sin\beta$.

$$\begin{aligned} \sin(-\beta) &= \cos[90^\circ - (-\beta)] \quad \textit{Proven in example 2.} \\ &= \cos(90^\circ + \beta) \\ &= \cos[\beta - (-90^\circ)] \\ &= \cos\beta\cos(-90^\circ) + \sin\beta\sin(-90^\circ) \\ &= \cos\beta \cdot 0 + \sin\beta(-1) \\ &= -\sin\beta \end{aligned}$$

Other identities involving sums and differences are listed below.

$\cos(\alpha + \beta)$
$\sin(\alpha + \beta)$
$\sin(\alpha - \beta)$
$\tan(\alpha + \beta)$
$\tan(\alpha - \beta)$

$$\cos(\alpha + \beta) = \cos\alpha\cos\beta - \sin\alpha\sin\beta$$

$$\sin(\alpha + \beta) = \sin\alpha\cos\beta + \cos\alpha\sin\beta$$

$$\sin(\alpha - \beta) = \sin\alpha\cos\beta - \cos\alpha\sin\beta$$

$$\tan(\alpha + \beta) = \frac{\tan\alpha + \tan\beta}{1 - \tan\alpha\tan\beta}$$

$$\tan(\alpha - \beta) = \frac{\tan\alpha - \tan\beta}{1 + \tan\alpha\tan\beta}$$

Proofs of the first two identities follow. The proofs of the other identities are left for you in the exercises.

Proof of $\cos(\alpha + \beta) = \cos\alpha\cos\beta - \sin\alpha\sin\beta$

$$\begin{aligned} \cos(\alpha + \beta) &= \cos[\alpha - (-\beta)] \\ &= \cos\alpha\cos(-\beta) + \sin\alpha\sin(-\beta) \\ &= \cos\alpha\cos\beta + \sin\alpha(-\sin\beta) \\ &= \cos\alpha\cos\beta - \sin\alpha\sin\beta \end{aligned}$$

Proof of $\sin(\alpha + \beta) = \sin\alpha\cos\beta + \cos\alpha\sin\beta$

$$\begin{aligned} \sin(\alpha + \beta) &= \cos[90^\circ - (\alpha + \beta)] \\ &= \cos[(90^\circ - \alpha) - \beta] \\ &= \cos(90^\circ - \alpha)\cos\beta + \sin(90^\circ - \alpha)\sin\beta \\ &= \sin\alpha\cos\beta + \cos\alpha\sin\beta \end{aligned}$$

Examples

3 **Evaluate $\cos 22^\circ \cos 8^\circ - \sin 22^\circ \sin 8^\circ$.**

$$\begin{aligned} \cos 22^\circ\cos 8^\circ - \sin 22^\circ\sin 8^\circ &= \cos(22^\circ + 8^\circ) \\ &= \cos 30^\circ \\ &= \frac{\sqrt{3}}{2} \end{aligned}$$

4 **If $\angle A$ and $\angle B$ are acute angles with $\sin A = \frac{3}{5}$ and $\sin B = \frac{\sqrt{5}}{3}$, find $\sin(A - B)$.**

Before using the formula to find $\sin(A - B)$, we need to find $\cos A$ and $\cos B$. Recall that $\sin^2 x + \cos^2 x = 1$ for any angle.

If $\sin A = \frac{3}{5}$, $\sin^2 A = \frac{9}{25}$.

$\cos^2 A = 1 - \sin^2 A$

$\cos^2 A = 1 - \frac{9}{25} = \frac{16}{25}$

$\cos A = \pm\sqrt{\frac{16}{25}}$ or $\pm\frac{4}{5}$

Since A is acute, we choose $\frac{4}{5}$.

If $\sin B = \frac{\sqrt{5}}{3}$, $\sin^2 B = \frac{5}{9}$.

$\cos^2 B = 1 - \sin^2 B$

$\cos^2 B = 1 - \frac{5}{9} = \frac{4}{9}$

$\cos B = \pm\sqrt{\frac{4}{9}}$ or $\pm\frac{2}{3}$

Since B is acute, $\cos B = \frac{2}{3}$.

Therefore, $\sin(A - B) = \frac{3}{5}\cdot\frac{2}{3} - \frac{4}{5}\cdot\frac{\sqrt{5}}{3} = \frac{6}{15} - \frac{4\sqrt{5}}{15} = \frac{6 - 4\sqrt{5}}{15}$.

Exercises

Exploratory **Give an example that shows each of the following is *false*.**

1. $\cos(\alpha + \beta) = \cos\alpha + \cos\beta$

2. $\sin(\alpha + \beta) = \sin\alpha + \sin\beta$

3. $\cos(\alpha - \beta) = \cos\alpha - \cos\beta$

4. $\sin(\alpha - \beta) = \sin\alpha - \sin\beta$

Choose the best answer.

5. $\cos\theta =$
a. $\sin\theta$ **b.** $-\cos\theta$ **c.** $\cos(-\theta)$ **d.** $-\sin\theta$

6. $\sin(-\theta) =$
a. $-\sin\theta$ **b.** $\sin\theta$ **c.** $\cos\theta$ **d.** $-\cos\theta$

7. If $\sin A = \cos B$, then
a. $A = B$ **b.** $A + B = 90°$ **c.** $A^2 + B^2 = 1$ **d.** $A - B = 90°$

8. Tan $(A + B)$ is undefined if
a. $\tan A = \tan B$ **b.** $\tan A + \tan B = 1$ **c.** $\tan A \tan B = 0$ **d.** $\tan A \tan B = 1$

9. If $\cos(90° - \theta) = \frac{3}{4}$, then $\sin\theta =$
a. $\frac{3}{4}$ **b.** $-\frac{3}{4}$ **c.** $\frac{\sqrt{3}}{2}$ **d.** none of these

10. If $\tan(-\theta) = 5$, then $\tan\theta =$
a. 5 **b.** -5 **c.** $\frac{1}{5}$ **d.** $-\frac{1}{5}$

Written **Use the formula for $\cos(\alpha - \beta)$ to evaluate each of the following.**

1. $\cos 15°$
2. $\cos 75°$
3. $\cos(-75°)$
4. $\cos 165°$
5. $\cos 195°$
6. $\cos 285°$
7. $\cos(-15°)$
8. $\cos 345°$

Find the exact value in simplest form for each of the following.

9. $\cos 105°$
10. $\cos 255°$
11. $\cos\left(\frac{2\pi}{3} + \frac{\pi}{6}\right)$
12. $\cos\left(\frac{5\pi}{6} + \frac{\pi}{4}\right)$
13. $\sin 15°$
14. $\sin 75°$
15. $\sin 105°$
16. $\sin 165°$
17. $\tan 75°$
18. $\tan 105°$
19. $\tan 15°$
20. $\tan 165°$
21. $\cos\frac{11\pi}{12}$
22. $\sin\frac{5\pi}{12}$
23. $\tan(-105°)$
24. $\cos 375°$
25. Show that $\sin(90° - \theta) = \cos\theta$ for all values of θ.

If A and B are acute angles of right $\triangle ABC$, show that each of the following is an identity.

26. $\sin A = \cos B$
27. $\tan A = \cot B$
28. $\sec A = \csc B$

In exercises 26–28, the pairs of functions are examples of *cofunctions*. Use the cofunction relationship to find the smallest positive value of A that satisfies each equation.

29. $\sin(A - 20°) = \cos 50°$
30. $\tan A = \cot(A + 28°)$
31. $\sec 3A = \csc 63°$
32. $\cos 2A = \sin(3A + 5°)$
33. $\cot\left(\frac{A - 10°}{2}\right) = \tan 2A$
34. $\cos(2A - 5°) = \sin(A - 4°)$

Evaluate each expression.

35. $\cos 96° \cos 6° + \sin 96° \sin 6°$
36. $\cos 45° \cos 15° + \sin 45° \sin 15°$
37. $\sin 110° \cos 20° - \cos 110° \sin 20°$
38. $\sin 20° \cos 25° + \cos 20° \sin 25°$
39. $\frac{\tan 65° - \tan 20°}{1 + \tan 65° \tan 20°}$
40. $\frac{\tan 10° + \tan 50°}{1 - \tan 10° \tan 50°}$
41. $\sin 283° \cos 13° - \cos 283° \sin 13°$
42. $\cos\frac{7\pi}{12}\cos\frac{5\pi}{12} - \sin\frac{7\pi}{12}\sin\frac{5\pi}{12}$

Prove each identity. Illustrate each statement with a sketch.

43. $\cos(180° - \theta) = -\cos\theta$
44. $\cos(270° - \theta) = -\sin\theta$
45. $\sin(90° + \theta) = \cos\theta$
46. $\tan(\theta + 180°) = \tan\theta$
47. $\cos(\pi + \theta) = -\cos\theta$
48. $\cos\left(\frac{\pi}{2} + \theta\right) = -\sin\theta$
49. $\sin(\pi + \theta) = -\sin\theta$
50. $\cos\left(\frac{3\pi}{2} + \theta\right) = \sin\theta$
51. $\tan(180° - \theta) = -\tan\theta$
52. $\tan(2\pi + \theta) = \tan\theta$

Express each of the following in terms of $\sin\theta$ or $\cos\theta$.

53. $\sin(\theta - \pi)$
54. $\cos\left(\frac{3\pi}{2} - \theta\right)$
55. $\sin\left(\frac{\pi}{4} - \theta\right)$
56. $\cos\left(\frac{\pi}{4} - \theta\right)$
57. $\sin\left(\frac{\pi}{12} - \theta\right)$
58. $\cos\left(\frac{\pi}{3} - \theta\right) - \sin\left(\frac{\pi}{6} - \theta\right)$

If $\sin x = \frac{4}{5}$, $\sin y = \frac{5}{13}$, and x and y are each in quadrant I, find the value of each of the following.

59. $\sin (x + y)$
60. $\cos (x + y)$
61. $\sin (x - y)$
62. $\cos (x - y)$
63. $\sin (y - x)$
64. $\cos (y - x)$
65. $\tan (x + y)$
66. $\tan (x - y)$

If x is in quadrant I, $\sin x = \frac{12}{13}$, y is in quadrant II, and $\cos y = -\frac{4}{5}$, find the value of each of the following.

67. $\cos (x + y)$
68. $\cos (90° - x)$
69. $\sin (x - y)$
70. $\cos (x - y)$
71. $\cos (-x)$
72. $\sin (x + y)$
73. $\tan (x + y)$
74. $\tan (x - y)$

If $\tan A < 0$, $\tan B < 0$, $\cos A = \frac{1}{3}$, and $\cos B = \frac{\sqrt{5}}{3}$, find the value of each of the following.

75. $\cos (A - B)$
76. $\cos (B - A)$
77. $\sin (A + B)$
78. $\sin (A - B)$
79. $\sin (-A)$
80. $\sin (-B)$
81. $\tan (A + B)$
82. $\tan (A - B)$

Derive each of the following formulas.

83. $\sin (\alpha - \beta) = \sin \alpha \cos \beta - \cos \alpha \sin \beta$
84. $\tan (-\theta) = -\tan \theta$
85. $\tan (\alpha + \beta) = \dfrac{\tan \alpha + \tan \beta}{1 - \tan \alpha \tan \beta}$
86. $\tan (\alpha - \beta) = \dfrac{\tan \alpha - \tan \beta}{1 + \tan \alpha \tan \beta}$

Challenge **Prove each of the following identities.**

1. $2 \sin \alpha \cos \beta = \sin (\alpha + \beta) + \sin (\alpha - \beta)$
2. $\sin 2A = 2 \sin A \cos A$
3. $\sin (x + y) \sin (x - y) = \sin^2 x - \sin^2 y$
4. $\cos 2A = 1 - 2 \sin^2 A$

Mathematical Excursions

Electricity

A generator consists of loops of wire placed in a magnetic field. As the loops are turned, current is induced in the wire. When the loop is in a horizontal position the maximum current is obtained. As it moves to a vertical position the current decreases. As the loop continues to turn, the direction of the current changes and moves toward a new maximum in that direction. This change occurs when the loop has turned 180°.

If this information is drawn on a graph, it yields a sine curve. Just as the sine curve repeats itself every 360°, so does the pattern of the currents produced in a generator.

9.5 Double and Half Angle Formulas

In advanced work with circular functions, it is often necessary to find alternate ways of representing various expressions. For example, the formula for $\sin(\alpha + \beta)$ can be used to derive a formula for $\sin 2\alpha$.

$$\begin{aligned}\sin 2\alpha &= \sin(\alpha + \alpha)\\ &= \sin\alpha\cos\alpha + \cos\alpha\sin\alpha\\ &= 2\sin\alpha\cos\alpha\end{aligned}$$

Double Angle Formula for Sine

$$\sin 2\alpha = 2\sin\alpha\cos\alpha$$

In the same manner, a formula for $\cos 2\alpha$ can be derived.

$$\begin{aligned}\cos 2\alpha &= \cos(\alpha + \alpha)\\ &= \cos\alpha\cos\alpha - \sin\alpha\sin\alpha\\ &= \cos^2\alpha - \sin^2\alpha\end{aligned}$$

Double Angle Formulas for Cosine

$$\begin{aligned}\cos 2\alpha &= \cos^2\alpha - \sin^2\alpha\\ \cos 2\alpha &= 2\cos^2\alpha - 1\\ \cos 2\alpha &= 1 - 2\sin^2\alpha\end{aligned}$$

The alternate formulas for $\cos 2\alpha$ listed above are obtained by using the basic identities $\sin^2\alpha = 1 - \cos^2\alpha$ and $\cos^2\alpha = 1 - \sin^2\alpha$.

$$\begin{aligned}\cos 2\alpha &= \cos^2\alpha - \sin^2\alpha\\ &= \cos^2\alpha - (1 - \cos^2\alpha)\\ &= \cos^2\alpha - 1 + \cos^2\alpha\\ &= 2\cos^2\alpha - 1\end{aligned}$$

$$\begin{aligned}\cos 2\alpha &= \cos^2\alpha - \sin^2\alpha\\ &= (1 - \sin^2\alpha) - \sin^2\alpha\\ &= 1 - 2\sin^2\alpha\end{aligned}$$

In the exercises you will be asked to derive the following double angle formula for the tangent.

Double Angle Formula for Tangent

$$\tan 2\alpha = \frac{2\tan\alpha}{1 - \tan^2\alpha}$$

Example

1 **If $\cos\alpha = -\frac{3}{5}$ and α is in quadrant II, find $\sin 2\alpha$.**

First find $\sin\alpha$.

$$\sin^2\alpha + \cos^2\alpha = 1$$

$$\sin^2\alpha + \left(-\frac{3}{5}\right)^2 = 1$$

$$\sin^2 \alpha = \frac{16}{25} \text{ or } \sin \alpha = \pm \frac{4}{5}$$

Since α is in quadrant II, $\sin \alpha > 0$. Therefore, reject $-\frac{4}{5}$.

Now use the formula for $\sin 2\alpha$.

$$\sin 2\alpha = 2 \sin \alpha \cos \alpha$$

$$= 2\left(\frac{4}{5}\right)\left(-\frac{3}{5}\right)$$

$$\sin 2\alpha = \frac{-24}{25}$$

The double angle formulas are used to derive the following half angle formulas.

Half Angle Formulas

$$\cos \frac{\theta}{2} = \pm \sqrt{\frac{1 + \cos \theta}{2}}$$

$$\sin \frac{\theta}{2} = \pm \sqrt{\frac{1 - \cos \theta}{2}}$$

$$\tan \frac{\theta}{2} = \pm \sqrt{\frac{1 - \cos \theta}{1 + \cos \theta}}$$

The derivation of the half angle formulas for sine and cosine are shown below. You will be asked to derive the half angle formula for tangent in the exercises.

Let $2\alpha = \theta$, so $\alpha = \frac{\theta}{2}$.

$$2 \cos^2 \alpha - 1 = \cos 2\alpha$$

$$2 \cos^2 \frac{\theta}{2} - 1 = \cos \theta$$

$$\cos^2 \frac{\theta}{2} = \frac{1 + \cos \theta}{2}$$

$$\cos \frac{\theta}{2} = \pm \sqrt{\frac{1 + \cos \theta}{2}}$$

$$1 - 2 \sin^2 \alpha = \cos 2\alpha$$

$$1 - 2 \sin^2 \frac{\theta}{2} = \cos \theta$$

$$\sin^2 \frac{\theta}{2} = \frac{1 - \cos \theta}{2}$$

$$\sin \frac{\theta}{2} = \pm \sqrt{\frac{1 - \cos \theta}{2}}$$

You should be careful to note that the choice of positive or negative depends on the quadrant in which the angle lies.

Examples

2 **Find the exact value of cos 105°.**

$$\cos 105° = \cos \frac{210°}{2} = \pm \sqrt{\frac{1 + \cos 210°}{2}}$$

$$= \pm \sqrt{\frac{1 + \left(-\frac{\sqrt{3}}{2}\right)}{2}} = \pm \sqrt{\frac{2 - \sqrt{3}}{4}} \text{ or } \pm \frac{\sqrt{2 - \sqrt{3}}}{2}$$

Since 105° is in Quadrant II, cos 105° is negative. Thus, $\cos 105° = -\frac{1}{2}\sqrt{2 - \sqrt{3}}$.

3 **Suppose z is an acute angle with $\cos z = \frac{3}{5}$. Find $\tan \frac{1}{2}z$.**

$$\tan \frac{1}{2}z = \sqrt{\frac{1 - \frac{3}{5}}{1 + \frac{3}{5}}} = \sqrt{\frac{5 - 3}{5 + 3}} = \sqrt{\frac{2}{8}} = \frac{1}{2}$$

Exercises

Exploratory **Complete each identity.**

1. $\sin 2\alpha = ?$
2. $\tan 2\alpha = ?$
3. $\sin \frac{\theta}{2} = ?$
4. $\cos \frac{\theta}{2} = ?$
5. $\tan \frac{\theta}{2} = ?$
6. Express $\cos 2A$ in terms of $\sin A$ and $\cos A$.
7. Express $\cos 2A$ in terms of $\cos A$.
8. Express $\cos 2A$ in terms of $\sin A$.
9. Show that $\cos 60° = \frac{1}{2}$ by completing $\cos 2(30°)$.
10. Show that $\sin 90° = 1$ by completing $\sin 2(45°)$.
11. Show that $\tan 120° = -\sqrt{3}$ by completing $\tan 2(60°)$.
12. If $\cos \theta = x$, express $\cos 2\theta$ in terms of x.
13. If $\sin \theta = t$, express $\cos 2\theta$ in terms of t.
14. If $\cos \theta = \frac{1}{8}$, and θ is in quadrant I, find $\cos \frac{\theta}{2}$.
15. If $\cos \theta = \frac{1}{5}$, and θ is in quadrant I, find $\tan \frac{\theta}{2}$.
16. If $\cos \theta = 0.5$ and θ is in quadrant I, find $\sin \frac{\theta}{2}$.

Written **Given the following values, find $\cos 2\theta$.**

1. $\sin \theta = \frac{3}{5}$
2. $\sin \theta = \frac{4}{5}$
3. $\cos \theta = \frac{1}{3}$
4. $\sin \theta = -\frac{2}{9}$
5. $\cos \theta = \frac{-\sqrt{5}}{3}$
6. $\sin \theta = -0.8$
7. $\cos \theta = -0.6$
8. $\cos \theta = \frac{\sqrt{3}}{4}$
9. $\tan \theta = \frac{3}{4}$
10. $\csc \theta = -\frac{3}{2}$

Given the following values, find $\sin 2\theta$.

11. $\sin \theta = \frac{4}{5}$, $0° < \theta < 90°$
12. $\sin \theta = \frac{5}{13}$, $0° < \theta < 90°$
13. $\sin \theta = \frac{1}{2}$, $0° < \theta < 90°$
14. $\cos \theta = \frac{3}{5}$, $0° < \theta < 90°$

15. $\cos\theta = -\frac{4}{5}$, $90° < \theta < 180°$

16. $\cos\theta = -\frac{5}{13}$, $90° < \theta < 180°$

17. $\cos\theta = -\frac{\sqrt{5}}{3}$, $180° < \theta < 270°$

18. $\sin\theta = -\frac{1}{4}$, $180° < \theta < 270°$

19. $\cos\theta = \frac{\sqrt{10}}{10}$, $270° < \theta < 360°$

20. $\sin\theta = \frac{-\sqrt{17}}{8}$, $270° < \theta < 360°$

Given the following values, find tan 2θ.

21. $\tan\theta = \frac{1}{4}$

22. $\tan\theta = \frac{1}{2}$

23. $\tan\theta = 4$

24. $\cos\theta = -\frac{5}{13}$, $90° < \theta < 180°$

25. $\sin\theta = -\frac{3}{5}$, $270° < \theta < 360°$

Use a half angle formula to find the exact value of each function.

26. tan 105°
27. tan 22.5°
28. sin 15°
29. cos 75°
30. sin 22.5°
31. cos 112.5°
32. cos 15°
33. tan 157.5°

Given the following values, find $\sin\frac{x}{2}$, $\cos\frac{x}{2}$, and $\tan\frac{x}{2}$.

34. $\cos x = \frac{7}{25}$, $0° < x < 90°$

35. $\cos x = \frac{1}{9}$, $0° < x < 90°$

36. $\cos x = \frac{17}{25}$, $0° < x < 90°$

37. $\cos x = \frac{3}{5}$, $0° < x < 90°$

38. $\cos x = -\frac{7}{18}$, $90° < x < 180°$

39. $\cos x = -\frac{1}{8}$, $180° < x < 270°$

40. $\sin x = -\frac{5}{13}$, $270° < x < 360°$

41. $\tan x = -\frac{\sqrt{5}}{2}$, $270° < x < 360°$

42. $\tan x = -2\sqrt{2}$, $270° < x < 360°$

43. $\sin x = \frac{\sqrt{35}}{6}$, $90° < x < 180°$

Evaluate each of the following without using trigonometric tables or a calculator.

44. $2\cos^2\frac{\pi}{12} - 1$

45. $2\sin\frac{11\pi}{12}\cos\frac{11\pi}{12}$

46. $\dfrac{2\tan\frac{5\pi}{12}}{1 - \tan^2\frac{5\pi}{12}}$

Derive each formula.

47. $\tan 2\alpha = \dfrac{2\tan\alpha}{1 - \tan\alpha}$

48. $\tan\frac{\theta}{2} = \pm\sqrt{\dfrac{1 - \cos\theta}{1 + \cos\theta}}$

Challenge

1. If $\sin A = \frac{2}{3}$, find $\cos 4A$.
2. If $f(x) = 2x$ and $g(x) = \cos x$, express $g(f(x))$ in terms of $\sin x$.
3. Find the value of $\log\left(20\sin\frac{5\pi}{6}\right)$.

9.6 Using Trigonometric Identities

You can use trigonometric identities to solve new kinds of equations and to prove additional identities. A summary of commonly used identities is on page 623.

Examples

1 **Prove that $\frac{\sin 2x}{2\sin^2 x} = \cot x$.**

$$\frac{\sin 2x}{2\sin^2 x} = \frac{2\sin x\cos x}{2\sin^2 x} \qquad \textit{\sin 2x = 2\sin x\cos x}$$

$$= \frac{\cos x}{\sin x} \text{ or } \cot x$$

2 **Prove that $1 + \cos 2y = \frac{2}{1+\tan^2 y}$.**

Work on both sides of the potential equality to prove them equal.

$$1 + \cos 2y \stackrel{?}{=} \frac{2}{1+\tan^2 y}$$

$$1 + (2\cos^2 y - 1) \stackrel{?}{=} 2\left(\frac{1}{\sec^2 y}\right) \qquad \textit{\cos 2y = 2\cos^2 y - 1}$$

$$2\cos^2 y = 2\cos^2 y$$

Therefore, $1 + \cos 2y = \frac{2}{1+\tan^2 y}$

3 **Find all values of x in the interval $0 \le x < 2\pi$ that satisfy $\sin 2x - \sin x = 0$.**

$$\sin 2x - \sin x = 0$$

$$2\sin x\cos x - \sin x = 0$$

$$\sin x(2\cos x - 1) = 0$$

$\sin x = 0$ or $2\cos x - 1 = 0$

$x = 0 \text{ or } \pi$ $\qquad$ $\cos x = \frac{1}{2}$

$x = \frac{\pi}{3} \text{ or } \frac{5\pi}{3}$

The solutions are $0, \frac{\pi}{3}, \frac{5\pi}{3}$, and π.

4 **Solve $\cos 2t + 2 = \sin t$ for all values of t.**

$$\cos 2t + 2 = \sin t$$
$$(1 - 2\sin^2 t) + 2 - \sin t = 0 \qquad \textit{cos } 2t = 1 - 2\sin^2 t$$
$$2\sin^2 t + \sin t - 3 = 0$$
$$(2\sin t + 3)(\sin t - 1) = 0$$

$2\sin t + 3 = 0$ or $\sin t - 1 = 0$

$\sin t = -\frac{3}{2}$ $\qquad$ $\sin t = 1$

No solution for t exists. $\qquad$ $t = \frac{\pi}{2}$

The solutions are $\frac{\pi}{2} + 2\pi n,\ n \in \mathbb{Z}$.

Check for $t = \frac{\pi}{2}$

$$\cos 2\left(\frac{\pi}{2}\right) + 2 \stackrel{?}{=} \sin \frac{\pi}{2}.$$
$$\cos \pi + 2 \stackrel{?}{=} 1$$
$$-1 + 2 \stackrel{?}{=} 1$$
$$1 = 1$$

5 **Solve $\sin 2\theta = -\frac{1}{2}$ if $0° \leq \theta < 360°$.**

It is not always necessary to use the double or half angle formulas when expressions such as $\sin 2\theta$ appear in an equation. In this case, 2θ names an angle whose sine is $-\frac{1}{2}$. This means that 2θ must be 210°, 330°, 570°, 690°, and so on. Thus $\theta = 105°$, 165°, 285°, or 345°.

There are other values of θ but not in the required interval.

6 **Solve $4 \sin x \cos x = -\sqrt{3}$ for all values of x.**

$$4\sin x \cos x = -\sqrt{3}$$
$$2(2\sin x \cos x) = -\sqrt{3}$$
$$\sin 2x = -\frac{\sqrt{3}}{2} \qquad \textit{2 sin x cos x = sin 2x}$$
$$2x = \frac{4\pi}{3} + 2n\pi \text{ or } \frac{5\pi}{3} + 2n\pi, \text{ where } n \in \mathbb{Z}$$

Therefore, $x = \frac{2\pi}{3} + n\pi$ or $\frac{5\pi}{6} + n\pi$, where $n \in \mathbb{Z}$.

Exercises

Exploratory **Solve each equation for x if $0° \leq x < 360°$.**

1. $\sin 2x + \sin x = 0$
2. $\sin 2x + \cos x = 0$
3. $2\sin 2x - 4\cos x = 0$
4. $\cos 2x + \sin x = 0$
5. $\sin 2x + \sqrt{2}\cos x = 0$
6. $\sin 2x + \sqrt{3}\sin x = 0$

Written **Solve for θ if $0° \le \theta < 360°$. Give answers to the nearest degree.**

1. $\cos 2\theta + \cos \theta = 0$
2. $\cos 2\theta - \cos \theta = 5$
3. $\cos 2\theta + 4 \cos \theta = -3$
4. $\cos 2\theta + 3 \cos \theta = 1$
5. $\cos 2\theta = \cos \theta$
6. $2 \sin \theta = 2 + \cos 2\theta$
7. $2 \sin \theta = \tan \theta$
8. $\frac{1}{2} \csc \theta = \sin \theta$
9. $\tan 2\theta = \cot \theta$
10. $\cos 2\theta = \cos \theta + 2$
11. $\sin \theta + 1 = \cos 2\theta$
12. $\cot \theta = \sin 2\theta$
13. $3 \cos 2\theta + \cos \theta + 2 = 0$
14. $\sec \theta = 1 + \cos \theta$
15. $2 \tan \theta = 5 + 6 \cot \theta$
16. $3 \cos 2\theta + \sin \theta - 2 = 0$
17. $\sin 2\theta = 1$
18. $2 \cos 2\theta = 1$
19. $2 \sin 2\theta = \sqrt{3}$
20. $\cos 2\theta = 1$
21. $\sin 3\theta = 1$
22. $\sin 5\theta = 0$
23. $\cos \frac{\theta}{2} = \frac{1}{2}$
24. $4 \sin^2 2\theta = 3$
25. $\sin 2\theta = \tan \theta$
26. $2 \cot \theta = \csc \theta$
27. $\cos \theta = -\sin \theta$
28. $\sin 2\theta = \cos \theta$
29. $\sin 3\theta = \cos \theta$
30. $\cos (\theta + 10°) = \sin \theta$
31. $\sin (\theta - 28°) = \cos \theta$
32. $\cos (\theta + 42°) = \sin \theta$
33. $\cos (2\theta - 3°) = \sin \theta$
34. $\sin (2\theta + 14°) = \cos (\theta - 23°)$
35. $\csc (\theta + 13°) = \sec (3\theta - 3°)$
36. $\sin \theta + \cos \theta = 1$
37. $\sin \theta - \cos \theta = 1$
38. $\cos 2\theta = 4 + 7 \cos \theta$
39. $16 \cos \theta \sin \theta = -4$

Prove each identity.

40. $\sin 2x = 2 \cot x \sin^2 x$
41. $(\sin x + \cos x)^2 = 1 + \sin 2x$
42. $\dfrac{1 + \cos 2\theta}{2 \cos \theta} = \cos \theta$
43. $\cot \theta = \dfrac{\cos 2\theta + 1}{\sin 2\theta}$
44. $\dfrac{\sin 2\theta}{1 - \cos 2\theta} = \cot \theta$
45. $\sin 2x = \dfrac{2 \tan x}{1 + \tan^2 x}$
46. $\tan x = \dfrac{\sin 2x}{1 + \cos 2x}$
47. $\csc 2x = \dfrac{\sec x}{2 \sin x}$
48. $\sec^2 x - \tan^2 x = \dfrac{4 \sin x \cos x}{2 \sin 2x}$
49. $\csc x \sec x = 2 \csc 2x$
50. $\dfrac{1 - \cos 2x}{1 + \cos 2x} = \tan^2 x$
51. $\dfrac{\cos 2x}{\cos x} - \cos x = \cos x - \sec x$
52. $\dfrac{\sec^2 \theta - 2 \tan^2 \theta}{1 + \tan^2 \theta} = \cos 2\theta$
53. $\sin^4 \theta - \cos^4 \theta = -\cos 2\theta$
54. $\cos^2 2x + 4 \sin^2 x \cos^2 x = 1$
55. $\dfrac{1}{\sin x \cos x} - \cot x = \tan x$
56. $\sin^2 x = \frac{1}{2}(1 - \cos 2x)$
57. $2 \cos^2 \frac{x}{2} = 1 + \cos x$
58. $\sin \left(\theta + \frac{\pi}{3}\right) - \cos \left(\theta + \frac{\pi}{6}\right) = \sin \theta$
59. $\sin (60° + \theta) + \sin (60° - \theta) = \sqrt{3} \cos \theta$
60. $\sin \left(x + \frac{\pi}{4}\right) + \cos \left(x + \frac{\pi}{4}\right) = \sqrt{2} \cos x$
61. $\cos (x + y) \cos (x - y) = \cos^2 y - \sin^2 x$

Challenge **Prove the following identity.**

$$\frac{2 \sin x}{\cos x - \sin x} + \frac{2 \cos^2 x}{\cos 2x} - \frac{\sin 2x + 2}{\cos 2x} = 0$$

Mixed Review

Solve. Round lengths to the nearest unit and angle measures to the nearest degree.

1. A deck of uniform width is to be built along both 22-foot ends of a rectangular swimming pool and along one 34-foot side of the pool. If the area of the deck is 752 square feet, find the width of the deck.

2. A rectangular rug is placed in an 18-foot by 20-foot room so that a strip of floor of equal width is uncovered along each edge of the rug. If two-thirds of the area of the room is left uncovered, what are the dimensions of the rug?

3. A painter works on a job for 8 days and is then joined by her assistant. Together they finish the job in 6 more days. Her assistant could have done this job alone in 27 days. How long would it have taken the painter to do this job alone?

4. The speed of the current in the Susquehenna River is about 4 miles per hour. A boat travels downstream 42 miles and returns in a total time of 10 hours. How long would this trip have taken if the boat had traveled in still water?

5. Martin invested $2670 at 7.8% annual interest. How much money must he invest at 6.8% annual interest in order to have the same income after one year?

6. A real estate agent made a commission of $4800 on the sale of a $90,000 house. At that rate, how much commission would he make on a $129,000 house?

7. Luisa's grandmother invested $150 at 5% interest compounded semiannually. When the account was recently given to Luisa, it contained $4740. How long ago did Luisa's grandmother invest the $150?

8. A bacteria culture grows from 40 to 4840 bacteria in 2 hours. How many bacteria will there be after 5 hours? *Hint: Use the population growth formula,* $y = y_0 (1 + r)^n$.

9. A large sheet of paper is 0.15 mm thick. Suppose it is torn in half and the two pieces are placed together and torn in half again. If this process is continued, after how many tears will the stack of papers be at least 1 cm thick?

10. The resultant of two forces has a magnitude of 190 lb. The angle between these forces is 63°. If the magnitude of one of the forces is 55 lb, find the magnitude of the other force.

11. A 24-foot ladder leans against a building. The ladder and the building form a 22° angle. How far is the foot of the ladder from the base of the building?

12. What is the minimum force required to pull a 500-pound trunk along a ramp that slopes up at 17° to the horizontal?

13. A ship sailed for 120 kilometers. It then changed course and sailed for another 85 kilometers. If the ship is 200 kilometers from its starting point, at what angle did the ship turn when it changed course?

14. The angle of elevation from a ship to the top of a lighthouse is 14°. If the ship sails in a straight line toward the foot of the lighthouse for 100 meters, then the angle of elevation is 17°. What is the height of the lighthouse?

15. A pilot, traveling from New York City to Seattle, wants to avoid a large area of severe weather. To do this, he changes his flight path by turning at an angle of 35° to his original path. He flies for 60 miles on this course and then turns back toward his original path at an angle of 50°. He travels on this course until he can turn back onto his original flight path. How many additional miles did this course change add to the flight?

9.7 Inverse Circular Functions

You have often worked with inverse functions. Here are some examples you may recall. The graph of each function is in black. The graph of the inverse is shown in red.

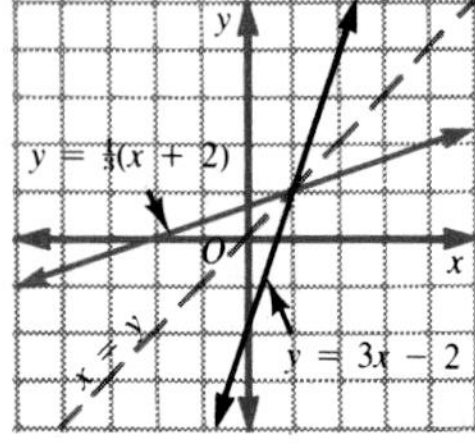

The inverse of $y = 3x - 2$, $y = \frac{1}{3}(x + 2)$, is the reflection of $y = 3x - 2$ in the line $x = y$.

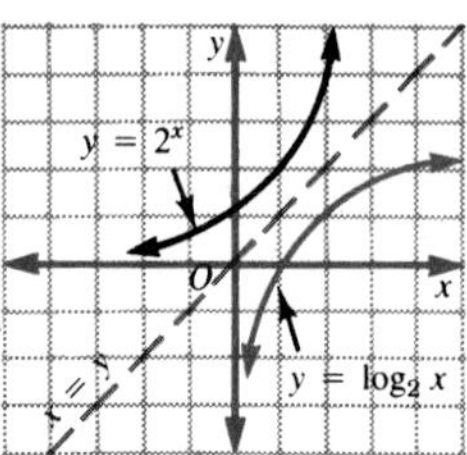

The inverse of $y = 2^x$ is $y = \log_2 x$. It is the reflection of $y = 2^x$ in the line $x = y$.

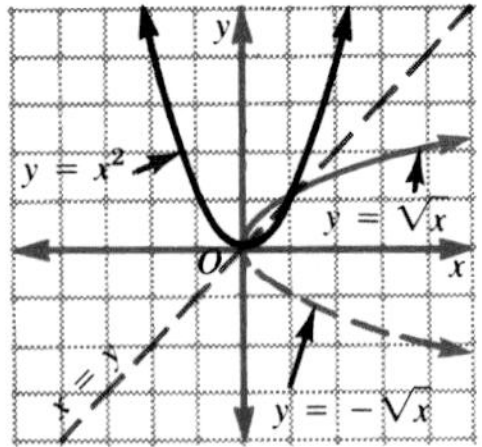

$x = y^2$ is the inverse relation to $y = x^2$. It is not a function, however. The "top" part, $y = \sqrt{x}$, is the inverse function of $y = x^2$, with the domain restricted to nonnegative values.

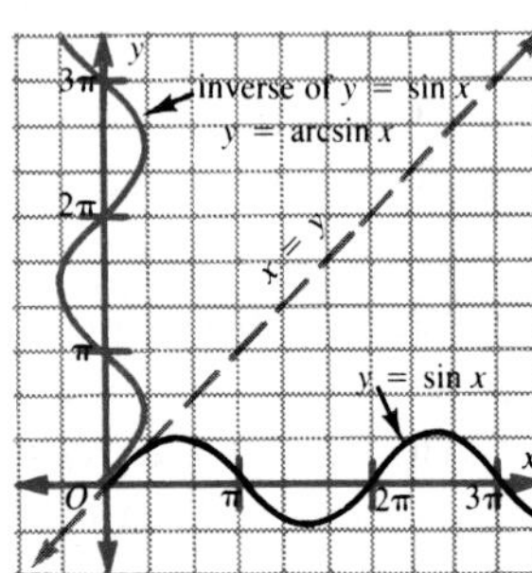

It is sometimes necessary to work with the inverses of the circular functions.

At the left the graph of the sine function has been reflected over the line $y = x$, producing the graph of the inverse of the sine function.

Notice that the inverse is *not* a function since it does not pass the vertical line test.

This inverse of the sine function is called the **arcsine** and is symbolized **arcsin** or $\mathbf{sin^{-1}}$. Thus, if $y = \sin x$, the inverse is expressed as $y = \sin^{-1} x$. You can think of this as meaning, "y is an angle whose sine is x."

Example

1 **Find arcsin $\frac{1}{2}$.**

In effect, this example is asking you to find the angle whose sine is $\frac{1}{2}$. There are at least two angles, measuring $\frac{\pi}{6}$ and $\frac{5\pi}{6}$, that have the appropriate sine value, but all such angle measures need to be given in the solution.

Therefore, $\arcsin \frac{1}{2} = \frac{\pi}{6} + 2\pi n$ or $\frac{5\pi}{6} + 2\pi n$, where $n \in \mathbb{Z}$.

The **arccosine** and **arctangent** relations can be defined in a similar way. The graphs of the arccosine and arctangent relations are shown below as well as their respective cosine and tangent functions.

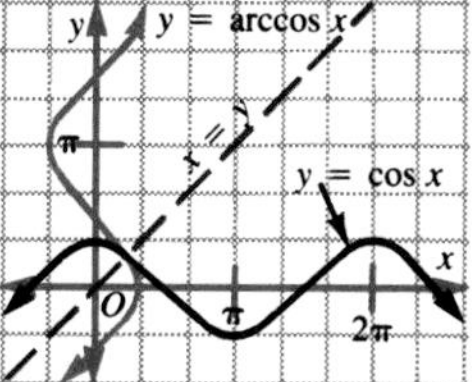

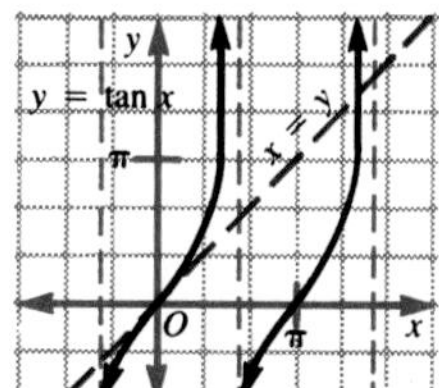

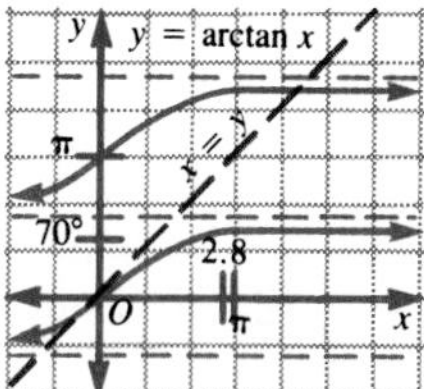

The graphs of y = tan x and y = arctan x are shown on separate coordinate planes to avoid confusion.

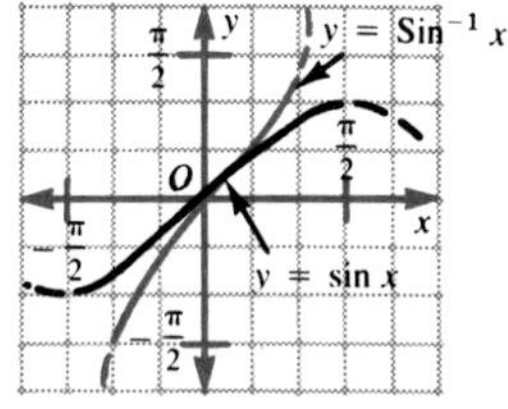

Restricted domain of $y = \sin x$ is $-\frac{\pi}{2} \leq x \leq \frac{\pi}{2}$. Its range is $-1 \leq y \leq 1$. The domain of $y = \text{Sin}^{-1} x$ is $-1 \leq x \leq 1$. Its range is $-\frac{\pi}{2} \leq y \leq \frac{\pi}{2}$.

Since the sine function is not one-to-one, its inverse, the arcsine relation, is not a function. To change this situation, we can restrict the domain of the sine function so that it is one-to-one. The special interval selected is $-\frac{\pi}{2} \leq x \leq \frac{\pi}{2}$. In this restricted domain, the inverse of the sine function is also a function. Capital letters are used to denote the inverse function. Thus, $y = \text{Sin}^{-1} x$ and $y = \text{Arcsin } x$ represent the *inverse sine function*.
The expressions $y = \text{Sin}^{-1} x$ and $y = \text{Arcsin } x$ are read as "y is the angle between $-\frac{\pi}{2}$ and $\frac{\pi}{2}$ whose sine is x."

A similar procedure is used for defining the inverse cosine and inverse tangent functions. The special intervals in which the cosine and tangent functions are one-to-one are listed below.

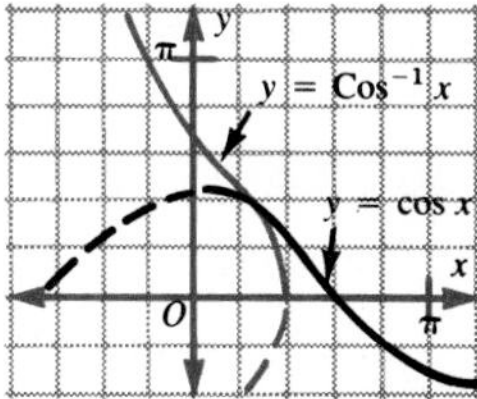

Restricted domain of $y = \cos x$ is $0 \leq x \leq \pi$. Its range is $-1 \leq y \leq 1$. The domain of $y = \text{Cos}^{-1} x$ is $-1 \leq x \leq 1$. Its range is $0 \leq y \leq \pi$.

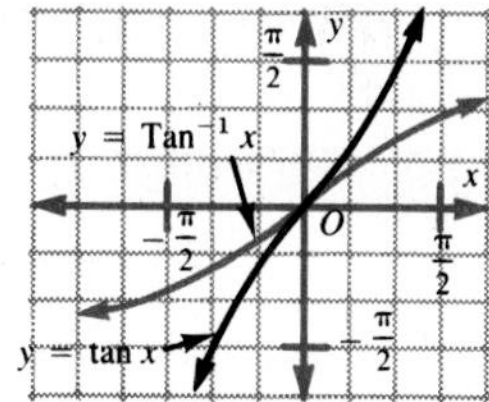

Restricted domain of $y = \tan x$ is $-\frac{\pi}{2} < x < \frac{\pi}{2}$. The range is $\mathcal{R}$, the reals. The domain of $x = \text{Tan}^{-1} x$ is $\mathcal{R}$. The range is $-\frac{\pi}{2} < y < \frac{\pi}{2}$.

Examples

2 **Find $\text{Cos}^{-1}\left(-\frac{\sqrt{3}}{2}\right)$.**

We are looking for θ such that $\cos\theta = -\frac{\sqrt{3}}{2}$ and $0 \leq \theta \leq \pi$. Since $\cos\frac{5\pi}{6} = -\frac{\sqrt{3}}{2}$, it follows that $\text{Cos}^{-1}\left(-\frac{\sqrt{3}}{2}\right) = \frac{5\pi}{6}$. *We could also write* $\text{Arccos}\left(-\frac{\sqrt{3}}{2}\right) = \frac{5\pi}{6}$.

3 **Find $\tan\left(\text{Sin}^{-1}\frac{1}{2}\right)$.**

Let $\theta = \text{Sin}^{-1}\frac{1}{2}$. Then $\sin\theta = \frac{1}{2}$ and $\theta = \frac{\pi}{6}$.

Therefore, $\tan\left(\text{Sin}^{-1}\frac{1}{2}\right) = \tan\frac{\pi}{6} = \frac{\sqrt{3}}{3}$.

4 **Find $\sin\left(\text{Arccos}\frac{2}{3}\right)$.**

Method 1

Let $\theta = \text{Arccos}\frac{2}{3}$.

Then, $\cos\theta = \frac{2}{3}$.

$$\sin^2\theta + \cos^2\theta = 1$$
$$\sin^2\theta + \frac{4}{9} = 1$$
$$\sin^2\theta = \frac{5}{9}$$
$$\sin\theta = \pm\frac{\sqrt{5}}{3}$$

Since $0 \leq \theta \leq \pi$ and $\cos\theta$ is positive, θ is in quadrant I.

Thus, $\sin\theta = \frac{\sqrt{5}}{3}$.

Therefore, $\sin\left(\text{Arccos}\frac{2}{3}\right) = \frac{\sqrt{5}}{3}$.

Method 2

Let $\theta = \text{Arccos}\frac{2}{3}$.

Sketch a right triangle with angle θ such that $\cos\theta = \frac{2}{3}$.

Now, use the Pythagorean Theorem to find that the other leg has length $\sqrt{5}$.

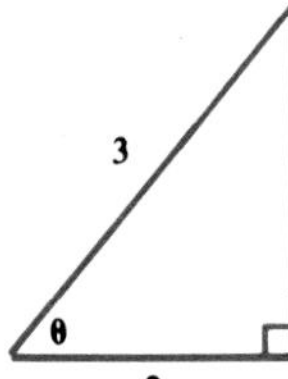

Thus, $\sin\theta = \frac{\sqrt{5}}{3}$.

Exercises

Exploratory **Answer each of the following.**

1. Explain the difference between $y = \sin x$ and $y = \sin^{-1} x$.
2. Explain the difference between $y = \sin^{-1} x$ and $y = \text{Sin}^{-1} x$.

Use arcsin, arccos, or arctan to write an equivalent statement for each of the following.

3. $\cos\theta = \frac{1}{2}$ **4.** $\tan\theta = 1$ **5.** $\sin\theta = \frac{\sqrt{2}}{2}$ **6.** $\cos\theta = 0.8$ **7.** $\tan\theta = -3.4$

Use sin θ, cos θ, or tan θ to write an equivalent statement for each of the following.

8. $\theta = \arcsin\frac{\sqrt{2}}{2}$ **9.** $\theta = \arccos\frac{1}{2}$ **10.** $\theta = \arctan\left(-\frac{\sqrt{3}}{3}\right)$ **11.** $\theta = \arccos(-1)$

State the domain and range of each of the following. Say if the equation defines a function.

12. $y = \sin x$ **13.** $y = \text{Arcsin}\, x$ **14.** $y = \cos x$
15. $y = \text{Cos}^{-1} x$ **16.** $y = \tan x$ **17.** $y = \text{Tan}^{-1} x$

Write an expression that represents each of the following.

18. $\arccos 1$ **19.** $\sin^{-1} 1$ **20.** $\tan^{-1} 1$ **21.** $\arccos 0$
22. $\sin^{-1}\frac{1}{2}$ **23.** $\cos^{-1}\frac{\sqrt{2}}{2}$ **24.** $\arcsin\frac{\sqrt{3}}{2}$ **25.** $\sin^{-1}\left(-\frac{\sqrt{3}}{2}\right)$
26. $\arccos\left(-\frac{1}{2}\right)$ **27.** $\arctan(-\sqrt{3})$ **28.** $\sin^{-1}(0.2588)$ **29.** $\tan^{-1}(-6.314)$

Written **Find each value if it exists. State values in both degree and radian measure.**

1. $\text{Arcsin}\frac{1}{2}$ **2.** $\text{Arccos}\, 0$ **3.** $\text{Arctan}\, 0$ **4.** $\text{Arccos}\frac{1}{2}$
5. $\text{Sin}^{-1}\frac{\sqrt{3}}{2}$ **6.** $\text{Cos}^{-1} 2$ **7.** $\text{Arcsin}\frac{\sqrt{2}}{2}$ **8.** $\text{Tan}^{-1}\sqrt{3}$
9. $\text{Tan}^{-1}(-1)$ **10.** $\text{Arcsin}\left(-\frac{\sqrt{3}}{2}\right)$ **11.** $\text{Arccos}\left(-\frac{\sqrt{2}}{2}\right)$ **12.** $\text{Sin}^{-1}(-1)$
13. $\text{Cos}^{-1}\left(-\frac{\sqrt{3}}{2}\right)$ **14.** $\text{Sin}^{-1}\left(\frac{3}{2}\right)$ **15.** $\text{Tan}^{-1}\frac{\sqrt{3}}{3}$ **16.** $\text{Tan}^{-1}\left(-\frac{\sqrt{3}}{3}\right)$
17. $\text{Arcsec}\, 2$ **18.** $\text{Arccsc}\frac{\sqrt{2}}{2}$ **19.** $\text{Cot}^{-1}(-\sqrt{3})$ **20.** $\text{Sec}^{-1}(-1)$

Evaluate

21. $\sin(\text{Arctan}\, 1)$ **22.** $\cos(\text{Arcsin}\, 0)$ **23.** $\tan(\text{Arccos}\, 1)$ **24.** $\tan(\text{Sin}^{-1} 0)$
25. $\cos\left(\text{Arcsin}\frac{1}{2}\right)$ **26.** $\cos\left(\text{Sin}^{-1}\frac{\sqrt{3}}{2}\right)$ **27.** $\sin\left(\text{Arctan}\frac{\sqrt{3}}{3}\right)$ **28.** $\sin(\text{Sin}^{-1} 1)$
29. $\cos\left(\text{Cos}^{-1}\frac{1}{2}\right)$ **30.** $\tan\left[\text{Cos}^{-1}\left(-\frac{\sqrt{2}}{2}\right)\right]$ **31.** $\sin\left[\text{Arccos}\left(\frac{\sqrt{3}}{2}\right)\right]$ **32.** $\tan\left[\text{Arccos}\left(-\frac{\sqrt{3}}{2}\right)\right]$
33. $\tan\left(\text{Sin}^{-1}\frac{3}{5}\right)$ **34.** $\cos\left(\text{Arcsin}\frac{4}{5}\right)$ **35.** $\sin\left(\text{Tan}^{-1}\frac{5}{12}\right)$ **36.** $\sin(\text{Sin}^{-1} 0.4823)$
37. $\tan\left(\text{Cos}^{-1}\frac{5}{13}\right)$ **38.** $\sin\left(\text{Arccos}\frac{15}{17}\right)$ **39.** $\tan\left[\text{Cos}^{-1}\left(-\frac{3}{5}\right)\right]$ **40.** $\cot(\text{Tan}^{-1} 1)$
41. $\csc\left(\text{Cos}^{-1}\frac{\sqrt{5}}{3}\right)$ **42.** $\csc\left(\text{Sin}^{-1}\frac{5}{8}\right)$ **43.** $\csc(\text{Tan}^{-1}\sqrt{3})$ **44.** $\sin(\text{Arccsc}\, 3)$

Challenge **Evaluate.**

1. $\sin^{-1}\left(\tan\frac{\pi}{4}\right)$ **2.** $\sin\left(\frac{\pi}{2} - \text{Tan}^{-1} 1\right)$ **3.** $\sin\left(2\,\text{Cos}^{-1}\frac{3}{5}\right)$

Problem Solving Application: **Working Backwards**

Sometimes we need to find the shortest or least expensive path between two points in order to solve a problem. This can always be accomplished by looking at *all* possible paths between the points, but that strategy can be extremely tedious and time-consuming if a large number of paths is involved. Often, by **working backwards,** the solution can be determined more quickly.

Example

1 **The figure at the right shows several city blocks in Pine City along with the time required, in minutes, to travel each block. Burchfield Freight must ship a package from point *A* to point *Z*. Which path should be used to complete the trip in the least amount of time?**

Explore Any path from A to Z must go along 4 blocks. Since 4-block paths must contain 2 horizontal and 2 vertical paths, there are $\frac{4!}{2!2!}$ or 6 different paths.

Plan Instead of finding travel times for the 6 paths by adding individual times, work backwards since that involves less computations. First, find the least travel times from C to Z, E to Z, and G to Z. Then, use these travel times to find the least travel times from B to Z and D to Z. Finally, use these last two travel times to find the least travel time from A to Z.

Solve There is one path from C to Z, which requires 11 minutes to travel. There is one path from G to Z, which requires 10 minutes to travel. There are two paths from E to Z. Since $3 + 8 > 5 + 4$, the least travel time from E to Z is 9 minutes on path EHZ. These times are shown in the figure at the right.

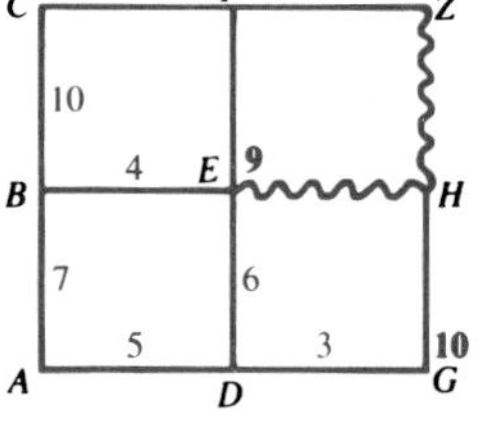

The path from B to Z must go through C or E. Since $10 + 11 > 4 + 9$, the least travel time from B to Z is 13 minutes on path $BEHZ$. The path from D to Z must go through E or G. Since $6 + 9 > 3 + 10$, the least travel time from D to Z is 13 minutes on path $DGHZ$. These times are shown in the figure at the right.

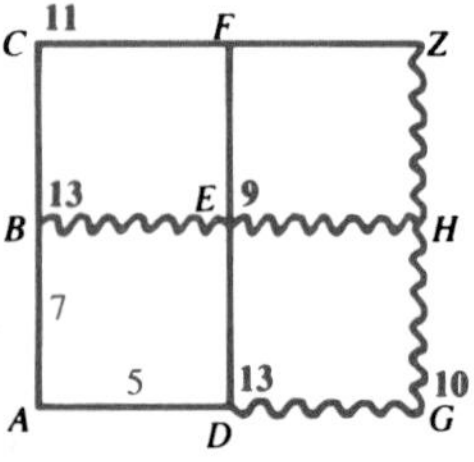

Finally, the path from A to Z must go through B or D. Since $7 + 13 > 5 + 13$, the least travel time from A to Z is 18 minutes on path $ADGHZ$.

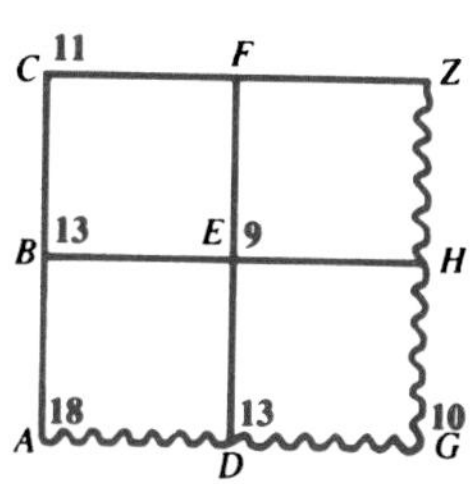

Examine You may want to compute the travel times, by adding individual times, for a few of the 6 possible paths to convince yourself that path $ADGHZ$ is the path with the least travel time.

Exercises

Written **In exercises 1 and 2, several city blocks in Beckville are shown, along with the time required, in minutes, to travel each block. Use the method shown in example 1 to find the shortest travel time from A to Z.**

1.

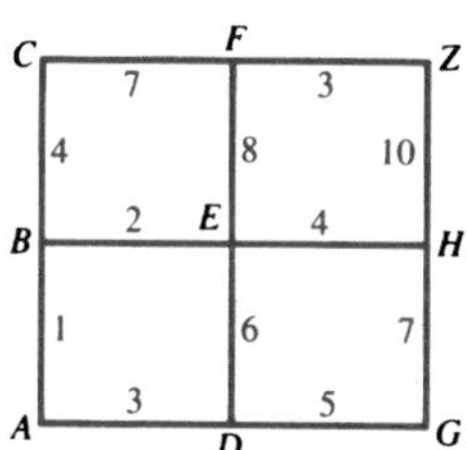

2.

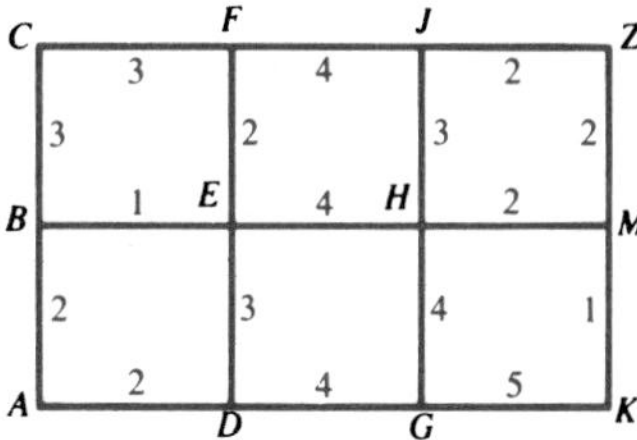

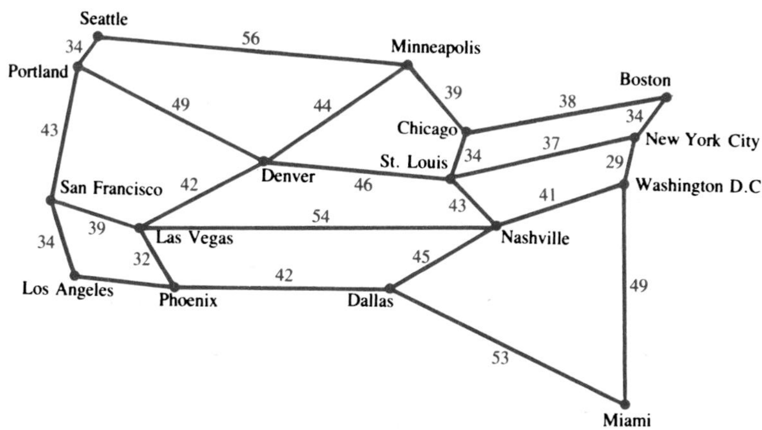

The figure at the left shows a communication's network along with the cost, in cents, to use each communication's path. Use the method shown in example 1 to find the least expensive path between each pair of cities. *Note: This figure is the same as a 3 x 3 grid.*

3. Las Vegas to Boston

4. New York City to Los Angeles

5. Seattle to Miami

Vocabulary

trigonometric identity
arcsine, Arcsine
arccosine, Arccosine
arctangent, Arctangent

Chapter Summary

1. Trigonometric identities enable us to simply expressions to facilitate solving trigonometric equations.
2. A list of commonly used trigonometric identities with page references appears on page 623.
3. Trigonometric equations are those in which the variable is a trigonometric expression.

Chapter Review

9.1 **Prove that each of the following is an identity.**

1. $\frac{\tan x + \cot x}{\csc x} = \sec x$
2. $\frac{\tan \theta}{\sec \theta + 1} = \frac{1 - \cos \theta}{\sin \theta}$

9.2 **Solve for all values of θ in the interval $0° \leq \theta < 360°$. State answers to the nearest degree.**

3. $2 \sin \theta = \sqrt{3}$
4. $\sec \theta = -\sqrt{2}$
5. $3 \cot \theta + \sqrt{3} = 0$
6. $3 \cos \theta + 4 = 9 - \cos \theta$
7. $4 \sin \theta \cos \theta = \sin \theta$

9.3

8. $\tan^2 \theta + \tan \theta = 0$
9. $2 \cos^2 \theta = 1 + \cos \theta$
10. $\sin^2 \theta + 3 \sin \theta + 1 = 0$
11. $2 \cos^2 \theta + \sin \theta = 1$
12. $\sin^2 \theta = \cos^2 \theta - 1$
13. $4 \cos^2 \theta = 1$

9.4 **If $\tan x = \frac{3}{4}$, $180° < x < 270°$, $\sin y = \frac{12}{13}$, $90° < y < 180°$, find the value of each of the following.**

14. $\sin (x + y)$
15. $\cos (x - y)$
16. $\cos (x + y)$
17. $\sin (x - y)$
18. $\tan (-y)$
19. $\cos (-x)$

If $\tan A = 8$ and $\tan B = -\frac{1}{2}$, find the value of each of the following.

20. $\tan (A + B)$
21. $\tan (A - B)$

Prove each identity.

22. $\tan\left(\frac{3\pi}{2} - x\right) = \cot x$

23. $\sin(\pi - x) = \sin x$

24. Find the smallest positive value of x such that $\sin 4x = \cos 50°$.

9.5 **If $\cos\theta = -\frac{2}{3}$ and $180° < \theta < 270°$, find each value.**

25. $\cos 2\theta$

26. $\sin 2\theta$

27. $\tan 2\theta$

If $\cos\theta = \frac{3}{8}$ and $270° < \theta < 360°$, find each value.

28. $\sin\frac{\theta}{2}$

29. $\cos\frac{\theta}{2}$

30. $\tan\frac{\theta}{2}$

31. If $\cos\frac{A}{2} = \frac{1}{5}$, find the value of $\cos A$.

9.6 **Prove that each of the following is an identity.**

32. $\frac{\cos\theta}{\sin 2\theta} = \frac{\csc\theta}{2}$

33. $\sin^2\theta = \frac{1 - \cos 2\theta}{2}$

34. $\cos 2A = \frac{1 - \tan^2 A}{1 + \tan^2 A}$

35. $\frac{1 + \cos 2x}{\cot x} = \sin 2x$

Solve for θ if $0° \leq \theta < 360°$. Give your answers to the nearest degree.

36. $\cos 2\theta + 13\cos\theta - 6 = 0$

37. $5\sin 2\theta + \frac{1}{2}\cos\theta = 0$

38. $\cos 2\theta = 8 - 15\sin\theta$

39. $8\sin\theta\cos\theta = 2\sqrt{3}$

9.7 **Evaluate.**

40. Arccos 1

41. Arcsin $\left(-\frac{\sqrt{2}}{2}\right)$

42. $\text{Tan}^{-1}(-\sqrt{3})$

43. $\sin\left(\text{Tan}^{-1}\frac{12}{5}\right)$

44. $\sec\left(\text{Arccos}\frac{2}{7}\right)$

45. $\csc[\text{Tan}^{-1}(-1)]$

46. $\cot[\text{Arctan}(-5)]$

Choose the best answer.

47. The expression $\cos(x + 90°)$ is equivalent to
   a. $-\sin x$ b. $\sin x$ c. $\cos x$ d. $\cos x + \sin x$

48. The expression $\cos 42° \sin 38° + \cos 38° \sin 42°$ is equivalent to
   a. $\sin 4°$ b. $\cos 4°$ c. $\sin 80°$ d. $\cos 80°$

49. Which is true for all values of x?
   a. $\cos x = \cos(90° + x)$ b. $\cos x = \cos(-x)$
   c. $\cos x = -\sin x$ d. $\sin x = \sin(180° + x)$

50. $1 - 2\sin^2\theta = ?$
   a. $\cos 2\theta$ b. $\cos^2\theta$ c. $\sin 2\theta$ d. $\sin\theta\cos\theta$

51. Which is defined for all values of x in the interval $90° < x < 270°$?
   a. $\tan x$ b. $\cot x$ c. $\csc x$ d. $\csc^2 x$

Chapter Test

Determine whether each statement is true or false. If it is false, write a related true statement.

1. $\sin(x + 90°) = \cos x$
2. $2\cos^2 x - 1 = \cos 2x$
3. $\cos x = \cos(180° + x)$
4. $\cos 12° \cos 4° + \sin 12° \sin 4° = \cos 16°$
5. $\cos(-\theta) = \cos(\theta)$
6. If $A = \text{Arccos}\,\frac{5}{13}$, then $\tan(90° - A) = \frac{12}{5}$.
7. $\tan y \sin y \csc y = \sec y$
8. $\sin(\pi + x) = \sin x$
9. $\sin(50° - 30°) = \sin 50° \cos 30° + \cos 50° \sin 30°$
10. $\sin(-\theta) = -\sin\theta$

11. If $\theta = \text{Arctan}\,1$, what is the measure of acute angle θ?
12. If x is the measure of an acute angle and $\sin x = \frac{3}{5}$, find the value of $\sin(180° - x)$.
13. If $\tan x = \frac{1}{2}$ and $\tan y = 1$, find the value of $\tan(x + y)$.
14. Find the positive value of $\tan\left(\text{Arcsin}\,\frac{3}{5}\right)$.
15. If x and y are measures of acute angles, $\sin x = \frac{3}{5}$, and $\sin y = \frac{1}{2}$, find the value of $\sin(x + y)$.
16. If x is in quadrant I and $\cos x = \frac{4}{5}$, find $\tan \frac{1}{2}x$.
17. If $\sin x = \frac{3}{5}$, find the value of $\cos 2x$.
18. If $\sin(A - 30°) = \cos 60°$, find the smallest positive value of A.
19. If x is the positive measure of an acute angle and $\cos x = \frac{5}{13}$, find $\sin 2x$.
20. If $\cos x = \frac{3}{5}$, find the positive value of $\sin \frac{1}{2}x$.

If $0° \le x < 360°$, solve each equation for x to the nearest degree.

21. $\cos x - 2\cos x \sin x = 0$
22. $3\sin^2 x + \sin x - 2 = 0$
23. $2\cos^2 x = \cos x$
24. $2\cos^2 x + 3\sin x = 3$
25. $\sin^2 x = 1 - \cos 2x$
26. $\sec^2 x + \tan x = 1$
27. $7\cos x + 1 = 6\sec x$
28. $\tan x - 2\cot x = 1$
29. $\sin 2x = \sqrt{3}\cos x$

Prove that each of the following is an identity.

30. $\sin 2x = \dfrac{2\tan x}{1 + \tan^2 x}$
31. $\dfrac{1 + \cos 2A}{\sin 2A} = \cot A$
32. $2 - \sec^2 x = (\cos 2x)(\sec^2 x)$
33. $\dfrac{\cos\theta + \cot\theta}{\cos\theta \cot\theta} = \tan\theta + \sec\theta$
34. $\cot x = \dfrac{\cos 2x + \cos x + 1}{\sin 2x + \sin x}$

More Work with Transformations

Chapter **10**

Reflections, translations, rotations, and dilations can be used in making various designs. Which of these transformations do you think could be used in making the design shown?

10.1 Composing Transformations

What happens when one transformation is performed followed by another?

Consider $\triangle ABC$ shown at the right. First, reflect $\triangle ABC$ over line ℓ, producing $\triangle A'B'C'$. Now, reflect $\triangle A'B'C'$ over line m, producing $\triangle A''B''C''$.

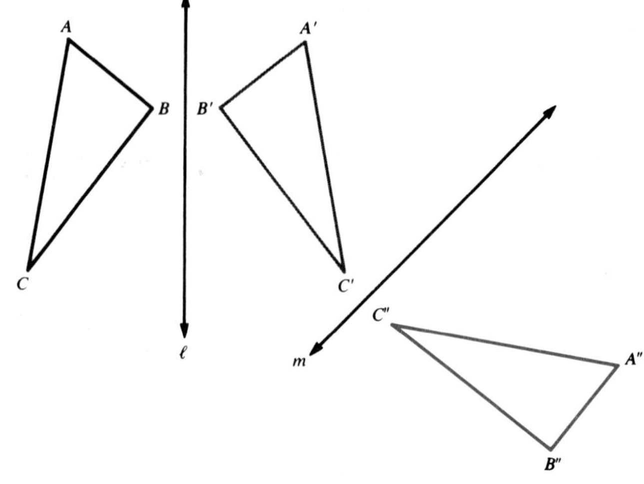

The operation which combines two transformations (or any two mappings) in this way is called a **composition.** The resulting transformation is called the **composite transformation.** In this case, the composite of R_m and R_ℓ is the transformation that maps $\triangle ABC$ to $\triangle A''B''C''$.

A notation for the composite of R_m and R_ℓ is $R_m \circ R_\ell$, which is read "R_m following R_ℓ". $R_m \circ R_\ell$ means to perform R_ℓ first, and then R_m. For $\triangle ABC$ shown above,

$$\triangle ABC \xrightarrow{R_\ell} \triangle A'B'C' \xrightarrow{R_m} \triangle A''B''C''.$$

Two alternative ways of writing this are shown below.

$$\triangle ABC \xrightarrow{R_m \circ R_\ell} \triangle A''B''C'' \qquad R_m(R_\ell(\triangle ABC)) = \triangle A''B''C''$$

Examples

1 **In the figure at the right, $\triangle KLM$ maps to $\triangle K'L'M'$ under R_ℓ. $\triangle K'L'M'$ maps to $\triangle K''L''M''$ under the half-turn $Rot_{P,180°}$. Thus the composite $Rot_{P,180°} \circ R_\ell$ maps $\triangle KLM$ to $\triangle K''L''M''$.**

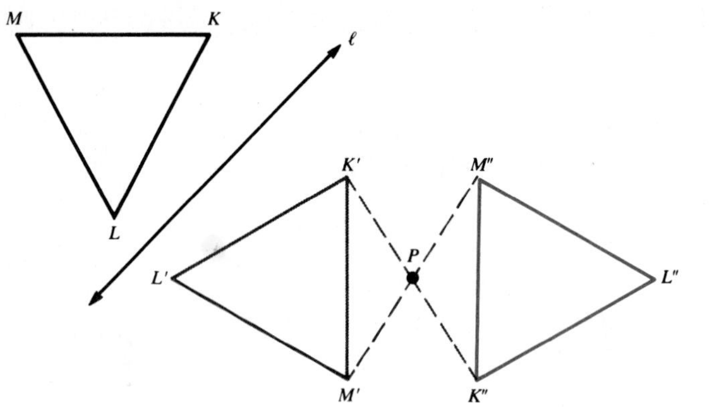

In symbols, we have

$$\triangle KLM \xrightarrow{Rot_{P,180°} \circ R_\ell} \triangle K''L''M''$$

or

$$Rot_{P,\ 180°}(R_\ell(\triangle KLM)) = \triangle K''L''M''.$$

2 **$\triangle ABC$ has been reflected over line ℓ, producing $\triangle A'B'C'$. Then $\triangle A'B'C'$ was translated 4 cm in a direction parallel to line ℓ.**

This composite is an example of a transformation called a *glide* reflection.

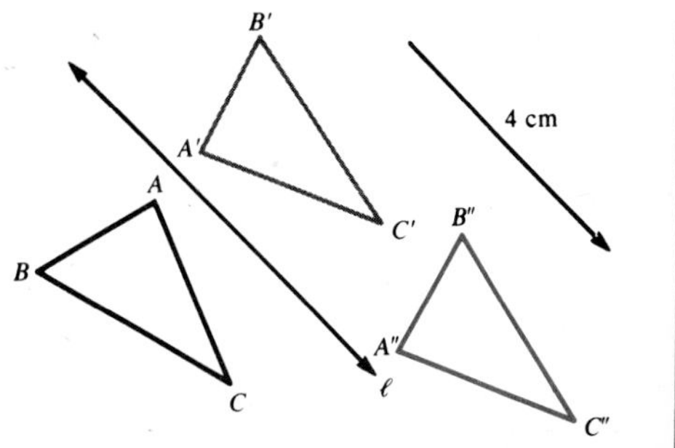

Glide Reflection

A **glide reflection** is a transformation which is the composite of a line reflection and a translation whose direction is parallel to the line of reflection.

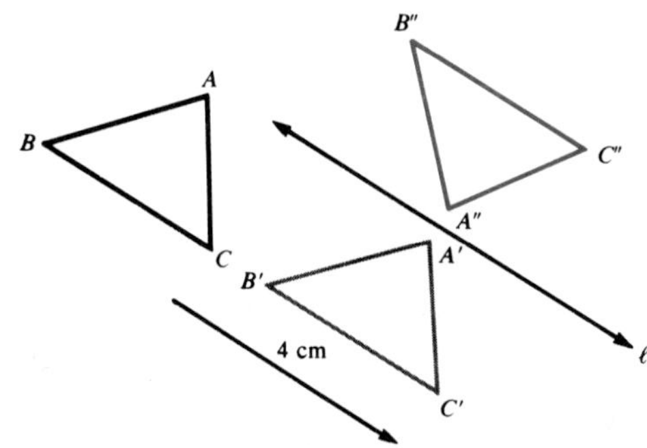

In example 2, $\triangle ABC$ was reflected over line ℓ first. Then the result was translated. The figure at the left suggests that it does not matter if the reflection or the translation is done first.

Examples

3 **Given the triangle with vertices $A(-5, 2)$, $B(-1, 4)$, and $C(-3, 7)$, find its image under $T_{6,-5} \circ R_{y\text{-axis}}$.**

First use the rule $(x, y) \xrightarrow{R_{y\text{-axis}}} (-x, y)$.

$A(-5, 2) \longrightarrow A'(5, 2)$

$B(-1, 4) \longrightarrow B'(1, 4)$

$C(-3, 7) \longrightarrow C'(3, 7)$

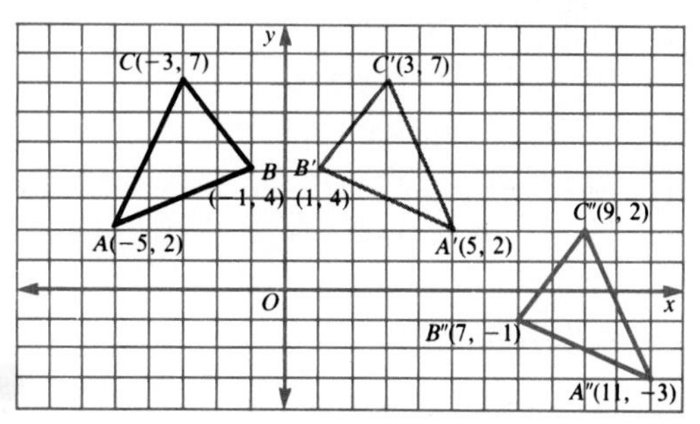

Then use $(x, y) \xrightarrow{T_{6,-5}} (x + 6, y - 5)$.

$A'(5, 2) \longrightarrow A''(11, -3)$ $\quad B'(1, 4) \longrightarrow B''(7, -1)$ $\quad C'(3, 7) \longrightarrow C''(9, 2)$

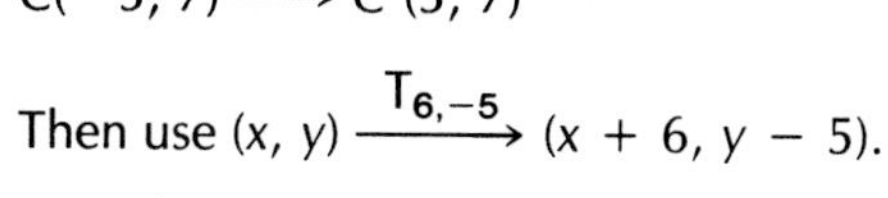

Thus, $\triangle ABC \xrightarrow{T_{6,-5} \circ R_{y\text{-axis}}} \triangle A''B''C''$.

4 **Using the triangle of example 3, find the image of $\triangle ABC$ under $R_{y\text{-axis}} \circ T_{6,-5}$.**

This time begin with the rule

$$(x, y) \xrightarrow{T_{6,-5}} (x + 6, y - 5).$$

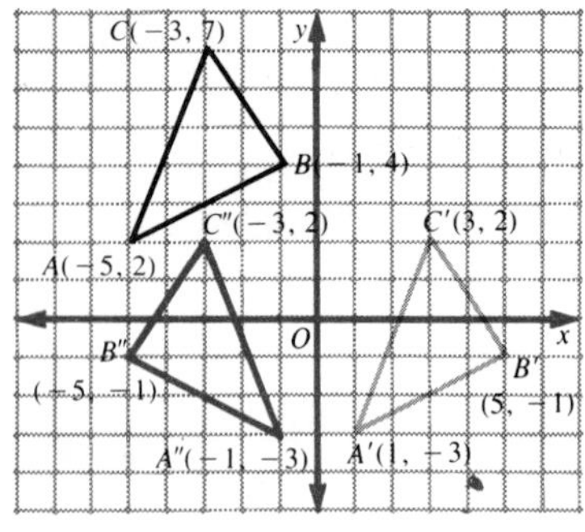

$A(-5, 2) \longrightarrow A'(1, -3)$
$B(-1, 4) \longrightarrow B'(5, -1)$
$C(-3, 7) \longrightarrow C'(3, 2)$

Then use $(x, y) \xrightarrow{R_{y\text{-axis}}} (-x, y)$.

$A'(1, -3) \longrightarrow A''(-1, -3)$
$B'(5, -1) \longrightarrow B''(-5, -1)$
$C'(3, 2) \longrightarrow C''(-3, 2)$

Thus, $\triangle ABC \xrightarrow{R_{y\text{-axis}} \circ T_{6,-5}} \triangle A''B''C''$.

Compare the results of examples 3 and 4. Clearly,

$$T_{6,-5} \circ R_{y\text{-axis}} \neq R_{y\text{-axis}} \circ T_{6,-5}.$$

Do you think composition of line reflections is commutative? How about composition of translations? Can you justify your answer?

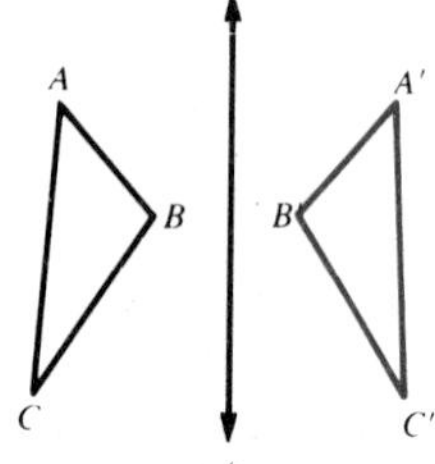

It is possible to form the composite of a transformation with itself. In the figure at the left, what is the effect of $R_\ell \circ R_\ell$?

You are right if you said that you end up where you started. That is, the composite transformation $R_\ell \circ R_\ell$ maps every point in the plane to itself.

Identity Transformation

A transformation that maps each point of the plane to itself is called an **identity transformation.**

The symbol I represents the identity transformation. Since $R_\ell \circ R_\ell$ leaves every point fixed, it follows that $R_\ell \circ R_\ell = I$.

The translation $T_{0,0}$ with rule $(x, y) \xrightarrow{T_{0,0}} (x, y)$ maps each point to itself. Therefore, translation $T_{0,0} = I$. Can you find other rules for the identity transformation?

Exercises

Exploratory **Explain the meaning of each term or symbol.**

1. transformation
2. line reflection
3. translation
4. rotation about O
5. half-turn with center (0, 0)
6. dilation
7. R_ℓ
8. $D_{O,\frac{1}{2}}$
9. $T_{4,-2}$
10. composition
11. identity transformation
12. glide reflection
13. $R_m(A) = B$
14. $R_\ell(R_m(C)) = D$
15. $Rot_{O,60°}$
16. In finding the image of a point under $R_\ell \circ Rot_{O,180°}$, which transformation is performed first?

In the figure at the right, use a single letter to name the following points.

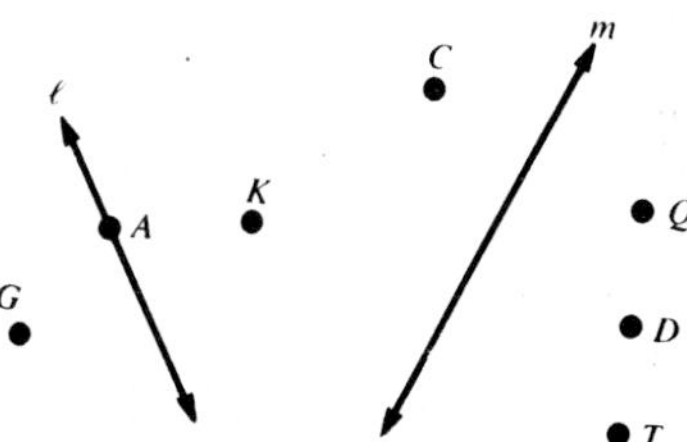

17. $R_\ell(K)$
18. $R_\ell(A)$
19. $R_m(K)$
20. $Rot_{D,180°}(T)$
21. $R_m(R_m(Q))$
22. $R_m(Rot_{D,180°}(T))$
23. $R_m(R_\ell(G))$
24. $R_m(Rot_{D,180°}(Q))$
25. $R_\ell(R_m(T))$

Written **Copy and enlarge the figure at the right. Find the image of point *P* under each composite.**

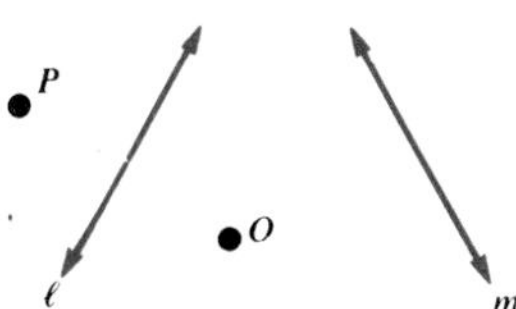

1. $R_m \circ R_\ell$
2. $R_\ell \circ R_m$
3. $R_m \circ R_m$
4. $Rot_{O,180°} \circ R_\ell$
5. $R_\ell \circ R_{O,180°}$
6. $R_m \circ Rot_{O,-90°}$
7. $D_{O,2} \circ R_m$
8. $R_\ell \circ D_{O,\frac{1}{2}}$
9. $D_{O,-3} \circ R_m$
10. $R_m \circ Rot_{O,90°} \circ D_{O,-\frac{1}{2}}$
11. $R_m \circ D_{O,\frac{1}{2}} \circ R_\ell$
12. $T_{0,5} \circ T_{0,-5}$

Graph the triangle with vertices $A(-6, 8)$, $B(-4, 2)$, and $C(-1, 4)$. Then find the image of $\triangle ABC$ under each composite.

13. $R_{y\text{-axis}} \circ R_{x\text{-axis}}$
14. $R_{x\text{-axis}} \circ R_{y\text{-axis}}$
15. $Rot_{O,180°} \circ R_{y\text{-axis}}$
16. $T_{-2,3} \circ T_{4,2}$
17. $R_{y\text{-axis}} \circ R_{y=x}$
18. $D_{O,\frac{1}{2}} \circ R_{x=3}$
19. $R_{x\text{-axis}} \circ D_{O,2}$
20. $D_{O,-2} \circ R_{y\text{-axis}}$
21. $Rot_{O,180°} \circ T_{0,8} \circ R_{y\text{-axis}}$
22. Copy and enlarge the figure. Find the image of $\triangle ABC$ under the glide reflection consisting of R_ℓ and a translation 3 cm in the indicated direction.

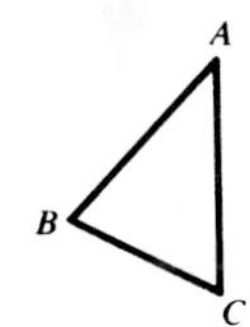

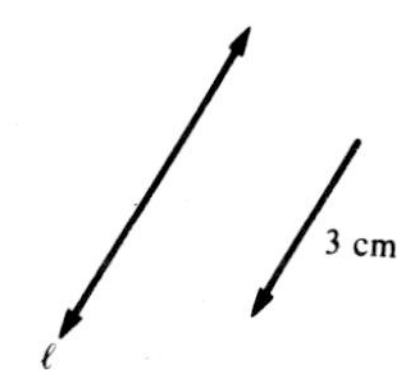

Answer the following questions.

23. Is composition of rotations with the same center commutative? Explain.
24. Is the composite of two translations equivalent to a single translation? Explain.
25. Is the composite of two rotations with the same center equivalent to a single rotation? Explain.
26. Is the composite of two line reflections equivalent to a single line reflection? Explain.

Which of the following are equivalent to the identity transformation? Justify your answer.

27. $R_\ell \circ R_\ell$
28. $Rot_{P,60°} \circ Rot_{P,120°}$
29. $Rot_{P,180°} \circ Rot_{P,180°}$
30. $Rot_{O,120°} \circ Rot_{O,240°}$
31. $T_{-2,3} \circ T_{2,-3}$
32. $R_\ell \circ R_m \circ R_m \circ R_\ell$

Copy and enlarge each figure. Find the image of the figure under $R_\ell \circ R_m$.

33.

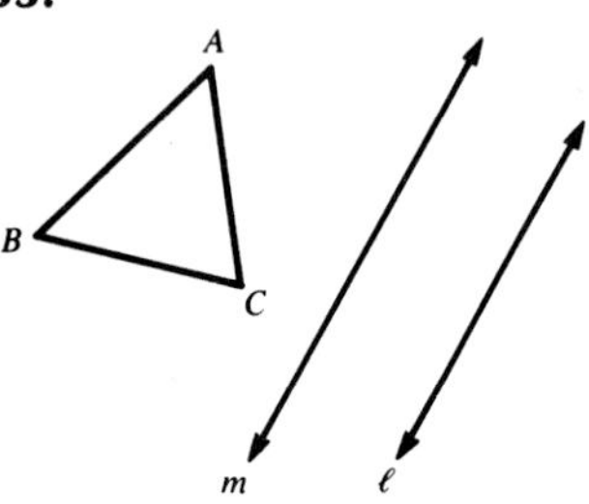

34.

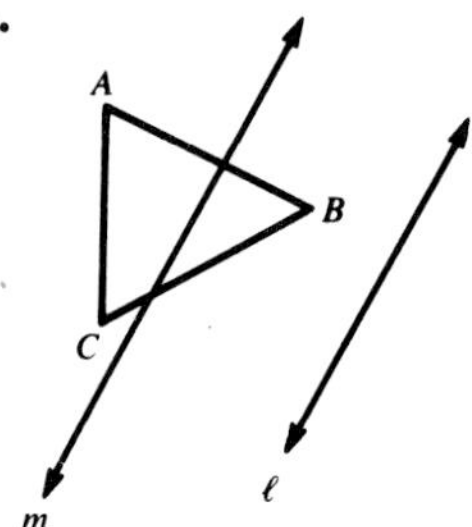

35.

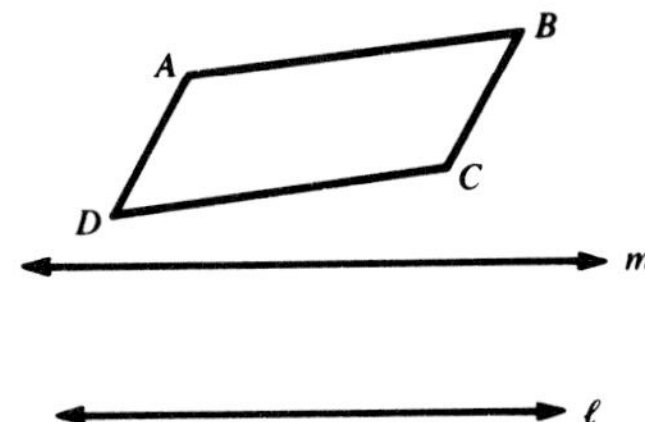

36.

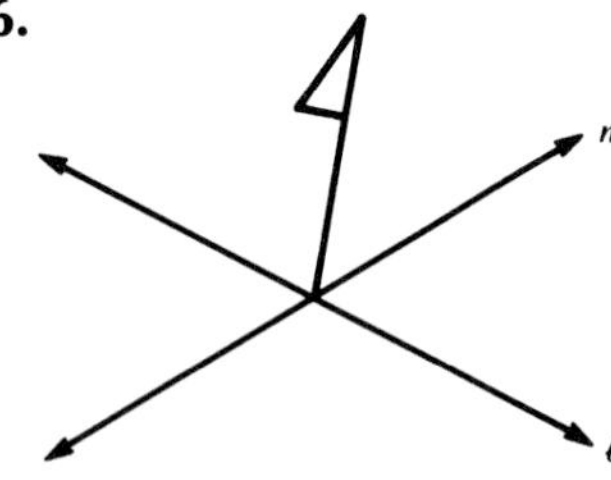

37.

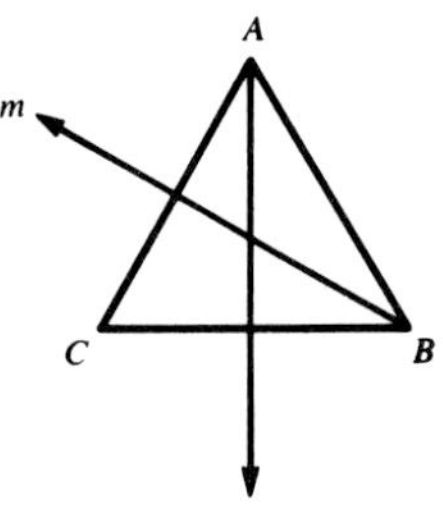

38.

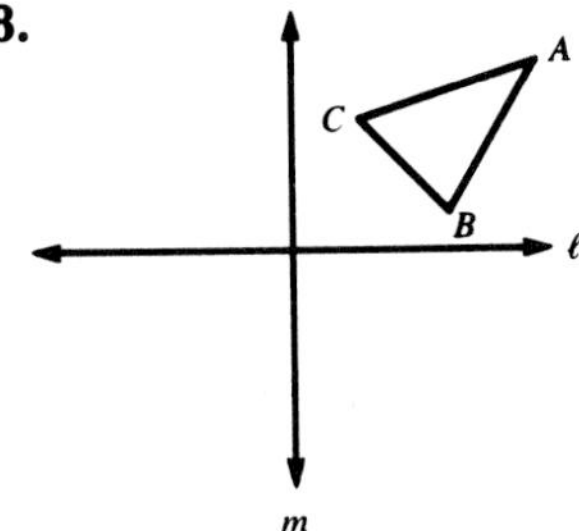

39. Look carefully at your results in exercises 33–35. How are ℓ and m related? What single transformation appears to map the original figure to the final image? Explain.
40. In exercises 36–38, what transformation is equivalent to $R_\ell \circ R_m$?
41. Let ℓ be a given line and P a point not on ℓ. Let G be a glide reflection with respect to ℓ. Suppose $G(P) = P'$. Show that ℓ bisects $\overline{PP'}$.

Recall that if function f is a one-to-one function, then its inverse is also a function. If possible, describe an inverse transformation for each of the following. Justify your answer.

42. $T_{4,5}$
43. R_ℓ
44. $Rot_{O,180°}$
45. $Rot_{O,50°}$

10.2 The Composite of Two Line Reflections

One thing you have probably noticed is that under any line reflection, the distance between the images of any two points is the same as the distance between the two original points. That is, line reflections preserve distance. Since there are other transformations that also preserve distance, it is convenient to introduce a new term to describe these transformations.

Definition of Isometry

An **isometry** is a transformation that preserves distance.

A line reflection is an isometry.

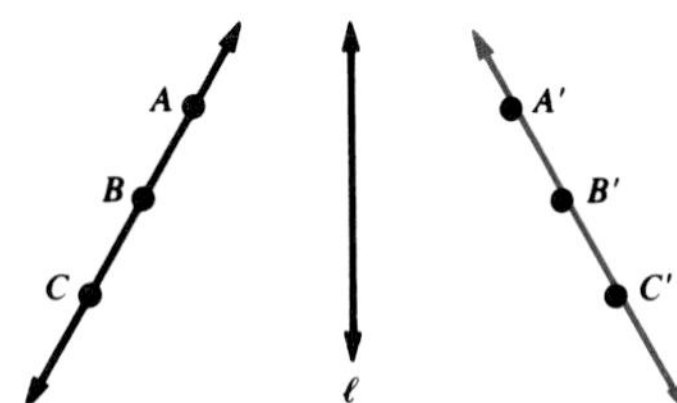

In the figure, if A, B, and C are *collinear,* and B is *between* A and C, then A', B', and C' are also *collinear* and B' is *between* A' and C'.

A line reflection preserves collinearity and betweenness of points.

Notice also that a line reflection maps lines to lines, segments to segments, rays to rays, and angles to angles.

A line reflection preserves angle measure.

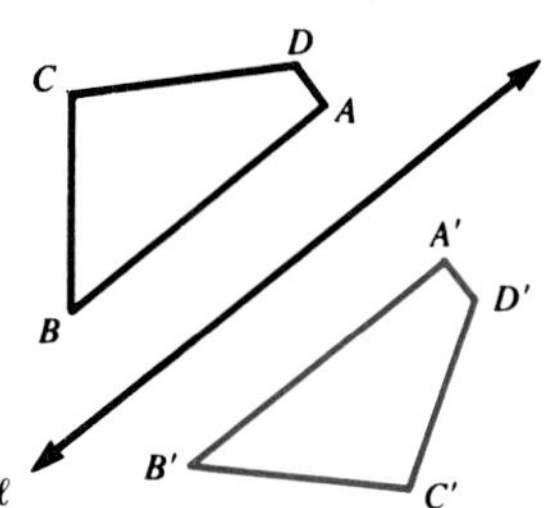

Can you think of a property not preserved by a line reflection? If you are having trouble answering this question, study quadrilateral $ABCD$ and its image under R_ℓ as shown at the left. Going from point A to point B, to point C, to point D, and back again to to point A, is moving in a clockwise direction. Thus, the orientation of quadrilateral $ABCD$ is said to be clockwise. Now, look carefully at quadrilateral $A'B'C'D'$. Its orientation is *counterclockwise.* We see that a *line reflection does not preserve orientation.* In fact, it *reverses* the orientation of the original figure.

Definition of Direct and Opposite Isometries

An isometry that preserves orientation is called a **direct isometry.** An isometry that reverses orientation is called an **opposite isometry.**

Line reflections are opposite isometries.

For the composite of two line reflections, there are three cases to consider. The reflecting lines may be two different parallel lines, the same line, or intersecting lines.

In the following experiment, the first case is considered. The second case is in the exercises and the third case is in the next section.

1. Copy the figure at the right. Note that $\ell \parallel m$.
2. Find the image of $\triangle ABC$ under $R_m \circ R_\ell$. Label the final image $\triangle A''B''C''$.
3. Compare $\triangle ABC$ and $\triangle A''B''C''$. Does there appear to be a single transformation that maps $\triangle ABC$ to $\triangle A''B''C''$?

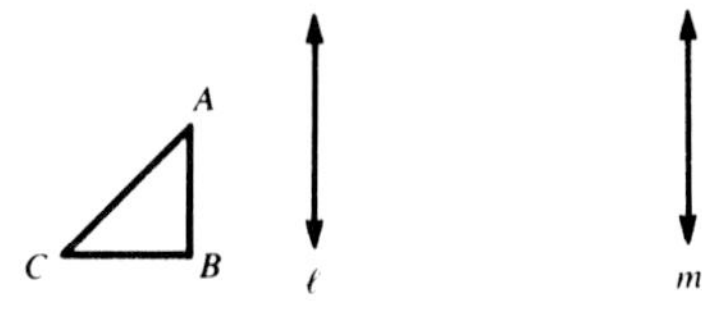

If you have been careful, your results should suggest that $\triangle A''B''C''$ is the image of $\triangle ABC$ under a translation, as shown at the right.

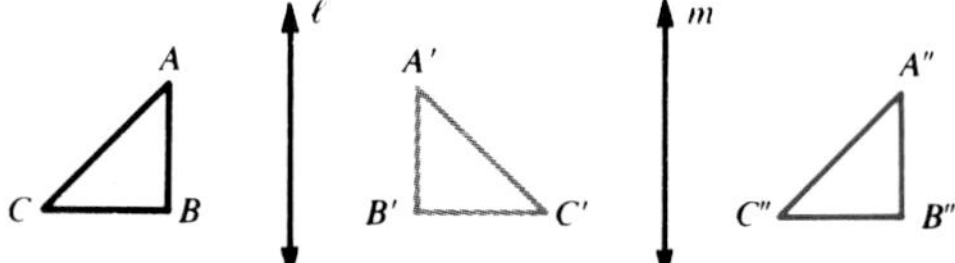

Now, using a ruler, measure the distance of the translation by measuring the distance between any point and its image. Then measure the distance between lines ℓ and m. What appears to be true? Finally, how is $\overline{AA''}$ related to both ℓ and m? The results of the investigation so far may be summarized as follows.

The composite $R_m \circ R_\ell$, where $\ell \parallel m$, is a translation. The distance of the translation is twice the distance between ℓ and m. The translation is in the direction from ℓ to m along a line perpendicular to both ℓ and m.

Do you think it is possible to express a translation as a composite of two line reflections? If so, how do you find the correct lines? These questions are answered by the following statement, which you will be asked to verify in the exercises.

Every translation can be expressed as the composite $R_m \circ R_\ell$, where

a. ℓ and m are parallel;

b. the distance from ℓ to m is half the distance of the translation; and

c. ℓ and m are both perpendicular to the direction of the translation.

Any property of a line reflection that is retained by the composition of line reflections must also be a property of a translation.

A translation is an isometry.
A translation preserves collinearity, betweenness, and angle measure.

Are translations direct or opposite isometries? Think carefully about this question. In the exercises, you will be asked to justify the following assertion.

A translation is a direct isometry.

Exercises

Exploratory **Tell whether each of the following is an isometry. Explain.**

1. line reflections
2. translations
3. dilations

Sketch an example to show that each of the following is an isometry.

4. The composite of two line reflections.
5. The composite of a line reflection and a translation.
6. The composite of two translations.

Answer the following.

7. Explain why the composite of any number of isometries must be an isometry.
8. Is a glide reflection an isometry? Explain.
9. Will the composite of any number of translations be a direct isometry? Explain.
10. Will the composite of any number of line reflections be an opposite isometry? Explain.
11. Explain why a glide reflection may be regarded as a composite of line reflections. Illustrate your answer with a sketch.
12. What does it mean to say that a property is "preserved" by a transformation? Give two examples.

Tell whether the following are direct or opposite isometries. Explain.

13. line reflections
14. translations
15. composite of two line reflections
16. composite of three line reflections

Written **Suppose $\ell \parallel m$ and $A \xrightarrow{R_\ell} B$, $B \xrightarrow{R_m} C$, $D \xrightarrow{R_\ell} E$, and $E \xrightarrow{R_m} F$. Tell whether the following are *always true*, *sometimes true*, or *never true*. Include a sketch.**

1. $AC = DF$
2. A and B are the same point.
3. $\overline{AB} \parallel \overline{DE}$
4. $\overline{AB} \perp \ell$
5. $AC = EF$
6. $ACFD$ is a parallelogram.

7. $R_\ell \circ R_m$ is a translation.

8. $R_\ell \circ R_\ell$ is a translation.

9. $\overline{AD} \parallel \overline{CF}$

10. $R_m \circ R_\ell$ is a line reflection.

Copy and enlarge the figure in which $m \parallel \ell$. Find the image of $\triangle ABC$ under each composite. Label the image $\triangle A''B''C''$.

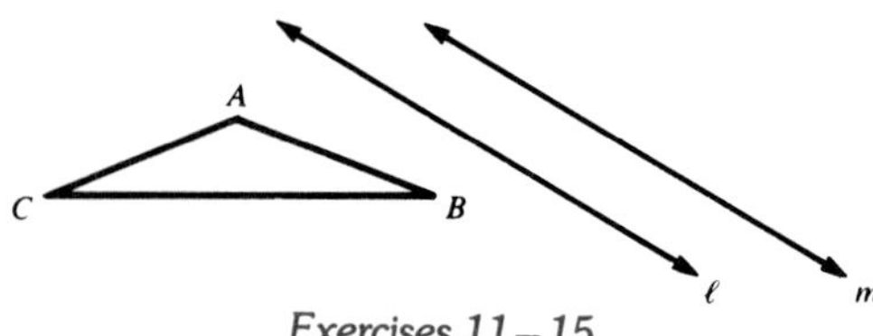

Exercises 11–15

11. $R_m \circ R_\ell$

12. $R_\ell \circ R_m$

13. Give the orientation of the original figure and its image in exercise 11.

Describe, as completely as you can, a single transformation equivalent to each composite.

14. $R_m \circ R_\ell$

15. $R_\ell \circ R_m$

Answer the following.

16. Copy and enlarge the figure. If $A \xrightarrow{R_\ell} A'$, locate line ℓ.

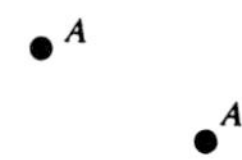

17. Under what circumstances is the composite of two line reflections a translation? Describe the translation.

18. Is every translation a composite of two line reflections? Explain.

19. Copy and enlarge the figure. If A maps to Q under a translation that is equivalent to $R_m \circ R_\ell$, locate line m.

20. Consider $R_\ell \circ R_\ell$ where ℓ is any line. Is this composite a translation? Explain.

21. Compare your results from exercise 20 with the results of the experiment done in the lesson.

22. What properties of line reflections are shared by translations? Which are not? Justify your answers.

In each case $\triangle ABC$ maps to $\triangle DEF$ under a translation. Copy each figure and find lines m and ℓ so that $R_m \circ R_\ell$ is equivalent to the translation.

23.

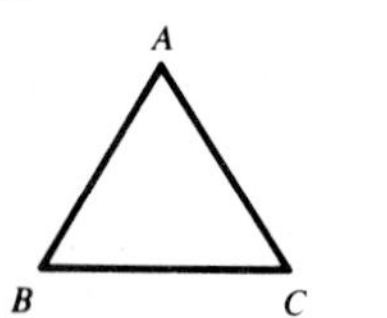

24.

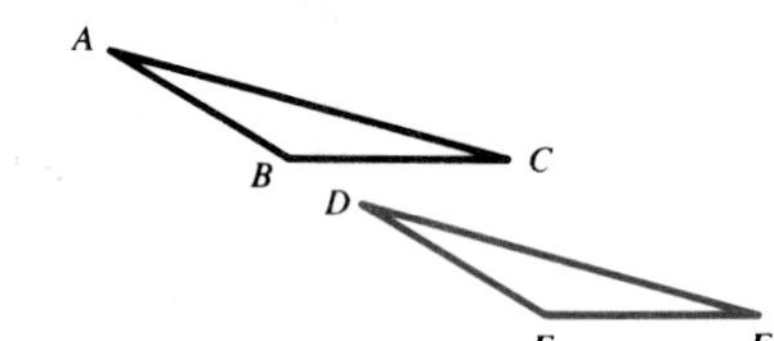

25. In exercises 23 and 24, are the two lines you found unique? Explain and illustrate with a figure.

26. DO SOME RESEARCH. Find out what you can about *strip,* or *frieze,* patterns which are used in many designs. How are these patterns related to the work of this chapter?

10.3 The Composite of Two Line Reflections—Continued

Now let us examine the composite of two line reflections in lines that intersect.

First copy and enlarge Figure 1. In this case ℓ and m intersect at X.

Then find the image of the flag under $R_m \circ R_\ell$. (See Figure 2.)

Now compare the original flag and the final image. What single transformation appears equivalent to $R_m \circ R_\ell$?

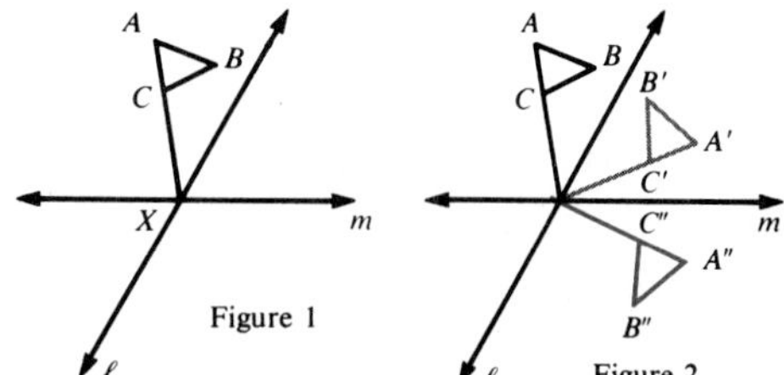

Figure 1

Figure 2

Another surprising result! It appears that the final image is a clockwise rotation of the original figure, as shown above.

In our investigations of translations, we found a relationship between the distance of the translation and the distance between the reflecting lines. Perhaps there is a similar relationship for a rotation and the two reflecting lines.

In the figure, $A \xrightarrow{R_\ell} A' \xrightarrow{R_m} A''$. The acute angle formed by ℓ and m measures 50°. Notice also that the angle of rotation measures 100°. It appears that the measure of the angle of rotation is twice the measure of the angle formed by the reflecting lines! The following statement summarizes our investigation.

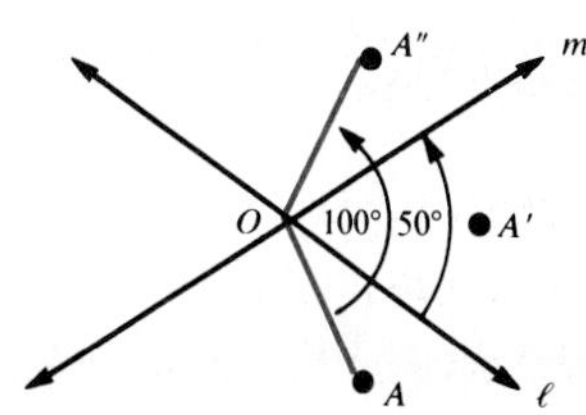

Let line ℓ and m intersect at O. The composite $R_m \circ R_\ell$ is a rotation about point O. The angle of rotation has a measure which is twice that of the acute or right angle formed by ℓ and m. The direction of the rotation is the same as the direction of the acute or right angle as measured from ℓ to m.

Suppose we are given a rotation. You have probably guessed that it is possible to express it as a composite of two line reflections. Compare the following result with our previous investigations of translations.

Every rotation $Rot_{O,2\theta}$ can be expressed as the composite $R_m \circ R_\ell$ where

a. ℓ and m intersect at O;

b. the angle formed by ℓ and m in the direction from ℓ to m has measure θ. If θ is positive, the angle is measured counterclockwise. If θ is negative, the angle is measured clockwise.

Example

1 **In the figure $C \xrightarrow{Rot_{O,-120°}} C'$. Find lines ℓ and m such that $R_m \circ R_\ell = Rot_{O,-120°}$.**

First draw any line through O. Label this line ℓ.

Then through O, draw line m such that the angle from ℓ to m measures $-60°$. You should verify that

$$C \xrightarrow{R_m \circ R_\ell} C'.$$

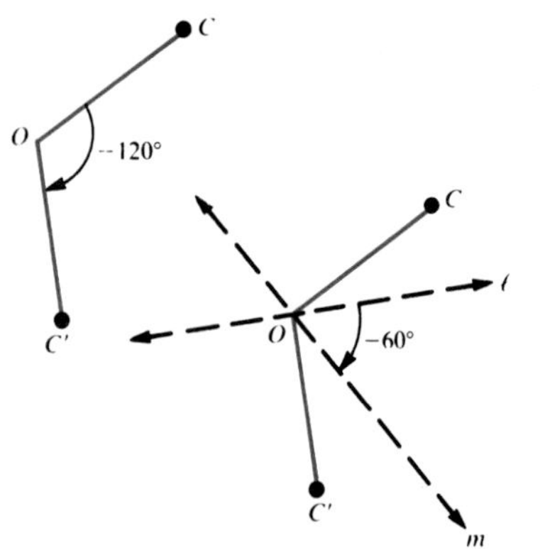

An important special case of a rotation is a rotation of 180° or half-turn. From the results so far, it is clear that a half-turn is the composite of two line reflections in perpendicular lines.

Since every rotation is a composite of two line reflections, we may conclude the following.

Rotations are direct isometries.

Rotations preserve collinearity, betweenness, and angle measure.

Exercises

Exploratory **Using the figure at the right, name each of the following images.**

1. $Rot_{O,60°}(A)$
2. $Rot_{O,-60°}(E)$
3. $Rot_{O,180°}(F)$
4. $Rot_{O,120°}(K)$
5. $Rot_{O,-120°}(O)$
6. $Rot_{O,-180°}(F)$

Under what circumstances will the composite of two line reflections be each of the following?

7. a translation
8. a rotation
9. a half-turn
10. the identity transformation
11. What properties are preserved by rotations? Which are not? Justify your answer.
12. Is a rotation an isometry? Explain.
13. Sara claims that the composite of a reflection and a rotation could not possibly be equivalent to a single translation or rotation. Do you agree? Explain.
14. Can the composite of two rotations about point P be equivalent to the identity transformation? Explain.

Tell whether the composite of each of the following is an isometry. If it is, classify it as direct or opposite. Justify your answer.

15. two rotations

16. a rotation and translation

17. a rotation and line reflection

18. two half-turns

19. a dilation and rotation

20. a rotation and glide reflection

21. three glide reflections

22. eight glide reflections

Written **Copy and enlarge each figure. Then find the image of the figure under $R_{\ell_2} \circ R_{\ell_1}$, and describe, as completely as you can, the effect of $R_{\ell_2} \circ R_{\ell_1}$.**

1.

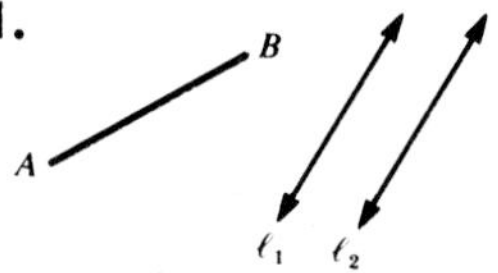

2.

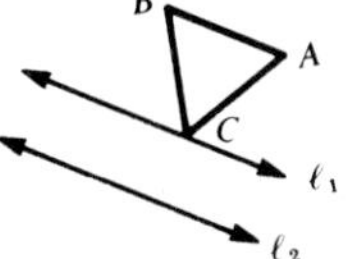

3.

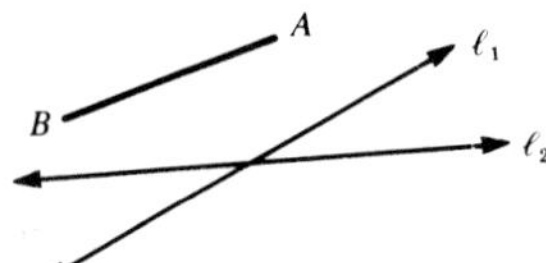

4.

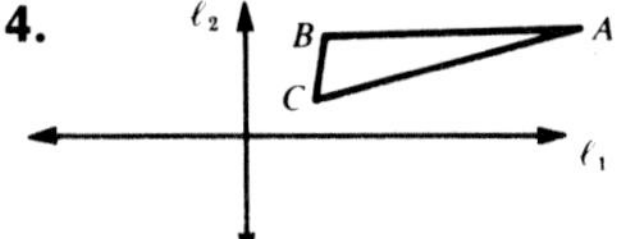

5.

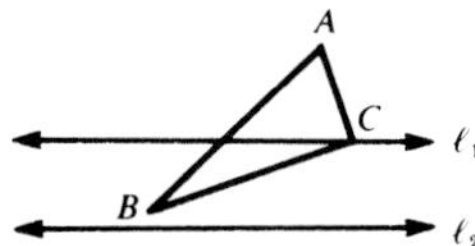

6.

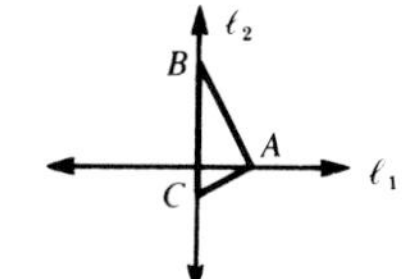

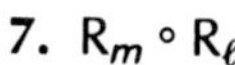

In the figure, lines ℓ and m form an angle of 30° and $A \xrightarrow{R_m \circ R_\ell} B$. Describe a single transformation equivalent to each of the following.

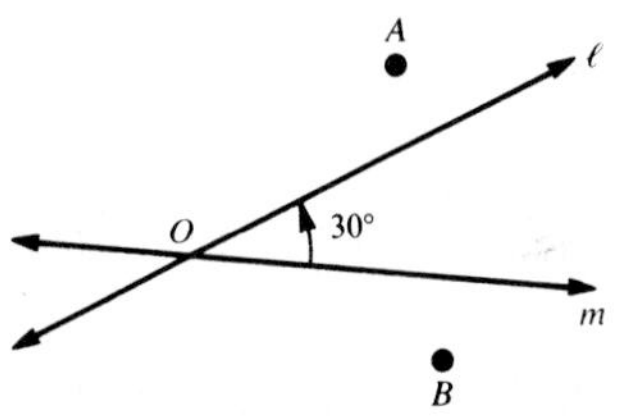

7. $R_m \circ R_\ell$

8. $R_\ell \circ R_m$

9. $R_\ell \circ R_m \circ R_m$

10. $R_m \circ R_\ell \circ R_\ell \circ R_m$

In the figure at the right, $\overline{AB} \xrightarrow{Rot_{O,80°}} \overline{A'B'}$. Copy and enlarge the figure.

11. Find two lines m and n such that $Rot_{O,80°} = R_m \circ R_n$.

12. Copy the figure again. Then find two lines different from the ones you found in exercise 11.

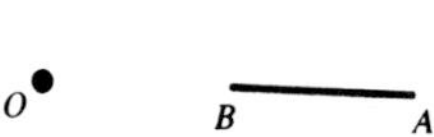

Determine whether each statement is *true* or *false*.

13. $R_{y=x} \circ R_{y\text{-axis}} = Rot_{O,90°}$

14. $R_{y=x} \circ R_{x\text{-axis}} = Rot_{O,90°}$

15. $R_{x\text{-axis}} \circ R_{y=1} = T_{0,-2}$

16. $R_{x=2} \circ R_{y\text{-axis}} = T_{2,0}$

17. $R_{x=1} \circ R_{y\text{-axis}} = R_{y\text{-axis}} \circ R_{x=1}$

18. $R_{x=1} \circ R_{x\text{-axis}} = R_{x\text{-axis}} \circ R_{x=1}$

19. If $\ell \perp m$, then $R_\ell \circ R_m = R_m \circ R_\ell$.

20. If $\ell \parallel m$, then $R_\ell \circ R_m = R_m \circ R_\ell$.

Find the image of each point under $Rot_{O,60°}$ without using a protractor.
Hint: Recall definitions for cos θ and sin θ.

21. (1,0)　**22.** $\left(\frac{1}{2}, -\frac{\sqrt{3}}{2}\right)$　**23.** (2,0)　**24.** (0,1)

25. $\left(\frac{1}{2}, \frac{\sqrt{3}}{2}\right)$　**26.** $(2, 2\sqrt{3})$　**27.** $\left(\frac{\sqrt{3}}{2}, \frac{1}{2}\right)$　**28.** $\left(\frac{\sqrt{2}}{2}, \frac{\sqrt{2}}{2}\right)$

Challenge **Each of the following composite transformations is equivalent to a glide reflection where the line of reflection is parallel to either the *x*-axis or the *y*-axis. Find the composite transformation of that glide reflection.**

> **Sample:** $T_{-2,4} \circ R_{x\text{-axis}}$
>
> $(x, y) \xrightarrow{R_{x\text{-axis}}} (x, -y) \xrightarrow{T_{-2,4}} (x - 2, 4 - y)$
>
> $R_{y=2}$ maps (x, y) to (x, 4 − y), and $T_{-2,0}$ maps (x, 4 − y) to (x − 2, 4 − y).
> Thus, $T_{-2,0} \circ R_{y=2}$ maps (x, y) to (x − 2, 4 − y).
> Therefore, $T_{-2,4} \circ R_{x\text{-axis}} = T_{-2,0} \circ R_{y=2}$.

1. $R_{x\text{-axis}} \circ T_{-2,4}$　**2.** $R_{y\text{-axis}} \circ T_{1,-1}$　**3.** $T_{6,3} \circ R_{x=-3}$

Mixed Review

Solve each equation. For exercises 6–8, 0° ≤ θ < 360°.

1. $20a^2 + a - 12 = 0$　**2.** $4y - 7 + y = 3$

3. $\frac{x-3}{2x} = \frac{x-2}{2x+1} - \frac{1}{2}$　**4.** $4^{2n-1} = \left(\frac{1}{8}\right)^{n+3}$

5. $\log_3 (r + 4) + \log_3 (r - 2) = 3$　**6.** $2 \sin^2 \theta - 1 = 0$

7. $9 \tan^4 \theta - 1 = 0$　**8.** $\cos 2\theta + \cos \theta = -1$

If $\sin x = -\frac{3}{5}$ and *x* is in the third quadrant, find each value.

9. $\cos 2x$　**10.** $\sin 2x$　**11.** $\cos \frac{x}{2}$　**12.** $\tan \frac{x}{2}$

Solve △*ABC*, if possible. Find the measure of sides to the nearest tenth and measures of angles to the nearest ten minutes.

13. $m\angle A = 70$, $m\angle B = 31$, $c = 17$　**14.** $m\angle A = 40$, $b = 8$, $c = 10$

Find the image of (−6, −2) under each composite transformation.

15. $T_{2,-5} \circ Rot_{O,270°}$　**16.** $D_{O,\frac{1}{2}} \circ R_{x\text{-axis}}$　**17.** $Rot_{O,180°} \circ T_{8,8} \circ R_{y=x}$

18. Find the area of parallelogram *ABCD* if $AB = 19$, $BC = 13$, and $m\angle C = 150$.

19. Express $\tan (45° - x)$ in terms of $\sin x$.

20. Evaluate $\cot [\text{Cos}^{-1} (-0.6)]$.

10.4 Three Line Reflection Theorem

Reflections, translations, and rotations are isometries. A glide reflection is the composite of a line reflection and a translation, both of which are isometries. So glide reflections are also isometries.

It appears that the isometries we have studied are line reflections or composites of line reflections. Is it the case that *every* isometry is a composite of line reflections?

Surprisingly, the answer turns out to be yes! Even more amazing is the fact that it never takes more than three line reflections to do the job. (Recall that a glide reflection may be regarded as the composite of three line reflections. Why?) This result is often stated as the **Three Line Reflection Theorem,** which can be proved in more advanced courses.

Three Line Reflection Theorem

Every isometry is equivalent to the composite of, at most, three line reflections.

Although we will not prove the theorem, it will be "tested out" in the examples.

Examples

1 **Suppose $\triangle A'B'C'$ is the image of $\triangle ABC$ under some isometry. Express the isometry as the composite of line reflections.**

Let ℓ be the perpendicular bisector of $\overline{AA'}$. Then

$$A \xrightarrow{R_\ell} A',\ B \xrightarrow{R_\ell} B_1,\ C \xrightarrow{R_\ell} C_1.$$

Let m be the perpendicular bisector of $\overline{C_1C'}$. Since $A'C_1 = A'C'$, it follows that A' is on line m.

A reflection in line m maps C_1 to C'. But in this case it happens that B_1 maps to B' also. So, we have

$$A \xrightarrow{R_m \circ R_\ell} A',\ B \xrightarrow{R_m \circ R_\ell} B',\ C \xrightarrow{R_m \circ R_\ell} C'.$$

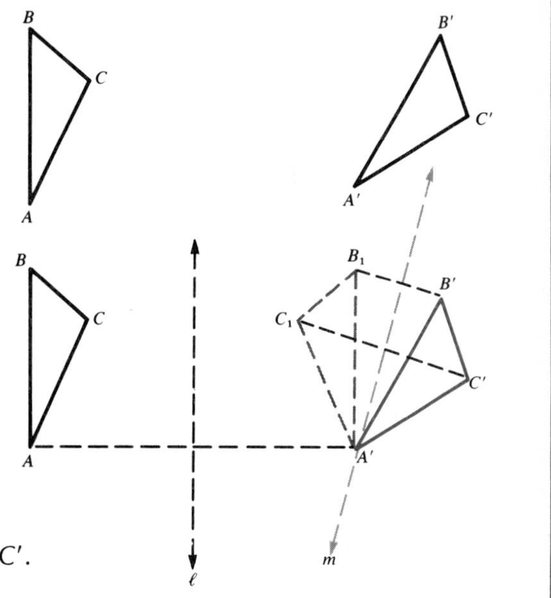

Therefore, the isometry is equivalent to $R_m \circ R_\ell$.

2 **Suppose $\triangle A'B'C'$ is the image of $\triangle ABC$ under some isometry. Express the isometry as the composite of line reflections.**

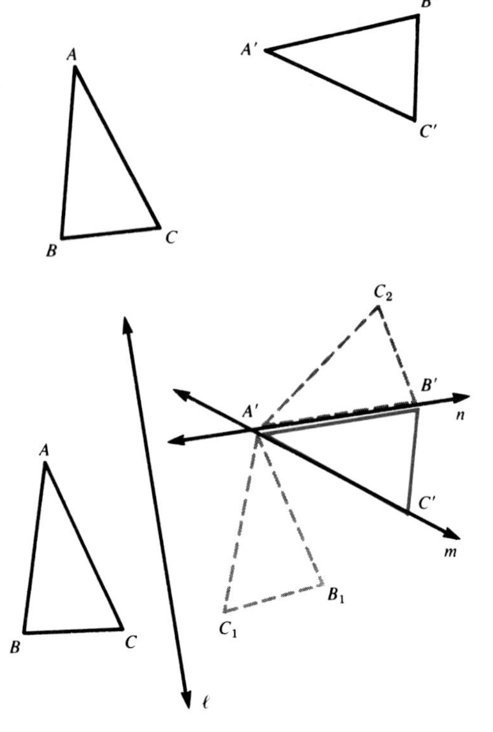

A procedure is outlined here. You should carry out the instructions on a separate piece of paper.

1. Let ℓ be the perpendicular bisector of $\overline{AA'}$. Reflect $\triangle ABC$ over ℓ, producing $\triangle A'B_1C_1$.

2. Let m be the perpendicular bisector of $\overline{B_1B'}$. Reflect $\triangle A'B_1C_1$ over m, producing $\triangle A'B'C_2$. Why will m contain A'?

3. Let n be the perpendicular bisector of $\overline{C_2C'}$. Reflect $\triangle A'B'C_2$ over n, producing $\triangle A'B'C'$. Why will n contain both A' and B'?

Thus, the composite $R_n \circ R_m \circ R_\ell$ maps $\triangle ABC$ to $\triangle A'B'C'$.

Therefore, the original isometry is equivalent to a composite of three line reflections.

You must wonder if there are any other isometries in addition to those already studied. The following corollary of the Three Line Reflection Theorem tells us the answer is no!

Every isometry is equivalent to one of the following: a line reflection, a rotation, a translation, or a glide reflection.

The Three Line Reflection Theorem and its corollary will be examined further in the exercises.

Exercises

Exploratory **In each case describe an isometry that appears to map one figure to the other.**

1.
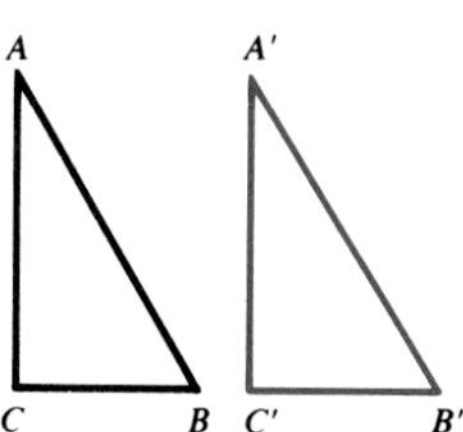

2.
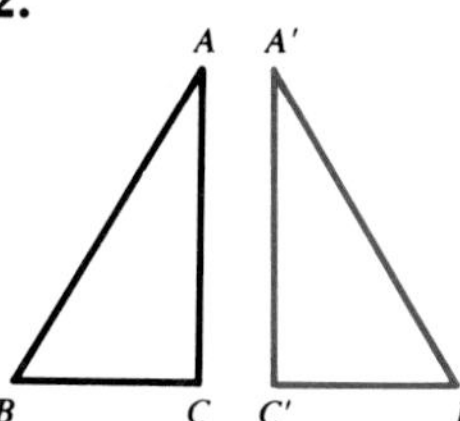

3.
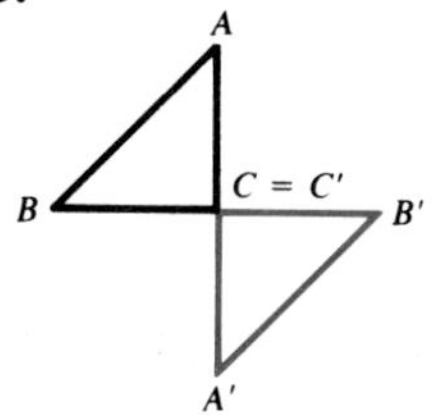

4.
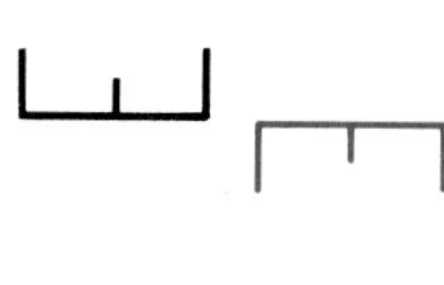

Complete the following.

5. Explain the Three Line Reflection Theorem.
6. Name four isometries.
7. Are there isometries other than those given in exercise 6? Explain.
8. Why can a glide reflection be regarded as the composite of three line reflections? Illustrate with a sketch.
9. An isometry that preserves order is called a direct isometry. Which transformations are direct isometries?

Suppose ℓ, *m*, and *n* are distinct lines. Tell if the statement is *always true, sometimes true,* or *never true*. Justify your answer.

10. $R_m \circ R_\ell$ is a rotation.
11. $R_\ell \circ R_n$ is a direct isometry.
12. $R_n \circ R_m$ is an opposite isometry.
13. $R_\ell \circ R_m$ is a line reflection.
14. $R_\ell \circ R_m$ is a translation.
15. $R_\ell \circ R_m \circ R_n$ is a line reflection.
16. $R_\ell \circ R_m \circ R_n$ is a glide reflection.
17. $R_\ell \circ R_m \circ R_m \circ R_n$ is a translation.

Written **Copy and enlarge each figure. Then verify the Three Line Reflection Theorem by showing that $\overline{AB}$ maps to $\overline{A'B'}$ by a composite of no more than three line reflections.**

1.
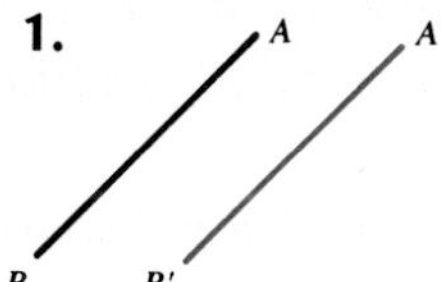

2.
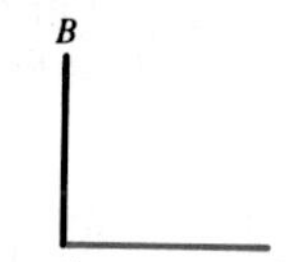

3.

4.

Copy and enlarge △*ABC* and △*DEF*. Then find a composite of line reflections that maps one triangle to the other.

5.

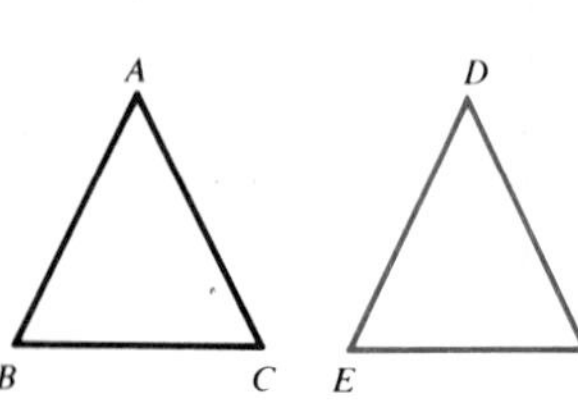

6.

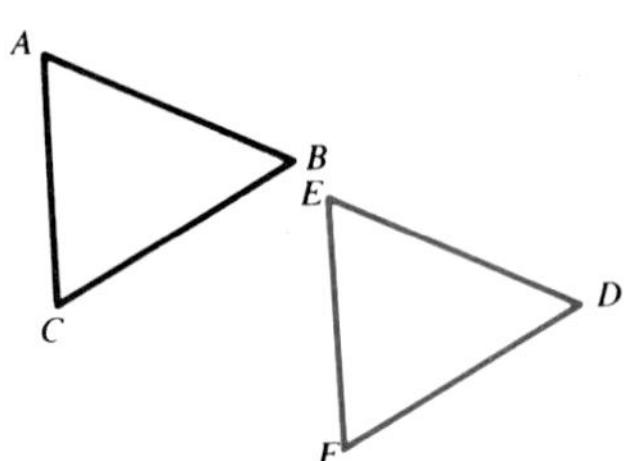

7.

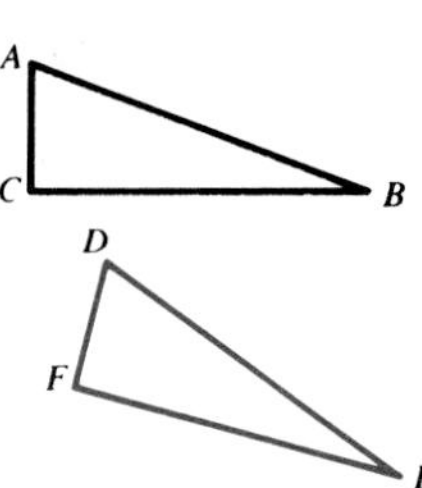

Complete each of the following.

8. On a piece of paper, draw any two congruent triangles in any position. Then find a composite of line reflections that maps one triangle to the other.

9. Draw an equilateral triangle. Name six different transformations that map the triangle to itself.

10. In the figure below, $\triangle A'B'C'$ is the image of $\triangle ABC$ under some isometry. Copy and enlarge the figure. Then find the images of *R* and *S* under the same isometry.

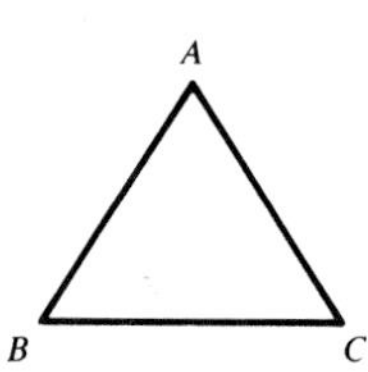

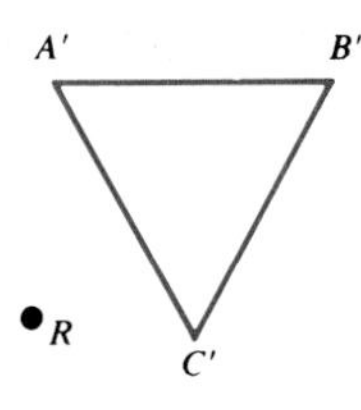

Graph △*ABC* with vertices *A*(0, 0), *B*(6, 1), and *C*(1, −2). Find the image of △*ABC* under the following. Then identify the transformation as a line reflection, translation, rotation, or glide reflection.

11. $R_{x\text{-axis}} \circ R_{y\text{-axis}}$

12. $Rot_{O,180°} \circ R_{y\text{-axis}}$

13. $T_{-2,4} \circ R_{x\text{-axis}}$

14. $T_{2,3} \circ T_{4,-2}$

15. $R_{y=x} \circ R_{x\text{-axis}}$

16. $T_{0,5} \circ R_{y\text{-axis}}$

17. $R_{y\text{-axis}} \circ T_{0,5}$

18. $Rot_{O,90°} \circ T_{4,2}$

19. $R_{x\text{-axis}} \circ T_{-2,4}$

20. $Rot_{O,90°} \circ Rot_{O,-30°}$

21. $T_{4,3} \circ Rot_{O,-60°}$

22. $Rot_{O,180°} \circ Rot_{O,180°}$

Challenge

1. Suppose *P* and *Q* are distinct points. Investigate a half-turn with center *Q* following a half-turn with center *P*. Do you think there is a single transformation equivalent to this composite? If so, describe it as completely as you can and justify your answer.

2. Draw any three distinct parallel lines. Find an equilateral triangle such that one vertex is on each of the three lines. Can you generalize your method?

10.5 Congruence and Similarity Revisited

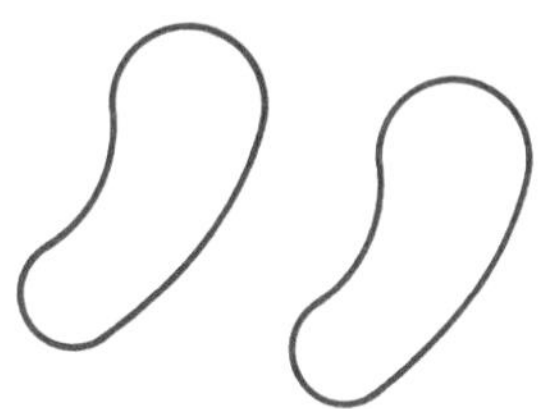

Most people would agree that the two figures at the left are congruent. In the past we have thought of congruent figures as having the same size and shape. Although this description is sufficient for many purposes, it is not a precise mathematical definition. The idea of an isometry can be used to give a more mathematical definition of congruence.

Definition of Congruence

Two figures are congruent if, and only if, there is an isometry that maps one figure onto the other.

In the figure at the left it appears that there is a translation which maps one fish onto the other. Since a translation is an isometry, the two figures are congruent.

Example

1 **Given:** $\overline{AC}$ and $\overline{DE}$ bisect each other at B.
Prove: $\triangle ABD \cong \triangle CBE$

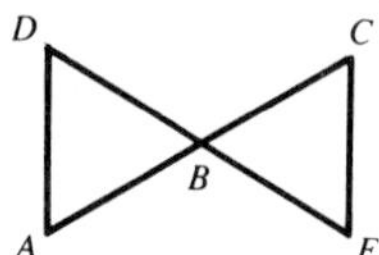

Informal Argument:

STATEMENTS	REASONS
1. $\overline{AB} \cong \overline{BC}$, $\overline{DB} \cong \overline{BE}$	1. Given, definition of bisector
2. B maps to itself, D maps to E, and A maps to C under the half-turn $Rot_{B,180°}$.	2. Definition of half-turn with center B
3. Under $Rot_{B,180°}$, we have $\overline{AB} \longrightarrow \overline{CB}$, $\overline{DB} \longrightarrow \overline{EB}$, and $\overline{AD} \longrightarrow \overline{CE}$. Therefore, $\triangle ABD \xrightarrow{Rot_{B,180°}} \triangle CBE$.	3. A rotation maps segments to segments.
4. $Rot_{B,180°}$ is an isometry.	4. A rotation is an isometry since it is the composite of two line reflections.
5. $\triangle ABD \cong \triangle CBE$	5. Two figures are congruent if there is an isometry that maps one figure onto the other.

Although a dilation is not an isometry, you know from your past experience that a dilation shares some properties with an isometry. Here is a summary of the properties of a dilation.

1. A dilation preserves collinearity and betweenness of points.
2. A dilation preserves angle measure, perpendicularity, and parallelism.
3. Under a dilation a line is parallel to its image.
4. A dilation preserves ratios of distances.

Example

2 **In the figure, $\triangle A'B'C'$ is the image of $\triangle ABC$ under a dilation with center O. If $AC = 6$, $A'B' = 16$, $A'C' = 8$, $BC = 9$, and $BB' = 6$, find $B'C'$, AB, and OB.**

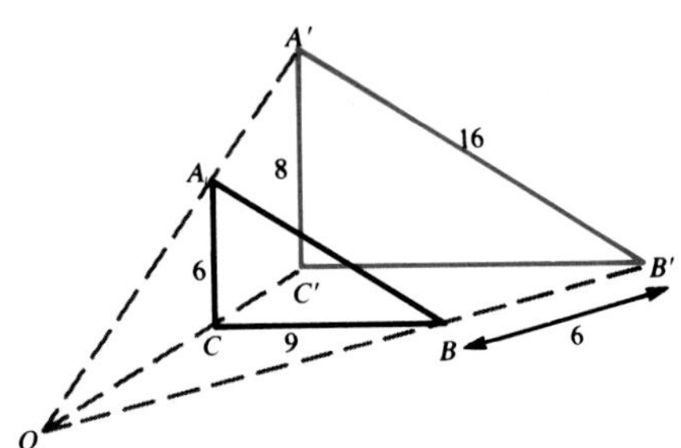

To determine $B'C'$, first find the scale factor k of the dilation $D_{O,k}$.

Since $\overline{AC} \xrightarrow{D_{O,k}} \overline{A'C'}$ it follows that $A'C' = k \cdot AC$, where k is the scale factor of the dilation. Then

$$8 = k \cdot 6.$$
$$k = \frac{8}{6} \text{ or } \frac{4}{3}$$

Since $B'C' = k \cdot BC$, it follows that

$$B'C' = \frac{4}{3} \cdot 9 \text{ or } 12.$$

Since $A'B' = k \cdot AB$, it follows that

$$16 = \frac{4}{3} \cdot AB.$$
$$\frac{3}{4} \cdot 16 = AB$$
$$12 = AB$$

We know that $OB' = \frac{4}{3}OB$. But $OB' = OB + BB'$ and $BB' = 6$. Therefore, $OB' = OB + 6$.

By equating the two expressions for OB', it follows that

$$OB + 6 = \frac{4}{3}OB.$$
$$6 = \frac{1}{3}OB$$
$$18 = OB$$

Just as we were able to use the idea of an isometry for a general definition of congruence, so can we use the idea of a dilation for a general definition of similarity. Before presenting this definition, though, let us consider what happens when we compose an isometry and a dilation. What properties will be preserved by such a transformation?

To help answer this question, consider $\triangle ABC$ shown at the right. First perform $D_{O,2}$, producing $\triangle A'B'C'$.

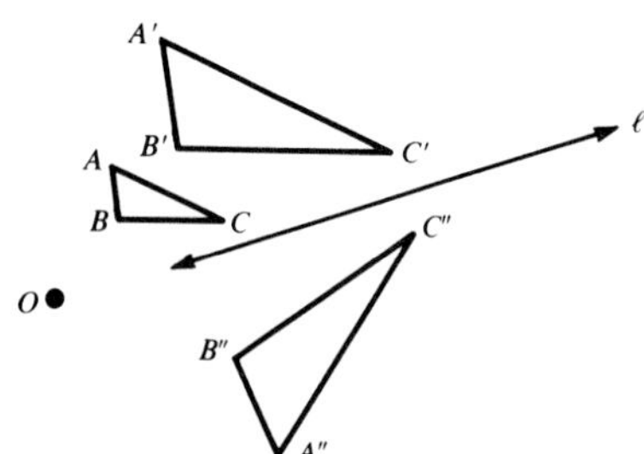

Now reflect $\triangle A'B'C'$ in line ℓ producing $\triangle A''B''C''$. Thus,

$$\triangle ABC \xrightarrow{R_\ell \circ D_{O,2}} \triangle A''B''C''.$$

It appears that one important feature has been retained by this composite: shape. The composite $R_\ell \circ D_{O,2}$ is an example of a transformation called a *similarity transformation*.

Definition of Similarity Transformation

A **similarity transformation** is a composite of isometries and dilations.

Is an isometry also a similarity transformation? *Yes. Why?*

We can now give a general definition of **similar figures.** Compare this definition with the definition of congruent figures.

Definition of Similar Figures

Two figures are similar if, and only if, there is a similarity transformation that maps one figure onto the other.

Exercises

Exploratory In each case, use the new definition of congruence to justify that each pair of figures are congruent.

1.

2.

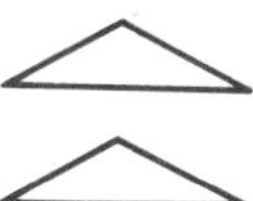

3.

4.

5.

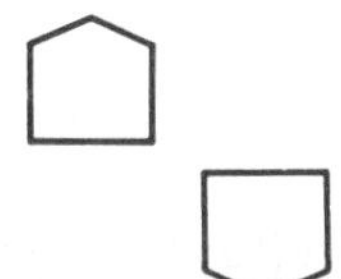

6.

Answer the following.

7. In a transformation approach to geometry, when are two figures congruent? Give an example.
8. How does the definition given in exercise 7 compare with the definition of congruent triangles you have previously encountered?
9. If two triangles are congruent according to the definition given last year, must they be congruent according to the definition given in this section?
10. Can a dilation ever be an isometry? Explain.
11. What is a similarity transformation?
12. Is an isometry a similarity transformation? Explain.
13. In the context of this section, when are two figures similar?
14. How does the definition given in exercise 13 compare with the definition of similar polygons given last year?
15. Is a dilation a similarity transformation? Explain.

Tell whether the following properties are preserved by a similarity transformation. Justify your answer.

16. collinearity
17. betweenness
18. distance
19. parallelism
20. angle measure
21. perpendicularity

Written **Draw two triangles such that one is the image of the other under the following transformation.**

1. line reflection
2. translation
3. rotation
4. glide reflection

Demonstrate with a sketch that a dilation preserves each of the following.

5. collinearity and betweenness
6. angle measure
7. perpendicularity
8. parallelism
9. ratios of distance
10. orientation

Under which of the following is a line parallel to its image?

11. line reflection
12. translation
13. rotation
14. half-turn
15. glide reflection
16. dilation

In each case, describe a dilation that maps $\overline{AB}$ to $\overline{A'B'}$. Assume $\overline{AB} \parallel \overline{A'B'}$.

17.

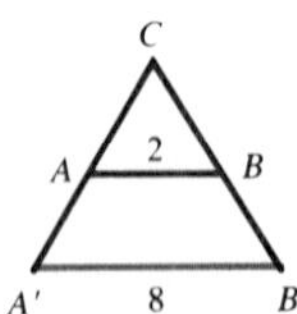

18.

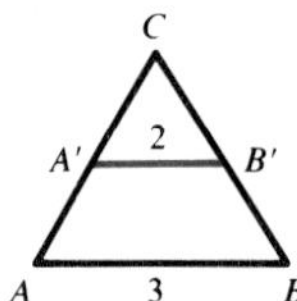

19.

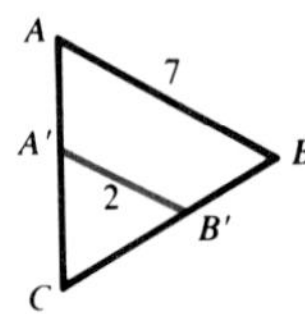

Copy and enlarge each figure. Find the image of each figure under $D_{P,\,2} \circ R_\ell$.

20.

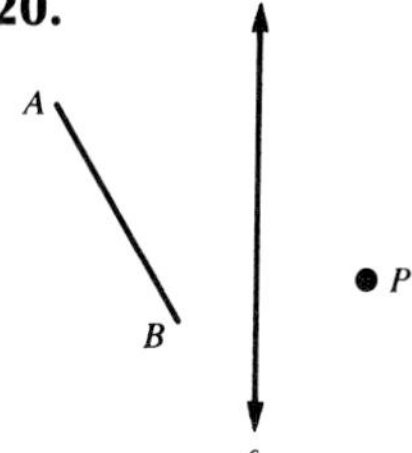

21.

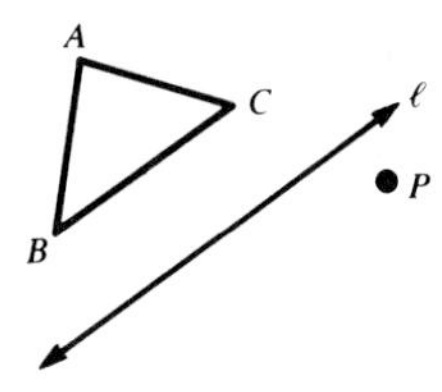

22.

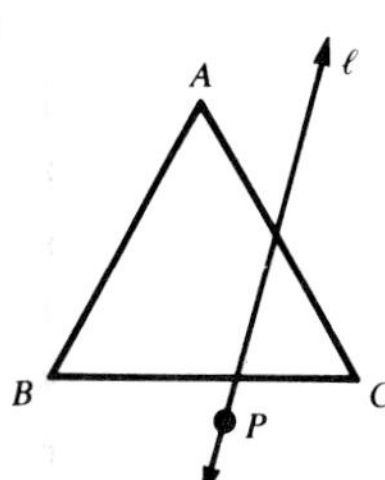

Sketch two figures such that one is the image of the other under the composition of a dilation with scale factor $\frac{1}{2}$ and the following transformation.

23. rotation of 120°
24. dilation, scale factor 2
25. translation
26. line reflection
27. rotation of 100°
28. glide reflection

Give an argument based on transformation geometry to support each conclusion.

29. Given: $\overline{AD} \perp \overline{BC}$, $\overline{BD} \cong \overline{DC}$
Conclusion: $\angle BAD \cong \angle CAD$

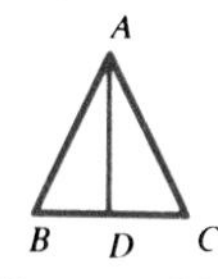

Exercise 29

30. Given: $\overline{CD}$ is the perpendicular bisector of $\overline{AB}$.
Conclusion: $\triangle ACD \cong \triangle BCD$

31. Given: $\overline{CD}$ is the perpendicular bisector of $\overline{AB}$.
Conclusion: $\triangle ACM \cong \triangle BCM$

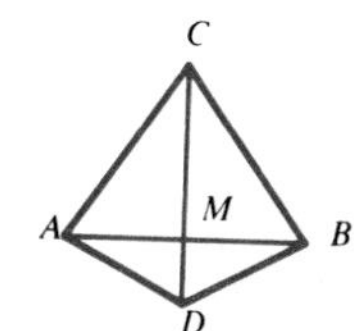

Exercises 30–31

Suppose $\triangle ABC$ and $\triangle DEF$ are images under the composite of a dilation with scale factor 3 and a translation. Then answer the following:

32. How are the perimeters of $\triangle ABC$ and $\triangle DEF$ related? Justify your answer.
33. How are the areas of $\triangle ABC$ and $\triangle DEF$ related? Justify your answer.

Tell whether any two of the given figures must be similar.

34. squares
35. trapezoids
36. equilateral triangles
37. rhombi
38. parallelograms
39. isosceles triangles
40. circles
41. regular pentagons
42. quadrilaterals
43. rectangles
44. regular polygons
45. semicircles

Without using a ruler or protractor, find the image of each point under $D_{O,\,2} \circ Rot_{O,\,30°}$.

46. (1, 0)
47. $\left(\frac{\sqrt{3}}{2}, \frac{1}{2}\right)$
48. $\left(\frac{\sqrt{2}}{2}, \frac{\sqrt{2}}{2}\right)$

10.6 A Group

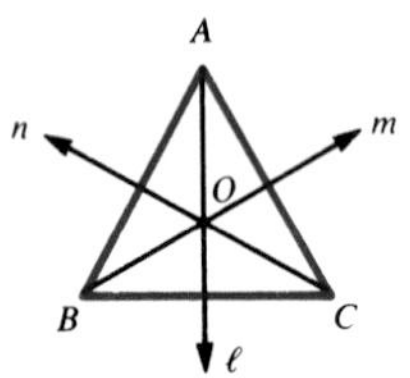

In previous sections you have investigated composites of various transformations. In this section, we will investigate the effect of six specific transformations on an equilateral triangle.

To begin, consider equilateral triangle *ABC*. An equilateral triangle has two types of symmetry, line symmetry and rotational symmetry. Five of the transformations are shown in the table below. Lines ℓ, m, and n are angle bisectors that intersect at point *O*.

Transformation	Symbol	Original Position	Final Position
Line Reflection in ℓ	R_ℓ		
Line Reflection in m	R_m		
Line Reflection in n	R_n		
Rotation of 120° about point *O*	$Rot_{O,120°}$		
Rotation of 240° about point *O*	$Rot_{O,240°}$		

The sixth transformation is the identity transformation, I, which leaves $\triangle ABC$ unchanged.

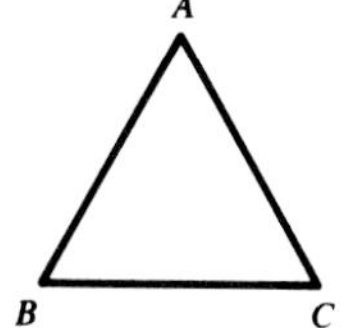

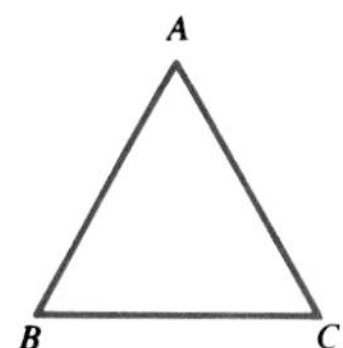

Examples

1 **Investigate the effect of $\text{Rot}_{O,240°} \circ \text{Rot}_{O,120°}$ on $\triangle ABC$.**

The effect of $\text{Rot}_{O,240°} \circ \text{Rot}_{O,120°}$ is shown below. Remember, do $\text{Rot}_{O,120°}$ first!

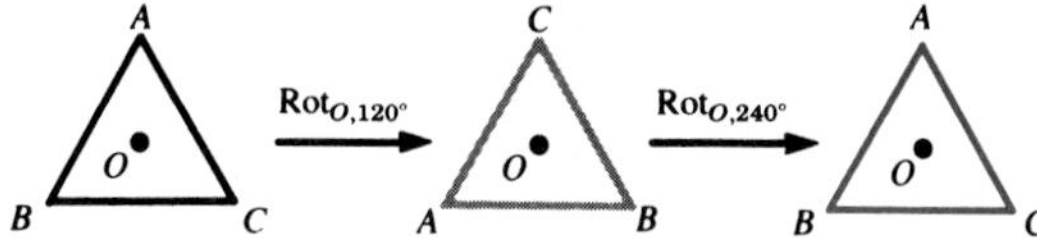

This composite leaves $\triangle ABC$ unchanged! Therefore, it is equivalent to the identity transformation.

We write $\text{Rot}_{O,240°} \circ \text{Rot}_{O,120°} = I$.

2 **What single transformation of $\triangle ABC$ has the same effect as $\text{Rot}_{O,120°} \circ R_\ell$?**

The composite $\text{Rot}_{O,120°} \circ R_\ell$ is shown below.

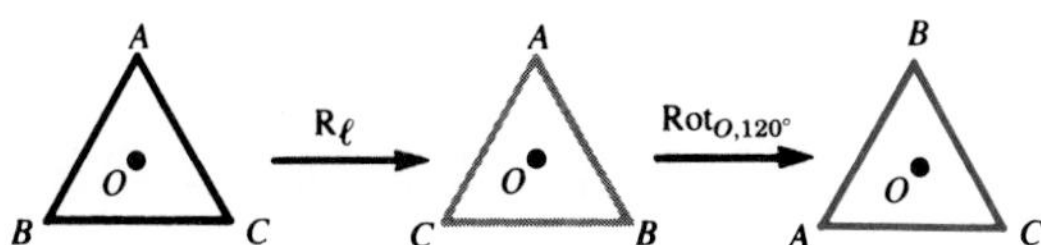

Now look at the table of transformations on the preceding page. Notice that R_n "takes" $\triangle ABC$ into the same position as the final one shown above. Therefore R_n has the same effect on $\triangle ABC$ as $\text{Rot}_{O,120°} \circ R_\ell$. That is, $\text{Rot}_{O,120°} \circ R_\ell = R_n$.

3 **Determine $\text{Rot}_{O,240°} \circ R_n$.**

First apply R_n and then $\text{Rot}_{O,240°}$, as shown below.

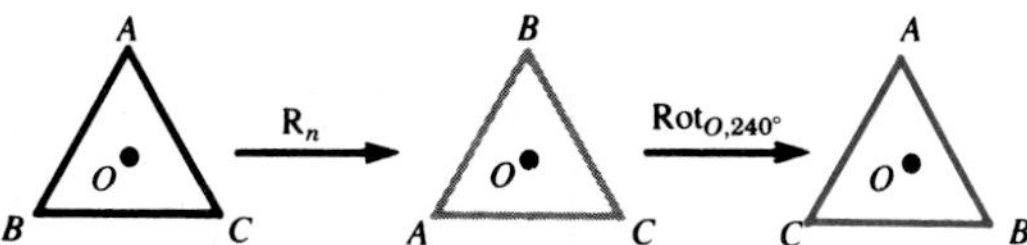

Using the table for transformations, we see that $\text{Rot}_{O,240°} \circ R_n = R_\ell$.

The results of Examples 1, 2, and 3 can be summarized conveniently in a table. Each entry in the following table is obtained by performing the transformation at the top first, and then the transformation at the left.

$\circ$	I	R_ℓ	R_m	R_n	$Rot_{O,120°}$	$Rot_{O,240°}$
I						
R_ℓ						
R_m						
R_n						
$Rot_{O,120°}$		R_n				
$Rot_{O,240°}$				R_ℓ	I	

In the exercises you will be asked to complete the table and show that, together with the operation composition, the set of six transformations forms a group. This group is called the **symmetry group** of an equilateral triangle.

Do other figures have symmetry groups? You will be asked to think about this question in the exercises.

Exercises

Exploratory For exercises 1–7, refer to the figure at the right. $\triangle ABC$ is equilateral and ℓ, m, and n are angle bisectors.

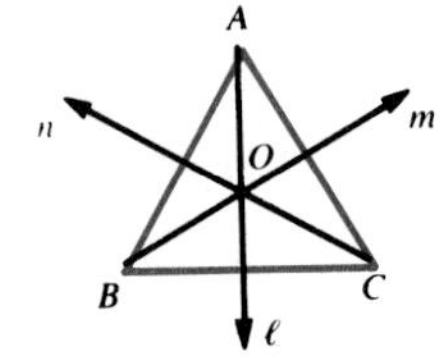

1. Explain why ℓ, m, and n are each lines of symmetry.
2. Explain why $\triangle ABC$ has rotational symmetry about point O.
3. Describe six transformations that map $\triangle ABC$ onto itself.

For each of the following draw a sketch of $\triangle ABC$ as it appears above. Then draw its image under each transformation.

4. R_ℓ
5. R_m
6. $Rot_{0,120°}$
7. $Rot_{0,240°}$

Written For each composite transformation below, draw a sketch of $\triangle ABC$ as it appears above. Then draw its image under each composite transformation.

1. $I \circ Rot_{O,120°}$
2. $R_\ell \circ R_m$
3. $R_\ell \circ Rot_{O,240°}$
4. $Rot_{O,240°} \circ R_m$

What single transformation of $\triangle ABC$ is equivalent to each of the following?

5. $I \circ R_\ell$
6. $R_\ell \circ R_m$
7. $Rot_{O,120°} \circ R_\ell$
8. $R_m \circ Rot_{O,240°}$
9. $R_\ell \circ Rot_{O,120°}$
10. $R_n \circ R_\ell$

Copy and complete the table at the top of the preceding page. Then answer the following.

11. Does the table define an operational system? Explain.
12. Verify that composition is an associative operation.
13. What is the identity element for the system?
14. Does every element in the system have an inverse? Explain.
15. Does the set of six transformations of $\triangle ABC$, together with composition, form a group? A commutative group? Explain.

For exercises 16–19, refer to rectangle *ABCD* at the right.

16. Explain why ℓ and m are lines of symmetry.
17. Explain why *ABCD* has point symmetry.

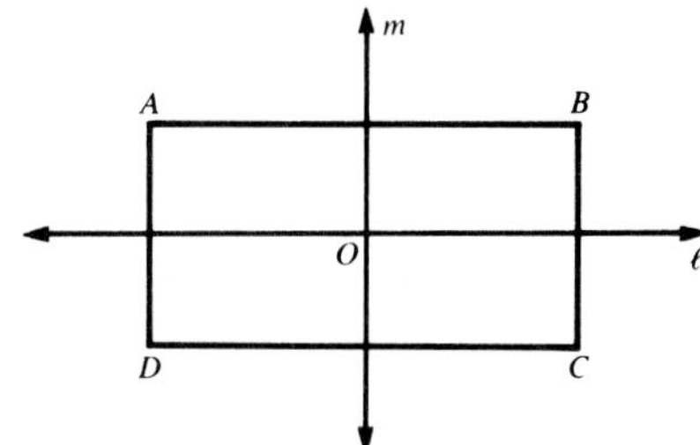

For exercises 18–19, refer to the table at the right.

18. Copy and complete the table.
19. Is the system a group? A commutative group?

$\circ$	I	R_ℓ	R_m	$Rot_{O,180°}$
I				
R_ℓ				
R_m				
$Rot_{O,180°}$				

Investigate the set of symmetries of each figure below for group properties. Include appropriate diagrams.

20. rhombus
21. isosceles triangle
22. square
23. parallelogram
24. regular pentagon
25. regular *n*-gon

Challenge **Do some research to answer the following.**

1. Examine the work of artist M. C. Escher. Find some examples related to the work of this chapter and discuss them from a "transformation" point of view.
2. How can the ideas of this chapter be generalized to three dimensions? For example, do you think the Eiffel Tower in Paris has any type of symmetry? Explain what you have in mind.

10.7 More About Groups

The set of symmetries of an equilateral triangle and the operation composition forms a group. In fact, transformation geometry abounds with groups. After some justification, it is convenient to accept the following.

Composition of transformations is an associative operation.

JUSTIFICATION: Suppose S, T, and W are any transformations. We must demonstrate that $(S \circ T) \circ W$ is the same transformation as $S \circ (T \circ W)$. To do this, let P be any point. Suppose

$$P \xrightarrow{W} P' \quad P' \xrightarrow{T} P'' \text{ and } P'' \xrightarrow{S} P'''.$$

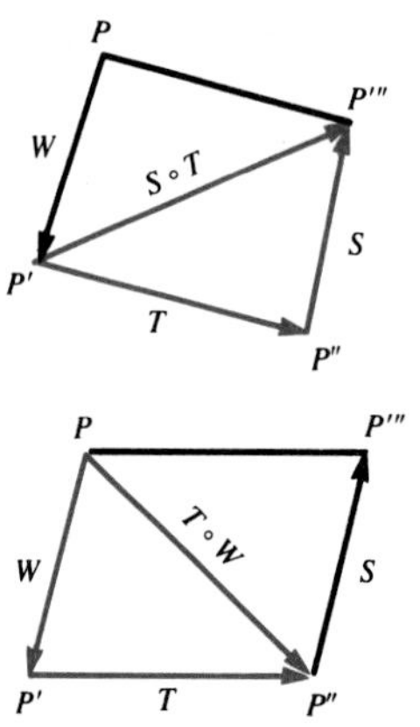

First consider the effect of $(S \circ T) \circ W$ on point P. The first figure at the left shows that $(S \circ T) \circ W$ maps P to P''', since W maps P to P' and $S \circ T$ maps P' to P'''.

Now consider the effect of $S \circ (T \circ W)$ on point P. The second figure shows that $S \circ (T \circ W)$ also maps P to P''', since $T \circ W$ maps P to P'' and S maps P'' to P'''.

Note that $(S \circ T) \circ W$ and $S \circ (T \circ W)$ both take point P to point P'''. We conclude, therefore, that $(S \circ T) \circ W$ and $S \circ (T \circ W)$ are the same transformation. This means that composition of transformations is an associative operation.

Examples

1 **Show that the set of all translations with the operation of composition is a commutative group.**

First assume that there is a coordinate system in the plane and that any translation $T_{a,b}$ has rule $(x, y) \rightarrow (x + a, y + b)$, where a and b are real numbers. Then show that the five properties of a commutative group are satisfied.

1. *Operational System (Closure Property)*

We must show that the composite of any two translations is a translation.

Let $T_{a,b}$ and $T_{c,d}$ be translations. Then

$$T_{a,b} \circ T_{c,d} \text{ has rule } (x, y) \rightarrow (x + c + a, y + d + b).$$

Since the rule is of the required form for a translation, we see that $T_{a,b} \circ T_{c,d}$ is a translation.

2. *Commutative Property.* Consider $T_{c,d} \circ T_{a,b}$ which has rule

$$(x, y) \rightarrow (x + a + c, y + b + d).$$

Compare this rule to the rule for $T_{a,b} \circ T_{c,d}$ in step **1.**

Since $x + a + c = x + c + a$ and $y + b + d = y + d + b$, the rules for $T_{a,b} \circ T_{c,d}$ and $T_{c,d} \circ T_{a,b}$ are the same.

We can conclude that composition of translations is commutative.

3. *Associative Property.* It has already been established that composition of transformations is associative. Since translations are transformations, it follows that composition of translations is associative.

4. *Identity Property.* The translation $T_{0,0}$ leaves every point fixed. Therefore, $T_{0,0}$ is the identity transformation. That is, $T_{0,0} = I$.

5. *Inverse Property.* Consider any translation $T_{a,b}$. We know $T_{-a,-b}$ exists since the real numbers a and b each have additive inverses in the reals. But $T_{a,b} \circ T_{-a,-b} = T_{a+(-a),b+(-b)} = T_{0,0}$. Therefore, every translation $T_{a,b}$ has an inverse translation $T_{-a,-b}$.

2 **Show that the set of all isometries with the operation of composition is a group.**

We will outline a solution. You should verify the results yourself.

1. Operational System. Is the composite of two isometries an isometry? Why?

2. Associativity. Isometries are transformations. Therefore, composition of isometries is associative.

3. Identity. Why is the identity transformation an isometry?

4. Inverses. Let T be an isometry. T is a one-to-one mapping of the plane onto the plane. Therefore, the inverse of T, T^{-1}, is also a one-to-one mapping of the plane onto the plane. Is T^{-1} also an isometry?

To answer this question suppose $A \xrightarrow{T^{-1}} B$ and $C \xrightarrow{T^{-1}} D$. Then, $B \xrightarrow{T} A$ and $D \xrightarrow{T} C$. Since T is an isometry, $BD = AC$. Therefore T^{-1} preserves distance. We have shown that every isometry has an inverse that is also an isometry.

We can now conclude that the set of all isometries with composition is a group. This group is called the *group of isometries,* or the **congruence group.**

The set of isometries is a subset of the set of all transformations of the plane. You will also be asked to show that the set of all transformations together with composition is a group. We say that the group of isometries is a **subgroup** of the group of transformations.

Definition of a Subgroup

Suppose (G, $*$) and (H, $*$) are groups and G is a subset of H. Then group (G, $*$) is called a subgroup of group (H, $*$).

From the results of Examples 1 and 2, and the fact that every translation is an isometry, it follows that the group of translations is a subgroup of the group of isometries.

Exercises

Written **Complete the rule $(x, y) \to$? for each of the following.**

1. $R_{x\text{-axis}}$
2. $R_{y\text{-axis}}$
3. $Rot_{O,180°}$
4. $R_{x=y}$
5. $T_{3,5}$
6. $D_{O,3}$
7. $R_{x\text{-axis}} \circ Rot_{O,180°}$
8. $R_{y\text{-axis}} \circ R_{x=y}$
9. $T_{-2,4} \circ Rot_{O,180°}$
10. $R_{x\text{-axis}} \circ T_{3,-5}$
11. $D_{O,-3} \circ T_{2,4}$
12. $(R_{x\text{-axis}} \circ R_{y\text{-axis}}) \circ R_{x=y}$
13. $R_{x\text{-axis}} \circ (R_{y\text{-axis}} \circ R_{x=y})$
14. $(T_{-2,3} \circ R_{x\text{-axis}}) \circ Rot_{O,180°}$
15. $T_{-2,3} \circ (R_{x\text{-axis}} \circ Rot_{O,180°})$
16. $(D_{O,-2} \circ T_{3,-4}) \circ R_{y\text{-axis}}$
17. $D_{O,-2} \circ (T_{3,-4} \circ R_{y\text{-axis}})$
18. Look carefully at your results in exercises 12–17. What property is suggested?
19. Give an argument which shows that, in general, composition of transformations is an associative operation.

Let S be the set of all rotations about a given point P. Then answer the following.

20. Give an argument that $(S, \circ)$ is an operational system.
21. Why is $\circ$ an associative operation in $(S, \circ)$?
22. Is there an identity in $(S, \circ)$? What is it?
23. Does every rotation in $(S, \circ)$ have an inverse? Justify your answer.
24. Give an argument to show that $(S, \circ)$ is a group.
25. Is $(S, \circ)$ a commutative group? Explain.

Consider the following sets and the operation composition to answer exercises 26–49.

W = {all isometries}
D = {all dilations with center (0, 0)}
L = {all line reflections}
R = {all rotations with center (0, 0)}
M = {all direct isometries}
T = {all transformations of the plane}
G = {all glide reflections}

Recall that $\cup$ means the union of two sets.

Which of the sets under composition has an identity?

26. W
27. D
28. L
29. $D \cup R$
30. M
31. T
32. G
33. $L \cup G$

Determine whether every element of the set has an inverse with respect to composition. *The identity need not be in the set.*

34. W
35. L
36. M
37. $L \cup G$
38. D
39. G
40. T
41. $D \cup R$

Which of the sets under composition forms a group? A commutative group?

42. W
43. L
44. T
45. $R \cup D$
46. G
47. D
48. M
49. $G \cup L$

50. Is the set of opposite isometries with composition a group? Explain.
51. Find three subgroup relationships in the real number system.
52. Show that the group of exercise 24 is a subgroup of the group of isometries.
53. Is the group of isometries a subgroup of a larger group? Explain.
54. Is the symmetry group of an equilateral triangle a subgroup of the group of isometries? Explain.

Challenge

Consider the set of rotations about point O through $30k$ degrees, where k is an integer. Is this set, together with composition, a subgroup of the group of all rotations about point O? Justify your answer.

Mathematical Excursions

Maurits Corneille Escher

The work on the opening page of this chapter is the work of the Dutch artist **Maurits Corneille Escher** (June 17, 1898–March 27, 1972). His work explored a strange world of optical illusion, visual puns, and distorted perspectives. In such prints as *Verbum* (1942) and *Metamorphosis* (1939–40), he dealt with the theme of metamorphosis and change using a gradual transformation of one shape into another. In the lithograph *Relativity* (1953), Escher created a visual paradox by combining three separate perspectives into a united, coherent whole. His work has become increasingly popular because of its unique combinations to project various visual deceptions.

Points in a plane can be identified using polar coordinates of the form (r, θ). A fixed point O in the plane is called the pole or origin. The polar axis is a ray whose initial point is the pole. The distance from the pole to a point P with polar coordinates (r, θ) is $|r|$.

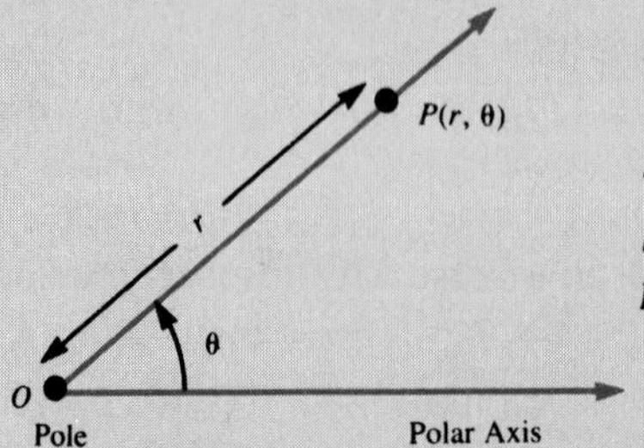

The polar axis is usually a horizontal line directed toward the right from the pole.

If r is positive, θ (theta) is the measure of any angle in standard position having $\overrightarrow{OP}$ as terminal side. If r is negative, θ is the measure of any angle having the ray opposite $\overrightarrow{OP}$ as terminal side. The angle can be measured in degrees or radians.

Polar transformations are sometimes used to simulate growth patterns or to alter the shape of an object. The following transformation formulas were used to alter the shape of a butterfly.

$$\theta' = \theta \qquad\qquad \theta'' = \theta$$

$$r' = r\left(1 + \frac{1}{3}(1 + \sin\theta)\right) \qquad r'' = r\left(1 + \frac{1}{2}(1 + \sin\theta)\right)$$

The original outline is shown as a solid black line. The (r', θ') transformation is shown as a colored line. The (r'', θ'') transformation is shown as a broken line.

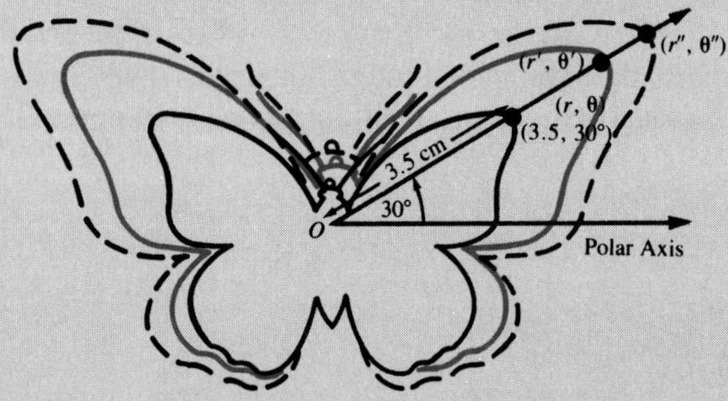

Consider the (r', θ') transformation. The image of any point (r, θ) on the black line is (r', θ') on the colored line. For example, the image, (r', θ'), of $(3.5, 30°)$ was determined as follows.

$$\theta' = \theta \qquad r' = r\left(1 + \frac{1}{3}(1 + \sin\theta)\right)$$

$$\theta' = 30° \qquad r' = 3.5\left(1 + \frac{1}{3}(1 + \sin 30°)\right)$$

The image of $(3.5, 30°)$ is $(5.25, 30°)$. The coordinates of the images of other points were determined in a similar manner.

Problem Solving Application: Using Transformations

In a plane, the shortest path between two points is a straight line. However, restrictions sometimes make a straight path impossible. Transformations can often be used in situations such as these to find the shortest path between two points.

Example

1 The state park officials have decided to build a footbridge from the park over the highway so that campers will have easy access to the nearby town of Troy. The bridge will be built perpendicular to the highway. Where should they build the bridge so that the walk from the park to Troy is the shortest?

Explore A straight line from the park to Troy would be the shortest path; however, the bridge must be perpendicular to the highway. If the highway could be represented by one line instead of two parallel lines, the straight line would provide the answer. We can use a composite of two line reflections to "move" the sides of the highway to one line.

Plan Draw a diagram of the situation. Lines m and n represent the two sides of the highway, point P is the park, and point T is Troy. We must find points X and Y so that the path from P to X to Y to T is the shortest possible.

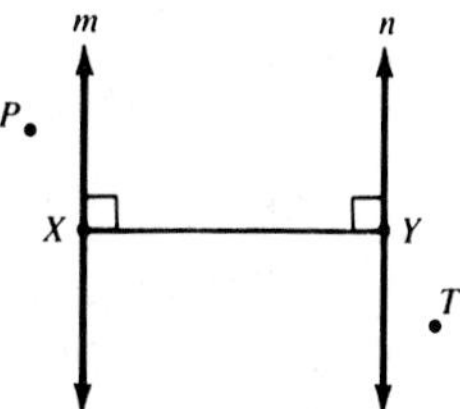

Solve Find line ℓ so that $\ell \parallel m$, $\ell \parallel n$, and line ℓ is equidistant from lines m and n. Find the reflection image of T with respect to line n and line ℓ, and label the point T'. We have "moved" line n onto line m. Draw the line segment from P to T'. The point X, which is the intersection of this line segment and line m, is the point where the bridge should start. The bridge should end at point Y, the point directly across from X.

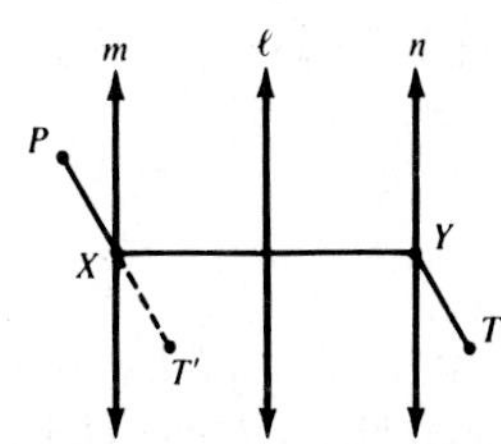

Examine $\overline{PT'}$ is the shortest path from P to T'. Because line reflections preserve distance, $\overline{PT'} \cong \overline{YT}$. Therefore, $PX + XY + YT$ is the shortest distance from P to T, and the bridge is placed correctly.

Exercises

Written Copy the diagram and solve each problem.

1. The cities of Springfield and Bloomington are on opposite sides of a river. The city officials would like to build a bridge over the river so that the drive between the cities is the shortest possible. Where should the bridge be built?

• Springfield

Bloomington •

2. A straight line drawn from Springfield to Bloomington would be 50 miles long. It is 11 miles from Springfield to the river and 7 miles from Bloomington to the river. The river is 12 miles wide. Use this information and the results of exercise 1 to find the approximate distance from Springfield across the bridge to Bloomington.

3. Use the results of example 1 and the following information to determine the length of the walk from the park to Troy. A straight line drawn from the park to Troy would be 1 mile long. The park is 1000 yards from the highway, and Troy is 500 yeards from the highway. The highway is 20 yards across.

For each of the following, copy the diagram and find the shortest path across the river. Assume that the banks are indicated by the parallel lines and the bridges must be perpendicular to the banks.

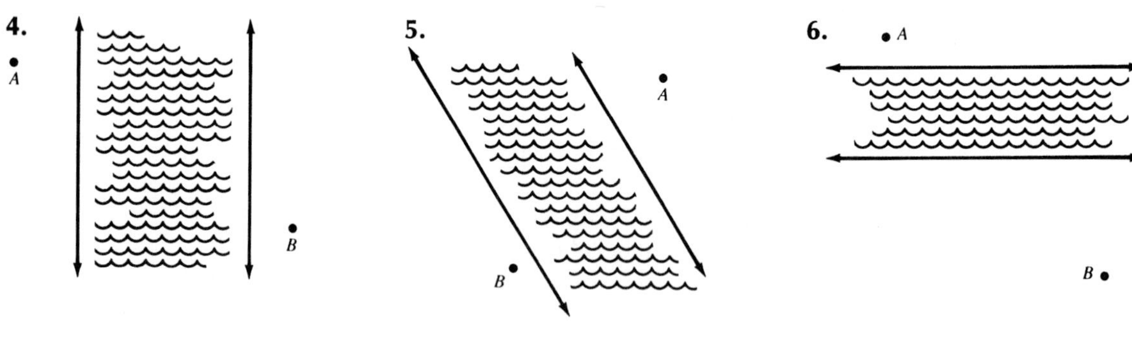

7. Copy the miniature golf hole shown at the right. Then use two or more line reflections to determine *two* different paths that could be used to score a hole-in-one for this hole.

Ball

Hole

Vocabulary

composition
composite transformation
glide reflection
identity transformation
isometry
direct isometry
opposite isometry
Three Line Reflection Theorem
congruence
similarity transformation
similar figures
symmetry group
congruence group
subgroup

Chapter Summary

1. The composite $R_m \circ R_\ell$ where $\ell \parallel m$, is a translation. The distance of the translation is twice the distance between ℓ and m. The translation is in the direction from ℓ to m along a perpendicular to both ℓ and m.
2. Every translation can be expressed as the composite $R_m \circ R_\ell$ where
   a. ℓ and m are parallel;
   b. the distance from ℓ to m is half the distance of the translation; and
   c. ℓ and m are both perpendicular to the direction of the translation.
3. A translation is an isometry and preserves collinearity, betweenness, and angle measure.
4. Suppose line ℓ and m intersect at O. The composite $R_m \circ R_\ell$ is a rotation about point O. The angle of rotation has a measure which is twice that of the acute or right angle formed by ℓ and m. The direction of the rotation is the same as the direction of the acute or right angle as measured from ℓ to m.
5. Every rotation $Rot_{O,2\theta}$ can be expressed as the composite $R_m \circ R_\ell$ where
   a. ℓ and m intersect at O; and
   b. the angle formed by ℓ and m in the direction from ℓ to m has measure θ.

   If θ is positive, the angle is measured counterclockwise. If θ is negative, the angle is measured clockwise.
6. Rotations are isometries and preserve collinearity, betweenness, and angle measure.
7. Three Line Reflection Theorem: Every isometry is equivalent to the composite of, at most, three line reflections.
8. Every isometry is equivalent to one of the following: a line reflection, a rotation, a translation, or a glide reflection.

9. Two figures are congruent if, and only if, there is an isometry that maps one figure onto the other.
10. A dilation preserves collinearity and betweenness of points, angle measure, perpendicularity, parallelism, and ratios of distances.
11. Under a dilation, a line is parallel to its image.
12. Two figures are similar if, and only if, there is a similarity transformation that maps one figure onto the other.
13. Composition of transformations is an associative operation.

Chapter Review

10.1 Explain the meaning of each of the following symbols.

1. R_ℓ
2. $D_{O,2}$
3. $R_m \circ D_{O,\frac{1}{2}}$
4. $T_{4,3}$
5. $Rot_{O,180°}$
6. $Rot_{O,180°} \circ T_{4,0}$

10.2 Given $\triangle ABC \cong \triangle A'B'C'$. Show that $\triangle ABC$ can be mapped onto $\triangle A'B'C'$ by a composite of each of the following transformations.

7. a reflection, a rotation and a translation
8. a reflection and a rotation
9. three reflections

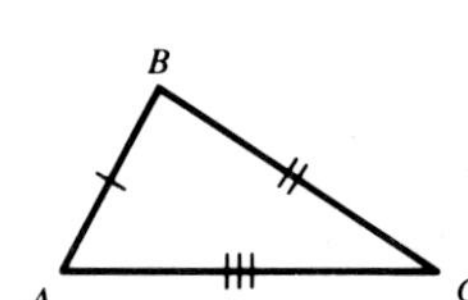

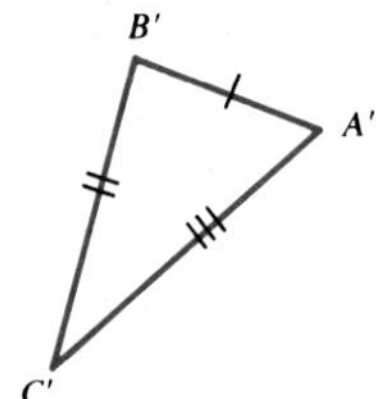

10.3 In the figure at the right, lines ℓ and m form an angle of 60° and A maps to B under $R_m \circ R_\ell$. Describe a single transformation equivalent to each of the following compositions.

10. $R_m \circ R_\ell$
11. $R_\ell \circ R_m$
12. $R_\ell \circ R_m \circ R_m$
13. $R_m \circ R_\ell \circ R_\ell \circ R_m$

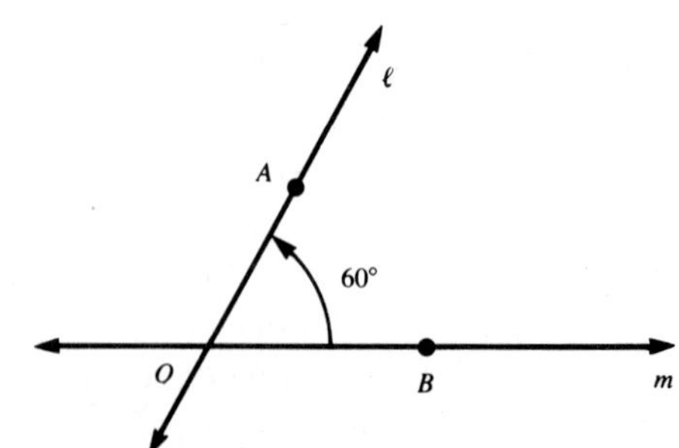

10.4 Graph $\triangle ABC$ with vertices $A(0, 0)$, $B(-5, 2)$, and $C(6, -1)$. Find the image of $\triangle ABC$ under the following. Then identify the transformation as a line reflection, translation, rotation, or glide reflection.

14. $R_{x\text{-axis}} \circ R_{y=x}$

15. $Rot_{O,180°} \circ R_{y=x}$

16. $T_{0,2} \circ Rot_{O,180°}$

17. $T_{2,3} \circ T_{0,2}$

18. $Rot_{0,90°} \circ T_{2,4}$

19. $Rot_{O,90°} \circ Rot_{O,-180°}$

10.5 Tell if each of the following is preserved by a similarity transformation.

20. parallelism
21. betweenness
22. perpendicularity
23. distance
24. collinearity
25. orientation
26. angle measure
27. ratio of distance
28. congruence

10.6 Give the inverse (under composition) of each transformation.

29. reflection over a line

30. translation mapping (0, 0) to (3, 4)

31. rotation of $-34°$ about M

32. rotation of 180° about M

10.7 Which of these sets contain the identity transformation?

33. set of all reflections
34. set of all translations
35. set of all rotations with same point as center
36. $\{Rot_{O,0°}, Rot_{O,90°}, Rot_{O,-90°}, Rot_{O,180°}\}$
37. set of all dilations with same point as center

38. When are two transformations inverses under composition?

With composition, does the given set form a group?

39. set of all rotations with center (2, 3)
40. set of all dilations with center $(0, \sqrt{5})$

Tell which group properties are satisfied by the given set and composition.

41. $\{Rot_{C,0°}, Rot_{C,120°}, Rot_{C,-120°}\}$
42. set of dilations with center M and integral scale factors
43. set of all reflections

44. Demonstrate that composition of transformations is not commutative.
45. Show that the set of all dilations is not closed under composition by describing $S \circ T$, where S is a dilation with scale factor 2 and center (0, 0), and T is a dilation of scale factor $\frac{1}{2}$ and center (10, 10).

Chapter Test

In the figure at right, each of the small triangles is equilateral. A translation T maps *O* onto *A*. Find each image.

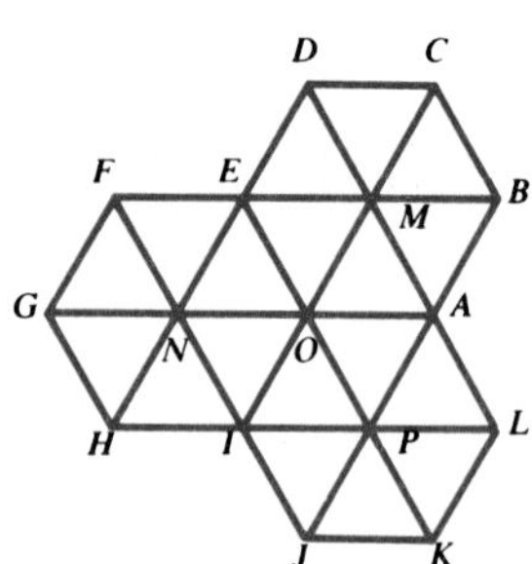

1. $T(P)$
2. $Rot_{O,60°}(I)$
3. $Rot_{O,60°}(\triangle POI)$
4. $T(\overline{FI})$
5. $(Rot_{O,60°} \circ Rot_{O,60°})(M)$
6. $(T \circ T)(E)$
7. $(T \circ Rot_{O,60°})(I)$
8. $(Rot_{O,60°} \circ T)(I)$
9. $(Rot_{O,60°} \circ T)(ENOM)$
10. $(Rot_{O,60°} \circ Rot_{O,60°})(\triangle NEF)$

Translations T_1 and T_2 are shown at the right. Draw the image of $\triangle ABC$ under each composite transformation.

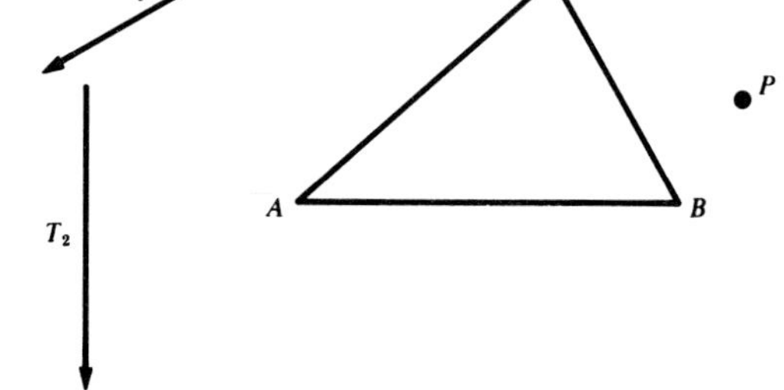

11. $Rot_{P,90°} \circ Rot_{P,90°}$
12. $T_1 \circ Rot_{P,90°}$
13. $T_2 \circ T_1$
14. $Rot_{P,90°} \circ T_1$
15. $T_1 \circ T_2$

Find the image of $\triangle XYZ$ for each composition.

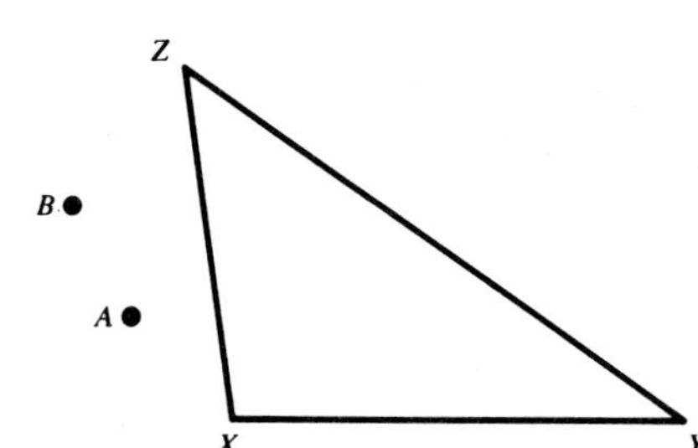

16. $Rot_{B,270°} \circ Rot_{A,90°}$
17. $Rot_{A,270°} \circ Rot_{B,90°}$
18. Find the image of $(2, -5)$ under $R_{y\text{-axis}} \circ T_{2,5} \circ D_{O,2}$.
19. Given the triangle with vertices $A(2, 0)$, $B(0, -2)$, and $C(-2, 4)$, find its image under $Rot_{O,90°} \circ R_{y=x}$. Is this transformation a line reflection, translation, rotation, or glide reflection?
20. Using the figure at the right, draw two reflecting lines a and b so that

    $(R_b \circ R_a)(\triangle MNO) = \triangle PQR$.
21. $\triangle PQR$ is the image of $\triangle MNO$ under a rotation. Locate the center of this rotation.

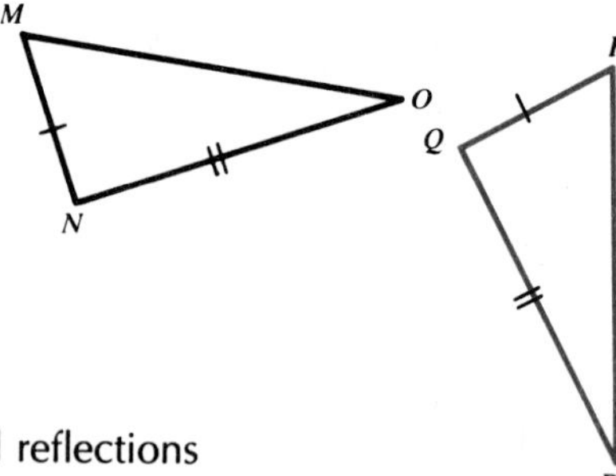

Which of these sets under composition forms a group?

22. set of all translations
23. set of all reflections
24. set of all rotations
25. set of all dilations

Probability and the Binomial Theorem

Chapter 11

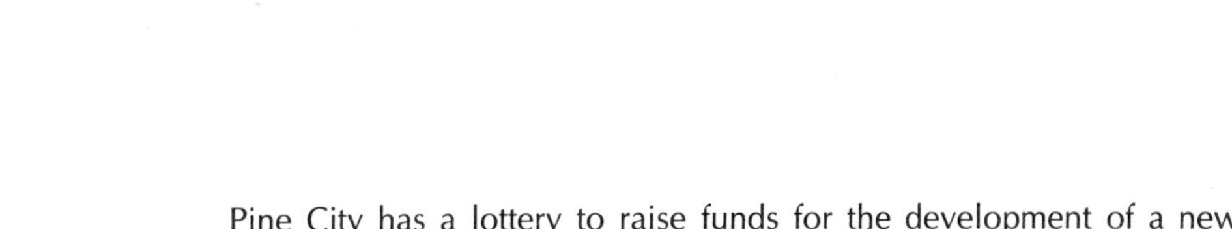

Pine City has a lottery to raise funds for the development of a new municipal park. Each week fifty balls, numbered 1–50, are placed in a hopper and six balls are chosen at random. You can use some of the basic laws of probability presented in this chapter to determine the probability of the winning combination.

11.1 The Basic Principles

In past mathematics courses, you studied some basic laws of probability. The most basic rule was that for an event E and an outcome set S,

$$P(E) = \frac{n(E)}{n(S)}.$$

In this notation, $n(E)$ represents the number of elements that make up the event E, and $n(S)$ the number of elements in the outcome set S. The event E is often considered a set itself. In this case E is necessarily a subset of the outcome set S.

Some familiar situations where the rule $P(E) = \frac{n(E)}{n(S)}$ can be applied concern tossing dice or coins, or selecting cards from a pack of cards. For example, the probability of picking the Queen of Spades when selecting a card at random from an ordinary deck of 52 is $\frac{1}{52}$, since there is one "success" and 52 possibilities altogether. The following tables about coins and dice should be familiar to you.

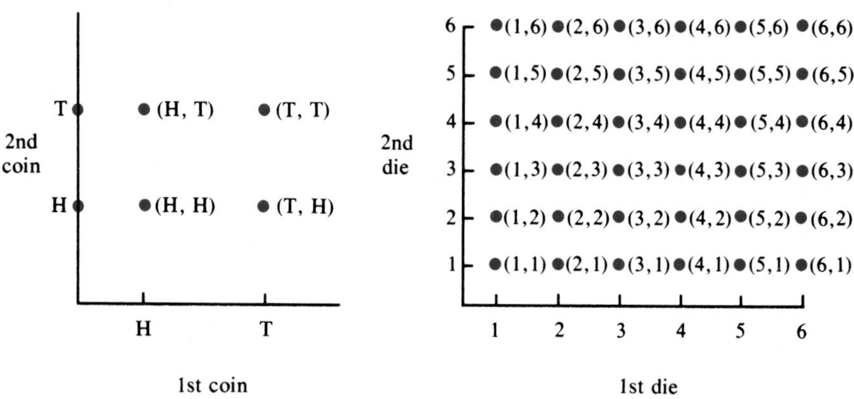

When two fair coins are tossed, the probability of obtaining two heads is $\frac{1}{4}$. When two fair dice are tossed, the probability of obtaining a sum of five is $\frac{4}{36}$ or $\frac{1}{9}$ since, out of 36 possible outcomes, four give a sum of five. They are the outcomes (2, 3), (3, 2), (4, 1), (1, 4).

The basic rule can also be used to derive the following facts.

Basic Rules of Probability

1. The probability of an impossible event is 0.
2. The probability of a certain event is 1.
3. For any event E, $0 \leq P(E) \leq 1$.

Here is another familiar idea about the probability of event A or event B.

$$P(A \text{ or } B) = P(A) + P(B) - P(A \text{ and } B)$$

Examples

1 What is the probability of rolling a six or a double when tossing two dice?

$$P(6 \text{ or Double}) = \frac{5}{36} + \frac{6}{36} - \frac{1}{36} = \frac{10}{36} \text{ or } \frac{5}{18}.$$

2 Find the probability of selecting an ace or a black card from a standard pack of 52.

$$P(\text{Ace or Black}) = \frac{4}{52} + \frac{26}{52} - \frac{2}{52} = \frac{28}{52} \text{ or } \frac{7}{13}.$$

Another very important principle you learned was the **Counting Principle.**

Counting Principle

Suppose one activity can occur in any of m ways. Another can occur in any of n ways. The total number of ways both activities can occur is given by the product mn.

When there are two events, neither of which affects the occurrence of the other, we say the events are *independent*. An alternate version of the Counting Principle may be used when the events in question are independent.

Counting Principle (Alternate Version)

Suppose the probability of one event E is r $(0 \le r \le 1)$. The probability of another event F is s $(0 \le s \le 1)$. Then the probability of both E and F occurring is the product rs.

Examples

3 Gwendoline has four skirts and five blouses. How many skirt-blouse outfits are possible?

Use the Counting Principle.

$$4 \cdot 5 = 20$$

4 One of Gwendoline's skirts is yellow with red dragons and one of her blouses is white. What is the probability she will pick that blouse and that skirt?

Use the Alternate Version of the Counting Principle.

$P(\text{skirt})$ is $\frac{1}{4}$ and $P(\text{blouse})$ is $\frac{1}{5}$.

$P(\text{skirt and blouse})$ is $\frac{1}{4} \cdot \frac{1}{5}$ or $\frac{1}{20}$.

5 **An urn contains seven red marbles and five white ones. A marble is drawn, not replaced, and another marble drawn. What is the probability both marbles are white?**

The solution to this problem is illustrated in the tree diagram shown at the right.

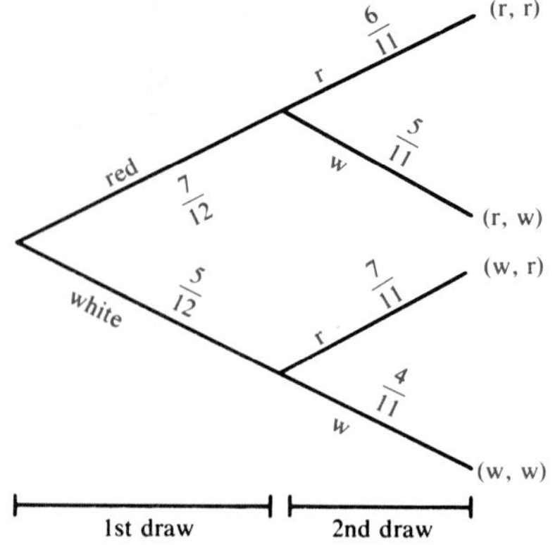

For the first draw, $P(\text{white}) = \frac{5}{12}$. If a white marble is drawn first, there remain in the urn eleven marbles of which four are white. Hence for the second draw, $P(\text{white}) = \frac{4}{11}$. We conclude that

$$P(\text{white, white}) = \frac{5}{12} \cdot \frac{4}{11} = \frac{20}{132} \text{ or } \frac{5}{33}.$$

Exercises

Exploratory **Use the Counting Principle—either version—to answer these problems.**

1. Juan has three shirts and four ties. How many shirt-tie combinations can he choose?
2. How many ways can you ferry across Pine Lake and back if there are four different ferry boats in service? How many ways can you make the trip across and then return on a different boat?
3. Mrs. Brayford owns nine dogs. She has five sheepdogs and four poodles. Every Wednesday, the dogs select one sheepdog and one poodle at random to go over and tear up Mr. Brullman's vegetable garden. How many sheepdog-poodle combinations are possible?

The outcome sets for the experiments of tossing two coins and tossing two dice appear on page 402. Draw similar illustrations for each of the following experiments.

4. tossing a coin and a die
5. tossing a die and picking a letter from C A T
6. picking a card from one of five cards marked A, E, I, O, U, and tossing a coin

Written **Two fair six-sided dice are tossed. Find the probability of each of these sums.**

1. 4 **2.** 5 **3.** 8 **4.** 7 **5.** 8 or 9

Two fair six-sided dice are tossed. One of the dice is green (*G*), the other red (*R*). Find each of the following.

6. $P(R \text{ is prime})$ **7.** $P(R = G)$ **8.** $P(R < G)$ **9.** $P(R = 3G)$

A card is drawn at random from a deck of 52. Find the probability of drawing each of the following.

10. ace
11. red card
12. king or queen
13. black six
14. card between 5 and 9, inclusive

An urn contains four red marbles and three blue ones. A marble is chosen. It is not replaced and then another is chosen. What is the probability of obtaining each of the following?

15. two red marbles
16. two blue marbles
17. one of each color

Answer questions 15–17 if the urn contains each of the following sets of marbles.

18. two red, five blue
19. two red, one green, four blue

20. A box contains red and blue disks. There are 5 less than twice as many red disks as blue. If the probability of drawing a red disk is $\frac{3}{5}$, how many red disks are in the box?

Mathematical Excursions

Predicting Outcomes with Computers

```
10 LET T2 = 0: LET T3 = 0
20 LET H2 = 0: LET H3 = 0
30 FØR I = 1 TØ 100
40 LET T = 0: LET H = 0
50 FØR N = 1 TØ 3
60 LET R = RND (1)
70 IF R < 0.5 THEN 110
80 PRINT "TAIL";" ";
90 LET T = T + 1
100 GØ TØ 130
110 PRINT "HEAD";" ";
120 LET H = H + 1
130 NEXT N
134 PRINT
140 IF H = 3 THEN 180
150 IF T = 3 THEN 200
160 IF H = 2 AND T = 1 THEN 220
170 IF H = 1 AND T = 2 THEN 240
180 LET H3 = H3 + 1
190 GØ TØ 250
200 LET T3 = T3 + 1
210 GØ TØ 250
220 LET H2 = H2 + 1
230 GØ TØ 250
240 LET T2 = T2 + 1
250 NEXT I
254 PRINT
260 PRINT "HHH: ";H3
270 PRINT "TTT: ";T3
280 PRINT "HHT: ";H2
290 PRINT "HTT: "; T2
300 END
```

Probability is used to predict the outcome in games of chance such as tossing coins. A computer can be used to do the actual counting in simulated situations where large samples are needed.

The program at the left simulates tossing three coins, prints the outcomes, and keeps totals of the possible outcomes of the 100 samples. Each time the program is run, the totals may be different due to the random selection of values in line 60.

The random numbers selected by the computer lie between 0 and 1. The program defines heads as numbers less than 0.5. Tails are defined as numbers greater than or equal to 0.5.

The output shows each toss of the coin and lists the totals for the four possible combinations: 3 heads, 3 tails, 2 heads and 1 tail, or 1 head and 2 tails.

Enter the program and run it several times. Notice the change in the totals. Average your results. Does the average come closer to the predicted outcomes?

A possible set of results from running the program is shown below. The symbols represent cumulative totals, not the order of heads and tails.

```
HHH: 11
 TTT: 19
HHT: 38
 HTT: 32
```

11.2 Permutations

The idea of a **permutation,** or arrangement, is an integral part of probability. Let us review the following principles involving permutations.

The number of permutations of n things taken all at a time is

$$n(n-1)(n-2)(n-3)\cdot\ldots\cdot 1.$$

This statement can be justified by reasoning that the first item may be chosen in any of n ways. Once this has been done, there are $n-1$ ways left of picking the second item, and so on. Finally, there is only one way to choose the last item.

The symbol for "number of permutations of n things taken all at a time" is ${}_nP_n$. Further, the symbol $n!$ is used to mean $n(n-1)(n-2)\cdot\ldots\cdot 1$. Hence

$${}_nP_n = n!$$

The number of permutations of n things taken r at a time, with r less than n, is given by the following rules.

Look at the rules for ${}_nP_n$ and ${}_nP_r$. Can you see why 0! is defined to be 1 in order to be consistent?

$${}_nP_r = \frac{n!}{(n-r)!}$$

$${}_nP_r = n(n-1)(n-2)\cdot\ldots\cdot(n-r+1)$$

This can be justified in a way similar to the previous argument. We note that if we have r places to fill with any of n objects, then the first can be chosen in n ways, the second in $n-1$ ways, the third in $n-2$ ways and the rth in $n-(r-1)$ or $n-r+1$ ways.

Examples

1 **How many ways can five people be seated in five seats?**

You need to determine the number of arrangements of five things taken all at a time. This is ${}_nP_n = 5! = 120$.

2 **How many three-letter "words" can be made from the letters X, B, R, V, T, L?**

There are six letters and we must decide the number of arrangements there are of any three. This is ${}_6P_3 = \frac{6!}{3!} = 120$.

3 **How many words can be made from the letters in GARBLE if the first and second letters must be vowels?**

There are two ways to choose the first letter (A or E) and then one way to choose the second. Then we fill the remaining four places with the remaining four letters. The final answer is $2 \times 1 \times 4 \times 3 \times 2 \times 1 = 48$.

4 **How many different arrangements are there of the letters in BUBBLE?**

If all the letters were different, there would be $6! = 720$ arrangements. But the letters are not different: for every arrangement of the three B's is one of six identical arrangements. **B B B U L E** is not a different arrangement from **B B B U L E.** Since three symbols can be arranged in $3! = 6$ ways, the final answer is $720 \div 6 = 120$.

B U B B L E	**B U B B L E**	**B U B B L E**
B U B B L E	**B U B B L E**	**B U B B L E**
B B B U L E	**B B B U L E**	**B B B U L E**
B B B U L E	**B B B U L E**	**B B B U L E**

Of course, there would be many others!

5 **How many arrangements are there of the letters in GIGGLING?**

For every one of the $4! = 24$ arrangements of the four G's, there are $2! = 2$ arrangements of the two I's. By the Counting Principle, this means there are $24 \times 2 = 48$ "apparent" arrangements for every "real" arrangement. The answer to the problem is $8! \div 48 = 840$.

Exercises

Exploratory **Explain each of the following symbols, phrases, or ideas.**

1. Events E and F are subsets of outcome set S.
2. $n(n - 1)(n - 2) \cdot \ldots \cdot 1$
3. arrangements of n things taken all at a time
4. arrangements of n things taken r at a time with $r < n$
5. permutations of seven things taken three at a time
6. $n!$
7. ${}_nP_n$
8. ${}_nP_r$
9. ${}_5P_3$
10. ${}_7P_7$
11. ${}_{18}P_2$

Written **Compute each of the following.**

1. ${}_{10}P_5$
2. ${}_9P_3$
3. $\frac{7!}{4!}$
4. $\frac{{}_5P_3}{{}_4P_3}$
5. $\frac{{}_8P_5}{{}_5P_5}$
6. $\frac{{}_{10}P_3}{{}_5P_3}$
7. How many terms are represented by the expression $n(n-1)(n-2)\cdot\ldots\cdot(n-r+1)$? What is the 6th term?
8. How many four-letter words can be made from the letters in OMSK?
9. How many four-letter words can be made from the letters XBOMT?

How many four-letter words can be made from the letters in BUNGLE under each condition?

10. there are no restrictions
11. the first letter must be B
12. the last letter must be E
13. the word must end in LE

How many six-letter words can be made from HUMANOID under each condition?

14. there are no restrictions
15. the first letter must be M
16. the first letter must be a vowel
17. the word must begin and end with a vowel

How many arrangements are there of the letters in each of the following?

18. MADRID
19. DULUTH
20. BARCELONA
21. PAMPLONA
22. BABBLE
23. GIGGLES
24. OHIO
25. XXXB
26. QQQXQ
27. QUIXOTE
28. BOOKKEEPER
29. MISSISSIPPI
30. Explain why the number of arrangements of the letters in BANANA is $\frac{6!}{3!2!}$.
31. In how many different orders can five different books be arranged on a shelf?
32. In how many ways can five students be seated in five chairs if one of the five has to sit in the center chair?
33. Mr. Brayton has the following books to place on a shelf: *Shakespeare's Histories, Shakespeare's Tragedies, Shakespeare's Comedies, Poems of Spenser, Poems of Chaucer, Poems of Milton,* and *Poems of Byron.* In how many different orders can Mr. Brayton shelve his books if the three Shakespeare volumes must be together in any order?

Tell how many five-letter "words" can be formed from the letter U, N, E, L, R, A, and D for each situation.

34. The first letter is D.
35. The second letter A.
36. The first two letters are U and N in that order.
37. The last two letters are L and E in any order.
38. The middle three letters are A, E, and R in any order.
39. How many six-digit numbers can be made using the digits from 966,349?
40. From all the five-digit numbers that can be formed from the digits 2, 3, 4, 4, 7, using each digit once, one is chosen at random. What is the probability it is divisible by 4?

Challenge **Six people can be seated in a row of 6 chairs in 6! ways. How many ways can these 6 people be seated around a circular table? One person can sit at any place, and then the other 5 can arrange themselves in 5! ways around the table.**

This is an example of a *circular permutation*. In general, if *n* objects are arranged in a circle, there are $(n - 1)!$ permutations of the *n* objects around the circle.

1. Find the number of ways in which five students can be seated around a circular table if two of the students insist on sitting next to each other.
2. Find the number of ways in which 4 boys and 4 girls can be seated around a circular table if they sit alternately.
3. Find the number of ways in which 7 tan chairs, 4 white chairs, and 1 gray chair can be placed around a circular table.

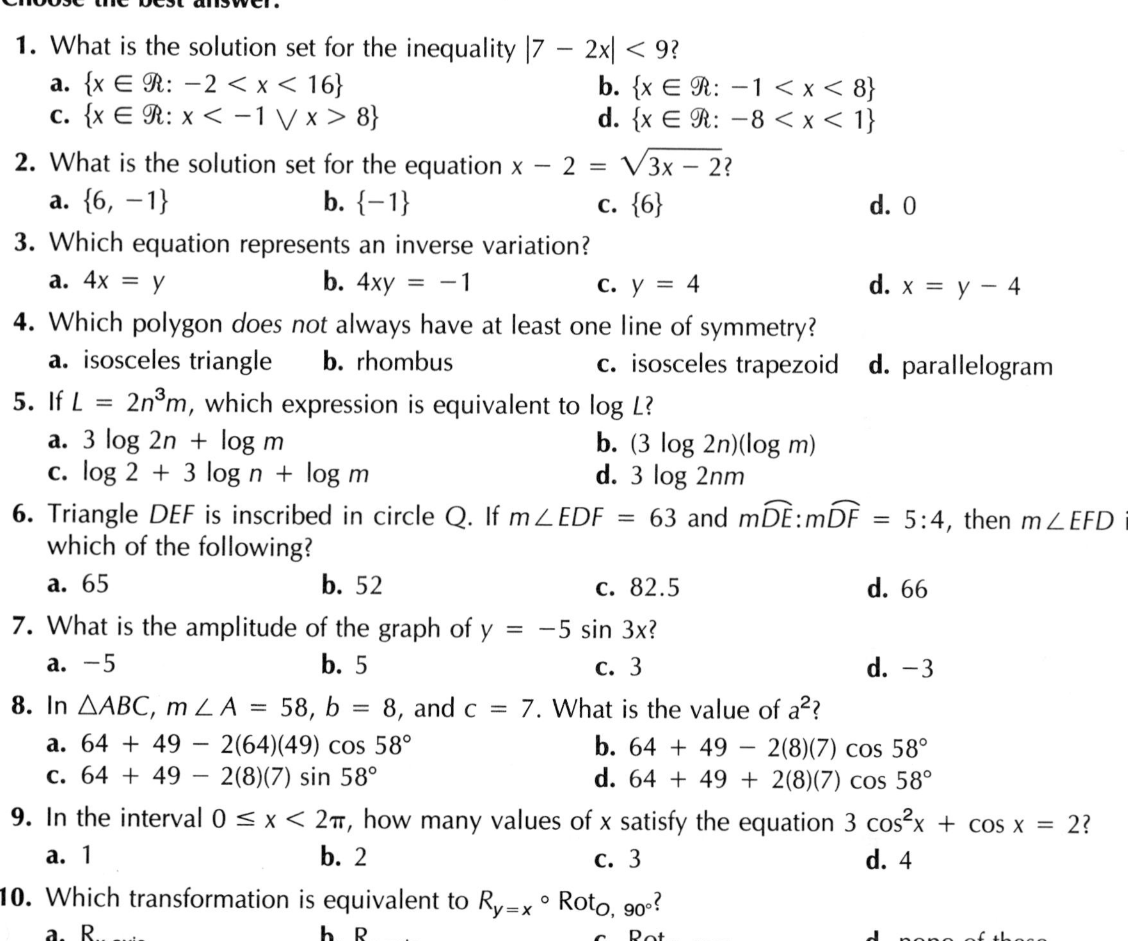

Mixed Review

Choose the best answer.

1. What is the solution set for the inequality $|7 - 2x| < 9$?
   a. $\{x \in \mathcal{R}: -2 < x < 16\}$ **b.** $\{x \in \mathcal{R}: -1 < x < 8\}$
   c. $\{x \in \mathcal{R}: x < -1 \vee x > 8\}$ **d.** $\{x \in \mathcal{R}: -8 < x < 1\}$
2. What is the solution set for the equation $x - 2 = \sqrt{3x - 2}$?
   a. $\{6, -1\}$ **b.** $\{-1\}$ **c.** $\{6\}$ **d.** 0
3. Which equation represents an inverse variation?
   a. $4x = y$ **b.** $4xy = -1$ **c.** $y = 4$ **d.** $x = y - 4$
4. Which polygon *does not* always have at least one line of symmetry?
   a. isosceles triangle **b.** rhombus **c.** isosceles trapezoid **d.** parallelogram
5. If $L = 2n^3m$, which expression is equivalent to $\log L$?
   a. $3 \log 2n + \log m$ **b.** $(3 \log 2n)(\log m)$
   c. $\log 2 + 3 \log n + \log m$ **d.** $3 \log 2nm$
6. Triangle DEF is inscribed in circle Q. If $m\angle EDF = 63$ and $m\widehat{DE}:m\widehat{DF} = 5:4$, then $m\angle EFD$ is which of the following?
   a. 65 **b.** 52 **c.** 82.5 **d.** 66
7. What is the amplitude of the graph of $y = -5 \sin 3x$?
   a. -5 **b.** 5 **c.** 3 **d.** -3
8. In $\triangle ABC$, $m\angle A = 58$, $b = 8$, and $c = 7$. What is the value of a^2?
   a. $64 + 49 - 2(64)(49) \cos 58°$ **b.** $64 + 49 - 2(8)(7) \cos 58°$
   c. $64 + 49 - 2(8)(7) \sin 58°$ **d.** $64 + 49 + 2(8)(7) \cos 58°$
9. In the interval $0 \leq x < 2\pi$, how many values of x satisfy the equation $3 \cos^2 x + \cos x = 2$?
   a. 1 **b.** 2 **c.** 3 **d.** 4
10. Which transformation is equivalent to $R_{y=x} \circ Rot_{O,\ 90°}$?
   a. $R_{x\text{-axis}}$ **b.** $R_{y\text{-axis}}$ **c.** $Rot_{O,\ 180°}$ **d.** none of these

11.3 Combinations

The idea of a **combination** was another important topic in probability. Do you remember the difference between a combination and a permutation?

You will recall that a helpful idea is to compare combinations to committees and permutations to "words." For example, the problem of finding the number of four-letter words which can be made from ten letters is a permutation problem because order must be considered. ABCD is a different word from ACDB. But in determining how many different four-person committees can be formed from ten people, we have a combination problem. Adams, Baker, Cohen, Danelli is the same committee as Adams, Cohen, Danelli, Baker. The membership of a committee does not depend on the order in which its members were chosen.

The number of combinations of n things taken r at a time is symbolized in this way, ${}_nC_r$.

If $r = n$, then we have ${}_nC_n$, which is the number of combinations of n things taken all at a time. Can you give an argument to show that ${}_nC_n = 1$?

If $r < n$, the plot becomes a bit thicker. In the exercises, you will be asked to justify the following formulas.

$${}_nC_r = \frac{{}_nP_r}{r!} \qquad {}_nC_r = \frac{n!}{r!(n-r)!}$$

$${}_nC_r = \frac{n(n-1)(n-2)\cdot\ldots\cdot(n-r+1)}{r(r-1)(r-2)\cdot\ldots\cdot 1}$$

Examples

1 **How many subcommittees of four people can be chosen from a committee of eight?**

You need to find ${}_8C_4$. This equals $\frac{8!}{4!4!} = 70$. There are seventy possible subcommittees.

2 **In the card game bridge, the entire deck of 52 cards is dealt to four players—13 cards each. How many different bridge hands are there?**

A hand at a card game does not depend on the order in which the cards were dealt. Therefore we have a combination problem. The answer is ${}_{52}C_{13} = \frac{52!}{13!\cdot 39!}$. This equals 635,013,559,600.

3 **The Pine City Opera Children's Chorus has five altos and seven sopranos. In how many ways can two altos and two sopranos be chosen to represent the chorus at a meeting of the town council?**

The altos can be chosen in any of ${}_5C_2$ ways and the sopranos in any of ${}_7C_2$ ways. Hence, by the Counting Principle, the total number of two-alto, two-soprano groups is ${}_5C_2 \cdot {}_7C_2 = 10 \cdot 21 = 210$.

4 **If two members of the chorus in example 3 are chosen at random, what is the probability they will both be sopranos?**

The number of two-soprano choices is ${}_7C_2 = 21$. The total number of possible choices is ${}_{12}C_2 = 66$. Using the fundamental principle of probability, we compute the number of "successes" divided by the total number of possibilities. In this case, we have $P(2 \text{ sopranos}) = \frac{21}{66} = \frac{7}{22}$.

Exercises

Exploratory

1. Explain the difference between combinations and permutations.

Tell whether each of the following represents a permutation or a combination problem.

2. Choosing three residents of Pine City to participate in a radio program.
3. Choosing three lottery tickets from 1500 in a hat for three equal prizes.
4. Choosing four half-hour television comedies from a possible ten to air between eight and ten o'clock, in any order.
5. Same as exercise 4, but *My Kid Tamara* must be shown first.

Explain the meaning of each of these symbols.

6. ${}_nC_r$
7. ${}_nC_n$
8. ${}_nC_0$
9. ${}_nC_{n-1}$
10. ${}_8C_5$
11. Give an argument to show that ${}_nC_n = 1$.

Written

1. Give an argument to show that ${}_nC_r = {}_nC_{n-r}$.
2. Show that the three formulas that follow are equivalent. Then state a combination problem and solve it using each of the formulas.

$${}_nC_r = \frac{{}_nP_r}{r!} \qquad {}_nC_r = \frac{n!}{r!(n-r)!} \qquad {}_nC_r = \frac{n(n-1)(n-2)\cdot \ldots \cdot (n-r+1)}{r(r-1)(r-2)\cdot \ldots \cdot 1}$$

Compute each of the following.

3. ${}_5C_5$ **4.** ${}_5C_4$ **5.** ${}_8C_5$ **6.** ${}_8C_3$ **7.** ${}_9C_6$

In how many ways can a subcommittee of three be chosen from each of the following?

8. 4 **9.** 5 **10.** 8 **11.** 12 **12.** 24

In how many ways can four people be selected from each of the following?

13. three people **14.** a dozen people **15.** fifteen people

16. How many different committees of three senators can be chosen from the United States Senate?

How many ways can five penguins be chosen from each of the following?

17. six penguins **18.** nine penguins **19.** thirteen penguins

Write expressions for the solutions to each of these problems. You need not evaluate these expressions.

20. How many committees of fifteen senators can be chosen in the Senate?

21. In the game of glorf, played in the Merglorf-Venglorf system, the deck consists of 44 snurls. Each of four players is dealt eleven. In how many different "hands" can this result?

22. Same as exercise 21, but the order in which the snurls are dealt counts.

The Pine City Opera Children's Chorus now has six altos and eight sopranos. In how many ways can the following selections of singers be made?

23. two altos, two sopranos

24. three altos, one soprano

If two people are selected at random from a fund-raising committee of four men and five women, what is the probability of choosing each of the following?

25. both men **26.** both women **27.** one man, one woman

Challenge

1. The tents manufactured by the Dick'n'Doug Tent Company have a maximum capacity of four persons. In how many different ways can six Boy Scouts occupy two tents? Think about this one!

2. Suppose five points are drawn on a piece of paper, no three in the same line. How many distinct lines can be drawn through these points?

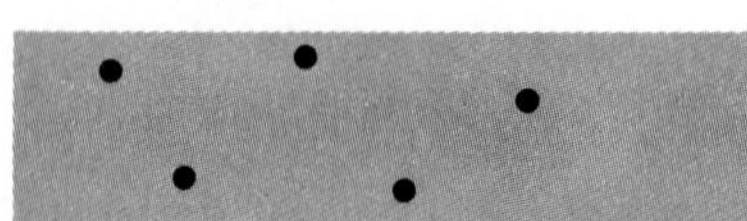

11.4 Exactly *r* Successes

In the experiment of tossing one fair die, you know that the probability of obtaining a 6 is $\frac{1}{6}$. For every trial of this experiment, $P(6) = \frac{1}{6}$. Thus, for five trials of this experiment, the probability of obtaining five 6's is $\frac{1}{6} \cdot \frac{1}{6} \cdot \frac{1}{6} \cdot \frac{1}{6} \cdot \frac{1}{6} = \left(\frac{1}{6}\right)^5 = \frac{1}{7776} \approx 0.00013.$

But suppose you wanted to know the probability of obtaining *just one* 6 in five trials. This problem requires further investigation.

First, be sure you understand the problem. You want to know the probability that any *one*, but *no more than one*, of the five tosses will land 6. For example, one possible "success" would be tosses of 3, 1, 3, 6, 4, rolled in that order.

The best way to solve the problem requires two steps. First, change the problem slightly and compute the probability that in five tosses of the die, the *first* one will land 6 and the others not-6. Since $P(6) = \frac{1}{6}$ and $P(\text{not-6}) = \frac{5}{6}$, we use the *Counting Principle* to conclude that

$$P(\text{6 first toss, not-6 the others}) = \frac{1}{6} \cdot \frac{5}{6} \cdot \frac{5}{6} \cdot \frac{5}{6} \cdot \frac{5}{6} = \frac{1}{6} \cdot \left(\frac{5}{6}\right)^4 = \frac{625}{7776} \approx 0.08$$

But in the original problem, *the 6 did not have to be first.* In fact, there are five different places where the 6 could occur. You conclude that the probability of exactly one 6 in five tosses of a fair die is given by $5 \cdot \frac{1}{6} \cdot \left(\frac{5}{6}\right)^4 = \frac{3125}{7776} \approx 0.40.$

Now suppose you wanted to know the probability of *exactly two 6's* in five tosses. If you changed the problem to two 6's on the *first two tosses only,* then you would have the following.

$$\begin{aligned} P(\text{6 first two tosses, not-6 the others}) &= \frac{1}{6} \cdot \frac{1}{6} \cdot \frac{5}{6} \cdot \frac{5}{6} \cdot \frac{5}{6} = \left(\frac{1}{6}\right)^2 \left(\frac{5}{6}\right)^3 \\ &= \frac{125}{7776} \approx 0.016 \end{aligned}$$

But the 6's do not have to be first. You must now answer the question, "In how many different ways can two 6's be distributed among five places?"

This is a combination problem very much like selecting two people from five. There are ${}_5C_2 = 10$ different ways of placing two 6's in five places, and so the final result is

6 6 X X X
6 X 6 X X
6 X X 6 X
6 X X X 6
and so on

$$P(\text{exactly two 6's in five tosses}) = {}_5C_2 \cdot \left(\frac{1}{6}\right)^2 \left(\frac{5}{6}\right)^3 = 10 \cdot \frac{125}{7776} \approx 0.16.$$

You can generalize this result to the following: when tossing a fair die five times, the probability of r 6's is ${}_5C_r\left(\frac{1}{6}\right)^r\left(\frac{5}{6}\right)^{5-r}$. And finally, the following generalization:

$$P(\text{exactly } r \text{ 6's in } n \text{ tosses of a fair die}) = {}_nC_r\left(\frac{1}{6}\right)^r\left(\frac{5}{6}\right)^{n-r}.$$

Example

1 **Find the probability of rolling exactly three 1's when a fair die is tossed ten times.**

$$P(\text{three 1's}) = {}_{10}C_3\left(\frac{1}{6}\right)^3\left(\frac{5}{6}\right)^7 = 120 \cdot \frac{1^3 \cdot 5^7}{6^{10}} = \frac{20 \cdot 5^7}{6^9} \approx 0.155$$

Exercises

Exploratory **Compute each of the following, expressing your result in fraction form, and in decimal form, accurate to 3 places.**

1. $\left(\frac{1}{6}\right)^3$ **2.** $\left(\frac{1}{6}\right)^4$ **3.** $\left(\frac{5}{6}\right)^4$ **4.** $\left(\frac{1}{6}\right)^2\left(\frac{1}{6}\right)^2$ **5.** $\left(\frac{1}{6}\right)^3\left(\frac{5}{6}\right)^2$

6. $\left(\frac{1}{6}\right)^4\left(\frac{5}{6}\right)^2$ **7.** $\left(\frac{1}{6}\right)\left(\frac{5}{6}\right)^5$ **8.** $\left(\frac{1}{6}\right)^3\left(\frac{5}{6}\right)^3$ **9.** $\left(\frac{1}{6}\right)^2\left(\frac{5}{6}\right)^6$

Written **A fair die is tossed four times. Find the probability of obtaining each of the following.**

1. 6 on the first toss only
2. 1 on first toss only
3. 1 on the third toss only
4. 5 on all four tosses
5. exactly one 6
6. exactly two 6's
7. exactly three 6's
8. no 5's

A fair die is tossed six times. Find the probability of obtaining each of the following.

9. 4 on the first toss only
10. 3 on the fourth toss only
11. exactly one 5
12. exactly two 5's
13. Inez says, "Since the probability of obtaining a 6 on any toss is $\frac{1}{6}$, then in six tosses of a die the probability of obtaining exactly one should be about 1." Comment on Inez's assertion.

The formula given at the top of page 414 can be rewritten as follows.

$$P(\text{exactly } r \text{ 6's in } n \text{ tosses of a fair die}) = {}_nC_r[P(6)]^r[P(not\ 6)]^{n-r}$$

Suppose a die is loaded so that $P(1) = \frac{1}{4}$, $P(2) = \frac{1}{15}$, $P(3) = \frac{1}{6}$, $P(4) = \frac{1}{10}$, $P(5) = \frac{1}{12}$, **and** $P(6) = \frac{1}{3}$. **If this die is tossed five times, find the probability of obtaining each result.**

14. 6 on the first toss only
15. exactly one 2
16. exactly two 3's
17. exactly three 4's
18. no 1's
19. at least four 5's
20. 6 on the first toss only and 1 on the last toss only.

Mixed Review

For the dartboard at the right, the side of the square is 25 cm, and the radius of the circle is 10 cm. A dart is thrown at random and hits the board. Find the probability of each result.

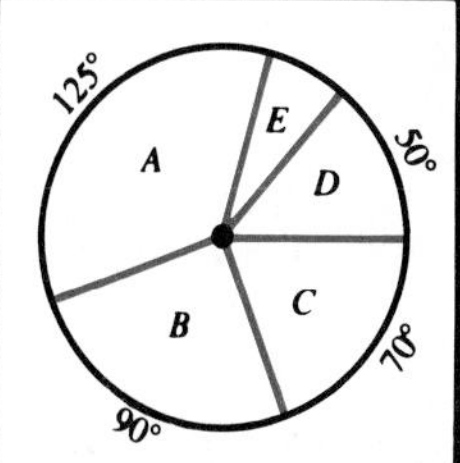

1. The dart is in region *A*.
2. The dart is in region *C*.
3. The dart is in region *B* or region *D*.
4. The dart is *not* in region *E*.
5. The dart is *not* in the circle.

Mathematical Excursions — Diagonals

The diagonal of a polygon is any segment connecting two non-adjacent vertices.

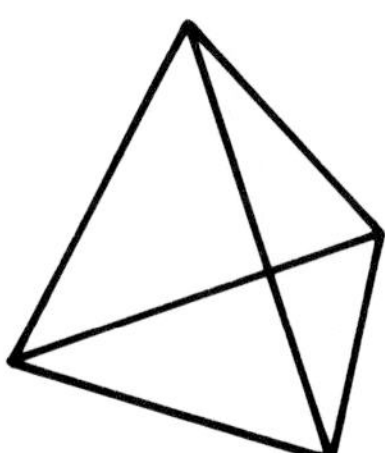

A quadrilateral has two diagonals.

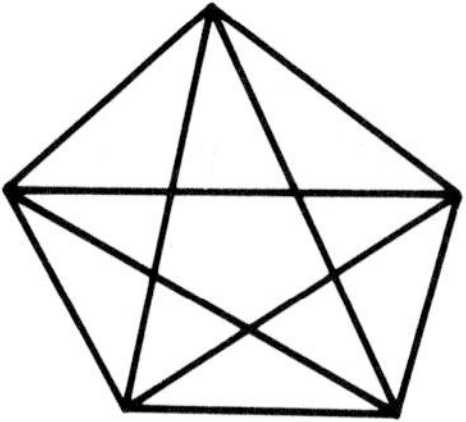

A pentagon has five diagonals.

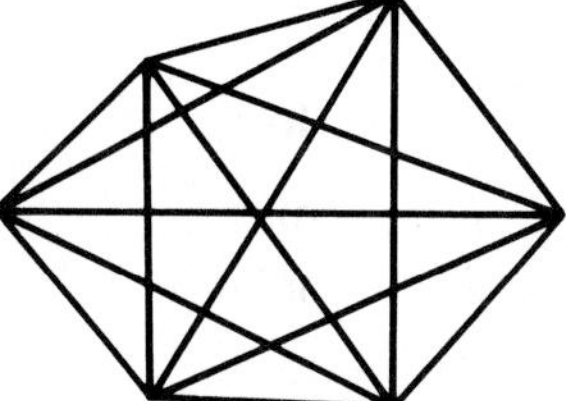

A hexagon has nine diagonals.

The problem of finding the number of diagonals of an n-sided polygon is similar to finding the number of line segments determined by n points, taken two at a time. Remember not to count the line segments that make up the polygon.

Exercises **Determine the number of diagonals of each polygon.**

1. octagon
2. decagon
3. dodecagon
4. 20-gon
5. n-gon

11.5 Exactly *r* Successes, Continued

In the last section, we saw how to compute the probability of obtaining exactly r successes in n trials of the experiment of tossing a die. Of course, there are other kinds of situations in which the same plan of reasoning would apply.

Example

1 **It is generally agreed that Beeville High has a better soccer team than Hornetville High. In fact, each time they play, it is accepted that the probability that Beeville will win is $\frac{3}{5}$. If they play four games, what is the probability that each team will win two?**

This is equivalent to asking, "What is the probability that Beeville will win exactly two games?" Use the reasoning of the previous section, as follows.

$$P(\text{Beeville wins exactly two}) = {}_4C_2\left(\frac{3}{5}\right)^2 \cdot \left(\frac{2}{5}\right)^2 = 6 \cdot \frac{9}{25} \cdot \frac{4}{25} = 6 \cdot \frac{36}{625} = \frac{216}{625} \approx 0.346$$

There are many situations in which an experiment has, or can be considered to have, only two outcomes. For example, a die can land "6" or "not-6." In a game of chess, player A can win or fail to win. On a given day, there can be 1 inch of rain, or not. In each case, there is a probability p for "success" and probability q for "failure." Furthermore, $q = 1 - p$. Can you explain why?

Experiments in which there are only two outcomes are called **Bernoulli experiments,** after the Swiss mathematician Jacques Bernoulli.

We can use our investigations of this and the previous section to come to the following important generalization.

Bernoulli experiment

Suppose there are n trials of a Bernoulli experiment with probability p for "success" and probability $q = 1 - p$ for "failure." Then the probability of obtaining exactly r successes in these n trials is

$${}_nC_r p^r q^{n-r}$$

Examples

2 **A coin is loaded so that $P(\text{heads}) = \frac{2}{3}$. In four tosses, what is the probability of obtaining exactly one tail?**

$P(\text{success}) = \frac{1}{3}$ and $P(\text{failure}) = \frac{2}{3}$.

$$P(\text{exactly one tail}) = {}_4C_1\left(\frac{1}{3}\right)^1\left(\frac{2}{3}\right)^3 = 4 \cdot \frac{1}{3} \cdot \frac{8}{27} = 4 \cdot \frac{8}{81} = \frac{32}{81}.$$

3 **A quiz has five multiple-choice questions. There are four possible choices for each question. If you choose answers at random on every question, what is the probability you will get at least one question right?**

"At least one" means "exactly one" or "exactly two" and so on, up to "exactly five." Furthermore, we know that $P(\text{success}) = \frac{1}{4}$ and $P(\text{failure}) = \frac{3}{4}$. Since "exactly one," "exactly two," and so on, are mutually exclusive, we can add the probabilities. We have

$$P(\text{at least one right}) = {}_5C_1\left(\frac{1}{4}\right)^1\left(\frac{3}{4}\right)^4 + {}_5C_2\left(\frac{1}{4}\right)^2\left(\frac{3}{4}\right)^3 + {}_5C_3\left(\frac{1}{4}\right)^3\left(\frac{3}{4}\right)^2 + {}_5C_4\left(\frac{1}{4}\right)^4\left(\frac{3}{4}\right)^1 + {}_5C_5\left(\frac{1}{4}\right)^5$$

However, there is an easier way. If it is true that you scored at least one right, then it is false that you scored zero right.

$$P(\text{at least one right}) = 1 - P(\text{none right})$$

This last quantity is much easier to compute!

$$P(\text{at least one right}) = 1 - {}_5C_0\left(\frac{3}{4}\right)^5 = 1 - 1 \cdot \left(\frac{3}{4}\right)^5 = 1 - \frac{243}{1024} = \frac{781}{1024} \approx 0.763$$

Exercises

Exploratory

1. Explain what is meant by the term Bernoulli experiment.
2. Give three ways in which the tossing of a fair die can be considered a Bernoulli experiment. *One way is to consider the outcome as "odd or even."*
3. In a Bernoulli experiment, if the probability of "success" is p and of failure q, why is $q = 1 - p$?

Describe a Bernoulli experiment in which the probability of success is each of the following.

4. $\frac{1}{2}$
5. $\frac{2}{3}$
6. $\frac{5}{26}$
7. $\frac{4}{52}$
8. $\frac{7}{19}$

Written

Assume that the probability that Beeville will beat Hornetville each time they play soccer is $\frac{2}{3}$. Find the probability of each of the following numbers of wins for Beeville.

1. exactly one win in three games
2. exactly one win in four games
3. at least one win in three games
4. no wins in five games
5. exactly three wins in five games
6. exactly four wins in five games

If the probability of Beeville's winning each game is $\frac{3}{4}$, find the probability of each of the following for Beeville.

7. exactly one win in three games
8. exactly one win in four games
9. exactly three wins in five games
10. exactly four wins in five games

A die is loaded so that $P(6) = \frac{1}{3}$. Find the probability of exactly these numbers of 6's in six tosses.

11. 0
12. 1
13. 2
14. 3
15. 4
16. 6

Using the die in exercises 11–16, find the probability of exactly the following number of 6's in eight tosses.

17. 1
18. 2
19. 3
20. 4
21. 5
22. 7

A fair coin is tossed the indicated number of times. Find the probability of the following.

23. at least one head in five tosses
24. at least one head in six tosses
25. at least two heads in five tosses
26. at least one head in eight tosses
27. at least one tail in ten tosses
28. at least two tails in seven tosses

29. What is the probability that a family with four children has at least one girl?

A multiple-choice test has four choices for each question, and eight questions.

30. If you guess at every item, what is the probability you will get exactly half of them right?
31. What is the probability you get at least one right?
32. What is *P*(at least two right)?

Write an expression for your solutions to each of the following problems, but do not try to evaluate them.

33. The probability of a certain conical (cone-shaped) drinking cup landing point-up is 0.43. What is *P*(exactly 43 points-up in 100 tosses)?
34. What is *P*(exactly twenty 5's) in 120 tosses of a fair die?
35. What is the probability of obtaining the sum twelve exactly ten times in 100 tosses of two fair dice?
36. Find the probability of at least one twelve in 50 tosses of two fair dice.

Challenge

1. What is the probability that the World Series will end in exactly five games, assuming the teams are evenly matched? Think about this one carefully.
2. If the probability that team A will win is $\frac{3}{5}$ for every game, what is the probability that team A will win the World Series in exactly five games?
3. *DO SOME RESEARCH.* Find out about the "Law of Large Numbers." Explain its relation to the material in this chapter.

11.6 The Binomial Theorem

The idea of combinations is an extremely useful one in mathematics. One of its uses is to help in a certain kind of algebraic procedure which would otherwise be very complicated. This procedure is raising a binomial to a higher power. The rule we are going to derive is called the Binomial Theorem.

Raising a binomial to a power is called *expanding the binomial.* The result is the expansion of the binomial. Let's begin by considering some simple expansions. Consider the binomial $(x + y)$ and raise it to the zero, the first and the second power.

$$(x + y)^0 = 1\text{, since } a^0 = 1 \text{ for all } a \neq 0.$$
$$(x + y)^1 = 1x + 1y$$
$$(x + y)^2 = 1x^2 + 2xy + 1y^2$$

All the coefficients are in red to make a special point, as you will see.

Further multiplication shows that

$$\begin{aligned}(x + y)^3 &= (x^2 + 2xy + y^2)(x + y)\\ &= x^3 + 2x^2y + xy^2 + x^2y + 2xy^2 + y^3\\ &= 1x^3 + 3x^2y + 3xy^2 + 1y^3.\end{aligned}$$

If we perform additional multiplications, we obtain the following results:

$$(x + y)^4 = 1x^4 + 4x^3y + 6x^2y^2 + 4xy^3 + 1y^4$$
$$(x + y)^5 = 1x^5 + 5x^4y + 10x^3y^2 + 10x^2y^3 + 5xy^4 + 1y^5$$
$$(x + y)^6 = 1x^6 + 6x^5y + 15x^4y^2 + 20x^3y^3 + 15x^2y^4 + 6xy^5 + 1y^6$$

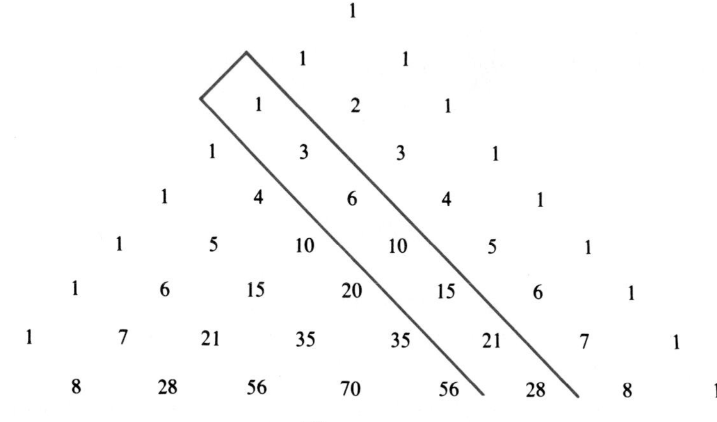

Figure 1

Now, consider just the coefficients (in red) of the x's and y's. Place them in a triangular array like the one shown here. (Figure 1)

This arrangement of numbers is sometimes called **Pascal's Triangle,** after the French philosopher and mathematician Blaise Pascal, who lived from 1623 to 1662.

What patterns can you see in the triangle?

There are quite a few. One pattern is that every entry is the sum of the two "parent" entries just above it. In the Challenge Exercises you will be asked to figure out why this is true.

Another interesting fact is that every entry in the table is equal to some value of ${}_nC_r$, for some correctly chosen n and r. For example, ${}_4C_2 = 6$, and 6 is the coefficient of x^2y^2 when $(x + y)$ is raised to the power 4.

Also, ${}_6C_4 = 15$ and 15 is the coefficient of x^2y^4 when $(x + y)$ is raised to the power 6.

The triangular array in Figure 2 shows the relation between the binomial coefficients and the combinatorial coefficients. Do not forget, ${}_nC_n$ and ${}_nC_0 = 1$ for any whole number n.

$$
\begin{array}{ccccccccccccc}
 & & & & & & {}_0C_0 & & & & & & \\
 & & & & & {}_1C_0 & & {}_1C_1 & & & & & \\
 & & & & {}_2C_0 & & {}_2C_1 & & {}_2C_2 & & & & \\
 & & & {}_3C_0 & & {}_3C_1 & & {}_3C_2 & & {}_3C_3 & & & \\
 & & {}_4C_0 & & {}_4C_1 & & {}_4C_2 & & {}_4C_3 & & {}_4C_4 & & \\
 & {}_5C_0 & & {}_5C_1 & & {}_5C_2 & & {}_5C_3 & & {}_5C_4 & & {}_5C_5 & \\
{}_6C_0 & & {}_6C_1 & & {}_6C_2 & & {}_6C_3 & & {}_6C_4 & & {}_6C_5 & & {}_6C_6
\end{array}
$$

Figure 2

We can now use this observation to state a rule for raising the binomial $(x + y)$ to any power. This rule allows us to expand $(x + y)^n$ without having to do the actual algebraic multiplication.

The following rule is called the **Binomial Theorem.**

Binomial Theorem

$$(x + y)^n = {}_nC_0x^n + {}_nC_1x^{n-1}y^1 + {}_nC_2x^{n-2}y^2 + \ldots + {}_nC_{n-1}x^1y^{n-1} + {}_nC_ny^n$$

Examples

1 **Expand $(a + 3)^4$ using the Binomial Thoerem and simplify the results.**

$$
\begin{aligned}
(a + 3)^4 &= {}_4C_0a^4 + {}_4C_1a^3(3)^1 + {}_4C_2a^2(3)^2 + {}_4C_3a(3)^3 + {}_4C_4(3)^4 \\
&= a^4 + 4a^3 \cdot 3 + 6a^2 \cdot 9 + 4a \cdot 27 + 81 \\
&= a^4 + 12a^3 + 54a^2 + 108a + 81
\end{aligned}
$$

2 **Expand $(2z - 1)^5$ and simplify.**

$$(2z - 1)^5 = {}_5C_0(2z)^5 + {}_5C_1(2z)^4(-1)^1 + {}_5C_2(2z)^3(-1)^2 + {}_5C_3(2z)^2(-1)^3 + {}_5C_4(2z)(-1)^4 + {}_5C_5(-1)^5$$
$$= 32z^5 + 5(16z^4)(-1) + 10(8z^3)(1) + 10(4z^2)(-1) + 5(2z)(1) + (-1)$$
$$= 32z^5 - 80z^4 + 80z^3 - 40z^2 + 10z - 1$$

Exercises

Exploratory

1. What does it mean to expand a binomial? Give two examples.
2. Which of the following is the correct expansion of $(x + 14)^2$?
   a. $x^2 + 196$ **b.** $x^2 + 14x + 28$ **c.** $x^2 + 28x + 28$ **d.** $x^2 + 28x + 196$
3. Write out the expansion of $(x + 2)^3$. First you will have to find $(x + 2)^2$, and then multiply by another factor of $(x + 2)$.

Write out each of the following expansions, multiplying each one out, "the long way."

4. $(x + 5)^3$
5. $(x + 2)^4$
6. $(2x + 1)^3$

7. Explain the value given to $(x + y)^0$ given in the text.
8. In the text, the expression $1x + 1y$ appears. How is this usually written? Why is it written differently here?

Written

Write out each of the following expansions, multiplying each one out, "the long way."

1. $(x + 9)^3$
2. $(3x + 2)^3$
3. $(x + \frac{1}{2})^3$
4. $(\frac{1}{2}w + 1)^3$
5. $(a - 2x)^3$
6. $(3x - 2)^3$
7. $(2x - 3)^4$
8. $(2x + 1)^4$

Carefully check each of the following expansions given in the text.

9. $(x + y)^4$
10. $(x + y)^5$
11. $(x + y)^6$

Find each of the following without using the Binomial Theorem. You may be able to use some of the previous exercises.

12. $(x + 2)^5$
13. $(3x + 2)^4$
14. $(x + \frac{1}{2})^5$
15. $(2x - 3)^5$

Exercises 16–20 refer to Pascal's Triangle. For convenience, call the line with the "1" only, the zero row, the line with "1 1" the first row, the one with "1 2 1" the second row, and so on.

16. Explain how to find each new entry in Pascal's Triangle.
17. Rewrite the triangle on your paper. Then write in the next two lines. How did you obtain them?

18. Which row in the table corresponds to $(x + y)^5$? to $(x + y)^7$?

19. To which power of $(x + y)$ does row 7 correspond? Write out the expansion of $(x + y)$ to that power.

20. If ${}_6C_3$ is the coefficient of a term in the expansion of $(x + y)^r$, tell what r is, and what the term is.

Answer exercise 20 for these coefficients.

21. ${}_7C_3$ **22.** ${}_6C_5$ **23.** ${}_6C_2$ **24.** ${}_8C_4$

25. Compare the entries in Pascal's Triangle, Figure 1, with those in Figure 2. Tell how you can find the value of an entry in Figure 2—say, ${}_{10}C_3$—by using Pascal's Triangle. Recall the agreement about naming the rows in Pascal's Triangle.

26. Often, the Binomial Theorem is written without the combinatorial coefficients—the C's. Such a statement of the theorem is begun below. Complete the statement.

$$(x + y)^n = x^n + \frac{n}{1}x^{n-1}y + \frac{n(n-1)}{1 \cdot 2}x^{n-2}y^2 + ?$$

27. Rewrite the Binomial Theorem in still another way—using factorial notation.

Expand and simplify using the Binomial Theorem.

28. $(x + 2)^4$ **29.** $(x + 3)^4$ **30.** $(x + 5)^3$ **31.** $(2x + 1)^3$

32. $(2x + 1)^5$ **33.** $(w - 11)^4$ **34.** $(2x - 1)^5$ **35.** $(x - 2y)^6$

36. $(\frac{1}{2}y + 3)^4$ **37.** $(2w + \frac{1}{3})^4$ **38.** $(2x - 3)^5$ **39.** $(4y - 3)^4$

40. $(x - 3)^4$ **41.** $(x + 2)^8$ **42.** $(2x + 3)^7$ **43.** $(x - \frac{1}{2}y)^5$

44. $\left(\frac{x}{y} + v\right)^5$ **45.** $(3v - \frac{1}{2}w)^5$ **46.** $(\frac{1}{2}a + \frac{2}{3}b)^5$ **47.** $(2a + b)^7$

Challenge

1. Describe any patterns you see in Pascal's Triangle which were not discussed in the text.

There is a sloping row marked on Pascal's Triangle on page 419. Starting at the top, what is the sum of each of the following?

2. its first and second numbers **3.** its second and third numbers

4. its third and fourth numbers **5.** its fourth and fifth numbers

6. The sums in exercises 2–5 are part of a certain number sequence. What sequence is it?

7. Show that ${}_nC_{r-1} + {}_nC_r = {}_{n+1}C_r$ *Hint: Use the factorial form of* ${}_nC_r$.

8. See if you can work out a convincing explanation for why the entries in the combinatorial table and in Pascal's Triangle correspond the way they do. Why should the coefficients of terms in the expansion of $(x + y)^n$ be the same as those of ${}_nC_r$?

9. *DO SOME RESEARCH:* find out about Blaise Pascal, the man after whom Pascal's Triangle is named. What were some of his contributions to mathematics, philosophy and religious thought?

11.7 The rth Term

Sometimes we need to find a specific term in the expansion of a binomial. For example, consider the third term in the expansion of $(x + y)^6$.

$$(x + y)^6 = x^6 + 6x^5y + 15x^4y^2 + 20x^3y^3 + 15x^2y^4 + 6xy^5 + y^6$$

The third term is indicated in red. Note that its coefficient is ${}_6C_2 = 15$ and that x is raised to the power 4 and y to the power 2. The fourth term of that expansion has coefficient ${}_6C_3 = 20$ and x is raised to the power 3, y to the power 3. We see that the

$$r\text{th term of } (x + y)^6 \text{ is } {}_6C_{r-1}x^{6-(r-1)}y^{r-1}.$$

And, in general, the following is true.

The rth term in a Binomial Expansion

The rth term of $(x + y)^n$ is ${}_nC_{r-1}x^{n-(r-1)}y^{r-1}$
and
the $(r + 1)$st term of $(x + y)^n$ is ${}_nC_rx^{n-r}y^r$.

Examples

1 **Find the fifth term of $\left(3a + \frac{1}{2}\right)^8$.**

Here, $n = 8$, and $r = 5$. The term we want will begin with ${}_8C_{5-1}$ or ${}_8C_4$. The power of $\frac{1}{2}$ will be 4 and of $3a$, $8 - 4$ or 4.

$${}_8C_4(3a)^4\left(\frac{1}{2}\right)^4 = 70(81a^4)\left(\frac{1}{16}\right) = \frac{2835}{8}a^4.$$

Thus, the fifth term of $\left(3a + \frac{1}{2}\right)^8$ is $\frac{2835}{8}a^4$.

2 **Find the coefficient of $a^{11}y^4$ in the expansion of $(2a - y)^{15}$.**

The *first* term of the expansion will have a^{15} in it, the *second* term $a^{14}y^1$, and so on. It is the *fifth* term which will contain $a^{11}y^4$.

$$\begin{aligned} \textit{fifth} \text{ term of } (2a - y)^{15} &= {}_{15}C_4(2a)^{11}(-y)^4 \\ &= \frac{15 \cdot 14 \cdot 13 \cdot 12}{4 \cdot 3 \cdot 2 \cdot 1} \cdot 2048a^{11}y^4 \\ &= 15 \cdot 7 \cdot 13(2048a^{11})(y^4) \quad \text{or} \quad 2{,}795{,}520\,a^{11}y^4 \end{aligned}$$

The coefficient of $a^{11}y^4$ is 2,795,520.

Exercises

Exploratory **Expand each of the following. Indicate carefully which term is first, which is second, and so on.**

1. $(x + 3)^3$
2. $(x + 2)^6$
3. $(5x + 1)^4$
4. $(y - 1)^8$

Written **Write the first and last terms of each of the following without using the rule.**

1. $(x + 1)^{50}$
2. $(2y - 3)^5$
3. $(0.5x + 2)^{10}$
4. $(8 + 3x)^5$

Find each of the following terms.

5. the second term of $(2x + 1)^4$
6. the second term of $(x + 3)^5$
7. the fourth term of $(2x + 1)^5$
8. the fourth term of $(3x - 1)^5$
9. the eighth term of $(2x + 3)^{11}$
10. the seventh term of $(x - 2y)^{10}$

Find the indicated coefficient in each of the following.

11. of x^5y in $(x + y)^6$
12. of $x^{11}y^2$ in $(2x + y)^{13}$
13. of x^4y^5 in $(x - 3y)^9$
14. of H^2K^8 in $(\frac{1}{2}H + K)^{10}$

Find each of the following.

15. the first four terms of $(x + y)^{16}$
16. the first four terms of $(x + y)^{20}$
17. the middle term of $\left(1 + \frac{1}{2}y^2\right)^{14}$
18. the twelfth term of $\left(4x - \frac{1}{3\sqrt{x}}\right)^{16}$

Find each of the following expansions and express in simplest form.

19. $(1 + 1)^5$
20. $(1 - 1)^4$
21. $(\frac{1}{2} + \frac{1}{2})^5$

Challenge

1. Show that ${}_nC_0 + {}_nC_1 + {}_nC_2 + \ldots + {}_nC_n = 2^n$. *Hint: Think about exercise 19.*

Expand each of these trinomials to the indicated power.

2. $(a + b - c)^3$ *Hint:* $a + b - c = (a + b) - c$
3. $(a + b + c)^4$

Find these values.

4. $(1.2)^4$ *Hint:* $1.2 = 1 + 0.2$
5. $(1.01)^4$
6. $(0.99)^4$

		1		
	$\frac{1}{2}$		$\frac{1}{2}$	
$\frac{1}{4}$		$\frac{2}{4}$		$\frac{1}{4}$

If *no* coins are tossed, the probability of *no* heads is 1. If one coin is tossed, $P(1H) = \frac{1}{2}$ and $P(0H) = \frac{1}{2}$. If two coins are tossed, $P(2H) = \frac{1}{4}$, $P(1H) = \frac{2}{4}$, and $P(0H) = \frac{1}{4}$. These entries have been noted in the table at the right.

7. Suppose three coins are tossed. Find $P(3H)$, $P(2H)$, $P(1H)$, and $P(0H)$. Enter your results in the table.
8. Continue your results for four coins and five coins.
9. Describe your results. What, for example, is the sum of the entries across one row? What else do you notice?

11.8 Conditional Probability

Suppose a friend draws a card from an ordinary deck of 52. What is the probability he picks a king? $P(\text{king}) = \frac{4}{52}$ since there are 4 kings and 52 cards altogether. But now, suppose that your friend draws the card and then informs you that he has drawn a picture card. With this new information, what is the probability he has drawn a king? Before this problem can be solved, some new terminology needs to be introduced.

"The probability of a king, given the fact of a picture card" will be symbolized as

$$P(\text{king}|\text{picture card}).$$

In general, we will use the symbol

$$P(A|B)$$

to mean, "the probability of A given B" or "the probability that event A will occur given the event B has already occurred," or "the probability of A under the condition of B." We call this the **conditional probability** of A, given B.

Now let us solve the original problem—that of finding $P(\text{king}|\text{picture card})$. We know that there are 12 picture cards and that four of these are kings. In a way, we have a new outcome set—the set of all picture cards in the deck. For this new set S', $n(S') = 12$. If the set K of kings is the set of successes, then $n(K) = 4$. Using the Fundamental Principle of Probability, we have

$$P(\text{king}|\text{picture card}) = \frac{n(K)}{n(S')} = \frac{4}{12} = \frac{1}{3}.$$

Consider another problem. Suppose your friend draws a card and announces that he has chosen a red card. Now, given this information, what is $P(\text{king})$? The "new outcome set" is the set of red cards. The "successes" have to be in this outcome set to start with (they have to be red) and, of course, they have to be kings. The successes are cards which are red and are kings—that is, red kings. There are two of these.

$$P(\text{king}|\text{red}) = \frac{n(\text{king and red})}{n(\text{red})} = \frac{2}{26} = \frac{1}{13}$$

If we let $R = \{\text{reds}\}$, and $K = \{\text{kings}\}$, then we have, using set notation,

$$P(K|R) = \frac{n(K \cap R)}{n(R)}.$$

By now we are ready to state a general rule. If we have sets A and B, the probability that A will occur, given that B has occurred, is given by the following rule.

The Probability of *A* given *B*

$$P(A|B) = \frac{n(A \cap B)}{n(B)}$$

Example

1 **At your school, a certain club is open only to seniors, and currently has five girls and five boys. It is decided to let juniors join the club and four new members are admitted: 3 boys and 1 girl. If, from the total new membership of 14, a new president is selected at random, what is the probability the new president is**

a. a girl?
b. a girl, given that the person is a junior?
c. a senior, given that a girl has been selected?

To solve these problems, we use basically the same process as used in the picture card situations which opened this section. Consider the following sets of club members.

$S = \{\text{seniors}\}$ $\quad J = \{\text{juniors}\}$ $\quad G = \{\text{girls}\}$

a. This is old business. There are 14 members of whom six are girls.

$$P(G) = \frac{6}{14} = \frac{3}{7}$$

b. We are looking for $P(G|J)$. According to the rule, $P(G|J) = \frac{n(G \cap J)}{n(J)}$. Now $n(J) = 4$ and $n(G \cap J) = 1$, since one member is a girl and a junior. Therefore, we have $P(G|J) = \frac{1}{4}$.

c. $P(S|G) = \frac{n(S \cap G)}{n(G)} = \frac{5}{6}$.

We have learned that $P(A|B) = \frac{n(A \cap B)}{n(B)}$. Now, suppose that sets A and B, and therefore set $A \cap B$, are contained in some outcome set S. From previous work we know that with respect to outcome set S,

$$P(A \cap B) = \frac{n(A \cap B)}{n(S)} \text{ and } P(B) = \frac{n(B)}{n(S)}.$$

Thus,

$$P(A|B) = \frac{n(A \cap B)}{n(B)} = \frac{\frac{n(A \cap B)}{n(S)}}{\frac{n(B)}{n(S)}} = \frac{P(A \cap B)}{P(B)}.$$

This final equality indicates that the probability of A, given B, can be computed not only from the *number of elements* in $A \cap B$ and in B. It can also be found directly from the probabilities of $A \cap B$ and B, if we know them. Sometimes this is all the information we have.

Example

2 **From a survey on ice cream preference conducted in front of the Pine City Ice Cream Parlor, the following information was obtained. Of the people questioned, 70% liked chocolate, 46% liked butter pecan, and 16% liked both. Now, if you know that a person likes chocolate, what is the probability that person also likes butter pecan?**

Let A = {people who like chocolate}. Let B = {people who like butter pecan}. This time we do not know the actual numbers involved, but we do know $P(A) = 0.70$, $P(B) = 0.46$ and $P(A \cap B) = 0.16$.

We have $P(B|A) = \frac{P(B \cap A)}{P(A)} = \frac{0.16}{0.70} = \frac{8}{35}$.

Note that in this example, there is one "extra" bit of information.

Here is another example, which makes use of permutations, as well as probability ideas. We will not provide the solution. It is left for you in the exercises.

Example

3 **A five-digit number is made from the digits 1, 2, 3, 4, 5. What is the probability that the number ends in the digits 52, given that it is even?**

First, determine which set represents A and which set represents B.

How many of the five digits, when used as the last digit, will produce an even number? How does this affect the number of elements in set B?

The solution is left for the exercises.

Exercises

Exploratory

1. Explain in your own words what is meant by conditional probability.

2. Explain the use of each of these symbols in the study of conditional probability. Give an example showing the use of each.

a. $P(A|B)$ **b.** $P(B|A)$ **c.** $P(A \cap B)$ **d.** $n(A)$ **e.** $n(B)$

3. Are $P(A|B)$ and $P(B|A)$ the same? Illustrate your answer.

4. Explain how the rule $P(A|B) = \frac{n(A \cap B)}{n(B)}$ can be justified.

5. Why is $\frac{n(A \cap B)}{n(B)}$ equal to $\dfrac{\frac{n(A \cap B)}{n(S)}}{\frac{n(B)}{n(S)}}$?
6. Justify the rule $P(A|B) = \frac{P(A \cap B)}{P(B)}$. Use your answers to exercise 4 and 5.
7. Which example in the text is an illustration of the rule given in exercise 6? Make up another problem like that one, and solve it using the rule.
8. If set A is a subset of set B, how does this affect the rules for $P(A \cap B)$? for $P(A|B)$?

A card is drawn from an ordinary deck of 52, and found to be red. In that event, find each of the following.

9. *P*(heart)
10. *P*(diamond)
11. *P*(ace)
12. *P*(three)
13. *P*(picture card)
14. *P*(jack or ten)
15. *P*(six of spades)
16. *P*(six of hearts)
17. *P*(red six)

Written **Suppose two fair dice are tossed and the result is $x \geq 9$. Given that information, find each of the following.**

1. *P*(8)
2. *P*(9 or 10)
3. *P*(double)
4. *P*(12)
5. *P*(even number)
6. *P*(double or even)

Draw a sample space for the experiment of tossing a red die and a green die. Then, for each of the following, find the indicated conditional probability. Also name the set of successes being discussed, using ordered pair notation.

7. *P*(red die shows 5, given that the sum is even)
8. *P*(red die shows 6, given a sum of 11)
9. *P*(red die shows 6, given sum 12)
10. *P*(green die shows 4, given sum 8 or 9)
11. *P*(red > green, given green 3)
12. *P*(red = green, given sum 7)

A survey of people's preferences for television situation comedies showed that 58% of the people polled liked *My Kid Tamara,* and 52% liked *Lil 'n' Mary*. Forty percent liked both shows. What is the probability that a person liked the following?

13. *My Kid Tamara* given that he liked *Lil 'n' Mary*
14. *Lil 'n' Mary* given that she liked *My Kid Tamara*
15. Can you express the information given in exercises 13–14 using *Venn diagrams?* What percentage of the people liked neither show?

A sports survey taken at Beeville High shows that 48% of the respondents liked soccer, 66% liked basketball, and 38% liked hockey. Also, 30% liked soccer and basketball, 22% liked basketball and hockey, and 28% liked soccer and hockey. Finally, 12% liked all three sports. Find the probability that a person chosen from those surveyed liked the following.

16. soccer given that he or she likes basketball
17. basketball given that he or she likes soccer
18. hockey given that he or she likes basketball
19. basketball given that he or she likes hockey
20. What is the probability a person likes hockey and basketball given that he likes soccer?
21. What is the probability she likes soccer and basketball, given she likes hockey?

22. Try to represent the information given in exercises 16–21 using a Venn diagram. What percentage of those interviewed like none of the sports discussed?

23. Answer the question in example 3 of the text. Explain every step in your solution.

In exercises 24–30, a five-digit number is to be formed from the digits 0, 2, 5, 6, 8. Zero as a first digit is permissible. What is the probability that the number meets the following conditions?

24. is even given that it ends in 0

25. ends in 0 given that it is even

26. ends in 0 given that it is divisible by 5

27. ends in 80 given that it is divisible by 5

28. ends in 625 given that it is divisible by 5

29. ends in 60 given that it is divisible by 4
Hint: remember the "test" for divisibility by 4.

30. ends in 65 given that it is divisible by 4

31. *A, B, C, D,* and *E* are running for election, and all are given equal chances of winning. Find *P*(*B* wins), and then *P*(*B* wins given that *E* drops out of the race).

Challenge

DO SOME RESEARCH. When was the study of probability, as a branch of mathematics, "born"? What were the circumstances? Find out about some of the people who contributed most notably to the early study of probability.

Mixed Review

Suppose the graphs of each pair of equations are drawn on the same set of axes. Determine a transformation that maps one graph onto the other.

1. $y = \tan\theta$
$y = \cot\theta$

2. $y = \sec\theta$
$y = \csc\theta$

3. $y = \cos\theta$
$y = \arccos\theta$

Write each expression in terms of the indicated circular function.

4. $\sin^2 x + \cos 2x$; $\cos x$

5. $\tan x$; $\sin x$

6. $\dfrac{2\cos 2x}{\cos 2x + \sin 2x + 1}$; $\tan x$

Suppose $\triangle ABC$ has vertices $A(0, 0)$, $B(3, 1)$, and $C(-3, -7)$. Find its image under each composite transformation. Then identify the transformation as a line reflection, translation, rotation, or glide reflection.

7. $R_{x=3} \circ T_{5,1}$

8. $Rot_{O,-90°} \circ T_{0,2}$

9. $T_{-3,3} \circ R_{y=x}$

10. A quiz has 5 true-false questions. If you choose your answers at random on every question, what is the probability you get at least four questions right?

11. In the simplification of the expansion $(2p - 3q)^6$, the exponent of p in a particular term is 2. What is the complete term?

12. An urn contains white and gold marbles. The probability of drawing a white marble from the urn is $\frac{9}{13}$. If 9 white marbles are taken out of the urn, then the probability of drawing a white marble is $\frac{3}{5}$. How many of each type of marble are in the urn?

Mathematical Excursions

Independent Events

The alternate version of the Counting Principle, given on page 403, is valid only when the occurrence of one of the events in question cannot affect the occurrence of the other. A more precise way of saying that event A cannot affect the occurrence of event B is to say that the probability of A is equal to the probability of A given that B has occurred. In symbols this becomes

$$P(A) = P(A|B).$$

We call A and B **independent events** if it is the case that

$$P(A) = P(A|B) \quad \text{and} \quad P(B) = P(B|A).$$

Another way of defining independence is to say that A and B are independent if

$$P(A \cap B) = P(A) \cdot P(B).$$

Exercises

1. Show that the two definitions for independent events, given above, are essentially equivalent.
2. Is each trial of a Bernoulli experiment—say, rolling a fair die—independent from each other trial? Explain.
3. In exercises 16–21 on page 428, is liking soccer independent of liking hockey? Explain your answer.
4. If two events are not independent, they are *dependent*. Show that in exercises 16–21, liking soccer and liking basketball are dependent. Rewrite the original problem to make them independent, if possible.
5. If A and B are subsets of S, and A and B are independent, show that A and $\overline{B}$ are also independent. (Recall that $\overline{B}$ means "the complement of B.")

Three events A, B, and C are independent if they are pairwise independent,

$$P(A \cap B) = P(A) \cdot P(B),\ P(A \cap C) = P(A) \cdot P(C), \text{ and } P(B \cap C) = P(B) \cdot P(C),$$

and if

$$P(A \cap B \cap C) = P(A) \cdot P(B) \cdot P(C).$$

6. A fair coin is tossed twice. Let A = {heads on first coin}, B = {heads on second coin}, and C = {heads on exactly one coin}. Determine if events A, B, and C are independent.

Problem Solving Application: **Look for a Pattern**

Many problems can be solved more easily by looking for a pattern. When using this method to solve problems, it is important to organize the information about the problem. Sometimes a pattern may be obvious, but in many instances, you will need to "play" with the information to find a pattern. Then, you can solve the problem by generalizing and applying the pattern.

Example

1 What is the greatest number of points of intersection of ten lines?

Explore There are ten lines. These lines must be drawn in a manner such that the points of intersection is the greatest number possible.

Plan Draw two lines. The greatest number of points of intersection is 1.

Draw another line. A third line would add the possibility of two more points of intersection.

Four lines can have a maximum of 6 points of intersection.

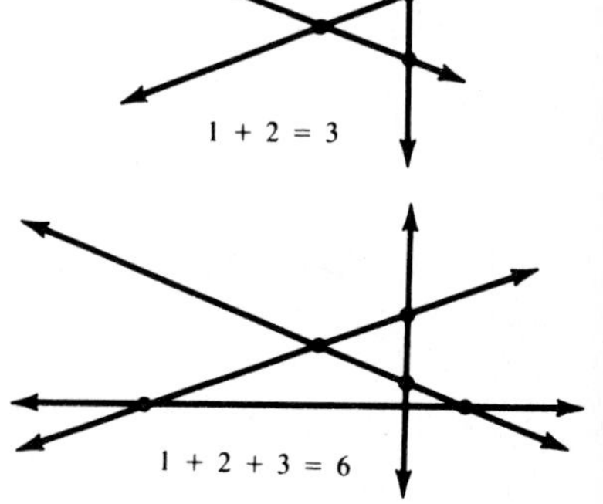

Use this pattern to find the greatest number of points of intersection of ten lines.

Solve $1 + 2 + 3 + 4 + 5 + 6 + 7 + 8 + 9 = 45$
The greatest number of points of intersection is 45.

Examine Look at the pattern. Draw 7 lines. Does the pattern hold true for 7 lines? Is it possible to have more than 45 intersecting points when 10 lines are drawn?

Exercises

Written **Write an equation showing the relationship between the variables in each chart. Then, copy and complete the table.**

1.

a	1	2	3	4	5	6
b	6	12	18			

2.

x	1	2	3	4	5	6
y	2	5	8			

3.

x	1	2	3	4	5	6
y	4	6	8			

4.

c	1	2	3	4	5	6
d	0	3	8			

Solve each problem.

5. If $1 \times 1 = 11$, $11 \times 11 = 121$, and $111 \times 111 = 12{,}321$, find $111{,}111{,}111 \times 111{,}111{,}111$ without using a calculator.

6. If $9 \times 9 = 81$, $99 \times 99 = 9801$, and $999 \times 999 = 998{,}001$, find $999{,}999 \times 999{,}999$ without using a calculator.

7. A sheet of paper is torn in half. Then, the pieces are placed on top of each other, and all of the pieces are torn in half. If this procedure is continued, how many pieces of paper will there be after 10 tears?

8. Shonté is saving money for a car. She saved $1 on April 1, $3 on April 2, and $5 on April 3. If Shonté continues this pattern throughout the month of April, how much will she have saved at the end of the month?

9. Imagine there are 250 children and 250 doors numbered 1 through 250. Suppose the first child opens each door. Then, the second child closes every second door. The third child changes the state of every third door (the child closes the open doors and opens the closed doors). This process continues until the 250th child changes the state of the 250th door. After this process is completed, which doors are open?

10. What symbol comes next? ___?___

Identify the correct pattern for the figure below.

11.

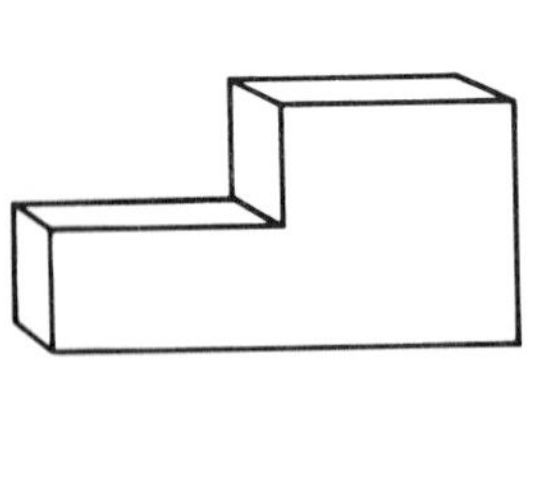

a.

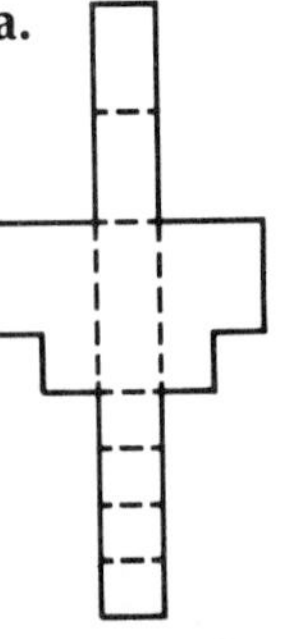

b.

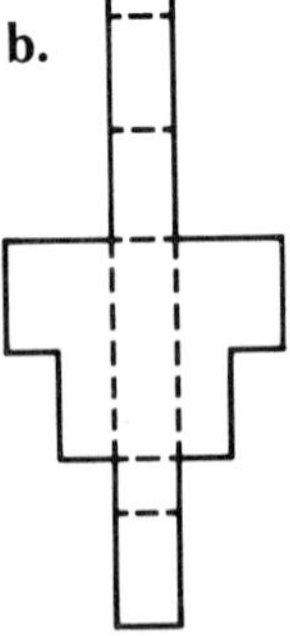

c.

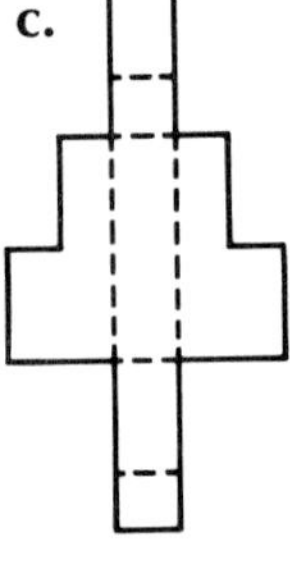

Vocabulary

Counting Principle
Counting Principle (Alternate Version)
permutation
combination
Bernoulli experiments
Pascal's Triangle
Binomial Theorem
Conditional Probability

Chapter Summary

1. For an event E and an outcome set S, $P(E) = \frac{n(E)}{n(S)}$.
2. The probability of an impossible event is 0.
3. The probability of a certain event is 1.
4. For any event E, $0 \leq P(E) = 1$.
5. For two events, A and B, $P(A \text{ or } B) = P(A) + P(B) - P(A \text{ and } B)$.
6. The symbol for "number of permutations of n things taken all at a time" is ${}_nP_n$. The value of ${}_nP_n$ is $n!$
7. The number of permutations of n things taken r at a time, with $r < n$, is given by the following rule.
$$ {}_nP_r = \frac{n!}{(n-r)!} \text{ or } n(n-1)(n-2)\ldots(n-r+1) $$
8. The number of combinations of n things taken r at a time is symbolized in this way: ${}_nC_r$.
9. The value of ${}_nC_n$ is 1.
10. To find the number of combinations of n things taken r at a time use the following formulas.
$$ {}_nC_r = \frac{{}_nP_r}{r!} \text{ or } {}_nC_r = \frac{n!}{r!(n-r)!} $$
or
$$ {}_nC_r = \frac{n(n-1)(n-2)\cdot\ldots\cdot(n-r+1)}{r(r-1)(r-2)\ldots 1} $$

11. $P(\text{exactly } r \text{ 6's in } n \text{ tosses of a fair die}) = {}_nC_r\left(\frac{1}{6}\right)^r\left(\frac{5}{6}\right)^{n-r}$

12. In a Bernoulli experiment, the probability of obtaining exactly r successes in n trials is

$$ {}_nC_r\, p^r q^{n-r} $$

where p is the probability for "success" and $q = 1 - p$ is the probability for "failure."

13. The rth term in a binomial expansion is ${}_nC_{r-1}x^{n-(r-1)}y^{r-1}$ and the $(r + 1)$st term is given by ${}_nC_r x^{n-r}y^r$.

Chapter Review

11.1 Two fair tetrahedral (four-sided) dice are tossed. Find the probability of each of the following sums. Make a diagram similar to those on page 402.

1. 2 **2.** 1 **3.** 5

4. 7 **5.** 8 **6.** an even number

Suppose the two tetrahedral dice are of different colors, one red, the other green. Find each of the following probabilities. ("$R = G$" means "the number of spots on the red die is equal to the number on the green die.")

7. $P(R = G)$ **8.** $P(R < G)$ **9.** $P(R = 2G)$ **10.** $P(R \text{ is odd})$

11.2 Show that the following two rules are equivalent.

11. ${}_nP_r = \dfrac{n!}{(n - r)!}$ ${}_nP_r = n(n - 1)(n - 2) \ldots (n - r + 1)$

11.3 State whether each of the following is a list of combinations, a list of permutations, or neither.

12. all possible ways of placing 6 people in a line for a driver's license

13. all possible football teams formed from a squad of 25 men, each of whom has the ability to play any position

14. all three-digit numbers in which no digits are repeated

15. all possible ways in which 2 ordinary dice can land in 1 toss

11.4 Compute each of the following.

16. ${}_6C_2\left(\frac{1}{6}\right)^2\left(\frac{5}{6}\right)^4$

17. ${}_8C_3\left(\frac{1}{6}\right)^3\left(\frac{5}{6}\right)^5$

18. ${}_6C_0\left(\frac{5}{6}\right)^6$

11.5 Write expressions for each of the following. Do not calculate.

19. the probability of obtaining exactly ten 6's in 100 tosses of a fair die
20. the probability of exactly fifteen 5's in 90 tosses of a die
21. the probability of exactly t twos in T tosses of a fair die

11.6 Use the Binomial Theorem to expand each expression, and then simplify.

22. $(x + 1)^8$

23. $(2x - 3)^5$

24. $\left(3a - \frac{1}{2}\right)^4$

11.7 In the simplification of the expansion of $(2a - b)^8$, the exponent of b in a particular term is 3.

25. Is this term the third or fourth in the expansion?
26. What is the complete term including its sign?
27. Write the term of the expansion which has 6 as the exponent of b.

Find each of the indicated terms.

28. the thirteenth term of $(x - y)^{17}$
29. the second term of $(x - 8k)^{21}$
30. the fifth term of $\left(3x - \frac{1}{2}\right)^{10}$
31. the sixth term of $(3x - 2y)^{13}$

Expand the following expression.

32. $(a + b + c)^3$ *Hint:* $a + b + c = (a + b) + c$

11.8 A small college has 100 students, each of whom plays at least one sport.
49 play football
35 run track
38 play basketball
9 play football and run track
7 play football and basketball
8 play basketball and run track
2 participate in all three
One student is picked at random. What is the probability that he or she plays the following?

33. at least one sport

34. only one sport

35. A card is drawn from a standard deck of 52 cards. Find $P(\text{ace}|\text{not a face card})$.
36. If bag 1 contains four red and four black marbles, bag 2 contains two red and six black marbles, and bag 3 contains seven red and one black marble, find the probability of having picked a marble from bag 1, given that a black marble has been picked.

Chapter Test

Find the value of each of the following.

1. $\frac{5!}{3!}$
2. ${}_6P_3$
3. $\frac{{}_7P_2}{{}_4P_4}$
4. $\frac{{}_{10}P_8}{{}_5P_4}$
5. ${}_6C_6$
6. ${}_7C_4$
7. ${}_{10}C_8$
8. ${}_7C_5\left(\frac{1}{6}\right)^4\left(\frac{5}{6}\right)^3$

Solve.

9. A card is drawn at random from a deck of 52. Find the probability of drawing a red seven.
10. Two fair six-sided dice are tossed. One of the dice is green, the other red. Find $P(R > G)$.
11. How many different committees of four people can be selected from a group of seven people?
12. How many five-letter words can be made from TOPICAL if the first letter must be a vowel?
13. Expand and simplify using the Binomial Theorem: $(2x - 4)^4$.
14. Find the first four terms of $(x + y)^{16}$.
15. Find the third term of $(3x + 1)^5$.
16. In the simplification of the expansion $(3a - 2b)^7$, the exponent of b in a particular term is 4. What is the complete term with its sign?
17. If the probability is $\frac{2}{3}$ that Beeville will defeat Hornetville each time they play soccer, what is the probability that Beeville will win exactly 2 games out of 4?
18. A fair coin is tossed five times. What is the probability of getting at least three heads?

Urn A contains 3 green and 7 white marbles. Urn B contains 4 green and 2 white marbles. A marble is selected at random from one of the urns. What is the probability of each event?

19. The marble selected was white.
20. The marble was selected from Urn A given that the marble selected was green.

Chapter 12

Statistics

Testing or surveying a large number of people can produce many interesting statistics. If the Scholastic Aptitude Test scores of a great number of high school students are collected and presented in a graph, the curve will most likely be the familiar bell shape. About 68% of the students who take the SAT score between 400 and 600.

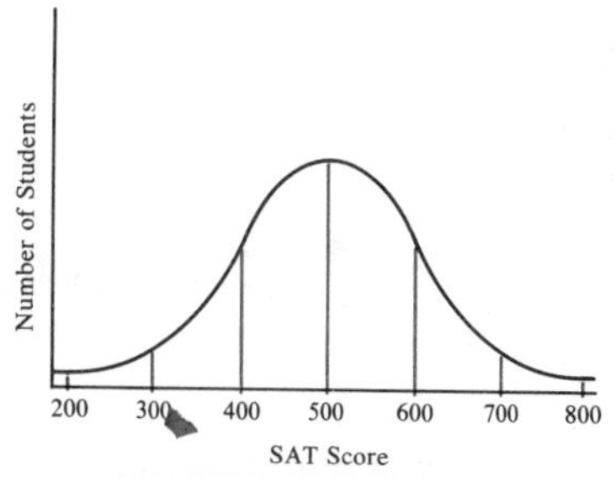

12.1 Measures of Central Tendency

Few of us need to be told about the importance of statistics in today's world. Indeed, there are certain people who say that statistics have become *too* important. Some people do not "trust" statistics. In fact, the science of statistics is misunderstood by many.

Statistics is a relatively new science. In the past, people had no efficient way of collecting the many small pieces of information which are used in statistics. But the introduction of modern computer technology has made possible the collection and analysis of great amounts of information.

You will recall that the individual pieces of information used in statistics are called **data.** The science of statistics may be defined as *the study of numerical data.*

You have already studied some important statistical measures, namely, the **mean,** the **median,** and the **mode.** These are *measures of central tendency.* As the name suggests, a measure of central tendency shows how numbers cluster around, or can be summarized by, one centrally located number.

Definition of Mean

> The **mean** of a set of n numbers is the sum of the numbers divided by n.

The mean is also sometimes referred to as the *arithmetic mean* or *numerical average.*

Examples

1 **Find the mean of the set {8, 6, 10, 6, 12, 12, 12, 10}.**

To find the mean, first find the sum.

$$8 + 6 + 10 + 6 + 12 + 12 + 12 + 10 = 76$$

Then divide the sum by 8, the number of measures. The mean is $\frac{76}{8}$ or $9\frac{1}{2}$.

2 **Brenda has the following grades on three Russian exams this term: 82, 74, 89. What grade must she get on her fourth exam in order to have an 85 average?**

Let x represent the minimum grade Brenda needs.

$$\frac{82 + 74 + 89 + x}{4} = 85$$
$$245 + x = 340$$
$$x = 95$$

Brenda needs at least a 95 to achieve her goal.

Definition of Median

The **median** of a set of n numbers is the one which lies in the middle when the numbers are arranged in order. If n is even, there are two middle numbers, and the median is their numerical average.

Examples

Find the median of each set.

3 {6, 8, 10, 12, 14, 16, 16, 17, 20}

The median is 14.

4 {6, 8, 10, 12, 14, 16, 16, 17, 20, 22}

The median is $\frac{14 + 16}{2}$ or 15.

Definition of Mode

The **mode** of a set of n numbers is the number which occurs most often. A set may have more than one mode. If it has two, it is called **bimodal.**

Examples

Find the mode of each set.

5 {4, 5, 6, 7, 7, 7, 8, 8, 14}

Since 7 occurs more frequently than any other number, the mode is 7.

6 {6, 8, 8, 8, 9, 10, 11, 12, 12, 12, 15}

This set has two modes, 8 and 12. The set is bimodal.

Exercises

Exploratory

1. Write a paragraph describing your feelings about statistics. Is your overall feeling positive or negative? Try to analyze your feelings, and give reasons for them.
2. Give an example of the use of statistics to deal with an issue important in your state or in the United States. Describe how the statistics were used.
3. Answer exercise **2** for a use of statistics in your own community.

Explain the meanings of these terms.

4. data
5. mean
6. numerical average
7. median
8. mode
9. bimodal

The following are terms from your previous study of statistics. Explain the meaning of each, and give an example, as your teacher directs.

10. bar graph
11. circle graph
12. percentile
13. quartile
14. upper quartile
15. histogram
16. frequency histogram
17. frequency polygon

Written

Find the mean of each set of numbers.

1. {6, 2, 10, 6, 10, 8, 11, 12}
2. {6.34, 8.02, 10.06, 4.70, 8.04}
3. $\left\{\frac{1}{4}, \frac{3}{8}, \frac{11}{16}, \frac{7}{8}, \frac{3}{4}, \frac{1}{8}, \frac{13}{16}, \frac{9}{16}\right\}$
4. {0.60, 0.002, 0.012, 0.088, 0.420}
5. Given the following set of numbers: 8, 6, 8, 8, 6, 10, 8, 8, 6, 6, 8, 8, 6, 6, 10, 10, 8, 6. Is there a shorter way to find the mean other than adding all the numbers and dividing?

Find the mode of each set of numbers. Is any bimodal?

6. {13, 11, 12, 13, 13, 11, 13, 13}
7. {15.06, 15.08, 15.08, 14.98, 15.08, 14.98, 14.98}
8. {8.08, 8.16, 8.02, 8.12, 8.12, 8.08, 8.16, 8.08, 8.16, 8.18, 8.16, 8.08}

Find the median of each set.

9. {11, 10, 13, 12, 12, 13, 15}
10. {50, 75, 65, 79, 55, 65, 50, 80}
11. $\{3n - 1, 2n, 3n + 1, 4n - 1, 3n, 3n + 2, 4n, 4n - 3\}$ if $n = 8$.

In each case, find the value of *x* which would make *M* the mean of the set.

12. $\{4, x, 6, 9\}$; $M = 7$
13. $\{14, 16, 10, 12, 20, 18, x\}$; $M = 15$
14. $\{3, 2, 3x - 1, 4\}$; $M = 4$
15. $\{15, 17, 18, 17.5, 4x + 3\}$; $M = 16$

Find the nonnegative number which makes *M* the mean of the set.

16. $\{3, 5, x^2 - 1, 8, 10\}$; $M = 8$
17. $\{10, 8, 6, x^2 - 3, 4\}$; $M = 5$
18. $\{11, 19, x^2 + 3, 14\}$; $M = 19$
19. $\{4, 4, 4, x^2 + 1, 5, 5\}$; $M = 4$

12.2 Introducing Σ

In this section we are going to study a new symbol. This symbol is used in many areas of mathematics, and especially in statistics. The symbol can simplify the process of dealing with large quantities of data.

The symbol is the Greek capital letter **sigma** and it looks like this.

$$\Sigma$$

The lower case sigma, σ, is used elsewhere.

The symbol means *the summation of* or *take the sum of what follows.* Thus,

$$\sum_{i=1}^{5} i$$

is read "the summation of i for $i = 1$ to 5".
It is computed this way.

$$\sum_{i=1}^{5} i = 1 + 2 + 3 + 4 + 5 = 15$$

In this case, 1 is the **lower limit of summation** and 5 is the **upper limit of summation.** The variable i is an example of a **dummy variable.** The symbol used for the variable is unimportant.

The following symbols all have the same meaning, the sum of the first ten natural numbers.

$$\sum_{i=1}^{10} i \qquad \sum_{j=1}^{10} j \qquad \sum_{k=1}^{10} k$$

Note that the sum of the first n natural numbers can be symbolized

$$\sum_{i=1}^{n} i$$

If n were equal to 5000, it would take a long time to write all the terms of that sum! The symbol Σ is a very efficient symbol indeed!

Examples

1 **Compute $\sum_{i=0}^{3} i^2$.**

This is "the summation of i^2 with $i = 0$ to 3."

$$\sum_{i=0}^{3} i^2 = 0^2 + 1^2 + 2^2 + 3^2 \quad \text{or} \quad 0 + 1 + 4 + 9 \text{ or } 14.$$

2 **Compute $\sum_{i=2}^{6}(i + 3)$.**

This is "the summation of $i + 3$ with $i = 2$ to 6."

$$\sum_{i=2}^{6}(i + 3) = (2 + 3) + (3 + 3) + (4 + 3) + (5 + 3) + (6 + 3) = 35$$

3 **Use sigma notation to express the mean of all the numbers from 10 to 50, inclusive.**

The mean is the sum of all the numbers divided by the number of numbers. The sum of the numbers from 10 to 50 can be symbolized $\sum_{i=10}^{50} i$. From 10 to 19 is ten numbers, to 29 is twenty, and to 49 is forty. Thus, there are 41 numbers from 10 to 50, inclusive.

The symbol is $\dfrac{\sum_{i=10}^{50} i}{41}$ or $\dfrac{1}{41}\sum_{i=10}^{50} i$.

4 **Use sigma notation to express the mean of the first n numbers.**

The symbol is $\dfrac{\sum_{i=1}^{n} i}{n}$ or $\dfrac{1}{n}\sum_{i=1}^{n} i$.

5 **Show that $\sum_{i=1}^{n} 4i = 4\sum_{i=1}^{n} i$.**

$$\sum_{i=1}^{n} 4i = 4(1) + 4(2) + 4(3) + \ldots + 4(n)$$

$$= 4(1 + 2 + 3 + \ldots + n) \qquad \textit{Distributive Law}$$

$$= 4\sum_{i=1}^{n} i \qquad \textit{Definition of } \textstyle\sum \textit{ symbol}$$

In general, $\sum_{i=1}^{n} ci = c\sum_{i=1}^{n} i$, where c is a constant.

Sometimes, when the limits of summation are clear, they can be omitted. In this case, the last rule might be written like this.

$$\sum ci = c\sum i$$

Exercises

Exploratory **Explain the meaning of each of the following symbols.**

1. $\sum$
2. $\sum_{i=1}^{4} i$
3. $\sum_{j=1}^{3} j$
4. $\sum_{i=3}^{6} i$
5. $\sum_{i=0}^{4} 3i$
6. $\sum_{i=1}^{2} i^2$
7. What is the difference between $\sum_{i=1}^{6} 2i$ and $\sum_{k=1}^{6} 2k$?

Compute each of the following. Show your work.

8. $\sum_{i=1}^{3} i$
9. $\sum_{i=1}^{6} i$
10. $\sum_{i=0}^{5} i$
11. $\sum_{i=1}^{4} 3i$

Written **Compute each of the following. Show your work.**

1. $\sum_{j=1}^{3} 2j$
2. $\sum_{k=0}^{8} k$
3. $\sum_{k=0}^{8} (k + 2)$
4. $\sum_{i=0}^{4} i^2$
5. $\sum_{i=1}^{8} \frac{i}{2}$
6. $\sum_{i=0}^{10} 5i$
7. $\sum_{i=1}^{4} (i^2 + 3)$
8. $\sum_{i=0}^{4} \frac{i^2}{3}$
9. $\frac{1}{6}\sum_{i=1}^{6} i$
10. $\frac{1}{10}\sum_{i=1}^{10} i^2$
11. $\sum_{i=5}^{10} (i + 1)$
12. $\sum_{k=12}^{20} \frac{k^3}{4}$
13. $\sum_{i=3}^{8} (i + 2)$
14. $\sum_{i=2}^{6} (i + 1)$
15. $2\sum_{i=1}^{5} i$
16. $10\sum_{i=0}^{5} \frac{i}{2}$
17. $3\sum_{j=0}^{4} j^2$
18. $6\sum_{k=1}^{3} k^3$
19. Explain why $\frac{1}{30}\sum_{i=1}^{30} i$ represents the mean of the first 30 natural numbers.

Use Σ-notation to write expressions for exercises 20–28.

20. the mean of the first 12 numbers
21. the sum of the first 12 squares
22. the mean of the first 100 numbers
23. the sum of the first *n* squares
24. the mean of the numbers from 25 to 75, inclusive
25. the sum of the first *r* squares
26. the sum of the cubes of whole numbers from 6 to 18, inclusive

27. the sum of the first 12 numbers of the form $\frac{1}{n}$ $\left(\text{The first three are } \frac{1}{1}, \frac{1}{2}, \frac{1}{3}.\right)$

28. the sum of the first n numbers of the form given in exercise 27

29. Show that $\sum_{i=0}^{6} 2i = 2\sum_{i=0}^{6} i$.

30. Show that $\sum_{i=1}^{n} 15i = 15\sum_{i=1}^{n} i$.

31. What does it mean to say that $\Sigma ci = c\Sigma i$? Why can c be "pulled out" of the summation in this way?

Do some research to find the contributions of each of the following to the study of statistics.

32. Gregor Mendel

33. Karl Pearson

34. Sir Francis Galton

Mixed Review

Find each value, if it exists.

1. $\text{Arccos}\left(-\frac{\sqrt{3}}{2}\right)$

2. $\tan(\text{Sin}^{-1} 1)$

3. $\csc\left(\text{Arctan}\frac{7}{24}\right)$

In the figure at the right, each of the small quadrilaterals is a square. Suppose translation T maps V onto Y. Find each image.

4. $\mathbf{T}(K)$

5. $R_{\overline{MR}}(Y)$

6. $(R_{\overline{BN}} \circ R_{\overline{LG}})(C)$

7. $(\text{Rot}_{Q,180°} \circ \text{Rot}_{H,-90°})(\overline{DF})$

8. $(\text{Rot}_{P,180°} \circ R_{\overline{VS}} \circ \mathbf{T})(OPIJ)$

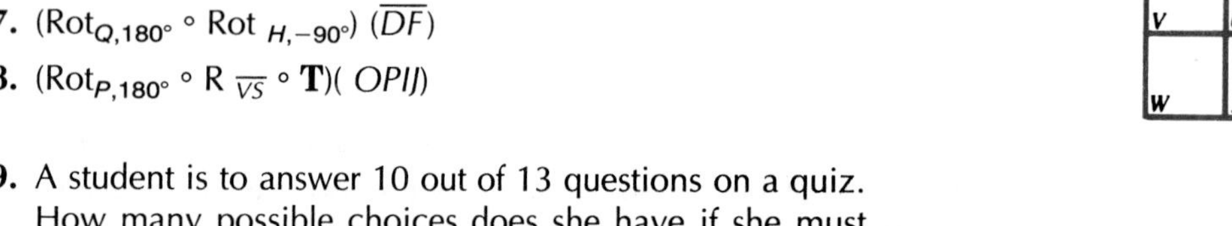

9. A student is to answer 10 out of 13 questions on a quiz. How many possible choices does she have if she must answer exactly 3 of the first 5 questions?

Thomas guesses on all 5 questions on a multiple-choice test. If each question has four choices, find the probability of the following.

10. exactly 3 correct

11. at least 3 correct

12. all incorrect

13. Find the sixth term of $(5x + 2y)^8$.

14. Expand $(3a - b)^5$.

15. Which event is more likely: obtaining at least one 6 when a fair die is tossed twice or obtaining at least one pair of 6's when a pair of fair dice is tossed 12 times?

12.3 Using Σ in Statistics

In this section we will see a special way in which the summation symbol Σ is used in many statistical situations.

Suppose a survey has produced the following ten measures.

10, 7, 7, 5, 7, 3, 8, 6, 12, 4

Let us agree to call the measures by the letter x and to use **subscripts** to indicate the order in which we have them.

Then, 10 is the first measure, indicated x_1 *read "x sub one"*
7 is the second measure, indicated x_2 *"x sub two"*
7 is the third measure, indicated x_3 *"x sub three"*
.
. and so on, down to
.
4, which is the tenth measure, or x_{10}. *"x sub ten"*

Thus, we have $x_1 = 10$, $x_2 = 7$, $x_3 = 7$, and so on to $x_{10} = 4$.

Note that the **indexing,** or assigning of subscripts to x, is according to order of appearance, not size. Thus, x_1 is the *first* measure, not necessarily the least. Similarly, x_8 is the eighth measure; it is not necessarily greater than x_5.

Now, the sum of these measures is

$$x_1 + x_2 + x_3 + \ldots + x_{10}.$$

Using Σ notation, this can be symbolized

$$\sum_{i=1}^{10} x_i.$$

In this case, the limits of summation refer to the *subscripts of x*. The symbol tells us to add the first measure, the second, the third, and so on, until we have come to the tenth.

Of course, there are times when you might want to rearrange your measures in "normal order." If we did that with the ten measures listed above, they would be in the following order.

3, 4, 5, 6, 7, 7, 7, 8, 10, 12

Then x_1 would equal 3, and x_2 would be 4, and so on, up to $x_{10} = 12$. In this case, too, the sum would be symbolized

$$\sum_{i=1}^{10} x_i$$

Again, x_i stands for the general, or ith, measure.

Examples

1 **Use Σ notation to write the mean of the numbers discussed on page 445.**

The sum is $\sum_{i=1}^{10} x_i = 69$. There are ten measures, so the mean is

$$\frac{1}{10}\sum_{i=1}^{10} x_i$$

The mean equals 6.9.

2 **Write an expression to name the mean of *n* given measures, $x_1, x_2, x_3, \ldots, x_n$.**

Their sum is $\sum_{i=1}^{n} x_i$. Thus, their mean can be written $\frac{1}{n}\sum_{i=1}^{n} x_i$.

Generally, we use the symbol $\bar{x}$ ("x bar") to name the mean of a set of measures named by x.

Thus, $\bar{x} = \frac{1}{n}\sum_{i=1}^{n} x_i$.

Another use of Σ is to restate the Binomial Theorem, which you learned in the previous chapter.

$$(x + y)^n = {}_nC_0x^n + {}_nC_1x^{n-1}y^1 + {}_nC_2x^{n-2}y^2 + \ldots + {}_nC_ny^n$$

Using Σ notation, the theorem can be expressed in this way.

$$(x + y)^n = \sum_{r=0}^{n} {}_nC_rx^{n-r}y^r$$

You can see that it is r which takes on values from 0 to n. If $n = 3$, for example, we have $(x + y)^3 = \sum_{r=0}^{3} {}_3C_rx^{3-r}y^r$. This symbolizes a sum of four terms, one for each r, where $r = 0, 1, 2, 3$.

We often make the following agreement if c is a constant.

$$\sum_{r=1}^{n} c = nc$$

Thus, $\sum_{i=1}^{5} 4 = 4 + 4 + 4 + 4 + 4$ or $5(4)$ or 20

and $\sum_{i=1}^{n} 4 = \underbrace{4 + 4 + 4 + \ldots + 4}_{n \text{ terms}}$ or $4n$.

Example

3 **Use the agreement in the preceding discussion to show that** $\sum_{i=1}^{n} (x_i + k) = \sum_{i=1}^{n} x_i + \sum_{i=1}^{n} k$.

$$\begin{aligned}\sum_{i=1}^{n}(x_i + k) &= (x_1 + k) + (x_2 + k) + \ldots + (x_n + k)\\ &= (x_1 + x_2 + \ldots + x_n) + \underbrace{(k + k + \ldots + k)}_{n \text{ terms}}\\ &= \sum_{i=1}^{n} x_i + nk\\ &= \sum_{i=1}^{n} x_i + \sum_{i=1}^{n} k\end{aligned}$$

Exercises

Exploratory

1. Explain the difference between Σi and Σx_i.
2. What does it mean to index a set of measurements?
3. Explain the use of subscripts in working with statistical data.
4. What does the symbol $\bar{x}$ mean? Find $\bar{x}$ if $x_1 = 2$, $x_2 = 5$, and $x_3 = 8$.

Find $\bar{x}$ for each of these sets of measures.

5. $x_1 = 2, x_2 = 7, x_3 = 8, x_4 = 9$
6. $x_1 = x_2 = x_3 = x_4 = 12.8$
7. $x_1 = 0, x_2 = 0.05, x_3 = 0.015, x_4 = 0$
8. $x_1 = n, x_2 = 2n - 1, x_3 = 3n, x_4 = n - 3, x_5 = n + 4$

Compute each of the following.

9. $\sum_{i=1}^{3} 4$
10. $\sum_{k=1}^{12} 2$
11. $\sum_{i=1}^{n} 15$

Written **For each of the following sets, index the measures in the order given, and find their sum and mean to the nearest tenth. Use Σ notation.**

1. {3, 0, 1, 2, 3, 4}
2. {15.2, 15.0, 15.0, 16, 15.8}
3. {4, 0, 2, 2, 3, 4}
4. {13.1, 13.0, 13.0, 14, 14.7}
5. {16.1, 16.2, 16.0, 16.0, 15.8, 16.2, 16.4, 16.2, 16.4, 16.2}
6. {18.1, 18.2, 18.0, 18.1, 17.9, 17.3, 17.5, 17.3, 18.4, 18.0}

In each of the following, find the value of x_i which makes the given value of $\bar{x}$ the mean of the set.

7. $x_1 = ?, x_2 = 2, x_3 = 5, \bar{x} = 4$
8. $x_1 = x_2 = x_3 = 1, x_4 = ?, \bar{x} = 1.1$
9. $x_1 = x_2 = 0, x_3 = 6, x_4 = ?, \bar{x} = 4$
10. $x_1 = 1, x_2 = ?, x_3 = 50, \bar{x} = 25$
11. $x_1 = 1, x_2 = 2, x_3 = 4, x_4 = ?, x_5 = 10, \bar{x} = 8$
12. $x_1 = \pi, x_2 = 2\pi, x_3 = ?, x_4 = \frac{5\pi}{2}, \bar{x} = 2\pi$

Given the following set of measures: {8, 8, 10, 9, 8, 11}, let $x_1 = 8, x_2 = 8, x_3 = 10$ and so on. Find the following.

13. x_3
14. $\sum_{i=1}^{6} x_i$
15. $\bar{x}$
16. $(\bar{x} - x_4)$
17. $(\bar{x} - x_4)^2$
18. $\sum_{i=1}^{6} (\bar{x} - x_i)$
19. $\sum_{i=1}^{6} (\bar{x} - x_i)^2$

Find n, $\bar{x}$, $\sum_{i=1}^{n} (\bar{x} - x_i)$, and $\sum_{i=1}^{n} (\bar{x} - x_i)^2$ for each of the following sets.

20. {12, 15, 18, 14, 14, 10, 13}
21. {6, 6, 8, 7, 6, 5, 4, 8, 9, 12, 4}
22. {12.2, 12.4, 12.6, 12.4, 11.8, 13.2, 12.4, 12.2, 12.2, 12.4}
23. {1.2, 1.2, 1.0, 1.8, 1.4, 1.4, 1.2, 1.6, 1.2, 1.2, 0.8}

Given: $x_1 = 6, x_2 = 4, x_3 = 4, x_4 = 12, x_5 = 8, x_6 = 10, x_7 = 12, x_8 = 14$. Find each of the following.

24. $\sum_{i=1}^{8} x_i$
25. $\bar{x}$
26. $(\bar{x} - x_5)^2$
27. $(\bar{x} - x_8)^2$
28. $\sum_{i=1}^{8} (\bar{x} - x_i)$
29. $|\bar{x} - x_8|$
30. $\sum_{i=1}^{8} |\bar{x} - x_i|$
31. $\sum_{i=1}^{8} (\bar{x} - x_i)^2$
32. Write the Binomial Theorem in Σ notation. Write the case for $n = 4$.
33. Write each term in the expansion of $(x + y)^5 = \sum_{r=0}^{5} {}_5C_r x^{5-r} y^r$.
34. Show that $\sum_{i=1}^{m} c = mc$.
35. Show that $\sum (x_i + 2) = \sum x_i + \sum 2$.

Let $x_1, x_2, x_3, \ldots, x_n$ and $y_1, y_2, y_3, \ldots, y_n$ be two sets of measures.

36. Show that $\sum_{i=1}^{n} (x_i + y_i) = \sum_{i=1}^{n} x_i + \sum_{i=1}^{n} y_i$.
37. Show that $\sum_{i=1}^{n} c(x_i + y_i) = c\sum_{i=1}^{n} x_i + c\sum_{i=1}^{n} y_i$.

12.4 Measures of Dispersion

In your statistics work so far, you have studied measures of central tendency. These measures are used often and give valuable information, but sometimes this information may be insufficient. Take the following situation.

Suppose that a certain test is given nationwide to ninth grade high school students. The test is marked in the usual way with grades from 0 to 100.

In one group of ninth graders, the mean grade is 80. What can we conclude about the test? Can we determine anything about the students who took it?

It is tempting to say, "Well, the average was about B−, which is as it should be. Everything seems OK."

But there are different ways in which the mean of 80 can be obtained! One way is that most people who took the test scored between 75 and 85. But another way is that half the students who took the test scored 60 and the other half scored 100. The mean is still 80, but these are two very different situations.

It is clear that we need another measure in addition to the mean. We need to have a measure which tells us how far the typical measure is from the mean, or how scattered the measures are. A measure which does this is called a *measure of dispersion*. It tells how a set of measures is *dispersed* rather than how it is *centered*. One measure of dispersion is the *range*.

Definition of Range

The **range** of a set of measures is the difference between the greatest and the least.

Example

1 **A survey has produced these measures: 2, 3, 1, 7, 6, 4, 6, 5, 5. Find the range.**

The range is the greatest (7) minus the least (1). Thus the range is 6.

Sometimes the range is useful, but it is easy to see cases where it does not give enough information. For example, the following two sets of measures both have the same mean and the same range. However, the first is clearly more dispersed or scattered than the second.

I. {2, 2, 2, 10, 10, 10} **II.** {2, 4, 5, 7, 8, 10}

It is tempting to try to develop another measure based simply on adding the differences of each measure from the mean. Here are two tables showing this operation for sets of numbers I and II.

I.

x_i	$\bar{x}$	$\bar{x} - x_i$
2	6	4
2	6	4
2	6	4
10	6	−4
10	6	−4
10	6	−4

The sum is $4 + 4 + 4 + (-4) + (-4) + (-4) = 0$

II.

x_i	$\bar{x}$	$\bar{x} - x_i$
2	6	4
4	6	2
5	6	1
7	6	−1
8	6	−2
10	6	−4

The sum is $4 + 2 + 1 + (-1) + (-2) + (-4) = 0$

In both cases the sum of the differences of each measure from the mean equals zero! Can you see why this is always the case? In symbols we write that for every set of measures,

$$\sum (\bar{x} - x_i) = 0.$$

A way out of this problem is to adopt a device which makes all the differences $\bar{x} - x_i$ *positive*. Then we can add them without fear that they all add to zero. The easiest way to do this is to take their *absolute value* $|\bar{x} - x_i|$.

We can then take the sum of all these positive differences and then divide by the number of measures. In symbols, we have

$$\frac{1}{n}\sum_{i=1}^{n} |\bar{x} - x_i|.$$

This is called the **mean absolute deviation.**

Example

2 **Find the mean absolute deviation for the following two sets of measures.**

I.

x_i	$\bar{x}$	$\lvert\bar{x} - x_i\rvert$
2	6	4
2	6	4
2	6	4
10	6	4
10	6	4
10	6	4

$$\frac{1}{n}\sum_{i=1}^{n} |\bar{x} - x_i| = \frac{1}{6} \cdot 24 = 4$$

II.

x_i	$\bar{x}$	$\lvert\bar{x} - x_i\rvert$
2	6	4
4	6	2
5	6	1
7	6	1
8	6	2
10	6	4

$$\frac{1}{n}\sum_{i=1}^{n} |\bar{x} - x_i| = \frac{1}{6} \cdot 14 = 2\tfrac{1}{3}$$

We conclude from the mean absolute deviations of sets **I** and **II** that the measures in set **I** are more scattered than those in set **II.**

Exercises

Exploratory

1. Explain in your own words the ideas of dispersion or scattering as used in this section.
2. Explain the difference between a measure of central tendency and a measure of dispersion.
3. Make up a specific example where measures of central tendency do not give adequate information about a set of numbers.
4. Give three different sets of six numbers each, all with a mean of 10.
5. What is the range of a set of measures?
6. Explain why the range is often an inadequate measure of dispersion.

Written

Give a set of five measures that satisfy each of the following conditions.

1. mean 50, mode 50
2. mean 50, mode 20
3. mean 50, median 60
4. mean 50, mode $1\frac{1}{2}$
5. mean 50, mode 20, median 40
6. If $x_1, x_2, x_3, \ldots, x_n$ is a set of n measures, indexed in the order in which they were obtained, is the range equal to $x_n - x_1$? Explain your answer.
7. For the set of measures in exercise 6, could the range be $x_8 - x_1$? Could it also be $x_8 - x_9$?
8. Give two different sets of measures with range 20 and mean 50.

Give a set of six measures satisfying each of these conditions.

9. range 10, mean 6
10. range 10, mean 100
11. range 30, median 4
12. range 30, mean 40, mode 42
13. range 100, median 61, mode 60
14. range 1, median 1.6, mean 1.7
15. range 100, mean 80, median 70
16. range 2, median 2.6, mean 2.7
17. If the mean of a set of r numbers is $\bar{x}$, what is the sum of the numbers? Why?

In exercises 18–25, let $x_1, x_2, x_3, \ldots, x_n$ be a set of measures. Explain what each of the following symbolizes.

18. $\bar{x} - x_4$
19. $x_4 - \bar{x}$
20. $|\bar{x} - x_8|$
21. $|x_{16} - \bar{x}|$
22. $(\bar{x} - x_1)^2$
23. $(x_{12} - \bar{x})^2$
24. $(x_2 - \bar{x})^4$
25. $\sum_{i=1}^{n} (\bar{x} - x_i)$
26. In general, is $\bar{x} - x_i = x_i - \bar{x}$?
27. Is $(\bar{x} - x_4)^2 = (x_4 - \bar{x})^2$? Why?
28. Is $|\bar{x} - x_4| = |x_4 - \bar{x}|$? Why?
29. Is $(\bar{x} - x_i)^2 = (x_i - \bar{x})^2$ for any i? Why?
30. Explain why $\sum |\bar{x} - x_i| = \sum |x_i - \bar{x}|$.

31. Give a set of six numbers, not all equal, with mean 8. Then show that for these numbers $\sum_{i=1}^{6} (\bar{x} - x_i) = 0$.

32. Answer exercise 31 for a set of ten numbers with mean 15.

33. Explain carefully why $\sum (\bar{x} - x_i) = 0$ for all sets of measures.

34. Explain how absolute value can help us out of the difficulty stated in exercise 33.

35. Explain the meaning of the symbol $\frac{1}{n}\sum_{i=1}^{n} |\bar{x} - x_i|$.

Compute the mean absolute deviation, $\frac{1}{n}\sum_{i=1}^{n} |\bar{x} - x_i|$, for each of the following sets of measures.

36. {2, 4, 6, 8, 10}

37. {2, 4, 6, 8, 20}

38. {4, 4, 4, 6, 12, 20}

39. {0, 2, 0, 4, 3, 1, 0}

40. {500, 450, 425, 525}

41. {6, 8, 6, 2, 10, 12, 2, 4, 4, 8}

42. {16.1, 16.0, 16.3, 15.4, 16.2}

43. {18.0, 17.0, 16.2, 17.1, 15.9}

44. {18.0, 17.0, 16.2, 17.1, 15.9, 14.0, 14.6, 15.2}

45. {1.4, 8.0, 1.6, 2.4, 6.2, 5.8, 5.5, 3.1, 0.8, 1.6}

46. {26.1, 22.7, 20.1, 20.2, 21.1, 19.8, 25.2, 24.3, 25.9}

Mathematical Excursions — Using Calculators

You can use a scientific calculator to compute the mean absolute deviation for a set of data. Follow these steps to find the mean absolute deviation for 3, 5, 19, 2, 1, 20, 7, 4, 10, 6, 11, 8.

First, find the mean and store it in memory.

ENTER: [(] 3 [+] 5 [+] 19 [+] 2 [+] . . . [+] 11 [+] 8 [)] [÷] 7 [=] [STO]

DISPLAY: 3 5 8 19 27 2 29 77 11 88 8 96 7 8

Then find the sum of the absolute value of the deviations.
Note that the [+/−] key is used only when the deviation is negative.

ENTER: [(] 3 [−] [RCL] [)] [+/−] [+] . . . [+] [(] 8 [−] [RCL] [)] [=] [÷] [RCL] [=]

DISPLAY: 3 8 −5 5 56 8 8 0 56 8 4.66666667

To the nearest tenth, the mean absolute deviation is 4.7. If your calculator has a statistical mode, consult your manual to help you use it. This will make computation easier.

Exercises Use a calculator to find the mean absolute deviation of Written Exercises 42–46.

12.5 Standard Deviation

In situations where large numbers of measurements are studied, most of the measures are quite close to the mean. Most students score close to the mean on standardized tests, such as the SAT. Most American males are close to the mean height for American males.

Most high school juniors are about $16\frac{1}{2}$ when they start their junior year. A few are closer to $15\frac{1}{2}$ or $17\frac{1}{2}$, but this is not usual. If we call $16\frac{1}{2}$ the mean, then $\left|16\frac{1}{2} - 15\frac{1}{2}\right| = 1$ and $\left|16\frac{1}{2} - 17\frac{1}{2}\right| = 1$. Are there any juniors who are $14\frac{1}{2}$ or $18\frac{1}{2}$ when they begin their junior year? Well, very few, surely. If their ages are tallied in a computation of the mean absolute deviation, we have $\left|16\frac{1}{2} - 14\frac{1}{2}\right| = 2$ and $\left|16\frac{1}{2} - 18\frac{1}{2}\right| = 2$.

But many people would say that a $14\frac{1}{2}$ year old junior is more than just "twice as rare" as a $15\frac{1}{2}$ year old. We would like to have a measure that gives greater weight to a measure the farther it gets from the mean.

For this and other reasons, we use the variance and standard deviation.

Definition of Variance

The **variance** is the sum of the squares of the differences of the measures from the mean, divided by the number of measures.

In symbols,

$$\text{var} = \frac{(\bar{x} - x_1)^2 + (\bar{x} - x_2)^2 + \ldots + (\bar{x} - x_n)^2}{n}$$
$$= \frac{1}{n}\sum_{i=1}^{n} (\bar{x} - x_i)^2.$$

Example

1 **Find the variance of this set of data: {5, 1, 3, 1, 2, 6, 6, 8}.**

Step 1. Find the mean. $\bar{x} = \frac{5 + 1 + 3 + 1 + 2 + 6 + 6 + 8}{8} = 4$

Step 2. Find the differences and then square them.

$(\bar{x} - x_1)^2 = (4 - 5)^2 = 1$ | $(\bar{x} - x_2)^2 = (4 - 1)^2 = 9$
$(\bar{x} - x_3)^2 = (4 - 3)^2 = 1$ | $(\bar{x} - x_4)^2 = (4 - 1)^2 = 9$
$(\bar{x} - x_5)^2 = (4 - 2)^2 = 4$ | $(\bar{x} - x_6)^2 = (4 - 6)^2 = 4$
$(\bar{x} - x_7)^2 = (4 - 6)^2 = 4$ | $(\bar{x} - x_8)^2 = (4 - 8)^2 = 16$

Step 3. Add these values. $1 + 9 + 1 + 9 + 4 + 4 + 4 + 16 = 48$

Step 4. Divide the sum by the number of measures. $\frac{48}{8} = 6$

The variance is 6.

As a partial correction to the fact that the differences $\bar{x} - x_i$ are squared in the variance, there is the standard deviation. This is the most important measure of dispersion.

Definition of Standard Deviation

The **standard deviation** is the square root of the variance.

In symbols,

$$\text{S.D.} = \sqrt{\frac{(\bar{x} - x_1)^2 + (\bar{x} - x_2)^2 + \ldots + (\bar{x} - x_n)^2}{n}}$$

$$\text{S.D.} = \sqrt{\frac{\sum_{i=1}^{n} (\bar{x} - x_i)^2}{n}} \quad \text{or} \quad \sqrt{\frac{1}{n}\sum_{i=1}^{n} (\bar{x} - x_i)^2}$$

Other symbols for standard deviation are s and σ (lower case sigma).

Examples

2 **Find the standard deviation of the measures in example 1.**

S.D. $= \sqrt{6} \approx 2.45$ *Unless otherwise specified, all measures will be computed to the nearest hundredth.*

3 **Find the standard deviation of this set of numbers: {2, 2, 3, 5, 8}.**

Step 1 S.D. $= \sqrt{\frac{1}{n}\sum_{i=1}^{n} (\bar{x} - x_i)^2}$ *Use a table to assist in compiling information.*

Step 2 Compute the mean and enter it in the second column.

$$\bar{x} = \frac{2 + 2 + 3 + 5 + 8}{5} = 4$$

Step 3 Find the differences, $\bar{x} - x_i$, and enter in the third column. Square the differences (fourth column).

x_i	$\bar{x}$	$\bar{x} - x_i$	$(\bar{x} - x_i)^2$
2	4	2	4
2	4	2	4
3	4	1	1
5	4	−1	1
8	4	−4	16

Step 4 Add the values in the fourth column. $\sum (\bar{x} - x_i)^2 = 26$

Step 5 Divide the sum by the number of measures. This gives the variance. var $= \frac{26}{5}$ or 5.2

Step 6 Take the square root. S.D. $= \sqrt{5.2} \approx 2.28$

Most practical uses of variance and standard deviation deal with many more measures than the ones in these examples. Also, the arithmetic is usually not so "neat." In most cases investigators use calculators or computers to assist them in their work.

The standard deviation and the mean are often used together to give a special kind of information. In many cases, the data produced by an experiment or survey cluster around the mean in such a way that if they are graphed to show the frequency of measures, the graph appears like the one in Figure 1.

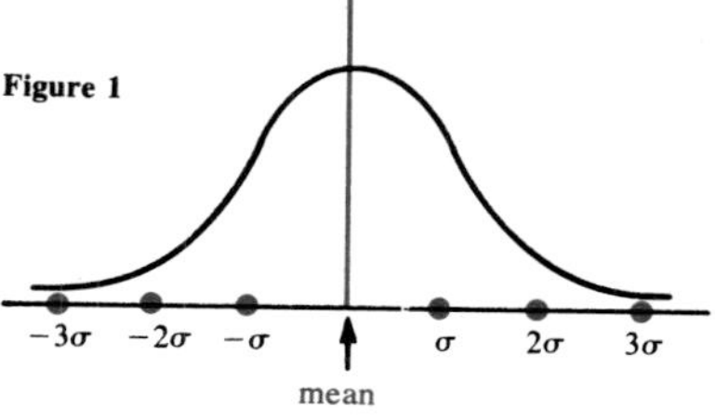

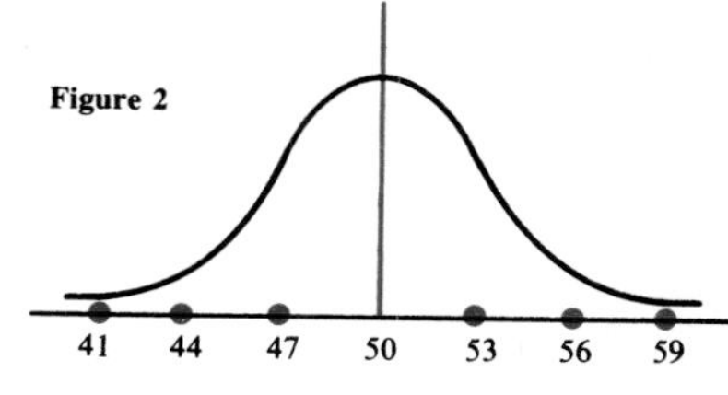

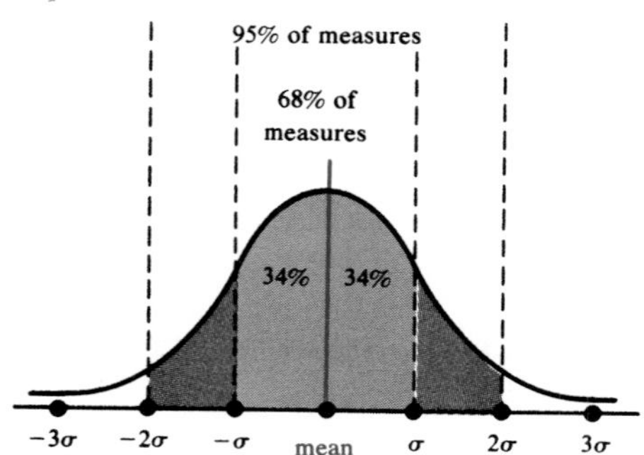

In these graphs, one unit on the horizontal axis is *one standard deviation*. For example, if the mean is 50 and the standard deviation is 3, the graph looks like the one in Figure 2.

Technically, such curves are called **normal curves.** Most people refer to them as **"bell curves."**

It can be shown that when data follow this bell-curve pattern, approximately 68% of all the measures involved fall within one standard deviation of the mean. (There are 34% on each side.) In addition, 95% fall within two S.D.'s of the mean and 99% within three S.D.'s.

Example

4 The approximate number of hours worked per week for 100 people in Hornetville is normally distributed. The mean is 40 hours per week and the standard deviation is 2 hours per week. About how many people work more than 42 hours per week?

This frequency distribution is shown by the following curve. The percentages represent the percentage of 100 people working the number of hours within the given interval.

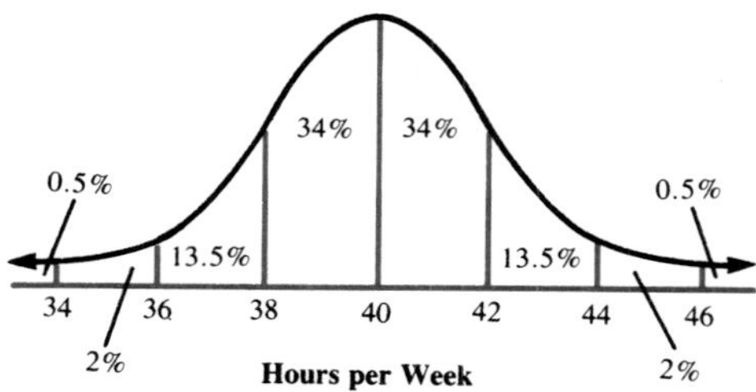

The percentage of people working more than 42 hours per week is 13.5% + 2% + 0.5% or 16%.

$$100 \times 16\% = 16$$

Thus, about 16 people work more than 42 hours per week.

Exercises

Exploratory

1. Explain one reason for introducing a measure in which the differences from the mean are squared.
2. Give a formula for the variance without using the Σ symbol. Tell what each part of the formula means.
3. Give a formula for variance with the Σ symbol. Explain each part of the formula.
4. What is the relationship between the variance and the standard deviation of a set of measures?
5. Give three symbols used for the standard deviation.
6. Answer exercise **2** for the standard deviation.
7. Answer exercise **3** for the standard deviation.
8. A very common symbol for variance is s^2. Why do you think this symbol was chosen?
9. The measures from survey S_1 have standard deviation 50. Those from survey S_2 have standard deviation 3. Binky says, "It appears that the measures from survey S_1 are more scattered." What do you think? Has she taken everything into consideration?
10. What is a normal curve? A "bell curve"?

Written **Compute the variance and standard deviation for each set of measures. In each case, show your work carefully and neatly.**

1. {1, 2, 4, 7, 11}
2. {17, 19, 17, 15, 16, 16, 19}
3. {3, 3, 4, 8, 12}
4. {500, 400, 500, 200, 400}
5. {2, 2, 4, 8, 14}
6. {250, 275, 325, 300, 200, 225, 175}
7. {2, 2, 8, 14, 6, 4}
8. {80, 82, 81, 81, 81, 81, 80, 82}
9. {2, 4, 0, 6, 0, 8, 0, 12}
10. {14, 12, 14, 16, 18, 18, 13}
11. {19.2, 19.6, 19.0, 18.6, 19.2, 18.8, 18.8}
12. {4, 6, 5, 2, 3, 4, 8, 5, 6, 2, 7, 7, 6}

In the following problems, suppose that a study has produced the data given. Compute the standard deviation in each case.

13. Life of light bulbs selected from factory assembly line, to the nearest 25 hours: 1050, 1175, 1075, 1025, 1100, 1125, 975, 1125, 1075, 1050.
14. Centiliters of chocolate consumed per week by teenagers in Cité du Pin: 120, 116, 104, 90, 82 108, 122, 88, 112, 96.

15. Hours of TV watched by families in Hornetville (per week): 34, 22, 28, 0, 17, 30, 18, 26, 28, 8, 30, 24, 24.

16. Heights of female juniors, Pine City High, 1989–1990, in centimeters: 132, 150, 138, 160, 133, 143, 148, 148, 148, 151, 148, 141.

17. Test scores on the BKTCS: 81, 80, 87, 97, 82, 86, 85, 82, 72, 80, 85, 84, 84, 63, 90, 82, 85, 79, 95, 81.

18. Test scores on the revised BKTCS: 90, 75, 87, 82, 79, 85, 85, 82, 82, 82, 89, 58, 87, 88, 82, 84, 86, 78, 91, 80, 94, 86, 88, 96.

Draw a bell curve which would describe a set of measures with these characteristics.

19. Mean = 14; S.D. = 2

20. Mean = 500; σ = 35

21. Mean = 0.06; σ = 0.005

22. Mean = 10,000; S.D. = 1850

23. Mean = 0; σ = 0.5

24. Mean = 0; σ = 1.3

25. Mean = 500; S.D. = 100

26. Mean = 6.16; S.D. = 2.7

A set of measures that follow a bell curve has mean 75 and standard deviation 3. Complete the following statements about these measures.

27. About _____% of the measures lie between 72 and 78.

28. About _____% of the measures lie between 72 and 75.

29. Between 75 and 81 there are about _____% of the measures.

30. The measure 69.8 is between _____ and _____ S.D.'s from the mean.

31. About _____% of the measures are between 78 and 81.

32. About _____% of the measures are less than 66 or greater than 84.

33. The measure 72.1 is larger than about _____% of the measures and smaller than about _____% of the measures.

The mean height of male juniors at Pine City High is 155 cm and the S.D. is 15. Compare the heights of each of these boys with their classmates'.

34. Bill, 155 cm

35. Troy, 140 cm

36. Mark, 171 cm

37. Bernie, 109 cm

38. Jim, 185 cm

39. Ben, 124 cm

40. On a standardized test, the mean is 61 and the S.D. is 3.2. Would you expect a score of 50, 56, 62, or 65 to occur less than 3% of the time?

Challenge

1. Discuss some situations where you think the measures involved could be described by a normal or bell curve.
2. Try to think of a situation which would *not* be describable by a bell curve.
3. Make up a set of measures which are clearly not distributed in the manner described by a bell curve.

12.6 Grouped Data

Very often when statistics are employed, there are a great number of measurements involved. Often some of the measures recur within the same set of data.

Example

1 **Find the mean of {2, 6, 2, 2, 2, 8, 6, 6, 6, 8, 6, 2, 2}.**

Of course, we could just add the numbers and divide by 13. We notice, however, that the 2's and the 6's repeat frequently, and there are two 8's. In fact, although there are 13 numbers, there are only 3 *different* numbers. We proceed as follows.

$$\text{SUM} = \underset{\substack{\uparrow \\ \text{number} \\ \text{of 2's}}}{6} \cdot 2 + \underset{\substack{\uparrow \\ \text{number} \\ \text{of 6's}}}{5} \cdot 6 + \underset{\substack{\uparrow \\ \text{number} \\ \text{of 8's}}}{2} \cdot 8$$

The numbers 6, 5, 2, in red, are the *frequencies* of 2, 6, 8, respectively.

The mean is equal to $\frac{6 \cdot 2 + 5 \cdot 6 + 2 \cdot 8}{6 + 5 + 2}$ or $\frac{12 + 30 + 16}{13}$ or $\frac{58}{13} \approx 4.46$.

The information from example 1 can be arranged in a table as shown at the right.

index	x_i	f_i	f_ix_i
1	2	6	12
2	6	5	30
3	8	2	16

In this table, the "typical" measure x_i occurs with corresponding frequency f_i. Thus, if $i = 1$, we have x_1, which is 2. Its corresponding frequency is f_1, which is 6. For $i = 2$, we have $x_2 = 6$ and $f_2 = 5$. The third measure is $x_3 = 8$ which occurs with corresponding frequency $f_3 = 2$. We can see that in every case, the product f_ix_i represents the same number you get by adding the particular x_i exactly f_i times.

Thus, the following two sums are equal.

$$\sum_{i=1}^{n} x_i = \sum_{i=1}^{k} f_ix_i$$

In the expression on the right side of the equation, k is the number of different measures. Note that in the example, $k = 3$.

Example

2 **Find the standard deviation of this set of numbers.**

{6, 6, 4, 6, 4, 2, 8, 8, 8, 2, 2, 6, 6}

Organize the information in the following table. Remember to round measures to the nearest hundredth.

index	x_i	f_i	f_ix_i	$\bar{x} = \frac{1}{n}\sum_{i=1}^{k} f_ix_i$	$\bar{x} - x_i$	$(\bar{x} - x_i)^2$	$f_i(\bar{x} - x_i)^2$
1	2	3	6	5.23	3.23	10.43	31.29
2	4	2	8	5.23	1.23	1.51	3.02
3	6	5	30	5.23	−0.77	0.59	2.95
4	8	3	24	5.23	−2.77	7.67	23.01

$k = 4$ $\quad \sum_{i=1}^{k} f_ix_i = 68$ $\quad \sum_{i=1}^{k} f_i(\bar{x} - x_i)^2 = 60.27$

$\bar{x} = \frac{1}{13} \cdot 68 \approx 5.23$ $\quad var = \frac{1}{13}\sum_{i=1}^{k} f_i(\bar{x} - x_i)^2 \approx 4.64$

$$\text{S.D.} = \sqrt{var} = \sqrt{\frac{1}{n}\sum_{i=1}^{n}(\bar{x} - x_i)^2} = \sqrt{\frac{1}{n}\sum_{i=1}^{k} f_i(\bar{x} - x_i)^2}$$

In this case, S.D. $= \sqrt{4.64} \approx 2.15$.

Note that the value 5.23 is an approximation for $\bar{x}$. In all computations in which we use this value (last three columns), our values will be subject to a slight rounding error.

Exercises

Exploratory **Which numbers recur in the following sets? How many times? Show this information in table form.**

1. {1, 3, 1, 1, 4, 8, 4, 1, 2, 2, 0, 1}
2. {3, 3, 2, 2, 1, 1, 5, 5, 4, 4}
3. {2, 6, 2, 2, 2, 8, 6, 6, 8, 2}
4. {1, 2, 5, 1, 2, 5, 2, 5}
5. {6, 3, 2, 2, 1, 1, 6, 3, 2}
6. {8, 2, 8, 8, 3, 5, 3, 8, 1, 1, 0}

Write a list of ten numbers in which the numbers indicated recur with the frequency indicated.

7. 6 with frequency 5
8. 1 with frequency 4, 0 with frequency 2
9. 10,500 with frequency 7
10. 76, 78, and 82 each with frequency 2
11. 0.2 with frequency 3 and 0.04 with frequency 2
12. 1.114 with $f = 3$ and 1.115 with $f = 6$

Let $x_1 = 4$, $x_2 = 6$, and $x_3 = 8$ recur with frequencies f_1, f_2, f_3, respectively. Find f_i for $i = 1, 2, 3$ (that is, find f_1, f_2, and f_3) for each of the following sets.

13. {4, 4, 4, 6, 8, 8, 8, 8}

14. {6, 4, 6, 4, 6, 6, 6, 8, 8, 8, 4}

15. {6, 6, 6, 6, 6, 8, 6, 8, 6, 8, 4, 8, 8, 4, 8, 4, 4}

Written **For each set, complete the accompanying table.**

1. {3, 3, 3, 5, 3, 5, 7, 7, 5, 8}

index	x_i	f_i	f_ix_i
1			
2			
3			

2. {1, 1, 3, 1, 3, 5, 1, 1, 3, 3, 11}

index	x_i	f_i	f_ix_i

3. 1, 0, 0, 2, 1, 3, 1, 3, 3, 3, 2 *Make your own table!*

4. Explain how $\frac{1}{n}\sum_{i=1}^{k} f_ix_i$ represents the mean of a set of measures.

Use the idea of grouped data and the symbolism of this section to find the mean of each of the following sets. Show your work in table form.

5. {6, 6, 6, 8, 6, 6, 8}

6. {6, 8, 7, 7, 8, 6, 7, 7, 7, 6}

7. {50, 40, 40, 44, 44, 42, 50, 42, 42, 42}

8. {8.1, 8.2, 8.0, 8.1, 8.1, 8.0, 8.2, 8.2, 8.2}

9. {35, 37, 36, 36, 37, 35, 35, 35, 35, 35, 35}

10. {8, 9, 8, 9, 7, 8, 9, 8, 6, 6, 7, 6, 5, 6, 5, 7, 7, 8, 6, 5, 6, 5, 9, 9, 6}

For exercises 11–14, set up a table with the headings shown below.

index	x_i	f_i	f_ix_i	$\bar{x}$	$\bar{x} - x_i$	$f_i\lvert\bar{x} - x_i\rvert$

Then use the information in the table to find the mean absolute deviation for each set.

11. the numbers in exercise 6

12. the numbers in exercise 7

13. the numbers in exercise 9

14. the numbers in exercise 10

Find the standard deviation for each of the following sets by setting up a table with the headings shown on page 459 and other information and then filling in the table for each set.

15. the numbers in exercise 5

16. the numbers in exercise 8

17. {14, 16, 18, 14, 12, 18, 12, 18, 12, 18, 14, 16, 16, 18, 16, 12, 14, 14}

18. {6, 8, 6, 10, 12, 6, 12, 6, 10, 8, 8, 10, 8, 8, 8, 8, 6, 8, 8, 10, 8}

19. {13, 14, 15, 15, 15, 13, 14, 17, 19, 17, 16, 15, 14, 13, 15, 19, 14}

20. {35, 34, 35, 35, 38, 39, 35, 34, 34, 34, 35, 38, 38, 38, 38, 38}

Mixed Review

1. Find the solution set for $|2x - 1| = 3$.
2. If $f(x) = 6 \log x$, find $f(10)$.
3. Evaluate $x^{\frac{2}{3}} + x^{-1}$ when $x = 8$.
4. Express 130° in radian measure.
5. Express cos 212° as a function of a positive acute angle.
6. A coin is tossed 4 times. What is the probability of getting exactly 3 heads?
7. Simplify $\frac{3}{2 - \sqrt{3}}$.
8. Chords $\overline{AB}$ and $\overline{CD}$ of circle O intersect at E. If $AE = 4$, $EB = 5$, and $CE = 2$, find ED.
9. If $\theta = \text{Arccos } \frac{\sqrt{3}}{2}$, find the degree measure of θ.
10. Simplify $\sqrt{-2} + \sqrt{-18}$.
11. Simplify $(2 - 2i)(2 + 2i)$.
12. Use the laws of logarithms to expand $\log \frac{a\sqrt{b}}{c}$.
13. If $f(x) = \sqrt{x - 2}$, find the domain of f.
14. If $\tan A = -\frac{5}{12}$ and $\cos A > 0$, find $\sin A$.
15. In $\triangle ABC$, $\sin A = 0.8$, $\sin B = 0.3$, and $a = 24$. Find b.
16. What kinds of symmetry does a rhombus have?
17. In a circle, a central angle of 3 radians intercepts an arc of 18 cm. Find the radius, in centimeters, of the circle.
18. Find the image of (1, 7) when it is reflected over the line $y = 3$.
19. If f is a linear function and $f(-2) = 5$ and $f(3) = -6$, find $f(x)$.
20. In $\triangle ABC$, $a = 2$, $b = 3$, and $c = 4$. Find the value of $\cos C$.
21. Express in simplest form the third term of $(a - 3b)^5$.
22. If $0° \leq \theta < 360°$, solve $3 \sin^2\theta + 3 \sin \theta - 2 = 0$ for θ.
23. On the same set of axes, sketch the graphs of $y = 2 \sin x$ and $y = \cos \frac{1}{2}x$ as x varies from 0 to 2π radians.
24. Using logarithms, solve $3^{2x} = 4$ for x to the nearest tenth.
25. Prove the identity $\tan \theta + \cot \theta = \sec \theta \csc \theta$.
26. Sketch the graph of $f = \{(x, y) : y = \log_2 x\}$ and then write an equation for f^{-1}, the inverse of f.
27. In the domain of complex numbers, solve $x^2 + 2x + 10 = 0$.
28. In trapezoid $ABCD$, $\overline{AB}$ and $\overline{DC}$ are parallel bases, $\overline{AD} \perp \overline{AB}$, $m\angle C = 128$, $m\angle DAC = 42$, and $AD = 26$. Find AB to the nearest whole number.
29. The vertices of $\triangle ABC$ have coordinates $A(2, 3)$, $B(9, 1)$, and $C(5, 5)$.
    a. On graph paper, draw and label $\triangle ABC$.
    b. Graph $\triangle A'B'C'$, the image of $\triangle ABC$ after a half-turn around the origin.
    c. Graph $\triangle A''B''C''$, the image of $\triangle A'B'C'$ after $R_{y=x}$.
30. A fair die is tossed six times. What is the probability of obtaining 2 or 6 on at least four rolls?

You have seen that the science of statistics deals with the collection and analysis of numerical data. In addition, statistical methods can be used to help make predictions.

Suppose, for example, you wish to predict the quantity of a food product sold based on its weekly selling price. The following table shows the quantity sold for each of the last ten weeks, and its selling price.

Quantity Sold (dozens)	30	47	38	28	49	23	47	46	39	42
Price (cents per dozen)	28	22	29	32	20	35	21	20	24	29

The graph on the left is a **scatter diagram** for the data. The scatter of dots suggests a straight line that slopes downward from the upper left corner to the lower right corner.

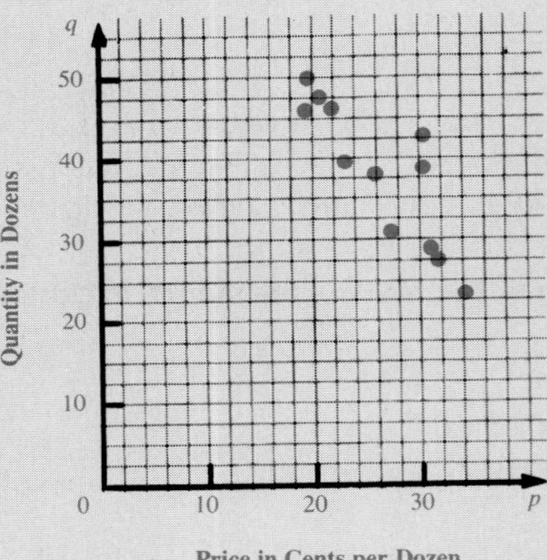

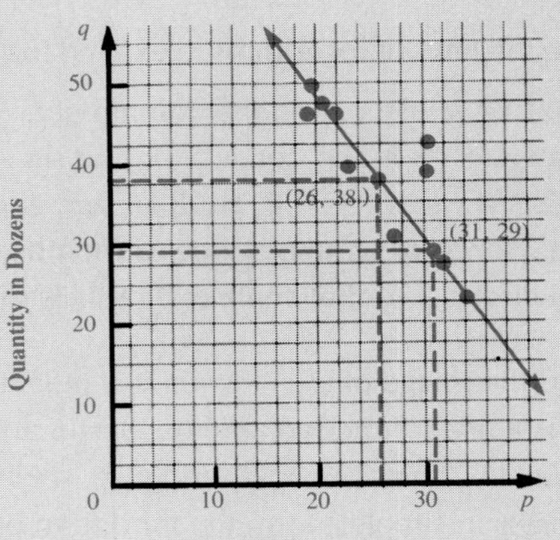

You can draw the line that is suggested by the dots. It represents the relationship between quantity and price. By choosing several points on the line, you can find the equation of the line. This equation is called the **prediction equation** for the relationship. *This procedure is dependent on your judgement. Statisticians normally use more precise procedures.*

slope $= \dfrac{38 - 29}{26 - 31}$ or -1.8

$q = -1.8p + b$ *q stands for quantity, and p stands for price*

$38 = -1.8(26) + b$

$84.8 = b$ The equation is $q = -1.8p + 84.8$.

Now, suppose that next week, the price of the food product will be 30 cents. Using the prediction equation, you can estimate that 30.8 dozen items will be sold.

$$q = -1.8p + 84.8 = -1.8(30) + 84.8 \quad \text{or} \quad 30.8$$

Problem Solving Application: Using Statistics

The *population* in a statistical study is all of the items or individuals in the group being considered. It rarely happens that 100% of the population is accessible as a source of data. Therefore, a random sample of the population must be selected that is representative of the population. The data from the random sample can then be used to make predictions about the entire population. For example, the mean of a random sample, called the *sample mean*, and its standard deviation can be used to estimate the mean of the entire population, called the *true mean*.

A statistic called the *standard error of the mean* is used to give a level of certainty, or confidence, about the sample mean.

Standard Error of the Mean

For a sample set of data with N numbers and standard deviation σ, the standard error of the mean, $\overline{\sigma}$, is $\frac{\sigma}{\sqrt{N}}$.

Sample means of various random samples of the same population are normally distributed about the true mean with the standard error of the mean as a measure of variation. Thus, the standard error of the mean behaves like the standard deviation and can be used to determine the probability that the true mean lies within a certain range of the sample mean.

Example

1 **The mean height of a random sample of 100 junior boys in Pine City is 67.4 inches with a standard deviation of 2.7 inches. Find the range of heights such that the probability is 68% that the mean height of the entire population of junior boys falls within it.**

For data that is normally distributed, 68% of all the data involved fall within one standard deviation of the mean. Therefore, there is a 68% probability that the true mean falls within *one standard error of the mean* of the sample mean.

$$\overline{\sigma} = \frac{\sigma}{\sqrt{N}} = \frac{2.7}{\sqrt{100}} \quad \text{or} \quad 0.27$$

$$67.4 - 0.27 = 67.13 \qquad 67.4 + 0.27 = 67.67$$

The probability is 68% that the true mean falls within the range 67.1 inches to 67.7 inches, to the nearest tenth of an inch.

A 32% level of confidence means there is less than a 32% chance that the true mean differs from the sample mean by a certain amount.

When the probability that the true mean falls within a certain range of the sample mean is 68%, we say that the *level of confidence* is 32%. The most commonly used levels of confidence are the 1%, the 5%, and the 10% levels. By looking at the normal curve, statisticians have determined that the following ranges correspond to these levels of confidence. *S.D. means standard deviation.*

1% level (99% probability) – within 2.58 S.D.'s of the mean
5% level (95% probability) – within 1.96 S.D.'s of the mean
10% level (90% probability) – within 1.65 S.D.'s of the mean

Example

2 **The mean height of a random sample of 144 senior boys in Beckville is 68.5 inches with a standard deviation of 2.9 inches. Find the range of the sample mean that has a 5% level of confidence.**

For a 5% level of confidence, the true mean must fall within 1.96 *standard errors of the mean* of the sample mean. *Recall that the standard error of the mean behaves like the standard deviation.*

$$\overline{\sigma} = \frac{2.9}{\sqrt{144}} \approx 0.24$$

$68.5 - 1.96(0.24) \approx 68.0$ $\qquad$ $68.5 + 1.96(0.24) \approx 67.0$

Thus, the range that has a 5% level of confidence is 67.0 inches to 68.0 inches.

Exercises

Written **In Mantorburg, a random sample of 100 families showed that the mean number of hours the television set was on each day was 4.6 hours. The standard deviation was 1.4 hours.**

1. Find the standard error of the mean to the nearest tenth.
2. Find the range about the sample mean that has a 32% level of confidence.
3. Find the range about the sample mean that has 10% level of confidence.
4. Find the range about the sample mean that has a 1% level of confidence.

A random sample of 50 acorns from an oak tree reveals a mean diameter of 16.2 mm with a standard deviation of 1.4 mm.

5. Find the standard error of the mean to the nearest tenth.
6. Find the range about the sample mean that has a 5% level of confidence.
7. Find the range about the sample mean that has a 1% level of confidence.

8. The standard deviation of the weights of 25 seven-year-olds in the United States is 10 pounds. What is the probability that the mean weight of the random sample will differ by more than 3.3 pounds from the mean weight of all seven-year olds?

Vocabulary

data
mean
median
mode
bimodal
sigma
limits of summation
dummy variable
subscripts
indexing
range
mean absolute deviation
variance
standard deviation
bell curve
normal curve

Chapter Summary

1. The science of statistics may be defined as the study of numerical data.
2. A measure of central tendency shows how numbers cluster around, or can be summarized by, one centrally located number. The mean, median, and mode are three common measures of central tendency.
3. The Greek capital letter sigma, used in connection with upper and lower limits, is a mathematical shorthand for expressing the sum of a series of data.
4. In general, $\sum_{i=1}^{n} ci = c\sum_{i=1}^{n} i$, where c is a constant.
5. The mean, $\overline{x}$, of a set of measures named by x can be expressed as the following summation.
$$\overline{x} = \frac{1}{n}\sum_{i=1}^{n} x_i$$
6. The Binomial Theorem can be restated in the following manner.
$$(x + y)^n = \sum_{r=0}^{n} {}_nC_r\, x^{n-r}y^r$$
7. If c is a constant, the following summation is accepted.
$$\sum_{r=1}^{n} c = nc$$
8. A measure of dispersion tells how far the typical measure is from the mean or how scattered the measures are. The range is a common measure of dispersion.

9. The mean absolute deviation equals $\frac{1}{n}\sum_{i=1}^{n}|\bar{x} - x_i|$.

10. The variance equals $\frac{1}{n}\sum_{i=1}^{n}(\bar{x} - x_i)^2$.

11. The standard deviation equals the square root of the variance. It can be symbolized as

$$\sqrt{\frac{1}{n}\sum_{i=1}^{n}(\bar{x} - x_i)^2}.$$

12. When data follow a bell curve pattern, approximately 68% of all measures fall within one standard deviation of the mean. Approximately 95% fall within two standard deviations of the mean and 99% fall within three standard deviations.

Chapter Review

12.1 **Find the mean, the median, and the mode of each of the following sets.**

1. {3, 4, 8, 6, 5, 5, 7, 2}
2. {8, 2, 2, 4, 6, 2}
3. {6, 1.5, 3, 8, 4, 3.5, 4, 1, 4.5}
4. {11, 10, 13, 12, 12, 13}

In each case, find the value of *x* which makes *M* the mean of the set.

5. $\left\{3x, 4, \frac{3}{2}, 1, 2\frac{1}{4}, \frac{5}{2}\right\}$; $M = \frac{5}{4}$
6. $\{x + 11, 26, 20, 19, 19, 21, 19, 22\}$; $M = 21$

12.2 **Compute each of the following expressions.**

7. $\sum_{i=1}^{7} i$
8. $\sum_{k=0}^{6} 4k$
9. $\frac{1}{8}\sum_{i=1}^{8} i^2$
10. $\sum_{j=1}^{4} \frac{j^3}{16}$
11. $\sum_{i=1}^{6} 8$
12. $\sum_{i=1}^{5} (i - 3)^2$

Use Σ notation to write expressions for the following. Then, compute each sum.

13. the sum of the squares of the whole numbers from 1 to 5, inclusive
14. the sum of the cubes of the integers from −3 to 3, inclusive
15. the mean of the first eight positive integers
16. the sum of the first n natural numbers
17. the sum of the absolute values of the differences of the first 5 natural numbers and their mean

12.3 **For each of the sets in exercises 18–22, index the measures in the order given and find their sum and mean. Use Σ notation.**

18. {5, 0, 3, 3, 4, 5}
19. {6, 1, 2, 1, 6, 3}

20. {13.2, 13.0, 13.0, 14, 13.8}

21. {12.1, 12.0, 12.0, 13, 13.6}

22. {17.1, 17.2, 17.0, 17.0, 16.8, 17.2, 17.4, 17.2, 17.4, 17.2}

Find n, $\bar{x}$, $\sum_{i=1}^{n} (\bar{x} - x_i)$ and $\sum_{i=1}^{n} (\bar{x} - x_i)^2$ for each of the following.

23. {11, 14, 17, 13, 13, 9, 12}

24. {5, 5, 8, 6, 5, 4, 3, 7, 8, 11, 3}

25. {11.1, 11.3, 11.6, 11.3, 10.7, 12.1, 11.3, 11.1, 11.1, 11.3}

12.4 **In exercises 26–32, tell whether each of the following sentences is *true* or *false*. Justify your answer in each case. Assume, where appropriate, that $x_1, x_2, \ldots, x_n$ is a set of measures.**

26. $\sum_{i=1}^{n} ci = c\sum_{i=1}^{n} i$

27. $\sum_{i=1}^{7} 8 = 15$

28. $\frac{1}{n}\sum_{i=1}^{n} x_i$ is the numerical average of a set of n measures.

29. $\sum_{i=1}^{n} (x_i + k) = \sum_{i=1}^{n} x_i + nk$

30. $\sum_{i=1}^{n} x_i = n\bar{x}$

31. $(\bar{x} - x_i)^2 = (x_i - \bar{x})^2$

32. $\bar{x} - x_i = x_i - \bar{x}$

Compute the mean absolute deviation for each of the following sets of measures.

33. {4, 4, 4, 6, 12, 20}

34. {0, 3, 0, 5, 3, 1, 0}

35. {1.2, 7.0, 1.4, 2.4, 6.2, 5.6, 5.4, 3.2, 0.6, 1.4}

12.5 **Compute the standard deviation to the nearest tenth for each set of measures.**

36. {1, 3, 5, 7, 9}

37. {15, 17, 15, 13, 14, 14, 18}

38. {70, 72, 71, 71, 71, 70, 70, 70, 72, 72, 71, 71}

39. {1.8, 1.8, 1.6, 1.6, 1.4, 2.0, 2.0, 2.0, 1.8, 1.4, 1.6, 2.0}

40. Roberta scored 92 on a standardized test which had a mean score of 78 and standard deviation of 7. Jane said, "You scored better than 95% of the people who took the test." Was Jane correct? Explain.

41. The Pine City High School juniors took the same test mentioned in exercise 40. There are 850 members of the junior class. How many juniors would you expect scored between 71 and 85?

12.6 **42.** Carefully copy the table and headings as shown below. Complete the table for this set of measures. (Find the standard deviation to the nearest tenth.)

1.8 1.8 1.6 1.6 1.4 2.0 2.0 2.0 1.8
1.4 1.6 2.0 1.6 1.6 2.0 1.8 1.4 1.4

index	x_i	f_i	$f_i x_i$	$\bar{x}$	$\bar{x} - x_i$	$(\bar{x} - x_i)^2$	$f_i(\bar{x} - x_i)^2$

Chapter Test

Find the mean, median, and mode of each of the following sets.

1. $\{1, 2, 3, 3, 2, 1\}$
2. $\{6, 5, 5, 4, 6, 3, 4\}$

In each case, find the value of x that makes M the mean of the set.

3. $\{4, x, 6, 9\}$; $M = 7$
4. $\{10, 8, 6, x^2 - 3, 4\}$; $M = 5$

Compute the following.

5. $\sum_{i=1}^{6} i$
6. $\sum_{k=0}^{4} 6k$
7. $\frac{1}{3}\sum_{i=1}^{3} i^2$

Use sigma notation to write expressions for the following. Then, compute each sum.

8. the mean of the first ten positive integers
9. the sum of the cubes of the whole numbers from 1 to 4, inclusive

Find n, $\bar{x}$, $\sum_{i=1}^{n} (\bar{x} - x_i)$ and $\sum_{i=1}^{n} (\bar{x} - x_i)^2$ for each of the following.

10. $\{10, 9, 11, 12, 12, 7, 11\}$
11. $\{3, 2, 5, 4, 2, 3, 2, 2\}$

Tell whether each of the following sentences is *true* or *false*. Justify your answer in each case. Assume, where appropriate, that $x_1, x_2, \ldots, x_m$ is a set of measures.

12. $\sum_{i=1}^{6} 7 = 14$
13. $\sum |\bar{x} - x_i| = \sum |x_i - \bar{x}|$
14. $\sum (\bar{y} - y_i)^2 = \sum (y_i - \bar{y})^2$

Compute the variance and the standard deviation for each set of measures.

15. $\{2, 4, 6, 10, 12\}$
16. $\{13, 14, 14, 14, 15, 13, 16\}$

Choose the best answer.

17. For a standardized test, the mean was 75 and the standard deviation was 5.8. Which of the following scores can be expected to occur less than 5% of the time?
    a. 90 **b.** 80 **c.** 70 **d.** 65
18. A standardized test has a mean of 55 and a standard deviation of 7.2. A student in the 60th percentile would have which score?
    a. 39 **b.** 57 **c.** 65 **d.** 72
19. Which set of numbers has a range of 30, a mean of 16, a mode of 12, and a median of 15?
    a. $\{1, 5, 12, 12, 18, 23, 27, 30\}$
    b. $\{0, 3, 12, 12, 20, 21, 22, 30\}$
    c. $\{5, 7, 8, 9, 15, 15, 34, 35\}$
    d. $\{1, 12, 12, 13, 17, 20, 22, 31\}$
20. Which set is more dispersed (scattered) based on the mean absolute deviation?
    a. $\{9, 14, 15, 18, 19, 33\}$
    b. $\{10, 12, 15, 19, 24, 28\}$
    c. $\{8, 14, 18, 20, 20, 28\}$
    d. $\{10, 13, 16, 20, 23, 26\}$

Sequences and Series

Chapter 13

After being released, a hot air balloon rises 24 meters in the first second, 18 meters in the second second, 13.5 meters in the third second, and so on. The number of meters that the hot air balloon rises during each second can be written as the following sequence.

$$24, 18, 13.5, 10.125, \ldots$$

Can you tell what the next number will be? What is the pattern?

In this chapter, you will learn more about sequences and series, and their applications in many different fields.

13.1 Sequences

Sequence and series are familiar words. They are usually used to relate the order in which something happens or an arrangement of related things in a row. But in mathematics a more precise meaning is given to each of these terms. Before examining a formal definition of sequence, consider the following example from science.

Example

1 **As shown in the figure, a freely-falling object, under perfect conditions, falls 16 feet during the first second, 48 feet in the next second, 80 feet in the next second, 112 feet in the next second, and 144 feet in the fifth second. This information can also be shown in a table.**

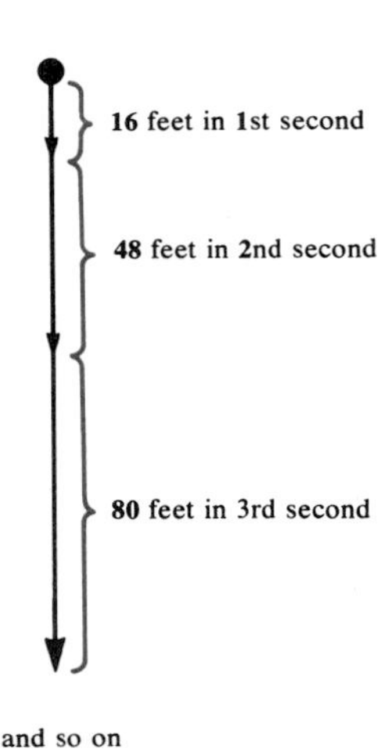

Second	1	2	3	4	5
Distance in feet	16	48	80	112	144

The table in Example 1 defines a function with domain {1, 2, 3, 4, 5} and range {16, 48, 80, 112, 144}. This function is an example of a special kind of function called a *sequence*.

Definition of Sequence

A function whose domain is a set of consecutive natural numbers that begins with 1 is called a **sequence.**

In a sequence, the members of the range are called the **terms of the sequence.** In Example 1, the terms are 16, 48, 80, 112, and 144.

In working with sequences, it is customary to use some special notation. Recall that if a denotes a function, then $a(1)$ denotes the member of the range that is paired with 1. If a is a sequence, however, we use

a_1 instead of $a(1)$,

a_2 instead of $a(2)$, and so forth.

The symbol a_1 is read "a-sub-one."

Therefore, the sequence of example 1 would be the following.

$$a_1 = 16 \qquad a_2 = 48 \qquad a_3 = 80 \qquad a_4 = 112 \qquad a_5 = 144$$

Since the domain of any sequence is always a set of consecutive natural numbers beginning with 1, a sequence can be defined by simply listing its terms as shown in the next example.

Examples

2 **The list 1, 4, 9, 16, 25 defines a sequence. Let b denote this sequence. Then the terms of the sequence can be written as follows.**

$$b_1 = 1 \qquad b_2 = 4 \qquad b_3 = 9 \qquad b_4 = 16 \qquad b_5 = 25$$

What is the rule for the nth term of the sequence? That is, what is b_n?

Since $1^2 = 1$, $2^2 = 4$, $3^2 = 9$, $4^2 = 16$, and $5^2 = 25$, a rule for the nth term is $b_n = n^2$.

3 **Write the first five terms of the sequence f, where $f_n = 2n + 1$.**

$$f_1 = 2 \cdot 1 + 1 = 3 \qquad f_2 = 2 \cdot 2 + 1 = 5 \qquad f_3 = 2 \cdot 3 + 1 = 7$$

$$f_4 = 2 \cdot 4 + 1 = 9 \qquad f_5 = 2 \cdot 5 + 1 = 11$$

The sequences considered so far have been *finite* sequences. Some sequences are *infinite*.

Definition of Infinite Sequence

A sequence whose domain is the set of all natural numbers is said to be an **infinite sequence.**

Example

4 **Consider the sequence $f_n = 2n + 1$ in Example 3. If the domain is $\mathcal{N}$, we obtain the following infinite sequence.**

$$3, 5, 7, 9, 11 \ldots$$

The three dots indicate that the terms go on forever.

Sometimes we list a few terms of a sequence and then the rule for the nth term. For example, the sequence in Example 4 could be written

$$3, 5, 7, \ldots, 2n + 1, \ldots$$

Example

5 **Find the twelfth term of the infinite sequence $\frac{1}{3}, \frac{1}{4}, \frac{1}{5}, \ldots, \frac{1}{n+2}, \ldots$.**

Denote the sequence by f. Then $f_n = \frac{1}{n+2}$. To find the twelfth term, let $n = 12$. Then, $f_{12} = \frac{1}{12+2} = \frac{1}{14}$.

Exercises

Exploratory **In exercises 1–8, *a* is the sequence 2, 4, 6, 8, 10, 12. Find each of the following.**

1. the domain of the sequence

2. the range of the sequence

3. a_1

4. a_3

5. a_6

6. a_5

7. What is a rule for the nth term?

8. Is the sequence finite or infinite? Explain.

Written **Write the first five terms of the sequence with the given *n*th term.**

1. $a_n = 2^n$

2. $a_n = n + 1$

3. $a_n = \frac{1}{n}$

4. $a_n = 2^{n-1}$

5. $a_n = 3n$

6. $a_n = 2^n - 1$

7. $a_n = 3(2^n)$

8. $a_n = 5n$

9. $a_n = 3^{n-1}$

10. $a_n = 3n - 2$

11. $a_n = \frac{(-1)^n}{n}$

12. $a_n = \frac{1}{n^2}$

13. $a_n = 1 + (n - 1)^2$

14. $a_n = 3\left(\frac{1}{2}\right)^{n-1}$

15. $a_n = 5 + (n - 1)^3$

16. $a_n = \frac{(-1)^n}{2^{n-1}}$ **17.** $a_n = 1 + \frac{1}{n}$ **18.** $a_n = \left(1 + \frac{1}{n}\right)^n$

19. The sequence a, where $a_n = 2$, is an example of a *constant* sequence. Can you think why? Give an example of another constant sequence.

20. Find the 2,403rd term of the sequence where $a_n = 1 + 3(n - 1)$.

Consider the sequence 1, 3, 5, ...

21. What do you think the fourth term is?

22. Do you think $a_n = 2n - 1$ is a rule for this sequence? Explain.

23. Is $a_n = (n - 1)(n - 2)(n - 3) + (2n - 1)$ also a rule for it? Why?

24. Using the rule in exercise 22, write the first 10 terms of the sequence.

25. Using the rule in exercise 23, write the first 10 terms of the sequence.

26. How do your results in exercises 24–25 show that listing a few terms of a sequence does not determine a unique sequence?

For each sequence, find a rule for the *n*th term and list the next three terms.

27. 2, 3, 4, 5, . . . **28.** 3, 6, 9, 12, . . . **29.** 1, 3, 9, . . .

30. $\frac{1}{2}, \frac{1}{4}, \frac{1}{8}, \ldots$ **31.** $1, -1, 1, -1, \ldots$ **32.** $-1, \frac{1}{3}, -\frac{1}{9}, \frac{1}{27}, \ldots$

Challenge **If $a_n = n^2 + n$ defines a sequence, then which term is 132?** *Note: This is not the same as asking for the 132nd term.*

Mathematical Excursions

Finite Differences

In many cases it is possible to determine the pattern in a sequence by studying the differences between consecutive terms. For the sequence 3, 8, 13, 18, . . . the difference between consecutive terms is 5, and the next term is found by adding 5 to the previous term. This method can be generalized as follows.

Consider the sequence 10, 16, 28, 46, 70, 100, . . . First, write the sequence and find the differences between terms. Then, find the differences of the differences. Continue this procedure until you arrive at a constant. When the differences are constant, continue the pattern as shown in red.

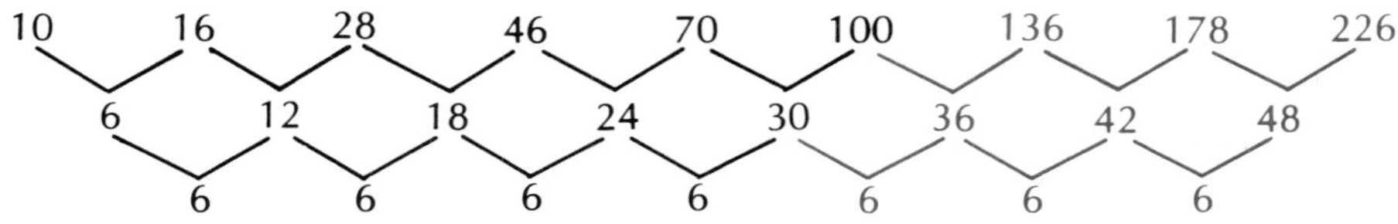

The next three terms of the sequence 10, 16, 28, 46, 70, 100, . . . are 136, 178, 226.

Exercises **Use the method of finite differences to find the next three terms in each sequence.**

1. 4, 18, 48, 100, 180, 294, . . . **2.** 1, 2, 4, 8, 16, 31, 57, . . .

13.2 Arithmetic Sequences

Many sequences have a definite pattern. For example, look carefully at the following sequence.

$$3, 8, 13, 18, 23$$

Notice that each term after the first is five more than the preceding term. This is an example of a certain type of sequence called an *arithmetic* (ar-ith-MET-ic) sequence.

Definition of Arithmetic Sequence

A sequence is an **arithmetic sequence** if and only if each term after the first is obtained by adding the same number d to the preceding term. The number d is called the **common difference.**

Sometimes an arithmetic sequence is called an *arithmetic progression* and its terms are said to be *in arithmetic progression.*

Examples

1 **Find the first five terms of the arithmetic sequence beginning with −6 and having a common difference of 3.**

$$-6, -3, 0, 3, 6$$

2 **Why is the sequence 1, 2, 4, 7, 11, 16 not arithmetic?**

No common difference exists between the terms.

3 **Let a denote the infinite arithmetic sequence.**

$$1, \frac{1}{2}, 0, -\frac{1}{2}, -1, \ldots$$

Find d for this sequence.

To find d, subtract the first term from the second term. Thus, $d = \frac{1}{2} - 1 = -\frac{1}{2}$.

Instead, the second term could have been subtracted from the third term, or the third term from the fourth, and so on.

4 **Use the value d and the sequence from example 3 to find a_6.**

Since $d = -\frac{1}{2}$ and the fifth term, a_5, is -1, we have

$$\begin{aligned} a_6 &= a_5 + d \\ &= -1 + \left(-\frac{1}{2}\right) \text{ or } -\frac{3}{2}. \end{aligned}$$

Let a be an arithmetic sequence with first term 4 and common difference 3. Here are the first few terms.

$$4, 7, 10, 13, 16, \ldots$$

Is it possible to find the 57th term without finding all the preceding terms? Look carefully at the following table. It illustrates a pattern that provides not only the 57th term, but the nth term of the sequence.

n	1	2	3	4	...	57	...	n
a_n	$4 + 0 \cdot 3$	$4 + 1 \cdot 3$	$4 + 2 \cdot 3$	$4 + 3 \cdot 3$	...	$4 + 56 \cdot 3$	...	$4 + (n - 1)3$

The nth term, or *general term,* of this sequence is

$$4 + (n - 1)3.$$

In general, the following is true.

nth Term of an Arithmetic Sequence

For any arithmetic sequence with first term a_1, and common difference d, *the* nth *term is given by*

$$a_n = a_1 + (n - 1)d$$

Examples

5 **If 8, 3, -2 are the first three terms of an arithmetic sequence, find the 46th term.**

First, find the common difference. $d = 3 - 8 = -5$

Then, use the formula $a_n = a_1 + (n - 1)d$.

In this case, $n = 46$, $a_1 = 8$, and $d = -5$.

$$\begin{aligned} a_{46} &= 8 + 45 \cdot (-5) \\ &= 8 + (-225) \\ &= -217 \end{aligned}$$

The 46th term is -217.

6 **Rosa's salary is currently \$15,500 per year. If her annual raise is \$750, how many years will it take for her salary to reach \$22,250?**

The first term of the arithmetic sequence is 15,500 and the common difference is 750. Find the term of the sequence which is 22,250. Use the formula $a_n = a_1 + (n - 1)d$ and solve the following equation for n.

$$\begin{aligned} 22{,}250 &= 15{,}500 + (n - 1)750 \\ 22{,}250 &= 15{,}500 + 750n - 750 \\ 7500 &= 750n \\ 10 &= n \end{aligned}$$

In 10 years, Rosa's salary will be \$22,250 per year.

The terms between any two terms of an arithmetic sequence are called **arithmetic means** between the two terms. In the arithmetic sequence,

$$18, 25, 32, 39, 46$$

25 and 32 are the two arithmetic means between 18 and 39.

Recall from statistics that the mean of two numbers is the average, or arithmetic mean, of the two numbers.

Example

7 **Find the four arithmetic means between 12 and 47.**

This means find the missing terms in the following sequence.

$$12, \underline{\quad ? \quad}, \underline{\quad ? \quad}, \underline{\quad ? \quad}, \underline{\quad ? \quad}, 47$$

This is an arithmetic sequence with $a_1 = 12$ and $a_6 = 47$. Again use the formula $a_n = a_1 + (n - 1)d$ and solve the following equation for d.

$$\begin{aligned} a_6 &= a_1 + (6 - 1)d \\ 47 &= 12 + 5d \\ 7 &= d \end{aligned}$$

Therefore, the arithmetic means are 19, 26, 33, and 40.

Exercises

Exploratory **Give the first five terms of each arithmetic sequence.**

1. $a_1 = 1, d = 3$
2. $a_1 = 4, d = 2$
3. $a_1 = -2, d = 3$
4. $a_1 = -12, d = 5$
5. $a_1 = \frac{1}{2}, d = \frac{1}{2}$
6. $a_1 = -5, d = -2$
7. $a_1 = 17, d = -3$
8. $a_1 = \frac{2}{3}, d = 1$
9. $a_1 = -\frac{3}{2}, d = -\frac{1}{4}$
10. $a_1 = \sqrt{5}, d = \sqrt{3}$
11. $a_1 = -\sqrt{5}, d = \sqrt{5}$
12. $a_1 = -\frac{3}{5}, d = \frac{1}{5}$

Tell if the sequence is arithmetic. If it is, find the common difference.

13. 1, 2, 3, 4, 5
14. 3, 6, 9, 12
15. 2, 4, 8, 16, 32
16. $1, \frac{1}{2}, 0, -\frac{1}{2}$
17. 32, 4, −24
18. $\sqrt{3}, \sqrt{3}, \sqrt{3}, \sqrt{3}$

Written **Write a formula for the *n*th term of each of the following arithmetic sequences.**

1. $a_1 = 2, d = 3$
2. $a_1 = 1, d = 4$
3. $a_1 = 5, d = 5$
4. $a_1 = -3, d = 2$
5. $a_1 = -2, d = 5$
6. $a_1 = -8, d = -3$
7. $a_1 = \frac{1}{2}, d = \frac{1}{2}$
8. $a_1 = \frac{1}{2}, d = \frac{1}{4}$
9. $a_1 = \frac{2}{3}, d = -\frac{1}{3}$
10. $a_1 = -\frac{3}{2}, d = \frac{1}{2}$
11. $a_1 = \sqrt{5}, d = \sqrt{3}$
12. $a_1 = -\frac{3}{5}, d = \frac{1}{5}$

In each case, the first three terms of an arithmetic sequence are given. Find the indicated term.

13. −2, 2, 6; 10th term
14. 4, 12, 20; 18th term
15. 10, −1, −12; 21st term
16. $\frac{1}{4}, \frac{3}{4}, \frac{5}{4}$; 12th term
17. $12, 6\frac{1}{2}, 1$; 35th term
18. $\frac{5}{6}, \frac{7}{6}, \frac{3}{2}$; 10th term

Find the missing terms in each arithmetic sequence.

19. 50, ___?___, ___?___, ___?___, 110
20. −10, ___?___, ___?___, 1
21. −9, ___?___, ___?___, ___?___, 3
22. 3, ___?___, ___?___, ___?___, ___?___, ___?___, 21
23. ___?___, −7, ___?___, ___?___, 14, ___?___
24. ___?___, 50, ___?___, ___?___, 29

For what values of *x* will each of the following form an arithmetic sequence?

25. 8, x, 18
26. 9, $2x + 3$, 14
27. $x + 1, 2x, x$
28. $\frac{1}{x + 3}, 2, x + 5$
29. $6x + 5, 2x - 7, 8x + 2$
30. $\frac{1}{x}, \frac{1}{x^2}, \frac{1}{x^3}$

Two terms of an arithmetic sequence are given. Find the indicated term.

31. $a_5 = 12, a_7 = 16, a_1 = ?$
32. $a_2 = 3, a_5 = 18, a_1 = ?$
33. $a_2 = -6, a_{10} = -22, a_1 = ?$
34. $a_1 = x, a_3 = y, a_2 = ?$
35. $a_4 = 8, a_{10} = 26, a_{12} = ?$
36. $a_6 = 13, a_4 = 28, a_3 = ?$

37. Is a constant sequence an arithmetic sequence? Explain.
38. If each term of an arithmetic sequence is multiplied by the same number, is the resulting sequence arithmetic? Explain.
39. If a and b are arithmetic sequences and a new sequence c is formed where $c_n = a_n + b_n$, is c also an arithmetic sequence? Explain.

Inez bought a car for $6800. Suppose the car depreciates $700 the first year and $400 every year after that.

40. How much will the car be worth after 8 years?
41. When will the car be worth 0 dollars?

42. In 1983, Kelvin earned $13,000. His salary increased by the same amount each year. If he earned $16,325 in 1990, what has been the amount of increase each year?
43. How many multiples of 7 are there between 11 and 391?
44. How many multiples of 13 are there between 29 and 258?
45. Sara's beginning salary in her new position is $23,450 per year. If she receives annual increases of $850, how long will it take for her salary to exceed $35,000?
46. Refer to exercise 45. If Sara remains in her new position for 10 years, what is the total amount she will have earned over this period?

Challenge **Ted has just won first prize in a local contest. He will receive $50 on June 1, $75 on June 2, $100 on June 3, and so on, forming an arithmetic sequence.**

1. How much will Ted receive on June 27?
2. At the end of June, how much money will Ted have received altogether?

Mixed Review

Solve each equation for θ in the interval $0° \leq \theta < 360°$.

1. $2 \cos^2 \theta - 5 \cos \theta = -2$
2. $\sin 2\theta = \sqrt{3} \cos \theta$
3. $\sin \theta = 1 + \cos \theta$

Compute each of the following. Show your work.

4. $\sum_{i=3}^{11} (i - 2)$
5. $\sum_{k=1}^{5} 2^k$
6. $\frac{1}{8} \sum_{n=0}^{8} (2n + 1)$

7. Find the positive value of x for which 14 is the mean of the set $\{x - 4, 2x + 3, 3x + 2, x^2\}$.
8. How many quadrilaterals can be drawn using 9 points in a plane if no three of the points lie on the same line?
9. Diane makes 8 out of every 10 free throws that she attempts. If she has 6 attempts in one game, what is the probability she will make at least 5?
10. A 24-foot ladder leans against the wall of a building. The top touches the wall 19 feet above the ground. Find, to the nearest degree, the measure of the angle formed by the ladder with the ground.

Mathematical Excursions

Fibonacci Sequence

An especially interesting sequence was discovered at the beginning of the thirteenth century by the Italian mathematician Leonardo of Pisa, also called Fibonacci (fi-boh-NAHT-chee).

The Fibonacci sequence first appeared in connection with the following problem.

A young pair of rabbits is placed in an enclosure. After waiting one month, this pair produces a pair of offspring in the second month and every month thereafter. In turn, each pair of offspring waits one month and produces a pair of offspring in the second month and each month thereafter. If none of the rabbits dies, how many *pairs* of rabbits will there be each month?

As you can see from the pictures of rabbits shown here, the following sequence, called the *Fibonacci sequence,* can be obtained.

1, 1, 2, 3, 5, 8, 13, 21, 34, . . .

Month	
1	
2	
3	
4	
5	

The sequence begins with two 1's. Each subsequent term is the sum of the two terms just before it. For example, 8 in the sequence is obtained by adding 3 and 5. Similarly, $13 = 5 + 8$, $21 = 8 + 13$, and so on.

What makes the Fibonacci sequence so fascinating is its frequent occurrence in nature. For example, the leaves or buds on a twig often grow in a spiral fashion down the twig. The distance between any two leaves is equal to the sum of the previous two distances between leaves.

Exercises

1. List the first 20 terms of a Fibonacci sequence.

2. A sequence may be defined *recursively* or by a *recursion formula* if the first term of the sequence is given and a relationship is defined for any term and its successor. For example, if $a_1 = 2$ and $a_{n+1} = 3a_n + 1$, then the first four terms are 2, 7, 22, 67. Write a_5 to a_8.

3. Write a recursion formula for the Fibonacci sequence.

4. DO SOME RESEARCH. Prepare a project that illustrates the Fibonacci sequence in nature. If possible, bring some examples to class.

13.3 Geometric Sequences

The sequence shown below is *not* an arithmetic sequence. But there is a pattern. What is it?

$$1, \frac{1}{2}, \frac{1}{4}, \frac{1}{8}, \frac{1}{16}$$

Each term after the first is *one-half times* the preceding term. This sequence is an example of a *geometric sequence*.

Definition of Geometric Sequence

A sequence is a **geometric sequence** if and only if each term after the first is obtained by multiplying the preceding term by the same number r. The number r is called the **common ratio.**

In the geometric sequence given above, the common ratio is $\frac{1}{2}$.

Sometimes a geometric sequence is called a geometric progression. *The terms are said to be* in geometric progression.

Example

1 If the following sequence is geometric, find the common ratio and the sixth term.

4, −8, 16, . . .

The common ratio can be found by dividing any term into the next term.

$$\frac{a_2}{a_1} = \frac{-8}{4} = -2 \quad \text{or} \quad \frac{a_3}{a_2} = \frac{16}{-8} = -2$$

Since the common ratio is -2,

$$a_4 = (-2)a_3 = -2(16) = -32.$$

$$a_5 = -2(-32) = 64$$

$$a_6 = -2(64) = -128$$

Therefore, the sixth term is -128.

As was the case with arithmetic sequences, there is a pattern in the formation of the terms of a geometric sequence. Suppose a_1 is the first term of a geometric sequence with common ratio r. Study the following table carefully. Notice how the formula for the nth term is found.

n	1	2	3	4	. . .	n
a_n	$a_1 \cdot r^0$	$a_1 \cdot r^1$	$a_1 \cdot r^2$	$a_1 \cdot r^3$	. . .	$a_1 \cdot r^{n-1}$

In general, the following is true.

nth Term of a Geometric Sequence

For any geometric sequence with the first term a_1 and common ratio r, the nth term is given by

$$a_n = a_1 r^{n-1}$$

Example

2 **Find the sixth term of the following geometric sequence.**

$$\mathbf{9, -6, 4, \ldots}$$

Find the common ratio: $r = \frac{-6}{9}$ or $-\frac{2}{3}$.

Use the formula $a_n = a_1 r^{n-1}$ with $n = 6$, $a_1 = 9$, and $r = -\frac{2}{3}$.

$$a_6 = 9\left(-\frac{2}{3}\right)^5$$

$$= 9\left(-\frac{32}{243}\right) = -\frac{32}{27}$$

The terms between any two terms of a geometric sequence are called **geometric means** between the two terms.

Example

3 **Find three geometric means between 3 and 243.**

We must find the missing terms in the sequence

$$3, \underline{\quad ? \quad}, \underline{\quad ? \quad}, \underline{\quad ? \quad}, 243.$$

Here, $a_1 = 3$ and $a_5 = 243$.

Using the formula $a_n = a_1 r^{n-1}$ with $n = 5$, $a_1 = 3$, and $a_5 = 243$, we have

$$243 = 3 \cdot r^4.$$

$$81 = r^4$$

$$\pm 3 = r$$

Since r can be either 3 or -3, there are two possible sequences.

When $r = 3$, the sequence is 3, 9, 27, 81, 243.

When $r = -3$, the sequence is 3, -9, 27, -81, 243.

Therefore, the geometric means are 9, 27, 81, or -9, 27, -81.

A familiar situation arises when a single geometric mean between two numbers is considered. Suppose b is the geometric mean between a and c. Then a, b, c forms a geometric sequence. It follows that

$$\frac{b}{a} = \frac{c}{b}.$$

Recall that b *is called the* mean proportional *between* a *and* c.

Example

4 **In $\triangle RST$, $\angle RST$ is a right angle and $\overline{SU} \perp \overline{RT}$. Find UT if $RU = 9$ and $US = 15$.**

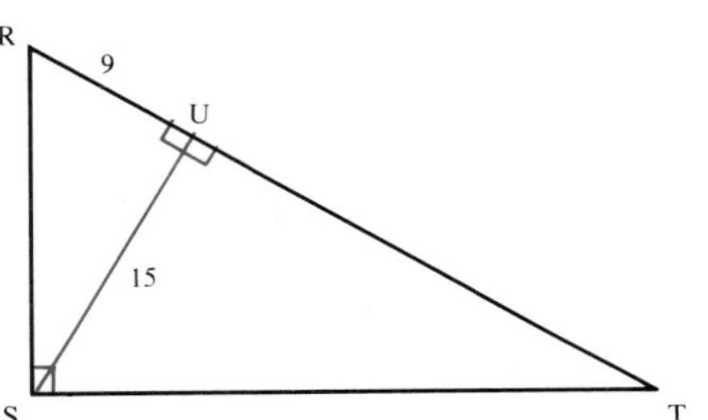

Recall that the length of the altitude drawn to the hypotenuse of a right triangle is the geometric mean between the lengths of the segments of the hypotenuse.

Therefore RU, US, UT forms a geometric sequence.

$$\frac{US}{RU} = \frac{UT}{US}$$

$$\frac{15}{9} = \frac{UT}{15}$$

$$9(UT) = 225$$

$$UT = 25$$

Exercises

Exploratory **Give the first five terms of each geometric sequence.**

1. $a_1 = 3, r = 2$

2. $a_1 = 2, r = -3$

3. $a_1 = 1, r = -2$

4. $a_1 = 4, r = \frac{1}{2}$

5. $a_1 = -5, r = \frac{1}{2}$

6. $a_1 = -2, r = -1$

7. $a_3 = -1, r = \frac{1}{4}$

8. $a_1 = 12, r = -\frac{1}{2}$

9. $a_1 = 100, r = 0.1$

In exercises 10–15, tell if the given sequence is arithmetic, geometric, both, or neither.

10. 2, 10, 50, 150

11. $9, 6, 4, \frac{8}{3}$

12. 7, 14, 21, 28

13. 3, 3, 3, 3 **14.** 5, 0, 0, 0 **15.** −1, 1, −1, 1

Written **The first three terms of a geometric sequence are given. Find the common ratio and the indicated term.**

1. 1, 2, 4; 8th term
2. $-6, 2, -\frac{2}{3}$; 7th term
3. 5, 15, 45; 5th term
4. −1, −4, −16; 6th term
5. 0.001, 0.01, 0.1; 28th term
6. 0.2, 0.22, 0.242; 11th term
7. $\frac{3}{2}, \frac{9}{4}, \frac{27}{8}$; 6th term
8. $\sqrt{3}, 3, \sqrt{27}$; 7th term

Find the terms between the two given terms of a geometric sequence.

9. $a_1 = 4, a_4 = 32$
10. $a_1 = 15, a_4 = 405$
11. $a_1 = \frac{1}{2}, a_5 = \frac{1}{32}$
12. $a_1 = \frac{1}{3}, a_5 = 27$
13. $a_1 = -5, a_6 = -1215$
14. $a_1 = -6, a_4 = 162$

Find the value(s) of *x* that make each sequence geometric.

15. 8, x, 2
16. 3, 6, $2x + 18$
17. $x + 1, x, x - 4$
18. $x + 1, 2x - 1, 4x - 3$

19. Can a sequence be both arithmetic and geometric? Explain.

20. If each term of a geometric sequence is multiplied by the same real number, is the resulting sequence geometric? Explain.

21. If $a_4 = 3$ and $a_7 = 24$ are terms of a geometric sequence, what is a_1?

22. The population of Pine City increases 10% each year. If it is now 20,000 and this rate of increase continues, what will it be after five years?

Challenge

1. Consider the geometric sequence $a_1, a_1r, a_1r^2, \ldots$. Now consider the logarithm of these terms: $\log a_1, \log (a_1r), \log (a_1r^2), \ldots$. Show that this sequence is arithmetic.

2. DO SOME RESEARCH. Find out what you can about the English economist Thomas R. Malthus (1766–1834). How were his views on population and food supply related to the mathematical ideas of this chapter?

3. The German physiologist E. H. Weber (1795–1878) performed pioneering experiments involving the sense of touch and the ability to perceive weights held in the hand. Prepare a report on Weber's research with special attention to his use of the mathematical ideas in this chapter.

13.4 Arithmetic Series

In many situations, you want to know the sum of the terms of the sequence. For example, suppose you were offered a salary of \$800 for the first month and a \$50 raise each month after that. Then your salary for a 12-month period would form the arithmetic sequence shown in the following table. *Why is the twelfth term of the sequence \$1350?*

Month	1	2	3	4	· · ·	12
Salary	800	850	900	950	· · ·	1350

One thing you would probably want to know is your *total* salary for the year. In other words, you would want to know the *sum* of the first 12 terms of the arithmetic sequence shown in the table.

Definition of Arithmetic Series

The indicated sum of the terms of an arithmetic sequence is called an **arithmetic series.**

The chart below contains examples of arithmetic sequences and their corresponding series.

Arithmetic Sequence	Arithmetic Series
2, 4, 6, 8, 10, 12	2 + 4 + 6 + 8 + 10 + 12
5, 2, −1, −4	5 + 2 + (−1) + (−4)
$\frac{5}{4}, \frac{9}{4}, \frac{13}{4}, \frac{17}{4}, \frac{21}{4}$	$\frac{5}{4} + \frac{9}{4} + \frac{13}{4} + \frac{17}{4} + \frac{21}{4}$
$a_1, a_2, a_3, a_4, \cdots, a_n$	$a_1 + a_2 + a_3 + a_4 + \cdots + a_n$

The symbol S_n represents the sum of the first n terms of a series. Thus, for the sequence 2, 4, 6, 8, 10, 12,

$$S_3 = 2 + 4 + 6 \quad \text{and} \quad S_6 = 2 + 4 + 6 + 8 + 10 + 12.$$

When n is large, it is not convenient to find the value of S_n by adding the terms. The following method, used to find the value of S_6 given above, is similar to the one used by the famous mathematician Karl Friedrich Gauss (1777–1855) when he was only 10 years old!

If $S_6 = 2 + 4 + 6 + 8 + 10 + 12$, then the terms of S_6 can be written in reverse order as follows.

$$S_6 = 12 + 10 + 8 + 6 + 4 + 2$$

Then, add the two equations.

$$2S_6 = 14 + 14 + 14 + 14 + 14 + 14$$
$$2S_6 = 6 \cdot 14 \text{ or } 84$$
$$S_6 = 42$$

Now, repeat this procedure for the general case.

$$\begin{aligned} S_n &= a_1 + (a_1 + d) + (a_1 + 2d) + (a_1 + 3d) + \cdots + a_n \\ + S_n &= a_n + (a_n - d) + (a_n - 2d) + (a_n - 3d) + \cdots + a_1 \\ \hline 2S_n &= (a_1 + a_n) + (a_1 + a_n) + (a_1 + a_n) + (a_1 + a_n) + \cdots + (a_1 + a_n) \end{aligned}$$

Notice that the right side of this equation contains the term $(a_1 + a_n)$ a total of n times. (You should be able to justify this.) Therefore,

$$2S_n = n(a_1 + a_n).$$
$$S_n = \frac{n}{2}(a_1 + a_n)$$

Sum of an Arithmetic Series

The sum of the first n terms of an arithmetic series is given by the following formula.

$$S_n = \frac{n}{2}(a_1 + a_n)$$

Example

1 **Tom's salary is \$800 for the first month. He receives a \$50 raise each month after that. Find the total salary for a 12-month period.**

The sequence is 800, 850, 900, . . . , 1350. Use the formula

$$S_n = \frac{n}{2}(a_1 + a_n).$$

In this case, $n = 12$, $a_1 = 800$, and $a_n = 1350$.

$$\begin{aligned} S_{12} &= \frac{12}{2}(800 + 1350) \\ &= 6(2150) \\ &= 12{,}900 \end{aligned}$$

Therefore, the total salary is \$12,900.

Recall that for an arithmetic sequence, $a_n = a_1 + (n - 1)d$. To find S_n in terms of a_1, substitute for a_n in the formula for S_n.

$$S_n = \frac{n}{2}(a_1 + a_n)$$
$$= \frac{n}{2}[a_1 + a_1 + (n - 1)d]$$
$$S_n = \frac{n}{2}[2a_1 + (n - 1)d]$$

Example

2 **Find the sum of the first 20 terms of the arithmetic series**

$$-6 + (-1) + 4 + \cdots$$

In this case, $d = 5$, $a_1 = -6$, and $n = 20$. Since the value of the 20th term is not given, substitute in the following formula.

$$S_n = \frac{n}{2}[2a_1 + (n - 1)d]$$
$$S_{20} = \frac{20}{2}[2(-6) + (20 - 1)5]$$
$$= 10(-12 + 95)$$
$$= 10(83) \text{ or } 830$$

Sometimes *sigma notation* is used to represent a series. For example, the arithmetic series $2 + 4 + 6 + 8 + 10 + 12$ can be represented as $\sum_{k=1}^{6} 2k$.

Example

3 **Find** $\sum_{k=1}^{200} (2k + 3)$.

$$\sum_{k=1}^{200} (2k + 3) = 5 + 7 + 9 + 11 + \cdots + 403$$

This is an arithmetic series with $a_1 = 5$, $d = 2$, $n = 200$, and $a_{200} = 403$. Therefore, you use the formula $S_n = \frac{n}{2}(a_1 + a_n)$.

$$S_{200} = \sum_{k=1}^{200}(2k + 3) = \frac{200}{2}(5 + 403)$$
$$= 100(408)$$
$$= 40,800$$

Exercises

Exploratory **Find the sum of each arithmetic series.**

1. $2 + 4 + 6 + 8 + 10 + 12$
2. $5 + 8 + 11 + 14 + 17 + 20 + 23$
3. $\frac{1}{2} + \frac{3}{4} + 1 + \frac{5}{4} + \frac{3}{2} + \frac{7}{4} + 2$
4. $-17 - 6 + 5 + 16 + 27$

Find S_n for each arithmetic series.

5. $a_1 = 10, a_n = 38, n = 14$
6. $a_1 = 2, a_n = 200, n = 100$
7. $a_1 = 50, n = 20, d = -4$
8. $a_1 = 4, n = 15, d = 7$
9. $a_1 = -20, n = 10, d = 17$
10. $a_1 = 9, n = 14, d = -6$
11. $a_1 = -3, n = 10, d = -4$
12. $a_1 = -11, n = 21, d = 3$
13. $a_1 = 3\frac{1}{2}, a_n = 11, n = 16$
14. $a_1 = 5, n = 13, d = 3\frac{1}{2}$
15. Rosa claims that a summation of the form $\sum_{k=1}^{n}(dk + b)$, where d and b are constants, is an arithmetic series. Is she right? Explain.

Written **Find the sum of the first 30 terms of each arithmetic series.**

1. $1 + 5 + 9 + \cdots$
2. $-5 - 2 + 1 + \cdots$
3. $-9 - 3 + 3 + \cdots$
4. $2 - 2 - 6 - \cdots$
5. $-3 + 1 + 5 + \cdots$
6. $-7 - 10 - 13 - \cdots$
7. The sum of an arithmetic series is 77. If its first term is 2 and its last term is 12, how many terms does it have?
8. Find the sum of the first 100 positive integers.
9. Find the sum of the first 100 positive even integers.
10. Find the sum of the first 100 positive odd integers.
11. The sum of an arithmetic series is 14,150. If the first term is -7 and the last term is 290, how many terms does it have?
12. Kelvin's starting salary is \$15,400. If he receives an annual raise of \$1200 for the next four years, what is his total salary for the five-year period?
13. Wanda plans to save \$825 by making fifteen weekly deposits in her savings account. Her first deposit is \$20 and she plans to deposit at least \$20 each week. By what amount must she increase each future deposit in order to reach her goal?

From Section 13.1, recall that a freely falling object falls 16 feet in the first second, 48 feet in the next second, and so on. The distance the object falls increases by 32 feet each second. How far does the object fall during each of the following seconds?

14. 6th second
15. 8th second
16. 10th second
17. 12th second

How far has the object fallen altogether after the following number of seconds?

18. 6 seconds **19.** 8 seconds **20.** 10 seconds. **21.** 12 seconds

Use sigma notation to express the following series.

22. $5 + 14 + 23 + 32$

23. $\frac{1}{5} + \frac{4}{5} + \frac{7}{5} + 2 + \frac{13}{5}$

24. $8 + 5 + 2 - 1 - 4$

25. $\frac{9}{2} + 7 + \frac{19}{2} + 12 + \frac{29}{2}$

26. $1 - 3 + 5 - 7 + 9$

27. $6 - 2 + \frac{2}{3} - \frac{2}{9} + \frac{2}{27}$

Find each sum.

28. $\sum_{k=1}^{100} k$

29. $\sum_{k=0}^{100} (2k + 1)$

30. $\sum_{k=1}^{51} (3k + 5)$

31. $\sum_{k=1}^{4} 4^k$

32. $\sum_{k=1}^{5} 4\left(\frac{1}{2}\right)^k$

33. $\sum_{k=1}^{4} \left(3^{k-1} + \frac{1}{2}\right)$

Challenge

1. Show that the sum of the first n positive integers is $\frac{n(n + 1)}{2}$.
2. Show that the sum of the first n positive odd integers is n^2.

Look carefully at the number of dots in the triangular arrays in the figure at the right. The numbers 1, 3, 6, 10 are the first four terms of the sequence of triangular numbers.

3. Find the next four numbers of this sequence.
4. Is this sequence arithmetic? Explain.
5. Show that the nth term of the sequence of triangular numbers is the sum of the first n positive integers.
6. Find the 40th triangular number.

Determine whether the given equations are *true* or *false*.

7. $\sum_{k=0}^{5} a^k + \sum_{n=6}^{10} a^n = \sum_{b=0}^{10} a^b$

8. $\sum_{r=3}^{7} 3^r + \sum_{a=7}^{9} 3^a = \sum_{j=3}^{9} 3^j$

9. $\sum_{n=1}^{10} (5 + n) = \sum_{m=0}^{9} (4 + m)$

10. $\sum_{r=2}^{8} (2r - 3) = \sum_{s=3}^{9} (2s - 5)$

11. $2 \sum_{k=3}^{7} k^2 = \sum_{k=3}^{7} 2k^2$

12. $3 \sum_{n=1}^{5} (n + 3) = \sum_{n=1}^{15} (n + 3)$

13.5 Geometric Series

In the preceding section, arithmetic series were formed from an arithmetic sequence. The indicated sum of the terms of a geometric sequence is called a **geometric series.**

Example

1 **Let 2, 4, 8, 16, 32 be a geometric sequence with $a_1 = 2$ and $r = 2$. Find the corresponding geometric series.**

$$2 + 4 + 8 + 16 + 32$$

As was the case with arithmetic series, there is a formula for the sum of the first n terms of a geometric series.

In the derivation given below, first write S_n in expanded form. Then multiply each side of the first equation by r. Subtracting the second equation from the first equation produces the third equation.

$$S_n = a_1 + a_1r + a_1r^2 + a_1r^3 + \cdots + a_1r^{n-1}$$

$$r \cdot S_n = a_1r + a_1r^2 + a_1r^3 + \cdots + a_1r^{n-1} + a_1r^n \quad \text{Multiply by } r \text{ and align like terms.}$$

$$S_n - rS_n = a_1 - a_1r^n \quad \text{Subtract.}$$

$$S_n(1 - r) = a_1 - a_1r^n \quad \text{Factor.}$$

$$S_n = \frac{a_1 - a_1r^n}{1 - r} \text{ or } \frac{a_1(1 - r^n)}{1 - r}, \text{ if } r \neq 1 \quad \text{Divide by } 1 - r. \text{ Then factor.}$$

Sum of a Geometric Series

The sum of the first n terms of a **geometric series** is given by

$$S_n = \frac{a_1 - a_1r^n}{1 - r} \text{ or } S_n = \frac{a_1(1 - r^n)}{1 - r} \textbf{ provided } r \neq 1.$$

Example

2 **Find the sum of the geometric series 3 + 6 + 12 + 24 + 48.**

In this case, $a_1 = 3$, $r = 2$, and $n = 5$.

$$S_n = \frac{a_1 - a_1r^n}{1 - r}$$

$$S_5 = \frac{3 - 3(2)^5}{1 - 2}$$

$$= \frac{3 - 3(32)}{-1}$$

$$= \frac{3 - 96}{-1} = \frac{-93}{-1} = 93$$

Thus, $S_5 = 93$.

For geometric sequences, recall that

$$a_n = a_1 r^{n-1}.$$

Multiplying both sides of the preceding equation by r produces

$$ra_n = a_1 r^{n-1} \cdot r^1.$$

$$ra_n = a_1 r^n$$

Substituting ra_n for $a_1 r^n$ in the formula for S_n gives another formula for S_n, shown below.

$$S_n = \frac{a_1 - a_1 r^n}{1 - r}$$

$$S_n = \frac{a_1 - ra_n}{1 - r}, \quad r \neq 1$$

Examples

3 **Find the sum of the geometric series with first term −2, last term −486, and common ratio 3.**

In this case, $a_1 = -2$, $a_n = -486$, and $r = 3$. Substitute in the following formula for S_n.

$$S_n = \frac{a_1 - ra_n}{1 - r}$$

$$S_n = \frac{-2 - 3(-486)}{1 - 3}$$

$$= \frac{-2 + 1458}{-2} = \frac{1456}{-2} = -728$$

4 **A company pays you a salary of \$800 for the first month with a raise of 5% each month. What is your total salary for twelve months?**

Since the salary increases by 5% each month, the salary for the second month is 800 + 0.05(800) or 1.05(800). The salary for the third month is

$$1.05(800) + (0.05)(1.05)(800) = (800)(1.05)^2.$$

Each month's salary is 105% of the preceding month's salary. Therefore, the monthly salaries form the following geometric sequence.

$$800, 800(1.05), 800(1.05)^2, 800(1.05)^3, \ldots$$

To find the total yearly salary, find the sum of the first 12 terms.

$$S_n = \frac{a_1 - a_1 r^n}{1 - r}$$

$$S_{12} = \frac{800 - 800(1.05)^{12}}{1 - 1.05} = \frac{800[1 - (1.05)^{12}]}{-0.05}$$

Using a calculator or logarithms, we find $S_n \approx 12{,}734$.

Exercises

Exploratory **Give an example of each type of series described below.**

1. arithmetic
2. geometric
3. both arithmetic and geometric
4. neither arithmetic nor geometric

Tell whether each of the following series is arithmetic, geometric, both, or neither.

5. $\frac{1}{4} + \frac{1}{2} + \frac{3}{4} + 1 + \frac{5}{4}$
6. $1 + \frac{1}{3} + \frac{1}{9} + \frac{1}{27}$
7. $2 + 2 + 2 + 2 + 2$
8. $1 + 2 + 4 + 7 + 11$
9. $1 - 3 + 5 - 7 + 9$
10. $-1 + 1 - 1 + 1$
11. $\sum_{k=1}^{4} \frac{1}{3}k$
12. $\sum_{i=1}^{5} (-2)^i$

Written **Find S_n for geometric series with the given values.**

1. $a_1 = 3, r = 2, n = 4$
2. $a_1 = 2, r = -3, n = 6$
3. $a_1 = 3, r = 2, n = 6$
4. $a_1 = 8, r = \frac{1}{2}, n = 6$
5. $a_1 = \frac{1}{9}, r = -3, n = 5$
6. $a_1 = 5, r = 3, n = 5$
7. $a_1 = 16, r = -\frac{1}{2}, n = 6$
8. $a_1 = 256, r = \frac{3}{4}, n = 5$
9. $a_1 = \frac{1}{2}, r = -\frac{1}{2}, n = 10$
10. $a_1 = 125, a_5 = \frac{1}{5}, r = \frac{1}{5}, n = 5$
11. $a_1 = 4, a_6 = \frac{1}{8}, r = \frac{1}{2}, n = 6$
12. $\frac{1}{4} - \frac{1}{12} + \frac{1}{36} - \cdots - \frac{1}{972}$
13. $1 + \frac{1}{2} + \frac{1}{4} + \cdots + \frac{1}{1024}$
14. $a_1 = -6, a_n = -93.75, r = 2.5$
15. $a_1 = 2, a_n = 10.125, r = 1.5$
16. In the formula for S_n, why can r not be equal to 1?
17. Give an example of a geometric series with r equal to 1. What is the sum of the first n terms? Generalize your results.

Since 1988, the number of auto accidents in Pine City has decreased by 10% each year. In 1988, there were 4000 accidents.

18. How many accidents were there in Pine City in 1991?
19. How many accidents were there from the beginning of 1988 to the end of 1991?

20. Ted begins a chain letter by writing to Sue, Zeb, and Rosa. Each of these three people write to three others, and so on, for a total of eight rounds. How many people will have received chain letters at the end of the eighth round?
21. Tax records show that the Dick and Doug Tool Company reported total profits of \$122,102 for the five-year period from 1987 to 1991. If their profits increased by 10% each year, what was the profit at the end of 1987?

22. Rosa has just won the Grand Prize in the Pine City Municipal Contest. She is given a choice of **A,** a flat grant of \$5 million, or **B,** 1¢ on the first day of a 30-day month, 2¢ on the second day, 4¢ on the third day, and so on, with the amount doubling each day during the month. Should Rosa choose **A** or **B** for the greater amount? Justify your answer.

23. A person has \$650 in a bank and is closing out the account by writing one check a week against the account. The first check is \$20, the second is \$25, and so on. Each check exceeds the previous one by \$5. In how many weeks will the account be closed if there is no service charge?

Write the following in expanded form and evaluate.

24. $\sum_{k=1}^{7} \left(\frac{1}{2}\right)^k$

25. $\sum_{i=1}^{6} 2^{i-4}$

26. $\sum_{j=1}^{4} \left(\frac{2}{3}\right)^j$

27. $\sum_{r=1}^{5} (-1)^r$

28. $\sum_{s=1}^{4} (-3)^s$

29. $\sum_{t=1}^{6} 2^{-t}$

Challenge

Suppose a side of the large square shown at the right is 10 cm. If the midpoints of consecutive sides are joined, an inscribed square is formed. If this process is continued to form a total of five squares, find the sum of the perimeters of the squares.

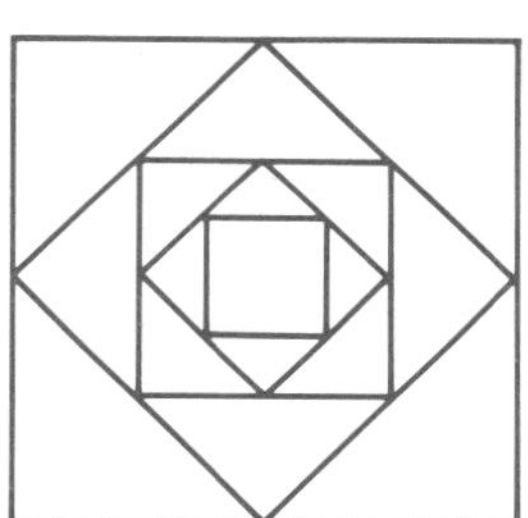

Mixed Review

1. Solve for x: $|x - 3| < 5$.
2. Write a quadratic equation in standard form with roots $1 \pm \sqrt{3}$.
3. Using logarithms, compute the value of $23\sqrt[3]{0.432}$ to the nearest tenth.
4. In circle O, chords $\overline{AB}$ and $\overline{CD}$ intersect at E. If $m\widehat{AC} = 30$ and $m\widehat{BD} = 90$, find $m\angle AEC$.
5. Solve for x: $\log x + \log (x - 1) = \log 2$.
6. If $f(x) = x^2 - x$, find $f(z + 2)$.
7. Simplify $\frac{2}{3 - \sqrt{3}}$.
8. Express $(2 - \sqrt{-5})^2$ in simplest $a + bi$ form.
9. Solve for x: $x^2 - 4x \leq 0$.
10. Solve for y: $\frac{3(y - 1)}{2 - y} + \frac{2y - 1}{y - 2} = 2$.

An infinite sequence has an unlimited number of terms. For example, the sequence $1, \frac{1}{2}, \frac{1}{3}, \frac{1}{4}, \frac{1}{5}, \cdots \frac{1}{n}, \cdots$ is an infinite sequence whose nth term is $\frac{1}{n}$. Several terms of this sequence are graphed below.

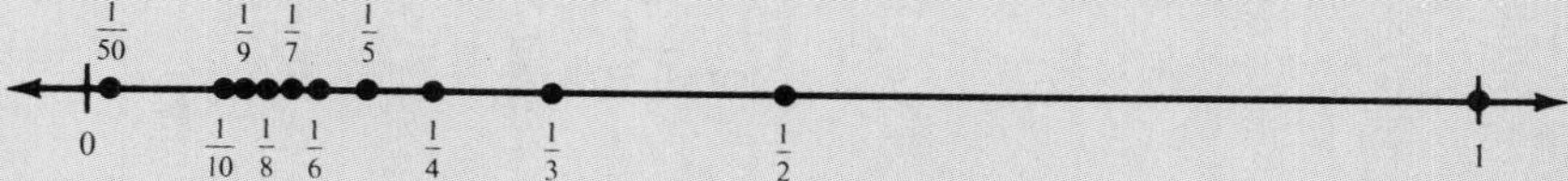

Notice that the terms approach zero as n increases in value. Zero is called the limit of the terms of this sequence. This can be expressed as follows.

$$\lim_{n\to\infty} \frac{1}{n} = 0 \qquad \textit{∞ is the symbol for infinity.}$$

This is read "the limit of 1 over n as n approaches infinity equals zero."

If a general expression for the nth term of a sequence is known, the limit can usually be estimated by substituting large values for n. For example, find the limit of the sequence $\frac{1}{2}, \frac{1}{4}, \frac{1}{8}, \frac{1}{16}, \cdots$

The nth term of this sequence is $\frac{1}{2^n}$.

The 50th term is $\frac{1}{2^{50}}$. The 100th term is $\frac{1}{2^{100}}$. Notice that the values approach zero.

Therefore, $\lim_{n\to\infty} \frac{1}{2^n} = 0$.

The form of the expression for the nth term of a sequence can be altered to make the limit easier to find. Consider the following:

Evaluate $\lim_{n\to\infty} \frac{3n + 1}{n}$.

Since $\frac{3n + 1}{n} = 3 + \frac{1}{n}$, $\lim_{n\to\infty} \frac{3n + 1}{n} = \lim_{n\to\infty}\left(3 + \frac{1}{n}\right)$.

Recall that $\lim_{n\to\infty} \frac{1}{n} = 0$. Therefore, $\lim_{n\to\infty}\left(3 + \frac{1}{n}\right) = 3 + 0$ or 3.

The limit is 3.

Limits do not exist for all sequences. If the terms of a sequence become arbitrarily large or approach two different values, the sequence has no limit.

Exercises **Evaluate each limit, or state that the limit does not exist.**

1. $\lim_{n\to\infty} \frac{n + 1}{n}$

2. $\lim_{n\to\infty} \frac{1}{3^n}$

3. $\lim_{n\to\infty} \frac{3n^2 + 4}{2n}$

4. $\lim_{n\to\infty} \frac{2n^2 - 6n}{5n^2}$

5. $\lim_{n\to\infty} \frac{2n^2 + 6n}{5n^2}$

6. $\lim_{n\to\infty} \frac{n^3 + 6n}{3n^3}$

13.6 Infinite Geometric Series

If a repeating decimal, such as $0.333\ldots$ or $0.\overline{3}$, is expressed as a sum, the following series is obtained.

$$0.\overline{3} = 0.3 + 0.03 + 0.003 + 0.0003 + 0.00003 + \cdots$$
$$\text{or } 0.\overline{3} = \frac{3}{10} + \frac{3}{100} + \frac{3}{1000} + \frac{3}{10{,}000} + \frac{3}{100{,}000} + \cdots$$

A few observations should be made about this series. First, the series is an **infinite series,** since an infinite number of fractions are being added. Second, the series is formed by adding the terms of the *infinite geometric sequence* $\frac{3}{10}, \frac{3}{100}, \frac{3}{1000}, \cdots$ where $a_1 = \frac{3}{10}$ and $r = \frac{1}{10}$. The series $\frac{3}{10} + \frac{3}{100} + \frac{3}{1000} + \cdots$ is an example of an *infinite geometric series.* Third, the sum of this series is $0.\overline{3}$ or $\frac{1}{3}$.

Not all infinite geometric series, though, have a sum. For example, it is clear that the infinite geometric series written below does not have a sum.

$$2 + 4 + 8 + 16 + \cdots + 2^n + \cdots$$

Under what circumstances, then, will an infinite geometric series have a sum? When there *is* a sum, is there a formula for obtaining it?

To investigate these questions, return to the original example.

$$\frac{3}{10} + \frac{3}{100} + \frac{3}{1000} + \cdots$$

The sum of the first n terms of this series is given by

$$S_n = \frac{a_1(1 - r^n)}{1 - r}.$$

What happens to S_n as larger values of n are taken?

$$\text{When } n = 1,\ S_1 = \frac{\frac{3}{10}\left(1 - \frac{1}{10}\right)}{1 - \frac{1}{10}} = \frac{1}{3}\left(1 - \frac{1}{10}\right) = 0.3.$$

Verify the calculations.

$$\text{When } n = 2,\ S_2 = \frac{\frac{3}{10}\left(1 - \frac{1}{10^2}\right)}{1 - \frac{1}{10}} = \frac{1}{3}\left(1 - \frac{1}{10^2}\right) = 0.33.$$

When $n = 3$, $S_3 = \dfrac{\frac{3}{10}\left(1 - \frac{1}{10^3}\right)}{1 - \frac{1}{10}} = \frac{1}{3}\left(1 - \frac{1}{10^3}\right) = 0.333.$

What is S_{10}? Do you see a pattern?

As the value of n increases, the corresponding values of S_n get closer and closer to $\frac{1}{3}$. This can also be seen by considering the general formula for S_n.

$$S_n = \frac{\frac{3}{10}\left(1 - \frac{1}{10^n}\right)}{1 - \frac{1}{10}} = \frac{\frac{3}{10}\left(1 - \frac{1}{10^n}\right)}{\frac{9}{10}}$$

$$S_n = \frac{1}{3}\left(1 - \frac{1}{10^n}\right)$$

Look carefully at the last formula. As n increases, the corresponding value of $\frac{1}{10^n}$ gets closer to 0. Therefore, $1 - \frac{1}{10^n}$ gets closer to 1. This means that S_n gets closer to $\frac{1}{3}$. Thus, the sum of the series is $\frac{1}{3}$. To summarize this discussion, we say that S_n approaches $\frac{1}{3}$ as n increases.

Now begin with a general infinite geometric series and repeat the argument given above.

$$a_1 + a_1r + a_1r^2 + a_1r^3 + \cdots + a_1r^{n-1} + \cdots$$

The sum of the first n terms is given by

$$S_n = \frac{a_1(1 - r^n)}{1 - r}, \ r \neq 1.$$

This formula can also be written in the following form.

$$S_n = \frac{a_1}{1 - r}(1 - r^n), \ r \neq 1$$

In the preceding example, you saw that for large values of n, r^n is close to 0 when $r = \frac{1}{10}$. Are there any other values of r for which this situation occurs? For example, suppose $r = \frac{1}{2}$. Does r^n get closer to 0 as n increases?

What about $r = 3$? $r = \frac{2}{3}$? Try these values before reading further.

You should conclude that r^n gets closer and closer to 0 *only when the value of r is between −1 and 1*. When r^n gets closer to 0, S_n approaches $\frac{a_1}{1-r}$. The expression $\frac{a_1}{1-r}$ is called the **sum of the infinite geometric series.**

Sum of an Infinite Geometric Series

If an infinite geometric series has common ratio r such that $|r| < 1$, then its sum S is given by

$$S = \frac{a_1}{1-r}.$$

The formula applies only when r *is between* −1 *and* 1*. For all other values of* r*, the series does* not *have a sum.*

Examples

1 **Find the sum of the infinite geometric series if it exists.**

$$2 + 1 + \frac{1}{2} + \frac{1}{4} + \cdots$$

In this case, $a_1 = 2$ and $r = \frac{1}{2}$. Use the formula for S.

$$S = \frac{2}{1 - \frac{1}{2}} = \frac{2}{\frac{1}{2}} = 4$$

The sum is 4.

2 **Find the sum of the infinite geometric series if it exists.**

$$1 + \frac{3}{2} + \frac{9}{4} + \frac{27}{8} + \cdots$$

Here, the common ratio is $\frac{3}{2}$ and $\frac{3}{2} > 1$, so this series has no sum.

3 **Express $0.\overline{82}$ in $\frac{a}{b}$ form where $a, b \in \mathbb{Z}$ and $b \neq 0$.**

$$0.\overline{82} = 0.82 + 0.0082 + 0.000082 + \cdots$$
$$= \frac{82}{100} + \frac{82}{10{,}000} + \frac{82}{1{,}000{,}000} + \cdots$$

$$= \frac{82}{100}\left(1 + \frac{1}{100} + \frac{1}{10{,}000} + \cdots\right)$$

$$= \frac{82}{100}\left(1 + \frac{1}{100} + \frac{1}{100^2} + \cdots\right)$$

The series in parentheses is an infinite geometric series with $a_1 = 1$ and $r = \frac{1}{100}$. Since $\frac{1}{100} < 1$, the sum of the series is found by using the formula for S with $a_1 = 1$ and $r = \frac{1}{100}$. Therefore,

$$0.\overline{82} = \frac{82}{100}\left(\frac{1}{1 - \frac{1}{100}}\right)$$

$$= \frac{82}{100}\left(\frac{1}{\frac{99}{100}}\right) = \frac{82}{100} \cdot \frac{100}{99}.$$

$$0.\overline{82} = \frac{82}{99}$$

Compare this method with the one given in section 1.1. of chapter 1.

Exercises

Exploratory

1. Give an example of an infinite geometric series that does not have a sum. Explain why it has no sum.
2. Explain what it means to say that $\frac{1}{3^n}$ approaches 0 as n increases.

If it exists, find the sum of each infinite geometric series with the given values of a_1 and r. If no sum exists, say so.

3. $a_1 = 6, r = \frac{1}{2}$
4. $a_1 = 8, r = \frac{1}{4}$
5. $a_1 = -3, r = \frac{1}{3}$
6. $a_1 = \frac{1}{2}, r = \frac{3}{2}$
7. $a_1 = -4, r = -\frac{1}{8}$
8. $a_1 = -12, r = -\frac{2}{3}$
9. $a_1 = -0.5, r = -1.25$
10. $a_1 = 3, r = -3$

What number does each of the following expressions get closer to as n increases? Explain.

11. $\frac{1}{2^n}$
12. $\frac{1}{3^n}$
13. $2 - \frac{1}{3^n}$
14. $8 - \left(\frac{2}{3}\right)^n$

15. Explain why $|r| < 1$ is equivalent to $-1 < r < 1$.

16. Under what circumstances does r^n get closer to 0 as n increases?

17. Explain how the formula $S = \frac{a_1}{1-r}$ is obtained from the formula $S_n = \frac{a_1(1-r^n)}{1-r}$.

Written **If it exists, find the sum of each geometric series. If no sum exists, explain why.**

1. $6 + 3 + \frac{3}{2} + \cdots$
2. $9 - 3 + 1 - \cdots$
3. $4 - 6 + 9 - \cdots$
4. $2 + 4 + 8 + 16 + \cdots + 128$
5. $\frac{1}{15} - \frac{1}{25} + \frac{3}{125} - \cdots$
6. $1 - \frac{1}{2} + \frac{1}{4} - \cdots$
7. $\frac{1}{25} + \frac{1}{35} + \frac{1}{49} + \cdots$
8. $-8 - 12 - 18 - \cdots - \frac{243}{4}$
9. $-2 + 3 - \frac{9}{2} + \cdots$
10. $\frac{1}{49} - \frac{1}{42} + \frac{1}{36} - \cdots$

Use the method given in Example 3 of this section to find a rational number in $\frac{a}{b}$ form for each repeating decimal.

11. $0.\overline{5}$
12. $0.\overline{2}$
13. $0.\overline{7}$
14. $0.\overline{1}$
15. $4.\overline{6}$
16. $0.\overline{36}$
17. $0.\overline{23}$
18. $0.\overline{231}$

19. If the sum of an infinite geometric series is 248 and its common ratio is -0.6, find the first term of the series.

20. If the sum of an infinite geometric series is $\sqrt{3}$ and its first term is $\sqrt{2}$, what is the common ratio for the series?

21. If the sum of an infinite geometric series is $-13\frac{1}{3}$ and its first term is -8, find the first four terms of the series.

22. If the sum of an infinite geometric series is $76\frac{4}{5}$ and its common ratio is $\frac{11}{16}$, find the first four terms of the series.

23. Each swing of a certain pendulum is 80% as long as the preceding swing. If the end of the pendulum travels 25 cm on the first swing, what is the total distance it travels before coming to rest?

24. A certain ball dropped from a height of 30 feet rebounds $\frac{2}{5}$ of the height from which it fell on each bounce. How far does it travel before coming to rest?

25. A hot air balloon rises 80 feet in its first minute of flight. If in each succeeding minute the balloon rises only 90% as far as in the previous minute, what will be the balloon's maximum altitude?

26. When dropped from a certain height, a silicon ball rebounds 91% of the height from which it fell on each bounce, and travels 500 feet before coming to rest. From what height was the ball dropped?

27. The end of a swinging pendulum traveled 385 cm before coming to rest. If each swing of the pendulum is $\frac{9}{11}$ as long as the preceding swing, how far did the end of the pendulum travel on the first swing?

28. A hot air balloon rises 48 meters in its first minute of flight. In each succeeding minute, the balloon rises a certain percent as far as in the previous minute until it reached its maximum altitude of 300 meters. Find this percent.

29. The Pine City Civic Association wishes to give a grant to Pine City College for a scholarship fund. A \$3000 scholarship will be awarded annually. Suppose that Pine City College invests the grant at 8% compounded semiannually. What is the amount of the grant the Pine City Civic Association must give to the college so that the annual scholarship can be given forever?

30. Infinite geometric series were considered in this section. Investigate infinite arithmetic series. Give a few examples. Will such a series ever have a sum? Explain.

Challenge **Refer to the sequence of inscribed squares given in the Challenge on page 492. Suppose the process of obtaining squares is continued forever.**

1. What is the sum of the perimeters of the squares?

2. What is the sum of the areas of all the squares?

3. A side of an equilateral triangle is 20 inches. The midpoints of its sides are joined to form an inscribed equilateral triangle. If this process is continued without end, find the sum of the perimeters of the triangles.

4. Find the sum of the areas of the series of triangles in the previous exercise.

Mixed Review

In the figure at the right, $\overline{PD}$ is tangent to $\odot O$. If $PD = 30$, $AB = 25$, $PE = 18$, $CF = 20$, $AD = 34$, and $AF > FD$, find each measure.

1. BP **2.** CE **3.** AF

If, in $\odot O$, $m\widehat{AC} = 92$, $m\widehat{BE} = 34$, and $m\angle EPD = 18$, find each measure.

4. $m\angle BPD$ **5.** mED **6.** $m\angle CFD$

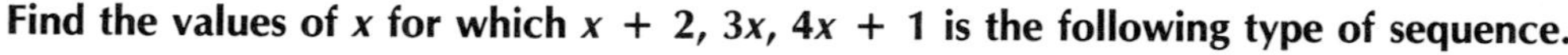

Find the values of x for which $x + 2$, $3x$, $4x + 1$ is the following type of sequence.

7. arithmetic **8.** geometric

9. Expand and simplify $(2x - 5)^5$ using the Binomial Theorem.

10. Find the pre-image of $(-1, 4)$ under $R_{y=x} \circ Rot_{O,-90°}$.

11. A flower bed is in the shape of an obtuse triangle. One angle is 45° and the opposite side is 7 meters long. The longest side is 9 meters long. Find the measures of the remaining side and angles.

12. The points scored by the winning team in each National Football League game over a three-day period were 17, 24, 23, 30, 21, 30, 24, 20, 34, 24, 23, 30, 40, and 31. Find the mean and the standard deviation of the scores.

13. The teaching staff of Beck High School informs its members of school cancellation by telephone. The principal calls 2 teachers, each of whom in turn calls 2 other teachers, and so on. This process must be repeated 6 times counting the principal's calls as the first time. How many teachers, including the principal, work at Beck High School?

Problem Solving Application: Using Geometric Sequences

Many different real-world problems can be modeled by using geometric sequences. For example, problems involving compound interest, appreciation, depreciation, and population growth can often be solved by determining a geometric sequence that describes the situation presented in the problem.

Examples

1 **The Sopher's paid \$87,000 for their house in 1990. If the value of the house increases at an annual rate of 5%, how much will it be worth in 10 years?**

Explore The problem can be represented by a geometric sequence.

1990	1991	1992	. . .	1999	2000
a_1	a_1r	a_1r^2	. . .	a_1r^9	a_1r^{10}
\$87,000	______	______	. . .	______	______

The pattern is a 5% increase each year for 10 years. Thus, for this geometric sequence, $a_1 = 87{,}000$, $r = 105\%$ or 1.05, and $n = 11$.

If the original price is a_1, then the value after 10 years will be a_{11}

Plan Use the formula for the *n*th term of a geometric sequence.

$$a_n = a_1r^{n-1}$$

Solve

$$a_{11} = 87{,}000(1.05)^{10}$$
$$\approx 141{,}713.83$$ *Use a calculator or logarithms to evaluate.*

The house will be worth approximately \$141,714 in 10 years.

Examine Five percent of \$87,000 is \$4350. If the house's value increased by only \$4350 each year, then it would be worth \$87,000 + 10(\$4350) or \$130,500 after 10 years. This value should be less than the actual value because the value increase is compounded annually. Since \$130,500 < \$141,714, the answer seems reasonable.

2 **Bryan has \$800 in a checking account. If he spends 15% of the balance each month, after how many months will the balance be less than \$1?**

Explore The pattern is a 15% decrease each month until the account balance is less than \$1. Thus, the problem can be represented by a geometric sequence with $a_1 = \$800$, $r = 85\%$ or 0.85 and $a_n = \$1$.

If $a_n = \$1$, then for the next integer greater than n, the balance will be less than \$1.

Plan Use the formula for the nth term of a geometric sequence.

$$a_n = a_1 r^{n-1}$$

Solve

$$1 = 800(0.85)^{n-1}$$
$$0.00125 = (0.85)^{n-1}$$
$$\log 0.00125 = \log (0.85)^{n-1}$$
$$\log 0.00125 = (n - 1) \log 0.85$$
$$\frac{\log 0.00125}{\log 0.85} = n - 1$$
$$41.13 \approx n - 1$$
$$42.13 = n$$

Use logarithm tables or a calculator to evaluate.

If the account contains \$1 after about 42.13 months, then the balance will be less than \$1 after 43 months. *Examine this solution.*

Exercises

Written Solve each problem.

1. The Kellar's paid \$80,000 for their house in 1990. If the value of the house increases at an annual rate of 7%, how much will it be worth in 5 years?
2. The number of bacteria in a culture increases at a rate of 100% every three hours. If there are 300 bacteria now, how many will there be after 24 hours?
3. Restaurant equipment valued at \$175,000 depreciates at an annual rate of 20%. How much will the equipment be worth in four years?
4. Michael has \$650 in a checking account. If he spends 8% of the balance each week, after how many weeks will the balance be less than \$5?
5. Frank invests \$1200 at 7.6% interest compounded quarterly (1.9% interest every 3 months). How long will it take for his investment to triple?
6. Ellie buys a new car for \$10,200. If the value of the car depreciates at an annual rate of 18%, after how many years will it be worth \$2000?
7. A new house purchased 20 years ago for \$32,000 is worth \$185,000 today. Assuming a steady growth rate, what was the annual rate of appreciation?
8. Printing equipment purchased 8 years ago for \$225,000 is worth \$80,000 today. What was the annual rate of depreciation, assuming the rate was the same each year?
9. A tank holds 5200 gallons of water. Each day, one half of the water in the tank is removed. After how many days will the tank contain less than 10 gallons of water?
10. In a game of chance, there are 25 squares. The first square is worth \$1 and each successive square is worth twice as much as the previous square. How much is the 24th square worth?
11. Computer components valued at \$350,000 appreciated for 10% per year for 4 years and the depreciated 15% per year after that. After how many years will their value have depreciated to less than \$150,000? *Hint: Use two geometric sequences.*

Vocabulary

sequence
terms of a sequence
infinite sequence
arithmetic sequence
common difference
arithmetic mean
geometric sequence
common ratio
geometric mean
arithmetic series
geometric series
infinite series
sum of an infinite geometric series

Chapter Summary

1. The domain of any sequence is always a set of consecutive natural numbers beginning with 1.
2. The symbol S_n represents the sum of the first n terms of a sequence.
3. The sum of the first n terms of an arithmetic series is given by the following formulas.

$$S_n = \frac{n}{2}(a_1 + a_n) \text{ or } S_n = \frac{n}{2}[2a_1 + (n - 1)d]$$

4. The sum of the first n terms of a geometric series is given by

$$S_n = \frac{a_1 - a_1 r^n}{1 - r} \text{ or } S_n = \frac{a_1(1 - r^n)}{1 - r} \text{ provided } r \neq 1.$$

5. Substituting ra_n for $a_1 r^n$ in the formula for S_n gives

$$S_n = \frac{a_1 - ra_n}{1 - r}, r \neq 1.$$

6. The sum of an infinite geometric series with common ratio r where $|r| < 1$ is given by $S = \dfrac{a_1}{1 - r}$.

Chapter Review

13.1 Give the first five terms of the sequence with the given nth term.

1. 3^n **2.** $2n + 1$ **3.** $4n$ **4.** 2^{n+1} **5.** $\frac{3n}{n^2}$

For each sequence, list the next three terms. Give a rule for the nth term.

6. 2, 4, 6, 8, . . . **7.** 2, 4, 8, 16, . . . **8.** $\frac{1}{3}, \frac{1}{6}, \frac{1}{9}, \frac{1}{12}, \ldots$

13.2 Give the first five terms of an arithmetic sequence with the given first term and common difference.

9. $a_1 = 1, d = 4$ **10.** $a_1 = 1, d = -4$ **11.** $a_1 = \frac{1}{2}, d = -\frac{1}{4}$

Give the nth term of the arithmetic sequence in each exercise.

12. $a_1 = 1, d = 4$ **13.** $a_1 = 1, d = -4$ **14.** $a_1 = \frac{1}{2}, d = -\frac{1}{4}$

Find the arithmetic means of the two terms given in each exercise.

15. $a_1 = 3, a_4 = 9$ **16.** $a_1 = 9, a_3 = 25$

13.3 Give the first five terms of the geometric sequence with the given first term and common ratio.

17. $a_1 = 1, r = 3$ **18.** $a_1 = -2, r = \frac{1}{2}$ **19.** $a_1 = 128, r = -\frac{1}{4}$

Find the geometric means between the given terms of these sequences.

20. 4, ___?___, ___?___, ___?___, 2500 **21.** 4, ___?___, ___?___, $\frac{1}{128}$

Tell whether each of the following sequences is arithmetic, geometric, both, or neither.

22. $\frac{1}{8}, \frac{1}{4}, \frac{1}{2}, 1$ **23.** $1, \frac{3}{4}, \frac{1}{2}, \frac{1}{4}$

24. 1, 3, 5 **25.** $4, -4, 4$

13.4 Find S_n for each arithmetic series given these values.

26. $a_1 = 8, a_n = 48, n = 9$ **27.** $a_1 = 9, n = 10, d = 3$

28. A clock strikes the hours from 1 to 12. How many strokes does it make in 24 hours?

13.5 Find S_n for each geometric series with these values.

29. $a_1 = 4, r = 2, n = 5$ **30.** $a_1 = 48, a_n = 3, r = -\frac{1}{2}$

Tell whether each of the following series is arithmetic, geometric, both, or neither.

31. $3 + 9 + 15 + 21 + 27$ **32.** $-\frac{1}{2} + \frac{1}{4} - \ldots + (-\frac{1}{2})^n$

33. $-2 - \frac{1}{2} + \frac{1}{2} + 2 + 3\frac{1}{2}$ **34.** $-6 + 12 - 24 + 48 - 96$

Give the sum of the series in each exercise.

35. $3 + 9 + 15 + 21 + 27$ **36.** $-\frac{1}{2} + \frac{1}{4} - \ldots + (-\frac{1}{2})^n$

37. $-2 - \frac{1}{2} + \frac{1}{2} + 2 + 3\frac{1}{2}$ **38.** $-6 + 12 - 24 + 48 - 96$

39. Kent invests \$2000 in his wife's business. At the end of each year he receives 10% of the money invested during the year and immediately reinvests this amount in the business. If he continues reinvesting in this way, how much does Kent have invested at the end of the fourth year?

13.6 Tell whether or not the following geometric series has a sum. If it has a sum, find it.

40. $\frac{7}{10} + \frac{7}{100} + \frac{7}{1000} + \cdots$ **41.** $3 + 1 + \frac{1}{3} + \frac{1}{9} + \cdots$

42. $1 + \frac{4}{3} + \frac{16}{9} + \frac{64}{27} + \cdots$ **43.** $\frac{1}{5} - \frac{2}{15} + \frac{4}{45} - \frac{8}{135} + \cdots$

Use the method given in example 3 of section 13.6 to find a rational number in $\frac{a}{b}$ form for each repeating decimal.

44. $0.\overline{8}$ **45.** $3.\overline{15}$ **46.** $0.8\overline{19}$

Chapter Test

Tell whether each of the following sequences is arithmetic, geometric, both, or neither.

1. $\frac{1}{2}, \frac{1}{4}, \frac{1}{8}, \ldots$ **2.** $-\frac{1}{3}, -\frac{1}{5}, -\frac{1}{7}, \ldots$ **3.** 6, 6, 6, . . . **4.** 4, 7, 10, . . .

5. Write the first five terms of the sequence defined by $a_n = \frac{(-1)^n}{2^n}$.

6. Find the 12th term of the arithmetic sequence $-3, 2, 7, \ldots$

7. Find the sum of the first 40 terms of an arithmetic sequence with first term 5 and 40th term 278.

8. Write the first four terms of the geometric sequence with first term -125 and common ratio $\frac{1}{5}$.

9. Find the 8th term of the geometric sequence $3, -6, 12, \ldots$

10. Find the sum of the first six terms of a geometric sequence with first term 5 and common ratio 3.

11. In the arithmetic sequence $\frac{7}{6}, \frac{4}{3}, \frac{3}{2}, \ldots$, find n if $a_n = 18$.

12. Find the sum of the first 50 terms of the arithmetic sequence with first term $\frac{1}{2}$ and common difference 1.

13. How many terms are there in a geometric sequence where $a_1 = 3$, $a_n = \frac{1}{243}$, and $r = -\frac{1}{3}$?

14. Find, in simplest radical form, the positive geometric mean between 2 and 10.

15. Find the value of x for which $5, 2x + 1, 17$ form an arithmetic sequence.

16. As n increases, what number does $3 - \frac{1}{2^n}$ approach?

Tell whether or not each of the following geometric series has a sum. If so, find it.

17. $5 + 1 + \frac{1}{5} + \frac{1}{25} + \cdots$ **18.** $2 + 2 + 2 + \ldots$ **19.** $\frac{5}{2} + \frac{5}{4} + \frac{5}{8} + \cdots$

Find the missing terms in each arithmetic sequence.

20. ____, 2, ____, ____, ____, 18 **21.** ____, $\frac{1}{3}$, ____, ____, ____, -1

22. Write the first four terms of an infinite geometric series that is equivalent to $0.\overline{37}$. Then, find the sum of the series.

23. Roberta received a starting salary of \$4000 per month. Each month she received an increase of \$50. What was her monthly salary at the end of a year? How much did she receive as salary during the year?

24. Juan's starting annual salary at Photon Inc. is \$18,672. At the end of each year, he will receive a raise of 6.5%. How much will he receive in salary during his first five years with Photon Inc.?

25. Each swing of a certain pendulum is 90% as long as the preceding swing. If the end of the pendulum travels 38 cm on its first swing, how far will it travel before coming to rest?

An Introduction to Matrices

Chapter 14

Many tables and charts involve the use of rectangular arrays of numbers. Some examples are pictured below. In this chapter, you will study arrays of numbers like these more carefully. You will be introduced to some new and interesting applications of such arrays in computer science and other fields.

TAX TABLE NO. 1

Taxpayers with Ohio taxable income of $55,000 or less may

If Ohio taxable income (Line 5) is: At least	But less than	The tax is:
1,950	1,975	14.58
1,975	2,000	14.77
2,000	2,025	14.95
2,025	2,050	15.14
2,050	2,075	15.32
2,075	2,100	15.51
2,100	2,125	15.70
2,125	2,150	15.88
2,150	2,175	16.07

NUMBER OF NIGHTS	ONE NIGHT	TWO NIGHTS	WEEKEND
ZONE II	$189	$249	$319
ZONE III	$209	$279	$349
ZONE IV	$219	$289	$369
ZONE V	$279	$349	$449

Use if your taxable income is less than $50,000. If $50,000 or more, use the Tax Rate Schedules.

Example: Mr. and Mrs. Brown are filing a joint return. Their taxable income on line 37 of Form 1040 is $25,300. First, they find the $25,300–25,350 income line. Next, they find the column for married filing jointly and read down the column. The amount shown where the income line and filing status column meet is $3,799. This is the tax amount they must write on line 38 of their return.

At least	But less than	Single	Married filing jointly	Married filing separately
25,200	25,250	4,652	3,784	5,051
25,250	25,300	4,666	3,791	5,065
25,300	25,350	4,680	3,799	5,079
25,350	25,400	4,694	3,806	5,093

If line 37 (taxable income) is— At least	But less than	And you are— Single	Married filing jointly	Married filing separately	Head of a household	If line 37 (taxable income) is— At least	But less than	And you are— Single	Married filing jointly	Married filing separately	Head of a household	If line 37 (taxable income) is— At least	But less than	And you are— Single	Married filing jointly	Married filing separately
$0	$5	$0	$0	$0	$0	1,400	1,425	212	212	212	212	2,700	2,725	407	407	407
5	15	2	2	2	2	1,425	1,450	216	216	216	216	2,725	2,750	411	411	411
15	25	3	3	3	3	1,450	1,475	219	219	219	219	2,750	2,775	414	414	414
25	50	6	6	6	6	1,475	1,500	223	223	223	223	2,775	2,800	418	418	418
50	75	9	9	9	9	1,500	1,525	227	227	227	227	2,800	2,825	422	422	422
75	100	13	13	13	13	1,525	1,550	231	231	231	231	2,825	2,850	426	426	426
100	125	17	17	17	17	1,550	1,575	234	234	234	234	2,850	2,875	429	429	429
125	150	21	21	21	21	1,575	1,600	238	238	238	238	2,875	2,900	433	433	433
150	175	24	24	24	24	1,600	1,625	242	242	242	242	2,900	2,925	437	437	437
175	200	28	28	28	28	1,625	1,650	246	246	246	246	2,925	2,950	441	441	441
200	225	32	32	32	32	1,650	1,675	249	249	249	249	2,950	2,975	444	444	444
225	250	36	36	36	36	1,675	1,700	253	253	253	253	2,975	3,000	448	448	448
250	275	39	39	39	39											
275	300	43	43	43	43	1,700	1,725	257	257	257	257	3,000				

14.1 Order, Addition

The tables and charts on the preceding page are examples of rectangular arrays of numbers, called *matrices*. The definition of a matrix is given below.

Definition of Matrix

> A **matrix** is a rectangular array of real numbers. A matrix has **order** $m \times n$ if it has m rows and n columns. A **square matrix** of order n has n rows and n columns.

Some samples of matrices are shown below.

$$\begin{bmatrix} 2 & 0 & 3 \\ -5 & 1 & 2 \end{bmatrix} \quad \begin{bmatrix} 3 & 1 \\ 6 & -1 \\ 2 & 2 \end{bmatrix} \quad \begin{bmatrix} 6 \\ \sqrt{2} \\ 1 \\ 2 \end{bmatrix} \quad \begin{bmatrix} 1 & 2 & 0 \\ 0 & 1 & 1 \\ 1 & -1 & -2 \end{bmatrix}$$

The first matrix shown above has order 2×3, the second has order 3×2, and the third has order 4×1. The fourth is a square matrix of order 3.

Capital letters are usually used to name matrices. Often, a matrix A of order 3×4 is denoted by $A_{3\times 4}$. If the matrix is square, only one number is used. For example, B_n names a square matrix B of order n.

We can now define an addition operation on matrices. To add two matrices, add their corresponding elements. Only matrices whose order is the *same* can be added. The sum of two matrices of different orders is *not defined*.

Examples

1 **Find the sum of** $\begin{bmatrix} \mathbf{3} & \mathbf{0} \\ \mathbf{2} & \mathbf{4} \\ \mathbf{1} & \mathbf{-7} \end{bmatrix}$ **and** $\begin{bmatrix} \mathbf{4} & \mathbf{0} \\ \mathbf{-3} & \mathbf{-3} \\ \mathbf{2} & \mathbf{7} \end{bmatrix}$.

Both matrices have order 3×2. Thus, they can be added.

$$\begin{bmatrix} 3 & 0 \\ 2 & 4 \\ 2 & -7 \end{bmatrix} + \begin{bmatrix} 4 & 0 \\ -3 & -3 \\ 2 & 7 \end{bmatrix} = \begin{bmatrix} 3+4 & 0+0 \\ 2+(-3) & 4+(-3) \\ 2+2 & -7+7 \end{bmatrix}$$

Add corresponding elements.

$$= \begin{bmatrix} 7 & 0 \\ -1 & 1 \\ 4 & 0 \end{bmatrix}$$

2 **Find the sum of** $\begin{bmatrix} 2 & 0 \\ 3 & -1 \end{bmatrix}$ **and** $\begin{bmatrix} -2 & 1 & 6 \\ 0 & 0 & 1 \end{bmatrix}$.

The matrices have different orders. Thus, there is no sum.

3 **Find** $A + B$ **if** $A = \begin{bmatrix} 2 & 0 & \frac{-6}{\sqrt{2}} \\ 3 & \pi & \end{bmatrix}$ **and** $B = \begin{bmatrix} -2 & 0 & \frac{6}{-\sqrt{2}} \\ -3 & -\pi & \end{bmatrix}$.

$$A + B = \begin{bmatrix} 2 & 0 & \frac{-6}{\sqrt{2}} \\ 3 & \pi & \end{bmatrix} + \begin{bmatrix} -2 & 0 & \frac{6}{-\sqrt{2}} \\ -3 & -\pi & \end{bmatrix}$$ *Add corresponding elements.*

$$= \begin{bmatrix} 0 & 0 & 0 \\ 0 & 0 & 0 \end{bmatrix}$$

Notice that in example 3, all the entries in the sum are zero. Such a matrix is called a **zero matrix.**

Definition of Zero Matrix

A zero matrix is a matrix, all of whose entries are zero. The symbol for a zero matrix is $\bar{0}$.

$\bar{0}_{2\times 3}$ denotes a zero matrix of order 2×3. $\bar{0}_n$ denotes a square matrix of order n.

When the symbol A is used to stand for a matrix, the corresponding lower case letter, a, is used to denote some element of the matrix. A "typical" element might be one in the ith row, jth column of the matrix and this element would be denoted by $a_{i,j}$. When there is no possibility of error, the comma can be omitted: a_{ij}. Another symbol for A would then be $[a_{ij}]$. Thus, A and $[a_{ij}]$ name *matrices,* but a_{ij} names a *real number.*

Using this notation, we can define addition of matrices more formally.

Addition of Matrices

If $A = [a_{ij}]$ and $B = [b_{ij}]$, then $A + B = [a_{ij} + b_{ij}]$.

That is, the element in the ith row, jth column of the sum is the sum of the elements in the ith row, jth column of each of the matrices being added.

We can also define the additive inverse of a matrix and a subtraction operation on matrices.

Additive Inverse of a Matrix

For the $m \times n$ matrix A, the additive inverse of A, written $-A$, is the $m \times n$ matrix whose entries are the additive inverses of the corresponding elements in A.

Subtraction of Matrices

The difference of $m \times n$ matrices A and B, written $A - B$, is $A + (-B)$. This is a definition of subtraction of matrices.

Example

4 Find $K - L$ if $K = \begin{bmatrix} 0 & 1 & -1 \\ -5 & 2 & 3 \end{bmatrix}$ and $L = \begin{bmatrix} 3 & 3 & 0 \\ 2 & -4 & 1 \end{bmatrix}$.

$$K - L = K + (-L)$$
$$= \begin{bmatrix} 0 & 1 & -1 \\ -5 & 2 & 3 \end{bmatrix} + \begin{bmatrix} -3 & -3 & 0 \\ -2 & 4 & -1 \end{bmatrix} \qquad -L = \begin{bmatrix} -3 & -3 & 0 \\ -2 & 4 & -1 \end{bmatrix}$$
$$= \begin{bmatrix} -3 & -2 & -1 \\ -7 & 6 & 2 \end{bmatrix}$$

Exercises

Exploratory **Explain the meaning of the following terms.**

1. 4×7 matrix
2. $A_{m \times n}$
3. B_n
4. $\bar{0}$
5. $\bar{0}_{2\times 3}$
6. a_{ij}
7. $[a_{ij}]$
8. $[b_{ij}]$
9. $-A$
10. $[a_{ij} + b_{ij}]$
11. ith row, jth column
12. subtraction of matrices

13. How many elements has a square matrix of order 7? Order 17? Order n?

In each of the following matrices, name a_{21} if it exists. Then name the matrix's additive inverse.

14. $\begin{bmatrix} 1 & 0 \\ 0 & 2 \end{bmatrix}$
15. $[1 \quad 2\sqrt{2} \quad \sqrt{2} \quad 0]$
16. $[9 \quad 0 \quad \sqrt{3}]$
17. $\begin{bmatrix} 0 \\ 1 \end{bmatrix}$

Name each of the following for matrix A.

$$A = \begin{bmatrix} 6 & 2 & 3 & -1 & -1 & 0 \\ 4 & 2 & 1 & 7 & 6 & 2 \\ 0 & 1 & 0 & -5 & -3 & 1 \\ 2 & 5 & -1 & 0 & 0 & -1 \end{bmatrix}$$

18. a_{11}
19. a_{12}
20. a_{13}
21. a_{24}
22. a_{32}
23. a_{46}
24. Can matrix A be added to $B_{6\times 4}$? Why or why not?

Written **Use the matrices below to find each additive inverse, sum, or difference, if it exists.**

$$A = \begin{bmatrix} 2 & 0 \\ -1 & 2 \end{bmatrix} \quad C = \begin{bmatrix} \sqrt{2} & -1 \\ 2 & 3 \end{bmatrix} \quad E = \begin{bmatrix} 2 & 0 \\ 1 & 2 \\ -3 & 0 \end{bmatrix} \quad F = \begin{bmatrix} -3 & 1 \\ 1 & 1 \\ 3 & 2 \end{bmatrix}$$

$$B = \begin{bmatrix} \sqrt{2} & 1 \\ 1 & -2 \end{bmatrix} \quad D = \begin{bmatrix} 2 & 0 & 1 \\ 3 & -1 & -1 \end{bmatrix}$$

1. $-A$
2. $-B$
3. $-C$
4. $-D$
5. $-E$
6. $-F$
7. $A + B$
8. $A + C$
9. $B + C$
10. $A + (-B)$
11. $A - B$
12. $B - C$
13. $A + D$
14. $B + E$
15. $D + E$
16. $E + F$
17. $E - F$
18. $-C + B$
19. $-E + F$
20. $B - E$
21. $C + \bar{0}_2$
22. $D - \bar{0}_{2\times 3}$
23. $E + \bar{0}_{2\times 3}$
24. $-F + \bar{0}_{3\times 2}$

14.2 Multiplication by a Scalar

In this section a special operation will be defined. This operation involves multiplying a matrix by a real number. Notice that we are *not* multiplying a matrix by another matrix.

Definition of Scalar Multiplication

The product of a real number k and a matrix A is the matrix kA, each of whose elements is the product of k and the corresponding element of A. That is, the product of the real number k and $A = [a_{ij}]$ is $kA = [ka_{ij}]$. This operation is called **scalar multiplication.** The real number k is called the **scalar.**

Some examples of scalar multiplication follow.

Examples

1 **Find** $2\begin{bmatrix} 3 & 0 & 1 \\ -1 & \sqrt{2} & \frac{1}{2} \end{bmatrix}$.

$$2\begin{bmatrix} 3 & 0 & 1 \\ -1 & \sqrt{2} & \frac{1}{2} \end{bmatrix} = \begin{bmatrix} 2(3) & 2(0) & 2(1) \\ 2(-1) & 2(\sqrt{2}) & 2(\frac{1}{2}) \end{bmatrix}$$

Multiply the scalar, 2, by each element of the matrix.

$$= \begin{bmatrix} 6 & 0 & 2 \\ -2 & 2\sqrt{2} & 1 \end{bmatrix}$$

2 **Find** $0\begin{bmatrix} 4 & 0 \\ \pi & 2(6) \end{bmatrix}$.

$$0\begin{bmatrix} 4 & 0 \\ \pi & 2(6) \end{bmatrix} = \begin{bmatrix} 0(4) & 0(0) \\ 0(\pi) & 0(2(6)) \end{bmatrix}$$

The scalar is 0.

$$= \begin{bmatrix} 0 & 0 \\ 0 & 0 \end{bmatrix} \text{ or } \bar{0}_{2\times 2}$$

3 **Find mB for $m = -2$ and $B = \begin{bmatrix} 2 \\ 0 \\ -6 \end{bmatrix}$.**

$$mB = -2\begin{bmatrix} 2 \\ 0 \\ -6 \end{bmatrix}$$

The scalar is -2.

$$= \begin{bmatrix} -4 \\ 0 \\ 12 \end{bmatrix}$$

4 **Show that for real numbers k and ℓ, $(k + \ell)A = kA + \ell A$.**

We will use the notation $A = [a_{ij}]$.

$(k + \ell)A = [(k + \ell)a_{ij}]$	*Definition of scalar multiplication*
$= [ka_{ij} + \ell a_{ij}]$	*The Distributive Law*
$= [ka_{ij}] + [\ell a_{ij}]$	*Definition of matrix addition*
$= k[a_{ij}] + \ell[a_{ij}]$	*Definition of scalar multiplication*
$= kA + \ell A$	*Notation*

5 **Solve the matrix equation $X + 3\begin{bmatrix} 1 & 2 & 4 \\ -1 & 0 & 1 \end{bmatrix} = \begin{bmatrix} 6 & 0 & 2 \\ 3 & 3 & 1 \end{bmatrix}$.**

To solve the equation, use the subtraction property for equations.

$$X + 3\begin{bmatrix} 1 & 2 & 4 \\ -1 & 0 & 1 \end{bmatrix} = \begin{bmatrix} 6 & 0 & 2 \\ 3 & 3 & 1 \end{bmatrix}$$

$$X + \begin{bmatrix} 3 & 6 & 12 \\ -3 & 0 & 3 \end{bmatrix} = \begin{bmatrix} 6 & 0 & 2 \\ 3 & 3 & 1 \end{bmatrix}$$ *Definition of scalar multiplication.*

$$X = \begin{bmatrix} 6 & 0 & 2 \\ 3 & 3 & 1 \end{bmatrix} - \begin{bmatrix} 3 & 6 & 12 \\ -3 & 0 & 3 \end{bmatrix}$$ *To subtract, add the additive inverse of the second matrix.*

$$X = \begin{bmatrix} 6 & 0 & 2 \\ 3 & 3 & 1 \end{bmatrix} + \begin{bmatrix} -3 & -6 & -12 \\ 3 & 0 & -3 \end{bmatrix}$$

$$X = \begin{bmatrix} 3 & -6 & -10 \\ 6 & 3 & -2 \end{bmatrix}$$

Exercises

Exploratory **Answer each of the following.**

1. Explain how to multiply a matrix A by a real number k.
2. Give an argument to show that $0A = \overline{0}$ for any matrix A.
3. Discuss the difference between scalar multiplication and regular multiplication.

Find each product.

4. $2\begin{bmatrix} 1 & 3 & -1 \\ 0 & 8 & \frac{1}{2} \end{bmatrix}$

5. $4\begin{bmatrix} 1 \\ 2 \\ 6 \end{bmatrix}$

6. $0\begin{bmatrix} 2 & \sqrt{2} \\ 0 & -1 \end{bmatrix}$

7. $2\begin{bmatrix} \sqrt{2} & -\sqrt{2} \\ 1 & 0 \end{bmatrix}$

8. $-3\begin{bmatrix} -2 & 6 \\ 4 & 7 \\ 1 & -5 \end{bmatrix}$

9. $-2\begin{bmatrix} 3 & -\frac{1}{2} & -1 \\ -2 & 6 & 0 \end{bmatrix}$

10. $-4\begin{bmatrix} \sqrt{5} \\ -3 \\ \frac{3}{4} \end{bmatrix}$

11. $-1\begin{bmatrix} 4 & -\frac{2}{3} \\ 0 & -8 \end{bmatrix}$

Written **Use the matrices and scalars below to perform the indicated operations, if possible. If not possible, explain why.**

$$A = \begin{bmatrix} 2 & -1 \\ 0 & 3 \end{bmatrix} \quad B = \begin{bmatrix} -3 & 1 \\ 4 & 1 \end{bmatrix} \quad C = \begin{bmatrix} 2 & 0 & -1 \\ 1 & 1 & 6 \end{bmatrix} \quad D = \begin{bmatrix} 0 & -1 & 1 \\ 1 & 2 & 1 \end{bmatrix}$$

$$k = 3 \qquad \ell = -2 \qquad m = \tfrac{1}{2}$$

1. kA
2. ℓA
3. $(k + \ell)A$
4. mC
5. mD
6. $(2\ell)B$
7. $(km)A$
8. $(\ell m)D$
9. $A + kB$
10. $A + kC$
11. $C + \ell D$
12. $C - \ell D$
13. $(\ell + m)B$
14. $6A + kB$
15. $k(A + B)$
16. $B - (8m)C$
17. $B - (8m)A$
18. $\sqrt{3}A + B$
19. $\sqrt{2}C + \sqrt{2}D$
20. $(k\ell)A + mB$
21. Check the solution to example 5.

Use matrices *A, B, C,* and *D* given above. Solve each equation for *X*.

22. $A + X = B$
23. $A + 2X = B$
24. $3A + X = B$
25. $-A + 2X = B$
26. $-3A + 3X = B$
27. $X - 3B = 2A$
28. $C - 3X = D$
29. $\frac{1}{2}C - X = D$
30. $\frac{1}{3}C - \frac{1}{4}X = \frac{2}{5}D$

Check the solutions for each of the following exercises.

31. exercise 22
32. exercise 23
33. exercise 24
34. exercise 25
35. exercise 26
36. exercise 27
37. exercise 28
38. exercise 29
39. exercise 30

Prove each of the following statements. Use the proof in example 4 as a guide. In each case, *A* and *B* are matrices, and *k* and *ℓ* are real numbers. Assume all matrices named are of appropriate order.

40. $(k\ell)A = k(\ell A)$
41. $k(A + B) = kA + kB$
42. $0A = \overline{0}$

Challenge

1. Let $M_{2\times3}$ = {all 2 × 3 matrices} and "+" be matrix addition. Show that ($M_{2\times3}$, +) is a group.
2. Is ($M_{2\times5}$, +) a subgroup of ($M_{3\times8}$, +)? Explain.
3. Let M_a = {all 2 × 2 matrices of the form $\begin{bmatrix} a & 0 \\ 0 & a \end{bmatrix}$, $a \in \mathcal{R}$}. Is (M_a, +) a group? Justify your answer.

14.3 Multiplying One Matrix by Another

In this section you are going to learn how to multiply two matrices together. The procedure is not the "obvious" one. However, it is a procedure that is adaptable to many uses in the sciences and in the business world.

To begin, consider the records of the Pine City Laser-Vision Company. The company produces three models of laser-vision: models A, B, and C. The number of tubes and speakers required for each of these models is shown in the parts-per-model matrix at the right. Also, the number of laser-visions of each type produced in January and February of 1990 is given by the models-per-month matrix shown at the right.

Parts per Model

Number of	Model A	Model B	Model C
Tubes	13	18	20
Speakers	2	3	4

Models per Month

	Jan.	Feb.
Model A	12	6
Model B	24	12
Model C	12	9

The manager of the company needed an answer to the following question: "How many tubes and speakers were required in all during January and February, 1990?

To answer this question, the computations shown below were completed.

	Model A		Model B		Model C	
Number of Tubes in January	13(12)	+	18(24)	+	20(12)	= 828
Number of Tubes in February	13(6)	+	18(12)	+	20(9)	= 474
Number of Speakers in January	2(12)	+	3(24)	+	4(12)	= 144
Number of Speakers in February	2(6)	+	3(12)	+	4(9)	= 84

The results are summarized in the parts-per-month matrix shown below.

	Jan.	Feb.
Number of Tubes	828	474
Number of Speakers	144	84

Note that while the order of the parts-per-model matrix is 2×3 and that of the models-per-month matrix is 3×2, the order of the parts-per-month matrix is 2×2.

The procedure used to obtain the parts-per-month matrix in this situation is an example of matrix multiplication.

Multiplication of Matrices

> Let $A_{m \times p}$ and $B_{p \times n}$ be matrices of order $m \times p$ and $p \times n$ respectively. Then the product AB of matrices A and B is the matrix C of order $m \times n$ of which the entry in the ith row, kth column is the sum of the products formed by multiplying entries of the ith row of A by corresponding elements in the kth column of B.

An illustration of how to find the element c_{34}, the element in the third row, fourth column, of the product C of matrices A and B is given below.

$$\text{3rd row} \rightarrow [a_{31} \quad a_{32} \quad \ldots \quad a_{3p}] \times \underbrace{\begin{bmatrix} b_{14} \\ b_{24} \\ \cdot \\ \cdot \\ \cdot \\ b_{p4} \end{bmatrix}}_{\text{4th column}} = c_{34}$$

$$c_{34} = a_{31}b_{14} + a_{32}b_{24} + a_{33}b_{34} + \ldots + a_{3p}b_{p4}$$

Examples

Given $A = \begin{bmatrix} 2 & 0 & 6 \\ -1 & -2 & 0 \end{bmatrix}$ **and** $B = \begin{bmatrix} 3 & 5 \\ 1 & 0 \\ 2 & -6 \end{bmatrix}$. **Find** AB. **Then find** BA.

1 $AB = \begin{bmatrix} 2 & 0 & 6 \\ -1 & -2 & 0 \end{bmatrix} \begin{bmatrix} 3 & 5 \\ 1 & 0 \\ 2 & -6 \end{bmatrix}$.

$$= \begin{bmatrix} 2(3) + 0(1) + 6(2) & 2(5) + 0(0) + 6(-6) \\ -1(3) + (-2)(1) + 0(2) & -1(5) + (-2)(0) + 0(-6) \end{bmatrix}$$

$$= \begin{bmatrix} 18 & -26 \\ -5 & -5 \end{bmatrix}$$

2 $BA = \begin{bmatrix} 3 & 5 \\ 1 & 0 \\ 2 & -6 \end{bmatrix} \begin{bmatrix} 2 & 0 & 6 \\ -1 & -2 & 0 \end{bmatrix}$.

$$= \begin{bmatrix} 3(2) + 5(-1) & 3(0) + 5(-2) & 3(6) + 5(0) \\ 1(2) + 0(-1) & 1(0) + 0(-2) & 1(6) + 0(0) \\ 2(2) + (-6)(-1) & 2(0) + (-6)(-2) & 2(6) + (-6)(0) \end{bmatrix}$$

$$= \begin{bmatrix} 1 & -10 & 18 \\ 2 & 0 & 6 \\ 10 & 12 & 12 \end{bmatrix}$$

3 **Find the product of** $\begin{bmatrix} 1 & 0 & 6 \\ \sqrt{2} & 5 & 1 \\ 2 & -5 & -5 \end{bmatrix}$ **and** $\begin{bmatrix} 1 & 0 & -6 \\ 2 & 3 & 4 \\ -1 & 7 & 0 \\ 0 & 0 & 3 \end{bmatrix}$.

There are three columns in the first matrix and four rows in the second matrix. Thus, the product does not exist.

4 **Given** $A = \begin{bmatrix} -1 & 2 \\ 3 & -4 \end{bmatrix}$ **and** $D = \begin{bmatrix} x \\ y \end{bmatrix}$. **Find** AD.

$$\begin{aligned} AD &= \begin{bmatrix} -1 & 2 \\ 3 & -4 \end{bmatrix}\begin{bmatrix} x \\ y \end{bmatrix} \\ &= \begin{bmatrix} -1(x) + 2(y) \\ 3(x) + (-4)(y) \end{bmatrix} \\ &= \begin{bmatrix} -x + 2y \\ 3x - 4y \end{bmatrix} \end{aligned}$$

Exercises

Exploratory **Answer each of the following.**

1. What is the difference between multiplying a matrix by a scalar and multiplying it by another matrix?
2. Under what circumstances is the product AB of the matrices A and B defined? When is it not defined?

State whether each of the following products exists. If it exists, find the product.

3. $\begin{bmatrix} 1 & 0 \end{bmatrix}\begin{bmatrix} 2 & 3 \\ 4 & -1 \end{bmatrix}$
4. $\begin{bmatrix} 1 \\ 0 \end{bmatrix}\begin{bmatrix} 2 & 3 \\ 4 & -1 \end{bmatrix}$
5. $\begin{bmatrix} 1 \\ 0 \end{bmatrix}\begin{bmatrix} 4 \\ 5 \end{bmatrix}$
6. $\begin{bmatrix} 1 \\ 0 \end{bmatrix}\begin{bmatrix} 2 & 3 \end{bmatrix}$
7. $\begin{bmatrix} 1 & 2 & 0 \end{bmatrix}\begin{bmatrix} 3 \\ 0 \\ 5 \end{bmatrix}$
8. $\begin{bmatrix} 2 & 3 \\ 4 & -1 \end{bmatrix}\begin{bmatrix} 1 & 0 \end{bmatrix}$
9. $\begin{bmatrix} 2 & 3 \\ 1 & 6 \end{bmatrix}\begin{bmatrix} 3 & 0 & 1 \\ 2 & 6 & 3 \\ -1 & 2 & 1 \end{bmatrix}$

Written

1. Compute AB and BA for $A = \begin{bmatrix} 1 & 2 \\ 4 & -3 \end{bmatrix}$ and $B = \begin{bmatrix} 1 & 0 \\ 2 & 1 \end{bmatrix}$. What can you conclude?
2. Is matrix multiplication commutative? Justify your answer.

For each of the indicated products, either give the order of the product or state that the product is not defined. Use $A_{4\times5}$, $B_{4\times9}$, $C_{1\times4}$, $D_{5\times6}$, and $E_{5\times5}$.

3. AB
4. BA
5. BC
6. CB
7. ED
8. EA
9. For matrix D and the real number scalar 3, what is the order of $3D$?

Use the matrices below to find each product. In the case of radicals, express your answers in simplest form. Define A^2 as $A \cdot A$.

$$A = \begin{bmatrix} 1 & 2 \\ 3 & 5 \end{bmatrix} \quad B = \begin{bmatrix} 2 & 0 \\ 3 & 1 \end{bmatrix} \quad C = \begin{bmatrix} 5 & 1 \\ 1 & 1 \end{bmatrix} \quad D = \begin{bmatrix} 1 & -2 \\ 3 & 4 \end{bmatrix} \quad E = \begin{bmatrix} -3 & -1 \\ 2 & -6 \end{bmatrix}$$

$$F = \begin{bmatrix} \sqrt{2} & 2 \\ 4 & 1 \end{bmatrix} \quad G = \begin{bmatrix} \sqrt{6} & 3 \\ 4 & \frac{1}{2} \end{bmatrix} \quad H = \begin{bmatrix} \frac{1}{3} & \frac{2}{5} \\ \frac{1}{4} & -1 \end{bmatrix} \quad K = \begin{bmatrix} \sqrt{10} & 3\sqrt{2} \\ -1 & -1 \end{bmatrix}$$

10. AB
11. BA
12. AC
13. CA
14. AD
15. DA
16. BD
17. DB
18. BC
19. CB
20. CD
21. DC
22. A^2
23. B^2
24. C^2
25. D^2
26. $(AB)C$
27. $A(BC)$
28. $(AB)D$
29. $A(BD)$
30. $(DE)C$
31. $D(EC)$
32. B^2E
33. $B(BE)$
34. A^2B^2
35. $(AB)^2$
36. $(DE)^2$
37. D^2E^2
38. AF
39. BF
40. CF
41. FG
42. GF
43. FK
44. KF
45. GK
46. KG
47. $(FG)K$
48. AH
49. BH
50. H^2
51. DH
52. $(HD)C$
53. Consider exercises 26–31. Do you think matrix multiplication is associative? Justify your answer.
54. Does $(AB)^2 = A^2B^2$ in matrix arithmetic? Explain. Give an example.
55. Does $A^2 - B^2 = (A + B)(A - B)$ in matrix arithmetic? Explain. Give an example.
56. Expand $(A + B)^2$ for matrices.
57. Investigate the existence of Distributive Laws for matrices. Does $A(B + C) = AB + AC$? What about $(B + C)A$?

Use the matrices below to find each product, if possible.

$$A = \begin{bmatrix} 2 & 0 & 3 \\ 1 & 1 & 2 \\ 2 & 1 & -1 \end{bmatrix} \quad B = \begin{bmatrix} 3 & 1 & 0 \\ 0 & 0 & 2 \\ 1 & -1 & -1 \end{bmatrix} \quad C = \begin{bmatrix} 0 & 0 & 1 \\ 2 & 1 & 0 \\ 1 & 0 & 1 \end{bmatrix} \quad D = \begin{bmatrix} 1 & 0 & 6 \end{bmatrix} \quad E = \begin{bmatrix} 2 \\ 1 \\ 1 \end{bmatrix}$$

58. AB
59. BA
60. CA
61. AC
62. BC
63. CB
64. AE
65. EA
66. BE
67. CE
68. EC
69. DE
70. ED
71. EB
72. B^2
73. A^2B
74. D^2
75. $A(B + C)$

14.4 Multiplicative Identity and Inverses

You have learned that the zero matrix, $\bar{0}$, acts as an additive identity. That is, for any matrix A, $A + \bar{0} = \bar{0} + A = A$. For example,

$$\begin{bmatrix} 1 & 2 \\ 2 & -3 \end{bmatrix} + \begin{bmatrix} 0 & 0 \\ 0 & 0 \end{bmatrix} = \begin{bmatrix} 1 & 2 \\ 2 & -3 \end{bmatrix}.$$

Furthermore, every matrix has an additive inverse, $-A$. Thus, $A + (-A) = \bar{0}$, and, for example,

$$\begin{bmatrix} 1 & 2 \\ 2 & -3 \end{bmatrix} + \begin{bmatrix} -1 & -2 \\ -2 & 3 \end{bmatrix} = \begin{bmatrix} 0 & 0 \\ 0 & 0 \end{bmatrix}.$$

In this section, we will investigate the possibility of a multiplicative identity and multiplicative inverses. To simplify computations, we will limit our discussion to 2×2 matrices.

Do you think there is a multiplicative identity in the set M_2 of 2×2 matrices? That is, do you believe that there exists a unique 2×2 matrix I such that $AI = IA$ for all $A \in M_2$?

Some might suggest $\begin{bmatrix} 1 & 1 \\ 1 & 1 \end{bmatrix}$ as a multiplicative identity. However, $\begin{bmatrix} 1 & 2 \\ 3 & 4 \end{bmatrix}\begin{bmatrix} 1 & 1 \\ 1 & 1 \end{bmatrix} = \begin{bmatrix} 3 & 3 \\ 7 & 7 \end{bmatrix}$ so that, if there is an identity for multiplication, $\begin{bmatrix} 1 & 1 \\ 1 & 1 \end{bmatrix}$ is not it! So, if there is an identity $I = \begin{bmatrix} p & q \\ r & s \end{bmatrix}$, then it must be the case that for $A = \begin{bmatrix} 1 & 2 \\ 3 & 4 \end{bmatrix}$,

$$\begin{bmatrix} 1 & 2 \\ 3 & 4 \end{bmatrix}\begin{bmatrix} p & q \\ r & s \end{bmatrix} = \begin{bmatrix} 1 & 2 \\ 3 & 4 \end{bmatrix} \text{ or } \begin{array}{ll} p + 2r = 1 & q + 2s = 2 \\ 3p + 4r = 3 & 3q + 4s = 4 \end{array}.$$

These systems of equations can be solved for the values p, q, r, and s as shown below.

$$\begin{cases} p + 2r = 1 \\ 3p + 4r = 3 \end{cases} \qquad \begin{cases} q + 2s = 2 \\ 3q + 4s = 4 \end{cases}$$

$$\begin{cases} 2p + 4r = 2 \\ 3p + 4r = 3 \end{cases} \qquad \begin{cases} 2q + 4s = 4 \\ 3q + 4s = 4 \end{cases}$$

$$-p = -1 \qquad -q = 0$$

$$p = 1 \qquad q = 0$$

Substitute 1 for p in p + 2r = 1 *Substitute 0 for q in q + 2s = 2*

$$1 + 2r = 1 \qquad 0 + 2s = 2$$

$$r = 0 \qquad s = 1$$

Thus, $p = 1$, $q = 0$, $r = 0$, and $s = 1$ or $I = \begin{bmatrix} 1 & 0 \\ 0 & 1 \end{bmatrix}$.

Thus, $\begin{bmatrix} 1 & 0 \\ 0 & 1 \end{bmatrix}$ is the candidate for the multiplicative identity in M_2. Additional investigation will show that this is indeed correct.

Now, do all the elements in M_2 have multiplicative inverses? That is, is it the case that for all $A \in M_2$, there exists an element B such that $AB = BA = I$?

The identity, $\begin{bmatrix} 1 & 0 \\ 0 & 1 \end{bmatrix}$ has an inverse, since $\begin{bmatrix} 1 & 0 \\ 0 & 1 \end{bmatrix}\begin{bmatrix} 1 & 0 \\ 0 & 1 \end{bmatrix} = \begin{bmatrix} 1 & 0 \\ 0 & 1 \end{bmatrix}$.

Thus, the inverse of $\begin{bmatrix} 1 & 0 \\ 0 & 1 \end{bmatrix}$ is itself.

If a matrix A has an inverse we will call it A^{-1}. However, if $A = \begin{bmatrix} 0 & 0 \\ 0 & 0 \end{bmatrix}$ then A^{-1} does not exist because, to find it, we would have to solve the following matrix equation.

$$\begin{bmatrix} 0 & 0 \\ 0 & 0 \end{bmatrix}\begin{bmatrix} p & q \\ r & s \end{bmatrix} = \begin{bmatrix} 1 & 0 \\ 0 & 1 \end{bmatrix}$$

This would require that $0p + 0r = 1$, which is clearly impossible.

To find out when a matrix in M_2 will have an inverse, choose a general matrix $A = \begin{bmatrix} a & b \\ c & d \end{bmatrix}$ and call its inverse, if it exists, $\begin{bmatrix} p & q \\ r & s \end{bmatrix}$. To find p, q, r, and s, solve the following matrix equation.

$$\begin{bmatrix} a & b \\ c & d \end{bmatrix}\begin{bmatrix} p & q \\ r & s \end{bmatrix} = \begin{bmatrix} 1 & 0 \\ 0 & 1 \end{bmatrix}$$

This is the same as solving the systems of equations below.

$$\begin{aligned} ap + br &= 1 \\ cp + dr &= 0 \end{aligned} \qquad \begin{aligned} aq + bs &= 0 \\ cq + ds &= 1 \end{aligned}$$

We will solve the first two equations for p and for r. You will be asked to solve the other equations for q and for s in the Written Exercises.

For p

$$\begin{cases} ap + br = 1 \\ cp + dr = 0 \end{cases}$$

$adp + dbr = d$	*Multiply by d.*
$bcp + dbr = 0$	*Multiply by b.*
$adp - bcp = d$	*Subtract.*
$p(ad - bc) = d$	*Factor.*
$p = \dfrac{d}{ad - bc}$	*Divide by $ad - bc$.*

For r

$$\begin{cases} ap + br = 1 \\ cp + dr = 0 \end{cases}$$

$acp + bcr = c$	*Multiply by c.*
$acp + adr = 0$	*Multiply by a.*
$bcr - adr = c$	*Subtract.*
$(bc - ad)r = c$	*Factor.*
$r = \dfrac{c}{bc - ad}$	*Divide by $bc - ad$.*
$r = \dfrac{-c}{ad - bc}$	*Multiply by $\dfrac{-1}{-1}$.*

Multiplicative Inverse of a 2 × 2 Matrix

For $A = \begin{bmatrix} a & b \\ c & d \end{bmatrix}$, the inverse, A^{-1}, is $\begin{bmatrix} \frac{d}{ad - bc} & \frac{-b}{ad - bc} \\ \frac{-c}{ad - bc} & \frac{a}{ad - bc} \end{bmatrix}$.

Since we cannot have a zero in the denominator, the matrix A in the definition above has an inverse if and only if $ad - bc$ is *not* zero. In other words, $ad \neq bc$.

Examples

1 **Given $A = \begin{bmatrix} 2 & 1 \\ 3 & 5 \end{bmatrix}$. Find A^{-1}, if it exists.**

Since $ad - bc = 2(5) - 1(3) = 10 - 3 = 7$, $A^{-1} = \begin{bmatrix} \frac{5}{7} & -\frac{1}{7} \\ -\frac{3}{7} & \frac{2}{7} \end{bmatrix}$.

To check this result, multiply A by A^{-1}. The product should be $\begin{bmatrix} 1 & 0 \\ 0 & 1 \end{bmatrix}$.

2 **Given $D = \begin{bmatrix} 4 & 10 \\ -2 & -5 \end{bmatrix}$. Find D^{-1}, if it exists.**

Since $ad - bc = 4(-5) - 10(-2) = -20 + 20 = 0$, D^{-1} does not exist.

Determinant of a 2 × 2 Matrix

For $A = \begin{bmatrix} a & b \\ c & d \end{bmatrix}$, the determinant of A, symbolized det A, or $|A|$, is the real number $ad - bc$.

Thus, the multiplicative inverse of $A = \begin{bmatrix} a & b \\ c & d \end{bmatrix}$ can be written

$$A^{-1} = \frac{1}{\det A}\begin{bmatrix} d & -b \\ -c & a \end{bmatrix}.$$

Example

3 **Given $K = \begin{bmatrix} 2 & 9 \\ 0 & 1 \end{bmatrix}$. Find det K and then use it to find K^{-1}.**

$$\det K = 2(1) - 0(9) = 2 - 0 = 2$$

Thus, $K^{-1} = \frac{1}{2}\begin{bmatrix} 1 & -9 \\ 0 & 2 \end{bmatrix} = \begin{bmatrix} \frac{1}{2} & -\frac{9}{2} \\ 0 & 1 \end{bmatrix}$.

Exercises

Exploratory **Answer each of the following.**

1. What must be true about a multiplicative identity in the set M_2 of 2 × 2 matrices? Why is $\begin{bmatrix} 1 & 1 \\ 1 & 1 \end{bmatrix}$ not the multiplicative identity?
2. Show that $\begin{bmatrix} 1 & 0 \\ 1 & 0 \end{bmatrix}$ is not a multiplicative identity for M_2.
3. Explain the method used on page 516 to find the identity I_2 for M_2. Repeat the method using $\begin{bmatrix} 3 & 4 \\ 6 & 2 \end{bmatrix}$ as the initial matrix.
4. Repeat the demonstration in Exploratory Exercise 3 for $\begin{bmatrix} 5 & -2 \\ -1 & 4 \end{bmatrix}$.
5. What do you think is the multiplicative identity in the set M_3 of 3 × 3 matrices? In M_4?

Find the multiplicative inverse of each of the following matrices, if it exists.

6. $\begin{bmatrix} 1 & 1 \\ 2 & 1 \end{bmatrix}$
7. $\begin{bmatrix} 0 & 0 \\ 3 & 0 \end{bmatrix}$
8. $\begin{bmatrix} 2 & 3 \\ 2 & 5 \end{bmatrix}$
9. $\begin{bmatrix} 4 & 5 \\ 3 & 4 \end{bmatrix}$

Written **Find the multiplicative inverse of each of the following matrices, if it exists. If it does not exist, explain why.**

1. $\begin{bmatrix} 1 & 2 \\ 1 & 3 \end{bmatrix}$
2. $\begin{bmatrix} 1 & 2 \\ 3 & 4 \end{bmatrix}$
3. $\begin{bmatrix} 5 & 4 \\ 6 & 5 \end{bmatrix}$
4. $\begin{bmatrix} 9 & -8 \\ -8 & 7 \end{bmatrix}$
5. $\begin{bmatrix} 6 & -3 \\ -8 & 4 \end{bmatrix}$
6. $\begin{bmatrix} 2 & 1 \\ -1 & 2 \end{bmatrix}$
7. $\begin{bmatrix} 0.5 & 1 \\ 1 & -2 \end{bmatrix}$
8. $\begin{bmatrix} \sqrt{2} & \sqrt{3} \\ \sqrt{3} & \sqrt{2} \end{bmatrix}$
9. $\begin{bmatrix} 0.02 & 0.1 \\ 0.03 & 0.3 \end{bmatrix}$
10. $\begin{bmatrix} k^2 & k^3 \\ k & k^2 \end{bmatrix}$
11. Amy says, "The multiplicative identity in M_n is the matrix $[e_{ij}]_n$ with $e_{ij} = 1$ if $i = j$ and $e_{ij} = 0$ otherwise." Explain what Amy means. Is she right?
12. Study the partial derivation of the rule for A^{-1} given on page 517. Complete the demonstration by finding values of q and s. Show every step.

Find the determinant of each of the following matrices. Then use the determinant to find the multiplicative inverse, if it exists.

13. $\begin{bmatrix} 1 & 0 \\ 0 & 1 \end{bmatrix}$
14. $\begin{bmatrix} 1 & 1 \\ 1 & 1 \end{bmatrix}$
15. $\begin{bmatrix} 1 & 2 \\ 3 & 4 \end{bmatrix}$
16. $\begin{bmatrix} -5 & 1 \\ 2 & 6 \end{bmatrix}$
17. $\begin{bmatrix} 5 & 6 \\ -8 & 2 \end{bmatrix}$
18. $\begin{bmatrix} 3 & 2 \\ 2 & \frac{1}{3} \end{bmatrix}$
19. $\begin{bmatrix} a_1 & a_2 \\ a_3 & a_4 \end{bmatrix}$
20. $\begin{bmatrix} \sqrt{2} & 1 \\ 1 & \sqrt{2} \end{bmatrix}$
21. $\begin{bmatrix} \sqrt{3} & 2\sqrt{2} \\ \sqrt{10} & 5\sqrt{3} \end{bmatrix}$
22. $\begin{bmatrix} a^2 & ab \\ b^2 & \frac{1}{a} \end{bmatrix}$
23. Give the rule for A^{-1} with "det" notation and without it. Illustrate each rule by finding $\begin{bmatrix} 2 & 1 \\ -3 & 4 \end{bmatrix}^{-1}$.

14.5 Systems of Equations

You have already learned how to solve systems of equations by certain algebraic and graphical methods, as shown below.

System of Equations	Algebraic solution		Graphical solution
$2x + 3y = 7$ $5x + 2y = 1$	$4x + 6y = 14$ $15x + 6y = 3$	*Multiply by 2.* *Multiply by 3.*	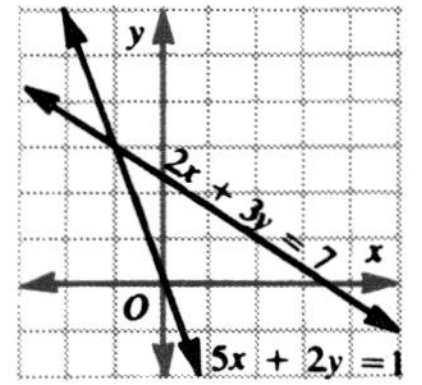
	$-11x = 11$	*Subtract.*	
	$x = -1$		
	$2(-1) + 3y = 7$	*Substitution*	
	$y = 3$		

Matrices can also be used to solve systems of equations. Consider the two expressions $\begin{cases} 2x + 3y \\ 5x + 2y \end{cases}$. Using matrix multiplication, this system can be rewritten as $\begin{bmatrix} 2 & 3 \\ 5 & 2 \end{bmatrix}\begin{bmatrix} x \\ y \end{bmatrix}$.

$\begin{bmatrix} 2 & 3 \\ 5 & 2 \end{bmatrix}$ is called the *coefficient matrix* of the system. $\begin{bmatrix} x \\ y \end{bmatrix}$ is a *column vector*. The two equations to be solved can be written as shown below.

$$\begin{bmatrix} 2 & 3 \\ 5 & 2 \end{bmatrix}\begin{bmatrix} x \\ y \end{bmatrix} = \begin{bmatrix} 7 \\ 1 \end{bmatrix}$$

To solve this matrix equation, solve for the column vector $\begin{bmatrix} x \\ y \end{bmatrix}$. Notice that the equation is of the form $AX = B$. Since the inverse of A is A^{-1}, multiply both sides of this equation by A^{-1}.

$$\begin{aligned} AX &= B \\ (A^{-1})AX &= (A^{-1})B \\ (A^{-1} \cdot A)X &= A^{-1} \cdot B \\ IX &= A^{-1}B \\ X &= A^{-1}B \end{aligned}$$

Thus, the solution is the matrix given by the product of A^{-1} and B.

Example

1 **Solve the system** $\begin{cases} 2x + 3y = 7 \\ 5x + 2y = 1 \end{cases}$ **using the matrix method.**

$$\begin{bmatrix} 2 & 3 \\ 5 & 2 \end{bmatrix}\begin{bmatrix} x \\ y \end{bmatrix} = \begin{bmatrix} 7 \\ 1 \end{bmatrix}$$

The equation is of the form $AX = B$, where $A = \begin{bmatrix} 2 & 3 \\ 5 & 2 \end{bmatrix}$ and $B = \begin{bmatrix} 7 \\ 1 \end{bmatrix}$. Since $X = \begin{bmatrix} x \\ y \end{bmatrix} = A^{-1}B$, we can find the product $A^{-1}B$.

$$A^{-1}B = \begin{bmatrix} -\frac{2}{11} & \frac{3}{11} \\ \frac{5}{11} & -\frac{2}{11} \end{bmatrix} \begin{bmatrix} 7 \\ 1 \end{bmatrix} \qquad A^{-1} = \frac{1}{-11}\begin{bmatrix} 2 & -3 \\ -5 & 2 \end{bmatrix}$$

$$= \begin{bmatrix} -\frac{14}{11} + \frac{3}{11} \\ \frac{35}{11} - \frac{2}{11} \end{bmatrix}$$

$$= \begin{bmatrix} -1 \\ 3 \end{bmatrix} \qquad \textit{Thus, } x = -1 \textit{ and } y = 3.$$

Of course, the method shown above works only if A^{-1} exists. If A^{-1} does not exist, then the system of equations either has no solutions or an infinite number of solutions.

Examples

2 **Solve** $\begin{cases} 2x + 3y = 2 \\ 4x + 6y = -3 \end{cases}$ **using the matrix method.**

The matrix equation is $\begin{bmatrix} 2 & 3 \\ 4 & 6 \end{bmatrix}\begin{bmatrix} x \\ y \end{bmatrix} = \begin{bmatrix} 2 \\ -3 \end{bmatrix}$.

However, the coefficient matrix does not have an inverse. Thus, the system does not have exactly one solution, but has either no solutions or an infinite number of solutions. To determine which of these is the case, go back to the original system. If the first equation is multiplied by 2 and the second equation is subtracted from the first, the result is $0 = 5$. This indicates that the system has no solutions. In other words, the equations represent lines in the same plane that are parallel.

3 **Solve** $\begin{cases} 2x + 3y = 5 \\ 4x + 6y = 10 \end{cases}$ **using the matrix method.**

Again, there is no inverse for the coefficient matrix. If the first equation is multiplied by 2 and the second equation is subtracted from the first, the result is $0 = 0$.

This indicates that there are an infinite number of solutions. In other words, the two equations name the same line in the plane.

A summary of the principles presented in this section is given below.

1. A system of two equations in two unknowns can be expressed by the matrix equation $AX = B$ where A is the coefficient matrix and X the column vector.
2. The system has exactly one solution if, and only if, A^{-1} exists. This solution can be determined by multiplying both sides of the equation by A^{-1}, obtaining $X = A^{-1}B$.
3. The system does not have exactly one solution if, and only if, A^{-1} does not exist. The system will have either no solutions or an infinite number of solutions.

Exercises

Exploratory **Answer each of the following.**

1. Briefly, describe the basic ideas of solving systems of two equations in two unknowns by the following methods: algebraic (addition–subtraction), algebraic (substitution), graphical, matrix.
2. What is a disadvantage of solving a system of equations graphically? Is this disadvantage taken care of by solving the system using a matrix?
3. What do you have to watch out for when using matrices to solve equations? Explain. How does this compare with an "ordinary" algebraic solution?
4. If the coefficient matrix of a system does not have an inverse, what does this tell you about the system? How would you investigate the system further?

Written **Solve each of the following systems of equations using the substitution method or the addition–subtraction method.**

1. $2x - 5y = 3$
   $x + 3y = 7$
2. $2x - 3y = 11$
   $3x - 2y = 14$
3. $2x - 5y = 1$
   $6x + 10y = 5$

Solve each of the following systems of equations graphically.

4. $x + y = 2$
$-2x + 3y = -2$

5. $y = -8x$
$7x + y = 1$

6. $2x + 5y = 3$
$4x - 5y = 1$

Solve the systems of equations in the following exercises using the matrix method, if possible.

7. exercise 1
8. exercise 2
9. exercise 3
10. exercise 4
11. exercise 5
12. exercise 6

Solve each of the following systems of equations using the matrix method, if possible. If not possible, use another method to find the solution.

13. $2x + y = 12$
$x + 2y = 9$

14. $3x + y = 13$
$2x + 3y = 4$

15. $x + y = -7$
$-4x + 7y = 6$

16. $2x + 3y = 11$
$-2x + y = 1$

17. $50x + 9y = -23$
$-22x - y = 19$

18. $x - 20y = 59$
$x + y = -4$

19. $x + 3y = 7$
$2x - 5y = -2$

20. $3y = 8$
$2x - y = 1$

21. $3x + 2y = 19$
$2x - 5y = 11$

22. $x - 3y = 8$
$5x + y = -1$

23. $23x + 19y = 14$
$-18x - 13y = -9$

24. $8x - y = 101$
$13x + 12y = 1$

25. Solve this problem using the matrix method. At Amy's Burger Pit, two burgers and three orders of french fries cost $4.45. Three burgers and two orders of fries total $5.05. What does Amy charge for one burger and one order of fries?

Mixed Review

The following daily high temperatures in degrees Fahrenheit were recorded during a cold spell that lasted 40 days in Ithaca, New York.

29	26	17	12	5	4	25	17	23	18
13	6	25	20	27	22	26	30	31	20
2	12	27	16	27	16	30	6	16	5
0	5	29	18	16	22	29	8	23	24

1. Find the median.
2. Find the mode.
3. Find the mean.
4. Find the range.
5. Find the standard deviation, to the nearest tenth.

6. A set has nine elements. How many subsets of six elements does it have?

7. Find the probability of selecting a face card or a red card from a deck of 52 cards.

8. Which term of the arithmetic sequence 7, 13, 19, . . . is 193?

9. What is the sum of the infinite geometric series $75 + 45 + 27 + 16.2 + \cdots$?

Solve each system of equations graphically or algebraically.

10. $y = x^2$
$y = -4x$

11. $y = x^2 - 6x + 2$
$4x + y = 5$

12. $x^2 + y^2 = 25$
$3y - x = 15$

14.6 Minors and 3 × 3's

You have learned that for $A = \begin{bmatrix} a & b \\ c & d \end{bmatrix}$, det $A = ad - bc$.

Another notation for det A is $|A|$. Thus, for $A = \begin{bmatrix} 1 & 2 \\ 3 & 4 \end{bmatrix}$, det $A =$ $\begin{vmatrix} 1 & 2 \\ 3 & 4 \end{vmatrix} = 4 - 6 = -2$.

This concept can be extended to 3 × 3 matrices.

Determinant of a 3 × 3 Matrix

For $A = \begin{bmatrix} a & b & c \\ d & e & f \\ g & h & i \end{bmatrix}$, det $A = |A| = \begin{vmatrix} a & b & c \\ d & e & f \\ g & h & i \end{vmatrix} =$
$aei + bfg + cdh - ceg - afh - bdi.$

The computation of the determinant of a 3 × 3 matrix can be simplified by a process known as *expanding by minors.*

Definition of Minor

The minor of an element of a determinant is the determinant of one lower order obtained by deleting the row and column of that element.

The minor of b in $\begin{vmatrix} a & b & c \\ d & e & f \\ g & h & i \end{vmatrix}$ is found by deleting row 1 and column 2.

Thus, the minor of b is $\begin{vmatrix} d & f \\ g & i \end{vmatrix}$.

Definition of Cofactor

The cofactor of an element a_{ij} of a determinant is the minor of a_{ij} if $i + j$ is even and the minor multiplied by -1 if $i + j$ is odd.

The cofactor of a_{22} in $\begin{vmatrix} a_{11} & a_{12} & a_{13} \\ a_{21} & a_{22} & a_{23} \\ a_{31} & a_{32} & a_{33} \end{vmatrix}$ is $\begin{vmatrix} a_{11} & a_{13} \\ a_{31} & a_{33} \end{vmatrix}$ since 2 + 2 is even.

The cofactor of a_{23} is $-1 \begin{vmatrix} a_{11} & a_{12} \\ a_{31} & a_{32} \end{vmatrix}$ since 2 + 3 is odd.

In more advanced work it can be shown that a determinant is equal to the sum of the products of the elements of any row or column and their corresponding cofactors. To illustrate this statement, use the first column of $\begin{vmatrix} a & b & c \\ d & e & f \\ g & h & i \end{vmatrix}$ and show that $\begin{vmatrix} a & b & c \\ d & e & f \\ g & h & i \end{vmatrix} = a\begin{vmatrix} e & f \\ h & i \end{vmatrix} - d\begin{vmatrix} b & c \\ h & i \end{vmatrix} + g\begin{vmatrix} b & c \\ e & f \end{vmatrix}$.

$$\begin{aligned} a\begin{vmatrix} e & f \\ h & i \end{vmatrix} - d\begin{vmatrix} b & c \\ h & i \end{vmatrix} + g\begin{vmatrix} b & c \\ e & f \end{vmatrix} &= a(ei - fh) - d(bi - ch) + g(bf - ce) \\ &= aei - afh - dbi + dch + gbf - gce \\ &= aei + bfg + cdh - ceg - afh - bdi \end{aligned}$$

Since $\begin{vmatrix} a & b & c \\ d & e & f \\ g & h & i \end{vmatrix} = aei + bfg + cdh - ceg - afh - bdi$, the statement is true.

Example

1 **Given** $B = \begin{bmatrix} 3 & -1 & 7 \\ 2 & 3 & 4 \\ 1 & -1 & 2 \end{bmatrix}$. **Find det** B.

Use row three. The cofactor of 1 is $\begin{vmatrix} -1 & 7 \\ 3 & 4 \end{vmatrix}$, the cofactor of -1 is $-1\begin{vmatrix} 3 & 7 \\ 2 & 4 \end{vmatrix}$, and the cofactor of 2 is $\begin{vmatrix} 3 & -1 \\ 2 & 3 \end{vmatrix}$.

$$\begin{aligned} \det B &= 1\begin{vmatrix} -1 & 7 \\ 3 & 4 \end{vmatrix} - 1\left(-1\begin{vmatrix} 3 & 7 \\ 2 & 4 \end{vmatrix}\right) + 2\begin{vmatrix} 3 & -1 \\ 2 & 3 \end{vmatrix} \\ &= 1(-1 \cdot 4 - 7 \cdot 3) + 1(3 \cdot 4 - 7 \cdot 2) + 2(3 \cdot 3 - 2 \cdot -1) \\ &= 1(-25) + 1(-2) + 2(11) \\ &= -5 \end{aligned}$$

The *adjoint* of a matrix can be used to find the inverse of any square matrix.

Definition of Adjoint

The **adjoint of** A—denoted **adj** A—is the matrix whose row elements are the cofactors of the corresponding elements in the columns of A.

Example

2 **Given** $A = \begin{bmatrix} 1 & 0 & 2 \\ 2 & -1 & 1 \\ 3 & 2 & 0 \end{bmatrix}$. **Find adj** A.

To find adj A, find the cofactor of *each* element of A. The cofactors of 1, 2, and 3 (the *first column* of A) are the elements of the *first row* of adj A, the cofactors of 0, -1, and 2 (the *second column* of A) are the elements of the *second row* of adj A, and the cofactors of 2, 1, and 0 (the *third column* of A) are the elements of the *third row* of adj A.

Thus, adj $A = \begin{bmatrix} -2 & 4 & 2 \\ 3 & -6 & 3 \\ 7 & -2 & -1 \end{bmatrix}$.

Multiplicative Inverse of a Square Matrix

For any square matrix A, $A^{-1} = \frac{1}{\det A} \cdot \operatorname{adj} A$.

Example

3 **Find A^{-1} for matrix A in example 2.**

$$\begin{aligned}\det A &= 1(-1)(0) + 0(1)(3) + (2)(2)(2) - 2(-1)(3) - 1(2)(1) - 0(2)(0)\\ &= 0 + 0 + 8 - (-6) - 2 - 0\\ &= 12\end{aligned}$$

$$A^{-1} = \frac{1}{\det A} \cdot \operatorname{adj} A = \frac{1}{12}\begin{bmatrix} -2 & 4 & 2 \\ 3 & -6 & 3 \\ 7 & -2 & -1 \end{bmatrix} = \begin{bmatrix} -\frac{1}{6} & \frac{1}{3} & \frac{1}{6} \\ \frac{1}{4} & -\frac{1}{2} & \frac{1}{4} \\ \frac{7}{12} & -\frac{1}{6} & -\frac{1}{12} \end{bmatrix}$$

Solving systems of three equations in three unknowns involves finding the multiplicative inverse of a 3×3 matrix.

Example

4 **Solve the system** $\begin{cases} x + 2z = 12 \\ 2x - y + z = 10. \\ 3x + 2y = 4 \end{cases}$

Use the methods introduced in section 14.5. First, rewrite the equations using matrices.

$$\begin{bmatrix} 1 & 0 & 2 \\ 2 & -1 & 1 \\ 3 & 2 & 0 \end{bmatrix}\begin{bmatrix} x \\ y \\ z \end{bmatrix} = \begin{bmatrix} 12 \\ 10 \\ 4 \end{bmatrix}$$

The coefficient matrix is the same as matrix A in examples 2 and 3.

Notice that this is a matrix equation in the form $AX = B$. To solve, multiply on the left by A^{-1}.

$$\begin{bmatrix} x \\ y \\ z \end{bmatrix} = \begin{bmatrix} -\frac{1}{6} & \frac{1}{3} & \frac{1}{6} \\ \frac{1}{4} & -\frac{1}{2} & \frac{1}{4} \\ \frac{7}{12} & -\frac{1}{6} & -\frac{1}{12} \end{bmatrix}\begin{bmatrix} 12 \\ 10 \\ 4 \end{bmatrix} = \begin{bmatrix} 2 \\ -1 \\ 5 \end{bmatrix}$$

The solution is $x = 2$, $y = -1$, and $z = 5$, or the ordered triple $(2, -1, 5)$.

Exercises

Exploratory **In exercises 1–5, explain the meaning of each of the following symbols or terms. For each term, also give an example.**

1. $\begin{vmatrix} a & b \\ c & e \end{vmatrix}$

2. $\begin{vmatrix} a_{11} & a_{12} \\ a_{21} & a_{22} \end{vmatrix}$

3. $\begin{vmatrix} a & b & c \\ d & e & f \\ g & h & i \end{vmatrix}$

4. the minor of an element in a determinant
5. the cofactor of an element in a determinant
6. Cofactors are sometimes called signed minors. Explain why.

Written **Given** $K = \begin{bmatrix} k & m & n \\ p & q & r \\ s & t & v \end{bmatrix}$. **Find the minors of the elements in each of the following.**

1. row 1
2. row 2
3. row 3
4. column 1
5. column 2
6. column 3

Find the cofactors of each of the following elements in matrix K above.

7. k
8. p
9. s
10. m
11. q
12. t
13. n
14. r
15. v
16. Show that if one row or one column of a matrix is multiplied by a constant k, then its determinant is multiplied by k.
17. Show that if two rows or two columns of a matrix are interchanged, then only the sign of its determinant is changed.
18. Show that if a matrix A has two rows or two columns identical, then $\det A = 0$. *Hint: Use Written Exercise 17.*

Solve each of the following systems of equations using the matrix method.

19. $$\begin{aligned} x + y + 2z &= 6 \\ 2x - y + z &= 5 \\ -x + 2y - z &= -3 \end{aligned}$$

20. $$\begin{aligned} x + y - z &= 5 \\ 3x \qquad + z &= 1 \\ 4x - 3y - z &= -14 \end{aligned}$$

21. $$\begin{aligned} 4x + y + z &= 5 \\ -2x + y + 7z &= 8 \\ -4x \qquad + z &= -1 \end{aligned}$$

22. $$\begin{aligned} x - 2y - 3z &= 13 \\ y - 8z &= 15 \\ -x + y + z &= -8 \end{aligned}$$

Challenge Develop a program of definitions and theorems similar to that in this section, but for 4×4 matrices.

14.7 Matrices and Transformations

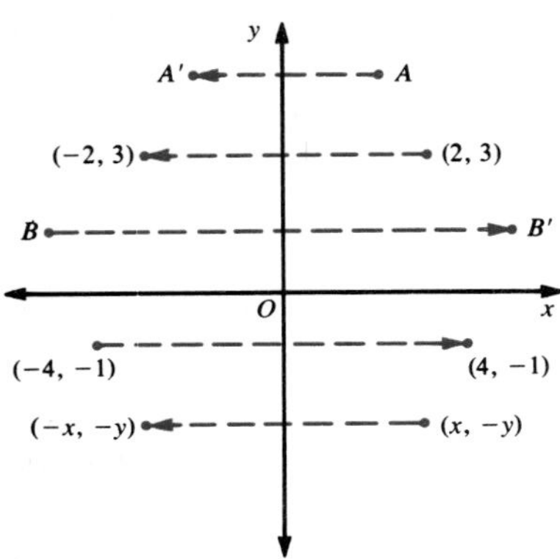

INVESTIGATION 1 Consider the line reflection $R_{y\text{-axis}}$. The transformation is shown at the left. Its rule is

$$(x, y) \xrightarrow{R_{y\text{-axis}}} (-x, y).$$

To determine if there is a matrix that can be used to denote this transformation, use the column vector $\begin{bmatrix} x \\ y \end{bmatrix}$. Find a 2 × 2 matrix, if one exists, that satisfies the equation $[?]\begin{bmatrix} x \\ y \end{bmatrix} = \begin{bmatrix} -x \\ y \end{bmatrix}$. Experimentation will show that $\begin{bmatrix} -1 & 0 \\ 0 & 1 \end{bmatrix}$ is the correct matrix. That is,

$$\begin{bmatrix} -1 & 0 \\ 0 & 1 \end{bmatrix}\begin{bmatrix} x \\ y \end{bmatrix} = \begin{bmatrix} -x \\ y \end{bmatrix}.$$

For example, the reflection of the point (5, −3) can be found by multiplication, $\begin{bmatrix} -1 & 0 \\ 0 & 1 \end{bmatrix}\begin{bmatrix} 5 \\ -3 \end{bmatrix}$. The image point is $\begin{bmatrix} -5 \\ -3 \end{bmatrix}$ or (−5, −3).

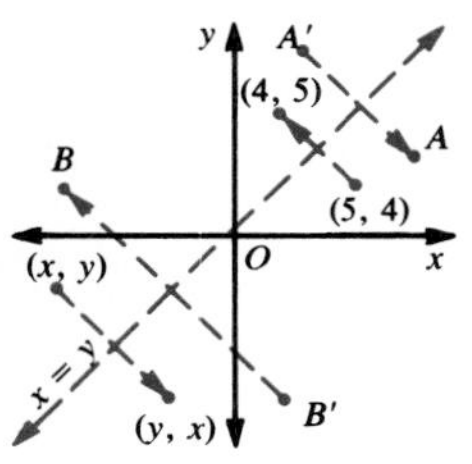

INVESTIGATION 2 Consider the transformation $R_{x=y}$. This is a reflection in the line $x = y$, and is shown at the left. Its rule is $(x, y) \xrightarrow{R_{x=y}} (y, x)$. Is there a matrix that denotes this transformation? Experimentation will show that $\begin{bmatrix} 0 & 1 \\ 1 & 0 \end{bmatrix}$ is such a matrix. That is,

$$\begin{bmatrix} 0 & 1 \\ 1 & 0 \end{bmatrix}\begin{bmatrix} x \\ y \end{bmatrix} = \begin{bmatrix} y \\ x \end{bmatrix}.$$

For example, the reflection of the point (5, −3) can be found by multiplication, $\begin{bmatrix} 0 & 1 \\ 1 & 0 \end{bmatrix}\begin{bmatrix} 5 \\ -3 \end{bmatrix}$. The image point is $\begin{bmatrix} -3 \\ 5 \end{bmatrix}$ or (−3, 5).

INVESTIGATION 3 Surely it is the case that $\begin{bmatrix} 1 & 0 \\ 0 & 1 \end{bmatrix}\begin{bmatrix} x \\ y \end{bmatrix} = \begin{bmatrix} x \\ y \end{bmatrix}$. But it also is the case that the square of the transformation $R_{x=y}$ sends every point in the plane back to itself. That is, it should be the case that

$$\begin{bmatrix} 0 & 1 \\ 1 & 0 \end{bmatrix}\begin{bmatrix} 0 & 1 \\ 1 & 0 \end{bmatrix}\begin{bmatrix} x \\ y \end{bmatrix} = \begin{bmatrix} x \\ y \end{bmatrix}.$$

You will be asked to show that this is true in the Written Exercises.

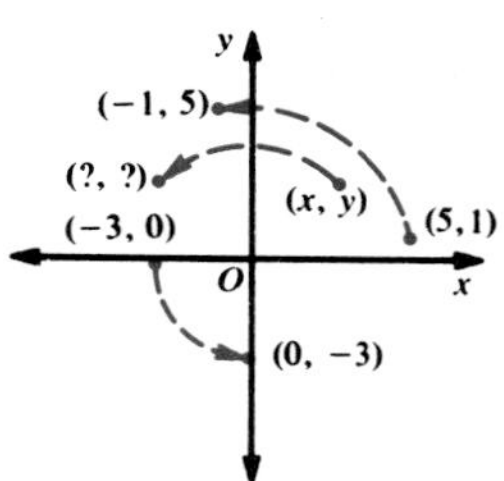

INVESTIGATION 4 Recall that the rule for a rotation of the plane 90° counterclockwise with center at the origin is $(x, y) \xrightarrow{Rot_{O,90°}} (-y, x)$. Is there a 2 × 2 matrix that denotes this transformation? Experimentation will show that the matrix $\begin{bmatrix} 0 & -1 \\ 1 & 0 \end{bmatrix}$ is such a matrix. That is,

$$\begin{bmatrix} 0 & -1 \\ 1 & 0 \end{bmatrix}\begin{bmatrix} x \\ y \end{bmatrix} = \begin{bmatrix} -y \\ x \end{bmatrix}.$$

For example, the rotation of the point (5, −3) can be found by multiplication, $\begin{bmatrix} 0 & -1 \\ 1 & 0 \end{bmatrix}\begin{bmatrix} 5 \\ -3 \end{bmatrix}$. The image point is $\begin{bmatrix} 3 \\ 5 \end{bmatrix}$ or (3, 5).

Rotations of 90°, or of multiples of 90°, are special, and also relatively easy to investigate. But what about rotations through a *general* angle of any number of degrees? Is there a matrix representation of a rotation of plane counterclockwise through an angle of 17°, or $\frac{2\pi}{3}$, or θ°?

The answer is yes. However, first consider the theorem about isometries that states that an isometry is uniquely determined by its action on three non-collinear points. This theorem will be used to help find the matrix we are looking for.

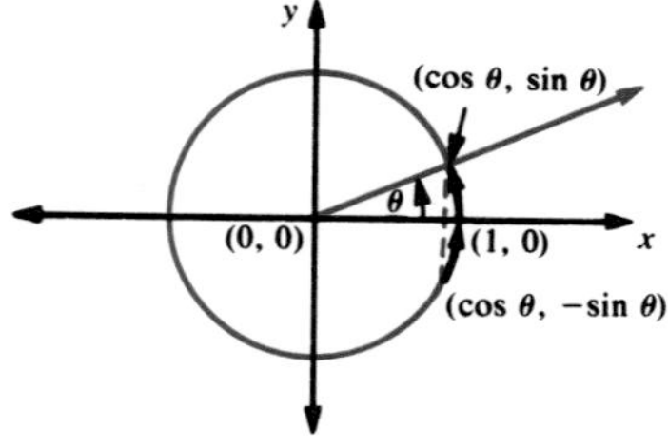

Study the figure at the left. It shows the effect of a rotation of θ° counterclockwise on three points: (0, 0), (1, 0), and (cos θ, −sin θ). This rotation takes

(0, 0) into (0, 0),

(1, 0) into (cos θ, sin θ), and

(cos θ, −sin θ) into (1, 0).

A matrix representation of such a rotation would have to satisfy each of the three equations shown below.

a. $[?]\begin{bmatrix} 0 \\ 0 \end{bmatrix} = \begin{bmatrix} 0 \\ 0 \end{bmatrix}$ **b.** $[?]\begin{bmatrix} 1 \\ 0 \end{bmatrix} = \begin{bmatrix} \cos\theta \\ \sin\theta \end{bmatrix}$ **c.** $[?]\begin{bmatrix} \cos\theta \\ -\sin\theta \end{bmatrix} = \begin{bmatrix} 1 \\ 0 \end{bmatrix}$

Experimentation will show that $\begin{bmatrix} \cos\theta & -\sin\theta \\ \sin\theta & \cos\theta \end{bmatrix}$ is the correct matrix. If you substitute this matrix for [?] in each equation above, you will find that it works in each case.

a. $\begin{bmatrix} \cos\theta & -\sin\theta \\ \sin\theta & \cos\theta \end{bmatrix}\begin{bmatrix} 0 \\ 0 \end{bmatrix} = \begin{bmatrix} 0 \\ 0 \end{bmatrix}$ **b.** $\begin{bmatrix} \cos\theta & -\sin\theta \\ \sin\theta & \cos\theta \end{bmatrix}\begin{bmatrix} 1 \\ 0 \end{bmatrix} = \begin{bmatrix} \cos\theta \\ \sin\theta \end{bmatrix}$

c. $\begin{bmatrix} \cos\theta & -\sin\theta \\ \sin\theta & \cos\theta \end{bmatrix}\begin{bmatrix} \cos\theta \\ -\sin\theta \end{bmatrix} = \begin{bmatrix} 1 \\ 0 \end{bmatrix}$

Example

1 **Find the image of the point $\left(-\frac{3}{2}, \sqrt{3}\right)$ under a rotation of 120° counterclockwise, with center (0, 0).**

Since the rotation is 120°, $\theta = 120°$.

So, $\cos\theta = -\frac{1}{2}$ and $\sin\theta = \frac{\sqrt{3}}{2}$.

Therefore, the rotation matrix is $\begin{bmatrix} -\frac{1}{2} & -\frac{\sqrt{3}}{2} \\ \frac{\sqrt{3}}{2} & -\frac{1}{2} \end{bmatrix}$.

Thus, the required image is given by $\begin{bmatrix} -\frac{1}{2} & -\frac{\sqrt{3}}{2} \\ \frac{\sqrt{3}}{2} & -\frac{1}{2} \end{bmatrix}\begin{bmatrix} -\frac{3}{2} \\ \sqrt{3} \end{bmatrix}$.

$$\begin{bmatrix} -\frac{1}{2} & -\frac{\sqrt{3}}{2} \\ \frac{\sqrt{3}}{2} & -\frac{1}{2} \end{bmatrix}\begin{bmatrix} -\frac{3}{2} \\ \sqrt{3} \end{bmatrix} = \begin{bmatrix} \frac{3}{4} - \frac{3}{2} \\ -\frac{3\sqrt{3}}{4} - \frac{\sqrt{3}}{2} \end{bmatrix}$$

$$= \begin{bmatrix} -\frac{3}{4} \\ -\frac{5\sqrt{3}}{4} \end{bmatrix}$$

or $\left(-\frac{3}{4}, -\frac{5\sqrt{3}}{4}\right)$

Exercises

Exploratory **Explain the meaning of each of the following terms and symbols.**

1. line reflection
2. $R_{x\text{-axis}}$
3. line of symmetry
4. transformation
5. isometry
6. rotation
7. $\sin\theta$
8. $\cos\theta$
9. $Rot_{O,90°}$

Find the image of the point *P* indicated under the transformation indicated. Include a sketch.

10. $P(2, 1)$; $R_{y\text{-axis}}$
11. $P(-3, 2)$; $R_{y\text{-axis}}$
12. $P(-5, -5)$; $Rot_{O,180°}$
13. $P(-4, 1)$; $R_{x=2}$
14. $P(6, 5.7)$; $R_{x=y}$
15. $P(6, 1)$; $Rot_{O,90°}$
16. $P(-3, 1)$; $Rot_{(0,1),90°}$
17. $P(3, 2)$; $Rot_{(0,0),270°}$
18. $P(6, 2)$; $D_{O,\frac{1}{3}}$

Written **Answer each of the following.**

1. In your own words, explain what it means to say that $\begin{bmatrix} -1 & 0 \\ 0 & 1 \end{bmatrix}$ is a matrix representation for a line reflection in the y-axis. As part of your discussion include the effect of this transformation on the point $Q(-1, 3)$.
2. Show that the square of the transformation $R_{x=y}$ sends every point in the plane back to itself. In other words, show that $\begin{bmatrix} 0 & 1 \\ 1 & 0 \end{bmatrix}\begin{bmatrix} 0 & 1 \\ 1 & 0 \end{bmatrix}\begin{bmatrix} x \\ y \end{bmatrix} = \begin{bmatrix} x \\ y \end{bmatrix}$. *See Investigation 3 on page 528.*
3. Suppose you square the matrix $\begin{bmatrix} 0 & -1 \\ 1 & 0 \end{bmatrix}$. What do you get? What would multiplication of this new matrix by $\begin{bmatrix} x \\ y \end{bmatrix}$ represent?

Find matrix representations for each of the following transformations. Justify your answer and include a sketch.

4. $R_{x\text{-axis}}$
5. $Rot_{O,180°}$
6. $Rot_{O,270°}$
7. $Rot_{O,-90°}$
8. $D_{O,2}$
9. $D_{O,7}$
10. $R_{x=-y}$
11. $Rot_{O,-180°}$

12. Can you think of two ways to find the matrix representation for a rotation of 270° counterclockwise?
13. What does it mean to say that an isometry is "uniquely determined by its action on three points"?
14. What are the three points and their images in the derivation of the general rotation matrix?

Use the general rotation matrix to find the images of each of the following points under a rotation of θ counterclockwise. Use degree or radian measure as indicated. If necessary, use the trig table on page 618 or a calculator to find sin θ and cos θ. Give all answers to three decimal places.

15. $\left(\frac{3}{2}, \sqrt{3}\right)$; $\theta = 120°$
16. $\left(\frac{1}{2}, \sqrt{3}\right)$; $\theta = 120°$
17. $\left(\frac{1}{2}, -\sqrt{3}\right)$; $\theta = 150°$
18. $(-1, -2\sqrt{2})$; $\theta = 45°$
19. $(-1, \sqrt{2})$; $\theta = 315°$
20. $\left(2, -\frac{\sqrt{3}}{2}\right)$; $\theta = \frac{2\pi}{3}$
21. $(0, 8)$; $\theta = 135°$
22. $(3.2, 1.8)$; $\theta = 122°40'$
23. $(1, \sqrt{2})$; $\theta = 4°$

Challenge

1. Interpret the matrix $\begin{bmatrix} 1 & 1 \\ 0 & 1 \end{bmatrix}$ in terms of geometric transformations. What is $\begin{bmatrix} 1 & 1 \\ 0 & 1 \end{bmatrix}\begin{bmatrix} x \\ y \end{bmatrix}$? $\begin{bmatrix} 1 & 1 \\ 0 & 1 \end{bmatrix}\begin{bmatrix} 2 \\ 5 \end{bmatrix}$? Illustrate what is happening with careful figures. *This transformation is called a shear.*
2. On your own, investigate other matrix representations of mappings in the plane. Be prepared to report in class. In each case, include careful figures.

You may recall the following identities from your work in chapter 9.

$$\cos(\alpha + \beta) = \cos\alpha\cos\beta - \sin\alpha\sin\beta$$
$$\sin(\alpha + \beta) = \sin\beta\cos\alpha + \sin\alpha\cos\beta$$

An alternate proof of these identities can be shown using the figure below and matrix multiplication.

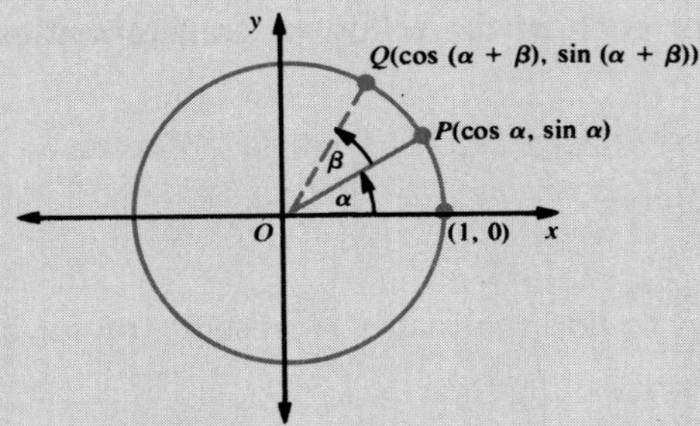

Consider point Q carefully. Its coordinates are

$$(\cos(\alpha + \beta), \sin(\alpha + \beta)).$$

But Q is also the image of $P(\cos\alpha, \sin\alpha)$ under a rotation of β counterclockwise. Therefore, it must be the case that

$$\begin{bmatrix} \cos\beta & -\sin\beta \\ \sin\beta & \cos\beta \end{bmatrix} \begin{bmatrix} \cos\alpha \\ \sin\alpha \end{bmatrix} = \begin{bmatrix} \cos(\alpha + \beta) \\ \sin(\alpha + \beta) \end{bmatrix}.$$

Recall that $\begin{bmatrix} \cos\beta & -\sin\beta \\ \sin\beta & \cos\beta \end{bmatrix}$ *is the general rotation matrix.*

Multiplying, we have

$$\begin{bmatrix} \cos\alpha\cos\beta - \sin\alpha\sin\beta \\ \sin\beta\cos\alpha + \sin\alpha\cos\beta \end{bmatrix} = \begin{bmatrix} \cos(\alpha + \beta) \\ \sin(\alpha + \beta) \end{bmatrix}$$

or, since these two matrices are equal,

$$\cos(\alpha + \beta) = \cos\alpha\cos\beta - \sin\alpha\sin\beta$$
$$\sin(\alpha + \beta) = \sin\beta\cos\alpha + \sin\alpha\cos\beta.$$

14.8 Transition Matrices

Many people move from the city to the suburbs each year. But because of the attractions of a large city, some people move from the suburbs back into the city. In the case of Pine City, 20% of the population moves from the central city to the suburbs each year. (Assume this means that 80% stay.) But at the same time, 10% of the people in the suburbs move back into Pine City. This yearly shift in population can be summarized with the matrix shown below. Note that "moving from the city to the city" means staying in the city.

		TO:	
		Pine City	**Suburbs**
FROM:	**Pine City**	**0.8**	**0.2**
	Suburbs	**0.1**	**0.9**

A matrix such as this is called a **transition matrix.** Note that the sum of the entries in each row is 1. Why?

This year the population of Pine City is 50,000 and the population of its suburbs 20,000. To find the populations of Pine City and its suburbs next year (that is, one year from now), perform the matrix multiplication shown below. For simplicity, do not consider changes in population due to births or deaths.

$$[50{,}000 \quad 20{,}000]\begin{bmatrix} 0.8 & 0.2 \\ 0.1 & 0.9 \end{bmatrix} = [42{,}000 \quad 28{,}000]$$

Thus in one year, the population of Pine City will be 42,000 and the population of its suburbs will be 28,000.

What will the population of Pine City be in two years? To answer this question, multiply the population matrix [42,000 28,000] by $\begin{bmatrix} 0.8 & 0.2 \\ 0.1 & 0.9 \end{bmatrix}$. However,

$$[42{,}000 \quad 28{,}000]\begin{bmatrix} 0.8 & 0.2 \\ 0.1 & 0.9 \end{bmatrix} = \left([50{,}000 \quad 20{,}000]\begin{bmatrix} 0.8 & 0.2 \\ 0.1 & 0.9 \end{bmatrix}\right)\begin{bmatrix} 0.8 & 0.2 \\ 0.1 & 0.9 \end{bmatrix}.$$

Since matrix multiplication is associative, this result would be equal to

$$[50{,}000 \quad 20{,}000]\left(\begin{bmatrix} 0.8 & 0.2 \\ 0.1 & 0.9 \end{bmatrix}\begin{bmatrix} 0.8 & 0.2 \\ 0.1 & 0.9 \end{bmatrix}\right) \text{ or } [50{,}000 \quad 20{,}000]\begin{bmatrix} 0.8 & 0.2 \\ 0.1 & 0.9 \end{bmatrix}^2.$$

Expanding upon this, if $A = [50{,}000 \quad 20{,}000]$, we find the respective population of city and suburbs in four years can be found by computing $A\begin{bmatrix} 0.8 & 0.2 \\ 0.1 & 0.9 \end{bmatrix}^4$ and in 16 years by computing $A\begin{bmatrix} 0.8 & 0.2 \\ 0.1 & 0.9 \end{bmatrix}^{16}$.

Transition matrices can also be used to describe probability situations. Here is an example.

Pine City is pretty much a Republican town. In general, a Republican has a better chance of being elected to city office than a Democrat. But, as in other places, there is a "power of the incumbency." That is, both parties have a better chance of winning the next election if their party is in power now. In Pine City, for example, if the present mayor is a Republican, the probability is $\frac{5}{6}$ that the next mayor will be a Republican (and $\frac{1}{6}$ that he or she will be a Democrat). If the present mayor is a Democrat, the probability is only $\frac{2}{3}$ that the next mayor will be a Republican (and $\frac{1}{3}$ a Democrat). This information is summarized in the following transition matrix.

		TO: R	TO: D
FROM:	R	$\frac{5}{6}$	$\frac{1}{6}$
	D	$\frac{2}{3}$	$\frac{1}{3}$

Note that this matrix of probabilities does not give any information about the party of the present mayor. *If* the present mayor is Republican, then the probabilities for the election of a Republican or Democrat in the next election are given by the row vector $\begin{bmatrix} \frac{5}{6} & \frac{1}{6} \end{bmatrix}$. *If* the current mayor is a Democrat, the probabilities are given by $\begin{bmatrix} \frac{2}{3} & \frac{1}{3} \end{bmatrix}$.

Now, suppose we know that the current mayor is a Republican. In this case the probability that the current mayor is Republican is 1 and the probability that he or she is a Democrat is zero. This can be represented by the *row vector* $[1 \quad 0]$. The multiplication

$$[1 \quad 0]\begin{bmatrix} \frac{5}{6} & \frac{1}{6} \\ \frac{2}{3} & \frac{1}{3} \end{bmatrix} = \begin{bmatrix} \frac{5}{6} & \frac{1}{6} \end{bmatrix}$$

yields $\begin{bmatrix} \frac{5}{6} & \frac{1}{6} \end{bmatrix}$ as the probability for the next election, as it should.

Note that in this example we have described a *chain of events where every outcome is dependent on the one just before it, and on nothing else*. Such a sequence of trials, together with the indicated probabilities, is called a *Markov Chain*. Each "link" in such a chain of events is a *Markov dependent Bernoulli trial*.

Exercises

Exploratory

1. What is a transition matrix? Give an example different from the ones in this section.

Use the transition matrix for the population of Pine City and suburbs to answer the following questions.

2. What will the population of Pine City and its suburbs be in two years? In three years? In four years?
3. Answer exercise 2 if the current population of Pine City is 60,000 and of its suburbs, 10,000.
4. How many computations would you have to perform to compute A^{16} for a matrix A?

Written **Given** $A = \begin{bmatrix} 0.8 & 0.2 \\ 0.1 & 0.9 \end{bmatrix}$. **Find each of the following, rounding to three decimal places each time. Use a calculator when necessary.**

1. A^2
2. A^4
3. A^8
4. A^{16}
5. A^{32}

In each case, write a fully labeled "FROM–TO" transition matrix to describe the situation. Then answer the question asked. In each case, assume that the only courses of action are those implied in the question.

6. Each year 30% of the population of Beeville moves to West Beeville, while 10% of West Beeville's population moves to Beeville. If the respective populations of Beeville and West Beeville are 20,000 and 12,000, what will the populations be in each of the following?
   a. one year **b.** two years **c.** four years
7. A certain chemical substance in a closed chamber changes from its liquid to its vapor state according to this rule: 3% of the liquid evaporates each hour while at the same time, 2% of the vapor condenses to liquid. An experimenter begins with 100 units of liquid and 20 units of vapor in the chamber. Find the amounts of liquid and vapor in the chamber at the end of the following periods.
   a. one hour **b.** two hours
   c. three hours **d.** four hours
8. Answer exercise 7 if 5% of the liquid evaporates per hour while 1% of the vapor condenses.
9. Answer exercise 7 if the experimenter started with each of the following.
   a. 100 units of liquid, no vapor
   b. no liquid, 100 units of vapor
   c. 50 units of each

10. Two coins have been "loaded" in the following way: for coin A, $P(H) = \frac{2}{3}$; for coin B, $P(H) = \frac{1}{4}$. A sequence of trials is as follows.

    After a head is tossed, coin A is tossed; after a tail, coin B.

    To start the experiment, a fair coin is tossed; if it lands heads, we start with coin A; if tails, B.

    What is the probability of a head after each of the following?
    a. the first trial **b.** the second trial **c.** the third trial

11. Answer exercise 10 if the starter-coin always comes up heads.

12. Answer exercise 10 if the sequences are started by drawing a letter from SNAKE. If a vowel is drawn, toss coin *A*; if a consonant, coin *B*.

13. If it is raining in Pine City today, it will be raining again tomorrow with probability $\frac{1}{2}$, and fair with probability $\frac{1}{2}$. If it is fair today it will rain tomorrow with probability $\frac{1}{4}$ and be fair with probability $\frac{3}{4}$. It is raining today. What are the probabilities of rain or fair tomorrow? In two days? Three days? Four days? Eight days?

14. If Amy is wearing her glasses today, then she will be wearing her glasses tomorrow with probability $\frac{1}{4}$, her contact lenses with $P = \frac{1}{2}$, and no lenses at all with $P = \frac{1}{4}$. If she is wearing her contacts today, she will not be wearing her glasses tomorrow under any circumstances, but will be wearing her contacts with $P = \frac{3}{4}$, and no lenses with $P = \frac{1}{4}$. If she has no lenses today, she will be wearing glasses tomorrow with $P = \frac{1}{2}$, will not be wearing contacts under any circumstances, and be without lenses again with $P = \frac{1}{2}$. Find the probability that Amy is wearing each of the following in two days, given that you saw Amy today and she was wearing glasses.

a. glasses **b.** contact lenses **c.** neither

15. Answer exercise 14 if Amy was wearing contact lenses.

16. Write a matrix expression for the probability that Amy will be wearing contact lenses thirty days from now, given that you saw her this morning squinting—no lenses on!

Challenge

Compare a Markov Chain with independent Bernoulli trials.

Mixed Review

1. Find the value of $\cos\left[\text{Arcsin}\left(-\frac{5}{9}\right)\right]$.

2. In square *ABCD*, diagonals $\overline{AC}$ and $\overline{BD}$ intersect at point *X*. Find $\text{Rot}_{X,90°}(R_{\overline{BD}}(A))$.

3. Suppose Marla and Elliott play 5 games of cribbage. The probability that Elliott wins a game is $\frac{2}{3}$ and that Marla wins is $\frac{1}{3}$. What is the probability that Marla will win at least 3 of the games?

4. On a standardized test, the mean is 68 and the standard deviation is 4.7. Would you expect a score of 59, 65, 70, 75, or 78 to occur less than 3% of the time.

5. Find the value of x for which 36, $5x - 1$, 16 form a geometric sequence.

6. Given $A = \begin{bmatrix} 8 & 5 \\ 3 & -1 \\ -2 & -4 \end{bmatrix}$ and $B = \begin{bmatrix} 4 & 5 & -9 \\ 2 & -2 & 1 \end{bmatrix}$, find *AB*.

Problem Solving Application: Using Matrices

Matrices are useful in the study of communication networks. The graph at the right shows the communication system linking four communication stations. Connected vertices represent the manner in which signals can be transferred and received by each communication station.

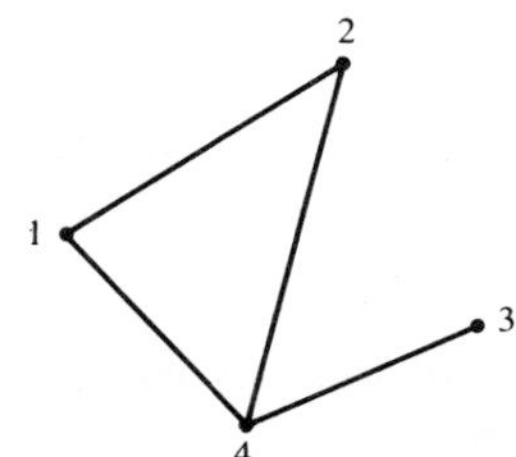

The figure above can be represented by the matrix at the right. The entry in position a_{12} is 1 because there is an edge between vertices 1 and 2. Similarly, the entry in position a_{23} is 0 because there is no edge between vertices 2 and 3.

$$M = \begin{bmatrix} 0 & 1 & 0 & 1 \\ 1 & 0 & 0 & 1 \\ 0 & 0 & 0 & 1 \\ 1 & 1 & 1 & 0 \end{bmatrix}$$

Example

1 **For the matrix M above, verify that M^2 represents the number of ways a message can be passed from one communication station to another using only one other communication station as a relay point.**

Explore To pass from vertex *i* to vertex *j* with exactly one relay point means that the message must pass from vertex *i* to vertex *j* by means of only one other vertex.

Plan Obtain M^2 by finding $M \times M$. Then, check the entries with the graph of the four communication stations.

Solve

$$M^2 = \begin{bmatrix} 2 & 1 & 1 & 1 \\ 1 & 2 & 1 & 1 \\ 1 & 1 & 1 & 0 \\ 1 & 1 & 0 & 3 \end{bmatrix}$$

Examine Check the entry in position a_{44}. The entry indicates there are 3 ways to send a message from communication station 4 back to communication station 4 using only one relay point. Using the graph, the three ways are $4 \rightarrow 1 \rightarrow 4$, $4 \rightarrow 2 \rightarrow 4$, and $4 \rightarrow 3 \rightarrow 4$. Compare other entries of M to the communications diagram to verify that the entries are correct.

Exercises

Written **Solve each problem.**

1. Using matrix M from Example 1, verify that M^3 represents the number of ways a message can be passed using exactly two relay points.

Not all networks have two-way communication between any connected members. The graph and matrix below represents a computer network where not all computers are connected, and some are only connected for one-way communication. Use the graph and matrix to complete exercises 2 and 3.

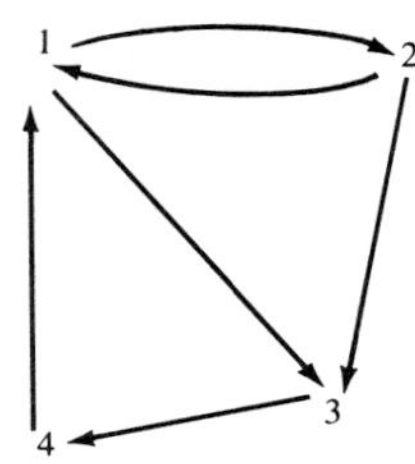

$$M = \begin{bmatrix} 0 & 1 & 1 & 0 \\ 1 & 0 & 1 & 0 \\ 0 & 0 & 0 & 1 \\ 1 & 0 & 0 & 0 \end{bmatrix}$$

2. Verify that $M + M^2$ represents the number of ways two computers are connected using no more than one relay point.

3. Verify that $M + M^2 + M^3$ represents the number of ways two computers are connected using no more than two relay points.

Find the matrix representing the communication network for each communications system.

4. 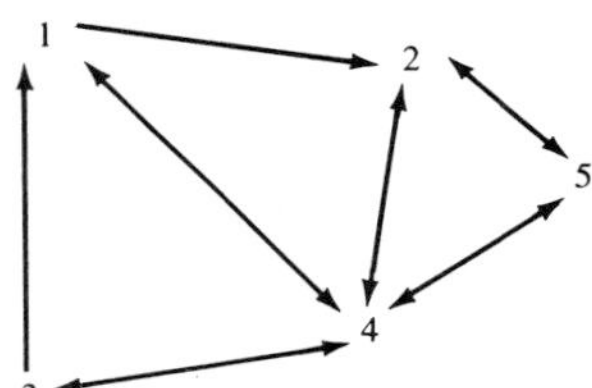

5. 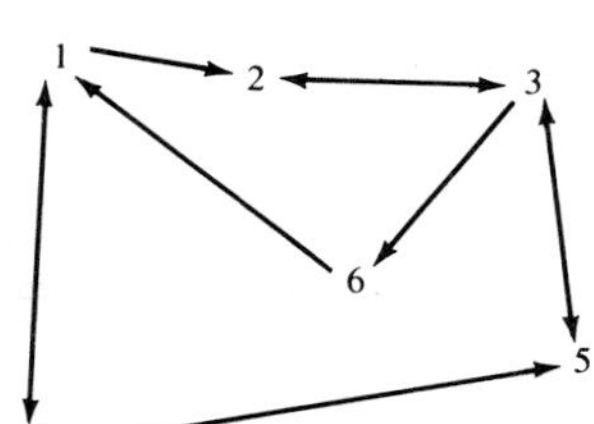

6. 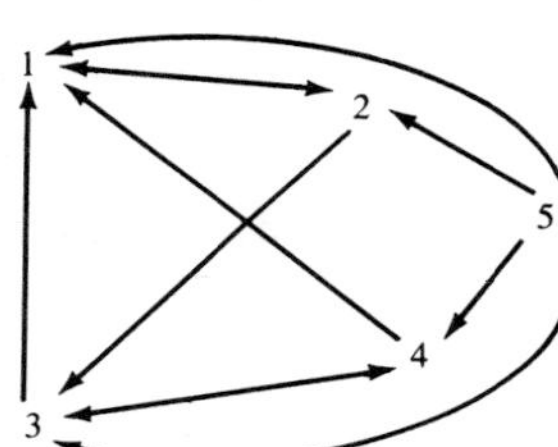

7. The graph at the right represents the passing of toxins through an ecological food chain. Ecologists follow the effects of this toxin through no more than two intermediate species. What species are affected under such an assumption if the toxin is applied to grass only?

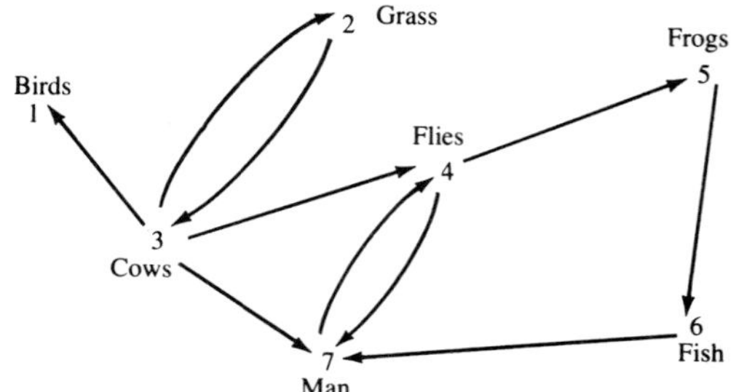

8. A museum is introducing "one-way" corridors for crowd control at popular exhibits. The proposed flow is shown. Decide whether it will be possible to visit all exhibits if you are allowed to start at any of the six exhibits.

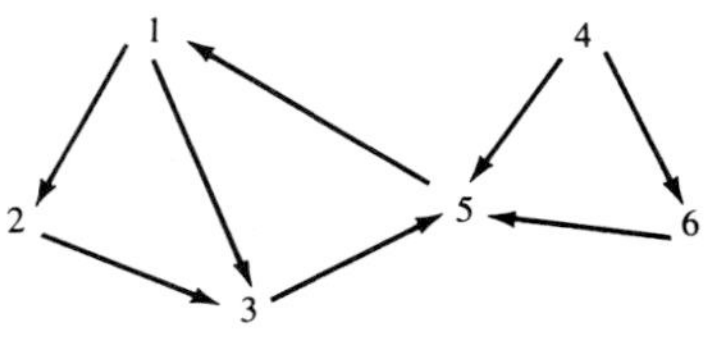

Vocabulary

matrix
order
square matrix
zero matrix
addition of matrices
additive inverse
subtraction of matrices
scalar
scalar multiplication
multiplication of matrices
multiplicative identity
multiplicative inverse
determinant of a 2×2 matrix
determinant of a 3×3 matrix
minor
cofactor
transition matrix

Chapter Summary

1. $A_{m \times n}$ denotes a matrix of order $m \times n$. That is, matrix A has m rows and n columns.
2. B_n denotes a square matrix of order n. That is, matrix B has n rows and n columns.
3. The sum of two matrices of different orders is not defined.
4. $\overline{0}_{m \times n}$ denotes a zero matrix of order $m \times n$. $\overline{0}_n$ denotes a square zero matrix of order n.
5. A and $[a_{ij}]$ can be used to name the same matrix. However, a_{ij} names an element in the matrix, specifically, the element in the ith row, jth column.
6. $-A$ denotes the additive inverse of matrix A.
7. The multiplicative identity of any 2×2 matrix, denoted by I, is $\begin{bmatrix} 1 & 0 \\ 0 & 1 \end{bmatrix}$.
8. The multiplicative inverse of a matrix $A = \begin{bmatrix} a & b \\ c & d \end{bmatrix}$, denoted by A^{-1}, can be written $\frac{1}{\det A}\begin{bmatrix} d & -b \\ -c & a \end{bmatrix}$.
9. Systems of two equations in two variables can be solved using the matrix method only if the multiplicative inverse of the coefficient matrix exists. If it does not exist, then the system either has no solutions or an infinite number of solutions.

10. Systems of three equations in three variables can be solved using the matrix method.

11. Matrices can be used to denote geometric transformations.

12. The general rotation matrix for a rotation of θ counterclockwise is $\begin{bmatrix} \cos\theta & -\sin\theta \\ \sin\theta & \cos\theta \end{bmatrix}$.

Chapter Review

14.1 Give the order of each of the following matrices.

1. $\begin{bmatrix} 2 & 3 \\ 4 & 1 \end{bmatrix}$

2. $\begin{bmatrix} 1 & 0 & \sqrt{2} & 1 \\ 0 & 0 & 3 & 0 \\ 6 & 2 & 0 & 0 \end{bmatrix}$

3. $\begin{bmatrix} 4 \\ 1 \\ 0 \\ -1 \end{bmatrix}$

4. $\begin{bmatrix} 3 & 0 & 0 & 1 & 2 \\ 0 & -1 & 2 & \frac{1}{2} & \frac{1}{4} \end{bmatrix}$

Name the additive inverses of the matrices in the following exercises.

5. exercise 1
6. exercise 2
7. exercise 3
8. exercise 4

Use the matrices below to find each sum or difference, if possible.

$A = \begin{bmatrix} 3 & 6 \\ 0 & -1 \end{bmatrix}$ $B = \begin{bmatrix} -1 & 4 \\ 2 & -1 \end{bmatrix}$ $C = \begin{bmatrix} 3 & 4 \\ -2 & 1 \\ 0 & 2 \end{bmatrix}$ $D = \begin{bmatrix} -3 & 7 \\ 2 & -1 \\ 5 & 6 \end{bmatrix}$

9. $A + B$
10. $A - B$
11. $B + C$
12. $C + D$
13. $C - D$
14. $D - C$
15. $A - C$
16. $B - A$

14.2 Find each product.

17. $3\begin{bmatrix} 7 & 4 & -3 \\ 2 & \frac{1}{2} & -1 \end{bmatrix}$

18. $3\begin{bmatrix} 2 \\ 6 \\ -2 \end{bmatrix}$

19. $-3\begin{bmatrix} 5 & -\frac{2}{3} \\ 0 & 2 \end{bmatrix}$

Solve each of the following matrix equations.

20. $\begin{bmatrix} 2 & 3 \\ -4 & 1 \end{bmatrix} + X = \begin{bmatrix} -7 & 3 \\ 5 & 2 \end{bmatrix}$

21. $X - 2\begin{bmatrix} 5 & -\frac{1}{2} \\ 3 & 2 \\ 1 & 0 \end{bmatrix} = \begin{bmatrix} -2 & 0 \\ 3 & 1 \\ 4 & -2 \end{bmatrix}$

14.3 Use the matrices below to find each product, if possible.

$X = \begin{bmatrix} 2 & 4 \\ -3 & 1 \end{bmatrix}$ $Y = \begin{bmatrix} -1 & 4 & 3 \\ 7 & 5 & -4 \end{bmatrix}$ $Z = \begin{bmatrix} -5 & 6 \\ 1 & -1 \end{bmatrix}$

22. XZ
23. YZ
24. X^2
25. Y^2
26. Z^2X

14.4 Find the multiplicative inverse of each of the following matrices, if it exists. If it does not exist, explain why.

27. $\begin{bmatrix} 4 & 2 \\ 3 & -1 \end{bmatrix}$ **28.** $\begin{bmatrix} -6 & 2 \\ -3 & 1 \end{bmatrix}$ **29.** $\begin{bmatrix} 7 & 4 \\ 3 & 2 \end{bmatrix}$ **30.** $\begin{bmatrix} -2 & 1 \\ 7 & \frac{1}{2} \end{bmatrix}$

14.5 Solve each of the following systems of equations using the matrix method, if possible. If not possible, use another method to find the solution.

31. $\begin{aligned} 2x + y &= 10 \\ x + y &= 7 \end{aligned}$

32. $\begin{aligned} -x + 4y &= 12 \\ -2x + 8y &= -8 \end{aligned}$

33. $\begin{aligned} x - 15y &= 37 \\ -3x + 5y &= -31 \end{aligned}$

14.6 Given $A = \begin{bmatrix} a & b & c \\ d & e & f \\ g & h & i \end{bmatrix}$. **Find the minors of the elements in each of the following.**

34. row 1 **35.** column 2 **36.** row 3

Find the cofactors of each of the following elements in matrix A above.

37. a **38.** c **39.** d **40.** f **41.** h

Solve each of the following systems of equations using the matrix method.

42. $\begin{aligned} 3x - y - z &= 6 \\ 3x + 2y + z &= 13 \\ -x + 5y - z &= 0 \end{aligned}$

43. $\begin{aligned} 2x + 3y - 4z &= 1 \\ -x + 2y + 3z &= -10 \\ 3x + y + z &= 19 \end{aligned}$

14.7 Find the image of the point P indicated under the transformation indicated. Include a sketch.

44. $P(-3, 2)$; $R_{y\text{-axis}}$ **45.** $P(7, -1)$; $\text{Rot}_{O,\,90°}$

46. $P(5, 4)$; $R_{x=y}$ **47.** $P(-4, -2)$; $R_{x\text{-axis}}$

Use the general rotation matrix to find the images of each of the following points under a rotation of θ counterclockwise.

48. $\left(\frac{1}{2}, \sqrt{3}\right)$; $\theta = 30°$ **49.** $(-1, \sqrt{2})$; $\theta = 45°$

14.8 Write a fully labeled "FROM–TO" transition matrix to describe the following situation. Then answer the question asked.

50. Each year 25% of the population of Cedarville moves to Yellow Springs, while 15% of Yellow Springs' population moves to Cedarville. If the respective populations of Cedarville and Yellow Springs are 10,000 and 5,000, what will the populations be in two years?

Chapter Test

Use the matrices below to find each sum or difference, if it exists.

$A = \begin{bmatrix} 2 & 3 \\ -4 & 1 \end{bmatrix}$ $B = \begin{bmatrix} 7 & -5 \\ 1 & 0 \end{bmatrix}$ $C = \begin{bmatrix} 2 & 1 \\ 3 & -1 \\ -4 & 5 \end{bmatrix}$ $D = \begin{bmatrix} 8 & -2 \\ 3 & 6 \\ 7 & 1 \end{bmatrix}$

1. $B + A$ **2.** $C + D$ **3.** $A + C$ **4.** $C - D$

Find each product, if possible.

5. $4\begin{bmatrix} 1 & 5 \\ -3 & 2 \\ -7 & 0 \end{bmatrix}$

6. $-5\begin{bmatrix} 4 & 3 & 2 \\ 1 & -4 & -1 \end{bmatrix}$

7. $2\begin{bmatrix} \frac{1}{2} & 3 \\ -\frac{1}{6} & 5 \end{bmatrix}$

8. $\begin{bmatrix} 3 & -1 \\ 4 & 2 \end{bmatrix}\begin{bmatrix} 7 & 3 \\ -6 & 1 \end{bmatrix}$

9. $\begin{bmatrix} -4 & 8 \\ 0 & 2 \\ 1 & 1 \end{bmatrix}\begin{bmatrix} 2 & 6 & -5 \\ 7 & -3 & 1 \end{bmatrix}$

10. $\begin{bmatrix} 2 \\ -8 \end{bmatrix}\begin{bmatrix} 5 & 3 \\ 2 & 0 \end{bmatrix}$

Solve each equation for *X*. Use matrices *A*, *B*, *C*, and *D* provided for exercises 1–4.

11. $X - A = B$ **12.** $4B + X = 3A$ **13.** $3C - 2X = -D$

Solve each of the following systems of equations using the matrix method.

14. $\begin{aligned} 3x - 2y &= -16 \\ x + y &= 2 \end{aligned}$

15. $\begin{aligned} -5x - 10y &= 50 \\ x - 2y &= 10 \end{aligned}$

16. $\begin{aligned} x + 3y - 3z &= 1 \\ 5x - 3y + 3z &= 11 \\ x + 2y - z &= 2 \end{aligned}$

17. $\begin{aligned} 2x - y - 5z &= 3 \\ x + 4y - 2z &= 3 \\ 5x + 3y + 2z &= 1 \end{aligned}$

Use the general rotation matrix to find the images of each of the following points under a rotation of θ counterclockwise.

18. $\left(\frac{3}{2}, \sqrt{3}\right); \theta = 30°$

19. $(2, -2\sqrt{2}); \theta = 45°$

20. Everyone in Beckville shopped at Janet's General Store until Don's Foodmart opened 2 years ago. Each year since then, 30% of Janet's customers switched to Don's Foodmart, while 10% of Don's customers switched back to Janet's General Store. If 20,000 people were shopping at Janet's General Store when Don's Foodmart first opened, use a "FROM-TO" transition matrix to determine how many people shop at each store now.

Polynomials, Complex Numbers, and DeMoivre's Theorem

Chapter 15

Polynomials and polynomial equations play an important role in the design of many different types of structures, such as the roller coaster shown below.

In this chapter you will have an opportunity to gain expertise in working with polynomials and solving polynomial equations. Complex numbers are examined and De Moivre's Theorem is introduced.

15.1 Polynomials

You have already worked with polynomials in one variable such as the ones shown below.

$$2x + 1 \qquad x^2 + 1 \qquad x^2 + 3x + 4$$
$$-x^3 + 3x - 10 \qquad 4x^5 - 3x^3 + 7x^2 - 2$$

Definition of Polynomial in One Variable

A **polynomial in one variable,** x, is an expression of the form $a_nx^n + a_{n-1}x^{n-1} + \ldots + a_2x^2 + a_1x + a_0$ where $a_n \neq 0$ and n is a whole number.

Since the **degree** of a polynomial in one variable, x, is the exponent of the greatest power of x, we see that the degree is n. The term a_0 is called the **constant term.** The term a_nx^n is called the **leading term.** The coefficient a_n is called the **leading coefficient.** A polynomial consisting of just zero is called the **zero polynomial.** The zero polynomial has no degree. The following chart gives some examples of polynomials.

Polynomial	Degree	Leading Coefficient	Constant Term	Name of Polynomial
4	0	4	4	Constant
$2x - 5$	1	2	−5	Linear
$4x^2 - 2x + 7$	2	4	7	Quadratic
x^3	3	1	0	Cubic
$8x^4 - 3x + 17$	4	8	17	Quartic

Definition of Polynomial Function

A function P such that $P(x) = a_nx^n + a_{n-1}x^{n-1} + \ldots + a_1x + a_0$, where $a_n \neq 0$ and n is a whole number, is called a **polynomial function.**

Examples

1 **The equation $y = 2x^2 - 3x + 5$ defines a polynomial function of degree 2 or, more simply, a quadratic function.**

2 **If $P(x) = 3x^3 - 4x + 5$, find $P(-2)$.**

$$\begin{aligned} P(-2) &= 3(-2)^3 - 4(-2) + 5 \\ &= 3(-8) + 8 + 5 \\ &= -24 + 8 + 5 \text{ or } -11 \end{aligned}$$

If a polynomial $P(x)$ is set equal to 0, the resulting equation is called a **polynomial equation.** For example, $x^4 - 8x^2 - 12 = 0$ is a polynomial equation of degree four. The **roots,** or *solutions,* of a polynomial equation are the values of x for which $P(x) = 0$. Such a value is also called a **zero of the polynomial.**

You have had a good deal of experience in finding roots of linear and quadratic polynomial equations. As you will recall, many quadratic equations can be solved by factoring. Sometimes this technique can be used with polynomial equations of higher degree. Here are some examples.

Examples

3 **Domain = $\mathscr{C}$. Solve $x^3 + 2x = 2x^2$.**

$$x^3 + 2x = 2x^2$$

$$x^3 - 2x^2 + 2x = 0 \quad \text{Express in standard form.}$$

$$x(x^2 - 2x + 2) = 0 \quad \text{Factor.}$$

$$x = 0 \text{ or } x^2 - 2x + 2 = 0 \quad \text{Set each factor equal to 0.}$$

$$x = \frac{2 \pm \sqrt{-4}}{2} \quad \text{Use quadratic formula.}$$

$$x = \frac{2 \pm 2i}{2} = 1 \pm i$$

Solution Set = $\{0, 1 + i, 1 - i\}$

4 **If $P(x) = x^4 - x^2 - 12$, find the zeros of P in the set of complex numbers.**

$$x^4 - x^2 - 12 = 0 \quad \text{Set } P(x) \text{ equal to 0.}$$

$$(x^2 - 4)(x^2 + 3) = 0 \quad \text{Factor polynomial.}$$

$$x^2 - 4 = 0 \quad \text{or } x^2 + 3 = 0 \quad \text{Set each factor equal to 0.}$$

$$x = \pm 2 \text{ or } \quad x = \pm i\sqrt{3}$$

The zeros of P are $-2, 2, i\sqrt{3}, -i\sqrt{3}$.

As you have probably noticed, the equations in examples 3 and 4 "work out" nicely because they are relatively easy to factor. But how do you deal with equations whose factors are not as obvious? We will begin our investigation of this question in the next section.

Exercises

Exploratory **Tell whether each of the following is a polynomial in one variable. If it is, give the degree, leading term, leading coefficient, and constant term.**

1. $3x - 4$
2. $4x^2 - 3x + 5$
3. $-2x^3 + 5x - 3$
4. $x^{-2} + 1$
5. 8
6. $ix^4 + 3x^5 - \sqrt{3}$
7. $x - 3x^2 + 7$
8. $\frac{3}{x^2}$
9. $x^2 - 2\sqrt{x} + 1$
10. $|x - 7|$
11. $\sin x$
12. $5x^3 - (2 + i)x^2 - \pi$

In exercises 1–12, identify polynomials of the following type.

13. constant
14. linear
15. quadratic
16. cubic

Written **If $P(x) = 2x^3 - 3x^2 + x + 6$, find each of the following.**

1. $P(1)$
2. $P(0)$
3. $P(2)$
4. $P(-i)$
5. $P(-1)$
6. $P(-2)$
7. $P(2i)$
8. $P(-2i)$
9. $P(2x + 1)$
10. $P(x + 2)$
11. $P(x + h)$
12. $P(x + i)$

Domain = $\mathscr{C}$. Solve each equation.

13. $2x - 3(7x - 4) = 3(2x - 1)$
14. $x^2 - 4x = 0$
15. $x^2 - 5x + 6 = 0$
16. $x^3 - x^2 = 0$
17. $x^3 = 9x$
18. $2x^2 - 7x + 5 = 0$
19. $x^3 - 5x^2 + 6x = 0$
20. $2x^3 + 4x^2 = 70x$
21. $x^3 + 3x = -2x^2$

Find the zeros of each polynomial.

22. $2x^2 + 32$
23. $3m^3 - 2m^2 + 4m$
24. $x^4 - 5x^2 + 6$
25. $t^4 + 4t^2 - 5$
26. $x^4 - 8x^2 + 12$
27. $x^5 - 9x^3 + 14x$
28. $4x^2 + 4x + 1$
29. $t^4 + t$
30. $8m^3 + 12m^2 + 9m$

Under what circumstances does a polynomial function of degree 2 have each of the following?

31. exactly one zero
32. two real zeros
33. two imaginary zeros

Challenge

1. Use sigma notation to represent a polynomial in one variable x of degree n.
2. Use sigma notation to represent the sum of two polynomials in one variable x, each with degree n.
3. Use sigma notation to represent the product of two polynomials in one variable x, each of degree 2.

15.2 Division of Polynomials

One way to determine whether 21 is a factor of 586 is to divide 586 by 21. If the remainder is 0, you can conclude that 21 is a factor of 586. In this case, the remainder is 19. Thus, 21 is not a factor of 586.

$$\begin{array}{r} 27 \\ 21\overline{)586} \\ \underline{42} \\ 166 \\ \underline{147} \\ 19 \end{array}$$

To check the division, verify the following identity:

$$\begin{array}{ccccccccc} 586 & = & 21 & \times & 27 & + & 19 \\ \text{dividend} & = & \text{divisor} & \times & \text{quotient} & + & \text{remainder} \end{array}$$

Now, suppose you wish to know if the polynomial $x - 3$ is a factor of the polynomial $x^3 - 8x^2 - 2x + 6$. The procedure for dividing polynomials is similar to the one used for integers.

Examples

1 **Divide $x^3 - 8x^2 - 2x + 6$ by $x - 3$.**

We arrange our work as shown at the right.

$$\begin{array}{r} x^2 - 5x - 17 \\ x - 3\overline{)x^3 - 8x^2 - 2x + 6} \\ \underline{x^3 - 3x^2} \\ -5x^2 - 2x \\ \underline{-5x^2 + 15x} \\ -17x + 6 \\ \underline{-17x + 51} \\ -45 \end{array}$$

Stop dividing when the degree of the remainder is less than the degree of the divisor, $x - 3$. In this case the remainder is -45, which has degree zero.

To check the division, you should verify the following identity by multiplying out the right side.

$$\underbrace{x^3 - 8x^2 - 2x + 6}_{\text{dividend}} = \underbrace{(x - 3)}_{\text{divisor}} \times \underbrace{(x^2 - 5x - 17)}_{\text{quotient}} + \underbrace{(-45)}_{\text{remainder}}$$

2 **Is $x + 2$ a factor of $x^3 + 3x^2 - 2x - 8$?**

$$\begin{array}{r} x^2 + x - 4 \\ x + 2\overline{)x^3 + 3x^2 - 2x - 8} \\ \underline{x^3 + 2x^2} \\ x^2 - 2x \\ \underline{x^2 + 2x} \\ -4x - 8 \\ \underline{-4x - 8} \\ 0 \end{array}$$

The remainder is the zero polynomial. Thus, $x + 2$ is a factor of $x^3 + 3x^2 - 2x - 8$.

3 **Divide $5x^2 + x + 2x^4 - 1$ by $x^2 - x + 1$.**

Arrange both polynomials in standard form. As you perform the division, you will see that it is convenient to have a cubic term, $0x^3$, in the dividend.

$$
\begin{array}{r}
2x^2 + 2x + 5 \\
x^2 - x + 1 \overline{)\,2x^4 + 0x^3 + 5x^2 + x - 1} \\
\underline{2x^4 - 2x^3 + 2x^2} \\
2x^3 + 3x^2 + x \\
\underline{2x^3 - 2x^2 + 2x} \\
5x^2 - x - 1 \\
\underline{5x^2 - 5x + 5} \\
4x - 6
\end{array}
$$

As a check, you should verify the following identity.

$$\underbrace{2x^4 + 0x^3 + 5x^2 + x - 1}_{\text{dividend}} = \underbrace{(x^2 - x + 1)}_{\text{divisor}} \times \underbrace{(2x^2 + 2x + 5)}_{\text{quotient}} + \underbrace{(4x - 6)}_{\text{remainder}}$$

dividend = divisor × quotient + remainder

Examples 1, 2, and 3 suggest the following principle, sometimes called the Division Algorithm.

Division Algorithm

If $P(x)$ is the dividend and $D(x)$ is the divisor, then it is possible to find polynomials $Q(x)$ and $R(x)$ such that $P(x) = D(x) \cdot Q(x) + R(x)$. $R(x)$ is either the zero polynomial or of degree less than the degree of $D(x)$.

When the divisor is a linear polynomial of the form $x - a$, a short cut, called synthetic division can be used. To divide $4x^3 - 2x^2 + x - 6$ by $x - 2$ using synthetic division, follow the steps shown below.

1. Write the coefficients of the dividend as shown. Since you are dividing by $x - 2$, place 2 in the upper left box. Then bring down the 4.

$$\begin{array}{c|cccc} 2 & 4 & -2 & 1 & -6 \\ & & & & \\ \hline & 4 & & & \end{array}$$

2. Multiply 4 by 2 and write the result under -2. Then add -2 and 8 to obtain 6.

$$\begin{array}{c|cccc} 2 & 4 & -2 & 1 & -6 \\ & & 8 & & \\ \hline & 4 & 6 & & \end{array}$$

3. Multiply 6 by 2 to obtain 12. Add 1 and 12 to get 13. Multiply 13 by 2 to get 26. Add -6 and 26 to get 20.

$$\begin{array}{c|cccc} 2 & 4 & -2 & 1 & -6 \\ & & 8 & 12 & 26 \\ \hline & 4 & 6 & 13 & 20 \end{array}$$

4. The quotient is → $4x^2 + 6x + 13$
   The remainder is → 20

Examples

4 **Divide $5x^3 - 2x^2 - 3x - 2$ by $x + 1$.**

Since you are dividing by $x + 1$ or $x - (-1)$, place -1 in the upper left box. Then proceed as in example 3.

−1	5	−2	−3	−2
		−5	7	−4
	5	−7	4	−6

The quotient is ⟶ $5x^2 - 7x + 4$
The remainder is ⟶ -6

5 **Divide $-x + x^3 + 3 + x^4$ by $x + 2$.** *Note $x + 2 = x - (-2)$.*

First write the dividend in standard form. Since no x^2 term appears, add one by writing $0x^2$. Then proceed as in example 4.

$$x^4 + x^3 + 0x^2 - x + 3$$

−2	1	1	0	−1	3
		−2	2	−4	10
	1	−1	2	−5	13

The quotient is ⟶ $1x^3 - 1x^2 + 2x - 5$
The remainder is ⟶ 13

Exercises

Exploratory **Use long division to find each quotient and remainder. Check your results.**

1. $(x^2 - 12x - 45) \div (x + 3)$
2. $(15y^2 + 14y - 8) \div (5y - 2)$
3. $(x^3 - 2x^2 + x - 5) \div (x + 4)$
4. $(6x^3 + 11x^2 - 4x - 4) \div (3x - 2)$
5. $(m^3 - 9m^2 + 27m - 28) \div (m - 3)$
6. $(6t^3 + 5t^2 + 9) \div (2t + 3)$
7. $(x^3 + 6x^2 + 12x + 12) \div (x + 2)$
8. $(2a^3 + 7a^2 - 29a + 29) \div (2a - 3)$
9. $(8b^3 - 22b^2 - 5b + 12) \div (4b + 3)$
10. $(x^3 - 8) \div (x - 2)$
11. $(x^3 - 27) \div (x - 3)$
12. $(2x^4 - 3x^3 + x + 4) \div (x + 2)$
13. $(x^3 + 6x^2 + 8x - 2) \div (x^2 - x + 1)$
14. $(2x^3 - 2x - 3) \div (x - 1)$
15. $(x^3 + 4x - 4) \div (x + 2)$
16. $(x^3 - 7x + 3x^2 - 21) \div (x^2 - 7)$
17. $(48x^3 - 15 + 6x^2 - 40x) \div (6x^2 - 5)$
18. $(x^4 - 2x^2 - 1) \div (x^2 - 1)$
19. $(8x^3 - 1) \div (2x - 1)$
20. $(x^4 - 3x^2 + 1) \div (x^2 + x - 1)$
21. $(x^4 - 4x^2 + 12x - 9) \div (x^2 + 2x - 3)$
22. $(y^4 + 4y^3 + 10y^2 + 12y + 9) \div (y^2 + 2y + 3)$

Written **For each pair of polynomials $P(x)$ and $D(x)$, find polynomials $Q(x)$ and $R(x)$ such that $P(x) = D(x)Q(x) + R(x)$ where R is the zero polynomial or its degree is less than the degree of D.**

1. $P(x) = 3x^2 - 2x + 4$; $D(x) = 2x - 6$
2. $P(x) = 4x^3 + 2x^2 - 3x - 4$; $D(x) = x^2 + x - 1$
3. $P(x) = 2x^3 + x - 3$; $D(x) = x - 4$
4. $P(x) = x^3 + 7x^2 - 18x + 18$; $D(x) = x - 3$
5. $P(x) = 3x^4 - 12x^2 - 3x + 6$; $D(x) = 3x - 6$
6. $P(x) = x^5 - 1$; $D(x) = x - 1$
7. $P(x) = x^4 + 1$; $D(x) = x + 1$
8. Can synthetic division be used to simplify $(x^3 - x^2 + 1) \div (x^2 + 1)$? Explain.

Use synthetic division to find each quotient and remainder.

9. $(2x^3 - 3x^2 + 3x - 4) \div (x - 2)$
10. $(x^3 + 6x^2 + 3x + 1) \div (x - 2)$
11. $(3x^3 - 2x^2 + 2x - 1) \div (x - 1)$
12. $(x^4 - 2x^3 + x^2 - 3x + 2) \div (x - 2)$
13. $(2y^3 - 5y + 1) \div (y + 1)$
14. $(a^3 + 3a^2 - 4a + 1) \div (a + 3)$
15. $(x^3 - 11x + 10) \div (x - 3)$
16. $(x^4 - 5x^3 - 13x^2 + 53x + 60) \div (x + 1)$
17. $(2y^4 - 5y^3 - 10y + 8) \div (y - 3)$
18. $(8x^3 - 27) \div (2x + 3)$
19. $(y^5 + 32) \div (y + 2)$
20. $(x^5 - 3x^2 - 20) \div (x - 2)$

Challenge

1. Is $3x - 2$ a factor of $6x^3 - x^2 - 5x + 2$? Explain.
2. Use the results of exercise 1 to solve the equation: $6x^3 - x^2 - 5x + 2 = 0$.
3. Show that 2 is a zero of $x^3 - 2x^2 - x + 2$. Find all other zeros.
4. Try to find the remainder $R(x)$ in the following equation without using long division or synthetic division.

$$x^3 - x^2 + 3x - 1 = (x - 2)Q(x) + R(x).$$

Mixed Review

Solve each equation. In exercise 6, $0° \le x < 360°$.

1. $3(x - 5) - 2(3 - 2x) = 4x$
2. $|4y + 5| = 23$
3. $2a^2 - 7a + 18 = 0$
4. $\sqrt{4n - 3} = n - 2$
5. $9^{4k-2} = 27^{1-2k}$
6. $2\cos^2 x + \sin^2 x = 2\cos x$
7. $\frac{4t}{3t-2} + \frac{2t}{3t+2} = 2$
8. $\log_9 (x - 1) + \log_9 (x - 3) = \frac{1}{2}$
9. When air is pumped into an automobile tire, the pressure required is inversely proportional to the volume. If the pressure is 30 pounds when the volume is 140 cubic inches, find the pressure when the volume is 100 cubic inches.
10. A stack of boxes in a warehouse is arranged so that there are 5 boxes in the top row, 7 boxes in the second row, 9 boxes in the third row, and so on. If the stack is 20 rows high, how many boxes are in the stack?

15.3 The Remainder and Factor Theorems

When the polynomial

$$P(x) = x^3 - 4x^2 + 2x - 3$$

is divided by $x - 2$, the remainder is -7.

$$\begin{array}{r|rrrr} 2 & 1 & -4 & 2 & -3 \\ & & 2 & -4 & -4 \\ \hline & 1 & -2 & -2 & -7 \end{array}$$

Now, let us compute $P(2)$.

$$\begin{aligned} P(2) &= 2^3 - 4 \cdot 2^2 + 2 \cdot 2 - 3 \\ &= 8 - 16 + 4 - 3 \\ P(2) &= -7 \end{aligned}$$

Notice that the remainder is $P(2)$. This is not a coincidence, but an application of the following theorem.

Remainder Theorem

If a polynomial $P(x)$ is divided by $x - a$, the remainder is $P(a)$. This can be symbolized as $P(x) = (x - a)Q(x) + P(a)$.

The proof is left as an exercise.

Examples

1 **Find the remainder when $P(x) = x^6 - 5x^4 + x^2 - 1$ is divided by $x - 1$.**

The Remainder Theorem tells us the remainder is $P(1)$.

$$P(1) = 1^6 - 5 \cdot 1^4 + 1^2 - 1 = -4$$

Thus, the remainder is -4.

2 **If $P(x) = x^3 - 3x^2 - 2x + 6$, find $P(3)$.**

To find $P(3)$ use direct substitution as shown in Method I or use the Remainder Theorem as shown in Method II.

Method I

$$\begin{aligned} P(3) &= 3^3 - 3 \cdot 3^2 - 2 \cdot 3 + 6 \\ &= 27 - 27 - 6 + 6 \\ P(3) &= 0 \end{aligned}$$

Method II

$P(3)$ is the remainder when $P(x)$ is divided by $x - 3$. Use synthetic division to find the remainder.

$$\begin{array}{r|rrrr} 3 & 1 & -3 & -2 & 6 \\ & & 3 & 0 & -6 \\ \hline & 1 & 0 & -2 & 0 \end{array}$$

Therefore, $P(3) = 0$.

In example 2, since $P(3) = 0$, it follows that $x - 3$ is a factor of $x^3 - 3x^2 - 2x + 6$. This is a special case of the *Factor Theorem*.

Factor Theorem

The linear polynomial $x - a$ is a factor of the polynomial $P(x)$ if and only if $P(a) = 0$.

The Factor Theorem is a biconditional statement. Therefore we must prove two conditionals.

Proof of Factor Theorem

1. Prove: If $P(a) = 0$, then $x - a$ is a factor of $P(x)$.
By the Remainder Theorem we know $P(x) = (x - a)Q(x) + P(a)$.
Since $P(a) = 0$, we have $P(x) = (x - a)Q(x) + 0$
or $P(x) = (x - a)Q(x)$.
The last equation shows that $x - a$ is a factor of $P(x)$.

2. Prove: If $x - a$ is a factor of $P(x)$, then $P(a) = 0$.
If $x - a$ is a factor of $P(x)$ there must be a polynomial $Q(x)$ such that
$$P(x) = (x - a)Q(x) \text{ for all } x.$$
Letting $x = a$, we obtain
$$\begin{aligned} P(a) &= (a - a)Q(a) \\ &= 0 \cdot Q(a) \\ P(a) &= 0. \end{aligned}$$

Examples

3 **Is $x - 1$ a factor of $P(x) = 2x^3 - x^2 - 13x - 6$?**

By the Factor Theorem, $x - 1$ is a factor of $P(x)$ if and only if $P(1) = 0$. To evaluate $P(1)$, use direct substitution or synthetic division.

$$\begin{aligned} P(1) &= 2 \cdot 1^3 - 1^2 - 13 \cdot 1 - 6 \\ &= 2 - 1 - 13 - 6 \\ P(1) &= -18 \end{aligned}$$

or

$$\begin{array}{r|rrrr} 1 & 2 & -1 & -13 & -6 \\ & & 2 & 1 & -12 \\ \hline & 2 & 1 & -12 & -18 = P(1) \end{array}$$

Since $P(1) \neq 0$, $x - 1$ is *not* a factor of $P(x)$.

4 **If 1 is a zero of $P(x) = x^3 - x^2 + 4x - 4$, find all the zeros.**

To solve $x^3 - x^2 + 4x - 4 = 0$, we attempt to factor the polynomial $P(x) = x^3 - x^2 + 4x - 4$. If 1 is a zero, then $P(1) = 0$. The Factor Theorem tells us that $x - 1$ must be a factor of $x^3 - x^2 + 4x - 4$. To find the other factor, use long or synthetic division.

$$\begin{aligned} x^3 - x^2 + 4x - 4 &= 0 \\ (x - 1)(x^2 + 4) &= 0 \end{aligned}$$

$$\begin{array}{r|rrrr} 1 & 1 & -1 & 4 & -4 \\ & & 1 & 0 & 4 \\ \hline & 1 & 0 & 4 & 0 \end{array}$$

$$\begin{aligned} x - 1 = 0 &\text{ or } x^2 + 4 = 0 \\ x = 1 &\text{ or } \quad x^2 = -4 \\ & \quad\quad\; x = \pm 2i \end{aligned}$$

The zeros are 1, $2i$, $-2i$.

Exercises

Exploratory **Use the Remainder Theorem to find the remainder when the first polynomial is divided by the second polynomial.**

1. $x^3 + 2x^2 - 2x + 1; x - 2$
2. $x^3 - x^2 + x - 3; x + 2$
3. $x^7 - 8x^6 - 3x^2 + 100; x - 1$
4. $2x^{12} - 3x^9 + 4x^5 - 26; x + 1$
5. $2x^4 - x^2 + 3x - 5; x - 3$
6. $x^3 + x^2 - 4x - 4; x + 2$
7. Can the Remainder Theorem be used to find the remainder when $x^3 - 2x + 1$ is divided by $x^2 - 1$? Explain.

Use the Factor Theorem to decide if the first polynomial is a factor of the second polynomial.

8. $x - 2; x^3 + 2x^2 - x + 1$
9. $x - 2; x^3 + x^2 - 4x - 4$
10. $x - 3; x^3 - 6x^2 + 11x - 6$
11. $x + 1; x^3 + 2x^2 - x - 2$
12. $x - 3; x^4 + 8x^3 + 22x^2 + 24x + 9$
13. $x - 2; x^5 - 32$
14. $x + 3; x^4 + x^3 - 13x^2 - 25x - 12$
15. $x - 1; x^{100} - 1$
16. $x + 1; x^{97} - 1$
17. $x - 3; x^3 - 27$
18. Can the Factor Theorem be used to determine if $x^3 - 1$ is a factor of $5x^7 - 2x - 3$? Explain.

Written **If $P(x) = x^4 - 3x^2 + x + 6$, use the Remainder Theorem to find each of the following. Verify your answer using direct substitution.**

1. $P(1)$
2. $P(2)$
3. $P(-1)$
4. $P(-3)$
5. $P(3)$
6. $P(-2)$
7. $P(5)$
8. $P\left(\frac{1}{2}\right)$
9. $P\left(-\frac{1}{2}\right)$
10. $P\left(-\frac{2}{3}\right)$

Find the values of k such that the first polynomial is a factor of the second polynomial.

11. $x - 1; x^2 + 4x + k$
12. $x - 1; x^2 + kx + 3$
13. $x + 1; 2x^2 - kx + k$
14. $x - 2; x^3 + 2x^2 - kx + 4$
15. $x + 3; x^3 + 2x^2 - kx - 6k$
16. $x + 4; x^3 + kx^2 - 3x + 1$

In each case, show that the given number is a zero of the polynomial. Find all other zeros.

17. $2x^3 - 5x^2 - 28x + 15; -3$
18. $x^3 - x^2 - 34x - 56; -2$
19. $2x^3 - 11x^2 + 12x + 9; -\frac{1}{2}$
20. $x^3 - 3x^2 + x - 3; 3$

Write a polynomial equation in standard form that has the following solution set.

21. $\{-1, 1, 2\}$
22. $\left\{\frac{1}{2}, -2, 1\right\}$
23. $\left\{\frac{2}{3}, -1, 1\right\}$
24. $\{1, 1 + \sqrt{2}, 1 - \sqrt{2}\}$
25. $\left\{-1, -\frac{1}{2}, \frac{1}{3}\right\}$
26. $\{2, i, -i\}$

Challenge

1. Why does the proof of the Factor Theorem have two parts?
2. Give an argument that justifies the Remainder Theorem.

15.4 Finding Rational Roots

Recall that the Factor Theorem can be used to find all the roots of a polynomial equation if you know one of its roots. The following theorem describes what rational numbers are *possible* roots of a polynomial equation.

Rational Root Theorem

Let $P(x) = a_nx^n + a_{n-1}x^{n-1} + \cdots + a_2x^2 + a_1x + a_0$ be a polynomial with integer coefficients. Let $\frac{p}{q}$ be a non-zero rational number in simplest form. If $\frac{p}{q}$ is a root of the polynomial equation $P(x) = 0$, then p is a factor of the constant term a_0 and q is a factor of the leading coefficient a_n.

Proof of Rational Root Theorem

We are given that $\frac{p}{q}$ is a root of $P(x) = 0$.

This means $a_n\left(\frac{p}{q}\right)^n + a_{n-1}\left(\frac{p}{q}\right)^{n-1} + \cdots + a_1\left(\frac{p}{q}\right) + a_0 = 0$.
Multiplying each side by q^n produces

$$a_np^n + a_{n-1}p^{n-1}q + \cdots + a_1pq^{n-1} + a_0q^n = 0.$$

Now subtract a_np^n from both sides,

$$a_{n-1}p^{n-1}q + \cdots + a_1pq^{n-1} + a_0q^n = -a_np^n.$$

Factoring out q on the left side, we obtain

$$q(a_{n-1}p^{n-1} + \cdots + a_1pq^{n-2} + a_0q^{n-1}) = -a_np^n.$$

Now, $P(x)$ has integer coefficients. Also, since $\frac{p}{q}$ is rational, both p and q are integers. Therefore, all the variables in the last equation are integers. Thus, the equation is of the following form.

$$q(\text{some integer}) = -a_np^n \text{ where } q \text{ and } -a_np^n \text{ are integers.}$$

This means q is a factor of $-a_np^n$. Since $\frac{p}{q}$ is in simplest form, q cannot be a factor of p. Consequently, q cannot be a factor of p^n. (Why?) We conclude that q must be a factor of a_n. Proving that p is a factor of a_0 is left as an exercise.

Examples

1 **Solve $x^3 - 4x^2 + x + 6 = 0$.**

Use the Rational Root Theorem. To find possible rational roots $\frac{p}{q}$, p must be a factor of 6 and q must be a factor of 1. Therefore, the *possible* rational roots $\frac{p}{q}$ are: ± 1, ± 2, ± 3, ± 6.

Now, check each possible root, using either direct substitution or synthetic division.

Check for 1:

$$\begin{array}{r|rrrr} 1 & 1 & -4 & 1 & 6 \\ & & 1 & -3 & -2 \\ \hline & 1 & -3 & -2 & 4 \end{array}$$

Since the remainder is not 0, 1 is not a root. We proceed to the next possibility.

Check for -1:

$$\begin{array}{r|rrrr} -1 & 1 & -4 & 1 & 6 \\ & & -1 & 5 & -6 \\ \hline & 1 & -5 & 6 & 0 \end{array}$$

The remainder is 0. Therefore -1 is a root.

By the Factor Theorem, $x + 1$ is a factor. From the synthetic division, you can see that the other factor is $x^2 - 5x + 6$.

$$x^3 - 4x^2 + x + 6 = 0$$
$$(x + 1)(x^2 - 5x + 6) = 0$$
$$x + 1 = 0 \text{ or } x^2 - 5x + 6 = 0$$
$$(x - 3)(x - 2) = 0$$
$$x = -1 \text{ or } x = 3 \text{ or } x = 2$$
$$\text{Solution set} = \{-1, 3, 2\}$$

2 **Find the zeros of P if $P(x) = 2x^3 - 5x^2 - 28x + 15$.**

To find the zeros of P, solve the following equation.

$$2x^3 - 5x^2 - 28x + 15 = 0$$

If $\frac{p}{q}$ is a rational root then p is a factor of 15 and q is a factor of 2.

$$p: \quad \pm 1, \pm 3, \pm 5, \pm 15$$
$$q: \quad \pm 1, \pm 2$$

Possible rational roots: $\pm 1, \pm\frac{1}{2}, \pm 3, \pm\frac{3}{2}, \pm 5, \pm\frac{5}{2}, \pm 15, \pm\frac{15}{2}$

You can verify that $\frac{1}{2}$ is a root. Thus, $x - \frac{1}{2}$ is a factor of $P(x)$.

$$2x^3 - 5x^2 - 28x + 15 = 0$$
$$\left(x - \frac{1}{2}\right)(2x^2 - 4x - 30) = 0$$
$$x = \frac{1}{2} \text{ or } 2x^2 - 4x - 30 = 0$$
$$x^2 - 2x - 15 = 0$$
$$(x - 5)(x + 3) = 0$$
$$x = 5 \text{ or } x = -3$$

The zeros are $\frac{1}{2}$, 5, and -3.

Exercises

Exploratory **Each of the following equations has at least one root from among the numbers -1, 2, $\frac{1}{2}$, and 3. Find all the roots of each equation.**

1. $x^3 - 4x^2 + x + 6 = 0$
2. $x^3 - 13x + 12 = 0$
3. $x^3 - 2x^2 + x = 2$
4. $3x^2 + 2x^3 = 1$
5. $2x^4 - 3x^3 - 3x^2 + 2x = 0$
6. $6x^4 + 11x^3 - 3x^2 = 2x$
7. Without doing any computation, Sally knows that 5 cannot possibly be a root of $x^7 - 3x^3 - 7x^2 + 6 = 0$. Explain.

Give the POSSIBLE rational roots of each equation.

8. $x^3 - 3x^2 + 5x - 3 = 0$
9. $x^3 - 4x + 10 = 0$
10. $x^3 + 7x^2 + 18 = 0$
11. $x^7 - 8x^2 - 12 = 0$
12. $2x^{10} + x^{11} - 7 + x^2 = 0$
13. $5x^8 - x^9 + 2x^3 = 7 + x^2$

Written **Find all rational roots, if any, of each equation.**

1. $6x^3 + 4x^2 - 14x + 4 = 0$
2. $2x^3 - 11x^2 + 12x + 9 = 0$
3. $3x^3 - 14x^2 + 11x - 2 = 0$
4. $4x^3 - 4x^2 - x + 1 = 0$
5. $6x^3 - 5x^2 + 7x + 2 = 0$
6. $x^4 - \frac{1}{2}x^3 - 2x^2 + \frac{1}{2}x + 1 = 0$
7. $x^4 - 3x^3 + x^2 - 3x = 0$
8. $3x^3 + 3x = 1 + x^2$

Each equation has at least one rational root. Find all the roots. Domain = $\mathscr{C}$.

9. $x^3 - 2x^2 - x + 2 = 0$
10. $2x^3 - x^2 - 4x + 2 = 0$
11. $3x^3 + x^2 - 6x - 2 = 0$
12. $2x^3 + x^2 + 8x + 4 = 0$
13. $x^3 - x^2 - 3x - 1 = 0$
14. $3x^3 - 5x^2 + 4x + 2 = 0$
15. $x^3 - 8 = 0$
16. $x^3 + 1 = 0$
17. $x^3 - 1 = 0$
18. $2x^4 + 5x^3 + 3x^2 - x - 1 = 0$
19. $2x^3 - 9x^2 + 6x = 1$
20. $2x^4 - 3x^3 + 17x^2 + 12x - 10 = 0$
21. $x^5 + 27x^2 - x^3 - 27 = 0$
22. Juan lists all the possible rational roots for exercise 12. He looks at the equation carefully and notices that he can eliminate one-half of the possibilities. What has Juan noticed?
23. Explain why one-half of the possible rational roots for exercise 19 can be eliminated immediately.
24. In solving $x^3 - x^2 + 4x + 5 = 0$, Jane claims that one of the roots must be ± 1, or ± 5, since these are the only factors of the constant term. Do you agree? Explain.

Challenge **Give an argument based on the Rational Root Theorem which shows that the following numbers are not rational.**

1. $\sqrt{2}$
2. $\sqrt{3}$
3. $\sqrt{5}$
4. $\sqrt[5]{100}$
5. Complete the proof of the Rational Root Theorem.

Mixed Review

Choose the best answer.

1. Which of the following is the simplest form of $\frac{x^2 + 16}{x^2 - 16} + \frac{4}{4 - x}$?

a. $\frac{x^2 - 4x + 32}{x^2 - 16}$ **b.** $\frac{x}{x + 4}$ **c.** $\frac{x - 8}{x - 4}$ **d.** $\frac{x + 8}{x + 4}$

2. What is the multiplicative inverse of $4 - 3i$?

a. $\frac{4 + 3i}{25}$ **b.** $\frac{4 - 3i}{25}$ **c.** $\frac{4 - 3i}{7}$ **d.** $\frac{4 + 3i}{7}$

3. If $f(x) = x^2 - 2x - 1$ and $g(x) = 4x - 3$, what is $(f \circ g)(x)$?

a. $-4x^2 + 8x + 7$ **b.** $16x^2 - 32x + 14$ **c.** $8x^2 - 16x + 7$ **d.** $16x^2 - 26x - 8$

4. Which word has *both* horizontal and vertical line symmetry (when written vertically)?

a. HOAX **b.** DOCK **c.** WHAT **d.** OHIO

5. Which expression is equivalent to $3 \log x - 2 \log 2x$?

a. $\log \frac{x}{2}$ **b.** $\log \frac{3}{4}$ **c.** $\log \frac{x}{4}$ **d.** $\log (x^3 - 4x^2)$

6. In circle O, chords $\overline{AB}$ and $\overline{CD}$ intersect at point X. If $AB = 17$, $DX = 12$, $CX = 6$, and $AX > BX$, then what is the value of BX?

a. 9 **b.** 8 **c.** 17 **d.** none of these

7. The graph of which of the following equations has amplitude 2 and period 120°?

a. $y = 3 \cos 2x$ **b.** $y = 2 \tan \frac{3}{2}x$ **c.** $y = 2 \sin \frac{1}{3}x$ **d.** $y = -2 \sin 3x$

8. What is the height, to the nearest meter, of a tree that casts a shadow 20 meters long when the angle of elevation to the sun is 21°40′?

a. 8 meters **b.** 7 meters **c.** 50 meters **d.** 19 meters

9. What is the exact value of 4 cos 15°?

a. $\sqrt{2} + \sqrt{6}$ **b.** $2\sqrt{2 + \sqrt{3}}$ **c.** both **a.** and **b.** **d.** none of these

10. Which transformation *cannot* be equivalent to the composite of two line reflections?

a. half-turn **b.** glide reflection **c.** translation **d.** rotation

11. If a fair coin is tossed 5 times, what is the probability of obtaining at least two heads?

a. $\frac{3}{16}$ **b.** $\frac{13}{16}$ **c.** $\frac{5}{32}$ **d.** $\frac{27}{32}$

12. The number of hours of television watched weekly by 3,000 people in Ocala is normally distributed with a mean of 22 hours and a standard deviation of 7.5 hours. About how many people watch between 7 and 29.5 hours of television per week?

a. 2040 people **b.** 2445 people **c.** 1425 people **d.** 2520 people

13. What is the 12th term of the arithmetic sequence $-5, -1, 3, \ldots$?

a. 43 **b.** 39 **c.** 47 **d.** none of these

14. If $A = \begin{bmatrix} -1 & 4 \\ 3 & 5 \end{bmatrix}$, what is the determinant of A?

a. 7 **b.** -19 **c.** -17 **d.** -23

15.5 The Fundamental Theorem

Some polynomial equations like $4x^2 - 1 = 0$ have no integral solutions. Some like $x^2 - 2 = 0$ have no rational solutions. Some like $x^2 + 1 = 0$ have no real solutions. Are there any polynomial equations that have no solutions in the set of complex numbers?

This question was first answered in the 18th century by the famous German mathematician Karl Friedrich Gauss, who proved the *Fundamental Theorem of Algebra* when he was 22 years old. The proof requires a knowledge of mathematics beyond this course.

Fundamental Theorem of Algebra

Every polynomial equation of degree greater than zero has at least one root in the set of complex numbers.

Although the Fundamental Theorem guarantees that a root exists, it does not tell us how to find the root. Nor does it tell us how many roots a polynomial equation has.

The following corollary to the Fundamental Theorem provides important information about the number of roots of a polynomial equation.

Corollary

Every polynomial equation of degree n, where $n > 0$, has n roots in the set of complex numbers.

When using this corollary, count a root as many times as it occurs. For example, consider solving the second degree equation shown below.

$$x^2 - 6x + 9 = 0$$
$$(x - 3)(x - 3) = 0$$
$$x = 3 \text{ or } x = 3$$

The root 3 appears twice. Thus, 3 is a **double root.**

When solving equations that have imaginary roots, you may have observed that imaginary roots always appear in conjugate pairs. This situation occurs for all polynomial equations having real coefficients.

Conjugate Root Theorem

Suppose a and b are real numbers with $b \neq 0$. If $a + bi$ is a root of a polynomial equation with real coefficients then $a - bi$ is also a root.

Examples

1 **If $1 + i$ is a root of $2x^3 - 3x^2 + 2x + 2 = 0$, find all the roots.**

Since $1 + i$ is a root, the Conjugate Root Theorem says that $1 - i$ must also be a root. The Factor Theorem says that $x - (1 + i)$ and $x - (1 - i)$ are both factors of the polynomial. Proceed as follows.

$$2x^3 - 3x^2 + 2x + 2 = 0$$
$$[x - (1 + i)][x - (1 - i)][?] = 0$$
$$[(x - 1) - i][(x - 1) + i][?] = 0 \quad \textit{Associative Property}$$
$$(x^2 - 2x + 2)(?) = 0$$

Use division to find the other factor.

$$\begin{array}{r} 2x \;\; + 1 \\ x^2 - 2x + 2\overline{)2x^3 - 3x^2 + 2x + 2} \\ \underline{2x^3 - 4x^2 + 4x \quad\;\;} \\ x^2 - 2x + 2 \\ \underline{x^2 - 2x + 2} \\ 0 \end{array}$$

$$(x^2 - 2x + 2)(2x + 1) = 0$$
$$2x + 1 = 0$$
$$x = \frac{1}{2}$$

The roots are $1 + i$, $1 - i$, $-\frac{1}{2}$. *How do you know there are no other roots?*

2 **Find a polynomial equation with real coefficients and lowest degree whose roots include $2 + i$ and -1.**

By the Conjugate Root Theorem, $2 - i$ must also be a root. Therefore, the polynomial equation can be written in the factored form shown below.

$$[x - (2 + i)]\,[x - (2 - i)](x + 1) = 0$$
$$[(x - 2) - i]\,[(x - 2) + i](x + 1) = 0$$

$x^3 - 3x^2 + x + 5 = 0$ is the required equation.

Information about the nature of the roots of a polynomial equation is provided by using a theorem developed by the famous French mathematician and philosopher *René Descartes*. Before stating the theorem, however, it is necessary to discuss the meaning of a variation in sign. Suppose the terms of a polynomial with real coefficients are arranged in descending powers of x. A **variation in sign** occurs when two consecutive coefficients have different *signs*. Terms with zero coefficients are deleted.

Examples

Find the number of variations in sign in each of the following polynomials.

3 $x^4 + 2x^3 - 5x^2 - 2x + 7$

$x^4 + 2x^3 - 5x^2 - 2x + 7$

There are two variations in sign.

4 $-5x^4 + x^2 - 7x + 3$

$-5x^4 + x^2 - 7x + 3$

There are three variations in sign.

Descartes' Rule of Signs

Let $P(x)$ be a polynomial in standard form with real coefficients and v variations in sign. Then the number of positive roots of $P(x) = 0$ is v or is less than v by a positive even number.

Example

5 **Give the possible number of positive roots of $P(x) = 0$ if $P(x) = x^3 - 3x^2 + x - 1$.**

$P(x) = x^3 - 3x^2 + x - 1$ has 3 variations in sign. The number of positive roots of $P(x) = 0$ is 3 or $3 - 2 = 1$. Therefore, $P(x) = 0$ has either 3 positive roots or just 1 positive root.

To find the possible number of negative roots of a polynomial equation, we make use of the following theorem.

If $P(x) = 0$ has roots $r_1, r_2, r_3, \ldots, r_n$ then the equation $P(-x) = 0$ has roots $-r_1, -r_2, -r_3, \ldots, -r_n$.

Examples

6 **Write an equation whose roots are the negatives of the roots of $P(x) = 0$ if $P(x) = x^4 - 2x^3 - 3x^2 - 4x + 2$.**

$$\begin{aligned} P(-x) &= (-x)^4 - 2(-x)^3 - 3(-x)^2 - 4(-x) + 2 \\ &= x^4 + 2x^3 - 3x^2 + 4x + 2 \end{aligned}$$

Therefore, $x^4 + 2x^3 - 3x^2 + 4x + 2 = 0$ is the required equation.

7 **Discuss the nature of the roots of $P(x) = 0$ if $P(x) = 2x^4 - 3x^2 + 2x - 1$.**

$P(x)$ has 3 variations in sign. Therefore, $P(x) = 0$ has either 3 positive roots or 1 positive root.

$P(-x) = 2x^4 - 3x^2 - 2x - 1$ has 1 variation in sign. Therefore, $P(-x) = 0$ has 1 positive root. But a positive root of $P(-x) = 0$ is a negative root of $P(x) = 0$. Therefore, $P(x) = 0$ has 1 negative root. Combining this information with the Fundamental Theorem produces the following summary.

Possible Number of:		
Positive Roots	**Negative Roots**	**Imaginary Roots**
3	1	0
1	1	2

Exercises

Exploratory **Tell how many roots each equation has in $\mathscr{C}$.**

1. $x^3 - 1 = 0$
2. $x^5 + 32 = 0$
3. $x^2 - 2x + 1 = 0$
4. $5x^4 - 2x^2 - \sqrt{2} = 0$
5. $x^5 - 3x^7 + i = 0$
6. The solution set of $x^2 - 8x + 16 = 0$ is $\{4\}$. Does this contradict the corollary to the Fundamental Theorem given in the text? Explain.

For exercises 7–9 consider the equation $x^2 - ix + x - i = 0$.

7. Show that i is a root.
8. Show that $-i$ is not a root.
9. Do the results of exercises 7 and 8 contradict the Conjugate Root Theorem? Explain.
10. If 1 and $2 + i$ are roots of $x^3 - 5x^2 - 9x - 5 = 0$, find the remaining root. How do you know there are no other roots?

Written **In exercises 1–8, one root is given. Find all the roots.**

1. $2x^3 - x^2 + 2x - 1 = 0;\ i$
2. $x^3 - 2x^2 + 4x - 8 = 0;\ 2i$
3. $x^3 - 3x^2 + x - 3 = 0;\ -i$
4. $x^3 - 10x^2 + 34x - 40 = 0;\ 3 - i$
5. $x^3 - 3x^2 + 9x + 13 = 0;\ 2 - 3i$
6. $x^3 + 2x^2 - 3x + 20 = 0;\ 1 + 2i$
7. $x^4 - x^3 + 6x^2 - x + 15 = 0;\ 1 - 2i$
8. $x^4 - 2x^2 + 9 = 0;\ i + \sqrt{2}$

Find a polynomial equation in standard form with real coefficients and lowest degree whose roots include the following.

9. $1, i$
10. $3, -i$
11. $-2, 1 + i$
12. $\frac{1}{2}, 3i$
13. $2i, 1 + i$
14. $3 - i, 1 + i$
15. i is a double root
16. 0 is a double root, i

Give the possible number of positive roots of each equation.

17. $x^4 - 3x^3 + x^2 - 2 = 0$

18. $2x^5 + 3x^2 - 4x + 1 = 0$

19. $x^{10} - 1 = 0$

20. $x^6 - 3x^5 + 4x^4 - 5x^3 + 3x^2 + 1 = 0$

21. If the coefficients of $P(x)$ are all positive, can $P(x) = 0$ have any positive roots? Explain.

Write a polynomial equation in standard form whose roots are the negative of the roots of the given equation.

22. $x^2 - x - 6 = 0$

23. $x^2 - 6x + 9 = 0$

24. $x^2 - 2x + 2 = 0$

25. $x^3 - x^2 + x - 3 = 0$

26. $2x^5 - 3x^3 + 2x^2 - x = 0$

27. $x^7 - 3x^5 + 6x^2 - 3 = 0$

Discuss the nature of the roots of each equation.

28. $x^3 - 1 = 0$

29. $x^3 - x^2 + x - 1 = 0$

30. $4x^3 - 6x + 1 = 0$

31. $2x^4 - 7x + 1 = 0$

32. $x^{10} - x^8 + x^6 - x^4 + x^2 - 1 = 0$

33. $4x^5 - x^2 + 1 = 0$

34. $x^3 - x = 0$

35. $x^4 - 16 = 0$

Challenge

1. Ted claims that $4x^3 + 5x^2 + 1 = 0$ must have one irrational root and two imaginary roots. Is he correct? Explain.

Mathematical Excursions

Using Computers to Find Real Roots

There are many ways to use a computer in working with polynomials. For example, the program at the right shows how a computer can solve the polynomial equation $y = x^3 + 11x^2 - 6x - 50$ by substitution.

```
10  FØR X=0 TØ 5
20    LET Y=X↑3+11*X↑2-6*X-50
30    PRINT X,Y
40  NEXT X
50  END
RUN
0    -50
1    -44
2    -10
3    58
4    166
5    320
```

Note that $-10 < 0 < 58$. Therefore, $y = 0$ for some value of x between 2 and 3. For greater accuracy, replace line 10 with the following line.

```
10 FØR X = 2 TØ 3 STEP .2
```

The output shows that the value of x is between 2 and 2.2. Replace line 10 as many times as necessary to find the root to the desired accuracy.

Exercise The polynomial equation in the program shown above has two more real roots. Change the program to find each of those roots to the nearest tenth.

15.6 Irrational Roots

Consider $P(x) = 0$ where $P(x) = x^3 + x - 4$. Since $P(x)$ has one variation in sign, *Descartes' Rule of Signs* tells us there is exactly one positive root. The Rational Root Theorem tells us that the only possible positive *rational* roots are 1, 2, and 4. Each of these "roots" is checked below.

$$P(1) = 1 + 1 - 4 = -2 \quad P(2) = 8 + 2 - 4 = 6 \quad P(4) = 64 + 4 - 4 = 64$$

We see that neither 1, 2, nor 4 are roots.

We conclude that the positive root must be *irrational*. Now we consider the problem of finding a rational approximation of this root.

To begin, look carefully at the values for $P(1)$ and $P(2)$ given above. They tell us that point $A(1, -2)$ and point $B(2, 6)$ are both on the graph of the function $P(x) = x^3 + x - 4$. If it is assumed that the graph of P is a smooth unbroken curve, then it must be the case that the graph crosses the x-axis between 1 and 2. That is, the positive irrational root of $x^3 + x - 4 = 0$ must lie between 1 and 2.

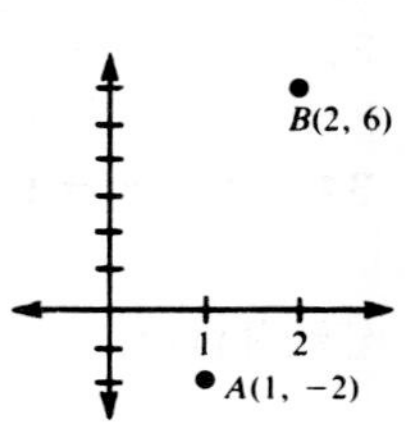

To "close in" on this root, check to see if the root lies between 1 and 1.5 or between 1.5 and 2. Since the calculations become a bit tedious, you might want to have a calculator handy or consider using the computer program from page 562. First compute $P(1.5)$.

$P(1.5) = (1.5)^3 + 1.5 - 4 = 0.875$. Since $P(1.5)$ is positive, you can conclude that the root is between 1 and 1.5.

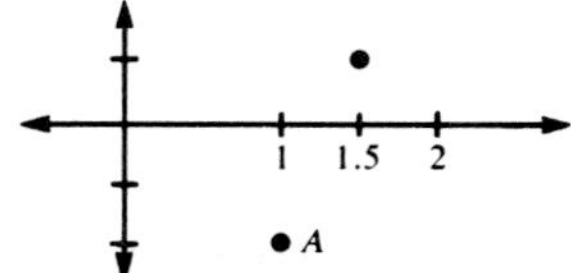

Checking 1.4, you should obtain $P(1.4) = 0.144$. Since $P(1.4)$ is also positive, the root must be between 1 and 1.4.

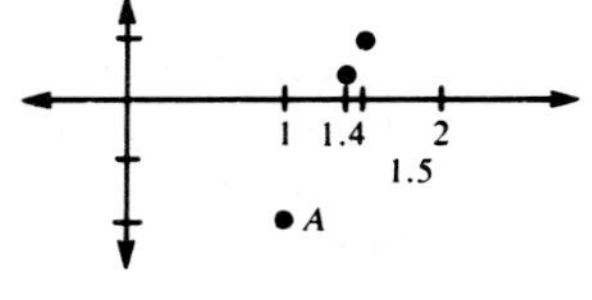

Checking 1.3, you should obtain $P(1.3) = -0.503$. Since $P(1.3)$ is negative, you can conclude that the root is between 1.3 and 1.4.

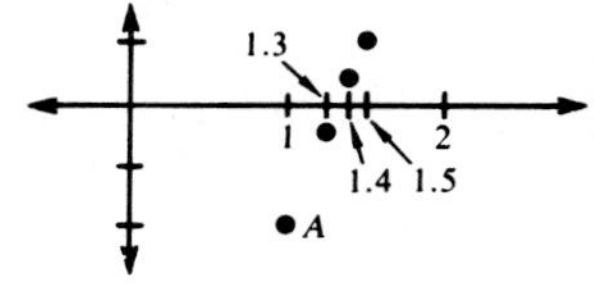

To determine whether the root is closer to 1.3 or 1.4, check the average of 1.3 and 1.4, or 1.35.

$$\begin{aligned} P(1.35) &= (1.35)^3 + 1.35 - 4 \\ &= -0.189625. \end{aligned}$$

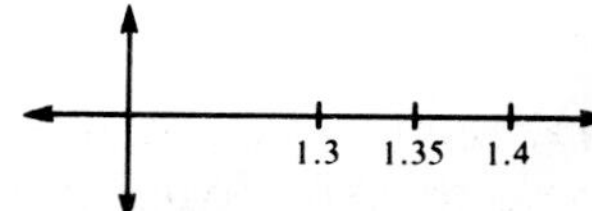

Since $P(1.35)$ is negative and $P(1.4)$ is positive, you can conclude that the graph of $P(x) = x^3 + x - 4$ crosses the x-axis between 1.35 and 1.4. Therefore, to the nearest tenth, a rational approximation of the positive root of $x^3 + x - 4 = 0$ is 1.4.

In the preceding example, it was assumed that the graph of a polynomial function is a smooth, unbroken curve. You are already familiar with the graphs of linear and quadratic polynomial functions and you may have graphed some cubics, such as $y = x^3$. The following examples show how knowledge of the zeros of a polynomial provides information about its graph.

Examples

1 **Sketch the graph of $P(x) = x^3 - 3x^2 + 2x$.**

$$P(x) = x(x^2 - 3x + 2) \quad \text{or} \quad x(x - 2)(x - 1)$$

We see that $P(x)$ has three zeros: 0, 2, 1. This means that $P(x)$ crosses the x-axis at (0, 0), (2, 0), and (1, 0). Next, we see that the following is true.

1. $P(x)$ is negative if $x < 0$ or $1 < x < 2$.

2. $P(x)$ is positive if $0 < x < 1$ or $x > 2$.

To complete the graph, set up a table of values and plot the points, connecting them with a smooth curve.

x	P(x)
-1	-6
0	0
$\frac{1}{2}$	$\frac{3}{8}$
1	0
$\frac{3}{2}$	$-\frac{3}{8}$
2	0
3	6

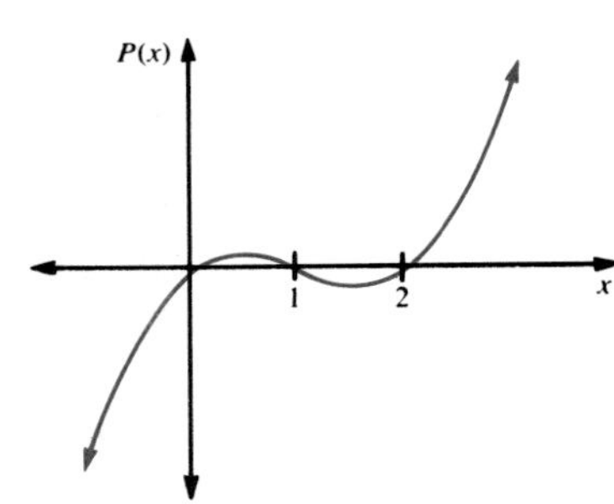

2 **Sketch the graph of $P(x) = -x^3 + 2x^2 + 4x - 8$.**

Using the Rational Root Theorem, we find that 2 and -2 are zeros. Therefore, we may write $P(x)$ in factored form: $P(x) = -(x - 2)^2(x + 2)$.

1. If $x < -2$ then $P(x)$ is positive.

2. If $x > -2$ then $P(x)$ is negative.

x	P(x)
-3	25
-2	0
-1	-9
0	-8
1	-3
2	0
3	-5

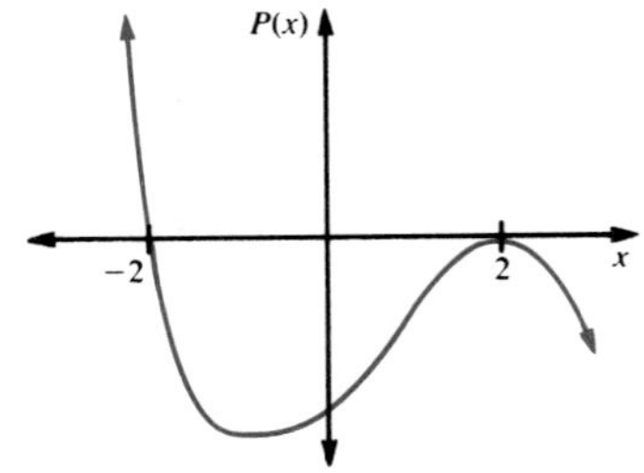

In general, the graph of a cubic polynomial is shaped like an *S* lying on its side.

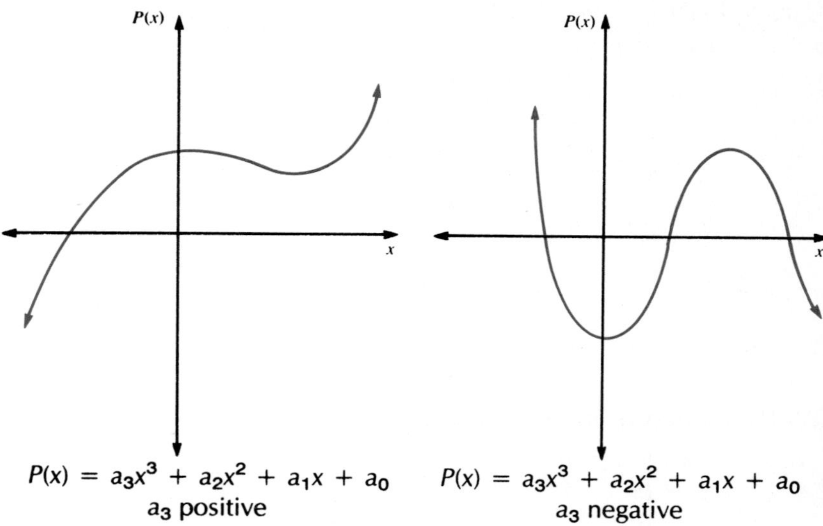

$P(x) = a_3x^3 + a_2x^2 + a_1x + a_0$
a_3 positive

$P(x) = a_3x^3 + a_2x^2 + a_1x + a_0$
a_3 negative

In approximating irrational roots, we assumed the following principle which can be proved in advanced mathematics.

The Location Principle

Suppose $P(x)$ is a polynomial with real coefficients and a and b are real numbers. If $P(a)$ is negative and $P(b)$ is positive, then the equation $P(x) = 0$ has at least one real root between a and b.

The Location Principle also applies when $P(a)$ is positive and $P(b)$ is negative, as shown in Figure 2.

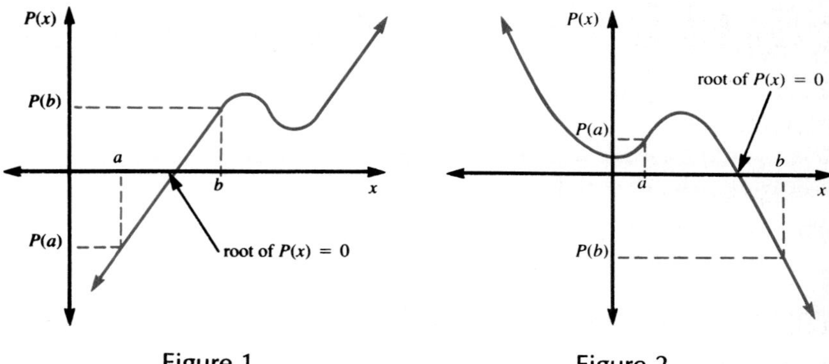

Figure 1

Figure 2

Figure 1 shows the graph of a polynomial, $P(x)$, where a and b are real numbers and $P(a) < 0$ and $P(b) > 0$. The graph of $P(x)$ must cross the x-axis at some value of x between a and b. At that point $P(x) = 0$ and x is a root.

In Figure 2, $P(a) > 0$ and $P(b) < 0$. Therefore, the polynomial has a root between a and b.

Exercises

Exploratory **Each of the following equations has a root between two consecutive integers. Find the two integers. Justify your answer.**

1. $x^3 - x - 3 = 0$
2. $x^2 - 2 = 0$
3. $x^4 - 4x^2 + 3 = 0$
4. $2x^5 + 3x - 2 = 0$
5. $x^3 - 12 = 0$
6. $2x^3 + 5 = 0$

For exercises 7–14, compute each of the following. Let $P(x) = x^4 - x^2 + 6$.

7. $P(-3)$
8. $P(-2)$
9. $P(-1)$
10. $P(0)$
11. $P(1)$
12. $P(2)$
13. $P\left(\frac{1}{2}\right)$
14. $P(-0.5)$
15. Discuss the nature of the roots of $x^4 - x^2 + 6 = 0$.

Written **Approximate at least one real root of each equation.**

1. $x^3 - 2x^2 + 6 = 0$
2. $2x^5 + 3x - 2 = 0$
3. $x^3 + 2x^2 - 3x - 5 = 0$
4. $x^3 - 5 = 0$
5. $x^3 - x^2 + 1 = 0$
6. $3x^2 - 8x + 1 = 0$
7. $x^5 - 6 = 0$
8. $x^3 - x - 3 = 0$
9. $x^3 - 2x^2 - x + 1 = 0$
10. $x^3 - 3x^2 + 7x - 11 = 0$

Sketch the graph of each function.

11. $P(x) = x^3$
12. $P(x) = x^4$
13. $P(x) = x^5$
14. $P(x) = x^6$
15. $P(x) = 4x^6$
16. $P(x) = 3x^5$
17. $P(x) = x^3 + 3$
18. $P(x) = x^4 - 2$
19. $P(x) = x^3 - x$
20. $P(x) = (x + 4)(x - 1)(x + 1)$
21. $P(x) = (x - 1)(x - 2)(x + 2)$
22. $P(x) = -x^3 - x$
23. $P(x) = -x^3 + 3$
24. $P(x) = x^4 - 81$
25. $P(x) = x^3 - 3x - 4$
26. $P(x) = x^4 + 2x^2 + 1$
27. $P(x) = x^3 - 3x + 4$
28. $P(x) = -x^3 - 13x - 12$

Use your knowledge of transformations to describe the relationship between the graphs of each of the following pairs of equations.

29. $y = x^2$; $y = (x - 3)^2$
30. $y = 2x^2$; $y = 2(x + 4)^2$
31. $y = x^3$; $y = (x - 2)^3$
32. $y = \frac{1}{2}x^2$; $y = \frac{1}{2}(x - 1)^2 + 2$
33. $y = \frac{4}{x}$; $y = \frac{4}{x - 3}$
34. $y = x^3$; $y = (x - 3)^3 - 2$

Challenge

1. Compute $\tan\left(\frac{\pi}{4}\right)$.
2. Compute $\tan\left(\frac{3\pi}{4}\right)$.
3. Do the results of Challenge Exercises 1 and 2 suggest that the graph of $y = \tan x$ crosses the x axis between $\frac{\pi}{4}$ and $\frac{3\pi}{4}$? Explain.
4. Determine values of k such that $x^3 - 4x^2 - 2x + k = 0$ has a root between -1 and -2.

15.7 Trigonometry and Complex Numbers

Suppose you wish to solve the equation $x^3 + 8i = 0$. None of the procedures you have learned so far can be used since the coefficient $8i$ is not a real number. In the next two sections, you will learn some techniques for dealing with such equations.

To begin, recall your work in trigonometry. In the diagram, point P has coordinates (x, y), $m\angle TOP = \theta$, and $OP = r$, where $r = \sqrt{x^2 + y^2}$. If S is the intersection of the unit circle and $\overline{OP}$, then $OS = 1$ and S has coordinates $(\cos\theta, \sin\theta)$. Point P is the image of point S under dilation $D_{O,r}$ with center at the origin. In symbols, we have

$$S \xrightarrow{D_{O,r}} P$$

$$\text{or } (\cos\theta, \sin\theta) \xrightarrow{D_{O,r}} (x, y).$$

But the rule for dilation D_r tells us that

$$(\cos\theta, \sin\theta) \xrightarrow{D_{O,r}} (r\cos\theta, r\sin\theta).$$

We conclude that

$$x = r\cos\theta \text{ and } y = r\sin\theta.$$

You can see, then, that the coordinates of point P can be expressed in two different forms.

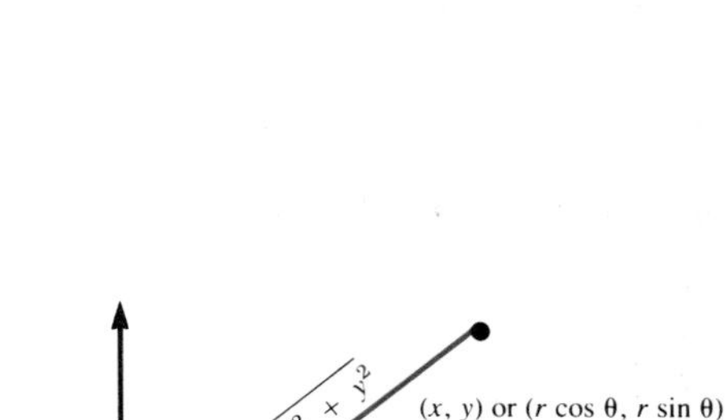

Rectangular Form: (x, y)

Trigonometric Form: $(r\cos\theta, r\sin\theta)$

Examples

1 **Express (1, 1) in trigonometric form.**

Plot (1, 1). Then find r and θ.

$$r = \sqrt{x^2 + y^2} = \sqrt{1^2 + 1^2} = \sqrt{2}$$

To find θ, we note that $x = r\cos\theta$ and $y = r\sin\theta$.

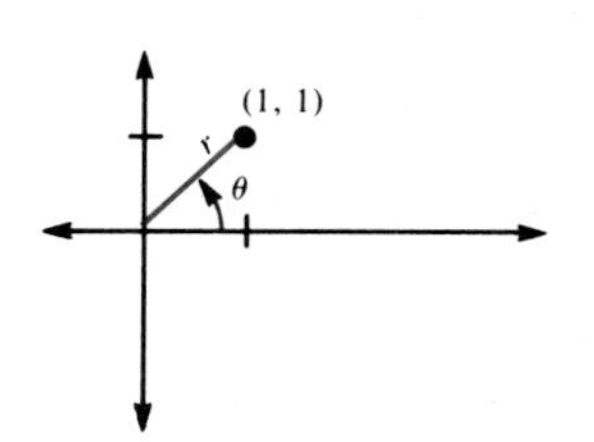

$$x = r\cos\theta \qquad y = r\sin\theta$$

$$1 = \sqrt{2}\cos\theta \qquad 1 = \sqrt{2}\sin\theta$$

$$\frac{\sqrt{2}}{2} = \cos\theta \qquad \frac{\sqrt{2}}{2} = \sin\theta$$

Since the sin and cos are both positive, $\theta = 45°$. Actually, we need use only one of the equations since the diagram tells us θ is in Quadrant I.

Therefore, the trigonometric form of (1, 1) is $(\sqrt{2}\cos 45°, \sqrt{2}\sin 45°)$.

2 **Express $(-\sqrt{3}, 1)$ in trigonometric form.**

Plot $(-\sqrt{3}, 1)$.

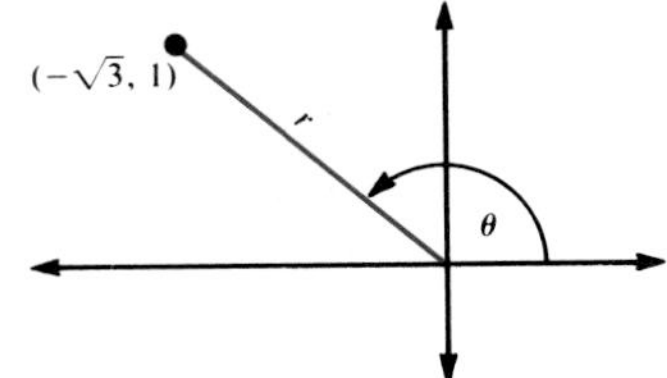

Find r.
$$r = \sqrt{x^2 + y^2}$$
$$r = \sqrt{3 + 1}$$
$$r = 2$$

Find θ.
$$x = r\cos\theta$$
$$-\sqrt{3} = 2\cos\theta$$
$$-\frac{\sqrt{3}}{2} = \cos\theta$$

The diagram tells us that θ is in Quadrant II. Therefore, $\theta = 150°$. We conclude that the trigonometric form of $(-\sqrt{3}, 1)$ is $(2\cos 150°, 2\sin 150°)$.

In previous work with complex numbers you saw that each complex number $x + yi$ corresponds to the point (x, y). But you have just seen that the point (x, y) may also be expressed in the form $(r\cos\theta, r\sin\theta)$, where $x = r\cos\theta$, $y = r\sin\theta$, and $r = \sqrt{x^2 + y^2}$.

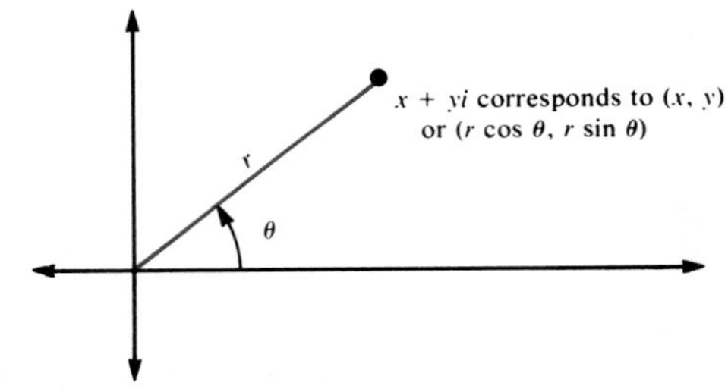

$$x + yi = r\cos\theta + (r\sin\theta)i$$
$$\text{or } x + yi = r(\cos\theta + i\sin\theta)$$

Thus, there are two forms of a complex number.

Rectangular Form of a Complex Number	$x + yi$
Trigonometric or Polar Form of a Complex Number	$r(\cos\theta + i\sin\theta)$

The number r is called the **modulus** (plural, **moduli**). The number θ is called the **amplitude.** The amplitude has the smallest positive measure.

Since the modulus is the distance of the point (x, y) from the origin, you may think of the modulus as the **absolute value of a complex number.** In symbols, this can be expressed as follows.

$$r = |x + yi| = \sqrt{x^2 + y^2}$$

Note that r is a non-negative real number.

Examples

3 **Express $1 - i$ in trigonometric form.**

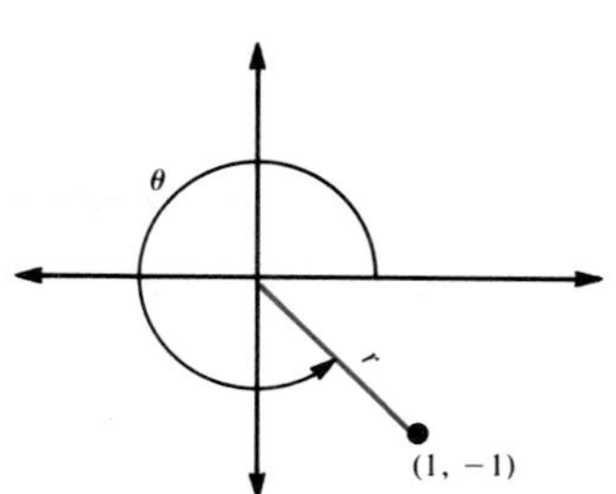

Plot point $(1, -1)$ which corresponds to $1 - i$.

Find r: $r = |1 - i| = \sqrt{1^2 + (-1)^2} = \sqrt{2}$

Find θ: $(1, -1)$ lies on the line $y = -x$, and θ is in quadrant IV.

Therefore, $\theta = 315°$.

$$1 - i = \sqrt{2}(\cos 315° + i \sin 315°)$$

4 **Express $-1 - i\sqrt{3}$ in trigonometric form.**

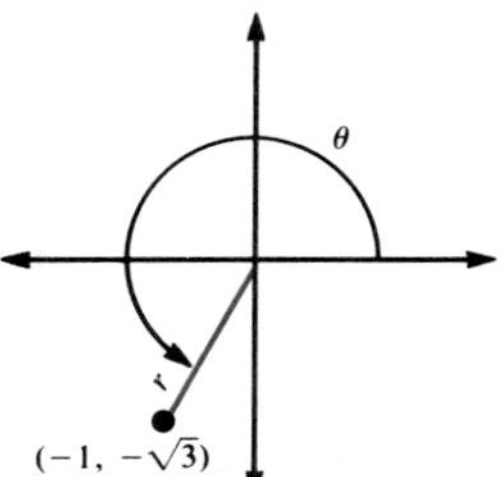

Graph $-1 - i\sqrt{3}$ by plotting $(-1, -\sqrt{3})$.

Find r. $\quad r = |-1 - i\sqrt{3}| = \sqrt{(-1)^2 + (-\sqrt{3})^2} = 2$

Find θ.

$$y = r \sin \theta$$
$$-\sqrt{3} = 2 \sin \theta$$
$$-\frac{\sqrt{3}}{2} = \sin \theta$$

Since θ is in Quadrant III, $\theta = 240°$.

Therefore, $-1 - i\sqrt{3} = 2(\cos 240° + i \sin 240°)$

5 **Express $5(\cos 120° + i \sin 120°)$ in rectangular form.**

$$5(\cos 120° + i \sin 120°) = 5\left(-\frac{1}{2} + i\frac{\sqrt{3}}{2}\right)$$
$$= -\frac{5}{2} + \frac{5i\sqrt{3}}{2}$$

Rectangular form is expressed as $x + yi$.

Exercises

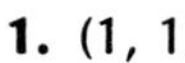

Exploratory **In the figure at the right, $r = OP$. Determine the values of r and θ if P has the following coordinates.**

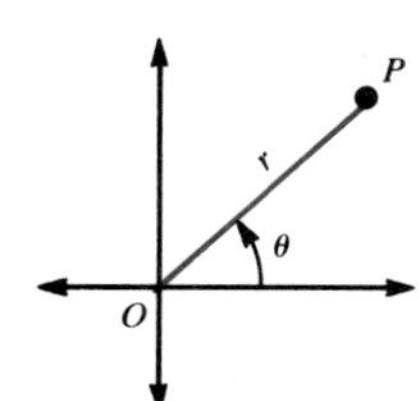

1. $(1, 1)$ **2.** $(2, 2)$ **3.** $(3, 3)$

4. $(1, \sqrt{3})$ **5.** $(\sqrt{3}, 1)$ **6.** $\left(\frac{\sqrt{2}}{2}, \frac{\sqrt{2}}{2}\right)$

7. $\left(\frac{\sqrt{3}}{2}, \frac{1}{2}\right)$ **8.** $\left(\frac{1}{2}, \frac{\sqrt{3}}{2}\right)$ **9.** $\left(\frac{3\sqrt{2}}{2}, \frac{3}{2}\right)$

Determine the coordinates of *P* if *r* and θ have the following values.

10. $r = 2; \theta = 30°$ **11.** $r = 1; \theta = 30°$ **12.** $r = \frac{1}{2}; \theta = 45°$

13. $r = 5; \theta = 60°$ **14.** $r = 10; \theta = 60°$ **15.** $r = 5; \theta = 45°$

Plot point *P* given the following values for *r* and θ.

16. $r = 1; \theta = 45°$ **17.** $r = 2; \theta = 45°$ **18.** $r = \frac{1}{2}; \theta = 60°$

Written **Plot each of the following points. Then give the rectangular coordinates.**

1. $(\cos 60°, \sin 60°)$ **2.** $(\cos 30°, \sin 30°)$ **3.** $(\cos 135°, \sin 135°)$
4. $(\cos 90°, \sin 90°)$ **5.** $(\cos 0°, \sin 0°)$ **6.** $(3 \cos 120°, 3 \sin 120°)$
7. $(2 \cos 90°, 2 \sin 90°)$ **8.** $(3 \cos 270°, 3 \sin 270°)$ **9.** $(6 \cos 330°, 6 \sin 330°)$
10. $\left(\frac{1}{2} \cos 315°, \frac{1}{2} \sin 315°\right)$ **11.** $\left(\frac{1}{2} \cos 45°, \frac{1}{2} \sin 45°\right)$ **12.** $(4 \cos 240°, 4 \sin 240°)$

Express each of the following in trigonometric form.

13. $(2, 2)$ **14.** $(3, 3)$ **15.** $(\sqrt{2}, -\sqrt{2})$ **16.** $(-5, 0)$
17. $\left(\frac{\sqrt{2}}{2}, \frac{\sqrt{2}}{2}\right)$ **18.** $(-4, 4)$ **19.** $\left(-\frac{\sqrt{3}}{2}, \frac{1}{2}\right)$ **20.** $(0, 4)$
21. $\left(-\frac{1}{2}, \frac{\sqrt{3}}{2}\right)$ **22.** $(2\sqrt{3}, -2)$ **23.** $(-3\sqrt{2}, -3\sqrt{2})$ **24.** $(-2, -2\sqrt{3})$

Graph each complex number. Then express the number in rectangular form and give the modulus and amplitude.

25. $\cos 90° + i \sin 90°$ **26.** $\cos 180° + i \sin 180°$
27. $\cos 270° + i \sin 270°$ **28.** $3(\cos 0° + i \sin 0°)$
29. $2(\cos 30° + i \sin 30°)$ **30.** $4(\cos 45° + i \sin 45°)$
31. $4(\cos 300° + i \sin 300°)$ **32.** $3\left(\cos \frac{\pi}{2} + i \sin \frac{\pi}{2}\right)$
33. $6(\cos 235° + i \sin 235°)$ **34.** $5\left(\cos \frac{3\pi}{2} + i \sin \frac{3\pi}{2}\right)$
35. $3(\cos 75° + i \sin 75°)$ **36.** $3(\cos 158° + i \sin 158°)$

Express each of the following in trigonometric form. Include a diagram.

37. $1 + i$ **38.** $2 + 2i$ **39.** $-1 - i$ **40.** $1 + i\sqrt{3}$
41. $-1 + i\sqrt{3}$ **42.** $\frac{\sqrt{3}}{2} - \frac{1}{2}i$ **43.** 10 **44.** $-\frac{1}{2} - \frac{i\sqrt{3}}{2}$
45. 1 **46.** -3 **47.** $-5i$ **48.** $3i$
49. $-8 - 8i$ **50.** $2 - 2i\sqrt{3}$ **51.** $-3\sqrt{3} - 3i$ **52.** $\frac{i}{\sqrt{2}}$

53. Explain why the modulus of a complex number cannot be a negative number.

Find the value of each of the following.

54. $|2 - 3i|$ **55.** $|-3 + 2i|$ **56.** $|-4 - 2i|$ **57.** $|1 - i|$
58. $|3i|$ **59.** $|-2i|$ **60.** $|-8 + 8i|$ **61.** $|i|$

Compute each product.

62. $(2 + 2i)(1 + i)$ **63.** $(\sqrt{3} + i)(1 + i\sqrt{3})$ **64.** $\left(\frac{3}{2} + \frac{3i\sqrt{3}}{2}\right)(1 + i\sqrt{3})$

65. In each of exercises 62–64, convert each factor and the product to trigonometric form. Can you find a pattern?

Mathematical Excursions — Geometry of Multiplying Complex Numbers

One of the most valuable tools in problem solving is the representation of abstract problems in concrete terms. Mathematicians strive to find the proper way to interpret formerly unsolvable problems, hoping to find a representation that will lead to better understanding of the problem and its solutions.

In 1797, Caspar Wessel (1745–1818), a Norwegian surveyor, published his geometric interpretation of complex numbers. Carl Friedrich Gauss (1777–1855), a German, and Jean Argand (1769–1822), a Frenchman, had each, independent of the others, been using a similar interpretation. This geometric interpretation helps us to describe and understand multiplication of complex numbers. The explanation is based on the "rotation and stretching" idea.

For example, the complex number $2 + 2i$ corresponds to the point (2, 2). The modulus, r, is $\sqrt{2^2 + 2^2}$ or $\sqrt{8}$ and the amplitude, θ, is 45°. When another complex number is multiplied by $2 + 2i$, that complex number is "stretched" by $\sqrt{8}$ units and "rotated" by 45°.

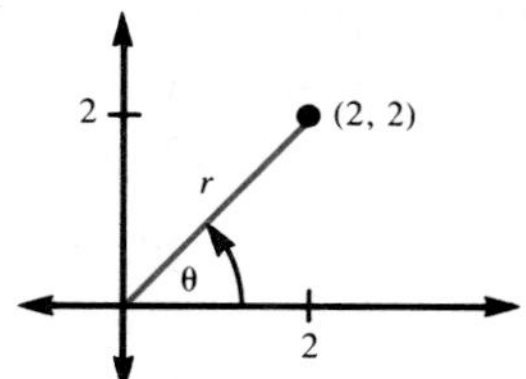

To illustrate this, multiply $2 + 2i$ by i. The number i corresponds to the point (0, 1), $r = 1$, and $\theta = 90°$. Multiplying i by $2 + 2i$ is interpreted as "stretching" it by $\sqrt{8}$ units and "rotating" it 45°. For the product, $r = \sqrt{8}$ and $\theta = 90° + 45°$ or 135°.

Verify this result by finding the trigonometric form of $(2 + 2i)(i) = -2 + 2i$.

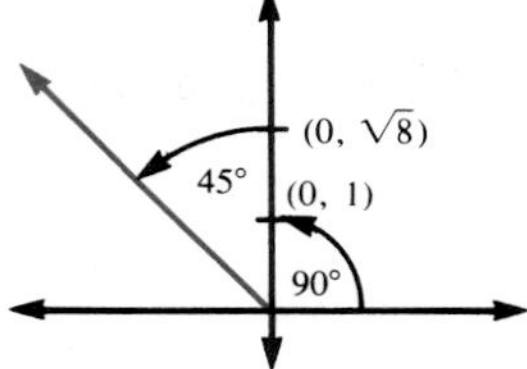

This geometric representation of multiplication of complex numbers serves as a basis for understanding and remembering De Moivre's Theorem given in the following section.

15.8 De Moivre's Theorem

One of the benefits of trigonometric form is a relatively simple method of multiplying complex numbers. The following example is written in both rectangular and trigonometric form.

Rectangular Form: $(\sqrt{3} + i) \times (2 + 2i\sqrt{3}) = 8i$

$\updownarrow \qquad \updownarrow \qquad \updownarrow$

Trigonometric Form: $2(\cos 30° + i \sin 30°) \times 4(\cos 60° + i \sin 60°) = 8(\cos 90° + i \sin 90°)$

In the trigonometric form, do you see any relationship between the modulus of the product and the modulus of each factor? Now compare the amplitude of the product with the amplitudes of the factors. What do you notice? Your answers should suggest the following theorem.

Product of Complex Numbers

The product of two complex numbers with moduli r_1 and r_2 and amplitudes θ_1 and θ_2 is the complex number with modulus r_1r_2 and amplitude $\theta_1 + \theta_2$. In symbols, we have

$$[r_1(\cos \theta_1 + i \sin \theta_1)][r_2(\cos \theta_2 + i \sin \theta_2)]$$
$$= r_1r_2[\cos(\theta_1 + \theta_2) + i \sin(\theta_1 + \theta_2)].$$

PROOF:

$[r_1(\cos \theta_1 + i \sin \theta_1)][r_2(\cos \theta_2 + i \sin \theta_2)]$

$= r_1r_2[(\cos \theta_1 \cos \theta_2 - \sin \theta_1 \sin \theta_2) + i(\sin \theta_1 \cos \theta_2 + \cos \theta_1 \sin \theta_2)]$

$= r_1r_2[\cos(\theta_1 + \theta_2) + i(\sin \theta_1 + \theta_2)]$ *Recall formula for $\cos(\theta_1 + \theta_2)$ and $\sin(\theta_1 + \theta_2)$*

Examples

1 **Express $4(\cos 70° + i \sin 70°)\ 3(\cos 80° + i \sin 80°)$ in trigonometric and rectangular forms.**

$4(\cos 70° + i \sin 70°)\ 3(\cos 80° + i \sin 80°)$

$= 12(\cos 150° + i \sin 150°)$ in trigonometric form

or $12\left(-\frac{\sqrt{3}}{2} + \frac{1}{2}i\right) = -6\sqrt{3} + 6i$ in rectangular form

2 **Simplify $[2(\cos 40° + i \sin 40°)]^3$**

$[2(\cos 40° + i \sin 40°)]^3$

$= \underbrace{[2(\cos 40 + i \sin 40°)][2(\cos 40° + i \sin 40°)]}\underbrace{[2(\cos 40° + i \sin 40°)]}$

$= \qquad [4(\cos 80° + i \sin 80°)] \qquad [2(\cos 40° + i \sin 40°)]$

$= 8(\cos 120° + i \sin 120°)$

Thus, $[2(\cos 40° + i \sin 40°)]^3 = 2^3(\cos 3 \cdot 40° + i \sin 3 \cdot 40°)$

In example 2 the product of complex numbers theorem was used to calculate the third power of the expression. Using the same technique, you can calculate the fourth power, the fifth power, the sixth power, and so on. This idea is expressed in the following theorem, named after Abraham De Moivre (1667–1754).

De Moivre's Theorem

If $r(\cos\theta + i\sin\theta)$ is a complex number and n is a real number, then

$$[r(\cos\theta + i\sin\theta)]^n = r^n(\cos n\theta + i\sin n\theta).$$

The proof of this theorem requires mathematical induction and is given in Chapter 16.

Examples

3 **Evaluate $(1 - i)^6$.**

$$(1 - i)^6 = [2^{\frac{1}{2}}(\cos 315° + i\sin 315°)]^6$$ *Trigonometric form*

$$= (2^{\frac{1}{2}})^6[\cos(6 \cdot 315°) + i\sin(6 \cdot 315°)]$$ *De Moivre's Theorem*

$$= 2^3(\cos 90° + i\sin 90°)$$
$$= 8i$$

4 **Evaluate $\left(-\frac{1}{2} + \frac{i\sqrt{3}}{2}\right)^3$.**

$$\left(-\frac{1}{2} + i\frac{\sqrt{3}}{2}\right)^3 = [1(\cos 120° + i\sin 120°)]^3$$
$$= 1^3[\cos(3 \cdot 120°) + i\sin(3 \cdot 120°)]$$
$$= \cos 360° + i\sin 360°$$
$$= 1$$

Think carefully about the results of example 3. Since $(1 - i)^6 = 8i$, you can say that $1 - i$ is a 6th root of $8i$. That is, one root of the equation $x^6 = 8i$ is $1 - i$. Similarly, by the results of example 4, $-\frac{1}{2} - i\frac{\sqrt{3}}{2}$ is a root of the equation $x^3 = 1$. In general, a non-zero complex number has n nth roots. The following examples use De Moivre's Theorem to find roots of complex numbers.

Example

5 Find the cube roots of $8i$.

The problem is equivalent to solving the equation $z^3 = 8i$ in the domain of complex numbers. If we let $z = r(\cos\theta + i\sin\theta)$ and express $8i$ in trigonometric form, our original equation becomes

$$[r(\cos\theta + i\sin\theta)]^3 = 8(\cos 90° + i\sin 90°).$$

Now, using De Moivre's Theorem on the left side produces

$$r^3(\cos 3\theta + i\sin 3\theta) = 8(\cos 90° + i\sin 90°).$$

If these two complex numbers are equal, their moduli are equal. Also, their amplitudes are equal or they differ by a multiple of 360°. Expressing these ideas in symbols, we have

$$\begin{aligned} r^3 &= 8 \text{ and } 3\theta = 90° + 360°k,\ k \text{ an integer.} \\ r &= 2 \qquad \theta = 30° + 120°k \\ &\text{If } k = 0,\ \theta = 30°. \\ &\text{If } k = 1,\ \theta = 150°. \\ &\text{If } k = 2,\ \theta = 270°. \end{aligned}$$

If $k = 3$, $\theta = 390°$ which is coterminal with 30°. In fact, choosing any other integral value of k will produce an angle coterminal with one of those obtained above.

Therefore, $8i$ has the following cube roots.

$$z_1 = 2(\cos 30° + i\sin 30°) = \sqrt{3} + i$$
$$z_2 = 2(\cos 150° + i\sin 150°) = -\sqrt{3} + i$$
$$z_3 = 2(\cos 270° + i\sin 270°) = -2i$$

Plotting the roots in the complex plane produces an interesting figure. Notice that the roots are equally spaced at intervals of 120° on a circle with radius 2. That is, the figure has 120° rotational symmetry about the origin.

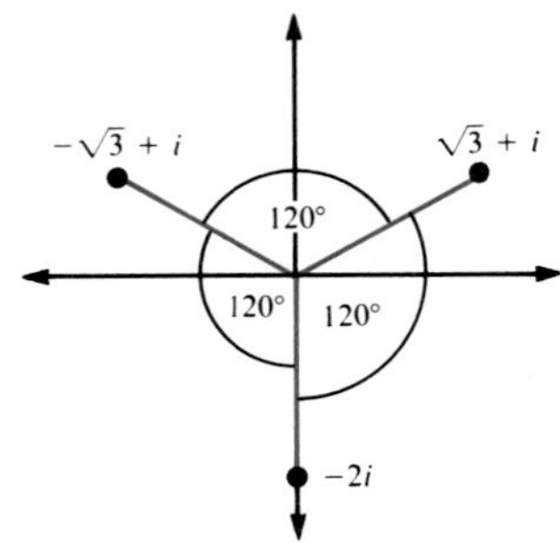

Example

6 Domain = $\mathscr{C}$. Solve $z^5 = -32i$. Graph the roots.

Expressing each side of the equation in trigonometric form produces

$$[r(\cos \theta + i \sin \theta)]^5 = 32(\cos 270° + i \sin 270°).$$
$$r^5(\cos 5\theta + i \sin 5\theta) = 32(\cos 270° + i \sin 270°)$$

Therefore,

$r^5 = 32$ and $5\theta = 270° + 360°k$, k an integer.

$r = 2$ $\qquad \theta = 54° + 72°k$

In this case, there are 5 roots. Therefore, compute θ for $k = 0, 1, 2, 3$, and 4.

The roots are as follows.

$z_1 = 2(\cos 54° + i \sin 54°)$

$z_2 = 2(\cos 126° + i \sin 126°)$

$z_3 = 2(\cos 198° + i \sin 198°)$

$z_4 = 2(\cos 270° + i \sin 270°)$

$z_5 = 2(\cos 342° + i \sin 342°)$

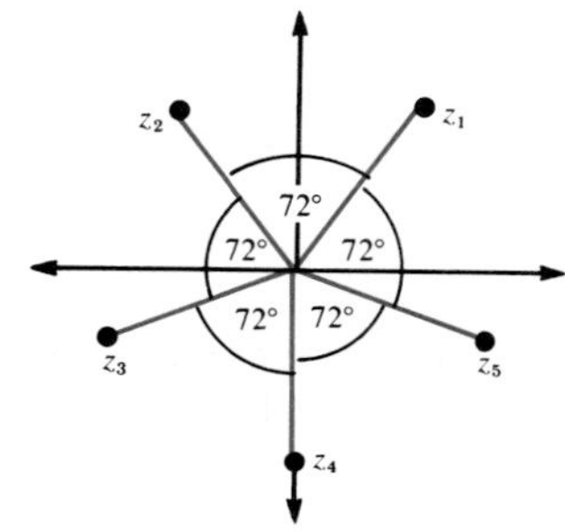

Perhaps you have noticed a pattern in finding the nth roots of a complex number. Here is a summary of the procedure.

$$r^{\frac{1}{n}}\left(\cos \frac{\theta + 360°k}{n} + i \sin \frac{\theta + 360°k}{n}\right)$$
for $k = 0, 1, 2, \ldots, n - 1$.

Exercises

Exploratory **Express each product in trigonometric form.**

1. $[3(\cos 30° + i \sin 30°)][4(\cos 70° + i \sin 70°)]$
2. $[5(\cos 145° + i \sin 145°)][2(\cos 20° + i \sin 20°)]$
3. $(\cos 160° + i \sin 160°)(\cos 40° + i \sin 40°)$
4. $\left(\cos \frac{\pi}{5} + i \sin \frac{\pi}{5}\right)\left(\cos \frac{2\pi}{5} + i \sin \frac{2\pi}{5}\right)$

Express each product in rectangular form.

5. $[2(\cos 20° + i \sin 20°)]\,[3(\cos 70° + i \sin 70°)]$
6. $(\cos 25° + i \sin 25°)(\cos 20° + i \sin 20°)$
7. $5(\cos 140° + i \sin 140°)(\cos 40° + i \sin 40°)$
8. $[6(\cos 10° + i \sin 10°)]\,[2(\cos 50° + i \sin 50°)]$
9. $[\sqrt{3}(\cos 190° + i \sin 190°)]\,[\sqrt{6}(\cos 20° + i \sin 20°)]$
10. $\left[2\frac{1}{3}(\cos 200° + i \sin 200°)\right]\left[5\frac{1}{4}(\cos 40° + i \sin 40°)\right]$

Written **Use De Moivre's Theorem to compute each of the following. Express your results in rectangular form.**

1. $[2(\cos 10° + i \sin 10°)]^6$
2. $(\cos 18° + i \sin 18°)^{10}$
3. $[3(\cos 30° + i \sin 30°)]^4$
4. $\left(\frac{1}{2} + i\frac{\sqrt{3}}{2}\right)^3$
5. $(1 - i)^4$
6. $(-1 - i)^3$
7. $(-2 + 2i)^4$
8. $(1 - i\sqrt{3})^8$
9. $(1 + i)^8$
10. $(-2i)^3$
11. $(-1 - i\sqrt{3})^{10}$

In exercises 12–21 express roots in rectangular form if the amplitude is a special angle. Otherwise, roots may be left in trigonometric form. Include a graph.

12. cube roots of 1
13. fourth roots of 1
14. cube roots of $27i$
15. sixth roots of 64
16. tenth roots of 1
17. fourth roots of i

Domain = $\mathscr{C}$. Solve each equation.

18. $z^3 + i = 0$
19. $x^5 - 32 = 0$
20. $x^6 - 64i = 0$
21. $z^2 = 4 - 4i$
22. Show that the equation $x^3 = 8$ can be solved by using either the Rational Root Theorem or De Moivre's Theorem.
23. Using De Moivre's Theorem, derive a formula for $\cos 3\theta$ and $\sin 3\theta$ in terms of $\cos \theta$ and $\sin \theta$. *Hint: Begin with* $(\cos \theta + i \sin \theta)^3$.

Suppose f is a mapping from $\mathscr{C}$ to $\mathscr{C}$ with the following rule: $f(z) = iz$. For each of the following, graph z and $f(z)$.

24. $z = 1 - i$
25. $z = -1 - i$
26. $z = (\cos 15° + i \sin 15°)$
27. Look at your results in exercises 24–26. Use your knowledge of transformations to describe the effect of f on each point of the complex plane.
28. If $f: \mathscr{C} \to \mathscr{C}$ and $f(z) = -iz$, what effect does f have on each point of the complex plane?
29. **Prove:** $\dfrac{r_1(\cos \theta_1 + i \sin \theta_1)}{r_2(\cos \theta_2 + i \sin \theta_2)} = \dfrac{r_1}{r_2}[\cos(\theta_1 - \theta_2) + i \sin(\theta_1 - \theta_2)]$.

Use the theorem in exercise 29 to compute each of the following. Check your answer by using the division procedure you learned in Chapter 2.

30. $\dfrac{1 - i}{1 + i}$
31. $\dfrac{-5 + 5i}{3 - 3i}$
32. $\dfrac{-2 + 2i}{1 + i}$
33. $\dfrac{-2\sqrt{3} - 2i}{4 + 4i\sqrt{3}}$
34. Prove the reciprocal of $r(\cos \theta + i \sin \theta)$ is $\dfrac{1}{r}(\cos \theta - i \sin \theta)$.
35. Find the reciprocal of $3(\cos 60° + i \sin 60°)$. Check your result by doing the computation in rectangular form.

Challenge

1. Investigate De Moivre's Theorem for the case where n is a negative integer. Justify any conclusions you reach.

Problem Solving Application: **Eliminating Possibilities**

Certain problems can be solved by using *matrix logic*. To solve such problems, a system of possibilities is set up using a matrix or grid. As each condition or clue for the problem is evaluated, the grid is marked to indicate logical choices or possibilities that can be eliminated. It is sometimes necessary to combine clues in order to reach a conclusion.

When working with the grid, a √ is often used to indicate a logical choice, or conclusion, and an **X** is used to indicate a possibility that is eliminated.

Example

1 **Brenda, Lynette, and Patti each participate in a different sport in their spare time. One plays volleyball (V), one softball (S), and one basketball (B). Their occupations are computer programmer (c), journalist (j), and teacher (t). Use the clues below to determine each person's sport and occupation.**

a. The basketball player and the teacher graduated from the same college.
b. The programmer, Lynette, and the volleyball player each bought a new car.
c. The journalist helps the softball player write articles for a newsletter.
d. Patti and the programmer used to live next door to each other.
e. Brenda beat both Patti and the basketball player in tennis.

We can list the possibilities using a grid, like the one shown at the right.

	c	j	t	V	S	B
Brenda						
Patti						
Lynette						
V						
S						
B						

Now we reason as follows.

1. From clues **a.** and **e.**, the basketball player is not the teacher, Patti or Brenda.
2. From clues **b.** and **d.**, the programmer is not the volleyball player, Lynette, or Patti.
3. From clue **b.**, the volleyball player is not Lynette.
4. From clue **c.**, the journalist is not the softball player.

The eliminated possibilities given above are indicated by *X*'s in the grid at the right. From this grid, we can conclude that Lynette is the basketball player and Brenda is the computer programmer.

	c	j	t	V	S	B
Brenda	√	X	X			X
Patti	X					X
Lynette	X			X	X	√
V	X					
S		X				
B			X			

We can now complete the grid.

5. Brenda, the computer programmer, is not the basketball player. Thus, the computer programmer, Brenda, is the softball player.
6. The basketball player, Lynette must be the journalist.
7. Patti must be the teacher and play volleyball.

	c	j	t	V	S	B
Brenda	√	X	X	X	√	X
Patti	X	X	√	√	X	X
Lynette	X	√	X	X	X	√
V	X	X	√			
S	√	X	X			
B	X	√	X			

Exercises

Written **Solve each problem.**

1. Ms. Davis and three of her associates work in a new shopping center. Each one is a supervisor in a different department of Plesich's General Store. Use these clues to determine each person's first and last name and the department that each person supervises.
   a. Mary and Mrs. Morris are friends of the clothing department supervisor.
   b. Cari started with the store before Ms. Wilson, who is the furniture department supervisor, and the jewelry department supervisor.
   c. Ann has lunch with Mrs. Davis every day.
   d. Paulette supervises the book department.
2. Dave, Denny, Dick, Don, and Drake were hired to do surveys to find out people's opinions on five different magazines. Each one asks people about only one magazine. Use these clues to determine each person's first and last name and the magazine each person surveyed.
   a. The five are Don, Drake, Mr. Chapman, Mr. Becker, and the person who surveys Auto.
   b. Drake, whose last name is not West, does not survey Sport.
   c. Mr. Chapman does not survey Teleview or Fashion Beat.
   d. Neither of the people who surveys Music Scene and Auto is Mr. Chapman or Drake.
   e. Dave's last name is not Jonson or West.
   f. Fashion Beat is not surveyed by Dave or Mr. Becker.
3. For the Gunthers' 25th wedding anniversary, their three children arranged a surprise party. Five of the people invited, including Mr. Gunther's boss, were unable to attend, but sent congratulations by telephone. Use these clues to determine the first and last name of the people that could not attend the party, along with their relationship to the Gunthers and the state from which they called.
   a. Frank, who is not Mr. Yates, called after Skip.
   b. The niece called from Ohio, the uncle from Kentucky, and the grandfather from New York.
   c. The call from California came before the call from Mr. Scott, who is not Andy.
   d. The college roommate called from Wisconsin, after the call from Mrs. Haar.
   e. Lynn's last name is not Carter or Haar.
   f. The three calls from relatives were from Wanda and Mr. Peterson and from New York.

Vocabulary

polynomial in one variable
constant term
leading coefficient
constant polynomial
polynomial function
degree
leading term
zero polynomial
cubic polynomial
roots
polynomial equation
zero of a polynomial
double root
variation in sign
modulus
amplitude

Chapter Summary

1. Division Algorithm: For polynomials $P(x)$ and $D(x)$, we can find $Q(x)$ and $R(x)$ such that $P(x) = D(x) \cdot Q(x) + R(x)$ with $R(x)$ either the zero polynomial or of degree less than the degree of $D(x)$.
2. Remainder Theorem: $P(x) = (x - a)Q(x) + P(a)$. The remainder when $P(x)$ is divided by $(x - a)$ is $P(a)$.
3. Factor Theorem: $(x - a)$ is a factor of $P(x)$ if and only if $P(a) = 0$.
4. Rational Root Theorem: $P(x) = a_nx^n + a_{n-1}x^{n-1} + \ldots + a_2x^2 + a_1x + a_0$ with a_i integers. If $\frac{p}{q}$ is a nonzero rational number in simplest form and a root of $P(x) = 0$, then p is a factor of a_0, and q is a factor of of a_n.
5. Fundamental Theorem of Algebra: Every polynomial equation of degree greater than zero has at least one root in the set of complex numbers.
6. Every polynomial equation of degree n, where $n > 0$, has n roots in the set of complex numbers.
7. Conjugate Root Theorem: Suppose a and b are real numbers with $b \neq 0$. If $a + bi$ is a root of a polynomial equation with real coefficients, then $a - bi$ is also a root.
8. Descartes' Rule: If $P(x)$ is a polynomial in standard form with real coefficients and v variations in sign, then the number of positive roots of $P(x) = 0$ is v or is less than v by a positive even number.
9. If $P(x) = 0$ has roots $r_1, r_2, r_3, \ldots, r_n$, then the equation $P(-x) = 0$ has roots $-r_1, -r_2, -r_3, \ldots, -r_n$.
10. Location Principle: For $P(x)$, a polynomial with real coefficients, and real numbers a and b, if $P(a) < 0$ and $P(b) > 0$ then $P(x) = 0$ has at least one real root between a and b.
11. Coordinates of a point can be expressed in rectangular form (x, y) and in trigonometric form $(r \cos \theta, r \sin \theta)$.

12. A complex number can be expressed in rectangular form $x + yi$ and in trigonometric or polar form $r(\cos\theta + i\sin\theta)$.
13. The absolute value of a complex number $x + yi$ is the distance of the point (x, y) from the origin. This is the modulus and is expressed as $r = |x + yi| = \sqrt{x^2 + y^2}$.
14. The product of two complex numbers with moduli r_1 and r_2 and amplitudes θ_1 and θ_2 is the complex number with modulus r_1r_2 and amplitude $\theta_1 + \theta_2$:
$$[r_1(\cos\theta_1 + i\sin\theta_1)][r_2(\cos\theta_2 + i\sin\theta_2)] = r_1r_2[\cos(\theta_1 + \theta_2) + i\sin(\theta_1 + \theta_2)]$$
15. De Moivre's Theorem: If $r(\cos\theta + i\sin\theta)$ is a complex number and n is a positive integer, then $[r(\cos\theta + i\sin\theta)]^n = r^n(\cos n\theta + i\sin n\theta)$.
16. If $r(\cos\theta + i\sin\theta)$ is a complex number, the nth roots of this complex number are $r^{\frac{1}{n}}\left(\cos\frac{\theta + 360°k}{n} + i\sin\frac{\theta + 360°k}{n}\right)$ for $k = 0, 1, 2, \ldots, n - 1$.

Chapter Review

15.1 Tell whether each of the following is a polynomial in one variable. If it is, give the degree, leading term, leading coefficient, and constant term. Identify the constant, linear, quadratic, and cubic polynomials.

1. $2x - 1$
2. $x^{-4} + 1$
3. $-3x^3 + 2x - 1$
4. 3
5. $ix^2 - 2x + 4$
6. $x^5 + \sqrt{2}\,x$
7. $|2x - 4|$
8. $3x^4 - (2 + i)x^3 - x^2$
9. $\sin x$

15.2 Perform the following divisions using synthetic division.

10. $x^4 - x^3 + 3x - 7$ by $(x - 2)$
11. $5x^4 - 10x^2 - 12x - 7$ by $(x - 4)$
12. $x^4 - x^2 + 3x - 7$ by $(x + 2)$
13. $3x^5 + 5x^4 - 2x^3 + x^2 - 10x + 3$ by $(x - 1)$
14. $4x^3 - 6x^2 + 5$ by $\left(x + \frac{3}{2}\right)$
15. $a_4x^4 + a_3x^3 + a_2x^2 + a_1x + a_0$ by $(x - r)$

15.3 Use the Remainder Theorem to find the remainder when the first polynomial is divided by the second polynomial.

16. $x^4 - 2x^3 + 7;\ x - 2$
17. $x^3 - 4x^2 + 2x - 3;\ x + 1$
18. $2x^5 - 3x^2 + x;\ x - 1$
19. $2x^4 - x^3 + 1;\ x + 3$

15.4 Find all the rational roots and if possible all the other roots of each equation.

20. $x^3 + 5x^2 + 8x + 6 = 0$
21. $x^3 + 2x^2 - x - 2 = 0$
22. $3x^3 - 13x^2 - 4x + 4 = 0$
23. $2x^3 - x^2 + 2x - 1 = 0$
24. $2x^3 - 5x^2 + 1 = 0$
25. $12x^3 + x^2 - 15 = 0$
26. $8x^5 - 27x^2 = 0$
27. $x^3 + x^2 + x + 6 = 0$
28. $x^4 - 4x^3 - 14x^2 + 36x + 45 = 0$
29. $2x^5 + x^4 - 4x^3 - 2x^2 - 16x - 8 = 0$

15.5 Form equations of the lowest possible degree, with integral coefficients, having the following roots.

30. $1, -1, 2$ **31.** $5, -2, -4$ **32.** $\frac{1}{2}, \frac{3}{2}, -1$

33. $\frac{2}{3}, \frac{1}{3}, -9$ **34.** $1 + \sqrt{3}, -\frac{1}{2}$ **35.** $2 - \sqrt{5}, -\frac{2}{5}$

36. $5, 2 - 3i$ **37.** $1, \frac{1 + \sqrt{3}\,i}{2}$ **38.** 2, a double root, and -1, a triple root

In exercises 39–42, one root is given. Find all the roots.

39. $x^3 + 2x^2 - 5x - 6 = 0; -1$ **40.** $x^3 - x^2 - 3x - 1; 1 - \sqrt{2}$

41. $8x^2 - 6x + 1 = 0; \frac{1}{2}$ **42.** $x^3 - 3x^2 - 9x - 5 = 0; -1$

15.6 Locate one real root between successive tenths of each of the following equations.

43. $x^3 + 3x - 2 = 0$ **44.** $x^3 + x^2 + x - 20 = 0$

45. $4x^4 - 6x^2 + 8x + 2 = 0$ **46.** $2x^3 + 5 = 0$

47. $x^4 - 14 = 0$ **48.** $x^4 + x - 7 = 0$

15.7 Express each of the following in rectangular form.

49. $7(\cos 30° + i \sin 30°)$ **50.** $6(\cos 180° + i \sin 180°)$

51. $8(\cos 20° + i \sin 20°)$ **52.** $2(\cos 90° + i \sin 90°)$

53. $4(\cos 45° + i \sin 45°)$ **54.** $3(\cos 120° + i \sin 120°)$

Express each of the following in trigonometric form.

55. $(-2, -2)$ **56.** $(-\sqrt{3}, 3)$ **57.** $1 - i$ **58.** $3 - i\sqrt{3}$

15.8 Express each product in trigonometric form.

59. $[5(\cos 30° + i \sin 30°)]\,[7(\cos 120° + i \sin 120°)]$

60. $[2(\cos 45° + i \sin 45°)]\,[3(\cos 45° + i \sin 45°)]$

61. $[2(\cos 240° + i \sin 240°)]\,[5(\cos 150° + i \sin 150°)]$

62. $[2(\cos 300° + i \sin 300°)]\,[8(\cos 120° + i \sin 120°)]$

63. $\left[\frac{1}{2}(\cos 70° + i \sin 70°)\right][6(\cos 110° + i \sin 110°)]$

Use De Moivre's Theorem to compute each of the following. State answers in polar and rectangular form.

64. $[3(\cos 45° + i \sin 45°)]^2$ **65.** $[2(\cos 72° + i \sin 72°)]^5$

66. $\left[\sqrt{2}\left(\cos \frac{5\pi}{4} + i \sin \frac{5\pi}{4}\right)\right]^4$ **67.** $\left[\sqrt{3}\left(\cos \frac{\pi}{6} + i \sin \frac{\pi}{6}\right)\right]^2$

68. $(1 + i)^5$ **69.** $(2 + 2i)^3$

Use De Moivre's Theorem to solve each of the following equations.

70. $x^3 = 1 + i$ **71.** $x^3 = i$ **72.** $x^6 = -1$

73. $x^2 = i$ **74.** $x^3 = -1$ **75.** $x^4 = -2 - 2i$

76. $x^{10} = -4i$ **77.** $x^5 = 2\sqrt{3} + 2i$ **78.** $x^2 = 4 - 4i$

Chapter Test

Tell whether each of the following is a polynomial. Identify the constant, linear, quadratic, or cubic polynomials.

1. $x^{-3} - 1$ **2.** $2x - 3x^2 + 3$ **3.** $\cos x$ **4.** $ix^3 + 2x^4 - \sqrt{2}$

5. Without performing the division, find the remainder when $x^3 - 4x^2 + 3x - 2$ is divided by $x - 1$.

6. Find the remainder when $3x^4 - 7x^3 + 12$ is divided by $x + 1$.

7. Find the remainder when $x^5 - 32$ is divided by $x - 2$.

Perform the following divisions using synthetic division.

8. $x^5 + 32 \div x + 2$ **9.** $x^3 - 9x^2 + 27x - 28 \div x - 3$

Locate one real root between successive tenths of each of the following equations.

10. $x^3 - 12 = 0$ **11.** $x^4 - 6x^2 + 8x = 0$

12. Form an equation of the lowest possible degree, with integral coefficients, having 2, 0, and $1 + i$ as roots.

Find all the rational roots and if possible all of the other roots of each of the following equations.

13. $x^4 + x - 78 = 0$ **14.** $x^3 - x^2 - 14x + 24 = 0$

15. One root of the equation $x^3 - 3x^2 + 9x + 13 = 0$ is $2 + 3i$. Find the others.

16. Show that $2 - \sqrt{3}$ is a root of the equation $x^4 - 12x^2 - 8x + 3 = 0$. Find the other roots.

17. If $3x^2 + 5x + D = Cx^2 + Ex + 5$ for $x = 0, 1, 2$, find the values of D, C, and E.

18. Solve the equation $2x^4 + 6x^3 + 11x^2 + 12x + 5 = 0$, given that -1 is a double root.

19. If $1 + \sqrt{2}$ is a root of $x^2 + Ax - 1 = 0$, find the value of A.

20. Given that $-5 + i$ is a root of the equation $x^4 + 5x^3 - 25x^2 - 140x - 26 = 0$, find the other roots.

21. Given that $-\frac{3}{4}$ is a root of the equation $4x^4 + 11x^3 + 26x^2 + 15x = 0$, find the other roots.

22. Given that $\frac{1}{2}$ is a root of $2x^3 - 5x^2 + 1 = 0$, find the other roots.

23. Solve the equation $x^3 + Ax^2 + Bx - 6 = 0$, given that A and B are integers, and that $2 - \sqrt{2}$ is a root.

Express each of the following in rectangular form.

24. $2(\cos 135° + i \sin 135°)$ **25.** $4(\cos 60° + i \sin 60°)$ **26.** $5(\cos 90° + i \sin 90°)$

Find the polar form of the product of each of the following pairs of complex numbers.

27. $[3(\cos 45° + i \sin 45°)][4(\cos 90° + i \sin 90°)]$

28. $[6(\cos 135° + i \sin 135°)][\frac{1}{2}(\cos 15° + i \sin 15°)]$

Use De Moivre's Theorem to compute each of the following. State answers in polar and rectangular form.

29. $[3(\cos 30° + i \sin 30°)]^4$ **30.** $(3 + 3i)^3$

Use De Moivre's Theorem to solve each of the following equations.

31. $x^2 = -9$ **32.** $x^3 = \sqrt{3} + i$ **33.** $x^2 = 4i$

Mathematical Induction

Chapter

16

Mathematical induction is an indirect method of proof which is used in cases where a direct method is either not possible or not convenient.

Numbers from the Fibonnaci sequence, 1, 1, 2, 3, 5, 8, 13, 21, 34, 55, . . . , are often found in nature. For example, the number of spirals in the core of a sunflower is usually one of the Fibonnaci numbers.

For many years, people have made various *observations* concerning patterns displayed by some Fibonacci numbers. Many of these observations can be confirmed for *all* Fibonacci numbers by mathematical induction.

16.1 Introduction

On several occasions in your mathematical career, you have used an **indirect method of proof.** Usually, in such cases, no convenient direct method was available to prove the particular assertion. One such method was the *Reductio ad Absurdum,* or RAA, method. Recall that in this method, you assume the *negation* of what you want to prove, and then show that *this* assumption is false. You may have used the RAA method to prove that there exists no rational number whose square is 2.

In this chapter we will study another indirect method of proof. This method is called **mathematical induction.** First, though, let us look at some situations where you might expect to use this kind of argument.

Very often in mathematics, people notice what appears to be a pattern. For example, a certain behavior appears to repeat in certain sets of numbers. However, *observation of the* **pattern** does not constitute a *proof.* We know that even many examples of something do not *prove* that the behavior in question always occurs.

Here are two examples of this kind of mathematical observation.

1. For hundreds of years, mathematicians have sought a formula which would generate prime numbers. That is, they looked for some expression, like $n^3 - 17n^2 - 1$, or like $2^n + 63$, such that for every replacement for n, the result is prime. One suggestion was that the expression

$$n^2 - n + 41$$

produces primes for every n. The chart at the right shows the results of the first 15 replacements. Note that in each case, the result is prime. But such a list, however many numbers it contains, does not mean that the formula always holds, though it certainly appears to do so for quite a few cases. Do you think that $n^2 - n + 41$ produces primes for every n? We will explore this further in the exercises.

n	$n^2 - n + 41$
1	41
2	43
3	47
4	53
5	61
6	71
7	83
8	97
9	113
10	131
11	151
12	173
13	197
14	223
15	251

2. Here is another observation. It appears that the sum of the first n odd numbers is equal to n^2. A geometric illustration of the observation appears at the right.

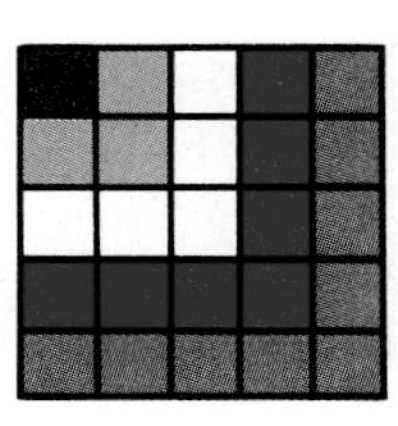

Here is a corresponding chart.

Number of Odd Numbers	Their Sum	The Square of n
1	1	$1^2 = 1$
2	$1 + 3 = 4$	$2^2 = 4$
3	$1 + 3 + 5 = 9$	$3^2 = 9$
4	$1 + 3 + 5 + 7 = 16$	$4^2 = 16$
5	$1 + 3 + 5 + 7 + 9 = 25$	$5^2 = 25$
and so on		

We will examine this pattern again in the next section.

Exercises

Exploratory

1. Explain the difference between indirect and direct proof. When are indirect proofs used?
2. Name a proposition, not given in the text, proved by an indirect method.
3. Use the symbols of logic to state the logical structure of the RAA proof. Explain your response, and include an example.

Written

Investigate the "formula" for primes, $n^2 - n + 41$, using the following steps:

a. Complete the chart for the indicated values of n.

b. In each exercise, are the results prime?

c. What conclusions can you reach for the "formula" from the results in each exercise?

1. 16–20
2. 21–25
3. 26–30
4. 31–35
5. 36–40
6. 41–45
7. Show that the assertion about the sum of the first n odd numbers is true for $n = 6, 7, \ldots, 10$.
8. Express the assertion mentioned in exercise 7 using Σ-notation.
9. Try to write a similar rule for the sum of the first n even numbers.

Challenge

1. Give a proof that there exists no rational number whose square is 2. Carefully explain the logic of your proof.
2. Answer Challenge Exercise 1 for a number whose square is 3.
3. Answer Challenge Exercise 1 for a number whose square is 5.
4. Try to prove that there is no rational number whose square is 4. Where does the proof break down?

16.2 The Method

In this section we shall use mathematical induction to prove that the sum of the first n odd numbers is equal to n^2. Other results shall be proved as well.

Many people have compared the method of proof by mathematical induction to a kind of "domino theory." When dominoes are lined up in the right way, if any one falls, then they all fall. Note that *two things are necessary for all the* dominoes *to fall.*

1. One domino has to fall.
2. The dominoes must be set up in such a way that if one falls, they all do.

Both of these requirements are necessary! Pushing over one domino will not do the job if the dominoes are not arranged properly. Arranging them properly is of no value if the initial one does not fall.

Induction is like an endless line of dominoes falling.

Principle of Mathematical Induction (PMI)

Let S be a set of natural numbers for which the following two conditions hold:

1. $1 \in S$
2. If the natural number k is in S, then $k + 1$ is also in S.

Then $S = \mathcal{N}$.

Recall that $\mathcal{N}$ names the set of natural numbers.

Example

1 **Prove that for all natural numbers n, the sum of the first n odd numbers is equal to n^2.**

Here is a representation of the first n odd numbers.

n	1	2	3	4	5	. . .	n
nth odd number	1	3	5	7	9	. . .	$2n - 1$

Thus, the nth odd number is $2n - 1$ and the sum of the first n odd numbers is

$$1 + 3 + 5 + 7 + \cdots + (2n - 1).$$

The assertion to be proved is

$$1 + 3 + 5 + 7 + \cdots + (2n - 1) = n^2.$$

PROOF:

First, define a set S.

Let S be the set of natural numbers for which it is true that

$$1 + 3 + 5 + 7 + \cdots + (2n - 1) = n^2.$$

Next, show that S satisfies Conditions 1 and 2 of PMI. Then we will be able to conclude that $S = \mathcal{N}$.

Step 1 Show that $1 \in S$ or that $1 = 1^2$. This is plainly true. Thus, $1 \in S$.

Step 2 Show that if $k \in S$, then $(k + 1) \in S$.

First assume $k \in S$. That is, for some natural number k, assume that it is true that

$$1 + 3 + 5 + 7 + \cdots + (2k - 1) = k^2.$$

This statement is called the **Induction Hypothesis.**

We must now show that $(k + 1) \in S$, or that the sum of the first $k + 1$ odd numbers is $(k + 1)^2$. That is, show that

$$1 + 3 + 5 + 7 + \cdots + (2k - 1) + (2k + 1) = (k + 1)^2.$$

Note that $2k - 1$ is the kth odd number. Then $2k + 1$ is the $(k + 1)$st odd number. This is true because $2(k + 1) - 1 = 2k + 1$.

Now we proceed by showing that the equation printed above is really true.

First, concentrate on the expression on the left of the equation. The first k terms of this expression, up to $2k - 1$, are equal to k^2, *according to the Induction Hypothesis.* Thus, the expression on the left of the equation is equal to $k^2 + (2k + 1)$. But note what happens when the expression on the right side of the equation is expanded. We have $(k + 1)^2 = k^2 + 2k + 1$ as well.

Thus, Step 2 of the proof is completed. We have shown that if k is assumed to be in S, then $k + 1$ must also be in S. Therefore, the set S as we have defined it is equal to the set of *all* natural numbers $\mathcal{N}$. This means for all natural numbers n, the sum of the first n odd numbers is n^2.

By now, the comparison of PMI with the "domino theory" should be clear to you. In establishing the first condition of PMI, the number 1 is the "first domino pushed over". Other times it might be necessary to begin with another number—say 14—and then adjust Condition 2 accordingly.

The first number can be called the **anchor.** It is *essential* that *some* natural number be established as belonging to S as Step 1 of an induction proof.

Condition 2 of PMI is a conditional statement of the form $p \rightarrow q$. To show that a conditional is true, we must assume that the antecedent p is true and then prove that the consequent q is true as well. That is, we assume some natural number k is in set S and then prove $k + 1$ is also in S. The assumption that k is in S is called the *Induction Hypothesis*. Sometimes Step 2 of PMI is called the **inductive step.** This step assures that "since the first domino fell, all the dominoes will fall."

Exercises

Exploratory

1. Explain, in your own words, the essential idea of a proof by mathematical induction. For what kinds of assertions is it possible, or desirable, to use such a proof? Describe the two vital steps in a proof using PMI.
2. Give an explanation of PMI using the idea of climbing a ladder.
3. Is it necessary that the "first domino to fall," or "the first rung of the ladder," be the number 1? Try to formulate a version of PMI which would allow for a different starting point.
4. Explain what is meant by the terms anchor and inductive step as used in this section.
5. Clem objects that a proof by mathematical induction is circular. "After all," he says, "you start off by assuming what you want to prove!" How would you answer Clem?
6. Refer to the Example in the text. Why is the kth odd number $2k - 1$? Why is the $(k + 1)$st odd number $2k + 1$?
7. Name the rth even number in terms of r. Name the $(r + 1)$st even number.

Written

Name the nth and the $(n + 1)$st terms in each of these sequences.

1. $1, \frac{1}{2}, \frac{1}{3}, \frac{1}{4}, \frac{1}{5}, \ldots$
2. $2, 5, 10, 17, 26, \ldots$
3. $3, 9, 27, 81, 243, \ldots$
4. $4, 7, 10, 13, 16, \ldots$
5. $\frac{1}{2}, \frac{2}{3}, \frac{3}{4}, \frac{4}{5}, \frac{5}{6}, \ldots$
6. $2, \frac{3}{2}, \frac{4}{3}, \frac{5}{4}, \frac{6}{5}, \ldots$

Prove each of the following for all $n \in \mathcal{N}$. Check them first for $n = 1, 2, 3$.

7. $2 + 2^2 + 2^3 + \cdots + 2^n = 2(2^n - 1)$
8. $1 \cdot 2 + 2 \cdot 3 + 3 \cdot 4 + \cdots + n(n + 1) = \dfrac{n(n + 1)(n + 2)}{3}$

16.3 The Method Applied

In chapter 13, certain statements about the terms, and the sum of terms, of arithmetic and geometric series and sequences were established. You are now able to prove similar statements using the Principle of Mathematical Induction.

Example

1 Show that the sum of the first n natural numbers is equal to $\frac{n(n+1)}{2}$.

It is necessary to prove that $1 + 2 + 3 + \cdots + n = \frac{n(n+1)}{2}$.

Step 1 Show $1 \in S$.

$$1 \stackrel{?}{=} \frac{1 \cdot (1+1)}{2}$$
$$1 \stackrel{?}{=} \frac{1 \cdot 2}{2}$$
$$1 = 1 \quad \text{Thus, } 1 \in S.$$

Step 2 Show that if $k \in S$, then $(k+1) \in S$.

Assume the statement is true for k; that is, assume that it is true that

$1 + 2 + 3 + \cdots + k = \frac{k(k+1)}{2}$. *This is the Induction Hypothesis.*

Now show that it is true that $k + 1$ is in S; that is, show that

$$1 + 2 + 3 + \cdots + k + (k+1) = \frac{(k+1)(k+2)}{2}.$$

First work on the left.

$$\underbrace{1 + 2 + 3 + \cdots + k}_{\frac{k(k+1)}{2}} + \underbrace{(k+1)}_{\frac{2(k+1)}{2}} = \frac{k(k+1) + 2(k+1)}{2}$$

By the Induction Hypothesis, the sum of the first k terms of this expression is equal to $\frac{k(k+1)}{2}$. Also $k + 1 = \frac{2(k+1)}{2}$. Thus, on the left, the expression is equal to $\frac{k(k+1) + 2(k+1)}{2}$. However, this is equal to $\frac{(k+2)(k+1)}{2}$, the expression on the right, by the Distributive Law, and the inductive step has been completed. We have shown that $1 \in S$ and that if k is in S then $k + 1$ is also in S. Thus, $S = \mathcal{N}$ and the original statement is true for all natural numbers n.

Exercises

Exploratory **Find the LCD (least common denominator) for each of the following pairs and then find their sum.**

1. $\dfrac{k+1}{k+2}, k-1$
2. $\dfrac{k+3}{2}, k+2$
3. $\dfrac{k+1}{k+2}, \dfrac{k}{k^2-4}$
4. $\dfrac{(k^2+2k+1)}{2}, \dfrac{k+1}{k+2}$
5. In example 1, why is $\dfrac{k(k+1)}{2} + k + 1 = \dfrac{k(k+1) + 2(k+1)}{2}$?

Written **Explain or justify each step in the following proof.**

Show that the sum of the cubes of three consecutive numbers is divisible by 9.

1. Restate the problem: Show that for $n \in \mathcal{N}$, $n^3 + (n+1)^3 + (n+2)^3$ is divisible by 9.
2. Let S equal the set of natural numbers for which the statement is true.
3. The statement is true for 1.
4. Assume the statement is true for $n = k$; that is assume that $k^3 + (k+1)^3 + (k+2)^3$ is divisible by 9. Show that $(k+1)^3 + (k+2)^3 + (k+3)^3$ is divisible by 9.
5. $(k+1)^3 + (k+2)^3 + (k+3)^3 = (k+1)^3 + (k+2)^3 + k^3 + 9k^2 + 27k + 27$.
6. This equals $k^3 + (k+1)^3 + (k+2)^3 + 9(k^2 + 3k + 3)$.
7. The expression printed in gray is divisible by 9; the expression in red is divisible by 9. Therefore, their sum is divisible by 9.
8. $k \in S \rightarrow (k+1) \in S$, and the assertion is proved.

In each of these exercises, first test to see if the assertion appears to be true. If it does, try to prove it by induction. If you determine that the assertion is false, explain.

9. $1^2 + 3^2 + 5^2 + 7^2 + \cdots + (2n-1)^2 = (3n-2)^2$
10. $2 + 4 + 6 + \cdots + 2n = n(n+1)$
11. $1 + \dfrac{1}{2} + \dfrac{1}{3} + \dfrac{1}{4} + \cdots + \dfrac{1}{n} = \dfrac{n^2}{n+1}$
12. $1 \cdot 2 \cdot 3 + 2 \cdot 3 \cdot 4 + \cdots + n(n+1)(n+2) = n(n^2-1)(n+1)$

Prove each assertion in exercises 13–23 for all $n \in \mathcal{N}$. Check them for $n = 1, n = 2, n = 3$ first.

13. $\dfrac{1}{1 \cdot 2} + \dfrac{1}{2 \cdot 3} + \dfrac{1}{3 \cdot 4} + \cdots + \dfrac{1}{n(n+1)} = \dfrac{n}{n+1}$
14. $\dfrac{1}{2} + \dfrac{1}{4} + \dfrac{1}{8} + \cdots + \dfrac{1}{2^n} = 1 - \dfrac{1}{2^n}$
15. $1^2 + 3^2 + 5^2 + \cdots + (2n-1)^2 = \dfrac{n(2n+1)(2n-1)}{3}$
16. The sum of the first n cubes is $\dfrac{n^2(n+1)^2}{4}$.
17. The sum of the first n squares is $\dfrac{n(n+1)(2n+1)}{6}$.

18. 2 is a factor of $n^2 + n$. *Do this in more than one way.*

19. The sum of the first n cubes equals the square of the sum of the first n natural numbers. *But think about this one first.*

20. $3 + 6 + 9 + \cdots + 3n = \frac{3(n + 1)n}{2}$

21. $4^n + 1$ is not divisible by 3.

22. $4^n - 1$ is divisible by 3.

23. $1 \cdot 2 + 3 \cdot 4 + 5 \cdot 6 + \cdots + (2n - 1)(2n) = \frac{n(4n^2 + 3n - 1)}{3}$

24. Suppose the number 5 is used as an "anchor" in a PMI proof. What has been proved? How can the proof be extended to include *all* natural numbers?

25. Reformulate the Principle of Mathematical Induction for the anchor, 5. Note that $S \neq \mathcal{N}$, but S equals a certain subset of $\mathcal{N}$.

Prove each of these relations about the arithmetic sequence $a_1, a_1 + d, a_1 + 2d, \ldots$

You may need to refer to chapter 13 for review.

26. The nth term is $a_1 + (n - 1)d$.

27. The sum of the first n terms is $\frac{n}{2}[2a_1 + (n - 1)d]$.

Prove each of these relationships about the geometric sequence $a_1, a_1r, a_1r^2, a_1r^3, \ldots$

28. The nth term is a_1r^{n-1}.

29. The sum of the first n terms is $\frac{a_1 - a_1r^n}{1 - r}$, $(r \neq 1)$.

Use Σ-notation to express the assertions in these exercises.

30. exercise 11 **31.** exercise 17 **32.** exercise 13

Mixed Review

One root is given for each polynomial equation. Find all the roots of the equation.

1. $3x^3 - 2x^2 - 5x - 6 = 0$; 2 **2.** $x^3 - 7x^2 + 17x - 15 = 0$; $2 + i$

3. Locate the positive real root between successive tenths for $x^3 - 3x^2 - 7x - 5 = 0$.

4. What are the nth and $(n + 1)$st terms of the sequence 3, 8, 15, 24, 35, . . . ?

5. A motorboat travels 36 miles downstream and then returns. The downstream part of the trip takes 6 hours less than the upstream part. If the speed of the motorboat had been doubled, then the downstream part of the trip would have only taken 1 hour less than the upstream part. What is the rate of the current for this stream?

16.4 Special Methods

There is no "fixed" way of proving theorems by mathematical induction. Often, each problem has its own special characteristics which require a different technique to complete the proof. You may have to multiply the numerator and denominator of an expression by the same quantity, or factor or expand an expression in a special way, or see an important relationship with the Distributive Law. That is, you have to "look around" to find the right approach. In this section, we will examine some different techniques for proving assertions by induction.

Examples

1 Show that for every natural number n, $5^n - 2^n$ is divisible by 3.

Another way of expressing this is to say that $\frac{5^n - 2^n}{3}$ is an integer. Either statement of the assertion can be used in the proof.

Begin by defining S. Let S be the set of natural numbers for which the indicated statement is true.

Step 1. Show $1 \in S$.

$$5^1 - 2^1 = 5 - 2 = 3$$

Since 3 is divisible by 3, $1 \in S$.

Step 2. Show $k \in S \rightarrow (k + 1) \in S$.

Induction Hypothesis: $\frac{5^k - 2^k}{3}$ is an integer.

Now, show that $\frac{5^{k+1} - 2^{k+1}}{3}$ is an integer.

Here are two ways to proceed.

Method 1. $5^{k+1} - 2^{k+1} = 5^{k+1} \underbrace{- 5 \cdot 2^k + 5 \cdot 2^k}_{} - 2^{k+1}$

subtracting, then adding back the same quantity, $5 \cdot 2^k$

$$= 5(5^k - 2^k) + 2^k(5 - 2)$$

$$\text{Thus, } \frac{5^{k+1} - 2^{k+1}}{3} = 5 \cdot \frac{5^k - 2^k}{3} + 2^k \frac{(5 - 2)}{3}$$

$$= \underbrace{5 \cdot \frac{5^k - 2^k}{3}}_{\text{an integer by Induction Hypothesis}} + \underbrace{2^k \cdot 1}_{\text{an integer}}.$$

Thus, $\frac{5^{k+1} - 2^{k+1}}{3}$ is the sum of two integers and is, therefore, an integer.

Method 2. If $\frac{5^k - 2^k}{3}$ is an integer b, then $5^k - 2^k = 3b$ for some $b \in \mathbb{Z}$. This means $5^k = 2^k + 3b$. Furthermore, $5 = 2 + 3$.

Now, multiply 5^k by 5.

$$5^k \cdot 5 = 5^{k+1} = (2^k + 3b)(2 + 3) = 2^{k+1} + 3 \cdot 2^k + 6b + 9b$$

Then, $5^{k+1} - 2^{k+1} = 3 \cdot 2^k + 6b + 9b$ or $3(2^k + 5b)$.

But this last expression is clearly 3 times an integer. Thus,

$5^{k+1} - 2^{k+1}$ is divisible by 3, or

$\frac{5^{k+1} - 2^{k+1}}{3}$ is an integer.

With either Method 1 or Method 2, we have shown that $k \in S$ implies $(k + 1) \in S$. Hence, the original assertion is proved.

2 **Show that $3^n \cdot 4^{2n} - 1$ is divisible by 47, for all $n \in \mathcal{N}$.**

Step 1 $1 \in S$ since $3^1 \cdot 4^{2(1)} - 1 = 3 \cdot 4^2 - 1 = 47$

Step 2 Assume $3^k \cdot 4^{2k} - 1$ is divisible by 47.

Show $3^{k+1} \cdot 4^{2(k+1)} - 1$ is divisible by 47.

First of all, $3^{k+1} \cdot 4^{2(k+1)} - 1 = 3^{k+1} \cdot 4^{2k+2} - 1$.

$$3^{k+1} \cdot 4^{2k+2} - 1 = 3 \cdot 4^2(3^k \cdot 4^{2k}) \underbrace{- 3 \cdot 4^2 + 3 \cdot 4^2}_{\text{Subtract } 3 \cdot 4^2 \text{ then add it back.}} - 1$$

$$= \underbrace{3 \cdot 4^2(3^k \cdot 4^{2k} - 1)}_{\substack{\text{divisible by 47 by} \\ \text{Induction Hypothesis}}} + \underbrace{3 \cdot 4^2 - 1}_{47}$$

Since $3^{k+1} \cdot 4^{2k+1} - 1$ is the sum of two quantities each shown to be divisible by 47, it is itself divisible by 47.

Hence, $k \in S \rightarrow (k + 1) \in S$, and the assertion is proved.

Exercises

Exploratory **Express each of these assertions using the term "divisible."**

1. $\frac{6^n - 2^n}{4}$ is an integer.
2. $\frac{2^k \cdot 5^{2k+1} - 1}{7}$ is an integer.
3. $5^{n+2} + 6^{2n+1} = 31t, t \in \mathbb{Z}$

Complete each of these statements.

4. $5^{k+1} = 5^k \cdot ?$
5. $5^{k+2} + 5^k = 5^k(? + ?)$
6. $3^{k+1} - 3 \cdot 2^k = 3(? - ?)$
7. $9 \cdot 17^m - 17^{m+1} = 17^m(? - ?)$

8. Is the assertion in exercise 2 true or false? Justify your answer.

Written **Explain or justify each step in the following proof.**

Show that $5^{n+2} + 6^{2n+1}$ is divisible by 31, for all $n \in \mathcal{N}$.

1. Let S be the set of natural numbers for which the assertion is true.

2. $1 \in S$

3. Assume $5^{k+2} + 6^{2k+1}$ is divisible by 31.

4. Consider $5^{k+3} + 6^{2k+3}$.

5. $5^{k+3} + 6^{2k+3} = 5^{k+3} + 5(6^{2k+1}) - 5(6^{2k+1}) + 6^{2k+3}$

6. Then, $5^{k+3} + 6^{2k+3} = 5(5^{k+2} + 6^{2k+1}) + 6^{2k+1}(6^2 - 5)$

7. But $5(5^{k+2} + 6^{2k+1})$ and $6^{2k+1}(6^2 - 5)$ are both divisible by 31.

8. Hence, their sum is divisible by 31.

9. Both requirements of PMI have been satisfied; the assertion is proved.

Prove each of the following assertions for all $n \in \mathcal{N}$.

10. $6^n - 2^n$ is divisible by 4.

11. $\dfrac{7^n - 2^n}{5}$ is an integer.

12. $9^n - 4^n$ is divisible by 5.

13. $17^n - 14^n$ is divisible by 3.

14. $6^{n+2} + 7^{2n+1}$ is divisible by 43.

15. $7^{n+2} + 8^{2n+1}$ is divisible by 57.

16. $(8^{n+2} + 9^{2n+1}) \div 73$ is an integer.

17. $2^{2n+1} + 3^{2n+1}$ is divisible by 5.

18. $2^{2n+1} + 5^{2n+1}$ is divisible by 7.

19. $3^{2n+1} + 4^{2n+1} = 7t$ where $t \in \mathcal{I}$.

20. $3^{2n} - 1$ is divisible by 8.

21. $5^{2n} - 1$ is divisible by 24.

22. $7^{2n} - 1$ is divisible by 48.

23. $4^n \cdot 5^{2n} - 1$ is divisible by 99.

Use your results from exercises 1–23 to answer the following.

24. What generalization can you make after completing exercises 1–9 and 14–16?

25. What generalization can you make after completing exercises 10–13?

26. Look at exercises 20–22. By what number do you think $4^{2n} - 1$ is divisible? Can you prove it? Use this result to show that $2^{4n} - 1$ is divisible by 15.

27. Answer exercise 26 for $3^{4n} - 1$; supply your own assertion!

Comment on each of the following versions of the Principle of Mathematical Induction. How is each similar to the version we have used in the text? How dissimilar? Under what circumstances might each be preferable?

28. Let $\mathcal{N}_t$ be the set of natural numbers $\geq t$. Let S be the subset of $\mathcal{N}_t$ for which a given statement about natural numbers is true. Then, $S = \mathcal{N}_t$ if the following conditions hold.

a. $t \in S$

b. $k \in S \rightarrow (k + 1) \in S$

29. Let P be a statement about natural numbers and let $P(n)$ mean that the statement is true for the natural number n. Then P is true for all $n \in \mathcal{N}$ if the following two conditions hold.

a. $P(1)$ is true.

b. $P(k) \rightarrow P(k + 1)$

16.5 Inequalities

In this section, induction shall be used to prove the truth of some assertions of inequalities. As before, the proofs will depend on algebraic technique, imagination, and, perhaps, some tricks.

Examples

1 **Show that for every natural number n, $2n \le 2^n$.**

Let S be the set of natural numbers for which the indicated statement is true.

Step 1. Show $1 \in S$. But, surely, $2 \cdot 1 \le 2^1 = 2$. (That is, $2 = 2$.)

Step 2. Show $k \in S \rightarrow (k + 1) \in S$.

Assume $2k \le 2^k$. Show that $2(k + 1) \le 2^{k+1}$.

$(2k \le 2^k) \rightarrow 2k + 2 \le 2^k + 2$ *Add 2 to both sides.*

$\rightarrow 2k + 2 \le 2^k + 2^k$ *Replace 2 by 2^k, which is surely equal to or greater than 2; use the transitive property of less than.*

$\rightarrow 2k + 2 \le 2 \cdot 2^k$

$\rightarrow 2k + 2 \le 2^{k+1}$

$\rightarrow 2(k + 1) \le 2^{k+1}$

Thus, $k \in S \rightarrow (k + 1) \in S$, and the inductive step of the proof is complete.

2 **Show that $n! > 2^n$ for any natural number $n \ge 4$.**

In this case, the "anchor" will be 4 rather than the usual 1. Use the form of the Principle of Mathematical Induction given in exercise 28 of Section 16.4. This problem also deals with factorials. Recall that since $n! = n(n - 1)(n - 2) \ldots 1$, it follows that $(n + 1) \cdot n! = (n + 1)!$

Let $\mathcal{N}_4$ be the set of natural numbers $n \ge 4$. Let S be the subset of $\mathcal{N}_4$ for which it is true that $n! > 2^n$. We must show that $S = \mathcal{N}_4$.

Step 1. (Note the anchor!) Show $4 \in S$. But this is true since

$$4! = 24 > 2^4 = 16.$$

Step 2. Show $k \in S \rightarrow (k + 1) \in S$.

Assume that $k! > 2^k$ and that $k \ge 4$.
It is required to show that $(k + 1)! > 2^{k+1}$.
Since $k \ge 4$, $(k + 1) \ge 5$.

$k! > 2^k \rightarrow (k + 1)k! \ge 5 \cdot 2^k > 2 \cdot 2^k$ *$5 \cdot 2^k$ is greater than $2 \cdot 2^k$.*

$(k + 1)! > 2^{k+1}$ *Transitivity of $>$*

Thus, in fact, $k \in S$ does imply $(k + 1) \in S$, and $S = \mathcal{N}_4$. The assertion is proved, for $n \ge 4$, as required.

The following inequality is very important in more advanced studies.

Example

3 Show that if $1 + k > 0$, then $(1 + k)^n \geq 1 + nk$, for $n \in \mathcal{N}$.

Let S be the set of natural numbers for which the indicated statement is true. Note that in this argument, the symbol k is kept constant. The induction will be "on" the letter n.

Step 1. $1 \in S$ since $(1 + k)^1 \geq 1 + 1k$. $\quad (1 + k = 1 + k)$

Step 2. Assume the statement true for n: $(1 + k > 0) \rightarrow [(1 + k)^n \geq 1 + nk]$.

Show that $(1 + k > 0) \rightarrow [(1 + k)^{n+1} \geq 1 + (n + 1)k]$.

Now, if $(1 + k)^n \geq 1 + nk$, then $(1 + k)^{n+1}$, which equals $(1 + k)^n(1 + k)$, is greater than $(1 + nk)(1 + k)$, since the substitution of $(1 + nk)$ for $(1 + k)^n$ has resulted in something smaller. By the Induction Hypothesis, $(1 + nk)$ is less than $(1 + k)^n$.

$$[(1 + k)^n \geq 1 + nk] \rightarrow (1 + k)^{n+1} = (1 + k)^n(1 + k)$$
$$\geq (1 + nk)(1 + k)$$
$$\geq 1 + (n + 1)k + nk^2 \quad \textit{Multiplication}$$

But $n > 0$ and $k^2 > 0$, so $nk^2 > 0$. The polynomial diminishes by subtracting nk^2.

$$(1 + k)^{n+1} \geq 1 + (n + 1)k$$

Exercises

Exploratory

1. How is the transitive property of inequality (in this case, less than) used in the proof given in example 1? Include a description of the property in your answer.
2. What is the "anchor" in example 2? Why? What would happen if we tried to prove the assertion given there for $n \geq 1$, as usual?

Explain where Clem will fail when he tries to prove each of these assertions.

3. $n^2 \leq 50n$ for $n \in \mathcal{N}_{50}$
4. $n^2 < 2^n$ for $n \in \mathcal{N}_3$
5. $(n + 2)(n + 1)! > (n + 2)!$ for $n \in \mathcal{N}$
6. $5^n < 4^{n+1}$ for $n \in \mathcal{N}$

Written **The product of the first n even numbers is equal to $2^n \cdot n!$ Prove each of the following assertions.**

1. $2^n \geq 1 + n$ for $n \geq 1$
2. $3^n \geq 1 + 2n$ for $n \geq 1$
3. $5^n \geq 1 + 4n$ for $n \geq 1$
4. $2^n \geq n$ for $n \geq 1$

5. $n^2 \geq n + 7$ for $n \geq 4$ *Hint: for $(k + 1)^2$, add $2k + 1$ to both sides.*

6. $2^n \geq n^2$ for $n \geq 4$

7. $n^2 \geq n + 30$ for $n \geq 6$

8. $4n^2 \geq 5n + 11$ for $n \geq 3$

9. $2^n \geq n^3$ for $n > 9$

10. $3^n > 2^{n+1}$ for $n \geq 2$

11. $4^n > 3^{n+1}$ for $n \geq 4$

12. $\frac{1}{1^2} + \frac{1}{2^2} + \frac{1}{3^2} + \cdots + \frac{1}{n^2} < 3 - \frac{1}{n}$ for $n \geq 1$

Challenge

1. Look carefully at Written Exercises 1–3. What conjecture can you make? Think of a plan by which you think your conjecture might be proved.
2. What conjecture can you make about Written Exercises 4, 6, and 9? Think of a plan by which you think your conjecture might be proved.

Mixed Review

1. Simplify $(x + 4)(x - 3)^{-1} - (x - 3)(x + 4)^{-1}$.
2. Simplify $(\sqrt{-18} - \sqrt{-12})(\sqrt{-3} + \sqrt{-2})$.
3. Find the inverse of $f(x) = 3x + 2$.
4. What transformation maps the graph of $x^2 + y^2 = 4$ onto the graph of $(x - 3)^2 + (y + 2)^2 = 4$?
5. Find y if $\log_4 (3y - 2) = 2$.
6. Points A, B, and C are on circle O such that $\overline{PC}$ is tangent to circle O at C and $\overline{PAB}$ is a secant. If $m\angle CPB = 30$ and $m\widehat{AC}:m\widehat{CB} = 2:5$, find $m\widehat{ACB}$.
7. If $\sec \theta = \frac{17}{15}$ and $\sin \theta < 0$, find $\cot \theta$.
8. A plane flew 900 kilometers due north. It then changed direction by turning 20° clockwise and flew for another 700 kilometers. How far was the plane from its starting point?
9. Solve $2 \tan^2 x - 5 \sec x + 4 = 0$ in the interval $0 \leq x < 2\pi$.
10. The vertices of $\triangle DEF$ have coordinates $D(-2, -1)$, $E(1, 3)$, and $F(4, -5)$. Find the image of $\triangle DEF$ under the composite transformation $Rot_{O,-90°} \circ T_{-2,-3}$.
11. A bag contains 4 red, 5 blue, and 11 green marbles. Five are selected at random. Find the probability that exactly two of the marbles are green.
12. The average points scored per game by the leading scorer in the NBA from 1966 to 1990 are given below. Find the mean and standard deviation, to the nearest tenth, for this data.

33.5	35.6	27.1	28.4	31.2	31.7	34.8	34.0	30.6	34.5	31.1	31.1	27.2
29.6	33.1	30.7	32.3	28.4	30.6	32.9	30.3	37.1	35.0	32.5	33.6	

13. Mrs. Dunston invested in computer equipment worth \$270,000. The equipment depreciates at a rate of 20% per year. What will be the value of her equipment at the end of four years?
14. Solve $\begin{cases} 3x + 5y = 22 \\ -2x + 3y = -15 \end{cases}$ using the matrix method.
15. Write a polynomial equation of lowest possible degree, with integral coefficients, having -2 and $4 - 3i$ as roots.

16.6 Polynomials; De Moivre's Theorem

In this section, some results about the factorization of polynomials will be proved. The section will conclude with a proof, by mathematical induction, of a famous theorem of mathematics called De Moivre's Theorem. This theorem is also discussed in Chapter 15.

It is often necessary to know when certain polynomials are divisible by others, particularly by linear polynomials. Your work with the Remainder Theorem and the Factor Theorem was evidence of this. Here is a theorem which carries this work a bit further.

Theorem 16–1

For all $n \in \mathcal{N}$, $x - 1$ is a factor of $x^n - 1$.

Proof by Induction:

Let S be the set of natural numbers for which the assertion is true.

Step 1 Show $1 \in S$.

Clearly, $x - 1$ is a factor of $x^1 - 1$ or $x - 1$.

Step 2 Show $k \in S \rightarrow (k + 1) \in S$.

We assume $x - 1$ is a factor of $x^k - 1$, and must show that $x - 1$ is a factor of $x^{k+1} - 1$.

But $x^{k+1} - 1 = x^{k+1} - x^k + x^k - 1 = x^k(x - 1) + x^k - 1$.

Surely $x - 1$ is a factor of $x^k(x - 1)$ and, by the Induction Hypothesis, $x - 1$ is a factor of $x^k - 1$. Therefore, $x - 1$ is a factor of $x^{k+1} - 1$. We have shown that $k \in S \rightarrow (k + 1) \in S$; therefore, the theorem is proved.

Note that the technique in this proof is similar to the technique developed in Section 16.4. A generalization of this theorem, to $x - y$ and $x^n - y^n$, will be considered in the exercises.

In your study of equations, complex numbers, and trigonometric representations of them, a very important theorem called **De Moivre's Theorem** was developed.

Theorem 16–2
De Moivre's Theorem

For all real numbers n and θ and for real numbers $r > 0$

$$[r(\cos \theta + i \sin \theta)]^n = r^n(\cos n\theta + i \sin n\theta)$$

Although the theorem is true for all real n, it will be proved true only for n, a natural number. In this case, "induction on n" will be used. The symbols r and θ will be considered "fixed", and the Principle of Mathematical Induction will be applied to n.

Proof:

Let S be the set of natural numbers for which the assertion is true.

Step 1 Show $1 \in S$. For $n = 1$,
$[r(\cos\theta + i\sin\theta)]^1 = r^1(\cos 1\theta + i\sin 1\theta)$. Thus, $1 \in S$.

Step 2 Assume $k \in S$. Show $(k+1) \in S$.
If $k \in S$, this means $[r(\cos\theta + i\sin\theta)]^k = r^k(\cos k\theta + i\sin k\theta)$.

We must show that

$$[r(\cos\theta + i\sin\theta)]^{k+1} = r^{k+1}[\cos(k+1)\theta + i\sin(k+1)\theta].$$

Begin with the left side of the preceding equation.

$[r(\cos\theta + i\sin\theta)]^{k+1}$

$= [r(\cos\theta + i\sin\theta)]^k \cdot [r(\cos\theta + i\sin\theta)]^1$

$= \underbrace{r^k(\cos k\theta + i\sin k\theta)} \cdot r(\cos\theta + i\sin\theta)$

by the Induction Hypothesis

$= r^{k+1}(\cos k\theta + i\sin k\theta)(\cos\theta + i\sin\theta)$

$= r^{k+1}[(\cos k\theta\cos\theta - \sin k\theta\sin\theta) + i(\sin k\theta\cos\theta + \cos k\theta\sin\theta)]$

multiplying out, using $i^2 = -1$, collecting terms, and factoring out i

$= r^{k+1}[\cos(k\theta + \theta) + i\sin(k\theta + \theta)]$

using $\left.\begin{array}{l}\cos(\alpha+\beta) = \cos\alpha\cos\beta - \sin\alpha\sin\beta \\ \sin(\alpha+\beta) = \sin\alpha\cos\beta + \cos\alpha\sin\beta\end{array}\right\}$ *with* $\left\{\begin{array}{l}\alpha = k\theta \\ \beta = \theta\end{array}\right.$

$= r^{k+1}[\cos(k+1)\theta + i\sin(k+1)\theta]$

With this final step, it has been shown that $k + 1 \in S$, and, therefore, De Moivre's Theorem is true for $n \in \mathcal{N}$.

Exercises

Exploratory

1. In the proof of the theorem that $x - 1$ is a factor of $x^n - 1$, why is $x^{k+1} - 1$ equal to $x^{k+1} - x^k + x^k - 1$?

As a brief review, use De Moivre's Theorem to find each of the following.

2. $(1 + i)^5$
3. $[\sqrt{3}(\cos 25° + i\sin 25°)]^6$
4. $(1 + i\sqrt{3})^3$

Written **Prove each of these theorems.**

1. For all $n \in \mathcal{N}$, $x - y$ is a factor of $x^n - y^n$.
2. For all $n \in \mathcal{N}$, $x + y$ is a factor of $x^{2n} - y^{2n}$.
3. For all $n \in \mathcal{N}$, $x + y$ is a factor of $x^{2n+1} + y^{2n+1}$.

4. Compare the following statement with either or both exercise 1 and exercise 2.

$$\text{If } n \in \mathcal{N} \text{ and } n \text{ is even, } x - y \text{ is a factor of } x^n - y^n.$$

5. **Prove:** If n is even, $a^n - 1$ is divisible by $a + 1$.
6. Does the *proof* of De Moivre's Theorem offered in the text hold for all real n? Explain.
7. Complete a proof of the theorem in greater detail than that offered in the text. In particular, expand the statements and reasons summarized by the red print on page 599.
8. Rewrite the proof of De Moivre's Theorem, but with s in place of r, ϕ in place of θ, and m in place of n.
9. **Prove:** $\cos n\pi = (-1)^n$ for $n \in \mathcal{N}$

Challenge

DO SOME RESEARCH. Find out what you can about the contributions of Abraham De Moivre to mathematics, in the field of probability as well as in trigonometry.

Mathematical Excursions — Problem Solving and Patterns

Many of the same mental processes used in mathematical induction can be used in other problem solving strategies, such as looking for a pattern. Sometimes a pattern may be obvious, but many times it is not. As with some induction strategies, you may have to "play" with the facts and figures until you arrive at a pattern you can use.

Example Write an equation showing the relationship between the variables in the chart.

x	1	2	3	4	5
y	17	21	25	29	33

Look for a pattern by finding differences between successive values of x and y. Notice that the changes in x are 1 while the changes in y are 4. Using our knowledge of slopes and points, we arrive at the following equation.

$$m = \frac{4}{1} \qquad (y - 17) = \frac{4}{1}(x - 1) \text{ or } y = 4x + 13$$

Any of the ordered pairs (x, y) will yield the same result.

Exercises Write an equation showing the relationship between the variables in each chart. Then complete the chart.

1.

x	1	2	3			6
y	8	16		32	40	

2.

a	1	2	3	4	5	6
b	5	7	9			

·3. The Worthington City Council has 12 members. After their last session, each member shook hands with each other member. How many handshakes were there in all?

4. At Dublin High School, there are 500 students and 500 lockers, numbered 1 through 500. Suppose the first student opens each locker. Then the second student closes every second locker. The third student changes the state of every third locker (the student closes the open lockers and opens the closed lockers). The fourth student changes the state of every fourth locker. This process continues until the five-hundredth student changes the state of the five-hundredth locker. After this process is completed, which lockers are open?

16.7 The Binomial Theorem

This section is devoted to a proof by induction of the **Binomial Theorem.**

$$(a + b)^n = {}_nC_0a^n + {}_nC_1a^{n-1}b + {}_nC_2a^{n-2}b^2 + \cdots + {}_nC_ra^{n-r}b^r + \cdots + {}_nC_nb^n$$

To prove the theorem, some familiar ideas about combinations will be necessary.

The two "triangles" given below name the same numbers.

$$\begin{array}{c} {}_0C_0 \\ {}_1C_0 \quad {}_1C_1 \\ {}_2C_0 \quad {}_2C_1 \quad {}_2C_2 \\ {}_3C_0 \quad {}_3C_1 \quad {}_3C_2 \quad {}_3C_3 \\ {}_4C_0 \quad {}_4C_1 \quad {}_4C_2 \quad {}_4C_3 \quad {}_4C_4 \\ {}_5C_0 \quad {}_5C_1 \quad {}_5C_2 \quad {}_5C_3 \quad {}_5C_4 \quad {}_5C_5 \\ {}_6C_0 \quad {}_6C_1 \quad {}_6C_2 \quad {}_6C_3 \quad {}_6C_4 \quad {}_6C_5 \quad {}_6C_6 \\ \text{and so on} \end{array}$$

Triangle of Combinatorial Coefficients

$$\begin{array}{c} 1 \\ 1 \quad 1 \\ 1 \quad 2 \quad 1 \\ 1 \quad 3 \quad 3 \quad 1 \\ 1 \quad 4 \quad 6 \quad 4 \quad 1 \\ 1 \quad 5 \quad 10 \quad 10 \quad 5 \quad 1 \\ 1 \quad 6 \quad 15 \quad 20 \quad 15 \quad 6 \quad 1 \\ \text{and so on} \end{array}$$

Pascal's Triangle

In both triangles, every entry is equal to the sum of the two "parent" entries immediately above it. In symbols, this fact is expressed as follows.

$${}_{n+1}C_r = {}_nC_r + {}_nC_{r-1}$$

The proof of this statement is left for you in the exercises.

Here is a proof of the Binomial Theorem, using induction.

Let S be the set of natural numbers for which the theorem is true.

Step 1 Show that $1 \in S$.

$(a + b)^1 \stackrel{?}{=} {}_1C_0a^1 + {}_1C_1a^{1-1}b^1$	*Substitute 1 for n.*
$a + b \stackrel{?}{=} 1 \cdot a + 1 \cdot a^0b^1$	${}_1C_0 = {}_1C_1 = 1$
$a + b = a + b$	$a^0 = 1$

Step 2 Show that if $k \in S$, then $k + 1 \in S$.

Assume $k \in S$. That is, assume the theorem is true for k.

$$(a + b)^k = {}_kC_0a^k + {}_kC_1a^{k-1}b + \cdots + {}_kC_ra^{k-r}b^r + \cdots + {}_kC_kb^k$$

We must show that the following statement, with $k + 1$ substituted for k is true.

$$(a + b)^{k+1} = \underbrace{{}_{k+1}C_0a^{k+1}}_{\text{first term}} + {}_{k+1}C_1a^kb + \cdots + \underbrace{{}_{k+1}C_ra^{k+1-r}b^r}_{\text{general term}} + \cdots + \underbrace{{}_{k+1}C_{k+1}b^{k+1}}_{\text{last term}}$$

Now, $(a + b)^{k+1} = (a + b)(a + b)^k$
$= a(a + b)^k + b(a + b)^k$. *Distributive Law*

Use the Induction Hypothesis *twice* to expand this sum. Previous study shows what $(a + b)^k$ equals.

$$a(a + b)^k = a({}_kC_0a^k + {}_kC_1a^{k-1}b + \cdots + {}_kC_ra^{k-r}b^r + \cdots + {}_kC_kb^k)$$
$$= {}_kC_0a^{k+1} + {}_kC_1a^kb + \cdots + \boxed{{}_kC_ra^{k+1-r}b^r} + \cdots + {}_kC_kab^k$$
$$b(a + b)^k = b({}_kC_0a^k + {}_kC_1a^{k-1}b + \cdots + {}_kC_ra^{k-r}b^r + \cdots + {}_kC_kb^k)$$
$$= {}_kC_0a^kb + {}_kC_1a^{k-1}b^2 + \cdots + \boxed{{}_kC_ra^{k-r}b^{r+1}} + \cdots + {}_kC_kb^{k+1}$$

Add the two rows of terms printed in red. Only like terms can be added; these may differ *only in coefficient.*

The first term in the sum is ${}_kC_0a^{k+1} = {}_{k+1}C_0a^{k+1}$, since ${}_nC_0 = 1$ for any n. (This first term comes only from the first red line.)

The last term in the sum is ${}_kC_kb^{k+1} = {}_{k+1}C_{k+1}b^{k+1}$. So far, the first and last terms are what we want them to be.

For the *general term,* however, we have to find the term in the expansion of $b(a + b)^k$ to add to ${}_kC_ra^{k+1-r}b^r$. The general term printed in the red rectangle in the second red line is not *like* it, so they cannot be added.

However, it is the term *before* ${}_kC_ra^{k-r}b^{r+1}$ (in the second red line) that is needed!

The reason for this is that this term will be as follows.

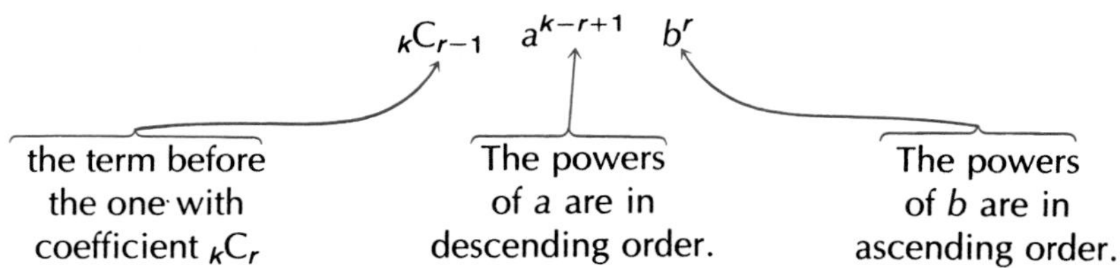

Since ${}_kC_ra^{k+1-r}b^r$ and ${}_kC_{r-1}a^{k+1-r}b^r$ are like terms, they can be added, obtaining $({}_kC_r + {}_kC_{r-1})a^{k+1-r}b^r$. But, by the theorem on combinatorial coefficients given at the beginning of this section, ${}_kC_r + {}_kC_{r-1} = {}_{k+1}C_r$. Thus, the sum of the two terms is

$${}_kC_ra^{k+1-r}b^r + {}_kC_{r-1}a^{k+1-r}b^r = {}_{k+1}C_ra^{k+1-r}b^r.$$

This is exactly what we wanted!

We have established that the first term, the last term, and the all-important general term for $(a + b)^{k+1}$ are what we wanted them to be.

So, $k \in S$ has implied $(k + 1) \in S$, and the assertion—the Binomial Theorem—is proved.

Exercises

Exploratory **The exercises in this section begin with a review of both combinatorial notation and the Binomial Theorem.**

1. Explain how to compute ${}_nC_r$, for $n \geq r$. What formulas can be used? Give a brief justification of each of the formulas.

Compute these values.

2. ${}_7C_4$
3. ${}_{10}C_3$
4. ${}_{16}C_{16}$
5. ${}_{20}C_5$
6. ${}_{20}C_{15}$
7. Justify this statement: ${}_nC_r = {}_nC_{n-r}$.

Use the Binomial Theorem to find each of these expansions.

8. $(x + y)^6$
9. $(2a + 3)^4$
10. $(x - \frac{1}{2}w)^6$
11. Explain the plan of the proof of the Binomial Theorem. Give the Induction Hypothesis, and the statement of what must be proved.

Written

1. Give reasons for each step in the following special case of the rule ${}_{n+1}C_r = {}_nC_r + {}_nC_{r-1}$. Show ${}_{10}C_4 = {}_9C_4 + {}_9C_3$.

$${}_9C_4 + {}_9C_3 = \frac{9!}{4!5!} + \frac{9!}{3!6!} = \frac{6 \cdot 9!}{6!4!} + \frac{4 \cdot 9!}{6!4!} = \frac{(6 + 4)9!}{6!4!} = \frac{10!}{6!4!} = {}_{10}C_4$$

Repeat the process for these values.

2. $n = 12, r = 6$
3. $n = 14, r = 8$
4. $n = 20, r = 5$
5. Explain how the rule referred to in exercises 1–4 is used in the proof of the Binomial Theorem.

Find each of the terms indicated.

6. the sixth, of $(a + 2b)^9$
7. the next-to-last, of $\left(3 - \frac{1}{x}\right)^{10}$
8. the term with x^8y^5, in the expansion of $(x + y)^{13}$

Expand each of the following.

9. $\sum_{r=0}^{3} {}_3C_r a^{3-r} b^r$
10. $\sum_{r=0}^{5} {}_5C_r a^{5-r} b^r$
11. $\sum_{j=2}^{4} {}_6C_j a^{6-j} b^j$
12. Give the first four terms of $\sum_{r=0}^{14} {}_{14}C_r a^{14-r} b^r$.

The following problems refer to the proof of the Binomial Theorem in the text.

13. Write a statement of the Binomial Theorem using Σ-notation.
14. Are the two terms added in the proof the rth terms of $a(a + b)^k$ and $b(a + b)^k$? Explain fully. Why are we not permitted to add the two terms?
15. In the inductive step, we are required to show that the first term of $(a + b)^{k+1}$ is ${}_{k+1}C_0 a^{k+1}$ and the last is ${}_{k+1}C_{k+1} b^{k+1}$. But we do not quite do this. The terms we do obtain appear different. Are they? Explain fully.

Simulation games are games that resemble real life processes. The game called LIFE, created by John Horton Conway, is one example. It shows, in a simple way, the evolution of a society of living organisms as it ages with time.

You can play LIFE using a grid of squares and counters in two different colors, say black and red. The counters are placed on the grid, one to a square. Then you change the positions of the counters according to the following *genetic laws*, or rules for births, deaths, and survivals.

SURVIVALS Every counter with 2 or 3 neighboring counters survives for the next generation. *Neighboring counters have common sides or corners.*

DEATH Every counter with 4 or more neighbors dies from overpopulation. Every counter with only one neighbor or none dies from isolation.

BIRTHS Every empty cell with exactly 3 neighbors is a birth cell. A counter is placed on the cell for the next generation.

To help eliminate mistakes in play, the following procedure is suggested.

1. Start with a pattern of black counters.
2. Put a red counter where each birth cell will occur.
3. Put a second black counter on top of each death cell. Ignore the red counters when determining death cells.
4. Check the new pattern. Then, remove all dead counters, and replace the newborn red counters with black counters.

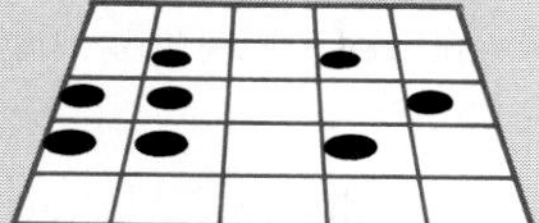

first generation

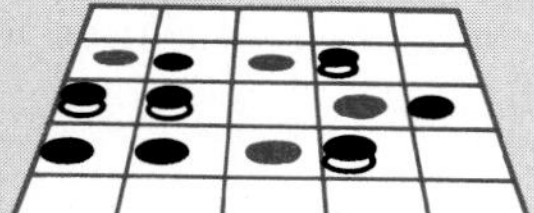

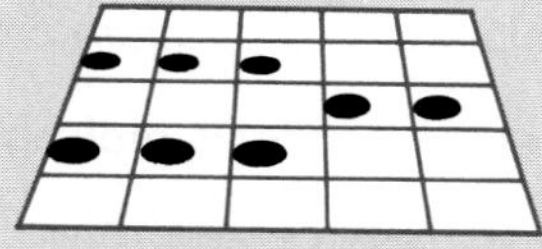

second generation

Notice that births and deaths occur simultaneously. Together, they are a move to the next generation.

When you play this game, you will find that many patterns either die, reach a stable repeating pattern or become blinkers, alternating between two patterns.

Exercises **Find the next six generations for each of the following.**

1.

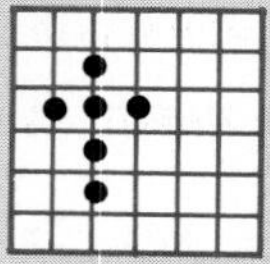

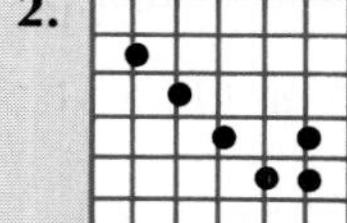

2.

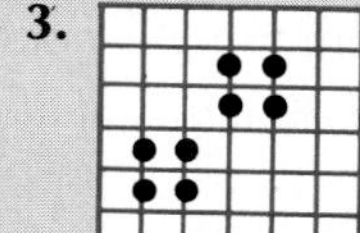

3.

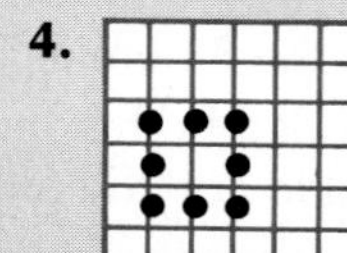

4.

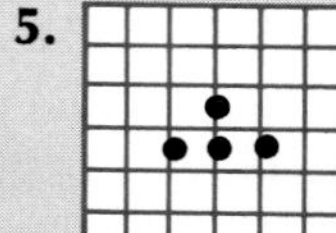

5.

16.8 Geometry

There are some interesting properties of points and lines in a plane, and of other geometric figures, which can be established by induction.

First, consider a problem which may be familiar. How many diagonals can be drawn for a polygon of n sides? Pictured below are some simpler cases.

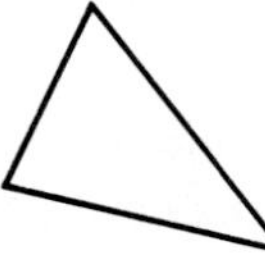

3 sides, 0 diagonals

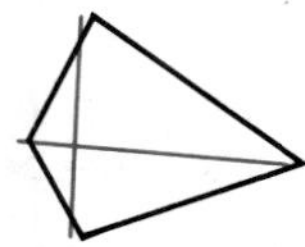

4 sides, 2 diagonals

5 sides, 5 diagonals

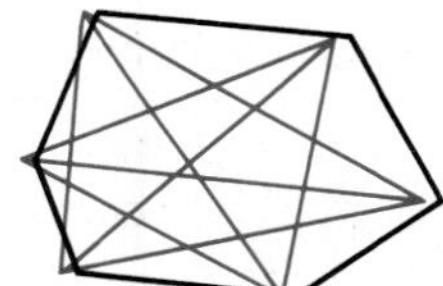

6 sides, 9 diagonals

Before continuing, try to figure out how many diagonals an n-gon will have. Of course, the answer should be in terms of n, and the special cases given above will have to satisfy any conjecture.

Upon careful investigation, you may have discovered the following assertion.

The number of diagonals of a convex polygon of n sides is $\frac{n(n-3)}{2}$, where $n \geq 3$.

Proof (by mathematical induction):

Let S be the set of natural numbers for which the assertion is true. Then $3 \in S$ since $\frac{3(3-3)}{2} = 0$, and a triangle has zero diagonals.

Suppose the assertion is true for $n = k$; that is, suppose a k-gon has $\frac{k(k-3)}{2}$ diagonals. To complete the problem, it must be shown that a $(k+1)$-gon has $\frac{(k+1)(k-2)}{2}$ diagonals.

Now, consider a $(k+1)$-gon. How many new diagonals are added with the addition of the $(k+1)$st vertex?

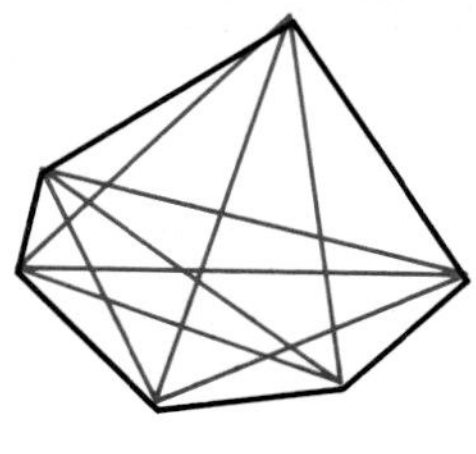

k-gon

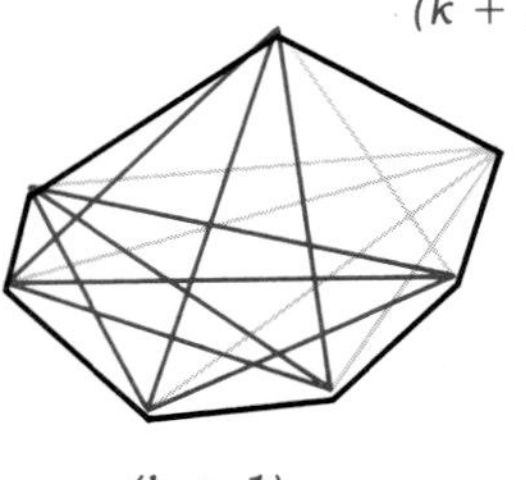

(k + 1)-gon

From the $(k+1)$st vertex, $k-2$ diagonals can be drawn to each of the $k-2$ nonneighboring vertices. (Verify this in the drawing.) Furthermore, an additional diagonal is drawn between the two neighboring vertices. (In the k-gon, this new diagonal was a side.) Thus, there are $k-2+1$ or $k-1$ new diagonals. The $(k+1)$-gon has $\frac{k(k-3)}{2}+k-1$ diagonals.

$$\begin{aligned}\frac{k(k-3)}{2}+k-1 &= \frac{k(k-3)}{2}+\frac{2(k-1)}{2}\\ &= \frac{k(k-3)+2(k-1)}{2}\\ &= \frac{k^2-3k+2k-2}{2}\\ &= \frac{k^2-k-2}{2}\\ &= \frac{(k+1)(k-2)}{2}\end{aligned}$$

Thus, $k \in S \rightarrow (k+1) \in S$, and the assertion is proved for $n \geq 3$.

Exercises

Written

1. Why must n equal or exceed 3 in the assertion discussed in this section?
2. How many diagonals has a dodecagon? a 15-gon? an 82-gon?
3. Draw some regular n-gons, and their diagonals, for different odd and even n. Try fairly large n, and draw careful diagrams. Can you come to any conclusions?

Check a number of cases of each of the following assertions to see if it appears true. If it does, prove it by mathematical induction. Careful! At least one of them is false!

4. Given a segment of length one unit, it is possible to construct (with straightedge and compass only) a segment of length $\sqrt{n}$, $n \in \mathcal{N}$.
5. The sum of the measures of the interior angles of a convex n-gon is $(n-2)180°$.
6. The number of regions formed when chords connect n points on a circle is 2^{n-1}, $n \geq 2$.
7. The number of chords connecting n points on a circle is $\frac{n(n-1)}{2}$.
8. The number of lines which can be obtained by connecting n distinct points in the plane, no three collinear, is $\frac{n(n-1)}{2}$, $n \geq 2$.

16.9 A Few Fundamentals

In this section, induction will be used to establish some very fundamental, and hard-to-get-at, concepts about our number system.

An **inductive** or **recursive definition** uses a kind of induction to *define* things.

We can define the use of exponents using a recursive definition.

1. $a^1 = a$

2. $a^{n+1} = a^n \cdot a$

Question: How must a^0 be defined in order to be consistent?

Here, as is usually the case, the number 1 is the "anchor." Part **2** of the definition is the inductive part. Is a^n defined clearly for all natural numbers n? Part **1** states specifically what a^n is when $n = 1$. Part **2** states that knowing how to find a^k implies knowing how to find a^{k+1}. Therefore, we see that the set S of natural numbers for which a^k is clearly defined is equal to $\mathcal{N}$.

Here is a recursive definition of "factorials."

1. $1! = 1$

2. $(n + 1)! = (n + 1) \cdot n!$

Question: How must 0! be defined in order to be consistent?

Another example goes back even further to the basic ideas of mathematics to consider the operation of addition. Addition is a binary operation. That is, addition is defined for exactly two numbers to be added. If the numbers are a_1 and a_2, we know how to find $a_1 + a_2$. But what of sums like $a_1 + a_2 + a_3 + a_4$? The following procedure can be used.

$a_1 + a_2 + a_3 = (a_1 + a_2) + a_3$

$a_1 + a_2 + a_3 + a_4 = (a_1 + a_2 + a_3) + a_4 = (a_1 + a_2) + a_3 + a_4$

and so on, for each new addend.

But, here is an inductive definition which takes care of such strings of addends for any finite-length string.

1. $a_1 = a_1$

2. $a_1 + \cdot \cdot \cdot + a_{n+1} = (a_1 + \cdot \cdot \cdot + a_n) + a_{n+1}$

Very often in your mathematical work you have used the Distributive Law, $a(b + c) = ab + ac$, in such a way which implied that the law worked for more than two numbers within the parentheses. That is, without even proving it, we assumed that

$$a(b_1 + b_2 + \cdots + b_n) = ab_1 + ab_2 + \cdots + ab_n.$$

A proof for this statement could be produced for any specific natural number n, say $n = 5$. A repeating, or *iterative*, method like the one outlined for addition would be used. But here is a proof by mathematical induction that the Distributive Law holds for any number of numbers "in parentheses." The previous definition of addition and the "ordinary" Distributive Law for two addends, $a(b_1 + b_2) = ab_1 + ab_2$, will be used.

Proof: Let S be the set of all natural numbers for which it is true that $a(b_1 + \cdots + b_n) = ab_1 + \cdots + ab$.

Step 1 Show $1 \in S$. If $n = 1$, $ab_1 = ab_1$. Thus, $1 \in S$.

Step 2 Assume $a(b_1 + \cdots + b_k) = ab_1 + \cdots + ab_k$. (Induction Hypothesis)

Show that $a(b_1 + \cdots + b_k + b_{k+1}) = ab_1 + \cdots + ab_k + ab_{k+1}$.

$a(b_1 + \cdots + b_k + b_{k+1})$

$= a[(b_1 + \cdots + b_k) + b_{k+1}]$	*Definition of addition*
$= a(b_1 + \cdots + b_k) + ab_{k+1}$	*Distributive Law for two addends*
$= (ab_1 + \cdots + ab_k) + ab_{k+1}$	*Induction Hypothesis*
$= ab_1 + \cdots + ab_k + ab_{k+1}$	*Definition of addition*

Thus, $k \in S \rightarrow (k + 1) \in S$, and our generalized Distributive Law is proved.

Generalized Associative and Commutative Laws can be proved in the same way.

Now we will prove a well-known law of exponents, namely,

$$a^m a^n = a^{m+n}.$$

The proof will be based on the definition given earlier for a^n. We must modify our induction plan to take account of the fact that there is more than one variable in the statement to be proved. Therefore, we will "hold m fixed" and perform the induction "on" the remaining variable n.

Proof Let S be the set of natural numbers for which it is true that $a^m a^n = a^{m+n}$.

Step 1 Show $1 \in S$. If $n = 1$, we have $a^m a = a^{m+1}$, which is precisely the definition of a^{m+1}. Therefore, $1 \in S$.

Step 2 Assume $k \in S$; that is, $a^m a^k = a^{m+k}$. We must now show that

$a^m a^{k+1} = a^{m+(k+1)}$. The proof is continued here horizontally. Follo each step carefully.

$a^m a^{k+1} = a^m(a^{k+1}) = a^m(a^k a) = (a^m a^k)a = (a^{m+k})a = a^{(m+k)+1} = a^{m+(k+1)}$

Thus, $k \in S$ implies that $(k + 1) \in S$, and the assertion is proved.

Exercises

Exploratory

1. Explain, in your own words, the difference between a recursive, or inductive, definition, and the kind of definition familiar to you.
2. Explain how $k!$ is actually defined for all $k \in \mathcal{N}$ by the definition given for factorials on page 607. Using this definition, find 7! directly.

Write the first eight terms of the sequences defined by the following.

3. $a_1 = 2;\ a_{n+1} = 2a_n$
4. $a_1 = 0;\ a_{n+1} = a_n + 5$
5. $a_1 = 2;\ a_n = a_{n-1} \cdot 2$
6. $a_1 = 3;\ a_2 = 4;\ a_{n+2} = a_{n+1} + a_n$

Written

Consider the sequence defined by $a_1 = 0.9;\ a_{n+1} = 0.9 \cdot a_n$.

1. Write out the first five terms of the sequence.
2. Do the terms of the sequence appear to be getting closer and closer to some number? Which number? How could you justify your answer?
3. Rewrite the following "horizontal" proof in a vertical format giving reasons at each step.

To Prove: $a_n < 1$ for all $n \in \mathcal{N}$.

Step 1 $1 \in S$.

Step 2 Suppose $a_k < 1$. Show $a_{k+1} < 1$. But $a_{k+1} = 0.9 \cdot a_k < 0.9 \cdot 1 = 0.9 < 1$.

4. Prove that $(a^m)^n = a^{mn}$ for all real m, and for $n \in \mathcal{N}$.
5. Prove that $1^n = 1$ for all $n \in \mathcal{N}$.
6. Complete this "definition" of Σ-notation for $n \in \mathcal{N}$.

$$\sum_{i=1}^{1} a_i = a \qquad \sum_{i=1}^{n+1} a_i = ?$$

Challenge

1. Recall that for mutually disjoint sets, $P(A \cup B) = P(A) + P(B)$. Use this result to prove that for any natural number n, $P(E_1 \cup E_2 \cup \cdots \cup E_n) = P(E_1) + P(E_2) + \cdots + P(E_n)$ for $E_1, E_2, \ldots, E_n$ mutually exclusive events.

Problem Solving Application: Trees

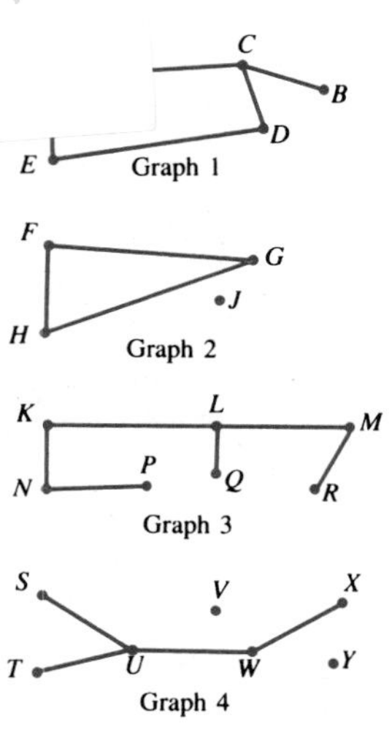

In the study of *graph theory*, a graph is defined as a set of points, called vertices, and a set of line segments, called edges, that connect pairs of vertices. The figure at the left shows four different graphs.

A graph is said to be *connected* if any vertex can be reached from any other vertex by traveling along the edges. Graphs 1 and 3 are connected while Graphs 2 and 4 are not.

A *cycle* is a path of edges in a graph that starts and ends at the same vertex but does not pass through any edge or vertex more than once. A cycle need not include every vertex in a graph. Graph 1 has a cycle, *A-E-D-C-A*. The cycle can start at any one of the four vertices. Graph 2 also has a cycle while Graphs 3 and 4 do not.

A *tree* is a connected graph with no cycles. Graph 3 is the only one of the four graphs that is a tree. Trees play an important role in graph theory and have many applications to real-world problems.

Examples

1 **Determine if each of the following graphs is a tree.**

a.

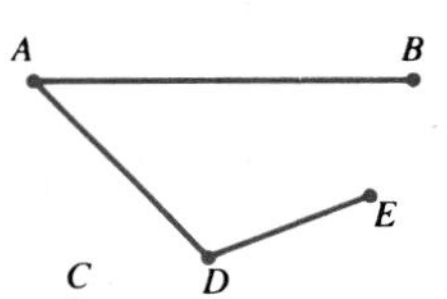

b.

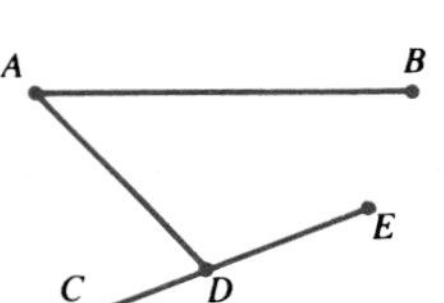

c.

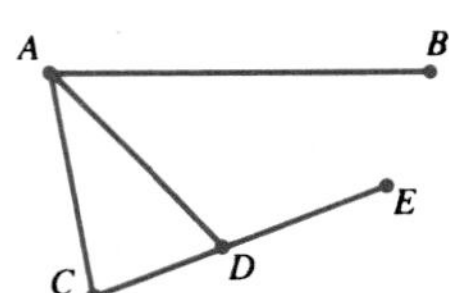

The graph in **a**. is not connected. The graph in **b**. is connected with no cycles so it is a tree. The graph in **c.** is connected but has a cycle, *A-D-C-A*.

2 **A telephone company intends to establish communications among 7 towns. Each town must be able to communicate with any other town, either directly or through other towns, and only certain towns can be linked as shown in the graph at the right. The numbers on each edge are estimates of the costs (in millions of dollars) of constructing links between towns. Determine what links should be used to minimize constructions costs.**

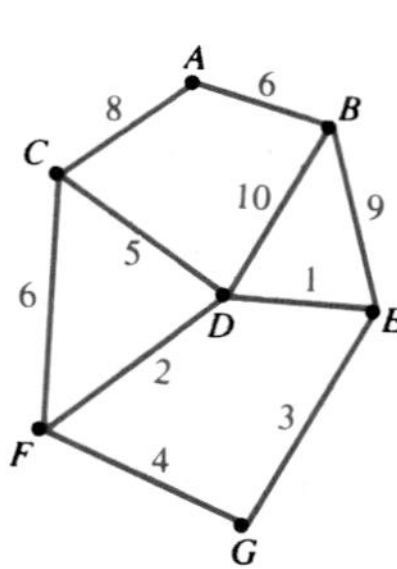

Explore The solution to this problem should be a tree that contains all of the vertices since there must be a path between any two vertices (connected) and since only one path is needed (no cycles).

Plan Use the following steps to find the tree with the least cost.

1. Choose the edge with the least cost.
2. Choose from among the remaining edges, the edge with the least cost that does not produce a cycle.
3. Repeat step 2 until all vertices are linked.

This process is called the Kruskal Algorithm.

Solve The edge with the least cost (1) is $\overline{DE}$. Now choose $\overline{DF}$ (2) and then $\overline{EG}$ (3). $\overline{FG}$ (4) cannot be chosen next since that would produce a cycle. Choose $\overline{DC}$ (5) instead. Either $\overline{AB}$ (6) or $\overline{CF}$ (6) could be chosen next, but $\overline{CF}$ produces a cycle, so choose $\overline{AB}$. Finally, choose $\overline{AC}$ (8) to complete the tree with the least cost, $1 + 2 + 3 + 5 + 6 + 8$ or 25 million dollars. This solution is shown at the right.

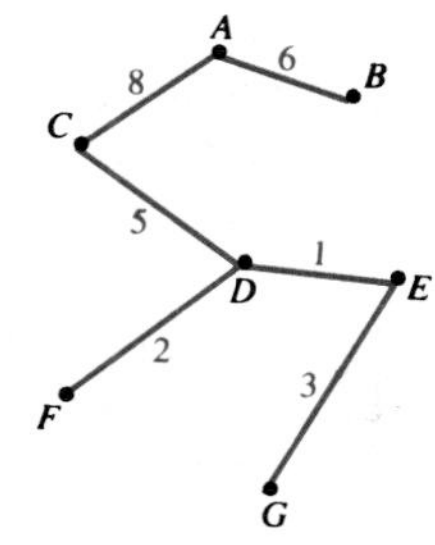

Examine You may want to find other graphs, and then compute their costs to convince yourself that this graph is the tree with the least cost.

Exercises

Written **Determine if each of the following graphs is connected, has a cycle, or is a tree.**

1.

2.

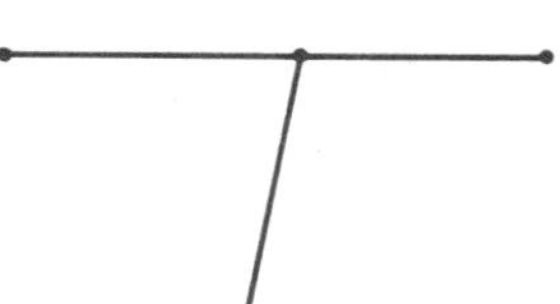

3.

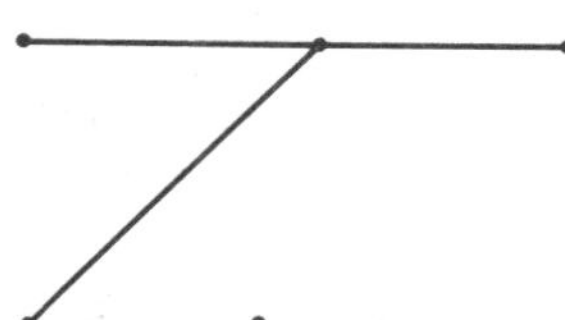

4. Five computers, labeled 1, 2, 3, 4, and 5, need to be connected by a cable system. Connecting an odd-numbered computer to an even-numbered computer is twice as expensive as connecting 2 even-numbered computers to two odd-numbered computers. Draw a graph to represent this situation. Then, find a tree with the least cost that connects the five computers.

5. Coffe Railways wants to construct a rail network of minimum length that connects 6 cities, *A, B, C, D, E,* and *F*. The distance between each pair of cities (in miles) is as follows.

$AB = 18$	$AC = 9$	$AD = 12$	$AE = 16$	$AF = 23$
$BC = 36$	$BD = 14$	$BE = 9$	$BF = 4$	$CD = 11$
$CE = 6$	$CF = 18$	$DE = 19$	$DF = 26$	$EF = 14$

Draw a graph to represent this situation. Then, find the rail network of minimum length that connects the cities such that any city is connected to any other city, either directly or through other cities.

Vocabulary

indirect proof
pattern
mathematical induction
PMI
Induction Hypothesis
anchor
inductive step
De Moivre's Theorem
Binomial Theorem
inductive definition

Chapter Summary

1. Proof by mathematical induction is often compared to a kind of "domino theory." When dominoes are lined up in the right way, if the first one falls, then they all fall.
2. The Principle of Mathematical Induction: Let S be a set of natural numbers for which the following two conditions hold.
   1. $1 \in S$
   2. If the natural number k is in S, then $k + 1$ is also in S.

   Then $S = \mathcal{N}$.
3. When establishing the first condition of PMI, the number 1 is the "first domino pushed over." The first number is sometimes called the anchor.
4. Condition 2 of PMI is a conditional statement in which some number k is assumed to be in set S. This assumption is called the Induction Hypothesis. It is then necessary to prove that $k + 1$ is also in set S. This step is sometimes called the inductive step.
5. Different techniques or special methods are sometimes required to complete a proof by mathematical induction.
6. An inductive or recursive definition uses a kind of induction to define things. Exponents, factorials, and addition can be defined using an inductive definition.

Chapter Review

16.1 1. Under what circumstances is PMI used?

16.2 **Prove each of the following assertions for $n \in \mathcal{N}$ by mathematical induction. Check them first for $n = 1$, $n = 2$, and $n = 3$.**

2. $4 + 8 + 12 + \cdots + 4n = 2n(n + 1)$

3. $\frac{1}{1 \cdot 3} + \frac{1}{3 \cdot 5} + \frac{1}{5 \cdot 7} + \cdots + \frac{1}{(2n - 1)(2n + 1)} = \frac{n}{2n + 1}$

4. Show that the assertion $1 + 3 + 5 + \cdots + (2n - 1) = n^2 + 1$ satisfies Step 2 of PMI but not Step 1.
5. Explain the idea of mathematical induction. Why is the Principle of Mathematical Induction often compared with the "domino theory"?

16.3 **Write the *n*th and $(n + 1)$st term in each of the following series.**

6. $\frac{1}{1 \cdot 2} + \frac{1}{2 \cdot 3} + \frac{1}{3 \cdot 4} + \frac{1}{4 \cdot 5} + \cdots$
7. $1 + 3 + 5 + 7 + \cdots$
8. $1 + x + \frac{x^2}{2!} + \frac{x^3}{3!} + \frac{x^4}{4!} + \cdots$

Find the first three terms of each series whose *n*th terms are given in each exercise.

9. $\frac{n + 1}{n!}$
10. $\frac{n}{3^n}$
11. $\frac{2(-1)^n \cdot n^2}{3^n}$

16.4 12. Explain why the statement "$6^n - 2^n$ is divisible by 4" means the same thing as the statement "$\frac{6^n - 2^n}{4}$ is an integer."

Prove each of the following assertions for $n \in \mathcal{N}$ by mathematical induction. Check them first for $n = 1$, $n = 2$, and $n = 3$.

13. $7^n + 5$ is divisible by 6.
14. $n^5 - n$ is divisible by 5. *Hint: Use the Binomial Theorem on $(n + 1)^5$.*

16.5 15. $6^n \geq 1 + 5n$ for $n \geq 1$.
16. $5^n > 4^{n+1}$, $n \geq 7$.

16.6 **Use De Moivre's Theorem to find each of the following values.**

17. $(1 + i)^3$
18. $(\cos 60° + i \sin 60°)^3$

16.7 19. Give a formula for computing ${}_nC_r$ for $n \geq r$.

Compute each of these values.

20. ${}_{10}C_0$
21. ${}_8C_4$
22. ${}_7C_2$

Find each of the following terms for the expansion of $(x + y)^5$.

23. second
24. third
25. fifth

16.8 26. "The numer of regions in which n lines, no two parallel and no three concurrent, divide the plane is $\frac{n(n + 1)}{2} + 1$." Check to see if this assertion is true. If it is, prove it by mathematical induction.

Chapter Test

1. Write the nth and $(n + 1)$st term of this series $\sqrt{2} + \sqrt[3]{2} + \sqrt[4]{2} + \sqrt[5]{2} + \cdots$.
2. Find the first three terms of $\sum_{n=1}^{100} \frac{(n + 1)^2}{n(n + 3)}$.

Prove the following assertions for $n \in \mathcal{N}$ are true by mathematical induction.

3. $1 + 5 + 9 + \cdots + (4n - 3) = n(2n - 1)$
4. $1 \cdot 2 + 2 \cdot 3 + 3 \cdot 4 + \cdots + n(n + 1) = \frac{1}{3}n(n + 1)(n + 2)$.
5. $5^n - 2^n$ is divisible by 3.

Compute each of these values.

6. ${}_5C_3$
7. ${}_{14}C_{14}$

Use De Moivre's Theorem to find each of the following values.

8. $(-3 + i\sqrt{3})^5$
9. $\left(-\frac{1}{2} + \frac{i\sqrt{3}}{2}\right)^2$

Find each of the following terms for the expansion of $(2a + b)^6$.

10. second
11. third
12. fifth

Write out the following sums.

13. $\sum_{i=1}^{5} a_i$
14. $\sum_{i=1}^{7} (2^i + 3)$
15. $\sum_{i=1}^{4} \frac{i + 3}{i!}$

Tell whether each statement is *true* or *false*. Justify your answers.

16. The sum of the first n natural numbers is $n^2 - 1$.
17. For $n \in \mathcal{N}$ and $1 \le n < 41$, $n^2 - n + 41$ is a prime number.
18. $\sum_{i=1}^{n} (2i - 1) = n^2$
19. $8^n - 5^n$ is divisible by 3.
20. $n! < 2^n$ for $n \ge 4$.
21. De Moivre's Theorem is true for all real n although mathematical induction proves it is true only for n, a natural number.
22. For all $n \in \mathcal{N}$, $x - y$ is a factor of $x^{2n} - y^{2n}$.
23. The expansion of $(a + x)^9$ contains nine terms.
24. The number of diagonals of a convex polygon of n sides is $\frac{n(n - 3)}{3}$, where $n \ge 3$.
25. The number of planes that pass through n distinct points in space, no four coplanar, is $\frac{n(2n - 5)}{3}$, where $n \ge 3$.

Symbols

$=$	is equal to
$\neq$	is not equal to
$>$	is greater than
$<$	is less than
$\geq$	is greater than or equal to
$\leq$	is less than or equal to
$\approx$	is approximately equal to
$*$	an operation in a group or a field
$\cdot$	times
$-$	negative
$+$	positive
$\pm$	positive or negative
$\overleftrightarrow{AB}$	line containing points A and B
$\overrightarrow{AB}$	ray with endpoint A passing through B
$\overline{AB}$	line segment with endpoints A and B
AB	measure of $\overline{AB}$
$\overset{\frown}{AB}$	arc with endpoints A and B
$m\overset{\frown}{AB}$	measure of arc AB
$\angle$	angle
$m\angle A$	measure of angle A
$\circ$	degree
$\triangle$	triangle
$\square$	parallelogram
$\odot P$	circle with center P
$\cong$	is congruent to
$\sim$	is similar to; negation
$\parallel$	is parallel to
$\perp$	is perpendicular to
π	pi
$\|a\|$	absolute value of a
$\sqrt{\ }$	principal square root
$a:b$	ratio of a to b
$\{\ \}$	set
$\in$	is a member of
$\cup$	union
$\cap$	intersection
$\wedge$	conjunction
$\vee$	disjunction
$\rightarrow$	conditional; is mapped onto
$n!$	n factorial
$x \xrightarrow{f} x^2$	f is the function that maps x to x^2
$f \circ g$	f following g
R_ℓ	reflection over line ℓ
$T_{2,3}$	translation 2 units to the right and 3 units up
$Rot_{A,65^\circ}$	rotation with center A through an angle of 65° counterclockwise
$D_{P,2}$	dilation with center P and a scale factor of 2
$\sum_{i=1}^{5} i$	summation of i for values of i from 1 to 5
$\mathcal{N}$	set of natural numbers
$\mathcal{W}$	set of whole numbers
$\mathcal{Z}$	set of integers
$\mathcal{Q}$	set of rational numbers
$\mathcal{R}$	set of real numbers

Common Logarithms

n	0	1	2	3	4	5	6	7	8	9
10	0000	0043	0086	0128	0170	0212	0253	0294	0334	0374
11	0414	0453	0492	0531	0569	0607	0645	0682	0719	0755
12	0792	0828	0864	0899	0934	0969	1004	1038	1072	1106
13	1139	1173	1206	1239	1271	1303	1335	1367	1399	1430
14	1461	1492	1523	1553	1584	1614	1644	1673	1703	1732
15	1761	1790	1818	1847	1875	1903	1931	1959	1987	2014
16	2041	2068	2095	2122	2148	2175	2201	2227	2253	2279
17	2304	2330	2355	2380	2405	2430	2455	2480	2504	2529
18	2553	2577	2601	2625	2648	2672	2695	2718	2742	2765
19	2788	2810	2833	2856	2878	2900	2923	2945	2967	2989
20	3010	3032	3054	3075	3096	3118	3139	3160	3181	3201
21	3222	3243	3263	3284	3304	3324	3345	3365	3385	3404
22	3424	3444	3464	3483	3502	3522	3541	3560	3579	3598
23	3617	3636	3655	3674	3692	3711	3729	3747	3766	3784
24	3802	3820	3838	3856	3874	3892	3909	3927	3945	3962
25	3979	3997	4014	4031	4048	4065	4082	4099	4116	4133
26	4150	4166	4183	4200	4216	4232	4249	4265	4281	4298
27	4314	4330	4346	4362	4378	4393	4409	4425	4440	4456
28	4472	4487	4502	4518	4533	4548	4564	4579	4594	4609
29	4624	4639	4654	4669	4683	4698	4713	4728	4742	4757
30	4771	4786	4800	4814	4829	4843	4857	4871	4886	4900
31	4914	4928	4942	4955	4969	4983	4997	5011	5024	5038
32	5051	5065	5079	5092	5105	5119	5132	5145	5159	5172
33	5185	5198	5211	5224	5237	5250	5263	5276	5289	5302
34	5315	5328	5340	5353	5366	5378	5391	5403	5416	5428
35	5441	5453	5465	5478	5490	5502	5514	5527	5539	5551
36	5563	5575	5587	5599	5611	5623	5635	5647	5658	5670
37	5682	5694	5705	5717	5729	5740	5752	5763	5775	5786
38	5798	5809	5821	5832	5843	5855	5866	5877	5888	5899
39	5911	5922	5933	5944	5955	5966	5977	5988	5999	6010
40	6021	6031	6042	6053	6064	6075	6085	6096	6107	6117
41	6128	6138	6149	6160	6170	6180	6191	6201	6212	6222
42	6232	6243	6253	6263	6274	6284	6294	6304	6314	6325
43	6335	6345	6355	6365	6375	6385	6395	6405	6415	6425
44	6435	6444	6454	6464	6474	6484	6493	6503	6513	6522
45	6532	6542	6551	6561	6571	6580	6590	6599	6609	6618
46	6628	6637	6646	6656	6665	6675	6684	6693	6702	6712
47	6721	6730	6739	6749	6758	6767	6776	6785	6794	6803
48	6812	6821	6830	6839	6848	6857	6866	6875	6884	6893
49	6902	6911	6920	6928	6937	6946	6955	6964	6972	6981
50	6990	6998	7007	7016	7024	7033	7042	7050	7059	7067
51	7076	7084	7093	7101	7110	7118	7126	7135	7143	7152
52	7160	7168	7177	7185	7193	7202	7210	7218	7226	7235
53	7243	7251	7259	7267	7275	7284	7292	7300	7308	7316
54	7324	7332	7340	7348	7356	7364	7372	7380	7388	7396

Common Logarithms

n	0	1	2	3	4	5	6	7	8	9
55	7404	7412	7419	7427	7435	7443	7451	7459	7466	7474
56	7482	7490	7497	7505	7513	7520	7528	7536	7543	7551
57	7559	7566	7574	7582	7589	7597	7604	7612	7619	7627
58	7634	7642	7649	7657	7664	7672	7679	7686	7694	7701
59	7709	7716	7723	7731	7738	7745	7752	7760	7767	7774
60	7782	7789	7796	7803	7810	7818	7825	7832	7839	7846
61	7853	7860	7868	7875	7882	7889	7896	7903	7910	7917
62	7924	7931	7938	7945	7952	7959	7966	7973	7980	7987
63	7993	8000	8007	8014	8021	8028	8035	8041	8048	8055
64	8062	8069	8075	8082	8089	8096	8102	8109	8116	8122
65	8129	8136	8142	8149	8156	8162	8169	8176	8182	8189
66	8195	8202	8209	8215	8222	8228	8235	8241	8248	8254
67	8261	8267	8274	8280	8287	8293	8299	8306	8312	8319
68	8325	8331	8338	8344	8351	8357	8363	8370	8376	8382
69	8388	8395	8401	8407	8414	8420	8426	8432	8439	8445
70	8451	8457	8463	8470	8476	8482	8488	8494	8500	8506
71	8513	8519	8525	8531	8537	8543	8549	8555	8561	8567
72	8573	8579	8585	8591	8597	8603	8609	8615	8621	8627
73	8633	8639	8645	8651	8657	8663	8669	8675	8681	8686
74	8692	8698	8704	8710	8716	8722	8727	8733	8739	8745
75	8751	8756	8762	8768	8774	8779	8785	8791	8797	8802
76	8808	8814	8820	8825	8831	8837	8842	8848	8854	8859
77	8865	8871	8876	8882	8887	8893	8899	8904	8910	8915
78	8921	8927	8932	8938	8943	8949	8954	8960	8965	8971
79	8976	8982	8987	8993	8998	9004	9009	9015	9020	9025
80	9031	9036	9042	9047	9053	9058	9063	9069	9074	9079
81	9085	9090	9096	9101	9106	9112	9117	9122	9128	9133
82	9138	9143	9149	9154	9159	9165	9170	9175	9180	9186
83	9191	9196	9201	9206	9212	9217	9222	9227	9232	9238
84	9243	9248	9253	9258	9263	9269	9274	9279	9284	9289
85	9294	9299	9304	9309	9315	9320	9325	9330	9335	9340
86	9345	9350	9355	9360	9365	9370	9375	9380	9385	9390
87	9395	9400	9405	9410	9415	9420	9425	9430	9435	9440
88	9445	9450	9455	9460	9465	9469	9474	9479	9484	9489
89	9494	9499	9504	9509	9513	9518	9523	9528	9533	9538
90	9542	9547	9552	9557	9562	9566	9571	9576	9581	9586
91	9590	9595	9600	9605	9609	9614	9619	9624	9628	9633
92	9638	9643	9647	9652	9657	9661	9666	9671	9675	9680
93	9685	9689	9694	9699	9703	9708	9713	9717	9722	9727
94	9731	9736	9741	9745	9750	9754	9759	9763	9768	9773
95	9777	9782	9786	9791	9795	9800	9805	9809	9814	9818
96	9823	9827	9832	9836	9841	9845	9850	9854	9859	9863
97	9868	9872	9877	9881	9886	9890	9894	9899	9903	9908
98	9912	9917	9921	9926	9930	9934	9939	9943	9948	9952
99	9956	9961	9965	9969	9974	9978	9983	9987	9991	9996

Trigonometric Ratios

Angle	Radians	Sin	Cos	Tan	Cot	Sec	Csc		
0°00′	0.0000	0.0000	1.0000	0.0000	—	1.000	—	1.5708	90°00′
10′	0.0029	0.0029	1.0000	0.0029	343.8	1.000	343.8	1.5679	50′
20′	0.0058	0.0058	1.0000	0.0058	171.9	1.000	171.9	1.5650	40′
30′	0.0087	0.0087	1.0000	0.0087	114.6	1.000	114.6	1.5621	30′
40′	0.0116	0.0116	0.9999	0.0116	85.94	1.000	85.95	1.5592	20′
50′	0.0145	0.0145	0.9999	0.0145	68.75	1.000	68.76	1.5563	10′
1°00′	0.0175	0.0175	0.9998	0.0175	57.29	1.000	57.30	1.5533	89°00′
10′	0.0204	0.0204	0.9998	0.0204	49.10	1.000	49.11	1.5504	50′
20′	0.0233	0.0233	0.9997	0.0233	42.96	1.000	42.98	1.5475	40′
30′	0.0262	0.0262	0.9997	0.0262	38.19	1.000	38.20	1.5446	30′
40′	0.0291	0.0291	0.9996	0.0291	34.37	1.000	34.38	1.5417	20′
50′	0.0320	0.0320	0.9995	0.0320	31.24	1.001	31.26	1.5388	10′
2°00′	0.0349	0.0349	0.9994	0.0349	28.64	1.001	28.65	1.5359	88°00′
10′	0.0378	0.0378	0.9993	0.0378	26.43	1.001	26.45	1.5330	50′
20′	0.0407	0.0407	0.9992	0.0407	24.54	1.001	24.56	1.5301	40′
30′	0.0436	0.0436	0.9990	0.0437	22.90	1.001	22.93	1.5272	30′
40′	0.0465	0.0465	0.9989	0.0466	21.47	1.001	21.49	1.5243	20′
50′	0.0495	0.0494	0.9988	0.0495	20.21	1.001	20.23	1.5213	10′
3°00′	0.0524	0.0523	0.9986	0.0524	19.08	1.001	19.11	1.5184	87°00′
10′	0.0553	0.0552	0.9985	0.0553	18.07	1.002	18.10	1.5155	50′
20′	0.0582	0.0581	0.9983	0.0582	17.17	1.002	17.20	1.5126	40′
30′	0.0611	0.0610	0.9981	0.0612	16.35	1.002	16.38	1.5097	30′
40′	0.0640	0.0640	0.9980	0.0641	15.60	1.002	15.64	1.5068	20′
50′	0.0669	0.0669	0.9978	0.0670	14.92	1.002	14.96	1.5039	10′
4°00′	0.0698	0.0698	0.9976	0.0699	14.30	1.002	14.34	1.5010	86°00′
10′	0.0727	0.0727	0.9974	0.0729	13.73	1.003	13.76	1.4981	50′
20′	0.0756	0.0756	0.9971	0.0758	13.20	1.003	13.23	1.4952	40′
30′	0.0785	0.0785	0.9969	0.0787	12.71	1.003	12.75	1.4923	30′
40′	0.0814	0.0814	0.9967	0.0816	12.25	1.003	12.29	1.4893	20′
50′	0.0844	0.0843	0.9964	0.0846	11.83	1.004	11.87	1.4864	10′
5°00′	0.0873	0.0872	0.9962	0.0875	11.43	1.004	11.47	1.4835	85°00′
10′	0.0902	0.0901	0.9959	0.0904	11.06	1.004	11.10	1.4806	50′
20′	0.0931	0.0929	0.9957	0.0934	10.71	1.004	10.76	1.4777	40′
30′	0.0960	0.0958	0.9954	0.0963	10.39	1.005	10.43	1.4748	30′
40′	0.0989	0.0987	0.9951	0.0992	10.08	1.005	10.13	1.4719	20′
50′	0.1018	0.1016	0.9948	0.1022	9.788	1.005	9.839	1.4690	10′
6°00′	0.1047	0.1045	0.9945	0.1051	9.514	1.006	9.567	1.4661	84°00′
10′	0.1076	0.1074	0.9942	0.1080	9.255	1.006	9.309	1.4632	50′
20′	0.1105	0.1103	0.9939	0.1110	9.010	1.006	9.065	1.4603	40′
30′	0.1134	0.1132	0.9936	0.1139	8.777	1.006	8.834	1.4573	30′
40′	0.1164	0.1161	0.9932	0.1169	8.556	1.007	8.614	1.4544	20′
50′	0.1193	0.1190	0.9929	0.1198	8.345	1.007	8.405	1.4515	10′
7°00′	0.1222	0.1219	0.9925	0.1228	8.144	1.008	8.206	1.4486	83°00′
10′	0.1251	0.1248	0.9922	0.1257	7.953	1.008	8.016	1.4457	50′
20′	0.1280	0.1276	0.9918	0.1287	7.770	1.008	7.834	1.4428	40′
30′	0.1309	0.1305	0.9914	0.1317	7.596	1.009	7.661	1.4399	30′
40′	0.1338	0.1334	0.9911	0.1346	7.429	1.009	7.496	1.4370	20′
50′	0.1367	0.1363	0.9907	0.1376	7.269	1.009	7.337	1.4341	10′
8°00′	0.1396	0.1392	0.9903	0.1405	7.115	1.010	7.185	1.4312	82°00′
10′	0.1425	0.1421	0.9899	0.1435	6.968	1.010	7.040	1.4283	50′
20′	0.1454	0.1449	0.9894	0.1465	6.827	1.011	6.900	1.4254	40′
30′	0.1484	0.1478	0.9890	0.1495	6.691	1.011	6.765	1.4224	30′
40′	0.1513	0.1507	0.9886	0.1524	6.561	1.012	6.636	1.4195	20′
50′	0.1542	0.1536	0.9881	0.1554	6.435	1.012	6.512	1.4166	10′
9°00′	0.1571	0.1564	0.9877	0.1584	6.314	1.012	6.392	1.4137	81°00′
		Cos	Sin	Cot	Tan	Csc	Sec	Radians	Angle

Trigonometric Ratios

Angle	Radians	Sin	Cos	Tan	Cot	Sec	Csc		
9°00′	0.1571	0.1564	0.9877	0.1584	6.314	1.012	6.392	1.4137	81°00′
10′	0.1600	0.1593	0.9872	0.1614	6.197	1.013	6.277	1.4108	50′
20′	0.1629	0.1622	0.9868	0.1644	6.084	1.013	6.166	1.4079	40′
30′	0.1658	0.1650	0.9863	0.1673	5.976	1.014	6.059	1.4050	30′
40′	0.1687	0.1679	0.9858	0.1703	5.871	1.014	5.955	1.4021	20′
50′	0.1716	0.1708	0.9853	0.1733	5.769	1.015	5.855	1.3992	10′
10°00′	0.1745	0.1736	0.9848	0.1763	5.671	1.015	5.759	1.3963	80°00′
10′	0.1774	0.1765	0.9843	0.1793	5.576	1.016	5.665	1.3934	50′
20′	0.1804	0.1794	0.9838	0.1823	5.485	1.016	5.575	1.3904	40′
30′	0.1833	0.1822	0.9833	0.1853	5.396	1.017	5.487	1.3875	30′
40′	0.1862	0.1851	0.9827	0.1883	5.309	1.018	5.403	1.3846	20′
50′	0.1891	0.1880	0.9822	0.1914	5.226	1.018	5.320	1.3817	10′
11°00′	0.1920	0.1908	0.9816	0.1944	5.145	1.019	5.241	1.3788	79°00′
10′	0.1949	0.1937	0.9811	0.1974	5.066	1.019	5.164	1.3759	50′
20′	0.1978	0.1965	0.9805	0.2004	4.989	1.020	5.089	1.3730	40′
30′	0.2007	0.1994	0.9799	0.2035	4.915	1.020	5.016	1.3701	30′
40′	0.2036	0.2022	0.9793	0.2065	4.843	1.021	4.945	1.3672	20′
50′	0.2065	0.2051	0.9787	0.2095	4.773	1.022	4.876	1.3643	10′
12°00′	0.2094	0.2079	0.9781	0.2126	4.705	1.022	4.810	1.3614	78°00′
10′	0.2123	0.2108	0.9775	0.2156	4.638	1.023	4.745	1.3584	50′
20′	0.2153	0.2136	0.9769	0.2186	4.574	1.024	4.682	1.3555	40′
30′	0.2182	0.2164	0.9763	0.2217	4.511	1.024	4.620	1.3526	30′
40′	0.2211	0.2193	0.9757	0.2247	4.449	1.025	4.560	1.3497	20′
50′	0.2240	0.2221	0.9750	0.2278	4.390	1.026	4.502	1.3468	10′
13°00′	0.2269	0.2250	0.9744	0.2309	4.331	1.026	4.445	1.3439	77°00′
10′	0.2298	0.2278	0.9737	0.2339	4.275	1.027	4.390	1.3410	50′
20′	0.2327	0.2306	0.9730	0.2370	4.219	1.028	4.336	1.3381	40′
30′	0.2356	0.2334	0.9724	0.2401	4.165	1.028	4.284	1.3352	30′
40′	0.2385	0.2363	0.9717	0.2432	4.113	1.029	4.232	1.3323	20′
50′	0.2414	0.2391	0.9710	0.2462	4.061	1.030	4.182	1.3294	10′
14°00′	0.2443	0.2419	0.9703	0.2493	4.011	1.031	4.134	1.3265	76°00′
10′	0.2473	0.2447	0.9696	0.2524	3.962	1.031	4.086	1.3235	50′
20′	0.2502	0.2476	0.9689	0.2555	3.914	1.032	4.039	1.3206	40′
30′	0.2531	0.2504	0.9681	0.2586	3.867	1.033	3.994	1.3177	30′
40′	0.2560	0.2532	0.9674	0.2617	3.821	1.034	3.950	1.3148	20′
50′	0.2589	0.2560	0.9667	0.2648	3.776	1.034	3.906	1.3119	10′
15°00′	0.2618	0.2588	0.9659	0.2679	3.732	1.035	3.864	1.3090	75°00′
10′	0.2647	0.2616	0.9652	0.2711	3.689	1.036	3.822	1.3061	50′
20′	0.2676	0.2644	0.9644	0.2742	3.647	1.037	3.782	1.3032	40′
30′	0.2705	0.2672	0.9636	0.2773	3.606	1.038	3.742	1.3003	30′
40′	0.2734	0.2700	0.9628	0.2805	3.566	1.039	3.703	1.2974	20′
50′	0.2763	0.2728	0.9621	0.2836	3.526	1.039	3.665	1.2945	10′
16°00′	0.2793	0.2756	0.9613	0.2867	3.487	1.040	3.628	1.2915	74°00′
10′	0.2822	0.2784	0.9605	0.2899	3.450	1.041	3.592	1.2886	50′
20′	0.2851	0.2812	0.9596	0.2931	3.412	1.042	3.556	1.2857	40′
30′	0.2880	0.2840	0.9588	0.2962	3.376	1.043	3.521	1.2828	30′
40′	0.2909	0.2868	0.9580	0.2994	3.340	1.044	3.487	1.2799	20′
50′	0.2938	0.2896	0.9572	0.3026	3.305	1.045	3.453	1.2770	10′
17°00′	0.2967	0.2924	0.9563	0.3057	3.271	1.046	3.420	1.2741	73°00′
10′	0.2996	0.2952	0.9555	0.3089	3.237	1.047	3.388	1.2712	50′
20′	0.3025	0.2979	0.9546	0.3121	3.204	1.048	3.356	1.2683	40′
30′	0.3054	0.3007	0.9537	0.3153	3.172	1.049	3.326	1.2654	30′
40′	0.3083	0.3035	0.9528	0.3185	3.140	1.049	3.295	1.2625	20′
50′	0.3113	0.3062	0.9520	0.3217	3.108	1.050	3.265	1.2595	10′
18°00′	0.3142	0.3090	0.9511	0.3249	3.078	1.051	3.236	1.2566	72°00′
		Cos	Sin	Cot	Tan	Csc	Sec	Radians	Angle

Trigonometric Ratios

Angle	Radians	Sin	Cos	Tan	Cot	Sec	Csc		
18°00′	0.3142	0.3090	0.9511	0.3249	3.078	1.051	3.236	1.2566	72°00′
10′	0.3171	0.3118	0.9502	0.3281	3.047	1.052	3.207	1.2537	50′
20′	0.3200	0.3145	0.9492	0.3314	3.018	1.053	3.179	1.2508	40′
30′	0.3229	0.3173	0.9483	0.3346	2.989	1.054	3.152	1.2479	30′
40′	0.3258	0.3201	0.9474	0.3378	2.960	1.056	3.124	1.2450	20′
50′	0.3287	0.3228	0.9465	0.3411	2.932	1.057	3.098	1.2421	10′
19°00′	0.3316	0.3256	0.9455	0.3443	2.904	1.058	3.072	1.2392	71°00′
10′	0.3345	0.3283	0.9446	0.3476	2.877	1.059	3.046	1.2363	50′
20′	0.3374	0.3311	0.9436	0.3508	2.850	1.060	3.021	1.2334	40′
30′	0.3403	0.3338	0.9426	0.3541	2.824	1.061	2.996	1.2305	30′
40′	0.3432	0.3365	0.9417	0.3574	2.798	1.062	2.971	1.2275	20′
50′	0.3462	0.3393	0.9407	0.3607	2.773	1.063	2.947	1.2246	10′
20°00′	0.3491	0.3420	0.9397	0.3640	2.747	1.064	2.924	1.2217	70°00′
10′	0.3520	0.3448	0.9387	0.3673	2.723	1.065	2.901	1.2188	50′
20′	0.3549	0.3475	0.9377	0.3706	2.699	1.066	2.878	1.2159	40′
30′	0.3578	0.3502	0.9367	0.3739	2.675	1.068	2.855	1.2130	30′
40′	0.3607	0.3529	0.9356	0.3772	2.651	1.069	2.833	1.2101	20′
50′	0.3636	0.3557	0.9346	0.3805	2.628	1.070	2.812	1.2072	10′
21°00′	0.3665	0.3584	0.9336	0.3839	2.605	1.071	2.790	1.2043	69°00′
10′	0.3694	0.3611	0.9325	0.3872	2.583	1.072	2.769	1.2014	50′
20′	0.3723	0.3638	0.9315	0.3906	2.560	1.074	2.749	1.1985	40′
30′	0.3752	0.3665	0.9304	0.3939	2.539	1.075	2.729	1.1956	30′
40′	0.3782	0.3692	0.9293	0.3973	2.517	1.076	2.709	1.1926	20′
50′	0.3811	0.3719	0.9283	0.4006	2.496	1.077	2.689	1.1897	10′
22°00′	0.3840	0.3746	0.9272	0.4040	2.475	1.079	2.669	1.1868	68°00′
10′	0.3869	0.3773	0.9261	0.4074	2.455	1.080	2.650	1.1839	50′
20′	0.3898	0.3800	0.9250	0.4108	2.434	1.081	2.632	1.1810	40′
30′	0.3927	0.3827	0.9239	0.4142	2.414	1.082	2.613	1.1781	30′
40′	0.3956	0.3854	0.9228	0.4176	2.394	1.084	2.595	1.1752	20′
50′	0.3985	0.3881	0.9216	0.4210	2.375	1.085	2.577	1.1723	10′
23°00′	0.4014	0.3907	0.9205	0.4245	2.356	1.086	2.559	1.1694	67°00′
10′	0.4043	0.3934	0.9194	0.4279	2.337	1.088	2.542	1.1665	50′
20′	0.4072	0.3961	0.9182	0.4314	2.318	1.089	2.525	1.1636	40′
30′	0.4102	0.3987	0.9171	0.4348	2.300	1.090	2.508	1.1606	30′
40′	0.4131	0.4014	0.9159	0.4383	2.282	1.092	2.491	1.1577	20′
50′	0.4160	0.4041	0.9147	0.4417	2.264	1.093	2.475	1.1548	10′
24°00′	0.4189	0.4067	0.9135	0.4452	2.246	1.095	2.459	1.1519	66°00′
10′	0.4218	0.4094	0.9124	0.4487	2.229	1.096	2.443	1.1490	50′
20′	0.4247	0.4120	0.9112	0.4522	2.211	1.097	2.427	1.1461	40′
30′	0.4276	0.4147	0.9100	0.4557	2.194	1.099	2.411	1.1432	30′
40′	0.4305	0.4173	0.9088	0.4592	2.177	1.100	2.396	1.1403	20′
50′	0.4334	0.4200	0.9075	0.4628	2.161	1.102	2.381	1.1374	10′
25°00′	0.4363	0.4226	0.9063	0.4663	2.145	1.103	2.366	1.1345	65°00′
10′	0.4392	0.4253	0.9051	0.4699	2.128	1.105	2.352	1.1316	50′
20′	0.4422	0.4279	0.9038	0.4734	2.112	1.106	2.337	1.1286	40′
30′	0.4451	0.4305	0.9026	0.4770	2.097	1.108	2.323	1.1257	30′
40′	0.4480	0.4331	0.9013	0.4806	2.081	1.109	2.309	1.1228	20′
50′	0.4509	0.4358	0.9001	0.4841	2.066	1.111	2.295	1.1199	10′
26°00′	0.4538	0.4384	0.8988	0.4877	2.050	1.113	2.281	1.1170	64°00′
10′	0.4567	0.4410	0.8975	0.4913	2.035	1.114	2.268	1.1141	50′
20′	0.4596	0.4436	0.8962	0.4950	2.020	1.116	2.254	1.1112	40′
30′	0.4625	0.4462	0.8949	0.4986	2.006	1.117	2.241	1.1083	30′
40	0.4654	0.4488	0.8936	0.5022	1.991	1.119	2.228	1.1054	20′
50′	0.4683	0.4514	0.8923	0.5059	1.977	1.121	2.215	1.1025	10′
27°00′	0.4712	0.4540	0.8910	0.5095	1.963	1.122	2.203	1.0996	63°00′
		Cos	Sin	Cot	Tan	Csc	Sec	Radians	Angle

Trigonometric Ratios

Angle	Radians	Sin	Cos	Tan	Cot	Sec	Csc		
27°00′	0.4712	0.4540	0.8910	0.5095	1.963	1.122	2.203	1.0996	63°00′
10′	0.4741	0.4566	0.8897	0.5132	1.949	1.124	2.190	1.0966	50′
20′	0.4771	0.4592	0.8884	0.5169	1.935	1.126	2.178	1.0937	40′
30′	0.4800	0.4617	0.8870	0.5206	1.921	1.127	2.166	1.0908	30′
40′	0.4829	0.4643	0.8857	0.5243	1.907	1.129	2.154	1.0879	20′
50′	0.4858	0.4669	0.8843	0.5280	1.894	1.131	2.142	1.0850	10′
28°00′	0.4887	0.4695	0.8829	0.5317	1.881	1.133	2.130	1.0821	62°00′
10′	0.4916	0.4720	0.8816	0.5354	1.868	1.134	2.118	1.0792	50′
20′	0.4945	0.4746	0.8802	0.5392	1.855	1.136	2.107	1.0763	40′
30′	0.4974	0.4772	0.8788	0.5430	1.842	1.138	2.096	1.0734	30′
40′	0.5003	0.4797	0.8774	0.5467	1.829	1.140	2.085	1.0705	20′
50′	0.5032	0.4823	0.8760	0.5505	1.816	1.142	2.074	1.0676	10′
29°00′	0.5061	0.4848	0.8746	0.5543	1.804	1.143	2.063	1.0647	61°00′
10′	0.5091	0.4874	0.8732	0.5581	1.792	1.145	2.052	1.0617	50′
20′	0.5120	0.4899	0.8718	0.5619	1.780	1.147	2.041	1.0588	40′
30′	0.5149	0.4924	0.8704	0.5658	1.767	1.149	2.031	1.0559	30′
40′	0.5178	0.4950	0.8689	0.5696	1.756	1.151	2.020	1.0530	20′
50′	0.5207	0.4975	0.8675	0.5735	1.744	1.153	2.010	1.0501	10′
30°00′	0.5236	0.5000	0.8660	0.5774	1.732	1.155	2.000	1.0472	60°00′
10′	0.5265	0.5025	0.8646	0.5812	1.720	1.157	1.990	1.0443	50′
20′	0.5294	0.5050	0.8631	0.5851	1.709	1.159	1.980	1.0414	40′
30′	0.5323	0.5075	0.8616	0.5890	1.698	1.161	1.970	1.0385	30′
40′	0.5352	0.5100	0.8601	0.5930	1.686	1.163	1.961	1.0356	20′
50′	0.5381	0.5125	0.8587	0.5969	1.675	1.165	1.951	1.0327	10′
31°00′	0.5411	0.5150	0.8572	0.6009	1.664	1.167	1.942	1.0297	59°00′
10′	0.5440	0.5175	0.8557	0.6048	1.653	1.169	1.932	1.0268	50′
20′	0.5469	0.5200	0.8542	0.6088	1.643	1.171	1.923	1.0239	40′
30′	0.5498	0.5225	0.8526	0.6128	1.632	1.173	1.914	1.0210	30′
40′	0.5527	0.5250	0.8511	0.6168	1.621	1.175	1.905	1.0181	20′
50′	0.5556	0.5275	0.8496	0.6208	1.611	1.177	1.896	1.0152	10′
32°00′	0.5585	0.5299	0.8480	0.6249	1.600	1.179	1.887	1.0123	58°00′
10′	0.5614	0.5324	0.8465	0.6289	1.590	1.181	1.878	1.0094	50′
20′	0.5643	0.5348	0.8450	0.6330	1.580	1.184	1.870	1.0065	40′
30′	0.5672	0.5373	0.8434	0.6371	1.570	1.186	1.861	1.0036	30′
40′	0.5701	0.5398	0.8418	0.6412	1.560	1.188	1.853	1.0007	20′
50′	0.5730	0.5422	0.8403	0.6453	1.550	1.190	1.844	0.9977	10′
33°00′	0.5760	0.5446	0.8387	0.6494	1.540	1.192	1.836	0.9948	57°00′
10′	0.5789	0.5471	0.8371	0.6536	1.530	1.195	1.828	0.9919	50′
20′	0.5818	0.5495	0.8355	0.6577	1.520	1.197	1.820	0.9890	40′
30′	0.5847	0.5519	0.8339	0.6619	1.511	1.199	1.812	0.9861	30′
40′	0.5876	0.5544	0.8323	0.6661	1.501	1.202	1.804	0.9832	20′
50′	0.5905	0.5568	0.8307	0.6703	1.492	1.204	1.796	0.9803	10′
34°00′	0.5934	0.5592	0.8290	0.6745	1.483	1.206	1.788	0.9774	56°00′
10′	0.5963	0.5616	0.8274	0.6787	1.473	1.209	1.781	0.9745	50′
20′	0.5992	0.5640	0.8258	0.6830	1.464	1.211	1.773	0.9716	40′
30′	0.6021	0.5664	0.8241	0.6873	1.455	1.213	1.766	0.9687	30′
40′	0.6050	0.5688	0.8225	0.6916	1.446	1.216	1.758	0.9657	20′
50′	0.6080	0.5712	0.8208	0.6959	1.437	1.218	1.751	0.9628	10′
35°00′	0.6109	0.5736	0.8192	0.7002	1.428	1.221	1.743	0.9599	55°00′
10′	0.6138	0.5760	0.8175	0.7046	1.419	1.223	1.736	0.9570	50′
20′	0.6167	0.5783	0.8158	0.7089	1.411	1.226	1.729	0.9541	40′
30′	0.6196	0.5807	0.8141	0.7133	1.402	1.228	1.722	0.9512	30′
40′	0.6225	0.5831	0.8124	0.7177	1.393	1.231	1.715	0.9483	20′
50′	0.6254	0.5854	0.8107	0.7221	1.385	1.233	1.708	0.9454	10′
36°00′	0.6283	0.5878	0.8090	0.7265	1.376	1.236	1.701	0.9425	54°00′
		Cos	Sin	Cot	Tan	Csc	Sec	Radians	Angle

Trigonometric Ratios

Angle	Radians	Sin	Cos	Tan	Cot	Sec	Csc		
36°00′	0.6283	0.5878	0.8090	0.7265	1.376	1.236	1.701	0.9425	54°00′
10′	0.6312	0.5901	0.8073	0.7310	1.368	1.239	1.695	0.9396	50′
20′	0.6341	0.5925	0.8056	0.7355	1.360	1.241	1.688	0.9367	40′
30′	0.6370	0.5948	0.8039	0.7400	1.351	1.244	1.681	0.9338	30′
40′	0.6400	0.5972	0.8021	0.7445	1.343	1.247	1.675	0.9308	20′
50′	0.6429	0.5995	0.8004	0.7490	1.335	1.249	1.668	0.9279	10′
37°00′	0.6458	0.6018	0.7986	0.7536	1.327	1.252	1.662	0.9250	53°00′
10′	0.6487	0.6041	0.7969	0.7581	1.319	1.255	1.655	0.9221	50′
20′	0.6516	0.6065	0.7951	0.7627	1.311	1.258	1.649	0.9192	40′
30′	0.6545	0.6088	0.7934	0.7673	1.303	1.260	1.643	0.9163	30′
40′	0.6574	0.6111	0.7916	0.7720	1.295	1.263	1.636	0.9134	20′
50′	0.6603	0.6134	0.7898	0.7766	1.288	1.266	1.630	0.9105	10′
38°00′	0.6632	0.6157	0.7880	0.7813	1.280	1.269	1.624	0.9076	52°00′
10′	0.6661	0.6180	0.7862	0.7860	1.272	1.272	1.618	0.9047	50′
20′	0.6690	0.6202	0.7844	0.7907	1.265	1.275	1.612	0.9018	40′
30′	0.6720	0.6225	0.7826	0.7954	1.257	1.278	1.606	0.8988	30′
40′	0.6749	0.6248	0.7808	0.8002	1.250	1.281	1.601	0.8959	20′
50′	0.6778	0.6271	0.7790	0.8050	1.242	1.284	1.595	0.8930	10′
39°00′	0.6807	0.6293	0.7771	0.8098	1.235	1.287	1.589	0.8901	51°00′
10′	0.6836	0.6316	0.7753	0.8146	1.228	1.290	1.583	0.8872	50′
20′	0.6865	0.6338	0.7735	0.8195	1.220	1.293	1.578	0.8843	40′
30′	0.6894	0.6361	0.7716	0.8243	1.213	1.296	1.572	0.8814	30′
40′	0.6923	0.6383	0.7698	0.8292	1.206	1.299	1.567	0.8785	20′
50′	0.6952	0.6406	0.7679	0.8342	1.199	1.302	1.561	0.8756	10′
40°00′	0.6981	0.6428	0.7660	0.8391	1.192	1.305	1.556	0.8727	50°00′
10′	0.7010	0.6450	0.7642	0.8441	1.185	1.309	1.550	0.8698	50′
20′	0.7039	0.6472	0.7623	0.8491	1.178	1.312	1.545	0.8668	40′
30′	0.7069	0.6494	0.7604	0.8541	1.171	1.315	1.540	0.8639	30′
40′	0.7098	0.6517	0.7585	0.8591	1.164	1.318	1.535	0.8610	20′
50′	0.7127	0.6539	0.7566	0.8642	1.157	1.322	1.529	0.8581	10′
41°00′	0.7156	0.6561	0.7547	0.8693	1.150	1.325	1.524	0.8552	49°00′
10′	0.7185	0.6583	0.7528	0.8744	1.144	1.328	1.519	0.8523	50′
20′	0.7214	0.6604	0.7509	0.8796	1.137	1.332	1.514	0.8494	40′
30′	0.7243	0.6626	0.7490	0.8847	1.130	1.335	1.509	0.8465	30′
40′	0.7272	0.6648	0.7470	0.8899	1.124	1.339	1.504	0.8436	20′
50′	0.7301	0.6670	0.7451	0.8952	1.117	1.342	1.499	0.8407	10′
42°00′	0.7330	0.6691	0.7431	0.9004	1.111	1.346	1.494	0.8378	48°00′
10′	0.7359	0.6713	0.7412	0.9057	1.104	1.349	1.490	0.8348	50′
20′	0.7389	0.6734	0.7392	0.9110	1.098	1.353	1.485	0.8319	40′
30′	0.7418	0.6756	0.7373	0.9163	1.091	1.356	1.480	0.8290	30′
40′	0.7447	0.6777	0.7353	0.9217	1.085	1.360	1.476	0.8261	20′
50′	0.7476	0.6799	0.7333	0.9271	1.079	1.364	1.471	0.8232	10′
43°00′	0.7505	0.6820	0.7314	0.9325	1.072	1.367	1.466	0.8203	47°00′
10′	0.7534	0.6841	0.7294	0.9380	1.066	1.371	1.462	0.8174	50′
20′	0.7563	0.6862	0.7274	0.9435	1.060	1.375	1.457	0.8145	40′
30′	0.7592	0.6884	0.7254	0.9490	1.054	1.379	1.453	0.8116	30′
40′	0.7621	0.6905	0.7234	0.9545	1.048	1.382	1.448	0.8087	20′
50′	0.7650	0.6926	0.7214	0.9601	1.042	1.386	1.444	0.8058	10′
44°00′	0.7679	0.6947	0.7193	0.9657	1.036	1.390	1.440	0.8029	46°00′
10′	0.7709	0.6967	0.7173	0.9713	1.030	1.394	1.435	0.7999	50′
20′	0.7738	0.6988	0.7153	0.9770	1.024	1.398	1.431	0.7970	40′
30′	0.7767	0.7009	0.7133	0.9827	1.018	1.402	1.427	0.7941	30′
40′	0.7796	0.7030	0.7112	0.9884	1.012	1.406	1.423	0.7912	20′
50′	0.7825	0.7050	0.7092	0.9942	1.006	1.410	1.418	0.7883	10′
45°00′	0.7854	0.7071	0.7071	1.000	1.000	1.414	1.414	0.7854	45°00′
		Cos	Sin	Cot	Tan	Csc	Sec	Radians	Angle

Trigonometric Formulas

Formulas for the Basic Identities

$\tan x = \dfrac{\sin x}{\cos x}$ $\cot x = \dfrac{\cos x}{\sin x}$ $\sec x = \dfrac{1}{\cos x}$ $\csc x = \dfrac{1}{\sin x}$

Formulas for the Pythagorean Identities

$\sin^2 x + \cos^2 x = 1$ $\tan^2 x + 1 = \sec^2 x$ $\cot^2 x + 1 = \csc^2 x$

Formulas for the Sum of Two Angles

$$\sin(\alpha + \beta) = \sin\alpha\cos\beta + \cos\alpha\sin\beta$$

$$\cos(\alpha + \beta) = \cos\alpha\cos\beta - \sin\alpha\sin\beta$$

$$\tan(\alpha + \beta) = \frac{\tan\alpha + \tan\beta}{1 - \tan\alpha\tan\beta}$$

Formulas for the Difference of Two Angles

$$\sin(\alpha - \beta) = \sin\alpha\cos\beta - \cos\alpha\sin\beta$$

$$\cos(\alpha - \beta) = \cos\alpha\cos\beta + \sin\alpha\sin\beta$$

$$\tan(\alpha - \beta) = \frac{\tan\alpha - \tan\beta}{1 + \tan\alpha\tan\beta}$$

Double-Angle Formulas

$$\sin 2\theta = 2\sin\theta\cos\theta$$

$$\cos 2\theta = \cos^2\theta - \sin^2\theta = 2\cos^2\theta - 1 = 1 - 2\sin^2\theta$$

$$\tan 2\theta = \frac{2\tan\theta}{1 - \tan^2\theta}$$

Half-Angle Formulas

$$\sin\frac{1}{2}\alpha = \pm\sqrt{\frac{1 - \cos\alpha}{2}}$$

$$\cos\frac{1}{2}\alpha = \pm\sqrt{\frac{1 + \cos\alpha}{2}}$$

$$\tan\frac{1}{2}\alpha = \pm\sqrt{\frac{1 - \cos\alpha}{1 + \cos\alpha}}$$

Some Special Relationships

$$\sin(-x) = -\sin x = \sin(360^\circ - x)$$

$$\cos(-x) = \cos x = \cos(360^\circ - x)$$

$$\tan(-x) = -\tan x = \tan(360^\circ - x)$$

$\sin(90^\circ - x) = \cos x$	$\sin(180^\circ - x) = \sin x$	$\sin(270^\circ - x) = -\cos x$
$\cos(90^\circ - x) = \sin x$	$\cos(180^\circ - x) = -\cos x$	$\cos(270^\circ - x) = -\sin x$
$\tan(90^\circ - x) = \cot x$	$\tan(180^\circ - x) = -\tan x$	$\tan(270^\circ - x) = \cot x$
$\sin(90^\circ + x) = \cos x$	$\sin(180^\circ + x) = -\sin x$	$\sin(270^\circ + x) = -\cos x$
$\cos(90^\circ + x) = -\sin x$	$\cos(180^\circ + x) = -\cos x$	$\cos(270^\circ + x) = \sin x$
$\tan(90^\circ + x) = -\cot x$	$\tan(180^\circ + x) = \tan x$	$\tan(270^\circ + x) = -\cot x$

Glossary

a_1 (471) A sub one, the first member of a sequence.

absolute value (10) The distance of a real number from 0 on the number line.

absolute value function (109) A function of the form $f(x) = |x|$.

addition of matrices (475) For $A = [a_{ij}]$ and $B = [b_{ij}]$, $A + B = [a_{ij} + b_{ij}]$.

additive inverse (507) For the $m \times n$ matrix A, the additive inverse of A, written $-A$, is the $m \times n$ matrix whose elements are the additive inverses of the corresponding elements in A.

amplitude (268) For functions of the form $y = a \sin bx$ and $y = a \cos bx$, the amplitude is $|a|$.

amplitude (568) The θ term in the trigonometric form of a complex number.

a_n (475, 481) The nth term of a sequence.

anchor (588) The number used in step 1 of an induction proof.

angle of depression (318) The angle between the horizontal and the line of sight when an object beneath the viewer is sighted.

angle of elevation (318) The angle between the horizontal and the line of sight when an object elevated from the viewer is sighted.

antilogarithm (190) The number whose common logarithm is a given value.

arc (211) A part of a circle.

arccosine (355) The inverse of the cosine function.

arcsine (354) The inverse of the sine function.

arctangent (355) The inverse of the tangent function.

arithmetic means (476) The terms between any two terms of an arithmetic sequence.

arithmetic progression (474) Another name for an arithmetic sequence.

arithmetic sequence (474) A sequence in which each term after the first is obtained by adding the same number to the preceding term.

arithmetic series (484) The indicated sum of the terms of an arithmetic sequence.

asymptote (274) The line on a graph at which point a function is undefined.

bell curve (455) See normal curve.

Bernoulli experiments (416) Experiments in which there are only two outcomes.

bimodal (439) A set with two modes.

binomial (14) A polynomial containing two terms.

Binomial Theorem (420) The rule that expands $(x + y)^n$ without having to do the actual algebraic multiplication.

center of the circle (210) A fixed point which is equally distant from all points on the circle.

central angle (211) An angle whose vertex is at the center of the circle.

change of base formula (197) If n, a, and b are positive, $a \neq 1$, $b \neq 1$, then $\log_b n = \dfrac{\log_a n}{\log_a b}$.

characteristic (190) The integer portion of a common logarithm.

chord (210) A segment whose endpoints are points of a circle.

circle (210) A set of coplanar points which are equally distant from a given fixed point.

circumscribed (211) To encircle a geometric figure so as to touch at as many points as possible.

coefficient (13) A numerical factor.

cofactor (524) The cofactor of an element a_{ij} of a determinant is the minor of a_{ij} if $i + j$ is even, and the minor multiplied by -1, if $i + j$ is odd.

combination (410) A subset of a set in which order does not matter.

common difference (474) The number added to each term in an arithmetic sequence to yield the next term of the sequence.

common external tangent (219) A common tangent that does not intersect a line segment joining the centers of the two circles.

common internal tangent (219) A common tangent that intersects the line segment joining the centers of the two circles.

common logarithm (125) Logarithms in base 10.

common ratio (480) The multiplier used to obtain a geometric sequence.

common tangent (219) A line tangent to each of two circles.

complex fractions (33) A fraction whose numerator or denominator contains one or more fractions.

complex number (66) A number that can be written in the form $a + bi$, where a and b are real numbers and $i = \sqrt{-1}$.

composite function (121) The combining of two functions, f and g, to produce a new function symbolized by $f \circ g$.

composite transformation (364) The transformation resulting from a composition.

composition (364) An operation which combines two transformations.

compound interest formula (201) If A is the amount accumulated, P is the principal, r is the interest rate per year, n is the number of interest periods per year, and t is the number of years, then $A = P\left(1 + \frac{r}{n}\right)^{nt}$.

concentric circles (210) Circles with the same center but not necessarily the same radii.

conditional probability (425) The probability that an event will occur given that another event has already occurred.

congruence (381) Two figures are congruent if, and only if, there is an isometry that maps one figure onto the other.

conic sections (119) Curves formed when a plane intersects a conical surface.

conjugate (57) A binomial pair of the form, $a + \sqrt{b}, a - \sqrt{b}$.

conjunction (6) An *and* statement. The symbol used is $\wedge$.

constant function (105) A function of the form $f(x) = b$ where b is a real number.

constant of proportionality (105) The slope in a direct variation.

constant of variation (105) The slope in a direct variation.

constant polynomial (544) A polynomial containing only the constant term.

constant term (533) The term a_0 in a polynomial.

continuously compounded interest formula (202) If the amount A is accumulated after t years at an annual interest rate r for the original principal P, then $A = Pe^{rt}$.

cosecant (276) consecant $\theta = \frac{1}{\sin \theta}$, when $\sin \theta \neq 0$.

cosine (257) The x-coordinate of the intersection of the terminal side of an angle, θ, in standard position with the unit circle.

cotangent (276) cotangent $\theta = \frac{\cos \theta}{\sin \theta}$, when $\sin \theta \neq 0$.

coterminal (251) Angles which have the same initial and terminal side but which differ in the number of complete rotations made around the origin.

Counting Principle (403) Suppose one activity can occur in any of m ways. Another can occur in any of n ways. The total number of ways both activities can occur is given by the product mn.

Counting Principle (alternate version) (403) Suppose the probability of one event E is r $(0 \leq r \leq 1)$. The probability of another event F is s $(0 \leq s \leq 1)$. Then the probability of both E and F occurring is the product rs.

cubic polynomial (544) A third degree polynomial.

data (438) Individual pieces of information used in statistics.

De Moivre's Theorem (598) For all real numbers n and θ, and for real numbers $r > 0$, $[r(\cos \theta + i \sin \theta)]^n = rn(\cos n\theta + i \sin n\theta)$.

degree (544) The exponent of the greatest power of the variable in a polynomial of one variable.

degree of a monomial (13) The sum of the exponents of the variables in a monomial.

degree of polynomial (14) The degree of the monomial, within the polynomial, with the greatest degree.

determinant of a 2 × 2 matrix (518) For a matrix $A = \begin{bmatrix} a & b \\ c & d \end{bmatrix}$, the determinant of A is the real number $ad - bc$.

diameter (210) A chord passing through the center of a circle.

difference of two squares (17) $(a + b)(a - b) = a^2 - b^2$

dilation (157) A transformation under which a plane is "stretched" or "shrunk" in relation to a fixed point.

direct isometry (369) An isometry that preserves orientation.

direct variation (105) A linear function defined by $f(x) = kx$ where $k \neq 0$.

direction (316) The effect of a force on a body determining the course it will move.

discriminant (76) The expression $b^2 - 4ac$ in the quadratic equation.

disjunction (6) An *or* statement. The symbol used is $\vee$.

domain (92) The set of all the first coordinates of the ordered pairs in a relation.

double root (557) A solution of a polynomial that appears twice.

ellipse (116) The curve formed when graphing an equation of the form $ax^2 + by^2 = c$.

equal circles (210) Circles with equal radii.

equivalent fractions (21) Different fractions that represent the same number.

exponential curve (173) The curve formed by the graph of an exponential function.

exponential function (173) A function of the form $f(x) = a^x$, where a is a positive real number other than 1.

externally tangent (219) Two circles which are tangent to the same line at the same point but where neither circle is inscribed in the other.

Fibonnaci sequence (479) An arithmetic sequence in which each term is the sum of the preceding two terms.

frequency (268) For functions of the form $y = a \sin bx$ and $y = a \cos bx$, the frequency is $|b|$.

function (96) A relation in which each member of the domain is paired with exactly one member of the range.

geometric means (481) The terms between any two terms of a geometric sequence.

geometric progression (480) Another name for a geometric sequence.

geometric sequence (480) A sequence in which each term after the first is obtained by multiplying the preceding term by the same number.

geometric series (489) The indicated sum of the terms of a geometric sequence.

glide reflection (365) A transformation which is the composite of a line reflection and a translation whose direction is parallel to the line of reflection.

greatest integer function (107) A function of the form $f(x) = [x]$ where $[x]$ stands for the greatest integer less than or equal to x.

half-turn (150) A rotation of 180° about a point.

horizontal line test (112) If no horizontal line intersects the graph of f in more than one point, then f is a one-to-one function.

hyperbola (118) The curve formed when graphing equations of the form $ax^2 - by^2 = c$.

identity function for composition (123) The function $I(x) = x$. For any function f, $(f \circ I)\,x = f(x)$ and $(I \circ f)(x) = f(x)$.

identity transformation (366) A transformation that maps each point of the plane to itself.

image (133) If A is mapped to A', then A' is called the image of A. A is the preimage of A'.

imaginary number (66) A complex number of the form $a + bi$ when $b \neq 0$.

imaginary unit (62) The number i, which is not a real number, where $i = \sqrt{-1}$.

index (50) n in the symbol $\sqrt[n]{y}$, read the nth root of y.

indexing (445) Assigning of subscripts to a variable.

Induction Hypothesis (587) A step in a mathematical induction proof where you assume that some natural number k is a member of the solution.

inductive definition (607) A type of induction used to define some fundamental concepts about the number system.

inductive step (588) Step 2 of an induction proof.

infinite sequence (471) A sequence whose domain is the set of all natural numbers.

infinite series (494) A series where an infinite number of terms are being added.

initial side (250) The ray of an angle that is fixed along the x-axis.

inscribed (211) To draw within a geometric figure so as to touch at as many points as possible.

inscribed angle (226) An angle whose vertex is on the circle and whose sides contain chords of the circle.

internally tangent (219) Two circles which are tangent to the same line at the same point and where one circle is inscribed in the other.

interpolation (194, 281) A method used to obtain more accurate approximations of logarithms or trigonometric values.

inverse of a function (111) An interchanging of the numbers of each ordered pair of a function. Symbolized by f^{-1}.

inverse variation (107) A function defined by $f(x) = \frac{k}{x}$ where $k \neq 0$.

irrational number (3) A number represented by a nonrepeating decimal.

isometry (369) A transformation that preserves distance.

laws of exponents (13) The rules that apply to multiplication and division of non-zero real numbers with integral exponents.

leading coefficient (544) The a_n term in a polynomial.

leading term (544) The term a_nx in a polynomial.

limits of summation (441) The lower and upper value assigned to a variable in a summation.

line of centers (219) The line joining the centers of two circles.

line of symmetry (135) A line, m, such that a figure is its own image after a line reflection over m.

line reflection (132) A transformation of the plane over line ℓ that maps each point P of the plane onto a point P', its image. If P is on line ℓ, then the image of P is P. If P is not on ℓ, then the image of P is the point P' such that ℓ is the perpendicular bisector of $\overline{PP'}$.

line reflections preserve distance (133) When a segment is reflected over a line, its image will be the same length as the original segment.

linear function (97) An equation in the form $y = mx + b$.

linear polynomial (15) A polynomial in one variable of the form $ax + b$ with $a \neq 0$.

logarithm (180) An exponent.

magnitude (316) The strength of a force acting on an object.

major arc (211) An arc which is more than half a circle.

mantissa (190) The decimal portion of a common logarithm.

mapping (96) A relation in which each member of the domain is paired with exactly one member of the range.

mathematical induction (583) A method in mathematics used to prove a statement indirectly.

matrix (506) A rectangular array of real numbers.

mean (438) The sum of a set of n numbers divided by n.

mean absolute deviation (450) For a set of n numbers, a measure of dispersion represented by $\frac{1}{n}\sum_{i=1}^{n}|\overline{x} - x_i|$.

mean proportional between *p* and *q* (482) A single geometric mean b, between two numbers p and q such that $\frac{b}{p} = \frac{q}{b}$.

median (439) The middle number when a set of n numbers are arranged in order. If n is even, the median is the average of the middle two numbers.

minor (524) The minor of an element of a determinant is the determinant of one lower order obtained by deleting the row and column of that element.

minor arc (211) An arc which is less than half a circle.

minute (280) One sixtieth of a degree.

mode (439) The number in a set of n numbers which occurs most often.

modulus (568) The r factor in the trigonometric form of a complex number.

monomial (13) A numeral, a variable or a product of a numerical factor and one or more variables.

negative exponent (168) For any real number r, except $r = 0$, and for any positive integer n, $r^{-n} = \frac{1}{r^n}$.

normal curve (455) The graph of a frequency distribution where approximately 68% of the measures fall within one standard deviation of the mean (34% on each side), 95% fall within two S.D.'s and 99% within three S.D.'s.

one-to-one function (112) A function in which no two ordered pairs have the same y-coordinate.

one-to-one mapping (132) A function that pairs each member of the domain with exactly one member of the range.

opposite isometry (369) An isometry that reverses orientation.

order (506) A matrix has order $m \times n$ if it has m rows and n columns.

Pascal's triangle (419) A triangular array of the coefficients resulting from the expansion of a binomial.

pattern (584) A certain behavior that appears to repeat in certain sets of numbers.

perfect square (49) A rational number whose square root is a rational number.

perfect trinomial square (17) $(a + b)^2 = a^2 + 2ab + b^2$ or $(a - b)^2 = a^2 - 2ab + b^2$

period (268) For functions of the form $y = a \sin bx$ and $y = a \cos bx$, their period is $\frac{2\pi}{|b|}$ or $\frac{360°}{|b|}$.

periodic function (266) A function f where there exists a non-zero number, a, such that for all x in the domain of f, $f(x) = f(x + a)$.

permutation (406) Arrangement

point symmetry (151) A figure that is its own image under some half-turn with center P is said to have point symmetry with respect to P.

polynomial (14) A monomial or the sum of two or more monomials.

polynomial equation (545) A polynomial set equal to zero.

polynomial function (544) A function P such that $P(x)$ equals a polynomial.

polynomial in one variable (544) An expression of the form $a_nx^n + a_{n-1}x^{n-1} + \cdots + a_2x^2 + a_1x + a_0$, where $a_n \neq 0$ and n is a whole number.

population growth formula (201) If y is the number of people, y_0 is the original population, r is the rate of growth, and n is the number of years, $y = y_0(1 + r)^n$.

pre-image (133) See image.

Principle of Mathematical Induction (PMI) (586) Let S be a set of natural numbers for which the following two conditions hold:
1. $1 \in S$
2. If the natural number k is in S, $k + 1$ is also in S.
Then $S = \mathcal{N}$.

property of equality for exponential functions (177) If $b > 0$ and $b \neq 1$, then $b^{x_1} = b^{x_2}$ if and only if $x_1 = x_2$.

pure imaginary number (63) Expressions with the form bi where b is a real number not equal to 0 and $i = 1$.

quadrantal angle (251) An angle in standard position that has its terminal side on one of the coordinate axes.

quadratic formula (19) If $ax^2 + bx + c = 0$, then $x = -b \pm \frac{\sqrt{b^2 - 4ac}}{2a}$.

quadratic function (106) A function of the form $f(x) = ax^2 + bx + c$ with $a \neq 0$.

quadratic polynomial (15) A polynomial of the form $ax^2 + bx + c$ with $a \neq 0$.

radian (253) The measure of the angle in standard position on a unit circle that intercepts an arc of length one unit.

radical equation (60) An equation in which a variable appears under a radical sign.

radical sign (48) The symbol $\sqrt{}$

radicand (48) The expression written underneath a radical sign.

radius (210) The distance, or an actual segment, from the center of the circle to any point on the circle.

range (92) The set of all the second coordinates of the ordered pairs in a relation.

range (449) The difference between the greatest and the least of a set of measures.

rational number (2) All numbers in the form of $\frac{a}{b}$, where a and b are integers and $b \neq 0$. The decimal representation of a rational number is always a repeating decimal.

real number (3) The set consisting of all rational and irrational numbers.

rectangular coordinates (290) The (x, y) coordinates of a point P.

reference angle (262) An angle in quadrant I used for determining trigonometric values for related angles in quadrants II, III and IV.

reflection over the line $y = x$ (139) If the point $P(x, y)$ is reflected over the line whose equation is $y = x$, its image is $P'(y, x)$.

reflection over the x-axis (138) If point $P(x, y)$ is reflected over the x-axis, its image is $P'(x, -y)$.

reflection over the y-axis (138) If point $P(x, y)$ is reflected over the y-axis, its image is $P'(-x, y)$.

relation (92) A set of ordered pairs.

resultant (316) The single force brought about as an effect of two forces acting on an object.

roots (545) Solutions of a polynomial.

rotation (146) A transformation that maps every point to an image point in the plane by rotating the plane around a fixed point.

rotational symmetry (147) A figure has rotational symmetry if it is the image of itself under some rotation. The angel of rotation must be greater than 0° and less than 360°.

scalar (509) A real number.

scalar multiplication (509) A real number times a matrix where each element of the matrix is multiplied by the real number to yield the corresponding element in the product.

scale factor (157) The multiple by which the size of a figure is to be increased or decreased.

secant (210) A line which intersects a circle in two distinct points.

secant (220) secant $\theta = \frac{1}{\cos \theta}$, when $\cos \theta \neq 0$.

second (280) One sixtieth of a minute.

sector (214) A region in a circle bounded by two radii and an arc of a circle.

semicircle (211) An arc which is half a circle.

sequence (467) A function whose domain is a set of consecutive natural numbers that begin with 1.

sigma (441) Σ, a Greek capital letter that indicates summation.

similar figures (383) Two figures are similar if, and only if, there is a similarity transformation that maps one figure into another.

similarity transformation (383) A composite of isometries and dilations.

sine (257) The y-coordinate of the intersection of the terminal side of an angle, θ, in standard position with the unit circle.

solving a triangle (307) The process of finding all the remaining parts of a triangle after being given values for some parts of that triangle.

square matrix (506) A matrix with equal numbers of rows and columns.

square root (48) A number x is called a square root of a number y if $x^2 = y$.

standard deviation (454) The square root of the variance.

standard position (250) An angle with its vertex at the origin and its initial side along the positive x-axis.

subscripts (445) The number assigned to the order of appearance of a value contained in a set of measures.

subtraction of matrices (508) The difference between $m \times n$ matrices A and B, written $A - B$, is $A + (-B)$.

tangent (210) A line which intersects a circle at exactly one point.

tangent (272) With θ being the measure of an angle in standard position, $\tan \theta = \frac{\sin \theta}{\cos \theta}$, when $\cos \theta \neq 0$.

term of a sequence (470) The members of the range of a sequence.

terminal side (250) The ray of an angle that is able to rotate around the unit circle.

Three Line Reflection Theorem (377) Every isometry is equivalent to the composite of, at most, three line reflections.

transformation (132) A one-to-one mapping whose domain and range are the set of all points in a given plane.

transition matrix (533) A matrix where the sum of the rows is 1.

translation (142) A transformation of the plane that "moves" every point in the plane the same distance in the same direction.

trigonometric coordinates (290) If P is a point r units from the origin and located on the terminal side of angle θ in standard position, that the coordinates of P are $(r \cos \theta, r \sin \theta)$.

trigonometric identify (330) Trigonometric statements that are true for all replacements of the variable for which the expression is defined.

trinomial (14) A polynomial containing three terms.

unit circle (253) A circle with center at the origin and radius of one unit.

variance (453) The sum of the squares of the differences of the measures from the mean, divided by the number of measures.

variation in sign (559) Two consecutive coefficients in a polynomial having different signs.

vector (316) A force that has both magnitude and direction.

vertical line test (98) If no vertical line intersects a graph of an equation more than once, the graph is the graph of a function.

zero exponent (168) For any real number r, except $r = 0$, $r^0 = 1$.

zero matrix (507) A matrix, all of whose entries are zero.

zero of polynomial (545) Solutions for the variable in a polynomial that allow the polynomial to equal zero.

zero polynomial (544) A polynomial consisting of just zero.

Selected Answers

CHAPTER 1 WORKING WITH POLYNOMIAL AND RATIONAL EXPRESSIONS

Pages 4–5 Lesson 1.1

Exploratory **1.** false **3.** true **5.** false **7.** Commutative property of addition **9.** Definition of multiplicative identity **11.** Commutative property of multiplication **13.** The (left) distributive law

Written **1.** 0.8 **3.** $0.\overline{4}$ **5.** 1.75 **7.** $0.1\overline{3}$ **9.** 0.04 **11.** $-\frac{18}{1}$ **13.** $\frac{9}{4}$ **15.** $\frac{635}{1000}$ **17.** $\frac{6}{9}$ **19.** $\frac{4}{9}$ **21.** $\frac{3}{99}$ **23.** $\frac{835}{990}$ **25.** $\frac{363}{90}$ **27.** −17 **29.** 1 **31.** 17 **33.** $-6\frac{1}{2}$ **35.** −15 **37.** −12 **39.** $-\frac{7}{12}$ **41.** 1 **43.** $12y + 4$ **45.** $4m + 1$ **47.** $-2z$ **49.** $12x - 3y + 1$ **51.** $12x + y$ **53.** $3x + 6y$ **55.** $8 - 2a$ **57.** $12 + 12x$ **59.** $-1.6m + 7.1n$ **61.** $x - 5$ **63.** $2(x + 7)$ **65.** $8 + 2x$ **67.** $\$1.68t$ **69.** $75h$ (miles) **71.** $\frac{y}{80}$ (hours) **73.** $\frac{1}{2}b(b + 4)$ **75.** $\frac{100 \text{ yd}}{x}$

Pages 8–9 Lesson 1.2

Exploratory **1.** true **3.** false **5.** true **7.** false

Written **1.** $x = 2$ **3.** $x = -2$ **5.** $x = 5$ **7.** $y = -7$ **9.** $x = 2$ **11.** $y = 10$ **13.** $x = 2$ **15.** $y = -4$ **17.** $x = -1$ **19.** $t = -4$ **21.** $x = -4$ **23.** $y = \frac{1}{2}$ **25.** $x = 7$ **27.** $x = \frac{5}{4}$ **29.** $x \le \frac{1}{4}$ **31.** $m > -3$ **33.** $m \ge -\frac{11}{3}$ **35.** $y < 2$ **37.** $x \le \frac{1}{6}$ **39.** $10 \le x$ **41.** no solution **43.** all x **45.** all x **47.** $x = \frac{3}{2}$ **49.** $-3 \le x < 8$ **51.** $y < 0$ or $y \ge 6$ **53.** $-7 < x < 1$ **55.** $x < -2$ or $x > 2$ **57.** $3 \le w \le 4$ **59.** all x **61.** $b < 3$ **63.** $y \ge 1$ **65.** 23 **67.** 7 **69.** −1, 0, 1 **71.** 92 m, 51 m **73.** When mileage for 6 days is less than 262 miles. **75.** 1 hour 48 minutes **77.** 54 mph **79.** No, consecutive means difference = 1. **81.** 16 cm **83.** Yes, 11 quarters, 14 nickels **85.** 50 **87.** no solution **89.** 6 hours **91.** 89 mph, 99 mph

Page 12 Lesson 1.3

Exploratory **1.** $(x = 7) \vee (x = -7)$ **3.** $(x < 10) \wedge (x > -10)$ **5.** $(x = 11) \vee (x = -11)$ **7.** $(x \ge 15) \vee (x \le -15)$

Written **1.** $\{-7, 7\}$ **3.** no solution **5.** $\{-8, 8\}$ **7.** $\{-5, 3\}$ **9.** $\{-1, 7\}$ **11.** $\{-25, 13\}$ **13.** no solution **15.** $\{-20, 12\}$ **17.** $\{11, 26\}$ **19.** $\{-8, 14\}$ **21.** $\left\{\frac{3}{2}\right\}$ **23.** $\emptyset$ **25.** $\{x \in \mathcal{R}: (x < -9) \vee (x > 9)\}$ **27.** $\{t \in \mathcal{R}: -7 < t < 7\}$ **29.** $\{m \in \mathcal{R}: (m \le -3) \vee (m \ge 3)\}$ **31.** $\{y \in \mathcal{R}: -4 < y < 4\}$ **33.** $\{t \in \mathcal{R}: (t \le -9) \vee (t \ge 9)\}$ **35.** $\{t \in \mathcal{R}\}$ **37.** $\{x \in \mathcal{R}: (x \le -5) \vee (x \ge 3)\}$ **39.** $\{y \in \mathcal{R}: 2 < y < 10\}$ **41.** $\{y \in \mathcal{R}: (y < 4) \vee (y > 14)\}$ **43.** $\{x \in \mathcal{R}: -9 \le x \le 18\}$ **45.** $\{x \in \mathcal{R}\}$ **47.** $\{p \in \mathcal{R}: (p < 1) \vee (p > 4)\}$ **49.** $\{n \in \mathcal{R}: 4 \le n \le 5\}$ **51.** $\{x \in \mathcal{R}: (x < -1) \vee (x > 6)\}$ **53.** $\emptyset$

Page 16 Lesson 1.4

Exploratory **1.** binomial, 4 **3.** binomial, 2 **5.** monomial, 6 **7.** no

Written **1.** $16x + 2y$ **3.** $-7a^2 - 2a$ **5.** $3a - 4ab$ **7.** $2b^2 - 5a$ **9.** $-2x - 4$ **11.** $-x^2$ **13.** $12a^2 - 11b^2$ **15.** x^2y^2 **17.** $8a^3$ **19.** $4 - a^2b$ **21.** $-125m^6n^3$ **23.** x^6y^2 **25.** $\frac{a}{2}$ or $\frac{1}{2}a$ **27.** $\frac{2a}{3x}$ **29.** $\frac{3}{4x}$ **31.** $32x^{11}y^{14}$ **33.** $4a^2 + 12a + 9$ **35.** $-q^2 - 8q + 7$ **37.** $4y^2 - 6y + 2$ **39.** $a^2 - a$ **41.** $2x^2 + 7x + 3$ **43.** $36a^2 - 4b^2$ **45.** $4y^2 - 31y + 42$ **47.** $2a^2 - ab - b^2$ **49.** $4q^2 - 28q + 49$ **51.** $9m^2 - 4$ **53.** $x^3 + 3x^2 + 5x + 3$ **55.** $2y^3 - 9y^2 - 7y + 24$ **57.** $2s^3 + 7s^2 - 5s - 4$ **59.** $a^3 + 27$ **61.** $a^3 + b^3$ **63.** $a^3 - 27$ **65.** $a^3 - b^3$

Pages 19–20 Lesson 1.5

Exploratory **1.** $8(x + 3)$ **3.** $x(3x + 1)$ **5.** $5(x^2 - 5)$ **7.** $(x + 3)(x - 3)$ **9.** $(m - 5)(m + 5)$ **11.** $(r + 7)(r - 7)$ **13.** $(x + 3)^2$ **15.** $(y - 2)(y - 5)$

Written **1.** $4y(1 + 3y)$ **3.** $3s(s + 3)(s - 3)$ **5.** $12 - 4y - 3x$ **7.** $(b + 12)(b - 12)$ **9.** $p(7m + 2p - 14x)$ **11.** $r^2(r^2 + rs + s^2)$ **13.** $3(4x^2 + xy - 3)$ **15.** $(p - 4)(p - 1)$ **17.** $2(a + 2)(a + 3)$ **19.** $(4x + 3)(x + 2)$ **21.** $xy(y + 3)(y - 3)$ **23.** $y(x + 3)^2$ **25.** $r(r - 6)(r + 5)$ **27.** $4(2 - m)(2 + m)$ **29.** $x(2x + 1)(x - 5)$ **31.** $4(4 - t)(4 + t)$ **33.** $9(x + y)(x - y)$ **35.** $2(3y - 1)(y + 2)$ **37.** $\{0, 4\}$ **39.** $\{0, 1\}$ **41.** $\{5, -6\}$ **43.** $\{4, -3\}$ **45.** $\left\{-\frac{3}{2}, -1\right\}$ **47.** $\left\{-\frac{1}{3}, 0, \frac{1}{3}\right\}$ **49.** $\left\{\frac{3 \pm \sqrt{17}}{2}\right\}$ **51.** $\{4 \pm \sqrt{2}\}$ **53.** $\left\{\pm\frac{\sqrt{30}}{2}\right\}$ **55.** −13, −14 or 13, 14 **57.** 7, 8, 9 or −7, −8, −9 **59.** 5 cm, 13 cm **61.** 8 m **63.** 10 meters **65.** 2 meters

Pages 23–25 Lesson 1.6

Exploratory 1. $x = 0$ **3.** $y = 0$ **5.** $x = 7$ **7.** $y = 0, -3$ **9.** $x = 0, -3, 3$ **11.** 18 **13.** $8my$ **15.** $3y + 12$ **17.** x^3 **19.** $x + 3$ **Written 1.** $\frac{1}{3}$ **3.** $\frac{4}{9}$ **5.** $\frac{37}{81}$ **7.** $x^3, x \neq 0$ **9.** $y, y \neq 0$ **11.** $x^6, x \neq 0$ **13.** $\frac{17}{21x^3}, x \neq 0$ **15.** $\frac{b}{a}, a, b \neq 0$ **17.** $\frac{7x}{3y^4}, y \neq 0$ **19.** $\frac{-y^3}{6x^3}, x, y \neq 0$ **21.** $\frac{7x^7}{3y^4}, y \neq 0$ **23.** $-2xy^2, x, y \neq 0$ **25.** $\frac{-3t^4}{4u}, t, u \neq 0$ **27.** $\frac{1}{2}, m \neq -5$ **29.** $\frac{1}{3}, y \neq -3$ **31.** $\frac{2y+1}{3y+2}, y \neq -\frac{2}{3}$ **33.** $\frac{1}{t-2}, t \neq \pm 2$ **35.** $\frac{1}{x}, x \neq 0, 6$ **37.** $\frac{1}{a-b}, a \neq \pm b$ **39.** $m, m \neq 1$ **41.** $\frac{u-10}{u+10}, u \neq -10$ **43.** $y + 3, y \neq 3$ **45.** $\frac{2x}{(x+3y)^2}, x \neq 0, -3y$ **47.** $\frac{a+2}{2a+1}, a \neq -\frac{1}{2}, 2$ **49.** $\frac{1}{y+4}, y \neq 3, -4$ **51.** $\frac{2x}{x-3}, x \neq 3, -4$ **53.** $\frac{8a}{a-2}, a \neq 2$ **55.** $\frac{1}{2-5m}, m \neq 3, \frac{2}{5}$ **57.** $-1, r \neq 1$ **59.** $-s^2, s \neq 1$ **61.** $\frac{1-x}{x+1}, x \neq -1$ **63.** $-\frac{r+1}{r+4}, r \neq \pm 4$ **65.** $m = \pm 4$ **67.** $t = 23$ **69.** $x = -3$ **71.** $y \in \{9, -1\}$ **73.** $y \in \{3, 5\}$ **75.** $m \in \{3, 5\}$ **Mixed Review 1.** $a = 1$ **2.** $n > -2$ **3.** $-\frac{1}{4} \leq x < 2$ **4.** $k = 0$ or 11 **5.** $b \leq -7$ or $b \geq -\frac{13}{3}$ **6.** $-\frac{4}{7} \leq x \leq 2$ **7.** 41° **8.** 47 mph **9.** 6 pounds of peaches, 8 pounds of bananas **10.** 10 feet

Pages 27–28 Lesson 1.7

Exploratory 1. $\frac{15}{8}$ **3.** $\frac{20}{21}$ **5.** $\frac{7}{5}$ **7.** $\frac{1}{50}$ **9.** $-\frac{7}{18}$ **11.** $\frac{63}{64}$ **13.** $\frac{x}{6}$ **15.** $\frac{3}{y}$ **Written 1.** $\frac{1}{9}$ **3.** 12 **5.** $3y^2$ **7.** $\frac{m^2}{nq}$ **9.** $\frac{2a^6b^2}{c}$ **11.** $\frac{x+3}{y}$ **13.** $\frac{4x+4}{3x}$ **15.** $\frac{5}{4}$ **17.** $\frac{10a}{5a+7}$ **19.** 4 **21.** -3 **23.** 4 **25.** $\frac{1}{2}$ **27.** $\frac{4y+6}{y}$ **29.** $\frac{12}{x-1}$ **31.** $\frac{y^2+3y}{2}$ **33.** a **35.** $\frac{-b^2a}{2}$ **37.** x **39.** $-\frac{t+6}{4t}$ **41.** $\frac{c+4}{b^2}$ **43.** $-\frac{y}{3}$ **45.** $-\frac{(4+x)}{2}$ **47.** $\frac{2c-2}{3c+6}$ **49.** $\frac{3-4t-4t^2}{4t+12}$ **51.** $\frac{x+1}{2x+3}$ **53.** $\frac{a(a-2)}{a+1}$

Pages 31–32 Lesson 1.8

Exploratory 1. $\frac{x(2x-1)}{y^2}$ **3.** $\frac{10}{x+3}$ **5.** $\frac{-y+7}{8}$ **7.** 1 **9.** $\frac{1}{x-1}$ **11.** 24 **13.** a^2 **15.** xy^2 **17.** $y^2 - 4$ **19.** $x^3 + 8x^2 - 65$ **21.** $96x^2(x + 9)$ **Written 1.** $\frac{4}{7}$ **3.** $\frac{5}{4}$ **5.** $\frac{8+3y}{9}$ **7.** $\frac{3r-14}{21}$ **9.** $\frac{13}{2x}$ **11.** $\frac{5x+1}{6}$ **13.** $\frac{4x+y}{7y}$ **15.** $\frac{ab+1}{b}$ **17.** $\frac{4a-14ab}{7b^2}$ **19.** $\frac{13+a}{15}$ **21.** $\frac{y-12}{14}$ **23.** $\frac{23a-20}{20}$ **25.** $\frac{x+4}{x+2}$ **27.** $\frac{b+5}{b}$ **29.** $\frac{3}{y-4}$ **31.** $x + y$ **33.** $\frac{1}{x(x+3)}$ **35.** $\frac{-2}{(b-3)(b-5)}$ **37.** $\frac{9m^2+4m+1}{3m+1}$ **39.** $\frac{m}{3+m}$ **41.** $\frac{2x-1}{(x-2)(x+1)^2}$ **43.** $\frac{1}{(3a-1)}$ **45.** $\frac{2a+1}{(6a+1)(a-3)}$

Pages 34–35 Lesson 1.9

Exploratory 1. $\frac{5}{6}$ **3.** $\frac{4}{15}$ **5.** -6 **7.** $\frac{x^2-x}{3}$ **9.** $\frac{7}{4}$ **Written 1.** $\frac{1}{x}$ **3.** $-\frac{x}{3}$ **5.** $\frac{2}{3}$ **7.** $\frac{38}{3}$ **9.** $x + 3$ **11.** $-\frac{3}{50}$ **13.** -1 **15.** $\frac{3x+1}{x}$ **17.** $\frac{b-a}{b+a}$ **19.** $\frac{2}{a^2}$ **21.** $\frac{4a-1}{4a+1}$ **23.** $\frac{1}{5(y-5)}$ **25.** $\frac{2}{2t-1}$ **27.** $x + 3$ **29.** $\frac{3}{3x-1}$ **31.** $y + 3$ **33.** $\frac{9}{20}$ **35.** $\frac{a-5}{a+4}$ **37.** $-x$ **39.** $\frac{y}{y+1}$ **41.** $-\frac{(x+y)}{y}$ **43.** $\frac{a+7}{a+2}$ **45.** $\frac{2}{m-2}$

Pages 38–39 Lesson 1.10

Exploratory 1. yes **3.** multiply by 5 **5.** multiply by x **7.** multiply by 3 **9.** the last; no fractions **Written 1.** $x = \frac{3}{10}$ **3.** $x = 9$ **5.** $y = 42$ **7.** $x = -6$ **9.** $y = 0$ **11.** $x = -3$ **13.** $y \in \{-6, 2\}$ **15.** $x \in \left\{\frac{5}{2}, -\frac{3}{2}\right\}$ **17.** $x = -1$ **19.** $r \in \left\{-3, \frac{1}{6}\right\}$ **21.** $y = -9$ **23.** $t = 2$ **25.** $q = 1$ **27.** $x = -5$ **29.** $y = 1500$ **31.** $y = 45$ **33.** $d \in \{-3, 2\}$ **35.** $x = 3$ **37.** $x = \frac{2}{3}$ **39.** $t = -2$ **41.** $q = 15$ **43.** $x \in \{-2, -6\}$ **45.** $x \in \{-6, 3\}$ **47.** $b = 0$ **49.** $t = 3$ **51.** no solution **53.** $y \in \{-6, 2\}$ **55.** $x \in \{0, -4\}$ **57.** 20° **59.** $x = -\frac{7}{3}$ **61.** $x = \frac{376}{685}$ **Mixed Review 1.** $\frac{3r^2+6r+18}{5r}$

2. $\frac{(x-3)^2}{(x-2)^2}$ **3.** $\frac{t+1}{t+2}$ **4.** $m+8$ **5.** $\frac{x+3}{x^2-1}$ **6.** $\frac{y-3}{y+2}$
7. $-x$ **8.** $-x$ **9.** $\frac{4(x-6)}{4x^2-6x+9}$ **10.** no solution
11. 8 or $\frac{3}{2}$ **12.** $r = 3$

Pages 40–41 Lesson 1.11

Written **1.** 90 **3.** 19 **5.** $\frac{7}{13}$ **7.** 3 hrs 45 min **9.** 4 and 12 **11.** $\frac{20}{9}$ hrs **13.** 2 hrs 24 minutes **15.** 52 mph **17.** 6 oz. **19.** $\frac{10}{9}$ liters **21.** 29 gal

Page 43 Problem Solving Application

1. $-18, -20$ or $18, 20$ **3.** 1 ft **5.** 18 ft by 24 ft **7.** 12.65, 37.95 **9.** 8 **11.** -2.07 or 0.87 **13.** 6 cm by 4 cm **15.** 0.618

Pages 44–45 Chapter Review

1. $\frac{13}{4}$ **3.** $\frac{7}{9}$ **5.** $-4y - 4xy$ **7.** $x = -1$ **9.** $-8 < x$ **11.** $(x \leq -4) \vee (x > 1)$ **13.** if more than 1150 kg, Acme **15.** no solution **17.** $-2 < y < 6$ **19.** $24m^5n^4$ **21.** $-7x^2 - 83x$ **23.** $x^3 - 5x^2 + 10x - 8$ **25.** $(n-4)(n+4)$ **27.** $5(p-2)(p+2)$ **29.** $(5h-6)(h-12)$ **31.** $y \in \left\{\frac{1}{2}, -5\right\}$ **33.** $y = 5$ **35.** $y \in \{-3, 5\}$ **37.** $x - 5$ **39.** -2 **41.** $\frac{5}{4}$ **43.** $\frac{x+5}{x+2}$ **45.** $\frac{2a}{3(a-2)}$ **47.** $x = 6$ **49.** $x = \frac{3}{2}$ **51.** 3

CHAPTER 2 WORKING WITH COMPLEX NUMBERS

Page 51 Lesson 2.1

Exploratory **1.** 3 **3.** -2 **5.** -11 **7.** 0.1 **9.** $\frac{1}{2}$ **11.** $\frac{6}{7}$ **13.** 2 **15.** 2 **17.** -3 **19.** -5 **21.** irrational **23.** neither **25.** rational **27.** rational **29.** rational **Written** **1.** -9 **3.** -15 **5.** x **7.** $3a$ **9.** x^4 **11.** $2m^6$ **13.** $\frac{y}{2}$ **15.** $-13y^7$ **17.** $2m^2$ **19.** y **21.** $8ab^4$ **23.** $-2m^3n$ **25.** $2x^2y^3z$ **27.** 9 **29.** 2 **31.** 1 **33.** 8 **35.** 4.6 **37.** 5.2 **39.** 16.9 **41.** $25, -25$ **43.** $4, -4$ **45.** $5, -5$ **47.** $\frac{1}{2}$

Pages 54–55 Lesson 2.2

Exploratory **1.** $\sqrt{10}$ **3.** $\sqrt{21}$ **5.** $4\sqrt{10}$ **7.** 2 **9.** $2\sqrt{2}$ **11.** $5\sqrt{3}$ **13.** $2\sqrt{5}$ **15.** $5\sqrt{2}$ **17.** $10\sqrt{7}$ **19.** $m\sqrt{m}$ **Written** **1.** $15\sqrt{6}$ **3.** $9\sqrt{y}$ **5.** $t^3\sqrt{t}$ **7.** $2x\sqrt{3y}$ **9.** $xy^2\sqrt{y}$ **11.** $4xy\sqrt{3x}$ **13.** $30\sqrt{2}$ **15.** 60 **17.** $2\sqrt[3]{2}$ **19.** $4\sqrt[3]{2}$ **21.** $x^5\sqrt{x^2}$ **23.** $2y\sqrt[3]{2}$ **25.** $3\sqrt[3]{2}$ **27.** $10y\sqrt[3]{2y^2}$ **29.** $\frac{\sqrt{3}}{3}$ **31.** $\frac{5\sqrt{3}}{3}$ **33.** $\frac{7\sqrt{3}}{3}$ **35.** $\sqrt{6}$ **37.** $\frac{\sqrt{2}}{3}$ **39.** $\frac{\sqrt{3}}{2}$ **41.** $2\sqrt{5}$ **43.** $\frac{2\sqrt{2}}{3}$ **45.** $\frac{\sqrt{15}}{5}$ **47.** $\frac{\sqrt{15}}{3}$ **49.** $\sqrt{2}$ **51.** $\frac{\sqrt{5}}{4}$ **53.** $\frac{5}{3}$ **55.** $\frac{5}{2}$ **57.** $\frac{\sqrt{6}}{10}$ **59.** $2\sqrt{2}$ **61.** 5 **63.** $\frac{7}{2}$ **65.** $\frac{16\sqrt{2}}{3}$ **67.** $2\sqrt[3]{4}$ **69.** $\frac{\sqrt[4]{54}}{3}$ **71.** $\frac{2\sqrt{14}}{7}$ **73.** $\frac{\sqrt[3]{3}}{2}$ **75.** $\frac{6\sqrt[3]{5}}{5}$

Pages 58–59 Lesson 2.3

Exploratory **1.** $6\sqrt{3}$ **3.** $2\sqrt{2}$ **5.** $23\sqrt{3}$ **7.** $\sqrt{5} + \sqrt{10}$ **9.** $2\sqrt{3}$ **11.** $4\sqrt{10}$ **Written** **1.** $-2\sqrt{2}$ **3.** $-26\sqrt{3}$ **5.** $-7\sqrt{3y}$ **7.** $\frac{3\sqrt{2}}{2}$ **9.** $-\sqrt{3}$ **11.** $\frac{\sqrt{5}}{5}$ **13.** $\frac{\sqrt{2}}{2}$ **15.** $6\sqrt{3}$ **17.** $5\sqrt[3]{3}$ **19.** $23\sqrt[4]{2}$ **21.** $(10m - 8n)\sqrt{mn}$ **23.** $6 - 3\sqrt{2}$ **25.** 12 **27.** $-4\sqrt{6} - 2$ **29.** $11 + 5\sqrt{5}$ **31.** $-3 - 3\sqrt{7}$ **33.** $31 - 8\sqrt{7}$ **35.** $28 - \sqrt{5}$ **37.** $27 + 10\sqrt{2}$ **39.** $35 - 10\sqrt{10}$ **41.** 18 **43.** 62 **45.** 50 **47.** 44 **49.** $-1 + \sqrt{3}$ **51.** $2 + \sqrt{2}$ **53.** $\frac{4+\sqrt{7}}{3}$ **55.** $\frac{3-\sqrt{5}}{4}$ **57.** $\frac{20-4\sqrt{15}}{5}$ **59.** $-3 - \sqrt{3}$ **61.** $\frac{1-2\sqrt{2}}{7}$ **63.** $\frac{9+5\sqrt{5}}{11}$ **65.** $\frac{-4-\sqrt{7}}{3}$ **67.** $-15 - 11\sqrt{2}$ **69.** $2 \pm \sqrt{3}$ **71.** $2 \pm 2\sqrt{3}$ **73.** $\frac{-5 \pm 3\sqrt{5}}{2}$ **75.** $\frac{11 \pm \sqrt{265}}{24}$ **77.** $x^2 - 3\sqrt{3}x + 6 = 0$ **79.** $x^2 - 4x + 1 = 0$ **81.** $x^2 - 8x + 13 = 0$

Page 61 Lesson 2.4

Exploratory **1.** 81 **3.** 4 **5.** 16 **7.** no solution **9.** no solution **Written** **1.** 12 **3.** 4 **5.** 7 **7.** 9 **9.** -7 **11.** no solution **13.** 5 **15.** 3 **17.** 4, 8 **19.** 25 **21.** 14 **23.** no solution **25.** 9 **27.** 21 **29.** 5 **31.** 7 **33.** 6 **35.** 0

Mixed Review **1.** $5 - 2\sqrt{3}$ **2.** $-\frac{\sqrt{10}}{25}$ **3.** $\frac{-12-4\sqrt{2}}{7}$ **4.** $\frac{-1-\sqrt{5}}{2}$ **5.** $2\sqrt{2}$ or $-\sqrt{2}$

6. no solution **7.** $t = \frac{6}{13}$ **8.** 3 or −4 **9.** 15 hits **10.** 800 ml of 25% and 200 ml of 50% **11.** 5 tons **12.** 1 hour 17 minutes

Page 64 Lesson 2.5

Exploratory **1.** $3i$ **3.** $6i$ **5.** $24i$ **7.** $-4i$ **9.** −2 **Written** **1.** $7i$ **3.** $i\sqrt{3}$ **5.** $\frac{i}{3}$ **7.** $-2i\sqrt{3}$ **9.** $3i\sqrt{2}$ **11.** $2i\sqrt{3}$ **13.** $-2i\sqrt{2}$ **15.** $-2i\sqrt{7}$ **17.** $-8i\sqrt{3}$ **19.** $14i\sqrt{2}$ **21.** $6i$ **23.** $\frac{i\sqrt{2}}{2}$ **25.** $-\frac{3i\sqrt{2}}{2}$ **27.** −2 **29.** −4 **31.** $-5\sqrt{3}$ **33.** $10i$ **35.** −12 **37.** 40 **39.** 15 **41.** $\frac{5}{2}$ **43.** $2i$ **45.** 4 **47.** $\frac{2\sqrt{5}}{5}$ **49.** $\frac{\sqrt{6}}{3}$ **51.** 1 **53.** i **55.** −1 **57.** 0 **59.** $2\sqrt{3}$

Pages 68–69 Lesson 2.6

Exploratory **1.** true **3.** true **5.** true **7.** false **9.** $x = 2, y = 3$ **11.** $x = 4, y = -5$ **13.** $x = 3, y = 0$ **15.** $-1 - i$ **17.** −12 **Written** **1.** $6 + 8i$ **3.** $-1 + 4i$ **5.** $-1 - 11i$ **7.** $-2 + 11i$ **9.** $1 + (3 - \sqrt{3})i$ **11.** $\frac{1}{8} - \frac{7i}{6}$ **13.** $5i$ **15.** $-13 + 13i$ **17.** $-26 - 7i$ **19.** $14 + 2i$ **21.** $-21 - 20i$ **23.** $1 - 2i\sqrt{2}$ **25.** $\frac{1}{2}$ **27.** $8 - i\sqrt{2}$ **29.** $26 + 7i$ **31.** 4 **33.** $14 - 17i$ **35.** $-22 + 43i$ **37.** $9 + 3i\sqrt{5}$ **39.** $-7 + i$ **41.** $-7 - 24i$ **43.** $22 + 7i$ **45.** $12 - 2i\sqrt{2}$ **47.** 4 **49.** $52 - 47i$ **51.** −13 **53.** $27 - 15i$ **55.** $4 + i$ **57.** −29 **59.** $-1 - i$ **61.** $-2i$ **63.** $1 - 3i$ **65.** $-1 - i$ **67.** $2 + 4i$ **69.** $\sqrt{2} - \sqrt{6} + (2\sqrt{3} - 6)i$ **71.** $63 + 54i$ **73.** 1 **75.** $-20 - 8\sqrt{6}$ **77.** $-\frac{77}{108} - \frac{5i\sqrt{3}}{24}$ **79.** 0 **81.** $i - 3$ **83.** $-19 - 73i$ **85.** $\sqrt{30} - 2\sqrt{5} + (\sqrt{6} - 3)i$ **87.** $11 - 18i$ **89.** $x = 2, y = 3$ **91.** $x = \frac{67}{11}, y = \frac{19}{11}$ **93.** $-a - bi$

Pages 71–72 Lesson 2.7

Exploratory **1.** $1 - i$ **3.** $-2 - 4i$ **5.** i **7.** 3 **9.** $-2 - i\sqrt{3}$ **11.** 2 **13.** 20 **15.** 1 **17.** 9 **19.** 7 **21.** $a + bi$ **Written** **1.** $-1 - i, \frac{1}{2} - \frac{1}{2}i$ **3.** $-2 - i, \frac{2}{5} - \frac{1}{5}i$ **5.** $1 + i, -\frac{1}{2} + \frac{1}{2}i$ **7.** $-1 + 4i, \frac{1}{17} + \frac{4}{17}i$ **9.** $2 - 4i, -\frac{1}{10} - \frac{1}{5}i$ **11.** $2i, \frac{1}{2}i$ **13.** $-1 - i\sqrt{3}, \frac{1}{4} - \frac{\sqrt{3}}{4}i$ **15.** $\pi + \pi i, -\frac{1}{2\pi} + \frac{1}{2\pi}i$ **17.** $2 - i$ **19.** $-\frac{2}{5} + \frac{1}{5}i$ **21.** $\frac{3}{5} + \frac{6}{5}i$ **23.** $\frac{1}{2} + \frac{1}{2}i$ **25.** $\frac{7}{13} + \frac{17}{13}i$ **27.** $\frac{5}{2} + \frac{1}{2}i$ **29.** $\frac{9}{41} + \frac{1}{41}i$ **31.** $\frac{1}{3} + \frac{2\sqrt{2}}{3}i$ **33.** $\frac{4\sqrt{3}}{7} - \frac{8}{7}i$ **35.** $-\frac{3}{11} - \frac{4\sqrt{7}}{11}i$ **37.** $\frac{16}{50} + \frac{63}{50}i$ **39.** $-\frac{1}{2} - \frac{1}{2}i$ **41.** $-\frac{1}{2} - \frac{1}{2}i$

Pages 77–78 Lesson 2.8

Exploratory **1.** $\pm 2i$ **3.** $\pm 7i$ **5.** $2 \pm i$ **7.** $3 \pm i$ **9.** $\frac{1}{2} \pm \frac{1}{2}i$ **11.** $\frac{1}{6} \pm \frac{1}{6}i$ **13.** $\frac{1}{3} \pm \frac{\sqrt{5}}{3}i$ **15.** $-\frac{1}{6} \pm \frac{\sqrt{7}}{6}$ **17.** $\frac{3}{2} \pm \frac{\sqrt{23}}{2}i$ **Written** **1.** −2, 2 **3.** −3, 2 **5.** −4, 1 **7.** 8 **9.** 33 **11.** 49 **13.** 100 **15.** −19 **17.** −11 **19.** 0 **21.** 61 **23.** $\frac{9}{16}$ **25.** −12 **27.** 29 **29.** $16 - 12\sqrt{3}$ **31.** 153 **33.** −37 **35.** $k < \frac{1}{2}$ **37.** no solution **39.** $k > \frac{2}{3}$ **41.** $-4 < k < 4$ **43.** $m - ni$

Pages 80–81 Lesson 2.9

Exploratory **1.** −6, −4 **3.** −5, −3 **5.** 2, −5, **7.** $-\frac{4}{5}, -\frac{7}{5}$ **9.** $\frac{1}{2}, -\frac{1}{4}$ **11.** −3, −6 **13.** 2, −8 **15.** $-\frac{4}{3}, -\frac{4}{3}$ **17.** $-\frac{\sqrt{3}}{3}, \frac{\sqrt{21}}{3}$ **19.** $1 + \sqrt{2}, 3 + 3\sqrt{2}$ **21.** $\frac{1}{15}, -\frac{4}{15}$ **Written** **1.** $x^2 - 5x + 6 = 0$ **3.** $x^2 + 7x + 5 = 0$ **5.** $x^2 + 9x + 10 = 0$ **7.** $3x^2 - 2x - 12 = 0$ **9.** $x^2 - 5x - 14 = 0$ **11.** $x^2 + 9x + 20 = 0$ **13.** $2x^2 - 7x + 3 = 0$ **15.** $4x^2 - 47x - 12 = 0$ **17.** $4x^2 - 1 = 0$ **19.** $4x^2 + 13x - 12 = 0$ **21.** $x^2 - 6x + 4 = 0$ **23.** $x^2 + 4 = 0$ **25.** $x^2 - 2x + 2 = 0$ **27.** $x^2 - 8x + 25 = 0$ **29.** $x^2 - 6x + 25 = 0$ **31.** $x^2 - 8x + 28 = 0$ **35.** $r_2 = 2, b = -8$ **37.** $r_2 = 2 + \sqrt{5}, c = -3$ **Mixed Review** **1.** $-2 + 4\sqrt{6}$ **2.** $\frac{18 - 48i}{25}$ **3.** $n = -7$ **4.** $x^2 - 2\sqrt{5}x + 1$ **5.** 2 or −6 **6.** 4 m

Page 85 Lesson 2.10

Exploratory **1.** c **3.** b **Written** **1.** $x < -4$ or $x > 4$ **3.** $x < -5$ or $x > 5$ **5.** $-6 \le x \le 0$ **7.** $-3 < x < 2$ **9.** $-1 \le x \le 5$ **11.** $-6 \le x \le 6$ **13.** $-\frac{3}{2} < x < 2$ **15.** $x < -2$ or $x > \frac{1}{3}$ **17.** $-\sqrt{3} \le x \le \sqrt{3}$ **Mixed Review** **1.** 8 **2.** 11 or 7 **3.** 2 **4.** $1800, 6% **5.** 11 free throws

Page 87 Problem Solving Application

1. 1–$100 bill, 9–$50 bills, 90–$5 bills

3. 3; \$1.07 **5.** 5 **7.** 12 **9.** 79 **13.** 82524 + 19722 + 106 = 102352 **15.** 12734 + 12734 = 25468 **17.** 92836 + 12836 = 105672

Pages 88–89 Chapter Review
1. −13 **3.** $\frac{1}{2}$ **5.** $6a$ **7.** $12x^2y$ **9.** −7 **11.** −3 **13.** $2x\sqrt{7x}$ **15.** $70\sqrt{2}$ **17.** 4 **19.** $3\sqrt{2}$ **21.** $-9\sqrt{2}$ **23.** $6\sqrt{5} - 15\sqrt{2}$ **25.** $-20 - 10\sqrt{3}$ **27.** $4 \pm \sqrt{3}$ **29.** 7 **31.** $-30\sqrt{2}$ **33.** −1 **35.** rational **37.** irrational **39.** $8 - 5i$ **41.** $76 + i$ **43.** $7 + 7i$ **45.** $-3 + 4i$ **47.** $2 - 2i$ **49.** $\frac{1}{17} + \frac{4}{17}i$ **51.** 0, 2 equal roots **53.** $x^2 + 9 = 0$ **55.** $x^2 - 2x + 26 = 0$ **57.** $x < -4$ or $x > 3$

CHAPTER 3 RELATIONS AND FUNCTIONS

Pages 94–95 Lesson 3.1
Exploratory **1.** {1}; {1, 3} **3.** {Mary, Alice}; {Bob, Tom} **5.** {a, b, c}; {a, b, c} **7.** {△, ~}, {⊥, ∥} **9.** {Alabama, Alaska, Arizona, Arkansas} or {states that begin with the letter A}; {Montgomery, Juneau, Phoenix, Little Rock} or {their capitals} **11.** {(5, 1), (5, 2), (5, 3), (5, 4), (10, 5), (10, 6), (10, 7), (10, 8)} **13.** {(2, 3), (4, 3), (5, 1)} **15.** {(9, 3), (9, −3), (4, 2), (4, −2)} **Written** **1.** $\{(x, y): y = 2x + 1\}$; {−2, 0, 1, 2, 10}; {−3, 1, 3, 5, 21} **3.** $\{(x, y): y = x^2\}$; {−3, 0, 1, 3, 5}; {0, 1, 9, 25} **5.** {(−1, 0), (0, −1), (0, 1), (3, −4)} **7.** {−1, 0, 3}; {−4, −1, 0, 1} **9.** none **11.** {−3, −2, −1, 0, 1, 2, 3} **13.** {0, 1, 4, 9} **15.** $\{0, \pm 1, \pm\sqrt{2}, \pm\sqrt{3}, \pm 2, \pm\sqrt{5}, \pm\sqrt{6}\}$ **17.** {−7, −6, −5, −4, −3, −2, −1} **19.** {−4} **21.** {0, 1, 4, 9} **23.** {0, 1, 2, 3, 4} **25.** {−3, −2, −1, 0} **27.** $x < 0$ **29.** $x < 3$ **31.** 0 **33.** $\frac{1}{2}$ **35.** ±2 **37.** −3, −1 **39.** ±5 **41.** 1 **43.** none **45.** $x = 7, x < 5$ **47.** $-\frac{1}{2}$, 0 **49.** $x \le 5$ **57.** $\mathcal{R}$ **59.** $\{x{:}x \ge 0\}$ **61.** $\mathcal{R} / \left\{\frac{7}{2}\right\}$ **63.** $\mathcal{R}$ **65.** $\mathcal{R}$ **67.** $\mathcal{R}$ **69.** $\{x{:}x \ge 0\}$ **71.** $\mathcal{R}$ **73.** $\mathcal{R}$ **75.** $\mathcal{R}$

Pages 99–100 Lesson 3.2
Exploratory **5.** {1, 2}; {4}; yes **7.** {1}, {8, 9}; no **9.** {1, 2}; {1, 2, 3, 4}; no **11.** {Bernie, Harry, Marie}; {78}; yes **13.** {2, 3}; {−3, −2, 2, 3}; no **15.** $\{x{:}|x| \le 2\}$; $\{y{:}|y| \le 2\}$; no **17.** 0 **19.** 10 **21.** 15 **23.** yes; $\{x{:}0 \le x \le \pi\}$; $\{y{:}0 \le y \le 1\}$ **25.** yes; $\mathcal{R}$; $\mathcal{R}$ **27.** yes; $\mathcal{R}$; $\{y{:}y \ge 0\}$ **29.** yes; $\mathcal{R}/\{3\}$; {−2, 4} **31.** yes; $\mathcal{R}/\{0\}$; $\mathcal{R}/\{0\}$ **33.** no; $\{x{:}x \ge 2\}$; $\mathcal{R}$ **Written** **1.** {−3, −1, 5} **3.** {−3, −2, 6} **5.** {−1, 0, 3} **7.** {1, 7} **9.** $\left\{0, \frac{1}{2}, \frac{3}{2}\right\}$ **11.** $y = -\frac{1}{2}x - \frac{1}{2}$ **13.** $y = -\frac{3}{5}x - \frac{38}{5}$ **15.** $y = 2x$ **17.** $y = 2x$ **19.** linear function; $\mathcal{R}$; $\mathcal{R}$; 5 **21.** not a function; $\{x{:}x \ge 0\}$; $\mathcal{R}$; ±5 **23.** function; $\mathcal{R}$; $\{y{:}y \le 0\}$; −25 **25.** absolute value function; $\mathcal{R}$; $\{y{:}y \ge -2\}$; 3 **27.** Temperature depends on the time of day, and for a given time, there is exactly one temperature. **29.** {−2, 3, 6} **31.** $\left\{-\frac{7}{2}, -1, \frac{1}{2}, 2, \frac{9}{2}\right\}$ **33.** {−5, −4, −3, −1, 0, 1} **35.** $y = 0.13x + 14.50$ **37.** $y = \frac{65}{16}x - \frac{407}{16}$

Pages 103–104 Lesson 3.3
Exploratory **1.** function h maps x to $3x$ **3.** function f maps 4 to −6 **5.** function h maps 2 to 7 **7.** {(−1, 1), (0, 0), (1, 1), (2, 4)}; {0, 1, 4} **9.** {(−1, −4), (0, −1), (1, 2), (2, 5)}; {−4, −1, 2, 5} **11.** $\left\{(-1, 0), \left(0, \frac{1}{3}\right), \left(1, \frac{2}{3}\right), (2, 1)\right\}; \left\{0, \frac{1}{3}, \frac{2}{3}, 1\right\}$ **13.** 0, 2, 3, 8 **15.** −54, −2, 0, 250 **17.** 0, 2, 3, 8 **19.** −144, −16, 0, −400 **21.** 12, 2, 0, 30 **Written** **1.** {(−2, −7), (−1, −4), (0, −1), (1, 2), (2, 5)}; {−7, −4, −1, 2, 5} **3.** $\left\{\left(-2, -\frac{1}{2}\right), (-1, -1), (1, 1), \left(2, \frac{1}{2}\right)\right\}; \left\{-1, -\frac{1}{2}, \frac{1}{2}, 1\right\}$ **5.** {(−2, 6), (−1, 5), (0, 4), (1, 3), (2, 2)}; {2, 3, 4, 5, 6} **7.** {(−2, 3), (−1, 3), (0, 5), (1, 9), (2, 15)}; {3, 5, 9, 15} **9.** 0 **11.** 7 **13.** −13 **15.** $\frac{1}{2}$ **17.** −59 **19.** $4\sqrt{2} - 5$ **21.** $2p^3 + 6p^2 + 6p - 3$ **23.** $f: x \to |x|$; $\mathcal{R}$ **25.** $f: x \to |x - 3|$; $\mathcal{R}$ **27.** $\{x{:}x \ge -4\}$ **29.** $\mathcal{R}/\{0\}$ **31.** $\{x{:}x \ge -2 \text{ and } x \ne 8\}$ **33.** $\mathcal{R}/\{-3, 3\}$ **35.** $\mathcal{R}$; $\{y{:}y \ge 0\}$ **37.** $\mathcal{R}/\{0\}$; $\mathcal{R}/\{0\}$ **39.** $\mathcal{R}$; $\{y{:}y \ge 1\}$ **41.** $\mathcal{R}$; $\{y{:}y \ge -4\}$ **43.** 1 **45.** 0 **47.** 0 **49.** $f(x) = -x + 3$ **51.** (2, 0) **53.** $(r, 0)$ **55.** $(b, f(a))$ **57.** $(t, f(a))$ **59.** $BE = \sqrt{(t - a)^2 + (f(a) - f(a))^2} = \sqrt{(t - a)^2} = t - a$

Pages 108–110 Lesson 3.4
Exploratory **1.** linear, direct variation **3.** constant **5.** step **7.** none of these **Written** **1.** linear, direct variation **3.** inverse variation **5.** linear **7.** linear **9.** $f(x) = 3x - 2$ **11.** $f(x) = -x + 5$ **13.** $f(x) = -5x + 44$ **15.** $g(x) = -\frac{11}{5}x + \frac{3}{5}$ **17.** $g(x) = \frac{6}{5}x - \frac{24}{5}$ **19.** 520 cc **21.** 120 cc **23.** 2 **25.** 9 **27.** −4 **41.** $f(s) = s^2$ **43.** $c = 4.99n$; 4.99 **45.** 4.25 cm **47.** 40 **49.** 640 **51.** $G(t) = 3^t$ **55.** 210 **57.** $nc = 24.99$; 24.99 **59.** 42 pounds

Pages 113–115 Lesson 3.5

Exploratory 1. The inverse of a function is the relation generated by interchanging the numbers of each ordered pair of the function. Thus the domain and range are interchanged. **3.** A function is a one-to-one function if each element of the range corresponds to exactly one element of the domain. **5.** yes **7.** no **9.** {(3, −2), (5, −1), (3, 4)}; no **11.** {(7, −3), (9, 4), (−3, 7), (5, 8)}; yes **13.** 18 **Written 1.** {(5, 1), (3, 2), (3, 4), (5, −1)}; no **3.** {(5, −2), (4, 3), (4, 4), (4, −1)}; no **5.** {(−3, −2), (−2, −1), (−1, 0), (1, 2), (2, 3)}; yes; {(−2, −3), (−1, −2), (0, −1), (2, 1), (3, 2)}; yes **7.** {(−3, −12), (−2, −8), (−1, −4), (1, 4), (2, 8)}; yes; {(−12, −3), (−8, −2), (−4, −1), (4, 1), (8, 2)}; yes **9.** {(−3, 9), (−2, 4), (−1, 1), (1, 1), (2, 4)}; no; {(9, −3), (4, −2), (1, −1), (1, 1), (4, 2)}; no **11.** {(−3, 0), (−2, 1), (−1, 2), (1, 4), (2, 5)}; yes; {(0, −3), (1, −2), (2, −1), (4, 1), (5, 2)}; yes **13.** $\{(-3, 0), (-2, 1), (-1, \sqrt{2}), (1, 2), (2, \sqrt{5})\}$; yes; $\{(0, -3), (1, -2), (\sqrt{2}, -1), (2, 1), (\sqrt{5}, 2)\}$; yes **15.** $\left\{(-3, -1), \left(-2, -\frac{3}{2}\right), (-1, -3), (1, 3), \left(2, \frac{3}{2}\right)\right\}$; yes; $\left\{(-1, -3), \left(-\frac{3}{2}, -2\right), (-3, -1), (3, 1), \left(\frac{3}{2}, 2\right)\right\}$; yes **17.** no; no **19.** no; no **21.** no; no **23.** 7 **25.** 8 **27.** $x = 2y$ **29.** $x = \frac{1}{2}y$ **31.** $x = -3y - 5$ **33.** $f^{-1}(x) = \frac{1}{4}x$ **35.** $f^{-1}(x) = -\frac{1}{3}x$ **37.** $f^{-1}(x) = -x + 5$ **39.** $f^{-1}(x) = 2x - 1$ **41.** $f^{-1}(x) = -4x + 10$ **43.** $f^{-1}(x) = -2x + \frac{8}{3}$ **45.** no; fails Horizontal Line Test **47.** yes **49.** yes **51.** no **53.** $\mathcal{R}$ **55.** $\{x:x > 0\}$ **57.** $\{x:x \geq 3\}$ or $\{x:x \leq -3\}$ **59.** $\{x:x \geq -3\}$ or $\{x:x \leq -3\}$ **61.** $\{x:x \geq 0\}$ or $\{x:x \leq 0\}$ **63.** $\{x:x \geq 0\}$ or $\{x:x \leq 0\}$ **Mixed Review 1.** c **2.** b **3.** c **4.** d **5.** a **6.** b

Pages 119–120 Lesson 3.6

Exploratory 1. circle **3.** line **5.** hyperbola **7.** hyperbola **9.** two points **11.** hyperbola **13.** hyperbola **15.** hyperbola **Written 1.** $\sqrt{6}, -\sqrt{6}$; 2, −2 **3.** −3, 9 **5.** −5, 5; −5, 5 **7.** $-\sqrt{7}, \sqrt{7}; -\sqrt{7}, \sqrt{7}$ **9.** none; 6 **11.** −1, 1; −4, 4 **13.** parabola **15.** hyperbola **17.** circle **19.** parabola **21.** hyperbola **23.** ellipse **25.** ellipse **27.** ellipse **29.** hyperbola **31.** parabola **33.** ellipse **35.** hyperbola

Pages 124–125 Lesson 3.7

Exploratory 1. {3, 4, 5, 6} **3.** {7, 9, 11, 13} **5.** 7 **7.** 11 **9.** {(7, 7), (9, 11), (11, 9), (13, 13)} **11.** {7, 9, 11, 13} **13.** {7, 9, 11, 13} **15.** 3 **17.** 4 **Written 1.** 9 **3.** 8 **5.** 12 **7.** −9 **9.** $\frac{25}{16}$ **11.** $-\frac{24}{25}$ **13.** $19 - 8\sqrt{3}$ **15.** $8 - 4\sqrt{5}$ **17.** $3x - 1$ **19.** yes, possibly **21.** $6x + 9$, $6x - 8$ **23.** $x^3 - 3x^2 + 3x - 1$; $x^3 - 1$ **25.** $|x|$, x **27.** no **29.** yes **31.** $\mathcal{R}$ **33.** $\mathcal{R}/\{0\}$ **35.** $\mathcal{R}/\{0\}$ **37.** $\{y:y > 0\}$ **Mixed Review 1.** 3 **2.** $\frac{39 + 3\sqrt{133}}{2}$ **3.** {(8, 3), (0, 4), (3, 6), (−1, 7)} **4.** {(6, 3), (6, 8), (−8, −1), (4, 0)} **5.** doesn't exist **6.** {(3, 6), (4, 4), (6, 6), (7, −8)} **7.** hyperbola **8.** $-5 + 3i$, $\frac{5 + 3i}{34}$ **9.** $10,560 **10.** 480 cm

Page 127 Problem Solving Application

1. 3780 calories **3.** 18°F **5.** $2040 **7.** 7.7% **9.** It triples.

Pages 128–129 Chapter Review

1. $\{y: -3 \leq y \leq 3\}$ **3.** $\{y:0 \leq y \leq 81\}$ **5.** $\{y:0 \leq y \leq 4\}$ **7.** $\frac{1}{2}$ **9.** $x < \frac{7}{2}$ **11.** no; fails Vertical Line Test; {4}; $\mathcal{R}$ **13.** yes; $\mathcal{R}$; $\{y:y \geq 0\}$ **15.** {(−2, −4), (−1, −1), (1, −1), (2, −4)}; {−4, −1} **17.** 1 **19.** $\frac{8}{3}$ **21.** $\{x:x \leq 6\}$ **23.** $\mathcal{R}$ **25.** $f(x) = \frac{3}{4}x - \frac{13}{2}$ **29.** {(−6, −3), (−4, −2), (−2, −1), (2, 1), (4, 2), (6, 3)}; yes **31.** {(6, −3), (5, −2), (4, −1), (2, 1), (1, 2), (0, 3)}; yes **33.** $f^{-1}(x) = \frac{x}{2} + \frac{5}{2}$ **35.** circle **37.** ellipse **39.** −3, 3; −2, 2 **41.** 12 **43.** −9 **45.** $x^2 - 2x + 1$ **47.** $x^2 - 1$

CHAPTER 4 AN INTRODUCTION TO TRANSFORMATION GEOMETRY

Pages 133–134 Lesson 4.1

Exploratory 1. no **3.** no **5.** Point *P* **7.** *RK* **9.** $\angle LNQ$ **11.** *LK* **13.** *K* **15.** *Q* **Written 11.** *B* is between *A* and *C*, *B* is between *A* and *D*, *C* is between *A* and *D*, and *C* is between *B* and *D*. Under R_ℓ, their images *A*′, *B*′, *C*′, and *D*′ maintain these betweenness relationships. **13.** yes

Pages 136–137 Lesson 4.2

Exploratory 3. A, H, I, M, O, T, U, V, W, X, Y **5.** O; any line through the "center" of the letter **7.** four **9.** infinitely many **11.** one **Written 15.** no **Mixed Review 1.** c **2.** a **3.** b **4.** d **5.** $0 \leq x \leq 2$; $\sqrt{4 - x^2}$

Pages 139–141 Lesson 4.3

Exploratory **1.** $(-x, y)$ **3.** (y, x) **5.** $(-6, -4)$ **7.** $(0, 3)$ **9.** $\left(23, 3\frac{3}{4}\right)$ **11.** $(6, 0)$ **13.** $(0, -5)$ **15.** $(3, 5)$ **17.** $(-8, 0)$ **19.** $(0, 0)$ **21.** $(0, 3)$ **23.** $(-8, 3)$ **25.** $(2, 7)$ **27.** $(0, 0)$ **29.** neither **31.** x-axis **33.** both **Written** **1.** $A'(-4, 2)$, $B'(-1, 5)$; $3\sqrt{2}$, $3\sqrt{2}$ **3.** $A'(-1, 4)$, $B'(-6, 2)$; $\sqrt{29}$, $\sqrt{29}$ **5.** $A'(2, -3)$, $B'(-5, 1)$; $\sqrt{65}$, $\sqrt{65}$ **7.** $A'(5, -6)$, $B'(4, -2)$; $\sqrt{17}$, $\sqrt{17}$ **9.** $A'(-1, -3)$, $B'(2, 1)$; 5, 5 **11.** $A'(1, -1)$, $B'(-1, 6)$; $\sqrt{53}$, $\sqrt{53}$ **13.** $A'(2, 4)$, $B'(-1, -3)$; $\sqrt{58}$, $\sqrt{58}$ **15.** $A'(-3, -2)$, $B'(-5, 0)$; $2\sqrt{2}$, $2\sqrt{2}$ **17.** $A'(-1, 4)$, $B'(-5, 3)$; $\sqrt{17}$, $\sqrt{17}$ **19.** $A'(-7, -2)$, $B'(0, 3)$ **21.** $A'(-9, -2)$, $B'(-2, 3)$ **23.** $(0, -3)$ **25.** $(1, -2)$ **27.** $(-y, -x)$ **29.** -1, -1 **31.** yes

Pages 144–145 Lesson 4.4

Exploratory **1.** L **3.** $\overline{FN}$ **5.** $\angle JML$ **7.** I **9.** P **11.** $\overline{JM}$ **13.** $\angle NMJ$ **15.** $T_{3,1}$ **17.** $T_{-4,2}$ **19.** $T_{-4,-19}$ **Written** **1.** $(0, 12)$ **3.** $(-12, 5)$ **5.** $(4, 9)$ **7.** $(9, -4)$ **9.** $T_{-4,3}$; $(-9, 9)$ **11.** $T_{-3,5}$; $(-8, 11)$ **13.** $T_{15,-14}$; $(10, -8)$ **15.** $(2, 2)$ **17.** $(-7, -1)$ **19.** $(-9, -15)$ **21.** $(\sqrt{3} - 3, \sqrt{5} - 6)$ **23.** $AB = 2\sqrt{5}$, $A'B' = 2\sqrt{5}$ **25.** $AA' = \sqrt{17}$, $BB' = \sqrt{17}$ **27.** because $\overleftrightarrow{AB} \parallel \overleftrightarrow{A'B'}$ and $\overleftrightarrow{AA'} \parallel \overleftrightarrow{BB'}$ **29.** $A'(-3, 1)$, $B'(0, -5)$, $C'(4, -2)$, $D'(1, 3)$ **31.** $A'(-10, -1)$, $B'(-7, 5)$, $C'(-3, 2)$, $D'(-6, -3)$ **33.** $A'(-1, -3)$, $B'(5, 0)$, $C'(2, 4)$, $D'(-3, 1)$ **35.** yes, $R_{x\text{-axis}}$ **37.** yes, $T_{-7,0}$ **39.** yes, $R_{y=x}$ **41.** $A'(5, 3)$, $B'(4, -2)$, $C'(1, 1)$ **43.** yes **45.** $A''(-7, 7)$, $B''(-4, 12)$, $C''(-2, 6)$

Pages 148–149 Lesson 4.5

Exploratory **1.** V **5.** J **7.** A **9.** H **11.** B **13.** E **15.** B **17.** no **19.** no **21.** no **23.** yes, 180 **25.** yes, 180 **27.** no **29.** yes, 180 **Written** **5.** $(-2, 4)$ **7.** $(-9, -6)$ **9.** $(5, 6)$ **11.** $(8, -8)$ **17.** $A'(-1, 1)$, $B'(-6, 1)$, $C'(-6, 5)$ **19.** $A'(-3, 4)$, $B'(-4, -1)$, $C'(1, -1)$ **21.** $A'(1, -3)$, $B'(1, 3)$, $C'(-1, 3)$ **23.** $E'(4, -1)$, $F'(4, -9)$, $G'(8, -9)$, $H'(8, -1)$ **25.** $E'(0, 1)$, $F'(-3, -2)$, $G'(0, -5)$, $H'(3, -2)$ **27.** $(y, -x)$ **Mixed Review** **1.** $(-3, -1)$ **2.** $(-1, 3)$ **3.** $(0, 2)$ **4.** $(1, 3)$ **5.** $(2, -5)$ **6.** $(-4, 5)$ **7.** $(-2, 4)$ **8.** $(5, -2)$ **9.** -29, $4x^2 + 4x$ **10.** $-1 \le x \le 2$ or $3 \le x \le 6$ **11.** $c = 0.06p + 2$

Pages 151–153 Lesson 4.6

Exploratory **1.** $(-2, -3)$ **3.** $(5, 0)$ **5.** $(-3, 5)$ **7.** $(6, -2)$ **9.** $(4, -5)$ **11.** $(-9, 11)$ **13.** $(4, 5)$ **Written** **5.** $(-3, -5)$ **7.** $(-7, 9)$ **9.** $(2, 3)$ **11.** $\left(-2, -\frac{3}{2}\right)$ **13.** $\left(-\frac{7}{2}, \frac{3}{2}\right)$ **15.** $AB = \sqrt{85}$, $A'B' = \sqrt{85}$ **17.** Slopes of $\overleftrightarrow{A'B'}$ and $\overleftrightarrow{AB}$ are $\frac{1}{8}$. **19.** Slopes of $\overleftrightarrow{AA'}$ and $\overleftrightarrow{BB'}$ are -5. **21.** $A'(-1, -1)$, $B'(-6, -1)$, $C'(-6, -4)$ **23.** $R_{x\text{-axis}}$ **25.** $A''(-1, -1)$, $B''(-6, -1)$, $C''(-6, -4)$ **27.** translation **29.** half-turn **31.** half-turn **33.** yes **35.** no **37.** yes **39.** yes **41.** yes

Pages 155–156 Lesson 4.7

Exploratory **1.** yes **3.** yes **5.** $f^{-1}(x) = \frac{1}{3}x$ **7.** $f^{-1}(x) = x - 3$ **9.** $f^{-1}(x) = \frac{x}{2} + \frac{1}{2}$ **11.** $f^{-1}(x) = -\frac{x}{5} + \frac{8}{5}$ **13.** $f^{-1}(x) = \pm\sqrt{4 - x^2}$, $0 \le x \le 2$ **Written** **1.** yes **3.** yes **5.** no **7.** no **9.** yes **11.** yes

Pages 159–160 Lesson 4.8

Exploratory **1.** $(10, 20)$ **3.** $(20, -30)$ **5.** $\left(7\frac{1}{2}, 11\right)$ **7.** $(10\sqrt{2}, -15\sqrt{2})$ **9.** 4 **11.** $\frac{1}{2}$ **13.** $-\frac{1}{2}$ **15.** yes, $D_{O,-3}$ **17.** yes, $D_{O,-\frac{1}{3}}$ **19.** no **Written** **7.** $A'(0, 12)$, $B'(4, 4)$, $C'(8, 4)$ **9.** $A'(0, -18)$, $B'(-6, -6)$, $C'(-12, -6)$ **11.** 3, -3 **13.** $\frac{2}{5}$, -2 **15.** 1, $\sqrt{3}$ **17.** $-\frac{11}{2}$ **19.** yes **21.** no **23.** yes **25.** false **27.** false **29.** true

Page 162 Problem Solving Application

1. 10 dresses, 20 suits **2.** 30 dresses, 17 suits **3a.** $A \ge 0$, $B \ge 0$, $4A + B \le 32$, $A + 6B \le 54$ **3c.** $A = 6$, $B = 8$ **4a.** $m \ge 0$, $m + t \le 20$, $3 \le t \le 8$ **4c.** $m = 12$, $t = 8$ **5a.** $c \ge 0$, $b \ge 0$, $c + b \le 250$, $\frac{c}{10} + \frac{b}{15} \le 20$ **5c.** $c = 100$, $b = 150$ **5d.** $c = 0$, $b = 250$

Pages 164–165 Chapter Review

1. D **3.** $\angle RVA$ **5.** $\overline{RV}$ **13.** $(6, 4)$ **15.** $(4, -6)$ **17.** $(14, 4)$ **19.** $T_{3,-4}$ **21.** $T_{11,-5}$ **23.** $A'(-3, 2)$, $B'(-2, 6)$, $C'(1, 5)$ **25.** $T_{4,2}$ **27.** $(4, -6)$ **29.** neither **31.** rotational, point **33.** rotational, point **35.** rotational, point **37.** rotational, point **39.** rotational, point **45.** Slopes of $\overline{ST}$ and $\overline{S'T'}$ are $\frac{5}{3}$. **47.** dilation **49.** rotation **51.** reflection **53.** rotation

CHAPTER 5 EXPONENTIAL AND LOGARITHMIC FUNCTIONS

Pages 170–171 Lesson 5.1

Exploratory 1. 1 **5.** $\frac{1}{81}$ **9.** $-\frac{5}{4}$ **11.** 8 **13.** 16 **15.** 16 **17.** 2 **19.** 2 **21.** 4 **23.** 27 **25.** −9 **27.** $\frac{1}{4}$ **29.** $\frac{1}{5}$ **31.** $\frac{1}{2}$ **33.** $\frac{2}{3}$ **35.** 8 **37.** $\frac{1}{a^4}$ **39.** $\frac{5}{x^7}$ **41.** $9x^2$ **43.** $6t^5$ **45.** $5^{\frac{1}{3}}y^{\frac{1}{3}}$ **47.** $r^{\frac{1}{3}}p$ **Written 1.** $\frac{1}{64}$ **3.** $\frac{1}{125}$ **5.** $\frac{27}{8}$ **7.** 243 **9.** $\frac{1}{27}$ **11.** $-\frac{1}{2}$ **13.** 1 **15.** $\frac{1}{32}$ **17.** $\frac{1}{3}$ **19.** $\frac{2}{3}$ **21.** $\frac{5}{2}$ **23.** 8 **25.** $\frac{1}{625}$ **27.** $\frac{1}{32}$ **29.** 2 **31.** 36 **33.** 1 **35.** $\frac{9}{4}$ **37.** 20 **39.** 9 **41.** $\frac{10}{9}$ **43.** $\frac{28}{9}$ **45.** −4 **47.** 28 **49.** $5y^3$ **51.** $\frac{2}{xy}$ **53.** $\frac{1}{4x^6y^2}$ **55.** $\frac{4}{x}$ **57.** $\frac{1}{x^3}$ **59.** $\frac{9y^2}{4x^2}$ **61.** $\frac{xy^3}{3}$ **63.** $\frac{3b^{\frac{1}{3}}}{a}$ **65.** 2 **67.** $\frac{2188}{19683}$ **69.** 1,000,000,100

Pages 175–176 Lesson 5.2

Exploratory 1. no **3.** yes **5.** They are reflections about the y-axis of each other. **7.** quadrants I and II **9.** yes, for each y value there is a unique x value **11.** reflect about the x-axis **Written 1.** $\frac{1}{16}$ **3.** 8 **5.** $\frac{1}{32}$

Pages 178–179 Lesson 5.3

Exploratory 1. $m = 2$ **3.** $m = \frac{-1}{3}$ **5.** 7^2 **7.** 3^3 **9.** 3^4 **11.** 10^{-2} **13.** 2^{-3} **Written 1.** $x = 8$ **3.** $x = 4$ **5.** $t = \frac{1}{16}$ **7.** $x = 8$ **9.** $x = \frac{1}{16}$ **11.** $x = 16$ **13.** $x = 1$ **15.** $x = \frac{1}{2}$ **17.** $x = 1$ **19.** $x \in \{-5, 3\}$ **21.** $x = 1$ **23.** $x = -1$ **25.** $x = 3$ **27.** $x \in \left\{1, \frac{1}{2}\right\}$ **29.** $x = \frac{8}{3}$ **31.** $x = \frac{12}{5}$ **33.** $x = \frac{2}{11}$ **35.** $x = \frac{-1}{3}$ **Mixed Review 1.** c **2.** d **3.** a **4.** d **5.** c **6.** c **7.** a **8.** d **9.** a **10.** d **11.** c **12.** c **13.** c **14.** b **15.** c

Pages 182–184 Lesson 5.4

Exploratory 1. $3 = \log_2 8$ **3.** $3 = \log_3 27$ **5.** $\frac{-1}{2} = \log_4 \frac{1}{2}$ **7.** $0 = \log_{17} 1$ **9.** $2 = \log_{\sqrt{5}} 5$ **11.** $\frac{3}{2} = \log_9 27$ **13.** $9 = 3^2$ **15.** $9 = 27^{\frac{2}{3}}$ **17.** $16 = \left(\frac{1}{2}\right)^{-4}$ **19.** $\frac{1}{9} = 3^{-2}$ **21.** $\frac{1}{10} = 10^{-1}$ **23.** $\frac{1}{25} = 5^{-2}$ **25.** I and II **27.** I and IV **29.** domain: all reals; range: all positive reals **31.** domain: all positive reals; range: all reals **33.** yes, by the definition of $\log_{11} x$ **Written 1.** 3 **3.** 0 **5.** 2 **7.** −2 **9.** $\frac{1}{5}$ **11.** −5 **13.** −3 **15.** $\frac{4}{3}$ **17.** true **19.** true **21.** true **23.** true **25.** $y = \log_3 x$ **27.** $y = \log_{10} x$ **29.** $y = -\log_8 x$ **31.** $y = \log_{\frac{3}{2}} x$ **33.** $x = 36$ **35.** $x = 1$ **37.** $x = \frac{1}{27}^{2}$ **39.** $y = 2$ **41.** $b = 6$ **43.** $b = 5$ **45.** $b = 18$ **47.** $b = \frac{1}{6}$ **49.** $b = 3$ **51.** $y = -3$ **53.** $x = 8$ **55.** $b = 7$ **57.** $x = 5$ **59.** $x = -1$ **61.** sometimes **63.** always **65.** $y = 4$ **67.** $x \in \{-8, 8\}$ **69.** $x \in \{-3, 2\}$ **71.** $x \in \{-2, 8\}$

Pages 187–188 Lesson 5.5

Exploratory 1. $\log_5 x + \log_5 y$ **3.** $2 \log_3 x$ **5.** $\log_5 a + \log_5 b$ **7.** $\frac{1}{3} \log_5 x$ **9.** $2 \log_b r - \log_b s$ **11.** $\frac{1}{4} \log_{10} x + \frac{1}{4} \log_{10} y$ **13.** $\log_4 ab$ **15.** $\log_4 \frac{x}{y}$ **17.** $\log_5 8x^3$ **19.** $\log_3 rst$ **21.** $\log_2 \frac{r^3}{s^2}$ **23.** $\log_{10} \sqrt{\frac{x}{y}}$ **Written 1.** 3.262 **3.** 2.893 **5.** 0.40775 **7.** $x = 21$ **9.** $x = 3$ **11.** $x = 8$ **13.** $r = 9$ **15.** $x = 63$ **17.** $m = 8$ **19.** $m = 5$ **21.** $x = 6$ **23.** $x = 2$ **25.** $x = 1$ **27.** $t = 5$ **29.** $x = 5$ **31.** $x = 4$ **33.** false **35.** false **37.** false **39.** true

Page 191 Lesson 5.6

Exploratory 1. 2 **3.** 0 **5.** −3 **7.** 2, 0.3819 **9.** −2, 0.8152 **11.** 3, 0.4193 **13.** 1 **15.** −1 **17.** 2 **19.** −5 **21.** 2.81 **Written 1.** 0.6794 **3.** 1.7300 **5.** 2.1673 **7.** 0.5465 − 2 **9.** 0.3201 **11.** 4.7973 **13.** 2.570 **15.** 7.130 **17.** 9,720,000 **19.** 0.090 **21.** 0.006 **23.** 3.380 **25.** 2.16 **27.** 8110 **29.** 0.674 **31.** 706 **33.** 42.35 **35.** 30,000 **37.** 0.03 **39.** 2.4132 **41.** 1.8264 **43.** 1.7416 **45.** 2 − 10.4496

Page 196 Lesson 5.7

Exploratory 1. 85.4 **3.** 1.783×10^{12} **5.** 15.748 **7.** 0.8662 **9.** 0.4276 **11.** 1803 **13.** $n = 49$ **15.** $n = 64$ **17.** $n = 35$ **Written 1.** 358.5 **3.** 0.05185 **5.** 1537 **7.** 2.008 **9.** 0.447 **11.** 5.609×10^{11} **13.** 59.25 **15.** 1.865 **17.** 0.7221 **19.** 0.6268 **21.** 2.7813 **23.** 0.8828 − 2 **25.** 3.6198 **27.** 0.2964 − 3 **29.** 3.553 **31.** 3757 **33.** 0.02164

35. 0.04374 **37.** 0.003904 **39.** 0.01841 **41.** 12.22 **43.** 1236

Pages 198–199 Lesson 5.8

Exploratory **1.** $x \log 2 = \log 3$ **3.** $x \log 5 = \log 61$ **5.** $3x = \log 191$ **7.** $5 \log x = \log 28$ **9.** $\frac{\log 8}{\log 3}$ **11.** $\frac{\log 3}{\log 5}$ **13.** $\frac{\log 60}{\log 20}$ **15.** $\frac{\frac{1}{2}\log 2}{\log \pi}$ **Written** **1.** 5.7286 **3.** 1.6131 **5.** 2.3325 **7.** 3.9835 **9.** 2.8444 **11.** 2.7736 **13.** 0.5674 **15.** 1.1674 **17.** 6.9004 **19.** 91.24 **21.** 1.0889 **23.** 1.4127 **25.** 1.277 **27.** 1.338 **29.** 2.387 **31.** 1.974 **33.** $x = -2$ **35.** $x \in \left\{1, \frac{-1}{2}\right\}$ **37.** $x = \pm 4$ **39.** 10 years 4 months **41.** 6.3% **Mixed Review** **5.** $x \in \{2, 7\}$ **6.** $\frac{\sqrt{3}}{3}$ **7.** -0.9 **8.** 2.1309 **9.** $x \in \{-1, 2\}$ **10.** 4 **11.** 2 **12.** no solution **13.** -2 **14.** $x \geq 0, \sqrt{x-4}$ **15.** 12 inches

Pages 202–203 Lesson 5.9

Exploratory **1.** \$125.97 **3.** \$126.82 **Written** **1.** \$3170.60 **3.** \$5520.10 **5.** \$2979.45 **7.** 15 years at 10% **9.** 6177 **11.** 11,060 **13.** \$127.14 **15.** 20 years **17.** 200 **19.** 5 days **21.** ~50% **23.** 930, 1395 **25.** 1,116

Page 205 Problem Solving Application

1. 20,000; 24 hours **3.** \$740.12 **5.** 6s, 3p; 7s, 1p; 7s, 2p; 8s; 8s, 1p **7.** 531,441 **9.** 6 hours

Pages 206–207 Chapter Review

1. 1 **3.** $\frac{1}{125}$ **5.** $\frac{1}{27}$ **7.** 4 **9.** $\frac{9}{4}$ **11.** $\frac{1}{4}$ **13.** 9 **15.** $\frac{10}{9}$ **17.** $6x^2$ **23.** $\frac{1}{9}$ **25.** $x = 4$ **27.** $x = -1$ **29.** $\frac{1}{2}$ **31.** $\frac{1}{4}$ **35.** $y = 10^x$ **37.** $x = 25$ **39.** $x = 3$ **41.** $x = 5$ **43.** $x = 7$ **45.** 3.6425 **47.** 35.498 **49.** 5.780×10^{-4} **51.** 3.781×10^{13} **53.** 2.6658 **55.** 0.07063 **57.** 1.7427 **59.** 2.8074 **61.** 0.7925 **63.** \$3987.16

CHAPTER 6 CIRCLES

Page 212 Lesson 6.1

Exploratory **5.** Any two of the following: $\overline{AE}$, $\overline{AB}$, $\overline{BD}$, $\overline{CD}$ **7.** $\overline{CD}$ and $\overline{AB}$ **9.** $\overleftrightarrow{AE}$ **13.** $\angle AOC$ **15.** no **17.** yes **19.** 73 **21.** no **Written** **1.** 80 **3.** 110 **5.** 70 **7.** 110 **9.** 180 **11.** 100 **13.** 330 **15.** 250

Pages 215–216 Lesson 6.2

Exploratory **1.** 10π **3.** π **5.** $6\sqrt{5}\pi$ **7.** 0.12π **9.** 3.016π **11.** π^2 **13.** 1.57, $\frac{11}{7}$ **15.** 26.64, $\frac{132\sqrt{2}}{7}$ **Written** **1.** 36π **3.** $\frac{\pi}{4}$ **5.** 18π **7.** 5 **9.** $\sqrt{10}$ **11.** $\sqrt{\pi}$ **13.** Length of the arc is $\frac{c}{360} \cdot 2\pi r$ **15.** 6π; 2π **17.** $\frac{\pi}{12}$; $\frac{\pi}{6}$ **19.** 20π; 4π **21.** $\frac{125}{12}\pi$; $\frac{25}{6}\pi$ **23.** $\frac{k}{2250}\pi$; $\frac{k}{450}\pi$ **27.** $8304 + 324\pi$; 9321.36 **29.** $144 - 36\pi$ **31.** $288 - 72\pi$ **33.** $50\pi - 100$ **35.** $169\pi - 240$ **37.** $2\pi + 32$; 38.3 **39.** $6\pi + 12$; 30.8

Pages 219–221 Lesson 6.3

Exploratory **1.** $\overleftrightarrow{AP}$, $\overleftrightarrow{BP}$ **3.** A, B **5.** sometimes **7.** never **9.** sometimes **11.** sometimes **13.** sometimes **Written** **1.** 90 **3.** 4 **5.** $3\sqrt{5}$ **7.** $\sqrt{157}$ **9.** $8\sqrt{3}$ **11.** 4, 10 **13.** $2\sqrt{2}$, $2\sqrt{23}$ **15.** 12, 15 **17.** $\frac{7}{2}$, 7 **19.** 12, $\sqrt{145}$ **21.** 9 **23.** $\overline{OK} \parallel \overline{CL}$ **25.** 41 **27.** 1.5 **29.** 48 **31.** $\sqrt{29}$

Pages 224–225 Lesson 6.4

Exploratory **1.** 5 **3.** 13 **5.** 8 **7.** 36 **9.** $8\sqrt{3}$ **11.** 16 **13.** 13 **Written** **1.** 6 **3.** 22 **5.** 41 **7.** They are converses of each other.

Pages 228–229 Lesson 6.5

Exploratory **1.** The vertex of a central angle is the center of the circle. The vertex of an inscribed angle is a point on the circle. **3.** $m\angle G = \frac{1}{2}m\angle H$ **5.** 90 **7.** 74 **9.** 82 **11.** 18 **13.** 37 **15.** 144 **17.** 63 **19.** 135 **21.** right **23.** 4 **25.** 15 **Written** **1.** 3 **3.** 15 **5.** $2\sqrt{2}$ **7.** $\frac{1}{2}$ or 2 **9.** 4 **11.** 12 **13.** 5 **15.** 3 **17.** right, $OC = AO = OB$ **Mixed Review** **1.** $(x - 5, y + 2)$ **2.** $y = 1$ **3.** $\frac{3}{8}$, $-\frac{3}{2}$ **4.** \$1089.97 **5.** $\log_b N = 3 - 2 \log_b a$ **6.** $\log_2 x$ **7.** 180° **8.** 17 cm

Pages 232–233 Lesson 6.6

Exploratory **1.** 40 **3.** 105 **5.** 145 **7.** 55 **9.** 50 **Written** **1.** 60 **3.** 75 **5.** 61 **7.** 110 **9.** 60 **11.** 49 **13.** 100 **15.** 100 **17.** 40 **19.** 40 **21.** 40 **23.** 16 **25.** 122 **27.** 6 **29.** 90

Pages 235–236 Lesson 6.7

Exploratory **1.** secants: $\overleftrightarrow{AB}$, $\overleftrightarrow{ED}$; tangents: none, $(AC)(BC) = (EC)(DC)$ **3.** 4 **5.** 4 **7.** 7 **9.** 9 **11.** 12 **Written** **1.** 14 **3.** 25 **5.** 4 **7.** 9 **9.** 16, 8 **11.** 14, 24 **13.** 9, 32 **15.** 2 **17.** $5\sqrt{7}$ **19.** 4 **21.** 24 **23.** 31 **25.** $6\sqrt{10}$

Pages 241–242 Lesson 6.8

Written 19. For any two points, there exists one and only one line containing them. **21.** Since $\angle OAB$ is a right angle, $\overleftrightarrow{OA} \perp \overleftrightarrow{BA}$ and if a line is perpendicular to a radius at a point on the circle, then the line is tangent to the circle at that point. **31.** We do not know if these segments will meet at a single point of intersection. **33.** So that TK is the distance from T to $\overline{AB}$, TL is the distance from T to $\overline{AC}$, and TM is the distance from T to $\overline{BC}$.

Page 244 Problem Solving Application

1. 250 m **3.** square DGFG: 576 cm^2; square KLMN: 288 cm^2 **5.** 28 **7.** maximum: between 55 and 56 miles; minimum: between 14 and 15 miles

Pages 246–247 Chapter Review

5. $\frac{275}{9}\pi$; $\frac{55}{9}\pi$ **7.** $\frac{5}{2}\pi$; $\frac{5}{6}\pi$ **9.** $\frac{20}{3}\pi$; $\frac{10}{3}\pi$ **11.** $\overleftrightarrow{ST}$, $\overleftrightarrow{WT}$ or $\overleftrightarrow{WS}$; $\overline{ST}$, $\overline{WT}$ or $\overline{WS}$; W **13.** $2\sqrt{3}$ **15.** $2\sqrt{34}$ **19.** 110 **21.** 100 **23.** 55 **25.** 7 **27.** 20 or 10 **29.** 84 **31.** 30 **33.** 45

CHAPTER 7 AN INTRODUCTION TO CIRCULAR FUNCTIONS

Page 252 Lesson 7.1

Written 1. I **3.** I **5.** IV **7.** II **9.** III **11.** I **13.** IV **15.** II **17.** −260°, 460° **19.** −150° **21.** 20° **23.** 210° **25.** 310° **27.** 240° **29.** 300°

Pages 255–256 Lesson 7.2

Written 1. $\frac{\pi}{2}$ **3.** $\frac{\pi}{3}$ **5.** $-\frac{\pi}{2}$ **7.** $\frac{5\pi}{6}$ **9.** $\frac{\pi}{4}$ **11.** $-\frac{3\pi}{4}$ **13.** $\frac{17\pi}{4}$ **15.** $\frac{7\pi}{3}$ **17.** 180° **19.** 45° **21.** −60° **23.** 270° **25.** 330° **27.** −315° **29.** −150° **31.** 373° **33.** 0° **35.** −90° **37.** 8 cm **39.** 5 cm **41.** 4 cm **43.** 2 radians **45.** $\frac{\pi}{2}$ **47.** $\frac{3\pi}{2}$ **49.** 83π **51.** $\frac{5\pi}{3}$ **55.** 57.3°

Mixed Review 1. 8 **2.** 26 **3.** 25 **4.** 24 **5.** 40 **6.** 130 **7.** 130 **8.** 75

Pages 259–260 Lesson 7.3

Exploratory 1. 1 **3.** −1 **5.** 0 **7.** 0 **9.** 0 **11.** −1 **13.** −1 **15.** 1 **17.** positive **19.** positive **21.** negative **23.** positive **25.** positive **27.** positive **Written 1.** I **3.** II **5.** $\frac{3}{5}$ **7.** $\frac{4}{5}$ **9.** 0.8 **11.** $-\frac{12}{13}$ **13.** −1 **15.** 1 **17.** 0 **19.** 0 **21.** −5 **23.** true **25.** true **27.** false **29.** false

Page 264 Lesson 7.4

Exploratory 1. $\frac{1}{2}$ **3.** $\frac{\sqrt{3}}{2}$ **5.** $\frac{1}{2}$ **7.** $\sqrt{2}$ **9.** 1

Written 1. $-\frac{1}{2}$ **3.** $\frac{\sqrt{2}}{2}$ **5.** $-\frac{\sqrt{3}}{2}$ **7.** $\frac{\sqrt{2}}{2}$ **9.** $-\frac{\sqrt{2}}{2}$ **11.** $-\frac{1}{2}$ **13.** $\frac{\sqrt{3}}{2}$ **15.** $\frac{1}{2}$ **17.** $\frac{1}{2}$ **19.** $-\frac{1}{2}$ **21.** $-\frac{\sqrt{3}}{2}$ **23.** $-\frac{\sqrt{2}}{2}$ **25.** $-\frac{1}{2}$ **27.** $-\frac{\sqrt{3}}{2}$ **29.** sin 50° **31.** −sin 50° **33.** cos 10° **35.** sin 24° **37.** −sin 76° **39.** −sin 35° **41.** sin 40° **43.** −sin 40° **45.** −sin 20° **47.** cos 70° **49.** $-\cos\frac{\pi}{7}$ **51.** $\sin\frac{\pi}{4}$ **53.** 1 **55.** $-\sqrt{3}$ **57.** $-\frac{3\sqrt{2}}{2}$ **59.** $-\frac{3}{4}$ **61.** $-2\sqrt{3}$

Pages 268–269 Lesson 7.5

Exploratory 1. 1; 2π; 1 **3.** 2; 2π; 1 **5.** 1; π; 2 **7.** 1; $\frac{\pi}{2}$; 4 **9.** 3; π; 2 **11.** $\frac{1}{2}$; π; 2 **13.** 5; 6π; $\frac{1}{3}$ **15.** 3; 2π; 1 **Written 1.** $\{y: -1 \le y \le 1\}$ **3.** $\{y: -3 \le y \le 3\}$ **5.** $\{y: -5 \le y \le 5\}$ **7.** $\{y: -3 \le y \le 3\}$ **9.** $\frac{\pi}{2}$; $\frac{3\pi}{2}$ **11.** $\frac{\pi}{2}$; $\frac{3\pi}{2}$ **13.** $\frac{\pi}{4}$; $\frac{5\pi}{4}$; $\frac{3\pi}{4}$, $\frac{7\pi}{4}$ **15.** π; 0, 2π **17.** $y = 2 \sin x$ **19.** $y = -\cos x$ **29.** point symmetry: none; line symmetry: $x = 0$

Page 271 Lesson 7.6

Exploratory 1. 1; π; $y = \cos 2x$ **3.** $\frac{1}{2}$; 4π; $y = \frac{1}{2}\cos\frac{1}{2}x$ **Written 13.** 2 **15.** 6 **17.** 5

Page 275 Lesson 7.7

Exploratory 1. I **3.** III **5.** III **7.** π **Written 1.** $\frac{\sqrt{3}}{3}$ **3.** $\sqrt{3}$ **5.** −1 **7.** $-\frac{\sqrt{3}}{3}$ **9.** −1 **11.** 1 **13.** 0 **15.** $-\frac{\sqrt{3}}{3}$ **17.** 0 **19.** $-\sqrt{3}$ **25.** 5 **27.** −tan 38° **29.** tan 20° **31.** tan 68° **33.** $-\tan\frac{\pi}{12}$ **35.** $\frac{3}{4}$ **37.** $-\frac{3}{4}$ **39.** $\frac{3}{4}$ **41.** −1 **43.** −2 **45.** $\frac{3\sqrt{2}}{2}$ **47.** 2 **49.** $-\frac{\sqrt{2}}{2}$

Pages 277–279 Lesson 7.8

Exploratory 1. $\frac{1}{\cos\theta}$ **3.** $\frac{\cos\theta}{\sin\theta}$ **5.** $\frac{\sin\theta}{\frac{1}{\sin\theta}}$; $\sin^2\theta$ **7.** $\frac{\frac{\cos\theta}{\sin\theta}}{\frac{1}{\sin\theta}}$; $\cos\theta$ **9.** II **11.** II **13.** II **15.** $-\frac{3\pi}{2}$, $-\frac{\pi}{2}$, $\frac{\pi}{2}$, $\frac{3\pi}{2}$ **17.** $-\frac{3\pi}{2}$, $-\frac{\pi}{2}$, $\frac{\pi}{2}$, $\frac{3\pi}{2}$ **19.** 2π **21.** π

23. π **25.** 4π **Written** **3.** $-\frac{2\sqrt{3}}{3}$ **5.** $-\frac{2\sqrt{3}}{3}$ **7.** $\sqrt{3}$ **9.** $-\frac{\sqrt{3}}{3}$ **11.** -1 **13.** 2 **15.** $\frac{7}{3}$ **17.** 0 **19.** 0 **21.** 2 **23.** $\sin\theta = \frac{\sqrt{3}}{2}$, $\tan\theta = \sqrt{3}$, $\cot\theta = \frac{\sqrt{3}}{3}$, $\sec\theta = 2$, $\csc\theta = \frac{2\sqrt{3}}{3}$ **25.** $\sin\theta = -\frac{2\sqrt{14}}{15}$, $\cos\theta = -\frac{13}{15}$, $\tan\theta = \frac{2\sqrt{14}}{13}$, $\cot\theta = \frac{13\sqrt{14}}{28}$, $\csc\theta = -\frac{15\sqrt{14}}{28}$ **27.** $\sin\theta = -\frac{2\sqrt{2}}{3}$, $\cos\theta = \frac{1}{3}$, $\tan\theta = -2\sqrt{2}$, $\cot\theta = -\frac{\sqrt{2}}{4}$, $\csc\theta = -\frac{3\sqrt{2}}{4}$ **29.** $\left\{\text{reals except odd multiples of } \frac{\pi}{2}\right\}$; $\{y: |y| \geq 1\}$ **31.** {reals except multiples of π}; $\mathscr{R}$ **33.** {reals except multiples of π}; $\left\{y: |y| \geq \frac{1}{2}\right\}$ **35.** $\left\{\text{reals except multiples of } \frac{\pi}{2}\right\}$; $\left\{y: |y| \geq \frac{3}{4}\right\}$ **37.** increasing in quadrants I and IV, decreasing in quadrants II and III **39.** increasing in all quadrants **41.** increasing in quadrants I and II, decreasing in quadrants III and IV **43.** point: (0, 0); line: none **45.** point: none; line: $x = 0$ **Mixed Review** **1.** {1} **2.** $\log 6 = A + B$ **3.** $\frac{1}{2}$ **4.** 70 **5.** $5 - 12i$ **6.** 8 **7.** 11 **8.** (7, −1) **9.** line, rotational **10.** {(2, −1), (5, 0), (−2, 3)} **11.** $f^{-1}(x) = \frac{x + 7}{2}$ **12.** I and II **13.** two conjugate complex roots **14.** 4 **15.** $\frac{ab}{b - a}$ **16.** $-\frac{1}{b + a}$ **17.** 6 **18.** 5 **19.** $\frac{2 - i}{5}$ **20.** $\frac{1}{4}$

Pages 282–283 Lesson 7.9
Exploratory **1.** 0.2419 **3.** 0.8225 **5.** 0.1794 **7.** 0.5735 **9.** 0.6248 **11.** 3.204 **13.** 8°50′ **15.** 63° **17.** 57°20′ **19.** 45°10′ **21.** 87°20′ **Written** **1.** sin 80°, 0.9848 **3.** −tan 50°, −1.192 **5.** −tan 40°, −0.8391 **7.** −sin 8°, −0.1392 **9.** −tan 20°, −0.3640 **11.** −cos 56°, −0.5592 **13.** −cos 48°, −0.6691 **15.** −tan 30°, −0.5774 **17.** sin 26°50′, 0.4514 **19.** −cos 36°10′, −0.08073 **21.** 0.5336 **23.** 0.4928 **25.** 0.9661 **27.** 0.3340 **29.** 0.4418 **31.** 7.3010 **33.** 0.8721 **35.** 0.4597 **37.** 59°7′ **39.** 17°28′ **41.** 70°32′ **43.** 48°35′ **45.** 44°25′ **47.** 64°, 116° **49.** 60°, 120° **51.** 76°, 284° **53.** Graph of $y = \sin x$ has point symmetry with respect to (180°, 0). **55.** Graph of $y = \cos x$ has line symmetry about line $x = \pi$. **57.** Graph of $y = \tan x$ has point symmetry with respect to $\left(\frac{\pi}{2}, 0\right)$. **59.** Graph of $y = \cos x$ has line symmetry about the y-axis and has period 360°. **61.** Graph of $y = \tan x$ has point symmetry with respect to the origin and has period π. **63.** Graph of $y = \sin x$ has line symmetry about the lines $y = \frac{3\pi}{2}$ and $y = \frac{\pi}{2}$. $\pi + x = \left(\frac{3\pi}{2} + x\right) - \frac{\pi}{2}$. $2\pi - x = \left(\frac{3\pi}{2} - x\right) + \frac{\pi}{2}$.

Page 285 Problem Solving Application
1. 3 **3.** 3

Pages 286–287 Chapter Review
1. 60° **3.** 35° **5.** $\frac{5\pi}{9}$ **7.** $\frac{5\pi}{2}$ **9.** 72° **11.** −270° **13.** 6 units **15.** $-\frac{5}{13}$ **17.** 0 **19.** 1 **21.** $\frac{\sqrt{3}}{2}$ **23.** 1 **25.** $\frac{3}{2}$ **27.** 3; π; $\{y: -3 \leq y \leq 3\}$ **29.** $\left\{\frac{1}{2}; \frac{2\pi}{3}; y: -\frac{1}{2} \leq y \leq \frac{1}{2}\right\}$ **31.** $-\frac{\sqrt{2}}{4}$ **33.** −1 **35.** $\sqrt{3}$ **37.** −2 **39.** −csc 20° **41.** −3 **43.** $-2\sqrt{3}$ **45.** 0.0552 **47.** 54°42′ **49.** 16°33′ **51.** 45°, 225°

CHAPTER 8 APPLICATIONS OF CIRCULAR FUNCTIONS

Pages 292–293 Lesson 8.1
Exploratory **1.** (0, 3) **3.** $(3\sqrt{2}, 3\sqrt{2})$ **5.** $(4\sqrt{2}, -4\sqrt{2})$ **7.** (0, 4) **9.** $(5\sqrt{3}, 5)$ **11.** $\left(-\frac{7}{2}, -\frac{7\sqrt{3}}{2}\right)$ **13.** $(-\sqrt{3}, 1)$ **15.** $(-3\sqrt{3}, -3)$ **17.** (1, −1) **Written** **1.** (−4, 0) **3.** (0, −5) **5.** $(-4\sqrt{3}, -4)$ **7.** (5 cos 90°, 5 sin 90°) **9.** (7 cos 180°, 7 sin 180°) **11.** (4 cos 60°, 4 sin 60°) **13.** (4 cos 120°, 4 sin 120°) **15.** (6 cos 45°, 6 sin 45°) **17.** $(4\sqrt{2}\cos 135°, 4\sqrt{2}\sin 45°)$ **19.** (8 cos 210°, 8 sin 210°) **21.** (10 cos 150°, 10 sin 150°) **23.** $\frac{3}{5}, \frac{4}{5}, \frac{3}{4}$ **25.** $-\frac{4}{5}, \frac{3}{5}, -\frac{4}{3}$ **27.** $\frac{5}{13}, -\frac{12}{15}, -\frac{5}{12}$ **29.** $-\frac{\sqrt{2}}{2}, -\frac{\sqrt{2}}{2}, 1$ **31.** $\frac{\sqrt{15}}{4}, -\frac{1}{4}, -\sqrt{15}$ **33.** $-\frac{\sqrt{3}}{2}, \frac{1}{2}, -\sqrt{3}$ **35.** $\sin\theta = \frac{5}{13}$, $\cos\theta = -\frac{12}{13}$,

$\cot\theta = -\frac{12}{5}$, $\sec\theta = -\frac{13}{12}$, $\csc\theta = \frac{13}{5}$
37. $\sin\theta = -\frac{4}{5}$, $\cos\theta = \frac{3}{5}$, $\tan\theta = -\frac{4}{3}$, $\sec\theta = \frac{5}{3}$, $\csc\theta = -\frac{5}{4}$ **39.** $\cos\theta = \frac{2\sqrt{2}}{3}$, $\tan\theta = -\frac{\sqrt{2}}{4}$, $\cot\theta = -2\sqrt{2}$, $\sec\theta = \frac{3\sqrt{2}}{4}$, $\csc\theta = -3$ **41.** $\sin\theta = -\frac{\sqrt{7}}{4}$, $\cos\theta = -\frac{3}{4}$, $\tan\theta = \frac{\sqrt{7}}{3}$, $\cot\theta = \frac{3\sqrt{7}}{7}$, $\csc\theta = -\frac{4\sqrt{7}}{7}$

Pages 296–297 Lesson 8.2

Exploratory **1.** 6 **3.** 7 **5.** $\sqrt{58}$, 7.6 **7.** $4\sqrt{6}$, 9.8 **9.** $\frac{1}{8}$ **11.** $-\frac{1}{4}$ **13.** $-\frac{11}{28}$ **Written** **1.** 4.0 **3.** 4.3 **5.** 1.4 **7.** 10.6 **9.** 6.3 **11.** 60° **13.** 47° **15.** 60° **17.** 36° **19.** 12 cm **21.** 11 cm **23.** 37 cm **25.** 24 cm **27.** 7.8 units, 4.6 units **29.** 8.7, 5.3 **31.** 32.9, 10.5 **33.** 9.4, 4.6 **35.** 73° **37.** 60 miles **39.** 89 miles **41.** 90 miles **43.** 111°, 69° **45.** 32°, 148° **47.** 124°, 56° **49.** 110 yards **53.** 87°, 53°, 40° **55.** 78°, 61°, 41° **57.** 97°, 53°, 30°

Page 299 Lesson 8.3

Exploratory **1.** $A = \frac{1}{2}(10)(17)\sin 46°$ **3.** $A = \frac{1}{2}(15)(30)\sin 90°$ **5.** $A = \frac{1}{2}(18)(26)\sin 142°$ **Written** **1.** 9 **3.** 110 **5.** 30 **7.** 72 **9.** 15 **11.** $25\sqrt{3}$ **13.** 55.2 **15.** 36.4 **17.** 218.2 **19.** 378.5 **21.** $\frac{1}{3}$ **23.** $\sqrt{3}$ units **25.** 30°

Pages 301–302 Lesson 8.4

Exploratory **1.** $\sin B$ **3.** $\frac{a}{c}$ **5.** $a \sin C$ **7.** 8 **9.** 30 **11.** 30 **Written** **1.** 6 **3.** 8 **5.** 10 **7.** $4\sqrt{6}$ **9.** 9 **11.** $18\sqrt{2}$ **13.** 22.3 **15.** 13° **17.** 18.1 cm **19.** 21.2 cm **21.** 93.2 cm **23.** 506.73 m **25.** 134.32 m **27.** 15.5 dm **29.** 15.4 dm

Pages 304–305 Lesson 8.5

Exploratory **1.** 0 **3.** 1 **5.** 2 **7.** 0 **9.** 1 **11.** 0 **13.** 0 **Written** **1.** right **3.** acute **5.** 2; 35° or 145° **7.** 0 **9.** 1; 70° **11.** 1; 11° **13.** 2; 77° or 103° **15.** no **Mixed Review** **1.** c **2.** c **3.** d **4.** a **5.** c **6.** c **7.** c

Pages 309–310 Lesson 8.6

Exploratory **5.** So you can be sure to get the angles and sides in proper correspondence, and so you can determine if the triangle is obtuse, right, or acute. **7.** $c = 11.5$ or 3.8 **Written** **1.** $a = 12.5$, $m\angle B = 44$, $m\angle C = 76$ **3.** $c = 12.7$, $m\angle A = 42$, $m\angle B = 93$ **5.** $b = 2.9$, $c = 8.0$, $m\angle A = 90$ **7.** $b = 7.0$, $m\angle A = 30$, $m\angle C = 50$ **9.** $m\angle A = 72$, $m\angle B = 58$, $m\angle C = 50$ **11.** No triangle. $\angle A$ is the largest angle, a is not the largest side. **13.** $a = 2.6$, $b = 2.8$, $m\angle C = 95$ **15.** $a = 7.2$, $b = 10.2$, $m\angle C = 105$ **17.** 1; $a = 10.1$, $m\angle B = 53$, $m\angle C = 85$ **19.** 1; $c = 9\sqrt{2}$, $m\angle B = 90$, $m\angle C = 45$ **21.** 1; $b = 7.5$, $m\angle A = 36$, $m\angle B = 88$ **23.** 0; $\angle B$ is the largest angle, b is not the largest side. **25.** 2; $b = 47.0$, $m\angle A = 58$, $m\angle B = 97$ or $b = 25.8$, $m\angle A = 122$, $m\angle B = 33$ **27.** 36° **29.** 36° **31.** 98°, 82°

Pages 313–315 Lesson 8.7

Exploratory **1.** 7.9 **3.** 3.2 **5.** 30°20′ **7.** 5.6 **9.** 62°30′ **11.** 1.8 **13.** $2b$, 60°, 30° **15.** 54° **17.** 72° **19.** 2 ft **21.** 5 ft **23.** 4 ft **Written** **1.** 41.3 ft **3.** 92.3 ft **5.** 4705 ft **7.** 2807.5 ft **9.** 1103.2 ft **11.** 173.2 ft **13.** 351.3 ft **15.** 3° **17.** 8° **19.** 205.1 m **21.** 676.2 m **23.** 1572.3 m **25.** 909.1 m **27.** 5.8 m **29.** 13.0 ft **31.** 188.9 ft **33.** 98.5 ft **35.** 580.9 ft **37.** 109.6 ft **39.** 129.5 cm

Pages 318–319 Lesson 8.8

Exploratory **1.** 250 mph, direction of plane **3.** 30° from lesser force, 315 lbs **5.** 29° from lower force, 350 lbs **Written** **1.** 197 kg **3.** 13 knots **5.** 49 lbs **7.** 10 lbs **9.** 45 kg **11.** 66.3 lbs **13.** 50.3 lbs **15.** 105°40′, 44°30′ **17.** 97°10′, 41°20′ **19.** 90°, 36°50′ **21.** 118°30′, 56°00′ **23.** 38°40′ **25.** 26°30′

Pages 322–323 Lesson 8.9

Exploratory **1.** 30.6 lbs, 25.7 lbs **3.** 36.4 lbs, 21 lbs **5.** 41.0, 112.8 **7.** 79.5 kg, 42.3 kg **Written** **1.** 115 lbs, 33 lbs **3.** 37 kg, 15 kg **5.** 29 kg, 14 kg **7.** 132.4 lbs **9.** 196.7 lbs **11.** 3359.1 lbs, 3066.8 lbs **13.** 5276.3 lbs, 3942.8 lbs

Page 325 Problem Solving Application

1. 22 miles **3.** 88 cm; $(225\pi - 535)$ cm^2 or 172 cm^2 **5.** 67 ft; 64 ft **7.** 173 km **9.** 24 m **11.** 41 m

Page 327 Chapter Review

1. 1, 0, undefined **3.** $-\frac{4}{5}, -\frac{3}{5}, \frac{4}{3}$ **5.** $-\frac{\sqrt{5}}{2}$ **7.** 14.5 cm **9.** 18 units^2 **11.** 36 units^2 **13.** 2 **15.** 0 **17.** $m\angle A = 36$, $m\angle B = 27$, $m\angle C = 117$ **19.** 56 m **21.** 44 m **23.** 125°

CHAPTER 9 TRIGONOMETRIC IDENTITIES AND EQUATIONS

Pages 332–333 Lesson 9.1

Exploratory **1.** A mathematical identity is an equation that is valid for any substitution of the variables, whenever the substitution makes sense. **3.** $\sin^2 x$ **5.** $\cot^2 x$ **7.** -1 **9.** If $x = 180°$, then $\sin 180° + \cos 180° = 0 + (-1) = -1 \neq 1$.
Written **1.** $\pm\frac{\sqrt{15}}{4}$ **3.** $\pm\frac{\sqrt{6}}{2}$ **5.** $\pm\sqrt{5.25}$
7. $-\cos^2 x$ **9.** $\cot\theta$ **11.** $2\sin^2\theta$ **13.** $\sec\theta$ **15.** 1 **17.** $\sin^2\theta$ **19.** $\cot^2\theta$ **21.** $\cos x$ **23.** $\csc\theta$ **25.** $\cos\theta$ **27.** $-\cos\theta$ **29.** $\cos\theta$ **31.** $\tan\theta$ **33.** $\cot^2 x$ **Mixed Review** **1.** 32 **2.** 44 **3.** 12 **4.** 78 **5.** 74 **6.** 56 **7.** 27 **8.** 12 **9.** 23 **10.** $11\frac{1}{3}$ **11.** 15 **12.** 17 **13.** 12.5 cm **14.** 454.3 miles **15.** 84°36′ **16.** 57.2 km

Pages 335–336 Lesson 9.2

Exploratory **1.** 30°, 150° **3.** 45°, 225° **5.** 210°, 330° **7.** 0°, 180° **9.** 225°, 315° **11.** 120°, 300° **Written** **1.** 60°, 240° **3.** 225°, 315° **5.** 90° **7.** 210°, 330° **9.** 135°, 315° **11.** no solution **13.** no solution **15.** 240°, 300° **17.** $\frac{4\pi}{3} + 2\pi n$ or $\frac{5\pi}{3} + 2\pi n$ **19.** $\frac{3\pi}{4} + \pi n$ **21.** $\frac{7\pi}{6} + 2\pi n$ or $\frac{11\pi}{6} + 2\pi n$ **23.** $\frac{3\pi}{4} + \pi n$ **25.** 104°, 256° **27.** 102°, 258° **29.** 27°, 207° **31.** 40°, 140° **33.** 21°, 339° **35.** 114°, 246° **37.** 0°, 120°, 180°, 240° **39.** 0°, 135°, 180°, 225° **41.** 0°, 60°, 180°, 300° **43.** 0°, 30°, 150°, 180° **45.** 90°, 270° **47.** 0°, 48°, 180°, 312°

Pages 338–339 Lesson 9.3

Exploratory **1.** 0°, 180° **3.** 90°, 270° **5.** no solution **7.** 45°, 135°, 225°, 315° **9.** 60°, 240° **11.** 63°, 135°, 243°, 315° **Written** **1.** 90° **3.** 45°, 210°, 225°, 330° **5.** 60°, 120° **7.** 90°, 270° **9.** 270° **11.** 180° **13.** 30°, 150°, 270° **15.** 19°, 161°, 270° **17.** no solution **19.** 27°, 108°, 207°, 288° **21.** 270° **23.** 11°, 63°, 191°, 243° **25.** 90°, 112°, 248°, 270° **27.** 14°, 166°, 194°. 346° **29.** 19°, 90°, 161° **31.** 48°, 312° **33.** 11°, 169° **35.** 47°, 133° **37.** 123°, 237° **39.** 90°, 270° **41.** 45°, 135°, 225°, 315° **43.** 0° **45.** 210°, 330° **47.** 63°, 135°, 243°, 315° **49.** 19°, 161°, 210°, 330° **51.** 26°, 154°, 230°, 310° **53.** 90°, 270° **55.** 45°, 90°, 135°, 225°, 270°, 315°

Pages 343–345 Lesson 9.4

Exploratory **5.** c **7.** b **9.** a **Written** **5.** c **7.** b **9.** a **Written** **1.** $\frac{\sqrt{6}+\sqrt{2}}{4}$ **3.** $\frac{\sqrt{6}-\sqrt{2}}{4}$ **5.** $\frac{-\sqrt{6}-\sqrt{2}}{4}$ **7.** $\frac{\sqrt{6}+\sqrt{2}}{4}$ **9.** $\frac{\sqrt{2}-\sqrt{6}}{4}$ **11.** $-\frac{\sqrt{3}}{2}$ **13.** $\frac{\sqrt{6}-\sqrt{2}}{4}$ **15.** $\frac{\sqrt{6}+\sqrt{2}}{4}$ **17.** $2+\sqrt{3}$ **19.** $2-\sqrt{3}$ **21.** $\frac{-\sqrt{6}-\sqrt{2}}{4}$ **23.** $2+\sqrt{3}$ **27.** $\tan A = \frac{\sin A}{\cos A} = \frac{\cos B}{\sin B} = \cot B$ **29.** 60° **31.** 9° **33.** 38° **35.** 0 **37.** 1 **39.** 1 **41.** -1 **53.** $-\sin\theta$ **55.** $\frac{\sqrt{2}}{2}(\cos\theta - \sin\theta)$ **57.** $\frac{\sqrt{6}+\sqrt{2}}{4}\cos\theta - \frac{\sqrt{6}+\sqrt{2}}{4}\sin\theta$ **59.** $\frac{63}{65}$ **61.** $\frac{33}{65}$ **63.** $-\frac{33}{65}$ **65.** $\frac{63}{16}$ **67.** $-\frac{56}{65}$ **69.** $-\frac{63}{65}$ **71.** $\frac{5}{13}$ **73.** $\frac{33}{56}$ **75.** $\frac{\sqrt{5}+4\sqrt{2}}{9}$ **77.** $\frac{-2\sqrt{10}-2}{9}$ **79.** $\frac{2\sqrt{2}}{3}$ **81.** $\frac{2}{3}(\sqrt{5}+\sqrt{2})$

Pages 348–349 Lesson 9.5

Exploratory **1.** $2\sin\alpha\cos\alpha$ **3.** $\pm\sqrt{\frac{1-\cos\Theta}{2}}$ **5.** $\pm\sqrt{\frac{1-\cos\Theta}{1+\cos\Theta}}$ **7.** $2\cos^2 A - 1$ **9.** $\cos 2(30°) = 1 - 2\sin^2 30° = 1 - 2\left(\frac{1}{2}\right)^2 = \frac{1}{2}$ **11.** $\tan 2(60°) = \frac{2\tan 60°}{1-\tan^2 60°} = \frac{2(\sqrt{3})}{1-(\sqrt{3})^2} = \frac{2\sqrt{3}}{1-3} = -\sqrt{3}$ **13.** $1 - 2t^2$ **15.** $\frac{\sqrt{6}}{3}$ **Written** **1.** $\frac{7}{25}$ **3.** $-\frac{7}{9}$ **5.** $\frac{1}{9}$ **7.** -0.28 **9.** $\frac{7}{25}$ **11.** $\frac{24}{25}$ **13.** $\frac{\sqrt{3}}{2}$ **15.** $-\frac{24}{25}$ **17.** $\frac{4\sqrt{5}}{9}$ **19.** $-\frac{3}{5}$ **21.** $\frac{8}{15}$ **23.** $-\frac{8}{15}$ **25.** $-\frac{24}{7}$ **27.** $\sqrt{3-2\sqrt{2}}$ **29.** $\frac{1}{2}\sqrt{2-\sqrt{3}}$ **31.** $-\frac{1}{2}\sqrt{2-\sqrt{2}}$ **33.** $-\sqrt{3-2\sqrt{2}}$ **35.** $\frac{2}{3}, \frac{\sqrt{5}}{3}, \frac{2\sqrt{5}}{5}$ **37.** $\frac{\sqrt{5}}{5}, \frac{2\sqrt{5}}{5}, \frac{1}{2}$ **39.** $\frac{3}{4}, -\frac{\sqrt{7}}{4}, -\frac{3\sqrt{7}}{7}$ **41.** $\frac{\sqrt{6}}{6}, -\frac{\sqrt{30}}{6}, -\frac{\sqrt{5}}{5}$ **43.** $\frac{\sqrt{21}}{6}, \frac{\sqrt{15}}{6}, \frac{\sqrt{35}}{5}$ **45.** $-\frac{1}{2}$

Pages 351–352 Lesson 9.6

Exploratory **1.** 0°, 120°, 180°, 240° **3.** 90°, 270° **5.** 90°, 225°, 270°, 315° **Written** **1.** 60°, 180°, 300° **3.** 180° **5.** 0°, 120°, 240° **7.** 0°, 60°, 180°, 300° **9.** 30°, 150°, 210°, 330° **11.** 0°, 180°, 210°,

330° **13.** 71°, 120°, 240°, 289° **15.** 74°, 138°, 254°, 318° **17.** 45°, 225° **19.** 30°, 60°, 210°, 240° **21.** 30°, 150°, 270° **23.** 120° **25.** 0°, 45°, 135°, 180°, 225°, 315° **27.** 135°, 315° **29.** 23°, 45°, 113°, 203°, 225°, 293° **31.** 59°, 239° **33.** 31°, 151°, 271° **35.** 20°, 110°, 143°, 200°, 290°, 323° **37.** 90°, 180° **39.** 105°, 165°, 285°, 345° **Mixed Review** **1.** 8 feet **2.** 10 × 12 feet **3.** 18 hours **4.** 8 hours 24 minutes **5.** $3062.65 **6.** $6880 **7.** 70 years **8.** 1.2628 **9.** 7 tears **10.** 158.6 lb **11.** 9 feet **12.** 146.2 lb **13.** 25°45′ **14.** 135.1 m **15.** 15.38 miles

Pages 356–357 Lesson 9.7

Exploratory **1.** They are inverses of each other. **3.** $\theta = \arccos \frac{1}{2}$ **5.** $\theta = \arcsin \frac{\sqrt{2}}{2}$ **7.** $\theta = \arctan(-3.4)$ **9.** $\cos \theta = \frac{1}{2}$ **11.** $\cos \theta = -1$ **13.** D:[−1, 1], R:$\left[-\frac{\pi}{2}, \frac{\pi}{2}\right]$, yes **15.** D:[−1, 1], R:[0, π], yes **17.** D:$\mathcal{R}$, R: $\left(-\frac{\pi}{2}, \frac{\pi}{2}\right)$, yes **19.** $\frac{\pi}{2} + 2\pi n$ **21.** $\frac{\pi}{2} + \pi n$ **23.** $\frac{\pi}{4} + 2\pi n, \frac{7\pi}{4} + 2\pi n$ **25.** $\frac{4\pi}{3} + 2\pi n, \frac{5\pi}{3} + 2\pi n$ **27.** $\frac{2\pi}{3} + \pi n$ **29.** $\frac{11\pi}{20} + \pi n$ **Written** **1.** 30°, $\frac{\pi}{6}$ **3.** 0°, 0 **5.** 60°, $\frac{\pi}{3}$ **7.** 45°, $\frac{\pi}{4}$ **9.** −45°, $-\frac{\pi}{4}$ **11.** 135°, $\frac{3\pi}{4}$ **13.** 150°, $\frac{5\pi}{6}$ **15.** 30°, $\frac{\pi}{6}$ **17.** 60°, $\frac{\pi}{3}$ **19.** −30°, $-\frac{\pi}{6}$ **21.** $\frac{\sqrt{2}}{2}$ **23.** 0 **25.** $\frac{\sqrt{3}}{2}$ **27.** $\frac{1}{2}$ **29.** $\frac{1}{2}$ **31.** $\frac{1}{2}$ **33.** $\frac{3}{4}$ **35.** $\frac{5}{13}$ **37.** $\frac{12}{5}$ **39.** $-\frac{4}{3}$ **41.** $\frac{3}{2}$ **43.** $\frac{2\sqrt{3}}{3}$

Page 359 Problem Solving Application

1. Path ABEFZ: 14 minutes **3.** Las Vegas–Denver–St. Louis–New York City–Boston: $1.57 **5.** Seattle–Portland–San Francisco–Las Vegas–Phoenix–Dallas–Miami: $2.43

Pages 360–361 Chapter Review

3. 60°, 120° **5.** 120°, 300° **7.** 0°, 76°, 180°, 284° **9.** 0°, 120°, 240° **11.** 90°, 210°, 330° **13.** 60°, 120°, 240°, 300° **15.** $-\frac{16}{65}$ **17.** $\frac{63}{65}$ **19.** $-\frac{4}{5}$ **21.** $-\frac{17}{6}$ **25.** $-\frac{1}{9}$ **27.** $-4\sqrt{5}$ **29.** $-\frac{\sqrt{11}}{4}$ **31.** $-\frac{23}{25}$ **37.** 90°, 183°, 270°, 357° **39.** 30°, 60°, 210°, 240° **41.** $-\frac{\pi}{4}$ **43.** $\frac{12}{13}$ **45.** $-\sqrt{2}$ **47.** a **49.** b **51.** a

CHAPTER 10 MORE WORK WITH TRANSFORMATIONS

Pages 367–368 Lesson 10.1

Exploratory **17.** *G* **19.** *T* **21.** *Q* **23.** *T* **25.** *G* **Written** **23.** yes **25.** yes **27.** yes **29.** yes **31.** yes **39.** $\ell \parallel m$. $R_\ell \circ R_m$ is equivalent to the translation with distance equal to twice the distance between ℓ and m and in the same direction as any ray that has its endpoint on m and perpendicular to ℓ. **41.** Suppose $R_\ell(P) = C$ and $T_{a,b}(C) = P'$ so that $G(P) = P'$. By definition of glide reflection, $\overline{CP'} \parallel \ell$. Thus, ℓ must divide $\overline{PC}$ and $\overline{PP'}$ proportionally. Since ℓ bisects $\overline{PC}$, ℓ must also bisect $\overline{PP'}$. **43.** R_ℓ **45.** $Rot_{O,310°}$

Pages 371–372 Lesson 10.2

Exploratory **1.** yes **3.** no **7.** Since any isometry preserves distance, the composite of isometries must also preserve distance. **9.** yes **11.** Since a glide reflection is the composite of a translation and a line reflection, and since a translation can be expressed as a composite of two line reflections, a glide reflection can be expressed as a composite of three line reflections. **13.** Line reflections reverse orientation. **15.** Composite of two line reflections preserves orientation. **Written** **1.** always **3.** sometimes **5.** sometimes **7.** always **9.** sometimes **13.** Both have clockwise orientations. **15.** $R_\ell \circ R_m$ is a translation whose distance is equal to twice the distance between lines ℓ and m and whose direction is the same as any ray that has its endpoint on line m and is perpendicular to ℓ. **17.** The lines of reflection must be either parallel or coinciding. **21.** The composite of two line reflections, where the lines are parallel or coincide, is a translation. **25.** no

Pages 374–375 Lesson 10.3

Exploratory **1.** *E* **3.** *K* **5.** *O* **7.** Lines are parallel or coinciding. **9.** Lines are perpendicular or coinciding. **11.** distance, collinearity, betweenness of points, angle measure, perpendicularity, parallelism, orientation **13.** yes **15.** Composite of two direct isometries **17.** Composite of a direct and an opposite isometry. **19.** Dilations do not always preserve distance. **21.** Composite of three direct isometries and three opposite isometries. **Written** **7.** $Rot_{O,-60°}$ **9.** R_ℓ **13.** true **15.** false **17.** false **19.** true **21.** $\left(\frac{1}{2}, \frac{\sqrt{3}}{2}\right)$ **23.** $(1, \sqrt{3})$ **25.** $\left(-\frac{1}{2}, \frac{\sqrt{3}}{2}\right)$ **27.** (0, 1) **Mixed Review** **1.** $\frac{3}{4}, -\frac{4}{5}$ **2.** 2

3. $\pm\frac{\sqrt{6}}{2}$ **4.** -1 **5.** 5 **6.** 45°, 135°, 225°, 315° **7.** 30°, 150°, 210°, 330° **8.** 90°, 120°, 240°, 270° **9.** $\frac{7}{25}$ **10.** $\frac{24}{25}$ **11.** $-\frac{\sqrt{10}}{10}$ **12.** -3 **13.** one; $a = 16.3$, $b = 8.9$, $m\angle C = 79$ **14.** two; $a = 6.4$, $m\angle B = 53°1'$, $m\angle C = 86°59'$, or $a = 6.4$, $m\angle B = 126°59'$, $m\angle C = 13°1'$ **15.** $(-7, 4)$ **16.** $(-3, 1)$ **17.** $(10, 2)$ **18.** 213.9 **19.** $\frac{1 - 2\sin x\sqrt{1 - \sin^2 x}}{1 - 2\sin^2 x}$ **20.** $-\frac{3}{4}$

Pages 379–380 Lesson 10.4

Exploratory 1. translation **3.** rotation **5.** Every isometry is either a line reflection, the composite of two line reflections, or the composite of three line reflections **7.** no **9.** rotations, translations, and dilations of scale factor 1 or -1 **11.** always **13.** never **15.** sometimes **17.** sometimes **Written 9.** Let the triangle be $\triangle ABC$ as drawn. Let ℓ, m, and n be the perpendicular bisectors of $\overline{BC}$, $\overline{AC}$, and $\overline{AB}$ respectively, and let O be the point of intersection of ℓ, m, and n. Then the following transformations map $\triangle ABC$ onto itself. R_ℓ, R_m, R_n, $Rot_{O,120°}$, $Rot_{O,240°}$, and I **11.** rotation **13.** glide reflection **15.** rotation **17.** glide reflection **19.** glide reflection **21.** rotation

Pages 383–385 Lesson 10.5

Exploratory 1. line reflection **3.** glide reflection **5.** glide reflection or rotation **7.** Two figures are congruent if and only if there is an isometry that maps one figure onto the other. **9.** yes **11.** A similarity transformation is a composite of isometries and dilations. **13.** Two figures are similar if and only if there is a similarity transformation that maps one figure onto the other. **15.** yes **17.** yes **19.** yes **21.** yes **Written 11.** Only if the line is parallel to the line of reflection **13.** Only if the line does not pass through the center of rotation and the rotation is equivalent to a half-turn. **15.** Only if the line is parallel to the line of reflection. **17.** $D_{C,4}$ **19.** $D_{C,\frac{2}{7}}$ **33.** The area of $\triangle DEF$ is nine times the area of $\triangle ABC$ since any base and height of $\triangle DEF$ is three times larger than the corresponding base and height of $\triangle ABC$. **35.** no **37.** no **39.** no **41.** yes **43.** no **45.** yes **47.** $(1, \sqrt{3})$

Pages 388–389 Lesson 10.6

Exploratory 1. Since $\triangle ABC$ is an equilateral triangle, ℓ, m, and n are the perpendicular bisectors of $\overline{BC}$, $\overline{AC}$, and $\overline{AB}$ respectively. Thus, $R_\ell(B) = C$, $R_\ell(C) = B$, $R_m(A) = C$, $R_m(C) = A$, $R_n(A) = B$, $R_n(B) = A$. Therefore, $\triangle ABC$ is its own image under R_ℓ, R_m, or R_n, which means ℓ, m, and n are each lines of symmetry. **3.** I, R_ℓ, R_m, R_n, $Rot_{O,120°}$, $Rot_{O,240°}$ **Written 5.** R_ℓ **7.** R_n **9.** R_m **11.** yes **13.** I, the identity transformation **15.** yes **19.** yes **21.** The perpendicular bisector of the base of the triangle is a line of symmetry. The line reflection corresponding to this symmetry, along with I, forms a commutative group under composition. **23.** The parallelogram has point symmetry about the point of intersection of the diagonals. The rotation corresponding to this symmetry, along with I, forms a commutative group under composition.

Pages 392–393 Lesson 10.7

Exploratory 1. $(x, -y)$ **3.** $(-x, -y)$ **5.** $(x + 3, y + 5)$ **7.** $(-x, y)$ **9.** $(-x - 2, -y + 4)$ **11.** $(-3x - 6, -3y - 12)$ **13.** $(-y, -x)$ **15.** $(-x - 2, y + 3)$ **17.** $(2x - 6, -2y + 8)$ **23.** yes **25.** yes **27.** yes **29.** yes **31.** yes **33.** no **35.** yes **37.** yes **39.** yes **41.** yes **43.** no, no **45.** yes, yes **47.** yes, yes **49.** no, no **53.** yes

Page 396 Problem Solving Application

3. 1743 yards

Pages 398–399 Chapter Review

1. A line reflection over line ℓ. **3.** A dilation with center O and scale factor $\frac{1}{2}$ followed by a line reflection over line m. **5.** A rotation, with center O, of 180°. Also known as a half-turn about the point O. **11.** $Rot_{O,120°}$ **13.** I **15.** line reflection **17.** translation **19.** rotation **21.** yes **23.** no **25.** no **27.** yes **29.** reflection over the same line **31.** $Rot_{M,34°}$ **33.** no **35.** yes **37.** yes **39.** yes **41.** closure, associativity, commutativity, identity element, inverse element for all elements **43.** associativity, inverse element for all elements **45.** $T(x, y) = \left(\frac{x + 10}{2}, \frac{y + 10}{2}\right)$, $S\left(\frac{x + 10}{2}, \frac{y + 10}{2}\right) = (x + 10, y + 10)$. Therefore, $(S \circ T)(x, y) = (x + 10, y + 10)$ which is not a dilation, but a translation. Thus, the set of all dilations is not closed under composition.

CHAPTER 11 PROBABILITY AND THE BINOMIAL THEOREM

Pages 404–405 Lesson 11.1

Exploratory 1. 12 **3.** 20 **Written 1.** $\frac{1}{12}$ **3.** $\frac{5}{36}$

5. $\frac{1}{4}$ **7.** $\frac{1}{6}$ **9.** $\frac{1}{18}$ **11.** $\frac{1}{2}$ **13.** $\frac{1}{26}$ **15.** $\frac{2}{7}$ **17.** $\frac{4}{7}$ **19.** $\frac{1}{21}$, $\frac{2}{7}$; 0

Pages 407–409 Lesson 11.2

Exploratory 1. Events E and F are possible outcomes out of the total collection of outcomes. **3.** The product $n(n-1)(n-2)\cdot\cdot\cdot 1$ or ${}_nP_n$ **5.** The product $7\cdot 6\cdot 5$ or 210 **7.** The number of arrangements of n things taken all at a time **9.** The number of arrangements of 5 things taken 3 at a time **11.** The number of arrangements of 18 things taken 2 at a time **Written 1.** 30, 240 **3.** 210 **5.** 56 **7.** r; $(n-5)$ **9.** 120 **11.** 60 **13.** 12 **15.** 2520 **17.** 4320 **19.** 360 **21.** 10,080 **23.** 840 **25.** 4 **27.** 5040 **29.** 34,650 **31.** 120 **33.** 720 **35.** $\frac{1}{7}$ **37.** $\frac{1}{42}$ **39.** $\frac{1}{35}$ **Mixed Review 1.** b **2.** c **3.** b **4.** d **5.** c **6.** a **7.** b **8.** b **9.** c **10.** b

Pages 411–412 Lesson 11.3

Exploratory 3. combination **5.** combination **7.** Number of combinations of n things taken all at a time **9.** Number of combinations of n things taken all but one at a time **11.** Since order does not matter with combinations, there is only one way to take n objects all at a time. **Written 3.** 1 **5.** 56 **7.** 84 **9.** 10 **11.** 220 **13.** 0 **15.** 1365 **17.** 6 **19.** 1287 **21.** $\frac{44!}{11!33!}$ **23.** 420 **25.** $\frac{1}{6}$ **27.** $\frac{5}{9}$

Pages 414–415 Lesson 11.4

Exploratory 1. $\frac{1}{216}$, 0.005 **3.** $\frac{625}{1296}$, 0.482 **5.** $\frac{25}{7776}$, 0.003 **7.** $\frac{3125}{46,656}$, 0.067 **9.** $\frac{15,625}{1,679,616}$, 0.009 **Written 1.** $\frac{125}{1296}\approx 0.096$ **3.** $\frac{125}{1296}\approx 0.096$ **5.** $\frac{500}{1296}\approx 0.386$ **7.** $\frac{20}{1296}\approx 0.015$ **9.** $\frac{3125}{46656}\approx 0.067$ **11.** $\frac{18750}{46656}\approx 0.402$ **13.** Inez is incorrect. **15.** $\frac{38416}{151875}$ **17.** $\frac{81}{10000}$ **19.** $\frac{7}{31104}$ **Mixed Review 1.** 0.1745 **2.** 0.0977 **3.** 0.1955 **4.** 0.0349 **5.** 0.4973

Pages 417–418 Lesson 11.5

Exploratory 1. A Bernoulli experiment is an experiment that has only two outcomes. **3.** We know the sum of the probabilities of all possible outcomes is 1. Since Bernoulli experiments have only two outcomes, if the probability of the outcomes are p and q, then $p+q=1$ or $q=1-p$. **Written 1.** $\frac{2}{9}\approx 0.22$ **3.** $\frac{26}{27}\approx 0.963$ **5.** $\frac{80}{243}\approx 0.329$ **7.** $\frac{9}{64}\approx 0.141$ **9.** $\frac{135}{512}\approx 0.264$ **11.** $\frac{64}{729}\approx 0.088$ **13.** $\frac{80}{243}\approx 0.329$ **15.** $\frac{20}{243}\approx 0.082$ **17.** $\frac{1024}{6561}\approx 0.156$ **19.** $\frac{1792}{6561}\approx 0.273$ **21.** $\frac{448}{6561}\approx 0.068$ **23.** $\frac{31}{32}\approx 0.969$ **25.** $\frac{13}{16}\approx 0.813$ **27.** $\frac{1023}{1024}\approx 0.999$ **29.** $\frac{15}{16}\approx 0.938$ **31.** $\frac{58975}{65536}\approx 0.900$ **33.** ${}_{100}C_{43}(.43)^{43}(.57)^{57}$ **35.** ${}_{100}C_{10}\left(\frac{1}{36}\right)^{10}\left(\frac{35}{36}\right)^{90}$

Pages 421–422 Lesson 11.6

Exploratory 1. To expand a binomial means to raise a binomial to a power. **3.** $x^3+6x^2+12x+8$ **5.** $x^4+8x^3+24x^2+32x+16$ **7.** $(x+y)^0=1$ as long as $x+y\neq 0$, since $a^0=1$ for all $a\neq 0$. **Written 1.** $x^3+27x^2+243x+729$ **3.** $x^3+\frac{3}{2}x^2+\frac{3}{4}x+\frac{1}{8}$ **5.** $a^3-6a^2x+12ax^2-8x^3$ **7.** $16x^4-96x^3+216x^2-216x+81$ **9.** $x^4+4x^3y+6x^2y^2+4xy^3+y^4$ **11.** $x^6+6x^5y+15x^4y^2+20x^3y^3+15x^2y^4+6xy^5+y^6$ **13.** $81x^4+216x^3+216x^2+96x+16$ **15.** $32x^5-240x^4+720x^3-1080x^2+810x-243$ **19.** 7th, $x^7+7x^6y+21x^5y^2+35x^4y^3+35x^3y^4+21x^2y^5+7xy^6+y^7$ **21.** 7, $35x^4y^3$ **23.** 6, $15x^4y^2$ **27.** $(x+y)^n=x^n+\frac{n!}{1!(n-1)!}x^{n-1}y^1+\frac{n!}{2!(n-2)!}x^{n-2}y^2+\cdots+\frac{n!}{(n-3)!3!}x^3y^{n-3}+\frac{n!}{(n-2)!2!}x^2y^{n-2}+\frac{n!}{(n-1)!1!}xy^{n-1}+y^n$ **29.** $x^4+12x^3+54x^2+108x+81$ **31.** $8x^3+12x^2+6x+1$ **33.** $w^4-44w^3+726w^2-5324w+14641$ **35.** $x^6-12x^5y+60x^4y^2-160x^3y^3+240x^2y^4-192xy^5+64y^6$ **37.** $16w^4+\frac{32}{3}w^3+\frac{8}{3}w^2+\frac{8}{27}w+\frac{1}{81}$ **39.** $256y^4-768y^3+864y^2-432y+81$ **41.** $x^8+16x^7+112x^6+448x^5+1120x^4+1792x^3+1792x^2+1024x+256$ **43.** $x^5-\frac{5}{2}x^4y+\frac{5}{2}x^3y^2-\frac{5}{4}x^2y^3+\frac{5}{16}xy^4-\frac{1}{32}y^5$ **45.** $243v^5-\frac{405}{2}v^4w+\frac{135}{2}v^3w^2-\frac{45}{4}v^2w^3+\frac{15}{16}vw^4-\frac{1}{32}w^5$ **47.** $128a^7+448a^6b+672a^5b^2+560a^4b^3+280a^3b^4+84a^2b^5+14ab^6+b^7$

Page 424 Lesson 11.7

Exploratory **1.** $x^3 + 9x^2 + 27x + 27$ **3.** $625x^4 + 500x^3 + 150x^2 + 20x + 1$ **Written** **1.** x^{50}; 1 **3.** $0.0009765625x^{10}$; 1024 **5.** $32x^3$ **7.** $40x^2$ **9.** $11547360x^4$ **11.** 6 **13.** −30618 **15.** $x^{16} + 16x^{15}y + 120x^{14}y^2 + 560x^{13}y^3$ **17.** $\frac{429}{16}y^{14}$ **19.** $1 + 5 + 10 + 10 + 5 + 1 = 32$ **21.** $\frac{1}{32} + \frac{5}{32} + \frac{10}{32} + \frac{10}{32} + \frac{5}{32} + \frac{1}{32} = 1$

Pages 427–429 Lesson 11.8

Exploratory **3.** no **9.** $\frac{1}{2}$ **11.** $\frac{1}{13}$ **13.** $\frac{3}{13}$ **15.** 0 **17.** $\frac{1}{13}$ **Written** **1.** 0 **3.** $\frac{1}{5}$ **5.** $\frac{2}{5}$ **7.** $\frac{1}{6}$; {(5, 1), (5, 3), (5, 5)} **9.** 1; {(6, 6)} **11.** $\frac{1}{2}$; {(4, 3), (5, 3), (6, 3)} **13.** $\frac{10}{13}$ **15.** 30% **17.** $\frac{5}{8}$ **19.** $\frac{11}{19}$ **21.** $\frac{6}{19}$ **25.** $\frac{1}{4}$ **27.** $\frac{1}{8}$ **29.** $\frac{1}{8}$ **31.** $\frac{1}{5}, \frac{1}{4}$ **Mixed Review** **1.** $R_{x=135°}$ **2.** $T_{90°,0}$ **3.** $Rot_{O,90°}$ **4.** $\cos^2 x$ **5.** $\frac{\sin x}{\sqrt{1 - \sin^2 x}}$ **6.** $1 - \tan x$ **7.** $A'(1, 1)$, $B'(-2, 2)$, $C'(7, -6)$; glide reflection **8.** $A'(2, 0)$, $B'(3, -3)$, $C'(-5, 3)$; rotation **9.** $A'(-3, 3)$, $B'(-2, 6)$, $C'(-10, 0)$; line reflection **10.** $\frac{3}{16}$ **11.** $\frac{1}{64}$ **12.** 27 white, 12 gold

Page 432 Problem Solving Application

1. 24, 30, 36; $b = 6a$ **3.** 10, 12, 14; $y = 2x + 2$ **5.** 12, 345, 678, 987, 654, 321 **7.** $2^{10} = 1024$ **9.** All lockers whose numbers are perfect squares. 1, 4, 9, 16, and so on. **11.** b

Pages 434–435 Chapter Review

1. $\frac{1}{16}$ **3.** $\frac{1}{4}$ **5.** $\frac{1}{16}$ **7.** $\frac{1}{4}$ **9.** $\frac{1}{8}$ **13.** combinations **15.** neither **17.** $\frac{175000}{1679616} \approx 0.104$ **19.** ${}_{100}C_{10}\left(\frac{1}{6}\right)^{10}\left(\frac{5}{6}\right)^{90}$ **21.** ${}_{T}C_{t}\left(\frac{1}{6}\right)^{t}\left(\frac{5}{6}\right)^{T-t}$ **23.** $32x^5 - 240x^4 + 720x^3 - 1080x^2 + 810x - 243$ **25.** fourth **27.** $112a^2b^6$ **29.** $-168x^{20}k$ **31.** $-270{,}208{,}224x^8y^5$ **33.** 1 **35.** $\frac{1}{10}$

CHAPTER 12 STATISTICS

Page 440 Lesson 12.1

Exploratory **5.** The mean of a set of n numbers is the sum of the numbers divided by n. **7.** The median of a set of n numbers is the one which lies in the middle when the numbers are arranged in order. If n is even, it is the numerical average of the two middle numbers. **9.** A set with two modes is bimodal. **Written** **1.** 8.125 **3.** $\frac{71}{128}$ **5.** Yes, since there are eight 8's, seven 6's, and three 10's, the mean is $\frac{8(8) + 7(6) + 3(10)}{18} = 7.5$ **7.** 15.08, 14.98; yes **9.** 12 **11.** 25.5 **13.** 15 **15.** 2.375 **17.** 0 **19.** 1

Pages 443–444 Lesson 12.2

Exploratory **7.** These symbols represent the same number. The only difference is that different dummy variables have been used. **9.** 21 **11.** 30 **Written** **1.** 12 **3.** 54 **5.** 18 **7.** 42 **9.** $\frac{7}{2}$ **11.** 51 **13.** 45 **15.** 30 **17.** 90 **21.** $\sum_{i=1}^{12} i^2$ **23.** $\sum_{i=1}^{n} i^2$ **25.** $\sum_{i=1}^{r} i^2$ **27.** $\sum_{i=1}^{12} \frac{1}{i}$ **Mixed Review** **1.** 150° **2.** undefined **3.** $\frac{25}{7}$ **4.** P **5.** E **6.** M **7.** $\overline{PX}$ **8.** $JKBC$ **9.** 80 **10.** $\frac{45}{512}$ **11.** $\frac{53}{512}$ **12.** $\frac{243}{1024}$ **13.** $224{,}000x^3y^5$ **14.** $243a^5 - 405a^4b + 270a^3b^2 - 90a^2b^3 + 15ab^4 - b^5$ **15.** obtaining at least one pair of 6s when a pair of fair six-sided dice are tossed 12 times

Pages 447–448 Lesson 12.3

Exploratory **5.** 6.5 **7.** 0.01625 **9.** 12 **11.** $15n$ **Written** **1.** 13; 2.2 **3.** 15; 2.5 **5.** 161.5; 16.2 **7.** 5 **9.** 10 **11.** 23 **13.** 10 **15.** 9 **17.** 0 **19.** 8 **21.** 11, $\frac{75}{11} \approx 6.82$, 0, $\frac{612}{11} \approx 55.64$ **23.** 11, $\frac{14}{11} \approx 1.27$, 0, $\frac{204}{275} \approx 0.74$ **25.** 8.75 **27.** 27.5625 **29.** 5.25 **31.** 103.5 **33.** $x^5 + 5x^4y + 10x^3y^2 + 10x^2y^3 + 5xy^4 + y^5$

Pages 451–452 Lesson 12.4

Exploratory **5.** The difference between the greatest and least measures **Written** **7.** yes, yes **17.** rx **19.** The difference between the fourth measure and the mean **21.** The absolute value of the difference between the sixteenth measure and the mean **23.** The square of the difference between the twelfth measure and the mean **25.** The sum of the differences between the mean and each measure **27.** yes **29.** yes **37.** 4.8 **39.** $\frac{66}{49} \approx 1.35$ **41.** 2.64 **43.** 0.632 **45.** 2.188

Pages 456–457 Lesson 12.5

Exploratory **3.** $\frac{1}{n}\sum_{i=1}^{n}(\overline{x} - x_i)^2$ **7.** $\frac{1}{n}\sqrt{\sum_{i=1}^{n}(\overline{x} - x_i)^2}$ **9.** no **Written** **1.** 13.2, 3.63 **3.** 12.4, 3.52

5. 20.8, 4.56 **7.** 17.33, 4.16 **9.** 17, 4.12 **11.** 0.10, 0.31 **13.** 54.14 **15.** 9.15 **17.** 7.05 **27.** 68% **29.** 47.5% **31.** 13.5% **33.** 16%, 84% **35.** taller than about 16% of his classmates **37.** taller than about 0.5% of his classmates **39.** taller than about 2.5% of his classmates

Pages 459–460 Lesson 12.6
Exploratory **13.** 3, 1, 4 **15.** 4, 7, 6 **Written** **5.** 6.57 **7.** 43.6 **9.** 35.55 **11.** 0.54 **13.** 0.70 **15.** 0.90 **17.** 2.23 **19.** 1.82 **Mixed Review** **1.** $\{-1, 2\}$ **2.** 6 **3.** $\frac{33}{8}$ **4.** $\frac{13\pi}{18}$ **5.** $-\cos 32°$ **6.** $\frac{1}{4}$ **7.** $6 + 3\sqrt{3}$ **8.** 10 **9.** 30° **10.** $4i\sqrt{2}$ **11.** 8 **12.** $\log a + \frac{1}{2}\log b - \log c$ **13.** $\{x : x \geq 2\}$ **14.** $-\frac{5}{13}$ **15.** 9 **16.** line and point **17.** 6 cm **18.** $(1, -1)$ **19.** $f(x) = -\frac{11}{5}x + \frac{3}{5}$ **20.** -0.25 **21.** $90a^3b^2$ **22.** 27° or 153° **24.** 0.6 **26.** $f^{-1}(x) = 2^x$ **27.** $x = -1 \pm 3i$ **28.** 44 **29.** $A'(-2, -3)$, $B'(-9, -1)$, $C'(-5, -5)$; $A''(-3, -2)$, $B''(-1, -9)$, $C''(-5, -5)$ **30.** $\frac{73}{729}$

Page 464 Problem Solving Application
1. 0.1 hours **3.** 4.44 to 4.77 hours **5.** 0.2 mm **7.** 15.68 to 16.72 mm

Pages 466–467 Chapter Review
1. 5; 5; 5 **3.** $\frac{71}{18} \approx 3.94$; 4; 4 **5.** $-\frac{5}{4}$ **7.** 28 **9.** 25.5 **11.** 48 **13.** $\sum_{i=1}^{5} i^2 = 55$ **15.** $\frac{1}{8}\sum_{i=1}^{8} i = 4.5$ **17.** $\sum_{i=1}^{5}\left|i - \frac{1}{5}\sum_{i=1}^{5} i\right| = 6$ **19.** 19; $\frac{19}{6}$ **21.** 62.7; 12.54 **23.** 7, $\frac{89}{7} \approx 12.71$, 0, $\frac{262}{7} \approx 37.43$ **25.** 10, 11.29, 0, 1.029 **27.** false **29.** true **31.** true **33.** $\frac{49}{6} \approx 5.11$ **35.** 2.088 **37.** 1.6 **39.** 0.2 **41.** 578

CHAPTER 13 SEQUENCES AND SERIES

Pages 472–473 Lesson 13.1
Exploratory **1.** $\{1, 2, 3, 4, 5, 6\}$ **3.** 2 **5.** 12 **7.** $2n$ **Written** **1.** 2, 4, 8, 16, 32 **3.** 1, $\frac{1}{2}$, $\frac{1}{3}$, $\frac{1}{4}$, $\frac{1}{5}$ **5.** 3, 6, 9, 12, 15 **7.** 6, 12, 24, 48, 96 **9.** 1, 3, 9, 27, 81 **11.** -1, $\frac{1}{2}$, $-\frac{1}{3}$, $\frac{1}{4}$, $-\frac{1}{5}$ **13.** 1, 2, 5, 10, 17 **15.** 5, 6, 13, 32, 69 **17.** 2, $\frac{3}{2}$, $\frac{4}{3}$, $\frac{5}{4}$, $\frac{6}{5}$ **19.** Every term in the sequence is the same number. **23.** yes **25.** 1, 3, 5, 13, 33, 71, 133, 225, 353, 523 **27.** $n + 1$; 6, 7, 8 **29.** 3^{n-1}; 27, 81, 243 **31.** $(-1)^{n+1}$; 1, -1, 1

Pages 477–478 Lesson 13.2
Exploratory **1.** 1, 4, 7, 10, 13 **3.** -2, 1, 4, 7, 10 **5.** $\frac{1}{2}$, 1, $\frac{3}{2}$, 2, $\frac{5}{2}$ **7.** 17, 14, 11, 8, 5 **9.** $-\frac{3}{2}$, $-\frac{7}{4}$, -2, $-\frac{9}{4}$, $-\frac{5}{2}$ **11.** $-\sqrt{5}$, 0, $\sqrt{5}$, $2\sqrt{5}$, $3\sqrt{5}$ **13.** yes, 1 **15.** no **17.** yes, -28 **Written** **1.** $2 + (n-1)3$ **3.** $5 + (n-1)5$ **5.** $-2 + (n-1)(5)$ **7.** $\frac{1}{2} + (n-1)\left(\frac{1}{2}\right)$ **9.** $\frac{2}{3} + (n-1)\left(-\frac{1}{3}\right)$ **11.** $\sqrt{5} + (n-1)\sqrt{3}$ **13.** 34 **15.** -210 **17.** -175 **19.** 65, 80, 95 **21.** -6, -3, 0 **23.** -14, 0, 7, 21 **25.** 13 **27.** $\frac{1}{2}$ **29.** $-\frac{21}{10}$ **31.** 4 **33.** -4 **35.** 32 **37.** yes **39.** yes **41.** after 17 years **43.** 54 **45.** Sara's salary for her 15th year will exceed $35,000. **Mixed Review** **1.** 60°, 300° **2.** 60°, 90°, 120°, 270° **3.** 90°, 180° **4.** 45 **5.** 62 **6.** 10.125 **7.** 5 **8.** 126 **9.** $\frac{6144}{15625}$ **10.** 52°

Pages 482–483 Lesson 13.3
Exploratory **1.** 3, 6, 12, 24, 48 **3.** 1, -2, 4, -8, 16 **5.** -5, $-\frac{5}{2}$, $-\frac{5}{4}$, $-\frac{5}{8}$, $-\frac{5}{16}$ **7.** -1, $-\frac{1}{4}$, $-\frac{1}{16}$, $-\frac{1}{64}$, $-\frac{1}{256}$ **9.** 100, 10, 1, 0.1, 0.01 **11.** geometric **13.** arithmetic, geometric **15.** geometric **Written** **1.** 2, 128 **3.** 3, 405 **5.** 10, 10^{24} **7.** $\frac{3}{2}$, $\frac{729}{64}$ **9.** 8, 16 **11.** $\frac{1}{4}$, $\frac{1}{8}$, $\frac{1}{16}$ or $-\frac{1}{4}$, $\frac{1}{8}$, $-\frac{1}{16}$ **13.** -15, -45, -135, -405 **15.** 4, -4 **17.** $-\frac{4}{3}$ **19.** yes **21.** $\frac{3}{8}$

Pages 487–488 Lesson 13.4
Exploratory **1.** 42 **3.** $\frac{35}{4}$ **5.** 336 **7.** 240 **9.** 565 **11.** -210 **13.** 116 **15.** yes **Written** **1.** 1770 **3.** 2340 **5.** 1650 **7.** 11 **9.** 10100 **11.** 100 **13.** $5 **15.** 240 ft **17.** 368 ft **19.** 1024 ft **21.** 2304 ft **23.** $\sum_{k=1}^{5} \frac{1}{5}(3k - 2)$ **25.** $\sum_{k=1}^{5} \frac{1}{2}(5k + 4)$ **27.** $\sum_{k=1}^{5} 6\left(-\frac{1}{3}\right)^{k-1}$ **29.** 10201 **31.** 340 **33.** 42

Pages 491–492 Lesson 13.5
Exploratory **5.** arithmetic **7.** both **9.** neither **11.** arithmetic **Written** **1.** 45 **3.** 189 **5.** $\frac{61}{9}$

7. $\frac{21}{2}$ **9.** $\frac{341}{1024}$ **11.** $\frac{63}{8}$ **13.** $\frac{2047}{1024}$ **15.** 26.375 **17.** $S_n = a_1 n$ **19.** 13756 **21.** $20,000 **23.** 13 **25.** $\frac{63}{8}$ **27.** −1 **29.** $\frac{63}{64}$ **Mixed Review** **1.** $-2 < x < 8$ **2.** $x^2 - 2x + 2 = 0$ **3.** 17.4 **4.** 60 **5.** 2 **6.** $z^2 + 3z + 2$ **7.** $\frac{3 + \sqrt{3}}{3}$ **8.** $-1 - 4i\sqrt{5}$ **9.** $0 \le x \le 4$ **10.** 0

Pages 497–498 Lesson 13.6
Exploratory **3.** 12 **5.** $-\frac{9}{2}$ **7.** $-\frac{32}{9}$ **9.** No sum exists. **11.** 0 **13.** 2 **Written** **1.** 12 **3.** No sum exists. **5.** $\frac{1}{6}$ **7.** $\frac{7}{50}$ **9.** No sum exists. **11.** 0 **13.** 2 **Written** **1.** 12 **3.** No sum exists. **5.** $\frac{1}{6}$ **7.** $\frac{7}{50}$ **9.** No sum exists. **11.** $\frac{5}{9}$ **13.** $\frac{7}{9}$ **15.** $\frac{14}{3}$ **17.** $\frac{23}{99}$ **19.** 396.8 **21.** $-8, -\frac{16}{5}, -\frac{32}{25}, -\frac{64}{125}$ **23.** 125 cm **25.** 800 feet **27.** 70 cm **29.** $36,764.71 **Mixed Review** **1.** 20 **2.** 32 **3.** 10 **4.** 47 **5.** 52 **6.** 108 **7.** 3 **8.** 2 or −0.2 **9.** $32x^5 - 400x^4 + 2000x^3 - 5000x^2 + 6250x - 3125$ **10.** (1, 4) **11.** 114°37′, 20°23′, 3.45 m **12.** 26.5, 5.99 **13.** 127

Page 501 Problem Solving Application
1. $112,204 **3.** $71,680 **5.** 14 years 8 months **7.** 9.17% **9.** 10 days **11.** 8 years

Pages 502–503 Chapter Review
1. 3, 9, 27, 81, 243 **3.** 4, 8, 12, 16, 20 **5.** $3, \frac{3}{2}, 1, \frac{3}{4}, \frac{3}{5}$ **7.** 32, 64, 128; 2^n **9.** 1, 5, 9, 13, 17 **11.** $\frac{1}{2}, \frac{1}{4}, 0, -\frac{1}{4}, -\frac{1}{2}$ **13.** $a_n = 1 + (n - 1)(-4)$ **15.** 5, 7 **17.** 1, 3, 9, 27, 81 **19.** $128, -32, 8, -2, \frac{1}{2}$ **21.** $\frac{1}{2}, \frac{1}{16}$ **23.** arithmetic **25.** geometric **27.** 225 **29.** 124 **31.** arithmetic **33.** neither **35.** 75 **37.** $3\frac{1}{2}$ **39.** $2928.20 **41.** $\frac{9}{2}$ **43.** $\frac{3}{25}$

CHAPTER 14 AN INTRODUCTION TO MATRICES

Page 508 Lesson 14.1
Exploratory **13.** 49, 289, n^2 **15.** no, a_{21}, $[-1 \quad -2\sqrt{2} \quad -\sqrt{2} \quad 0]$ **17.** 1, $\begin{bmatrix} 0 \\ -1 \end{bmatrix}$ **19.** 2 **21.** 7 **23.** −1 **Written** **1.** $\begin{bmatrix} -2 & 0 \\ 1 & -2 \end{bmatrix}$ **3.** $\begin{bmatrix} -\sqrt{2} & 1 \\ -2 & -3 \end{bmatrix}$ **7.** $\begin{bmatrix} 2 + \sqrt{2} & 1 \\ 0 & 0 \end{bmatrix}$ **11.** $\begin{bmatrix} 2 - \sqrt{2} & -1 \\ -2 & 4 \end{bmatrix}$ **13.** not possible **15.** not possible **17.** $\begin{bmatrix} 5 & -1 \\ 0 & 1 \\ -6 & -2 \end{bmatrix}$ **21.** $\begin{bmatrix} \sqrt{2} & -1 \\ 2 & 3 \end{bmatrix}$ **23.** not possible

Pages 510–511 Lesson 14.2
Exploratory **1.** Multiply each element of matrix A by k. **5.** $\begin{bmatrix} 4 \\ 8 \\ 24 \end{bmatrix}$ **7.** $\begin{bmatrix} 2\sqrt{2} & -2\sqrt{2} \\ 2 & 0 \end{bmatrix}$ **9.** $\begin{bmatrix} -6 & 1 & 2 \\ 4 & -12 & 0 \end{bmatrix}$ **Written** **1.** $\begin{bmatrix} 6 & -3 \\ 0 & 9 \end{bmatrix}$ **3.** $\begin{bmatrix} 2 & -1 \\ 0 & 3 \end{bmatrix}$ **7.** $\begin{bmatrix} 3 & -\frac{3}{2} \\ 0 & \frac{9}{2} \end{bmatrix}$ **11.** $\begin{bmatrix} 2 & 2 & -3 \\ -1 & -3 & 4 \end{bmatrix}$ **15.** $\begin{bmatrix} -3 & 0 \\ 12 & 12 \end{bmatrix}$ **17.** $\begin{bmatrix} -11 & 5 \\ 4 & -11 \end{bmatrix}$ **21.** $\begin{bmatrix} 3 & -6 & -10 \\ 6 & 3 & -2 \end{bmatrix}$ **25.** $\begin{bmatrix} -\frac{1}{2} & 0 \\ 2 & 2 \end{bmatrix}$ **27.** $\begin{bmatrix} -5 & 1 \\ 12 & 9 \end{bmatrix}$

Pages 514–515 Lesson 14.3
Exploratory **3.** yes, [2 3] **5.** no **7.** yes, [3] **9.** no **Written** **1.** Matrix multiplication is not commutative. **3.** not defined **5.** not defined **7.** 5 × 6 **9.** 5 × 6 **11.** $\begin{bmatrix} 2 & 4 \\ 6 & 11 \end{bmatrix}$ **13.** $\begin{bmatrix} 8 & 15 \\ 4 & 7 \end{bmatrix}$ **15.** $\begin{bmatrix} -5 & -8 \\ 15 & 26 \end{bmatrix}$ **19.** $\begin{bmatrix} 13 & 1 \\ 5 & 1 \end{bmatrix}$ **23.** $\begin{bmatrix} 4 & 0 \\ 9 & 1 \end{bmatrix}$ **27.** $\begin{bmatrix} 42 & 10 \\ 110 & 26 \end{bmatrix}$ **29.** $\begin{bmatrix} 14 & -8 \\ 36 & -22 \end{bmatrix}$ **33.** $\begin{bmatrix} -12 & -4 \\ -25 & -15 \end{bmatrix}$ **37.** $\begin{bmatrix} 145 & -385 \\ -75 & 475 \end{bmatrix}$ **41.** $\begin{bmatrix} 8 + 2\sqrt{3} & 1 + 3\sqrt{2} \\ 4 + 4\sqrt{6} & \frac{25}{2} \end{bmatrix}$ **45.** $\begin{bmatrix} 2\sqrt{15} - 3 & 6\sqrt{3} - 3 \\ 4\sqrt{10} - \frac{1}{2} & 12\sqrt{2} - \frac{1}{2} \end{bmatrix}$ **49.** $\begin{bmatrix} \frac{2}{3} & \frac{4}{5} \\ \frac{5}{4} & \frac{1}{5} \end{bmatrix}$

59. $\begin{bmatrix} 7 & 1 & 11 \\ 4 & 2 & -2 \\ -1 & -2 & 2 \end{bmatrix}$ **63.** $\begin{bmatrix} 1 & -1 & -1 \\ 6 & 2 & 2 \\ 4 & 0 & -1 \end{bmatrix}$

65. not possible **69.** [8] **71.** not possible

73. $\begin{bmatrix} 33 & 7 & 3 \\ 24 & 4 & 3 \\ 18 & -6 & -9 \end{bmatrix}$ **75.** $\begin{bmatrix} 12 & -1 & 2 \\ 9 & 0 & 3 \\ 6 & 4 & 4 \end{bmatrix}$

Page 519 Lesson 14.4

Exploratory **1.** If I is the multiplicative identity of M_2, then $AI = IA = A$ for all A in M_2. **7.** no inverse **9.** $\begin{bmatrix} 4 & -5 \\ -3 & 4 \end{bmatrix}$ **Written** **1.** $\begin{bmatrix} 3 & -2 \\ -1 & 1 \end{bmatrix}$

5. no inverse **7.** $\begin{bmatrix} 1 & \frac{1}{2} \\ \frac{1}{2} & -\frac{1}{4} \end{bmatrix}$ **11.** She is correct.

13. 1, $\begin{bmatrix} 1 & 0 \\ 0 & 1 \end{bmatrix}$ **15.** −2, $\begin{bmatrix} -2 & 1 \\ \frac{3}{2} & -\frac{1}{2} \end{bmatrix}$ **19.** $a_1a_4 - a_2a_3$,

$\begin{bmatrix} -\frac{a_4}{a_1a_4 - a_2a_3} & -\frac{a_2}{a_1a_4 - a_2a_3} \\ \frac{a_3}{a_1a_4 - a_2a_3} & \frac{a_1}{a_1a_4 - a_2a_3} \end{bmatrix}$ **23.** $\begin{bmatrix} \frac{4}{11} & -\frac{1}{11} \\ \frac{3}{11} & \frac{2}{11} \end{bmatrix}$

Pages 522–523 Lesson 14.5

Exploratory **3.** The inverse of the coefficient matrix must exist. **Written** **1.** {(4, 1)} **3.** $\left\{\left(\frac{7}{10}, \frac{2}{25}\right)\right\}$

5. {(−1, 8)} **13.** {(5, 2)} **15.** {(−5, −2)}

17. {(−1, 3)} **19.** $\left\{\left(\frac{29}{11}, \frac{16}{11}\right)\right\}$ **21.** $\left\{\left(\frac{117}{19}, \frac{5}{19}\right)\right\}$

23. $\left\{\left(-\frac{11}{43}, \frac{45}{43}\right)\right\}$ **25.** \$1.25, \$0.65 **Mixed Review**

1. 19 **2.** 16 **3.** 18.175 **4.** 31 **5.** 8.94 **6.** 84 **7.** $\frac{8}{13}$ **8.** 32 **9.** 187.5 **10.** (0, 0) or (−4, 16) **11.** (3, −7) or (−1, 9) **12.** (0, 5) or (−3, 4)

Page 527 Lesson 14.6

Exploratory **5.** The minor of that element multiplied by 1 if $i + j$ is even or −1 if $i + j$ is odd, given that the element is in the *i*th row and the *j*th column. **Written** **1.** $\begin{vmatrix} q & r \\ t & v \end{vmatrix}, \begin{vmatrix} p & r \\ s & v \end{vmatrix}, \begin{vmatrix} p & q \\ s & t \end{vmatrix}$

3. $\begin{vmatrix} m & n \\ q & r \end{vmatrix}, \begin{vmatrix} k & n \\ p & r \end{vmatrix}, \begin{vmatrix} k & m \\ p & q \end{vmatrix}$

5. $\begin{vmatrix} p & r \\ s & v \end{vmatrix}, \begin{vmatrix} k & n \\ s & v \end{vmatrix}, \begin{vmatrix} k & n \\ p & r \end{vmatrix}$ **7.** $\begin{vmatrix} q & r \\ t & v \end{vmatrix}$

9. $\begin{vmatrix} m & n \\ q & r \end{vmatrix}$ **11.** $\begin{vmatrix} k & n \\ s & v \end{vmatrix}$ **13.** $\begin{vmatrix} p & q \\ s & t \end{vmatrix}$ **15.** $\begin{vmatrix} k & m \\ p & q \end{vmatrix}$

19. $\left\{\left(\frac{5}{3}, \frac{1}{3}, 2\right)\right\}$ **21.** $\left\{\left(\frac{1}{2}, 2, 1\right)\right\}$

Pages 530–531 Lesson 14.7

Exploratory **11.** (3, 2) **13.** (8, 1) **15.** (−1, 6) **17.** (2, −3) **Written** **3.** $\begin{bmatrix} 1 & 0 \\ 0 & -1 \end{bmatrix}$ A rotation about the origin or a half-turn about the origin.

5. $\begin{bmatrix} -1 & 0 \\ 0 & -1 \end{bmatrix}\begin{bmatrix} x \\ y \end{bmatrix} = \begin{bmatrix} -x \\ -y \end{bmatrix}$ **7.** $\begin{bmatrix} 0 & 1 \\ -1 & 0 \end{bmatrix}\begin{bmatrix} x \\ y \end{bmatrix} =$ [

9. $\begin{bmatrix} 7 & 0 \\ 0 & 7 \end{bmatrix}\begin{bmatrix} x \\ y \end{bmatrix} = \begin{bmatrix} 7x \\ 7y \end{bmatrix}$ **11.** $\begin{bmatrix} -1 & 0 \\ 0 & -1 \end{bmatrix}\begin{bmatrix} x \\ y \end{bmatrix} =$ [

15. $\left(-\frac{9}{4}, \frac{\sqrt{3}}{4}\right)$ **17.** $\left(\frac{\sqrt{3}}{4}, \frac{7}{4}\right)$ **19.** $\left(\frac{2 - \sqrt{2}}{2}, \frac{2 + \sqrt{2}}{2}\right)$

21. $(-4\sqrt{2}, -4\sqrt{2})$ **23.** (0.899, 1.481)

Pages 535–536 Lesson 14.8

Exploratory **3.** 41,300, 28,700; 35,910, 34,090; 32,137, 37,863 **Written** **1.** $\begin{bmatrix} 0.66 & 0.34 \\ 0.17 & 0.83 \end{bmatrix}$

3. $\begin{bmatrix} 0.371 & 0.629 \\ 0.314 & 0.686 \end{bmatrix}$ **5.** $\begin{bmatrix} 0.333 & 0.667 \\ 0.333 & 0.667 \end{bmatrix}$

7. a. 97.4 units, 22.6 units **b.** 94.93 units, 25.07 units **c.** ≈92.584 units, ≈27.416 units **d.** ≈90.354 units, ≈29.646 units **9. a.** 97 units, 3 units; 94.15 units, 5.85 units; 91.4425 units, 8.5575 units; ≈88.870 units, ≈11.130 units. **b.** 2 units, 98 units; 3.9 units, 96.1 units; 5.705 units, 94.295 units; ≈7.420 units, ≈92,580 units **c.** 49.5 units, 50.5 units; 49.025 units, 50.975 units; ≈48.574 units, ≈51.426 units; ≈48.145 units, ≈51.855 units **11. a.** $\frac{2}{3} \approx 0.667$ **b.** $\frac{19}{36} \approx 0.528$ **c.** $\frac{203}{432} \approx 0.470$ **13.** $\frac{1}{2}, \frac{1}{2}; \frac{3}{8}, \frac{5}{8}; \frac{11}{32}, \frac{21}{32}; \frac{43}{128}, \frac{85}{128}; \frac{171}{512}, \frac{341}{512}$ **15. a.** $\frac{1}{8}$ **b.** $\frac{9}{16}$ **c.** $\frac{5}{16}$ **Mixed Review**

1. $\frac{2\sqrt{14}}{9}$ **2.** *B* **3.** $\frac{17}{81}$ **4.** 78 **5.** 5 or −4.6

6. $\begin{bmatrix} 42 & 30 & -68 \\ 10 & 17 & -28 \\ -16 & -2 & 14 \end{bmatrix}$

Page 538 Problem Solving Application

7. All except fish

Pages 540–541 Chapter Review

1. 2×2 **3.** 4×1 **5.** $\begin{bmatrix} -2 & -3 \\ -4 & -1 \end{bmatrix}$ **9.** $\begin{bmatrix} 2 & 10 \\ 2 & -2 \end{bmatrix}$

11. not possible **13.** $\begin{bmatrix} 6 & -3 \\ -4 & 2 \\ -5 & -4 \end{bmatrix}$ **15.** not possible

19. $\begin{bmatrix} -15 & 2 \\ 0 & -6 \end{bmatrix}$ **23.** not possible **27.** $\begin{bmatrix} \frac{1}{10} & \frac{1}{5} \\ \frac{3}{10} & -\frac{2}{5} \end{bmatrix}$

31. $\{(3, 4)\}$ **33.** $\{(7, -2)\}$ **37.** $\begin{vmatrix} e & f \\ h & i \end{vmatrix}$

39. $-\begin{vmatrix} b & c \\ h & i \end{vmatrix}$ **43.** $\{(7, -3, 1)\}$ **47.** $(-4, 2)$

CHAPTER 15 POLYNOMIALS, COMPLEX NUMBERS, AND DEMOIVRE'S THEOREM

Page 546 Lesson 15.1

Exploratory **1.** yes; 1, $3x$, 3, -4 **3.** yes; 3, $-2x^3$, -2, -3 **5.** yes; 0, 8, 8, 8 **7.** yes; 2, $-3x^2$, -3, 7 **9.** no **11.** no **13.** exercise 5 **15.** exercises 2, 7 **Written** **1.** 6 **3.** 12 **5.** 0 **7.** $18 - 14i$ **9.** $16x^3 + 12x^2 + 2x + 6$ **11.** $2x^3 + 6x^2h + 6xh^2 + 2h^3 - 3x^2 - 6xh - 3h^2 + x + h + 6$ **13.** $\left\{\frac{3}{5}\right\}$ **15.** $\{2, 3\}$ **17.** $\{-3, 0, 3\}$ **19.** $\{0, 2, 3\}$ **21.** $\{0, -1 \pm i\sqrt{2}\}$ **23.** $0, \frac{1 \pm i\sqrt{11}}{3}$ **25.** ± 1, $\pm i\sqrt{5}$ **27.** $0, \pm\sqrt{2}, \pm\sqrt{7}$ **31.** $b^2 - 4ac = 0$

Pages 549–550 Lesson 15.2

Exploratory **1.** $x - 15$; 0 **3.** $x^2 - 6x + 25$; -105 **5.** $m^2 - 6m + 9$; -1 **7.** $x^2 + 4x + 4$; 4 **9.** $2b^2 - 7b + 4$; 0 **11.** $x^2 + 3x + 9$; 0 **13.** $x + 7$; $14x - 9$ **15.** $x^2 - 2x + 8$; -20 **17.** $8x + 1$; -10 **19.** $4x^2 + 2x + 1$; 0 **21.** $x^2 - 2x + 3$; 0 **Written** **1.** $\frac{3}{2}x + \frac{7}{2}$; 25 **3.** $2x^2 + 8x + 33$; 129 **7.** $x^3 - x^2 + x - 1$; 2 **11.** $3x^2 + x + 3$; 2 **15.** $x^2 + 3x - 2$; 4 **19.** $y^4 - 2y^3 + 4y^2 - 8y + 16$; 0 **Mixed Review** **1.** 7 **2.** -7 or 4.5 **3.** $\frac{7 \pm i\sqrt{95}}{4}$ **4.** 7 **5.** 0.5 **6.** 0° **7.** -2 **8.** 4 **9.** 42 pounds **10.** 480 boxes

Page 553 Lesson 15.3

Exploratory **1.** 13 **3.** 90 **5.** 157 **7.** no **9.** yes **11.** yes **13.** yes **15.** yes **17.** yes **Written** **1.** 5 **3.** 3 **5.** 63 **7.** 561 **9.** $\frac{77}{16}$ **11.** -5 **13.** -1 **15.** -3 **17.** $\frac{1}{2}$, 5 **19.** 3 **21.** $x^3 - 2x^2 - x + 2 = 0$ **25.** $6x^3 + 7x^2 - 1 = 0$

Page 556 Lesson 15.4

Exploratory **1.** $-1, 2, 3$ **3.** $2, \pm i$ **5.** $-1, 0, \frac{1}{2}, 2$ **9.** $\pm 1, \pm 2, \pm 5, \pm 10$ **11.** $\pm 1, \pm 2, \pm 3, \pm 4, \pm 6, \pm 12$ **13.** $\pm 1, \pm 7$ **Written** **1.** $-2, \frac{1}{3}, 1$ **3.** $\frac{2}{3}$ **5.** none **7.** 0, 3 **9.** $-1, 1, 2$ **11.** $-\frac{1}{3}, \pm\sqrt{2}$ **13.** $-1, 1 \pm\sqrt{2}$ **15.** $-2, -1, \pm i\sqrt{3}$ **17.** $1, \frac{-1 \pm i\sqrt{3}}{2}$ **19.** $\frac{1}{2}, 2 \pm\sqrt{3}$ **21.** $-3, -1, 1, \frac{3 + 3i\sqrt{3}}{2}$ **Mixed Review** **1.** b **2.** a **3.** b **4.** d **5.** d **6.** b **7.** d **8.** a **9.** c **10.** b **11.** b **12.** b **13.** b **14.** c

Pages 561–562 Lesson 15.5

Exploratory **1.** 3 **3.** 2 **5.** 7 **9.** no **Written** **1.** $-i, \frac{1}{2}$ **3.** $3, i$ **5.** $2 + 3i, -1$, **7.** $1 + 2i$, $-\frac{1}{2} \pm \frac{\sqrt{11}}{2}i$ **9.** $x^3 - x^2 + x - 1 = 0$ **11.** $x^3 - 2x + 4 = 0$ **13.** $x^4 - 2x^3 + 6x^2 - 8x + 8 = 0$ **15.** $x^4 + 2x^2 + 1 = 0$ **17.** 3 or 1 **19.** 1 **21.** no **23.** $x^2 + 6x + 9 = 0$ **25.** $x^3 + x^2 + x + 3 = 0$ **27.** $x^7 - 3x^5 - 6x^2 + 3 = 0$ **29.** 3 positive, or 1 positive and 2 imaginary **31.** 2 positive and 2 imaginary, or 4 imaginary **33.** 2 positive, 1 negative, and 2 imaginary, or 1 negative and 4 imaginary **35.** 1 positive, 1 negative, 2 imaginary

Page 566 Lesson 15.6

Exploratory **1.** 1, 2 **3.** $-2, -1$; 1, 2 **5.** 2, 3 **7.** 78 **9.** 6 **11.** 6 **13.** $\frac{93}{16}$ **15.** 4 imaginary roots **Written** **1.** -1.3 **3.** $-2.4, -1.3, 1.7$ **5.** -0.8 **9.** $-0.8, 0.6, 2.2$ **29.** $T_{3,0}$ **31.** $T_{2,0}$ **33.** $T_{3,0}$

Pages 569–571 Lesson 15.7

Exploratory **1.** $\sqrt{2}$, 45° **3.** $3\sqrt{2}$, 45° **5.** 2, 30° **7.** 1, 30° **9.** $\frac{3\sqrt{3}}{2}$, $\approx 35°$ **11.** $\left(\frac{\sqrt{3}}{2}, \frac{1}{2}\right)$ **13.** $\left(\frac{5}{2}, \frac{5\sqrt{3}}{2}\right)$ **15.** $\left(\frac{5\sqrt{2}}{2}, \frac{5\sqrt{2}}{2}\right)$ **17.** $(\sqrt{2}, \sqrt{2})$ **Written** **1.** $\left(\frac{1}{2}, \frac{\sqrt{3}}{2}\right)$ **3.** $\left(-\frac{\sqrt{2}}{2}, \frac{\sqrt{2}}{2}\right)$ **5.** $(1, 0)$ **7.** $(0, 2)$ **9.** $(3\sqrt{3}, -3)$ **11.** $\left(\frac{\sqrt{2}}{4}, \frac{\sqrt{2}}{4}\right)$ **13.** $(2\sqrt{2} \cos 45°, 2\sqrt{2} \sin 45°)$ **15.** $(2 \cos 315°,$

2 sin 315°) **17.** (cos 45°, sin 45°) **19.** (cos 150°, sin 150°) **21.** (cos 120°, sin 120°) **23.** (6 cos 225°, 6 sin 225°) **25.** i; 1, 90° **27.** $-i$; 1, 270° **29.** $\sqrt{3} + i$; 2, 30° **31.** $2 - 2i\sqrt{3}$; 4, 300° **33.** $\approx -3.44 - 4.91i$; 6, 235° **35.** $\approx 0.78 + 2.90i$; 3, 75° **37.** $\sqrt{2}$ (cos 45° + i sin 45°) **41.** 2(cos 120° + i sin 120°) **43.** 10(cos 0° + i sin 0°) **47.** 5(cos 270° + i sin 270°) **51.** 6(cos 210° + i sin 210°) **55.** $\sqrt{13}$ **57.** $\sqrt{2}$ **59.** 2 **61.** 1 **63.** $4i$

Pages 575–576 Lesson 15.8

Exploratory **1.** 12(cos 100° + i sin 100°) **3.** cos 200° + i sin 200° **5.** $6i$ **7.** -5 **9.** $-\frac{3\sqrt{6}}{2} - \frac{3\sqrt{2}}{2}i$ **Written** **1.** $32 + 32i\sqrt{3}$ **3.** $-\frac{81}{2} + \frac{81\sqrt{3}}{2}i$ **5.** -4 **7.** -64 **9.** 16 **11.** $-512 - 512i\sqrt{3}$ **13.** $\pm 1, \pm i$ **15.** ± 2, $1 \pm i\sqrt{3}$, $-1 \pm i\sqrt{3}$ **17.** cos 22.5° + i sin 22.5°, cos 112.5° + i sin 112.5°, cos 202.5° + i sin 202.5°, cos 292.5° + i sin 292.5° **21.** $\{2\sqrt[4]{2}$(cos 157.5° + i sin 157.5°), $2\sqrt[4]{2}$(cos 337.5° + i sin 337.5°)$\}$ **25.** $f(z) = 1 - i$ **31.** $-\frac{5}{3}$ **33.** $-\frac{\sqrt{3}}{4} + \frac{1}{4}i$ **35.** $\frac{1}{6} - \frac{\sqrt{3}}{6}i$

Page 578 Problem Solving Application

1. Mary Davis–jewelry; Paulette Morris–books; Cari Sullivan–furniture; Ann Wilson–clothing **3.** Frank Carter–grandfather–New York; Lyn Yates–boss–California; Wanda Haar–niece–Ohio; Andy Pinnow–uncle–Kentucky; Skip Scott–roommate–Wisconsin

Pages 580–581 Chapter Review

1. yes; 1; $2x$, 2, -1, linear **3.** yes; 3; $-3x^3$, -3, -1, cubic **5.** yes; 2; ix^2, i, 4, quadratic **7.** no **9.** no **11.** $5x^3 + 20x^2 + 70x + 268$, 1065 **13.** $3x^4 + 8x^3 + 6x^2 + 7x - 3$, 0 **15.** $a_4x^3 + (a_3 + a_4r)x^2 + (a_2 + a_3r + a_4r^2)x + (a_1 + a_2r + a_3r^2 + a_4r^3)$, $a_4r^4 + a_3r^3 + a_2r^2 + a_1r + a_0$ **17.** -10 **19.** 190 **21.** $-2, -1, 1$ **23.** $\frac{1}{2}$, $\pm i$ **25.** no rational roots **27.** -2, $\frac{1 \pm i\sqrt{11}}{2}$ **29.** -2, $-\frac{1}{2}$, 2, $\pm i\sqrt{2}$ **31.** $x^3 + x^2 - 22x - 40 = 0$ **33.** $9x^3 + 72x^2 - 79x + 18 = 0$ **35.** $5x^3 - 18x^2 - 13x - 2 = 0$ **37.** $x^3 - 2x^2 + 2x - 1 = 0$ **39.** -3, 2 **41.** $\frac{1}{4}$ **43.** 0.5, 0.6 **45.** $-1.6, -1.5$; $-0.3, -0.2$ **47.** $-2, -1.9$; 1.9, 2

51. $\approx 7.52 + 2.74i$ **53.** $2\sqrt{2} + 2i\sqrt{2}$ **55.** ($2\sqrt{2}$ cos 225° + $2\sqrt{2}$ sin 225°) **57.** $\sqrt{2}$ (cos 315° + i sin 315°) **59.** 35(cos 150° + i sin 150°) **61.** 10(cos 30° + i sin 30°) **63.** 3(cos 180° + i sin 180°) **65.** 32(cos 360° + i sin 360°), 32 **67.** $3\left(\cos\frac{\pi}{3} + i \sin\frac{\pi}{3}\right)$, $\frac{3}{2} + \frac{3\sqrt{3}}{2}i$ **71.** $\left\{\frac{\sqrt{3}}{2} + \frac{i}{2}, -\frac{\sqrt{3}}{2} + \frac{i}{2}, -i\right\}$ **73.** $\left\{\frac{\sqrt{2}}{2} + \frac{\sqrt{2}i}{2}, -\frac{\sqrt{2}}{2} - \frac{\sqrt{2}i}{2}\right\}$ **77.** $\{\sqrt[5]{4}$ (cos 6° + i sin 6°), $\sqrt[5]{4}$ (cos 78°, i sin 78°, $\sqrt[5]{4}\left(-\frac{\sqrt{3}}{2} + \frac{i}{2}\right)$, $\sqrt[5]{4}$ (cos 222° + i sin 222°), $\sqrt[5]{4}$ (cos 294° + i sin 294°)$\}$

CHAPTER 16 MATHEMATICAL INDUCTION

Page 585 Lesson 16.1

Exploratory **3.** To prove p is true, first we assume $\sim p$ is true. Next we show this assumption leads to a contradiction. From this we can conclude $\sim p$ is false. Since $\sim p$ is false, we can then conclude $\sim(\sim p)$ or p is true. **Written** **1. a.** 281, 313, 347, 383, 421 **b.** yes **c.** no conclusions **3. a.** 691, 743, 797, 853, 911 **b.** yes **c.** no conclusions **5. a.** 1301, 1373, 1447, 1523, 1601 **b.** yes **c.** no conclusions **7.** $1 + 3 + 5 + 7 + 9 + 11 = 36 = 6^2$, $36 + 13 = 49 = 7^2$, $49 + 15 = 64 = 8^2$, $64 + 17 = 81 = 9^2$, $81 + 19 = 100 = 10^2$ **9.** $\sum_{i=1}^{n} 2i = n^2 + n$

Page 588 Lesson 16.2

Exploratory **3.** no **5.** no **7.** $2r$, $2r + 2$ **Written** **1.** $\frac{1}{n}, \frac{1}{n+1}$ **3.** 3^n, 3^{n+1} **5.** $\frac{n}{n+1}, \frac{n+1}{n+2}$

Pages 590–591 Lesson 16.3

Exploratory **1.** $k + 2$, $\frac{k^2 + 2k - 1}{k + 2}$ **3.** $k^2 - 4$, $\frac{k^2 - 2}{k^2 - 4}$ **5.** Since $\frac{k(k+1)}{2} + (k + 1) = \frac{k(k + 1}{2} + \frac{2(k+1)}{2} = \frac{k(k+1) + 2(k+1)}{2}$ **Written** **1.** n, $n + 1$, and $n + 2$ are three consecutive integers, and $n^3 + (n + 1)^3 + (n + 2)^3$ is the sum of their cubes. **3.** $1^3 + 2^3 + 3^3 = 36 = 9 \cdot 4$. Thus, $1 \in S$ **5.** Multiplication **7.** Using induction hypothesis and definition of divisibility **9.** false **11.** false

31. $\sum_{i=1}^{n} i^2 = \frac{n(n+1)(2n+1)}{6}$ **Mixed Review**
1. $\frac{-2 \pm i\sqrt{5}}{3}$, 2 **2.** $2 \pm i$, 3 **3.** 4.7 and 4.8
4. $n^2 + 2n$, $n^2 + 4n + 3$ **5.** 8 mph

Pages 593–594 Lesson 16.4
Exploratory **1.** $6^n - 2^n$ is divisible by 4 **3.** $5^{n+2} + 6^{2n+1}$ is divisible by 31 **5.** 5^2, 1 **7.** 9, 17 **Written** **1.** Definition of the set for which the assertion is true **3.** Statement of the Induction Hypothesis **5.** Adding, then subtracting away the same quantity, $5(6^{2k+1})$ **7.** $5^{k+2} + 6^{2k+1}$ is divisible by 31 by the Induction Hypothesis, and $6^{2k+1}(6^2 - 5)$ is divisible by 31 since $6^2 - 5 = 31$. **9.** From exercises 2, 3, and 8 **25.** $a^n - b^n$ is divisible by $a - b$.

Pages 596–597 Lesson 16.5
Exploratory **3.** $n^2 \leq 50n$ for $n \leq 50$, not $n \geq 50$. **5.** $(n+2)(n+1)! = (n+2)!$ for $n \in \mathcal{N}$. Thus, $(n+2)(n+1)! \not> (n+2)!$ **Mixed Review**
1. $\frac{7(2x+1)}{(x-3)(x+4)}$ **2.** $-\sqrt{6}$ **3.** $\frac{x-2}{3}$ **4.** $T_{3,-2}$
5. 6 **6.** 140 **7.** $-\frac{15}{8}$ **8.** 1576 km **9.** 60°, 300°
10. $D'(-4, 4)$, $E'(0, 1)$, $F'(-8, -2)$ **11.** $\frac{693}{40000}$
12. 31.9, 2.5 **13.** \$110,592 **14.** $\left(7\frac{8}{19}, -\frac{1}{19}\right)$
15. $x^3 - 6x^2 + 9x + 50 = 0$

Pages 599–600 Lesson 16.6
Exploratory **1.** Because $-x^k + x^k = 0$ and $x^{k+1} + 0 - 1 = x^{k+1} - 1$ **3.** $-\frac{27\sqrt{3}}{2} + \frac{27}{2}i$

Page 603 Lesson 16.7
Exploratory **3.** 120 **5.** 15504 **9.** $16a^4 + 96a^3 + 216a^2 + 216a + 81$ **Written** **7.** $-\frac{30}{x^9}$ **9.** $a^3 + 3a^2b + 3ab^2 + b^3$ **11.** $15a^4b^2 + 20a^3b^3 + 15a^2b^4$
13. $(a+b)^n = \sum_{i=0}^{n} {}_nC_i\, a^{n-1}b^i$ **15.** no

Page 606 Lesson 16.8
Exploratory **1.** Because all convex polygons have at least three sides. **5.** true **7.** true

Page 609 Lesson 16.9
Exploratory **3.** 2, 4, 8, 16, 32, 64, 128, 256 **5.** 2, 4, 8, 16, 32, 64, 128, 256 **Written** **1.** 0.9, 0.81, 0.729, 0.6561, 0.59049

Page 611 Problem Solving Application
1. connected, cycle, not a tree **3.** not connected, no cycle, not a tree **5.** BF, CE, AC, BE, CD; length = 39 miles

Pages 612–613 Chapter Review
1. PMI is used in cases where a direct method of proof is either not possible or not convenient.
3. $n = 1$: $\frac{1}{3} = \frac{1}{2 \cdot 1 + 1}$. Thus, $1 \in S$. $n = 2$: $\frac{1}{3} + \frac{1}{15} = \frac{6}{15} = \frac{2}{5} = \frac{2}{2 \cdot 2 + 1}$. Thus, $2 \in S$. $n = 3$: $\frac{1}{3} + \frac{1}{15} + \frac{1}{35} = \frac{45}{105} = \frac{3}{7} = \frac{3}{2 \cdot 3 + 1}$. Thus, $3 \in S$.
Assume $k \in S$. That is, $\frac{1}{1 \cdot 3} + \frac{1}{3 \cdot 5} + \frac{1}{5 \cdot 7} + \cdots + \frac{1}{(2k-1)(2k+1)} = \frac{k}{2k+1}$. Now, $\frac{1}{1 \cdot 3} + \frac{1}{3 \cdot 5} + \frac{1}{5 \cdot 7} + \cdots + \frac{1}{(2k-1)(2k+1)} + \frac{1}{(2k+1)(2k+3)} = \frac{k}{2k+1} + \frac{1}{(2k+1)(2k+3)} = \frac{k(2k+3)+1}{(2k+1)(2k+3)} = \frac{2k^2+3k+1}{(2k+1)(2k+3)} = \frac{(2k+1)(k+1)}{(2k+1)(2k+3)} = \frac{k+1}{2(k+1)+1}$. Thus, $k + 1 \in S$.
Therefore, $S = \mathcal{N}$. **5.** Mathematical induction is used to prove assertions about the natural numbers. First, we define S to be the set of all natural numbers for which the assertion is true. Next, we show that $n \in S$ for some natural number n (for PMI, $n = 1$, but n can be some other natural number). Then we assume $k \in S$ and prove $k + 1 \in S$. Completion of these steps means that $S = \mathcal{N}$. PMI is often compared to the "domino theory" because all the dominos will fall only if one domino has to fall (show $1 \in S$) and the dominos are set up in such a way that if one falls, they all do (assume $k \in S$, prove $k + 1 \in S$). **7.** $2n - 1$, $2n + 1$ **9.** $2 + \frac{3}{2} + \frac{2}{3}$ **11.** $-\frac{2}{3} + \frac{8}{9} - \frac{2}{3}$ **17.** $-2 + 2i$
19. ${}_nC_r = \frac{n!}{r!(n-r)!}$ **21.** 70 **23.** $5x^4y$ **25.** $5xy^4$

Index

A

B

C

D

E

F

G

H

I

N

O

P

Q

T

U

Z